FUNDAMENTALS OF FOOD ENGINEERING

Fundamentals of FOOD ENGINEERING

Second Edition

D.G. Rao
Former Scientist and Head
CSIR-Central Food Technological Research Institute (CFTRI)
Resource Centre
Habsigude, Hyderabad

PHI Learning Private Limited
Delhi-110092
2026

*In fond memory of **Shri Asoke K. Ghosh** (October 1942 – February 2024), Founder Chairman and Managing Director of PHI Learning, whose vision endlessly inspires.*

The Legacy Continues....

Published by Pushpita Ghosh, PHI Learning Private Limited, Rimjhim House, 111, Patparganj Industrial Estate, Delhi-110092 and Printed by Multi Colour Services, 92, DSIDC Sheds, Okhla Industrial Area, Phase-I, New Delhi-110020.

₹1195.00

FUNDAMENTALS OF FOOD ENGINEERING, Second Edition
D.G. Rao

ISBN-978-81-963789-2-9 (Print Book)
ISBN-978-81-963789-4-3 (e-Book)

The export rights of the book are vested solely with the publisher.

To my beloved parents
Late Mrs. Dubasi Savitri and Late Dubasi Balakrishna Dora
who provided me with every opportunity in life which made me what I am today

Contents

Foreword

India is one of the leading producers of agricultural produce in the world. However, exploitation of the processing capabilities of the harvested produce is one of the lowest in the world causing a huge loss of agro-produce, which otherwise could have been usefully utilized for feeding millions of our people and improving their nutrition. The post harvest losses can be mitigated by proper application of various post harvest technological operations which are a part of food technology. The processing operations at small scale level may not be adequate to handle large amounts of agricultural produce. The rapid growth in population adds an additional burden to food technology to make it extend its horizons. The rapid industrialization has its sway on food processing, which metamorphosed it into industrial ventures by encompassing a host of engineering operations known as **unit operations.**

Probably the first industrial food processing operation was canning of food which was invented by Donkin and Hall in England in the year 1812 which was a thermal way of degradation of microorganisms and enzymes that spoil the foods. This marvelous invention has ushered in the era of modern food technology. Design and fabrication of equipment to carry out the food technological operations and to successfully run such equipment have necessitated the coupling of engineering principles with food technology to process a host of raw materials which are essentially biological in nature, and are susceptible to spoilage. Thus, wedding of food technology with engineering operations has given birth to a new field in food processing known as **food engineering** and constitutes the subject matter of this book.

Food engineering encompasses various other peripheral disciplines like mathematics, physics, chemistry and biological sciences including microbiology and human nutrition. These basic sciences in combination with chemical engineering and mechanical engineering lay the foundation for food engineering.

I am glad to note that Dr. D.G. Rao has a long experience in teaching food engineering in various universities. It is apt that he has authored the book on *Fundamentals of Food Engineering* which is a long felt need. The bringing out of this book is very timely when most of the Indian universities are starting courses in food technology both in engineering and science streams offering B.Tech. (Food Technology) and M.Sc. (Food Technology). Dr. D.G. Rao has presented the text lucidly in a simple language with a good number of worked examples so that the subject matter is comprehensible even to non-mathematics students who are novice to engineering discipline. I am sure that the book is useful not only to students as a textbook who are pursuing their academic degrees, but also to practising food technologists and engineers as reference material in their professional pursuit.

I wish the author all the best, and the book all success.

Dr. M.V. Rao
President, Agri Biotech Foundation
Acharya N.G. Ranga Agricultural University (ANGRAU)
Hydrabad
Former Special Director General, ICAR
and
Former Vice Chancellor, ANGRAU

Preface

The motivation to revise this edition of the book has mainly emerged from various fruitful discussions I had with some students, colleagues, and friends in the last one decade who are using this text. Accordingly, a new chapter on Fermentation (Chapter 25) is added in view of its importance in food processing. The treatment of the subject matter in this chapter is not very elaborate. Hence, for any further details on Fermentation, the readers are advised to refer to any standard books on Biotechnology. However, the mathematical component on thermal death kinetics, application of Michaelis-Menten equation, etc. is all deliberately avoided keeping in view the heterogeneous nature of readers, and the limited need of such elaborate calculations of fermentation in food processing. The mixing chapter is enlarged by adding a section on 'Emulsification' because of its imminent need in food processing particularly for making functional foods and convenience foods. The chapters on 'Thermodynamics' and 'Crystallization' are dropped in the new edition because of their limited application to food engineers. Problem 11.7 and Problem 21.8 have been suitably corrected, which escaped my attention during the first edition. Some typographic errors have also been attended to in the revised edition.

I express my sincere thanks to the editorial and production departments of PHI Learning, New Delhi for their unstinted support and efforts to bring the new edition in the present form. I am also thankful to various authors, publishers and industries who have accorded copyright permissions to utilize their published material in the book. My sincere thanks are also due to my wife, Dr. D. Sai Leela for her support and ungrudging endurance during the revision work of this edition.

I believe that there is always a scope of further improvement in the book. Hence, any further specific suggestions and positive criticisms for improving the quality of the book are always welcome. Users of this book can contact me at dgrao1950@rediffmail.com or share their feedback at phi@phindia.com

D.G. Rao

Preface to the First Edition

This book essentially deals with the fundamentals of food engineering and more precisely with unit operations in food processing which the students of food technology study in their undergraduate level in engineering stream and in their postgraduate level in science streams. The motivation for writing this book has come to me from my interaction with a number of young students when I was teaching food engineering as a scientist at CFTRI, Mysore during 1988–90, and subsequently as a visiting faculty in a number of universities in Andhra Pradesh. The fruitful discussions I had with a number of students in the last two decades have motivated me to pen a book on food engineering which is comprehensible to students who are novice to engineering discipline. In the process, I have consulted the syllabi of various food technology curriculum, and imbibed them in this book.

The book is divided into four parts viz., (i) General Introduction, (ii) Basic Engineering Principles, (iii) Unit Operations, and (iv) Food Industry Management. In the first part, a brief introduction about food processing is given, and various food preservation techniques are highlighted. In the second part basic engineering principles are covered which include steam generation and utilization, refrigeration principles, and humidification operations. Also a brief introduction is given on various process control methods.

The third part covers various unit operations and is the core subject of the book. The unit operations include various transport processes like fluid mechanics, heat transfer and mass transfer. Rheology of foods is dealt with in Chapter 10. Heat transfer is covered in five chapters to include heat transfer by various modes. A separate chapter (Chapter 15) is devoted to microwave heating which is an important emerging concept in food processing.

Various mass transfer operations such as dehydration, distillation, gas absorption, and extraction, and so on are dealt with separately. Mechanical separation processes, size reduction and grinding, and materials handling equipment are covered separately.

The last part on food plant management is covered in three chapters which also includes a chapter on food process economics.

A number of worked-out examples are included in many chapters. Excessive mathematical derivations have been deliberately avoided keeping in view the heterogeneous readership of the book. It has also been my endeavour throughout to keep the language of the book simple and lucid. The contents have been made pragmatically crisp for easy comprehension. All efforts have been made to judiciously blend all food engineering operations in one single book, and make it as comprehensive as possible. Still I feel that the contents in the book are by no means exhaustive.

Hence, the enthusiastic students are advised to refer to the original reference materials given at the end of each chapter. However, the basic principles discussed throughout the book are well applicable for day-to-day processing operations.

The book is amply illustrated with figures, photographs, diagrams and tables which will purportedly help students comprehend the subject matter. Review questions and numerical problems provided at the end of each chapter will help the students in testing their understanding of the subject.

The book is ideally suited for students pursuing their B.Tech. (Food Technology) or M.Sc. (Food Technology) courses. It also caters to the needs of students of Biotechnology and Chemical Engineering as all the unit operations have been adequately covered.

I am thankful to all my senior colleagues and friends for their good wishes and moral support. I am grateful to Padma Shri (Dr.) M.V. Rao, former Special Director General of ICAR and a renowned agricultural scientist, for readily agreeing to write the Foreword for this book. I would also like to acknowledge the help and cooperation I have received from my colleagues at CFTRI-Resource Centre. I appreciate the fruitful discussions I had with them.

My thanks are also due to a few of my relatives and friends who have invisibly encouraged me in this endeavour.

I also express my sincere thanks to the staff of editorial and production departments of PHI Learning, New Delhi, for their unstinted support and efforts to bring the book in the present form. I am also thankful to various authors, publishers and industries who have accorded copyright permissions to utilize their published material in the book.

I also wish to thank my young daughters Asha Latha and Priyanka who have cheerfully typed the text from my illegibly hand-written manuscripts. Last but not the least, I wish to thank my wife, Dr. Sai Leela for her ungrudging endurance during the two-year period of writing this book.

The views/information expressed in this publication are entirely my own, and are not connected with my official capacity in CSIR/CFTRI.

This book is an honest attempt by me to introduce engineering principles, and the importance of engineering calculations to food technologists and chemical technologists in the day to day operations of process equipment. Hence, I feel that I am amply rewarded if the book serves at least partly to that end. Any exhaustive book of this nature is not free from some mistakes, omissions and commissions. Any suggestions and positive criticisms for improving the quality of the book are always welcome. Users of this book can contact me at dgrao1950@rediffmail.com

D.G. Rao

Annam Rakshati Rakshitam

(Food protects if protected)

Part I

General Introduction

CHAPTER 1

Introduction

Food processing is probably one of the oldest avocations man has been involved with from time immemorial, and evolved into various types of activities finally to convolute into a generic term what is today called **food technology**. The incessant and intensive need of man to satiate the biological requirement of hunger by consuming food has resulted into multifarious activities of processing of a multitude of raw materials which nature has bestowed on man. Since the bounty of nature cannot be unlimited, it also necessitated man to go relentlessly in search of greener pastures. Thus, the mobility of man has started from one region to other, which has demanded accumulation of a large number of gadgets and paraphernalia. Since the basic requirement of food is to satiate hunger which is a biological need, the ingenuinity and greed of man knew no bounds to accumulate as much quantity of food as possible. Thus, in a way we can say *the history of man is nothing but the history of food technology*. Imagine a world without hunger! the world which does not care for food, a food which does not require processing, a processing which does not involve an activity, the activity which is sedentary! probably there would not have been any progress in the society had it been free of hunger.

1.1 HISTORICAL BACKGROUND

The origin of food technology perhaps is cooking, and cooking more food to make it in more quantities in a more palatable form, and extend shelf-life considerably. Roasted meat was found to stay for longer periods as compared to raw meat. Thus, invention of fire and its effects may possibly be considered to be a prelude for food technology. Use of fire for cooking dates back to at least 250,000 years in the pre-neolithic period (Stewart and Amerine 1973). Subsequently, the cooking process has gone through various stages. As a technological process, perhaps food processing originated as baking of bread in Egypt as early as 4000 BC which involved operations, such as kneading, mixing, leavening and baking. Wine making was known to Romans in 7000 BC. Thus, bakers and brewers were the forerunners of current food industries. Table 1.1 shows the developments in agriculture and food processing through various ages (Stewart and Amerine 1973).

Table 1.1 Development of Agriculture, Food Processing and Food Preservation

Period	*Dates*[1]	*Agriculture*	*Foods*	*Processing and preservation techniques*	*Examples of science and technology*
Palaeolithic (old stone age)	Before 15,000 BC	None	Eggs, fruits, nuts, seeds, roots, insects, fish, honey, small animals[2]	Roasting, pounding, drying	Bags, baskets, clothes, stone and bone implements, made fire, painting, sculpture, language
Mesolithic (middle stone age)	15,000 BC	None	Greater variety, stored wild fruits and berries	Dried fish, boiling, food storage, smoking, steaming	Bow and arrow, dog, goat, reindeer and sheep probably domesticated, clay covered baskets
Neolithic (new stone age) (villages)	9000 BC or earlier	Seasonal culture of cereals, horticulture, plowing, permanent fields, pruning	Domesticated animals[3], milk, butter, cheese, gruel, dates, olives, grapes, beer, vinegar, wine	Alcoholic fermentation, acetification, salting, baking, bread-making, sieving, primitive processing, seasoning	Pottery wheel, spinning, weaving, wood, flint and bone sickles, saddle quern, mortar, fishing with hooks and nets
Bronze (cities)	3500 BC	Irrigation[4], horse and ox-drawn plows, much local and long-distance trade	Soybean, figs, rice, olive oil, vegetables, lentils, cabbage, cucumbers, onions	Filtration, lactic acid fermentation, more types of flavouring, flotation, leavened bread, sausage making, frying, sophisticated and complicated pressing, clarification	Architecture, smelting, wheeled carts, ships, writing, bronze tools, mathematics, rotary millstones, bronze weapons, astronomy, shadufs, medicine, chemistry
Iron	1500 BC	Land and sea trade common, heavier plows	Fruits, spices, beans, artichokes, lettuce, sauces	Refinement of flavouring and of cookery	Pulleys, glass, improved and cheaper tools and weapons, currency
Roman	600 BC–400 AD	Reaping machines, legume rotation, plows on wheels, food trade extensive	Sugar cane in West, apples, asparagus, beets, orange	Food adulteration common	Water mills, donkey mills, wooden cooperage

[1]The dates indicate only the beginnings in the main centres of origin. They appeared much later in other areas, may not have developed at all, or may even have retrogressed. Tasmania was discovered in 1642, and the evidence up to Cook's visit in 1777, was that the people of Tasmania had retrogressed from the Neolithic to Palaeolithic period. Stone implements continued to be used long after the old stone age. It is important to remember that a food processing operation may have originated in one region long before another. Also, bronze age implements continued to be used for a long time into the iron age.

[2]Big-game hunts occurred in areas of cliffs about 400,000 BC when fire and axes and spears were used. Pit-hunting and use of knives appeared about 75,000 BC.

[3]The order of domestication is unknown but goats, yak, buffalo, pig and cattle were domesticated early in this period but not the horse or camel. The horse, camel, ass, elephant and poultry were domesticated towards the end of this period.

[4]Irrigation existed prior to 3500 BC but its widespread use about this time is believed to account for the spectacular increase in the population of Mesopotamia.

Source: Reprinted from *Introduction to Food Science and Technology*, Stewart, G.F. and Amerine, M.A., pp. 2–3, Copyright (1973), with the permission from Elsevier.

However, the modern food technology started as early as 1812 when the successful canning operation of foods was carried out in England. The stress at that point of time was more on inactivating the microorganisms by sterilization (thermal processing) which were responsible for spoilage of foods. From then onwards, the food technology has undergone several modifications until the development of recent technique of aseptic packaging of fruit pulps in the year 1960.

Various other food preservation methods are given in Table 1.2. They are ancient, traditional and modern methods. Most of the traditional methods are followed in modern food processing (may be except smoking) with the application of engineering skills to make commercially viable high volume produce.

Table 1.2 Various Processing Methods to Preserve the Food Products

1. Salting	Aseptic packaging
2. Pickling in	10. Modified Atmospheric Packaging (MAP)
3. acid	11. Controlled Atmospheric Packaging (CAP)
4. oil	Chemical preservation in
5. Fermentation	Potassium meta bisulphite
6. Roasting and grinding	Benzoate
7. Drying/dehydration	Acids
8. Freezing	Natural preservation
9. Cold storage	Salt
Canning	Sugar
Packaging	Spices
Packing and wrapping	Fermentation

Dehydration and pickling were known to Indians from time immemorial. They are still being used in the modern food technology. The reduction in moisture content reduces water activity, and makes water not available for microorganisms to grow and pathogens to survive. Subsequently, it was realized that reduction in moisture content alone was not adequate. Proper packing techniques are to be adopted, so that the food does not get rehydrated. Thus, modern food technology heavily relies upon packaging techniques and packaging materials, not only as a source of providing body and shape for the foods, but also as an effective tool in combating rehydration, and intrusion of outside pathogens. Indeed packaging itself is a wide subject in food technology.

The modern food technology is a saga of application of engineering principles to process high volume of raw materials to preservable products. In fact, modern food industry does not change the taste of food materials from the traditional practices; the industry adopts singularly the modern engineering practices only for ease and convenience of processing.

1.2 IMPORTANCE OF FOOD PROCESSING

The importance of food processing cannot be overemphasized for the reasons best known. The post harvest losses are enormous. The production of agricultural produce is seasonal. Their availability during the season (which hardly lasts for 2–3 months) is enormous. It is humanly impossible to consume all of them during the season itself. They will not be available subsequently. Hence processing them to safer moisture levels is essential so that the food produced can be preserved and made available from the seasons of glut to gloomy days. Thus, food processing may be viewed to have the following important attributes:

1. By imbibing the knowledge of scientific understanding of the spoilage of foods, it has virtually transformed craft-based industry to a science-based industry.
2. It reduces the spoilage of food by various post harvest methods, and makes it available during off-season also, which tantamounts to increasing production.

3. By processing large quantities of agricultural produce at reasonable price, it provides a justifiable return to the farming community which is a fulfillment of our social obligation.
4. By processing large quantities of foods and making them available in a Ready To Eat (RTE)/Ready To Cook (RTC)/Heat And Eat (HAE) forms, it reduces the kitchen drudgery to women.
5. It has a special advantage, unlike any other production sectors, that the Multiplication Factor (MF) for food processing is 2.4; i.e., for every one rupee of wealth it makes directly, it creates an additional 2.4 rupees of wealth indirectly in the allied areas like transportation, packaging, cold storages, etc. (Anonymous 1997).
6. Since the food processing industries are based on the locally available agricultural produce which is grown in rural areas, it helps establish the processing units in rural areas which not only provides rural employment, but also reduces the urban migration which is the root cause for most of the social evils.

1.3 CATERING TECHNOLOGY VS FOOD TECHNOLOGY

For many a people, both elite and illiterate, food technology is synonymous with the catering technology. Even though the difference is very subtle, it is very distinct. Catering technology is not meant for preserving the food materials for a longer period of time. It is meant for consumption either on the spot or within a day's time in most of the cases (of course, there are certain exceptions like some of the Indian traditional foods which last for two-three weeks of time). Food technology is meant for processing the foods so that they can be preserved for very long periods anywhere between three months to one year (of course, there are certain exceptions in it also like bread which stays for only 96 hours, but still bakery is considered as a component of food technology).

Generally the volumes of production are not considered as a major characteristic feature to distinguish between food technology and catering technology. For example, Tirupathi Tirumala Devasthanam (TTD) feeds food to about 10,000–1,000,000 devotees in each session of the day under *Nityanna dana pathakam.* They also use most of the modern cooking gadgets like boilers, autoclaves, grinders, mixers, trolleys for conveying or transportation. Still it is considered as an activity under catering technology because the basic requirement here is not storage or preservation of foods by processing. Similarly, a processing unit manufacturing even 100 kg per day of toffees or chocolates or 100 kg of pickles per day is considered as a food processing unit, and is categorized as an activity under food technology.

1.4 FOOD ENGINEERING AS A DISTINCT DISCIPLINE

Food engineering is considered as that branch of food technology which deals with design, fabrication, maintenance and operation of food processing equipment. It could be a boiler, a pump, an evaporator, an extraction unit or a dryer. Thus, it merges with the other allied branches of engineering, viz., chemical engineering and mechanical engineering. Apparently we can define food engineering as the study of engineering operations in which the food materials/agricultural produce is involved as raw material or product or a material which is used in food processing like the packaging materials and food-related synthetic chemicals, viz., food colours, food flavours, food acidulants or food preservatives. Thus, the fluid dynamic studies using milk or a fruit juice flowing through a pipe or a conduit are considered as a part of food engineering. Extraction (or leaching) of oil from the oil seeds by solvent extraction process is a part of food engineering even though it is almost akin to chemical engineering studies.

Then why is it that food engineering is considered a distinct discipline?†
The reasons for it could be many. An attempt is made here to identify a few:

1. Food materials change their characteristic features during processing. For example, milk on concentration becomes *khoa*, a semi-solid mass which has different rheological and heat transfer characteristics as compared to the raw material, and needs to change the process parameters during processing.
2. Same type of raw materials have different characteristic features in terms of their size, moisture content, acid to sugar ratio, flowability, etc. depending upon the source and time of harvesting, extent of maturity at the time of harvesting, etc. Hence, they need a separate treatment; and very interestingly most of the food materials are not amenable to engineering operations. Imagine how difficult it is to shape a *ladoo* using machine or make a *bonda* using co-extrusion technique.
3. Food materials contain various types of components, viz.,
    - Proteins
    - Vitamins
    - Enzymes
    - Minerals
    - Carbohydrates
    - Sugars
    - Fats
    - Various micro-nutrients

    Their processing characteristics are varied and demand different processing conditions. Engineering operations which are done at scaled-up level need special attention, which are not typical of other engineering operations.
4. Grinding, crushing and milling are very common size reduction operations in food processing like in metallurgical engineering operations; but the crushing strength of food materials is much low compared to stones and ores, and hence require different types of gadgets and treatment.
5. Since food materials are used for edible purposes, hygienic processing of them is very essential. Hence, good quality control and Good Manufacturing Practices (GMP) have to be strictly followed. The results of any deviation from these may be sometimes disastrous. Maintaining high quality of aseptic conditions is a must in any food processing industry. Hence, food engineer needs to reassure contamination-free environment.
6. Some process operations, particularly dehydration can be done by a variety of dryers like cabinet tray dryer, vacuum shelf dryer, tunnel dryer or a freeze dryer. A proper choice of dryer is to be made not purely on technical considerations, but mainly based on the capacity and value of the product. Hence, good engineering judgement is required.

†University of Mysore at one time even recognized food engineering as a separate discipline.

1.5 NATIONAL SCENARIO

India is essentially an agricultural country and the economy is basically agrarian in nature. More than 70 per cent of the population lives in rural areas; and out of them 80 per cent depend on agriculture for employment and livelihood. For an agrarian economy, rural population can be considerably benefited by food technology atleast in the following three ways (Reuter 1977):

1. Instant foods, energy foods and baby foods can be produced from the locally available raw materials which will reduce child-malnutrition.
2. Integrated food management for storage, transportation and distribution.
3. Application of food technology practices for processing traditional foods by way of drying, pickling, salting and smoking (traditional foods are a rich heritage of India and cater to the nutritional requirements of rural people substantially).

Thus, food processing is vital to India's prosperity. It is very pertinent to note here the introductory remarks of FAIDA report (Anonymous 1995).

> *There is no country in the world which is to lose more than India if it does not succeed in food processing. If food succeeds, India succeeds.*
>
> [FAIDA report, p. 1 (1995)]

India's progress in the agricultural production is envious. The food grains production has increased from a meagre 50 million tons in the year 1947 to a staggering 323.5 million tons today. India stands second in the world next to China, in terms food production with a net value $ 382.2 billion (965 million tons excluding sugar cane) today which is almost comparable to that of US (https://www.investopedia.com/articles/investing/100615/4-countries-produce-most-food.asp). The food materials production statistics are given in Table 1.3. The spectacular progress in the agricultural production could be made by government's **Green revolution** programme and **Save grain** campaign during 1960s. But processing capabilities are very low; as low as less than 2.5 per cent of horticultural produce, whereas in other developing and developed countries the situation is different. For example, the processing levels are: Malaysia (83 per cent), Brazil (70 per cent), Phillippines (78 per cent) and USA (70 per cent) (Patnaik 1993).

Table 1.3 National Production Statistics (million tons)*

Food grains	323.5
Oil seeds	40.01
Legumes	27.8
Horticultural produce	333.25
Milk	207.0
Meat	4.5
Fish	16.3
Sugar cane	468.8

* Ref: https://pib.gov.in/PressReleaseIframePage.aspx?PRID=1899193

1.6 PROCESS INDUSTRIES BASED ON RAW MATERIALS

Food industries are essentially raw material-based industries and may be categorized into a large number (Table 1.4). The industries are more or less resource-based, and hence, the project economics very much rely upon the quality and cost of raw materials. Hence, supply-chain management, volumes of production and good engineering practices play a vital role in the success of the food processing industries.

Table 1.4 Different Varieties of Food Processing Industries Based on the Raw Materials

1. Animal products	4. Cereals and pulse-based products
2. Bakery products	5. Convenience foods
3. Beverage products	6. Fruits and vegetables-based products
Alcoholic	7. Microbiology and fermented foods
Non-alcoholic	8. Spices and condiments
Carbonated	9. Proteins and proteinaceous foods
Non-carbonated	10. Packaging materials

REVIEW QUESTIONS

1.1 *The history of man and history of food processing are synonymous*—justify the statement with suitable examples.

1.2 Outline a brief history of food technology.

1.3 What is the importance of food processing in development of India?

1.4 Distinguish between catering technology and food technology with suitable examples.

1.5 What are various traditional and modern methods of food preservation?

1.6 How do you justify food engineering to be a distinct discipline?

1.7 What are various types of food processing industries we come across?

1.8 Classify the food industries based on raw materials.

REFERENCES

Anonymous (1995), Food and Agriculture Integrated Development and Action (FAIDA)—Modernising the Indian Food Chain, Mc-Kinsey & Co., New Delhi.

Patnaik, G. (1993), Development of Food Processing Industries in Eastern India, *Indian Food Packer*, Nov.–Dec. issue, pp. 54–73.

Reuter, F.W. (1977), Food Technology: Past, Present and Future in *Food Technology*, Desrosier, N.W. (Ed.), AVI Publishing Co., West Port, p. 4.

Stewart, G.F. and Amerine, M.A. (1973), *Introduction to Food Science and Technology*, Academic Press, New York, pp. 1–27.

CHAPTER

2

Food Preservation Methods

The whole of food science, food technology and food engineering revolve around a singular objective of preserving the food materials from the days of glut to gloomy days, so that an incessant food chain is maintained throughout the year. The main objective in food preservation is to preserve the foods over longer periods of time keeping intact the nutrients *as far as possible*, or at least causing as little damage to the nutrients as possible. The whole purpose of food is to meet the metabolic needs of living beings so that the body cells receive adequate amount of energy, nutrients, and strength for rebuilding the damaged or dead cells.

Thus, the foods for energy should be rich in carbohydrates, fats and proteins; foods for strength should be rich in proteins, minerals and vitamins while the nutritious food should be rich in minerals, vitamins, essential amino acids and fibre.

An incessant and intensive search has always been going on to devise various methods to preserve the foods so that their storage life is extended without causing any physical, chemical and microbiological damage to the food stuffs to the maximum possible extent. One of the major culprits found to be responsible for spoilage of foods is water content (which was later defined as water activity (a_w) rather than water content). Another responsible factor was noticed to be temperature. The room temperatures of 30 °C– 40 °C were found to be most conducive to most of the pathogens and microorganisms to thrive, multiply and be active in causing spoilage of foods. Hence, methods have been devised from time immemorial to reduce the water content in foods, and reduce/increase the storage temperature of foods.

2.1 TRADITIONAL METHODS

The traditional methods were mostly based on reducing the moisture content or activity of water so that the food spoiling microorganisms could not survive. This would obviously cause some changes or damage to the physical structure of the foods, viz., contraction in volume and surface area and hence contraction in cellular structure, etc. Some of the commonly used traditional methods were shown in Table 1.2.

2.1.1 Dehydration by Sun Drying

This method is indeed considered as the grand-mother's technique for preserving foods. The food material on dehydration loses moisture, and thus, water is not available for microorganisms to grow and survive. The material also loses weight, and hence, is easy to handle and transport. In the olden days the food materials, may it be grains, pulses, meat, fish, fruits or vegetables, were kept in yards under sun-shine to get dried. It was an excellent method in terms of energy saving and convenience. However, it has the following disadvantages:

(i) Chances of contamination by outside air
(ii) Pilferage of the material
(iii) Attack by pets and birds
(iv) Browning and/charring of the food materials
(v) Uncertainty of the sun-light
(vi) Requirement of large drying areas for spreading the materials in yards

In spite of the above disadvantages, it is still considered as an excellent method, and is still followed in rural areas.

2.1.2 Salting

Salt (sodium chloride) is used as a preservative in food processing. It also works on the principle that it does not allow moisture to be available for microorganisms for their survival and growth. In fact, salt is considered as a number one natural preservative. There is no limit for its addition except for taste. A combination of both salting and drying is also used for preservation of fish, meat and some vegetables.

2.1.3 Pickling

The effect of pickling is also akin to salting, which reduces the water activity. Pickling is done both by acids and by oil. The food materials like fish, meat, prawns, some vegetables, etc. are initially dried (or cooked in case of animal-based products to reduce the water content and also to enhance the taste by avoiding the raw taste). Later, the dry material is pickled in acids (like acetic acid (vinegar)) or in vegetable oils. Some of the Indian pickles made out of fruits/vegetables have a shelf-life of one year and more.

2.1.4 Fermentation

Fermentation is the name given to the biochemical changes brought about by microorganisms (like yeast or bacteria) in organic compounds, usually known as **substrate**. In fact, the microorganisms decompose the organic matter to generate energy for their own survival. In the process, the organic matter (food material) gets processed and develops some new flavours or tastes. Fermentation does not extend the shelf-life of products to great extent, it only extends marginally. For example, milk at room temperature is spoiled quickly, whereas curds stay for more than a day and cheese stays much more. More than extending the shelf-life, fermentation is known to enhance the organoleptic and nutritional qualities of products. A detailed discussion on fermentation is available in Chapter 25.

2.2 THERMAL PROCESSING METHODS

Thermal processing of foods to preserve them over a period of time is a common practice. Thermal processing takes place by two methods. It could be by heating or cooling.

2.2.1 Heat Processing

The commonly used heating processes are canning, sterilization, pasteurization, deep fat frying, roasting, baking, etc. Some of them are discussed here.

Canning: Canning is a heat sterilization process for preserving food materials like fruits, vegetables, meat, fish, etc.

Canning is done at high temperatures of the order of the boiling point of water/boiling point of the syrup. Typically a canning unit consists of can reformer, can flanger, exhaust box, double seamer, and retorting line/autoclave.

The food materials like fruit pulp/sugar syrup/brine are heated to boiling and filled into pre-sterilized cans. The filled cans pass through an exhaust box in which there is a continuous jet of steam. The exhausting takes about 10 min time. Later the cans are closed in a double seamer and then put to retorting/autoclaving. The retorting is done at the boiling temperature of water (100 °C), whereas the autoclaving is done at higher temperatures (~120 °C). The material inside the cans is not directly exposed to heat. The heat transfers from the sealed cans by conduction and convection. The time duration of retorting/autoclaving is dependent upon the type of food materials being canned, and also acidic or non-acidic nature of the food materials. Less heating may not sterilize the material, whereas overheating may cook the material, and hence, affects its textural quality.

Except for the consumption of large quantities of heat and can metal (usually tin metal is used), canning is an excellent process for preservation.

Sterilization (bottling): Most of the RTS beverages, nectars and juices are preserved in bottles after sterilization. The liquid materials are heated up to their boiling point (beyond 100 °C along with all additives, viz., colours, flavours, essences and emulsifying agents, etc.), and are filled into pre-sterilized bottles. Later the bottles are crown corked. By this method, the liquids can be preserved up to a period of one year. The RTS beverages are usually served in chilled conditions.

Pasteurization: Pasteurization is a thermal method by which most of the pathogenic microorganisms are destroyed, but not all. Temperatures used are less than 100 °C, and is usually of the order of 76–82 °C. Pasteurization is done quickly by raising the temperature to the desired level, and later it is cooled. The material is not generally kept at higher temperatures as is done in sterilization. Pasteurization of milk at 72 °C is a very classical example. Pasteurization has the advantage that it does not destroy the nutrients of the food as the food material is subjected to lower heating and is cooled quickly. Other microorganism which are not destroyed during pasteurization are generally kept under control by other preservative methods like refrigeration or drying, etc.

Obviously, pasteurization method can be applied only if the spoilage organisms are not very heat resistant.

Deep fat frying: Deep fat frying can also be considered as a kind of thermal processing only. Here the heat is transferred to the food material by hot oil; i.e., hot oil acts as the heat transfer medium. The solid foods in the form of flours are made into dough by adding known quantity of water, which in turn is shaped into desired forms. The wet material is fried in a hot oil which is normally at 180 °C to 200 °C. At this high temperature most of the water diffuses out of the food material, and the cavities are filled with oil. The mechanism of moisture diffusion and oil penetration during deep fat frying of *papads* was explained by Math, et al. (2004). The preservation mainly takes place because of the reduction in moisture content. Most of the snack foods and some traditional foods are prepared by deep fat frying. This technique is mostly used for processing of cereals and pulse based solid products or a combination of cereal- and pulse based-solid foods.

The deep fat fried foods contain as much as 50 per cent of oil or fat, which imparts good organoleptic characteristics to the food. Except for the consumption of oil, it is an excellent method of preservation for shorter duration, which is of the order of few days to a month. The embedded oil gets rancid over a period of time which is an ineluctable nuisance with the deep fat-fried foods. To reduce the problem of rancidity, some permissible antioxidants are added to the frying oil in few ppm level. In the recent years, a method is devised to remove some portion of oil from the food products by centrifugation. This makes the process economical as well as reduces the problem of rancidity considerably.

2.2.2 Cold Processing

The commonly applied cooling processes are refrigeration, freezing, freeze drying, etc. Some of them are given here.

Refrigeration: Refrigeration is normally resorted to storing food materials for a limited period which varies from 2–10 days. The refrigeration temperatures are usually of the order of 0 °C to 5 °C. Food materials, whether solids or liquids or pastes or doughs, can be stored either with some amount of preprocessing or in the raw form. Refrigerated preservation is generally used for retail vending of processed/unprocessed foods. Frozen foods like ice creams or frozen shrimp are transported in refrigerated vans, which operate at –18 °C. The frozen foods are also stored at –18 °C until they are transported to retail selling.

Freezing: Freezing is an excellent process for preservation of food materials without virtually losing any nutrients. The texture of the food material is also not lost during freezing. The food materials are initially preprocessed to remove most of the unwanted materials and then frozen at –18 °C to –25 °C, later they are stored at –18 °C temperature until they are transported for retail vending. Ice creams, fruit pulps, prawns, shrimps, processed meat are preserved by freezing. Particularly, the pulp of custard apple fruit is preserved by freezing alone as it is not amenable to canning.

Individual Quick Freezing (IQF) techniques are also used in which virtually each piece of the food material is individually frozen quickly. Later they are all packed in containers. Generally prawns and shrimps are preserved by IQF technique.

Except for consumption of high energy for freezing, it is an excellent method for preservation.

2.3 PRESERVATION BY DEHYDRATION

Preservation by dehydration is a very common practice. The food materials in the form of solids (like fruits, vegetables, nuts or food grains, etc.) and pastes like fruit pulps are dried by using mechanical dryers after certain amount of preprocessing like blanching, etc. The mechanical dryers operate at 60 °C–80 °C except in some cases. Hot air is generally used to dry in the convective dryers. The heat source could be electrical heating, steam, burning of agro wastes or furnace oils. Various types of dryers generally used in food processing are given in Table 2.1. Of them, the cabinet tray dryer is very popular and is shown in Figure 17.12. Further details on the dehydration along with the criteria for selection of dryers will be dealt in Chapter 17.

Table 2.1 Various Types of Dryers

1. Cabinet tray dryer	11. Pneumatic dryer
2. Tunnel dryer	12. Ring dryer
3. Bin dryer	13. Fluidized bed dryer
4. Vacuum shelf dryer	14. Moving bed fluidized dryer
5. Agitated pan dryer	15. Falling particle dryer
6. Continuous belt dryer	16. Drum dryer
7. Pneumatic conveyor dryer	17. Rotary dryer
8. Screw conveyor dryer	18. Batch freeze dryer
9. Belt conveyor dryer	19. Continuous freeze dryer
10. Spray dryer	

2.4 PRESERVATION BY CHEMICALS

A good number of chemical preservatives are used in food processing with the approval of law, and are considered as class II preservatives. They are in addition to the natural preservatives like salt and spices. Some of the chemical preservatives are: potassium metabisulphite, sodium benzoate, benzoic acid, sorbic acid, acetic acid, propionic acid, and citric acid.

Gaseous sterilants mostly used in food processing units are methyl bromide (permitted to be used under ''Controlled use'' category), ethylene oxide, propylene oxide.

Fruit acids, viz., citric acid and tartaric acid are also used as acidulants and preservatives.

The chemical preservatives act as antimicrobial agents. They prevent spoilage by eliminating the ineffective and toxinogenic microorganism from food. Table 2.2 highlights some of the chemical preservatives (Stewart and Amerine 1973).

Ethylene Diamine Tetra Acetic acid (EDTA) and its salts are especially useful in arresting the catalysis of rancidity of fats by metal ions like copper and iron. Citric acid, phosphoric acid and their salts are also used for this purpose. Antioxidative agents like Butylated Hydroxyanisole (BHA) and propyl gallate are also used in deep fat frying to reduce the problems of rancidity in the residual oils of the deep fat fried foods, as well as to protect the unused oil to a great extent for successive use. Earlier Butylated Hydroxytoluene (BHT) was also being used as an antioxidizing agent; but of late its use is banned.

The use of chemicals is generally discouraged because of possible health hazards. There is lack of information on this, and there are contradicting views on the possible health hazards. While food industry claims and boasts about the usage of chemicals because of their obvious advantages, nutritionists caution about their indiscriminate usage which is likely to cause health

Table 2.2 Chemical Additives Used in Food Processing for Antimicrobial Action

Chemical preservative	*Food products*										
	Soft drinks	*Fruit juices*	*Alcohol drinks*	*Fruit concentrates*	*Cheese products*	*Baked goods*	*Cured meat*	*Fish products*	*Salads*	*Dehydrated F and V*	*Sauerkraut (pickles)*
Ethyl benzoate	*	*	*	*		*			*	*	*
Propyl benzoate	*	*	*	*		*			*	*	*
Sorbates	*	*	*	*	*		*	*	*	*	*
Propianates					*	*					
Sulphites	*	*	*	*						*	*
Acetates						*		*			*
Diacetates						*		*			*
Nitrites							*				
Nitrates							*				
Ethylene oxide										*	
Propylene oxide										*	

hazards on continuous usage, the data on which is scanty. Some unscrupulous people using hazardous chemicals indiscriminately have created fear and distrust in the minds of the consumers. The use of chemicals for preservation or processing should not violate the basic understanding which is termed as *chemicals should not be used to conceal inferiority, to substitute for the real thing or to create human health problems.*

2.5 PRESERVATION BY EXTRUSION

Extrusion is a process in which the food material in the form of dough (usually cereal based products) is passed through a screw conveyor under high pressure. Figure 2.1 shows a typical extruder. The high pressure in the screw chamber raises the temperature to as high as a 150 °C and above, and gelatinises the starch. This also removes the moisture. The material under high pressure is passed through openings of different shapes during which period it puffs. Thus, the process offers versatility in product range and reduces the cost of production, particularly for production of snack foods in which cooking takes place by deep fat frying. Extrusion offers a special economic advantage by reducing the cost of fats. The food materials can be preserved because of both low moisture content and low fat content. The storage life of the products can be of the order of 3–4 months depending upon packaging conditions. Extrusion cooking is a HTST (high temperature and short time) process. The high temperature reduces the microbial contamination and inactivates enzymes.

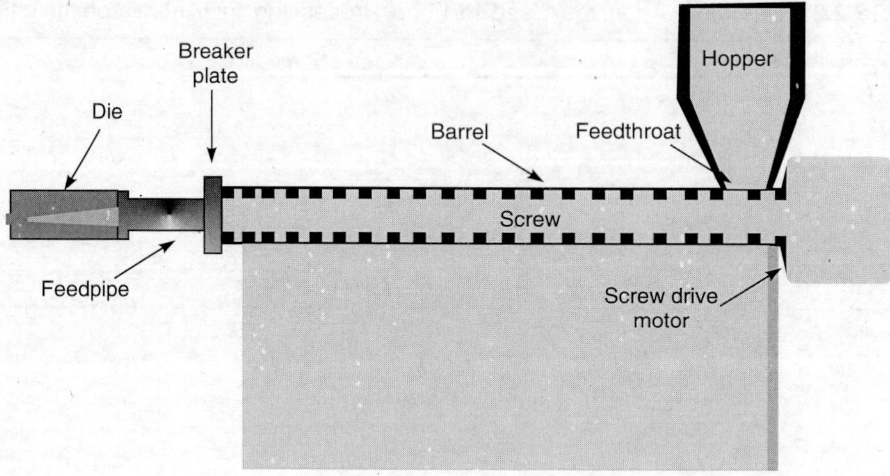

Figure 2.1 Extruder.

REVIEW QUESTIONS

2.1 What are various traditional methods of food preservation?

2.2 Explain briefly what are various thermal processing methods for preservation of food materials.

2.3 Describe dehydration as an effective method for preservation of foods.

2.4 Write a note on the use of chemicals for preservation of food materials.

2.5 Describe extrusion processing of foods.

REFERENCES

Math, R.G., Velu, V., Nagender, A. and Rao, D.G. (2004), Effect of frying conditions on moisture, fat and density of papad, *J. Food Engg.*, **64,** pp. 429–434.

Stewart, G.F. and Amerine, M.A. (1973), *Introduction to Food Science and Technology,* Academic Press, New York, p. 198.

Part II

Basic Engineering Principles

CHAPTER

3

Basic Principles

Food engineer does industrial operations in which the raw food materials have to be processed with the intention of preservation, transportation or handling. The raw material could be in the form of vapours, gases, liquids, pastes, doughs, suspensions or solids. During processing, the matcrials need to be mixed, cooled or heated, and the products need to be separated from the unwanted or undesired so that they could be consumed conveniently. The food engineer must develop, design and engineer the technology and equipment, by choosing the right raw materials of specified physical structure, operate the plant efficiently, safely and economically. At the same time, he has to ensure that the product material meets the requirements of the consumers.

Because of the necessity to operate the plant safely and economically keeping the objective of consumers' satisfaction and convenience in mind, food engineer may not be able to go strictly with scientific principles, but may have to use his judgement. Thus, *food engineering is both an art and science.* He uses the science to the extent it is required to solve his problems; he uses judgement which he gained from experience to circumvent a problem to the extent desired when science does not give a complete and satisfactory answer. Thus, *an engineer should not be dogmatic, he should be more pragmatic.*

Technologies for processing of foods have already been set in. What we are seeing today as developments in food processing are only engineering marvels. Consumer accepts any new technology only with hesitation, but he readily agrees any engineering contributions to the processing. For example, the Genetically Modified (GM) foods are not readily accepted; people have hesitation to take irradiated foods. An extruded food is readily accepted. Extrusion is an engineering marvel to translate a scientific invention.

A food technologist wants the food to be cooked, starch to be gelatinized; a microbiologist wants food to be sterilized at high temperature to avoid all pathogenic microorganisms; a nutritionist demands food should be nutritious and wholesome with as little of fat being used as possible; a consumer desires a snack food in a variety of shapes and colours; the answer for all these an engineer gives is extrusion cooking, which is a combination of high pressure physics, thermodynamics of heat transfer and chemistry of starch gelatinization with a consumer's satisfaction of fancy-shaped snack foods.

3.1 BASIC PRINCIPLES OF PHYSICS AND CHEMISTRY

Physics is defined as *the study of accurate measurements.* In addition to accurate measurements, it encompasses various physical processes like fluid flow, heat, light, sound, etc. The fluid flow in its totality has given birth to *fluid mechanics* or *fluid dynamics.* Heat is studied as heat transfer or thermodynamics, which is based on classical *Carnot cycle.* The whole study of light has given birth to optics, whereas sound to acoustics.

The modern electronics is the offspring of applied physics. Most of the physical processes or physical changes taking place in any systems are the outcome of physics; could it be unit operations in any industrial process. Unit operations encompass a whole gamut of food engineering, viz.,

- Fluid flow
- Heat transfer
- Mass transfer
  - diffusional operations
  - equilibrium processes
- Mixing operations
- Mechanical separations

Unit operations are so named because there may be a large number of operations, which all can be broken down and simplified; the individual operations have a common technique and are based more or less on the same scientific principles. For example, it could be distillation or gas absorption or liquid-liquid extraction, all of them follow the same equilibrium operations. Thus if the operations are simplified, their treatment could be unified. They are all amenable to simple mathematical expressions. If the process operations are considered at micro level, they can be reduced to simple differential equations which on integration will unify the whole processing operations. Hence, the name *unit operations.*

Chemistry is the study of molecular structure and molecular transformations. Most of the food materials are organic in nature, and they consist of carbon, hydrogen, nitrogen and oxygen in some form or the other. All the unit processes taking place in food processing are the outcome of chemistry; could it be oxidation, hydrogenation, esterification/trans esterification, dehydration or gelatinization of starch. Most of the synthetic food colours, food flavours, food acidulants or food preservatives are the outcome of chemical transformations. An understanding of chemistry is very much essential for proper comprehension of food technology. Now food chemistry has evolved as a distinct discipline in view of its importance in food processing. Indeed most of the food technologists in the olden days were chemists.

3.2 IDEAL GAS LAW AND PVT RELATIONSHIPS

Equations of State (EOS) are an important concept in engineering design for finding out the pressure/volume/temperature of gases and vapours. Based on the volume, the vessels holding the vapours can be designed.

Boyle's Law states that at constant temperature, the volume of a definite mass of the gas is inversely proportional to the pressure.

$$V \propto \left(\frac{1}{p}\right) \text{ at constant } T \tag{3.1}$$

Charles' Law (which is also known as the Gay–Lussac's law) states that at constant pressure, the volume of any gas expands by same fraction of its volume at 0 °C for every 1 °C rise in temperature. Thus,

$$V_t - V_0 = V_0 \, \alpha_t \, T \tag{3.2}$$

where V_t is the volume of gas at temperature t °C and V_0 is that at 0 °C. α_t is the coefficient of volumetric expansion. From the value of α_t which is approximately constant for all gases as 0.0033609; it leads us to conclude

$$\frac{V_t}{T_t} = \frac{V_0}{T_0} \tag{3.3}$$

where T_t is temperature in degrees Kelvin and $T_t = t + 273.16$.

Combining the above two laws with Avogadro's law which states that *equal volumes of all gases contain equal number of molecules, under same conditions of temperature and pressure* leads us to the concept of ideal gas. It is defined as the one whose molecules occupy a very negligible volume compared to the volume of the gas, and intramolecular attractions are very small. This leads to write Eq. (3.3) as follows:

$$pv = RT \tag{3.4}$$

or

$$pV = nRT \tag{3.5}$$

where p is pressure, v is molar volume, V is volume occupied by the gas, n is no. of moles of the gas, T is temperature in degree Kelvin and R is the universal gas constant.

R has different values based on the units of V and T. pV in turn has the units of energy. R is 1.9872 (if energy unit is calorie); R is 8.3144 (if energy unit is joule); R is 82.057 (if energy unit is cm^3·atm.); R is 0.082057 (if energy unit is lit · atm.) and R is 0.000082057 (if energy unit is m^3·atm.). The other values of R can be derived based on the units of p and V.

Ideal gas has also one characteristic feature that the molar volume (volume occupied by one mole of the gas) of any ideal gas at standard conditions (p = 1 atm.pr. and T = 273 °K) is same.

1 g mole occupies 22.4 lit. or 1 kg mole occupies 22.4 m^3, i.e., 2 grams of hydrogen (one mole) will occupy 22.4 litres under standard conditions and 28 grams of nitrogen (one mole) will also occupy 22.4 litres under standard conditions.

Virtually all gases and vapours can be assumed to obey ideal gas behaviour as pressure tends to zero. At least under low pressures, ideal gas law is a reasonable approximation. Ideal gas law behaviour is frequently assumed for gases and vapours for design purposes.

Subsequently, ideal gas law is modified with some approximations to account for real gases. A large number of equations of state has been proposed from time to time. One of the most important of them is van der Waals' equation of state, which accounts for the volume occupied by the real gas molecules, and accordingly the volume term is corrected. Similarly the intramolecular forces have also taken into account, and accordingly the pressure term is corrected. Thus, van der Waals' equation can be written as:

$$\left\{p+\left(\frac{a}{v^2}\right)\right\}(v-b)=RT \tag{3.6}$$

where a and b are van der Waals' constants, the values of which are available in chemical engineering literature†. For example, for air a is equal to 135.653 kNm^4/kmol., b is equal to 0.037 m^3/kmol. and R is equal to 8.3144 kJ/kmol.K.

Similarly for water vapour, they are as a is 553.667 kNm^4/kmol., b is 0.3 m^3/kmol. and R is 8.3144 kJ/kmol.K.

For the above, p is in N/m^2 and v is in m^3/kmol.

PROBLEM 3.1 The density of air at 25 °C is reported to be 1.1843 kg/m^3 at 1 atm. pressure. Compare the value with those calculated using ideal gas law and van der Waals' equation of state.

Solution Ideal gas law:

Air contains approximately 21 per cent oxygen and 79 per cent nitrogen in gaseous phase.

Volumetric composition = molar composition for gases and vapours.

Hence, composition of air = 21 per cent O_2 + 79 per cent N_2.

Molecular weight of air = 0.21 × (Molecular weight of oxygen) + 0.79 × (Molecular weight of nitrogen)

$$= (0.21 \times 32) + (0.79 \times 28)$$

$$= 6.72 + 22.12 = 28.84 \cong 29$$

One kg mole of the gas occupies 22.4 m^3 at 273 K.

∴ Volume occupied by 29 kg of gas at (273 + 25)K $= 22.4 \times \dfrac{298}{273} = 24.45$ m^3

∴ $$\text{Density} = \frac{29}{24.45} = 1.186 \text{ kg/m}^3$$

van der Waals' equation:

$$\left(p+\frac{a}{v^2}\right)+(v-b)=RT$$

The van der Waals' constants for air are: a is 135.653 kNm^4/kmol and b is 0.037 m^3/kmol.

$$\text{Energy unit} = pv = \frac{N}{m^2} \times \text{m}^3 = \text{Nm} = \text{Joule}$$

$$R = 8.3144 \text{ Joules/gmol.K}$$

$$= 8314.4 \text{ Nm/kmol. K}$$

†See Chandrasekharan, K.D. and Venketeswarlu, D. (1974), *SI Units in Chemical Engineering and Technology*, Chemical Engineering Education Development Centre, IIT Madras, p. 110.

Hence, van der Waals' equation for air can be written as:

$$\left(10132.5+\frac{135.635}{v^2}\right)\times (v-0.037) = 8314.4 \times 298 = 2{,}477{,}700 \tag{3.7}$$

In Eq. (3.7) v can be solved easily by trial and error. As a first approximation, we can take it as being equal to that given by ideal gas law (Table 3.1).

Table 3.1

v, m^3	*LHS of Eq.* (3.7), joules
24.45	2473700
24.0	2428100
25.0	2429430
24.4	2468640
24.5	2478770
24.48	2476740
24.49	2477750

Hence, the vol. of one kg mole = 24.49 m^3

$$\text{Density} = \frac{29}{24.49} = 1.1842 \text{ kg/m}^3$$

as compared to 1.1868 kg/m^3 by ideal gas law, whereas the reported density is 1.1843 kg/m^3.

3.3 MATERIAL BALANCE

Material balance, popularly known as mass balance, is a very powerful tool in the analysis of engineering systems, and for design of equipment. It is an outcome of conservation of mass, i.e., mass can neither be destroyed nor created; it can only be transformed. Generally we write material balance equation for closed systems.[†] The material balance equation states that, whatever be the material entering into a closed system or a closed vessel (Figure 3.1) will eiter get

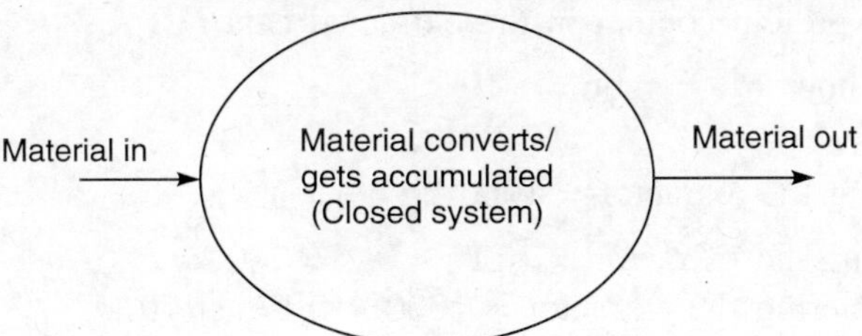

Figure 3.1 Material balance in a closed system.

converted into some useful products or gets accumulated in the system or will come out of the system unaffected. Thus, it can be written in the form of Eq. (3.8) as follows:

[†]A closed system is one which has definite boundaries.

$$\begin{bmatrix}\text{Material going into}\\ \text{the system}\end{bmatrix} = \begin{bmatrix}\text{Material converts into}\\ \text{some other products}\end{bmatrix} + \begin{bmatrix}\text{Material accumulates}\\ \text{in the closed system}\end{bmatrix} + \begin{bmatrix}\text{Material comes out of the}\\ \text{closed system unaffected}\end{bmatrix} \quad (3.8)$$

If Eq. (3.8) is written in terms of rate expressions,

$$\begin{bmatrix}\text{Rate of material}\\ \text{entering}\end{bmatrix} = \begin{bmatrix}\text{Rate of}\\ \text{conversion}\end{bmatrix} + \begin{bmatrix}\text{Rate of accu-}\\ \text{mulation}\end{bmatrix} + \begin{bmatrix}\text{Rate of material}\\ \text{leaving}\end{bmatrix} \quad (3.9)$$

Both Eq. (3.8) and (3.9) are same except that Eq. (3.9) is written in terms of rate expressions, whereas Eq. (3.8) is written in quantitative terms. The equations can be written as a whole for the entire mass or can also be written for any component in the whole mass. In either case, they hold good, i.e.,

$$M_i = M_o + \frac{dM}{dt} + r_M \quad (3.10)$$

or

$$M_{A,i} = M_{A,o} + \frac{dM_A}{dt} + r_{M,A} \quad (3.11)$$

In the above equations, subscripts, i and o stand for inlet and outlet conditions, subscript A stands for component A, dM/dt and dM_A/dt stand for rate of accumulation of M and M_A, whereas r_M and $r_{M,A}$ stand for rate of disappearance of M and component A respectively.

In case, there is any formation of component in the closed system, the conversion term will come with negative sign in the RHS or with a positive sign on the left hand side.

PROBLEM 3.2 Ground nut seeds containing 20 per cent oil and 17 per cent protein, are used for extraction of oil using hexane in a continuous extractor. The equilibrium oil content in the solvent leaving the extractor is in the ratio of 1:10. The extracted cake contains 0.5 per cent of unextracted oil. What should be the flow rate of the solvent to the continuous extraction unit processing 1000 kg of seeds per hour. The extracted cake retains 0.1 per cent of solvent with it. What will be protein content in the extracted cake?

Solution Basis = 1 hour of operation

Inlet to the extractor:

Raw seeds entering the extractor = 1000 kg

(a) Oil entering the extractor = 1000 × 0.2 = 200 kg
Protein entering the extractor = 1000 × 0.17 = 170 kg

(b) The other material = 1000 – 200 – 170 = 630 kg
Solvent entering the extractor = x kg

Outlet of the extractor:

Other material leaving the extractor = 630 kg
Protein leaving the extractor = 170 kg
Extracted cake leaving the extractor = 630 + 170 kg = 800 kg

Oil leaving the extractor along with the cake = 800 × 0.05 = 40 kg

∴ Oil extracted in the extractor = 200 – 40 = 160 kg

$$\text{Solvent required to extract this oil} = \frac{160}{0.1} = 1600 \text{ kg}$$

Solvent retained in the extracted cake = 0.1 per cent

∴ Solvent in the cake = (800 + 40) × 0.01 = 8.4 kg

Solvent entering the extractor = solvent required to extract oil + solvent retained in the cake

∴ $x = 1600 + 8.4 = 1608.4$ kg

The material balance for seeds, cake, solvent, protein and cake can be shown in Figure 3.2 and Tale 3.2.

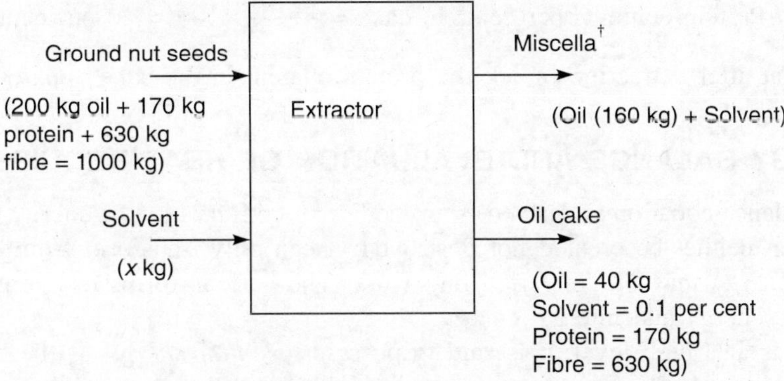

Figure 3.2 Material balance of Problem 3.2.

†Miscella–The solvent containing extracted oil is known as *miscella*. More details are available in Chapter 19.

Table 3.2

Material entering (kg)		*Material leaving* (kg)	
Oil in seeds	= 200	Oil in miscelle	= 160
		Solvent in miscelle	= 1600
Protein in seeds	= 170	Protein in cake	= 170
Fibre in seeds	= 630	Fibre in cake	= 630
Solvent	= 1608.4	Solvent in cake	= 8.4
		Oil in cake	= 40
Total	= **2608.4**	**Total**	= **2608.4**

Material balance on oil:

Oil entering = 20 per cent of 1000 kg seeds	Oil in the cake	= 40 kg
= 200 kg	Oil in the miscelle	= 160 kg
	Total	= **200 kg**

Material balance on solvent:

Solvent entering = 1608.4 kg	Solvent in miscelle (10%)	=	1600 kg
	Solvent in cake (0.1%)	=	8.4 kg
	Total	=	**1608.4 kg**

Composition of extracted cake:

Fibre in cake	=	630 kg
Protein in cake	=	170 kg
Solvent in cake	=	8.4 kg
Oil in cake	=	40 kg
Total weight of cake	=	**848.4 kg**

$$\text{Protein content (per cent) in cake} = \frac{170}{848.4}\times 100 = 20 \text{ per cent}$$

This shows that after extraction of oil, the protein content in the cake apparently increases.

3.4 ENERGY BALANCE AND EVALUATION OF HEAT REQUIREMENTS

The energy balance equation is derived from the law of conservation of energy which stipulates that energy can neither be created nor destroyed; it can only transform from one form to the other, or the energy may be converted into work. The various forms of energy are: potential energy, kinetic energy and internal energy.

If a body is at a high level, it is said to possess *potential energy*. If the body is dropped from that height, the body attains momentum, and thus, it will have some velocity. This means that the potential form of energy is getting converted to *kinetic* form of energy. If the body is heated, it receives the energy. The received energy is converted to raise the temperature of the body which we call as the *internal energy* of the body. Similarly, when steam at high pressure is entered into cylinder with piston, the pressure energy of the steam is destroyed by moving the piston; i.e., the pressure energy is converted into doing some external work by pushing the piston.

Now we can write a generalized energy balance equation for a closed system which is almost similar to Eq. (3.8). For a closed system,

$$\begin{bmatrix}\text{Total energy entering}\\ \text{the system}\end{bmatrix} = \begin{bmatrix}\text{Energy utilized}\\ \text{in the system}\end{bmatrix} + \begin{bmatrix}\text{Energy dissipated}\\ \text{from the system}\end{bmatrix} + \begin{bmatrix}\text{Energy lost out of the}\\ \text{system in the outlet}\end{bmatrix} \quad (3.12)$$

If some external work is done on the system, this will also appear on the LHS of Eq. (3.12). Equation (3.12) can also be written in terms of rate expression as follows:

$$\begin{bmatrix}\text{Rate of energy}\\ \text{entering into system}\end{bmatrix} = \begin{bmatrix}\text{Rate of utilization of}\\ \text{energy in the system}\end{bmatrix} + \begin{bmatrix}\text{Rate of energy dissipation}\\ \text{from the system to the surroundings}\end{bmatrix} + \begin{bmatrix}\text{Rate of loss of energy}\\ \text{from the system in the outlet}\end{bmatrix} \quad (3.13)$$

Energy entering the system and leaving the system are also dependent upon the quantity of the material. Thus, energy entering into the system for M_i moles or kg of the material

$$= M_i (u + E_k + E_p + pv)_i \tag{3.14}$$

where u is the internal energy per mole or kg, E_k is the kinetic energy associated per unit mole or kg of the material, E_p is the potential energy associated with unit mole or kg of the material, pv is the energy associated with mole or kg of the material entering the system in which p is the pressure and v is specific volume or molar volume.

Similarly, for the outlet stream also, we can write $M_o (u + E_k + E_p + pv)_o$ (3.15)

Thus, the generalized energy balance equation can be written as:

$$M_i (u + E_k + E_p + pv)_i + W_s = M_o (u + E_k + E_p + pv)_o + Q + \Delta E \tag{3.16}$$

where W_s is the shaft work done on the system, Q is the heat energy lost to the environment from the system and ΔE is the energy accumulated in the system. We can also write

$$h = u + pv \tag{3.17}$$

$$H = U + pV \tag{3.18}$$

where H is defined as the enthalpy and h is the molar enthalpy. It can be understood as the heat content of the system. In other words, we can say that the heat supplied to a system will result either in increasing the internal energy of the system or in doing some external work against pressure p or both. Hence, Eqs. (3.17) or (3.18) are customarily written in terms of:

$$\Delta H = \Delta u + p\,(\Delta V) \tag{3.19}$$

where ΔH stands for enthalpy change, Δu stands for internal energy change and ΔV stands for volume change.

Generally, the kinetic energy term and potential energy term are negligible. Hence, Eq. (3.16) can be written as:

$$M_i h_i + W_s = M_o h_o + Q + \Delta E \tag{3.20}$$

Under adiabatic conditions[†] $Q = 0$, under steady state conditions[‡] $\Delta E = 0$. Equation (3.20) can be simplified as:

$$M_i h_i - M_o h_o + W_s = 0 \tag{3.21}$$

Equation (3.21) can be used to calculate heat requirements or work to be done on a system for heating conditions with terms $M_i h_i$ and $M_o h_o$ indicating energy entering the system and energy leaving the system. Obviously, they are above a reference temperature.

The concept of specific heat is very useful for calculating the heat content of a system over a reference temperature, or heat required to raise the temperature of a system from an initial temperature to a final temperature. We define here the term specific heat or heat capacity,

---

[†]Adiabatic conditions mean neither heat is supplied to the system nor heat is taken away from the system.

[‡]Steady state conditions mean that the process conditions do not change with time; hence, accumulation term becomes zero.

usually denoted by C_p (to indicate specific heat capacity at constant pressure). *Specific heat* is the quantity of heat required to raise the temperature of unit mass of substance by one degree. It has the units of calories/g°C or kcal/kg°C or more appropriately in SI units J/kg°C or kJ/kg°C. It also leads us to define a unit called *calorie* which is the quantity of heat required to raise the temperature of one gram of water by one degree centigrade (usually from 14.5 °C to 15.5 °C); in SI units,

$$1 \text{ calorie} = 4.18 \text{ joules}$$

If Q is the quantity of heat required to raise the temperature of M kg of material by ΔT °C whose specific heat is C_p,

$$Q = MC_p\ (\Delta T) \tag{3.22}$$

Equation (3.22) has emanated from the definition of specific heat, and is a very useful design equation for calculating the heat requirements of the process systems.

PROBLEM 3.3 Calculate the quantity of heat required to heat 100 litre of water initially at 38 °C to a final temperature of 90 °C. The specific enthalpy of water at 33 °C is 137.8 kJ/kg and that at 90 °C is 136.8 kJ/kg (Appendix 1B).

Solution It is a simple application of Eq. (3.21)

$$M_i h_i - M_o h_o + W = 0$$

$$M_i = M_o = 100 \times 10^{-3} \times 10^3 \text{ kg/m}^3 = 100 \text{ kg}$$

Assuming that the density of water does not change considerably in the temperature range and is constant at 1000 kg/m^3

$$h_i = 137.8 \text{ kJ/kg}$$

$$h_o = 376.8 \text{ kJ/kg}$$

$$100\ (137.8 - 376.8) + W = 0$$

where W is the work done on the system or energy supplied to the system.

$$-23{,}900 + W = 0$$

$$\therefore \qquad W = 23{,}900 \text{ kJ}$$

Alternatively, the problem can also be solved by using Eq. (3.22). Assuming C_p of water does not change in the temperature range, and is constant at 1 cal/g°C.

Hence, $$C_p = 4.18 \text{ kJ/kg°C}$$

$$\therefore \qquad Q = MC_p\ (\Delta T) = 100 \times 4.18 \times (90 - 33) = 23{,}826 \text{ kJ}$$

PROBLEM 3.4 100 litre of water is heated in an insulated electrical heating kettle with 30 kW heater for ½ hour. Find the quantity of water converted to steam. The specific enthalpy of water and steam at 100 °C is, 417.54 and 2675.4 kJ/kg, respectively (Appendix 1B).

Solution The difference in the specific enthalpies of steam (h_g) and water (h_l) is equal to the latent heat of vaporization (λ)

Thus, $$\lambda = h_g - h_l = 2675.4 - 417.54 = 2257.86 \text{ kJ/kg}$$

Latent heat of vaporization is the quantity of heat required to be supplied to convert the liquid into vapour at that particular temperature.

In the present problem, the heating process goes like this:

Step I: Initially, all the 100 kg of water is heated from 25 °C to 100 °C.

Step II: Some quantity of water at 100 °C is converted to steam depending upon the quantity of heat supplied to the system.

Quantity of heat supplied is determined by the time for which the electrical heater is used.

Thus, Q = 30 kW × 1800 s = 54,000 kWs = 54,000 kJ

Quantity of heat required to heat 100 kg of water from 25 °C to 100 °C

$$= 100 \times 4.18 \times (100 - 25) = 31{,}350 \text{ kJ}$$

∴ Quantity of heat utilized for converting a part of water at 100 °C to steam at 100 °C

$$= 54{,}000 - 31{,}350 = 22{,}650 \text{ kJ}$$

Let x kg of water is converted to steam

$$\therefore \qquad 22{,}650 = 2257.86 \times x$$

$$\therefore \qquad x = \frac{22{,}650}{2257.86} = 10.03 \cong 10 \text{ kg}$$

Thus, 10 per cent of the water is converted to steam.

PROBLEM 3.5 In the above problem, instead of using an electric heater, the heat source is by condensation of steam at 0.2 MN/m^2 pressure. Calculate the quantity of steam.

Solution From Appendix 1B, the latent heat of steam at 0.2 MN/m^2 pressure is

$$\lambda_s = 2706.3 - 504.7 = 2201.6 \text{ kJ/kg.}$$

The total quantity of heat Q = 54,000 kJ

$$\therefore \qquad \text{Steam required} = \frac{54{,}000}{2201.6} = 24.53 \text{ kg.}$$

This assumes that there is no loss of heat either by radiation or by any other means.

Symbols

a: van der Waals' constant

b: van der Waals' constant

C_p: specific heat (kJ/kg°C)

E_k: kinetic energy

E_p: potential energy

H: enthalpy (J)

h: molar enthalpy (J/mole)

M: moles or mass of substance

n: no. of moles

p: pressure (N/m^2)

Q: quantity of heat (J)

R: universal gas constant

r: rate of conversion

T: temperature (K)

t: temperature in °C

U: internal energy (J)

u: molar internal energy (J/mole)
V: volume (m^3)
v: molar volume (m^3/mole)
W_S: shaft work done on the system (J)

Subscripts

A: component A
g: gas phase
i: inlet
l: liquid phase
m: whole mass
o: outlet
s: steam
t: at time *t*

Greek Symbols

λ: latent heat of vaporization (kJ/kg)
α_t: coefficient of volumetric expansion

REVIEW QUESTIONS

3.1 Describe the scope of food engineering.
3.2 Describe the application of basic principles of physics and chemistry in food processing.
3.3 What is meant by ''unit operations''? What are various unit operations?
3.4 Define Boyle's law, Charles's law and Avogadro's law.
3.5 Describe the concept of ideal gas. What is the ideal gas law?
3.6 What are the reasons for deviations from ideal behaviour of gases?
3.7 Describe van der Waals' equation of state.
3.8 Write a generalized material balance equation.
3.9 Write a generalized energy balance equation.
3.10 What is meant by specific heat of a substance?
3.11 Write the equation to find out the heat required to raise the temperature of a body by ΔT °C.

NUMERICAL PROBLEMS

3.1 The outlet flue gases coming out of a boiler has the following volumetric per cent composition at 125 °C.

CO_2 : 56
CO : 0.2
N_2 : 35
O_2 : 5
H_2O : 3.8

Find the weight of water vapour in 1 m^3 of flue gases. Assume that the gaseous mixture follows ideal gas law. (**Ans:** 22 grams)

3.2 A bakery dough contains 40 per cent moisture. 1.4 kg of dough is kept in a chamber where there is a constant air flow rate of 2 lpm containing 10 per cent moisture in the inlet air, and 22 per cent moisture in the outlet air at 30 °C. Assuming constant rate of evaporation find the moisture content in the dough after ½ hour. During this period, 2 per cent of moisture is bound to the dough and is not available for analysis. (**Ans:** 37.9 per cent)

3.3 Tomato soup with an initial solids concentration of 10 per cent is heated in an open kettle by using steam. The soup has a specific heat of 1.1 cal/g °C. 10 kg of soup at an initial temp of 28 °C is heated for one hour. The steam flow rate at 0.15 MN/m^2 pressure is 1.4 g/s. Only the latent heat of vaporization of steam alone is available for heating. Find the final concentration of tomato soup assuming that the soup boils at 104 °C. (**Ans:** 15.25 per cent)

3.4 Canning of peas is carried out in brine solution which has specific heat 1.05 cal/g°C, and specific gravity 1.0. The can is filled with 800 ml of hot brine at 100 °C into which 100 g of peas at 28 °C is added. Assuming that there is no heat loss, find the equilibrium temperature if the specific heat of peas is 1.4 cal/g°C. (**Ans:** 89.7 °C)

CHAPTER

4

Steam Generation and Utilization

Steam has a very specific role to play in food processing. It is mainly used as a source of heat in thermal processing of foods. One of the best advantages with steam is that the process heat temperature can be adjusted and controlled upto 150 °C without much pressure of the steam. Mostly we require the temperatures of processing approximately 110–125 °C as it is only desired to vaporize the moisture from food stuffs, or to sterilize the food stuffs. Hence, steam is a very convenient heating source. The advantages of using steam can be summarized as follows:

- Since the temperatures are moderate with the usage of steam as heating sources, charring of the food materials does not normally take place.
- The temperatures of processing can be adjusted conveniently upto 150 °C without resorting to much higher steam pressures (0.5 MN/m^2 which is equal to 5 atm. pr.).
- Since steam is pure moisture, any chances of contamination of steam with food materials do not affect the quality of food material.
- In case of canning operations, exhausting can be done with steam alone.
- Since the properties of steam are well studied and documented in the form of steam tables, design of equipment is easy.

Concept of normal boiling point: Let us take any liquid in a container with the top surface open to atmosphere [Figure 4.1(a)]. The liquid vaporizes slowly depending upon the temperature, and the vapours mix with the surrounding air. Thus, slowly the liquid evaporates. Instead, if the vessel is closed with a lid as shown in Figure 4.1(b), the liquid slowly vaporizes and exerts certain vapour pressure above the surface of the liquid. This would raise the pressure of the vapours, and they slowly start condensing back into liquid state. Thus, the vapours and liquid exist in a state of dynamic equilibrium. Now let us see, what happens if the liquid is heated from the bottom as shown in Figure 4.1(c). The liquid starts exerting more pressure in the vapour phase. That is, the vapour pressure exerted by the liquid increases. However, the lid is not displaced as the atmospheric pressure is acting on the top of the lid. If the temperature is further increased, a stage comes where the vapour pressure exerted by the vapours is equal

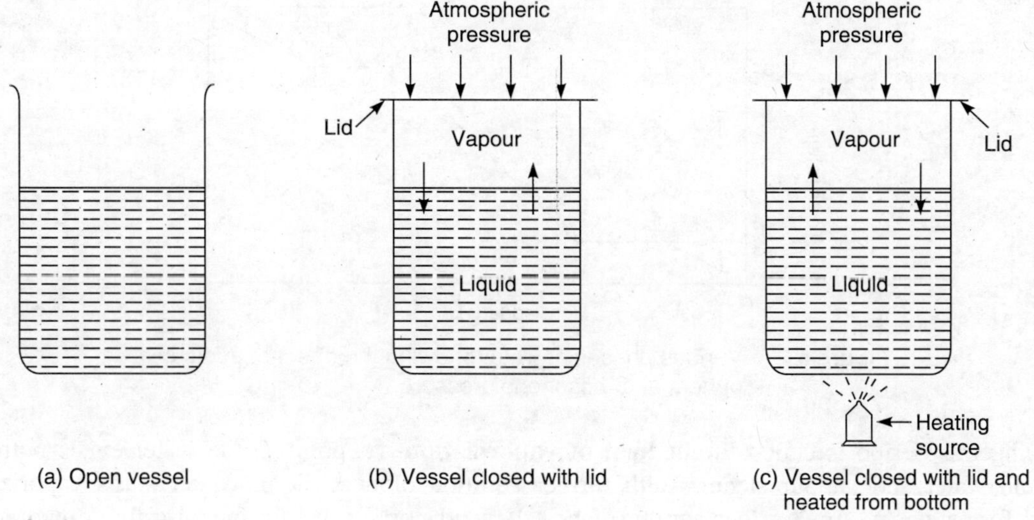

Figure 4.1 Vapour and liquid existing in a state of dynamic equilibrium.

to the atmospheric pressure, and hence, the lid is displaced. The temperature of the liquid corresponding to this stage is called the **normal boiling point**. Thus, the normal boiling point of a liquid is defined as *that temperature of the liquid at which the vapour pressure exerted by the vapours is equal to the normal atmospheric pressure.* This is the process of boiling which is applicable to any liquid including water.

4.1 PROPERTIES OF STEAM

Steam is obtained by boiling water. In fact, water exists in all the three phases, viz.,

- **Solid state (ice):** The temperature is normally below or equal to 0 °C at normal atmospheric pressure.
- **Liquid state (water):** The temperature is normally between 0–100 °C at atmospheric pressure.
- **Vapour state (steam):** The temperature is normally equal to or above 100 °C.

On further heating, the steam forms superheated steam, which almost behaves like a hot gas. The process of formation of steam is represented in Figure 4.2.

Ice at point P on heating increases the temperature and reaches a value of 0 °C. At this temperature, any further supply of heat does not increase the temperature, but builds up heat in the ice. When a quantity of approximately 303.8 kJ of heat is added to one kg of ice, ice starts changing the phase from solid state to liquid state at point R as in Figure 4.2. The heat supplied during QR does not manifest in raise in temperature and is remaining latent in the ice, this quantity of heat is called **latent heat of fusion**. At point R, water exists both in solid state and liquid state. From point R onwards, any further addition of heat would raise the temperature upto point S at 100 °C. At point S, water exists in liquid state. From this point onwards, any heat supplied will be latent in the liquid upto point T. The quantity of heat supplied between points S and T is equal to approximately 2257 kJ per kg of liquid water. The heat supplied

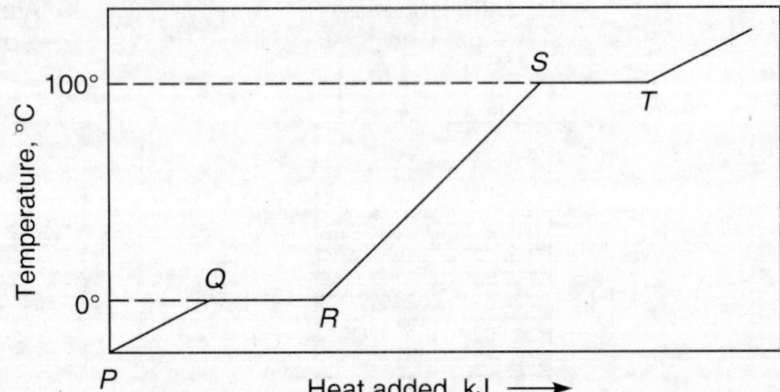

Figure 4.2 Various phases of water showing the effect of addition of heat at atmospheric pressure.

during this period is called **latent heat of vaporization**. At point *T*, the water exists both as liquid water and vapour steam. With further addition of heat, liquid water starts vaporizing and forms steam. Any further supply of heat beyond point *T* would increase the temperature of steam, and makes it a saturated steam, and still further addition of heat makes steam a superheated steam.

4.1.1 Pressure–Enthalpy Diagram

The transformation of water from liquid state to vapour state to form steam can also be well represented thermodynamically by drawing a plot of specific enthalpy and pressure relationships. The pressure–enthalpy diagram is shown in Figure 4.3, which is a bell-shaped curve. It essentially contains three zones

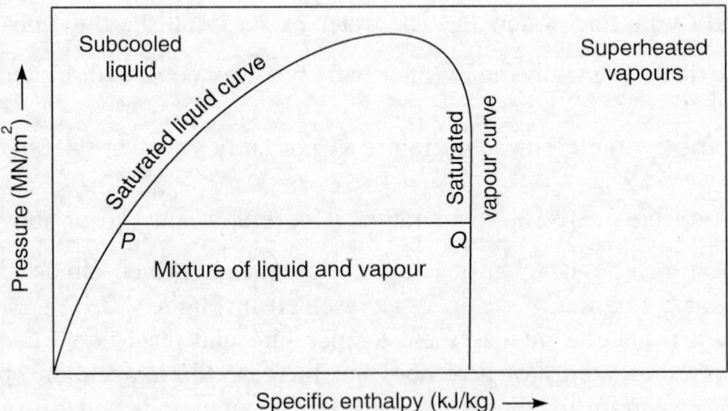

Figure 4.3 Pressure–enthalpy diagram for steam–water system.

(i) On the left-hand side of the bell is the subcooled liquid.

(ii) On the right-hand side of the bell is the saturated vapour which is superheated steam.

(iii) Under the bell is a system which consists of a mixture of liquid and vapour, i.e., water and steam.

The bell curve boundaries indicate saturated liquid curve and saturated vapour curve. Let us take any two points *P* and *Q* on the curve. The point *P* indicates the position of a saturated liquid, i.e., if point *P* is at 100 °C, it indicates water in the liquid state at 100 °C. Any further addition of heat does not raise the temperature, but we get a mixture of liquid water and vapour steam until we reach point *Q*, which is also at 100 °C. At point *Q*, the liquid disappears, and we have saturated steam at 100 °C. Any further addition of heat raises the temperature at the same pressure, and makes super-heated steam. Further details of Figure 4.3 can be found in Singh and Heldman (2001).

Similarly if we draw a similar plot on volume–pressure data, we get a bell-shaped curve as shown in Figure 4.4. At point *P*, water is in liquid state. Further addition of heat at this point at constant pressure increases the volume of the system upto point *Q* where whole water converts to steam. The diagram is essentially important in drawing our attention to point out that how the volume increases significantly as the liquid water vaporizes to steam at constant pressure.

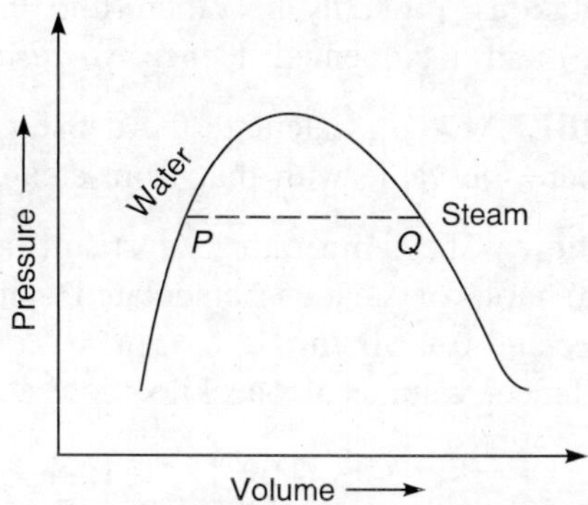

Figure 4.4 Pressure–volume diagram for water–steam system during phase change at constant temperature.

4.1.2 Dryness Fraction of Steam

As we have seen in Figure 4.3, the steam and water mixture under the bell-shaped zone is a mixture of steam and water. If we are close to RHS (Right Hand Side) of the bell, we will be having mixture which is more in steam and less in water. Similarly, if the mixture is close to LHS (Left Hand Side) of the bell, it will be richer in water and poorer in steam. The extent to which the phase change has progressed towards the steam zone is defined as the **steam quality**. The dryness fraction (x) is the quantity of steam (W_v) divided by the total quantity of steam (W_v) plus the quantity of water (W_l) present in it. Thus,

$$x = \frac{W_v}{W_v + W_l} = \frac{W_v}{W} \tag{4.1}$$

For a dry steam, $x = 1$ since $W_l = 0$. It is the region on the RHS of the bell (Figure 4.3).

4.2 STEAM TABLES AND THEIR APPLICATION

The data in Figure 4.3 are also presented in the form of a table and is popularly known as **steam tables**. The data are shown in Appendix 1 (Keenam, et al., 1969). The tables show a relationship between temperature and pressure for the liquid water. The data indicate that water can boil at any temperature provided we maintain corresponding pressure (vacuum). The tables also show the data on specific enthalpy, specific entropy and specific volume of the water and vapour. Therefore, obviously the difference in the specific enthalpies of vapour and liquid states gives the latent heat of vaporization at that temperature or pressure. Thus, we find at 100 °C,

$$h_g = 2676 \text{ kJ/kg}$$
$$h_l = 419 \text{ kJ/kg}$$

The difference 2257 kJ/kg is the latent heat of vaporization of water at 100 °C, which is equal to 540 cal/g.

Since the data of steam related to pressure, temperature, specific heat, specific enthalpy, specific entropy and specific volume are frequently required in process calculations and process operations, these data have been experimentally evaluated for steam and tabulated. These data are known as **steam tables**. These data are frequently used by mechanical engineers in process calculations. Probably no other fluid in scientific literature is as much studied as water (and steam) and documented. Hence, *steam tables* are unique in that respect.

PROBLEM 4.1 Calculate the density of steam at 0.5 MN/m^2 pressure by ideal gas law, and compare the value with that from steam tables.

Solution The temperature of steam corresponding to 0.5 MN/m^2 is 151.84 °C. Volume of 1 kilo mole of steam is calculated using ideal gas law. One kilo mole occupies 22.4 m^3 at 273 K and 0.1 MN/m^2.

Hence, volume of one kilo mole at 151.84 °C is:

$$\frac{22.4 \times (151.84 + 273)}{273} = 34.87 \text{ m}^3 \text{ at } 0.1 \text{ MN/m}^2$$

$$\text{Volume at } 0.5 \text{ MN/m}^2 = 34.7 \times \frac{0.1}{0.5} = 6.97 \text{ m}^3$$

One kilo mole of steam weighs 18 kg

$$\therefore \quad \text{density} = \frac{18}{6.97} = 2.57 \text{ kg/m}^3$$

Specific volume of steam from steam tables = 0.3818 m^3/kg

$$\therefore \quad \text{density} = \frac{1}{\text{Specific volume}} = \frac{1}{0.3818} = 2.62 \text{ kg/m}^3$$

PROBLEM 4.2 A 2 m^3 capacity autoclave is filled with saturated steam upto a pressure of 0.5 MN/m^2. Later the autoclave is drained off. Find the quantity of condensate. What is the quantity of heat released in the autoclave.

Solution The specific volume of steam at 0.5 MN/m^2 = 0.3818 m^3/kg

$$\therefore \quad \text{Quantity of steam entered into 2 m}^3 \text{ autoclave} = \frac{2}{0.3818} = 5.24 \text{ kg}$$

The steam condenses to give condensate

Hence, quantity of condensate drained = 5.24 kg

Enthalpy of steam at 0.5 MN/m^2 = 2747.5 kJ/kg

Enthalpy of water at atm. pr. = 417.54 kJ/kg

Heat released by kg of steam = 2747.5 – 417.54 = 2330 kJ

∴ Total quantity of heat realized = 2330 × 5.24 = 12,209 kJ.

PROBLEM 4.3 Find the enthalpy of steam (1 kg) which is 85 per cent dry and is at 0.15 MN/m^2 pressure.

Solution 85 per cent dry steam contains 0.85 kg of steam + 0.15 kg of water.

Specific enthalpy of water at 0.15 MN/m^2 = 467.13 kJ/kg

Specific enthalpy of 85 per cent dry steam = (0.85 × 2693.4) + (0.15 × 467.13) = 2359.5 kJ/kg

4.3 BOILERS

Generation of steam in process operations is a very common activity to use as a heating medium or to participate in process operations. The equipment used for generation of steam is known as **boiler**. Even though the word boiler literally means one that boils, invariably the boiling medium is conceived as water, the vapour is steam, and the generator of steam is boiler. The boiler essentially consists of two zones.

- Heat generation unit by burning/combusting the fuel.
- Water heating unit for generation of steam.

Thus, the boiler is expected to

(a) generate steam in a safe environment,
(b) utilize the heat from the fuel to the maximum possible extent,
(c) have adequate provision to minimize the heat loss from the heat generation source,
(d) contain minimum number of joints and bends in view of the safety considerations, and
(e) have an automatic control provision to regulate the heating source based on the output demand of steam.

4.3.1 Classification of Boilers

Classification of boilers is based on various factors. However, it is to be noted that the classification is for the sake of convenience only, even though all of them generate heat. The classification is shown in Figure 4.5.

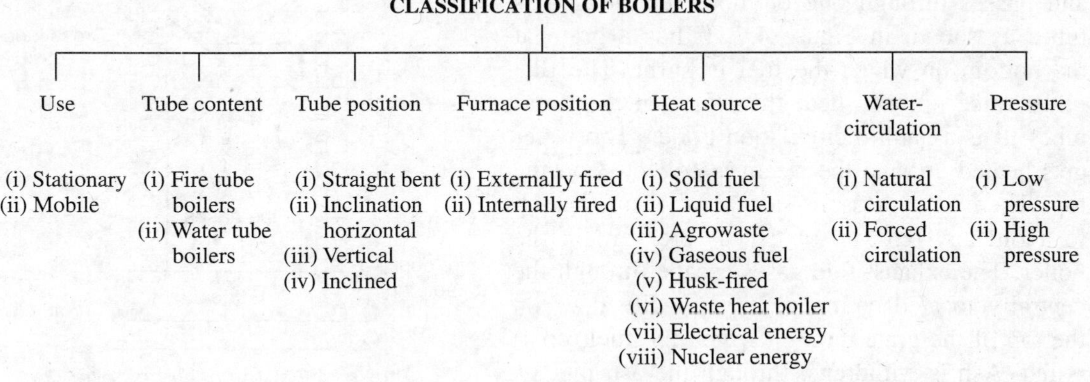

Figure 4.5 Classification of boilers.

4.3.2 Types of Boilers

Based on the classification shown in Figure 4.5, there are various types of boilers as follows:

(i) Locomotive boilers—straight, horizontal, mobile.
(ii) Cornish boilers—low pressure, single horizontal tube.
(iii) Lancashire boilers—stationary, fire-tube, horizontal straight tubes, internally-fired and natural circulation.
(iv) Vertical Cochran boiler.
(v) Babcock–Wilcox boiler—natural circulation.
(vi) Stirling boiler.
(vii) Lamont boiler—high pressure, forced circulation, water-tube.
(viii) Velox boiler.
(ix) Benson boiler—forced circulation, single water-tube, high pressure.

Some of them will be described in the following subsections briefly. A detailed discussion on boilers is available in any standard textbook on thermal engineering (Ballaney 1984).

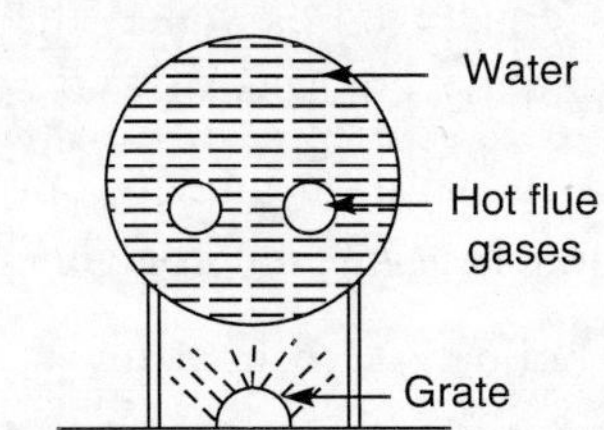

Figure 4.6 Lancashire boiler.

Horizontal tube boiler: For a typical horizontal tube boiler, we describe Lancashire boiler. It consists of two horizontal tubes through which hot flue gases pass, and water is contained outside the hot tubes (Figure 4.6). The firing grate† is at the bottom on which the fuel burns. The hot flue gases pass through the fire tubes.

Vertical tube boiler: It is positioned in a vertical position. The water is around the heating zone, and passes through one or more inclined cross tubes as shown in Figure 4.7. It has a grate at the bottom on which the fuel is burnt. The flue gases raise up, and heat the water in the cross tubes. It is a natural circulation boiler. The water gets heated up, and the steam is collected on the top. There is a pressure safety valve on the top to avoid any built-up of excess pressure in the boiler. The exhaust flue gases escape through the centrally located chimney. It has a fire door on the top of the grate through which the fuel (coal) is fed. Ash is withdrawn through the ash pit.

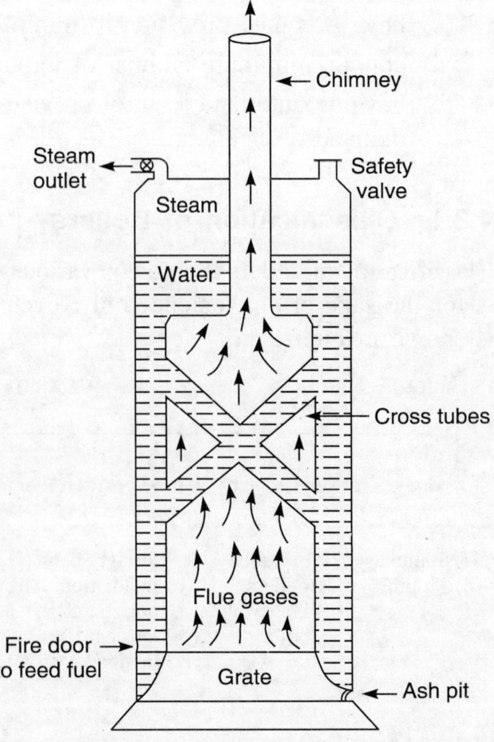

Figure 4.7 Vertical tube boiler.

†The place where fuel is burnt is known as **grate**.

Multitubular fire tube boiler: Figure 4.8 shows one typical multitubular, horizontal, fire tube boiler. It consists of a large number of tubes through which the hot flue gases pass. Water surrounds the hot tubes. The steam forms on the top. There is a provision for the steam to come in contact with the hot flue gases so that some superheated steam can be produced. Multitubular fire tube boilers are similar to Shell-and-Tube Heat Exchangers (See Figure 12.5 and Figure 12.6).

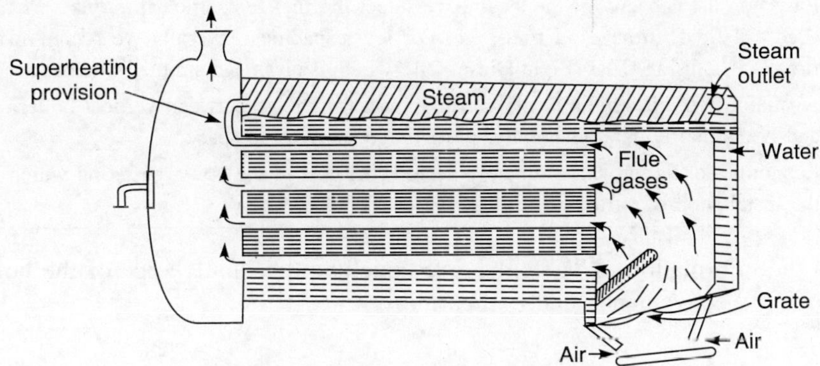

Figure 4.8 Multitubular fire tube boiler.

Multitubular water tube boiler: Babcock–Wilcox boiler is a typical multitubular water tube boiler. A schematic diagram is shown in Figure 4.9. The inclined tubes contain water, and hot water moves by natural circulation. Baffles are provided to deflect the hot flue gases for their effective utilization.

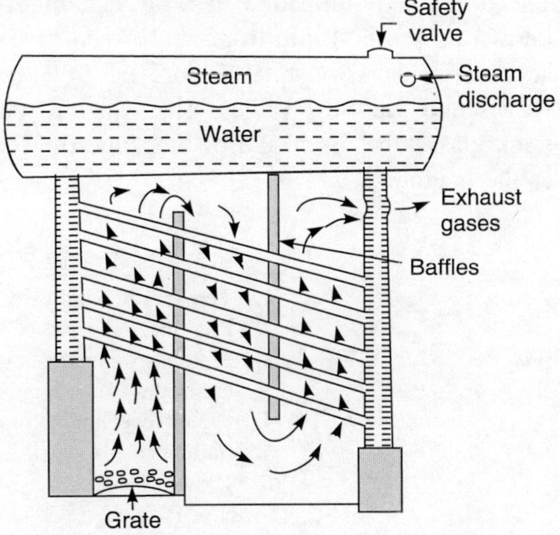

Figure 4.9 Multitubular water tube boiler.

4.4 CRITERIA FOR CHOICE OF A BOILER

The boiler should be able to generate adequate quantity of steam at the desired steam pressure as early as possible. The economics of the boiler in terms of its initial cost, operating costs should be lowest. The heat lost from the boiler surface to the surroundings should be least. Hence, we need to make adequate quantity of brickwork. The boiler should be versatile, and should be able to meet the steam load requirements. The boiler should be constructed with adequate number of the accessories so that the operation of the boiler should be safe and trouble-free. The boiler should be mounted in such a way that it should be easily approachable for repairs and maintenance. The joints, bends, elbows and valves should be as minimum as possible.

Considering the above points, the criteria for the selection of a boiler for any heating purpose are given in Table 4.1.

Table 4.1 Guidelines for the Choice of a Boiler

The capacity of the boiler (kg of steam/hour) should be based on the highest load requirements of the process.
The pressure of the steam is based on process conditions and temperature of the process requirements.
Based on the capacity of the steam, we select the fuel for combustion, viz., coal fired, oil fired, furnace oil fired, etc. For low capacities generally we prefer oil fired boilers using LDO (Light Diesel Oil), which is clean and simple in operation.
Availability of space is a prime consideration for choice between vertical boilers and horizontal boilers. Generally, the latter occupy more space.
Economics of the project ultimately decides the cost of the boiler based on which the installation of various safety gadgets is decided.

Based on the information in Table 4.1, we conclude the final choice of the boiler is based on: cost, capability, convenience and compatibility.

4.5 MAINTENANCE OF A BOILER AND BOILER HOUSE

The boiler is equipped with a large number of accessories for safe operation. They are also known as **boiler mountings** and given in Table 4.2. Routine maintenance of all these boiler accessories is a must, as failure of any one of them could be fatal, particularly the safety valves and blow-off valves. The water level indicator is usually made out of a high pressure thick glass tube. Breakage or blockage of it may result in spilling of hot water causing injuries to the operators.

Table 4.2 Boiler Accessories/Mountings

1. Temperature indicators	7. Steam valves
2. Pressure gauges	8. Water softener
3. Pressure indicator-cum-controller	9. Water feed valve
4. Water level indicators	10. Air-inlet valve
5. Safety valves	11. Fire put-off horn
6. Blow-off cocks	12. Steam line

All the pipe lines, valves, bends and elbows in the steam line should be thoroughly checked to avoid any blocking. On many occasions, the scale formation in the pipe lines may cause blockages which results in explosion. The scale formation in the water tubes or outside of the flue-gas tubes will result in loss of heat transfer coefficient,† since the scale formation is responsible for an additional resistance to the heat transfer. Supply of softened water to the boiler should be ensured. For this purpose, the routine maintenance of the water-softener unit should be undertaken at regular intervals.

†The description of heat transfer coefficient will be dealt in Chapter 11.

The boiler consists of a continuous flame onto which the fuel oil is spread. The spray of fuel oil burns and generates heat which in turn is used to heat water. For any reason, if this flame is put out, we have a continuous flow of fuel oil without being burnt; to avoid this kind of a situation, the boiler is provided with a horn system which gives alarm whenever the flame is put out.

The boiler is provided with an air inlet through an air filter. Frequently the filter is chocked/ blocked in which case, the air flow rate for combustion gets reduced which results in poor performance of the boiler. Hence, the cleaning of the air filter should be taken up routinely.

In addition to the routine maintenance of the boiler accessories/mounting, it is also equally essential to maintain the *boiler house*. The boilers are generally housed at a far off place from the processing hall, and the steam is transported through steam lines. Hence, the boiler house maintenance needs a special mention. The house is normally dirty with the ash that comes out of the boiler or with the spillover of furnace oil, or other oils. They need to be properly maintained. A pump is used for pumping the fuel oils, usually the pumps are reciprocating pumps and consist of V-belts. These pumps are in the vicinity of hot temperatures in the boiler house, and hence, are susceptible to failure or spoilage. The coal fired boilers are fed with coal, and the fly-ash that comes out of the grate needs to be regularly removed and cleaned. These operations are usually done manually by operators, and hence, is susceptible to human error. Thus, the boiler house needs special care and maintenance.

4.6 UTILIZATION STEAM IN FOOD PROCESSING

The utilization of steam in food processing operations is versatile and multifarious. They may be for the following varieties of operations:

(i) Thermal processing of food materials.
(ii) Drying of food grains/solid foods/pastes.
(iii) Sterilization of the foods.
(iv) Sanitization of process equipment.

We will see some of them in brief details.

4.6.1 Thermal Processing

It includes the following:

(i) Canning of fruit pulps, fruit slices, meat, fish or chicken products.
(ii) Blanching of fruits, vegetables, etc. before going for process operations or drying.
(iii) Concentration of fruit juices or fruit pulps to make concentrates.
(iv) Evaporation of sugar juice for making crystal cane sugar.
(v) Autoclaving of food products for making products like *idlis*, etc.
(vi) Autoclaving/retorting in canning operations.
(vii) Cooking of food stuffs in steam heated jacketed vessels.
(viii) Cooking of fruit pulps with all ingredients to make fruit jams/jellies.
(ix) Roasting of grains for puffing or for generation of flavour.

(x) Gelatinization of starch in the grains for cooking purpose.
(xi) Heating of food materials or water for processing operations.
(xii) Concentration of milk or cane sugar juice in open pans to make *khoa* or jaggery, respectively.

4.6.2 Drying

(i) Drying of cereals, pulses or millets or minor food grains to reduce the moisture content to improve the shelf-life of the grains.
(ii) Drying of fruit juices/pulps to make fruit bars and fruit toffees.
(iii) Vacuum drying of fruit juices to make powders.
(iv) Drum drying of pasty foods to make baby foods, weaning foods and RTE (Ready To Eat) foods.
(v) Fluidized bed drying of grains.
(vi) Pneumatic drying of coconut grits to make dessicated coconut powder.
(vii) Vibratory drying of wet solid grains.
(viii) Rotary drying of wet grains or seeds including oil seeds to bring down the moisture content for safe storage.

4.6.3 Pasteurization/Sterilization

(i) Sterilization of the media for biochemical operations.
(ii) Pasteurization of milk for preservation.
(iii) Sterilization of fruit juices or milk for aseptic packaging.
(iv) Sterilization of cans, bottles and other packaging materials before filling.

4.6.4 Sanitization of Process Equipment

(i) Sterilization of all the dairy equipment with live steam after processing.
(ii) Sterilization with live steam the process line and accessories like valves for sanitizing the process equipments.
(iii) Sterilization of the individual gadgets used in the processing operation like spoons, ladels, stirring equipments, etc.
(iv) Sanitization of the entire process equipment with live dry steam used for processing of animal-based products.
(v) Sterilization of the entire process equipment with live steam after processing is over in biochemical processes.

Symbols

h: specific enthalpy (kJ/kg)
x: Dryness fraction
W: Quantity (kg)

Subscripts

l: liquid phase (water) *v*: vapour phase (steam)

REVIEW QUESTIONS

4.1 Describe the importance of steam in food processing.

4.2 Describe the process of generation of steam from water.

4.3 Describe the pressure–enthalpy diagram for water.

4.4 What is meant by steam tables? Describe the importance of steam tables in process design.

4.5 What is meant by dryness fraction of steam?

4.6 What is meant by a boiler? Describe the characteristics of a good boiler.

4.7 How do you classify the boilers?

4.8 Describe the working of a horizontal tube boiler with a neat diagram.

4.9 Describe a vertical tube boiler.

4.10 What are the criteria you follow while choosing a boiler for a food processing operation?

4.11 What are various components of a boiler?

4.12 Describe the importance of maintenance of a boiler.

4.13 Narrate various applications of steam in food processing.

NUMERICAL PROBLEMS

4.1 An autoclave is used to cook 500 idlis each weighing approximately 50 g. The batter contains 60 per cent moisture and the cooked idli contains 40 per cent moisture. Calculate the quantity of steam to be supplied to the autoclave from a boiler operating at 0.1 MN/m^2 gauge. For want of data the specific heat of idli batter is taken as that of water, and idli takes another 10 per cent extra heat for gelatinization of the solids (cereal and pulse). Assume a total of 20 per cent heat loss from the autoclave to the surroundings, and batter is initially at 28 °C. (**Ans:** 10.85 kg)

4.2 A small 1.2 litre flask is heated with an electrical coil wound outside for sometime. The total energy supplied is 2.3 kJ. The flask contains 1.0 g of water at 28 °C initially. Find the dryness fraction of the steam in the flask when the pressure inside the flask went upto 0.15 MN/m^2. Assume that there is no heat loss to the surroundings. (**Ans:** 0.82)

REFERENCES

Ballaney, P.L. (1984), *Thermal Engineering*, 13th ed., Khanna Publications, New Delhi.

Keenam, J.H., Keyes, F.G., Hill, P.G., and Moore, J.J. (1969), *Steam Tables: Thermodynamic Properties of Water Including Vapour, Liquid and Solid Phases*, John Wiley & Sons, Inc., New York.

Singh, R.P. and Heldman, D.R. (2001), *Introduction to Food Engineering*, 3rd ed., Academic Press, Inc., p. 176.

CHAPTER

5

Refrigeration

Thermal processing plays a major role in preservation of food materials. It could be by heating or cooling.

Cold storage of food materials is known as **refrigeration technique**. Food materials are cooled much below the Room Temperature (RT) which is probably of the order of 4 °C or 0 °C or even –18 °C. Refrigeration techniques are used for preservation of mostly fruits, vegetables, meat and meat products, marine produce, milk and dairy products, enzymes used in food processing.

That is, in a way we can say that refrigeration technique is mostly used for preservation of those categories of food items which are spoiled by microbial attack and/enzymatic degradation. The effect of refrigeration (or low temperatures) is obvious in reducing the spoilage of foods. Spoilage is a biochemical reaction. Low temperatures reduce the rates of biochemical degradation. That is how the food materials are preserved. As is already mentioned, the biochemical degradation can only be reduced, but not avoided. The degradation/deterioration rates are reported to reduce by half for every 10 °C lowering in temperature (Singh and Heldman 2001).

5.1 REFRIGERATION AND BASIC CONCEPTS

Refrigeration is a process of pumping of heat from a high temperature source to a lower temperature sink. This process which otherwise would have been contradictory to the second law of thermodynamics, is achieved by employing a compressor to compress and expand a material which is known as **refrigerant**. The refrigeration materials (refrigerants) have a very peculiar property of changing the phase (gas to liquid and vice versa) with changes in pressure of the system as a whole. A vapour-compression refrigeration cycle is shown in Figure 5.1. The refrigerant in vapour phase entering at h into the compressor at temperature T_4 and pressure p_2, is compressed, and it leaves the compressor at a. The pressure and temperature at a are T_1 and p_1 which are higher than T_4 and p_2. The phase changes to liquid phase. At these conditions, the refrigerant enters the condenser where heat is removed by air cooling or by circulating cold water. Hence, the temperature falls to T_2 which is less than T_1, but pressure remains constant at p_1 and in liquid phase. The refrigerant enters an expansion valve at d which is at higher pressure p_1 and is almost at room temperature T_2. It gets expanded in the expansion valve and leaves at e, where its pressure falls to p_2, and the phase changes to vapour phase. The phase change takes place with absorption of heat, which in turn results in lowering the temperature to T_3. The lowering of temperature in the expansion value is due to *Joule–Thomson effect*

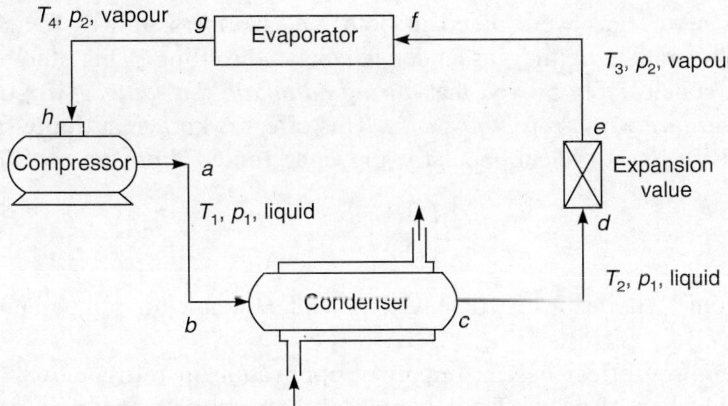

Figure 5.1 Schematic representation of a vapour-compression refrigeration cycle.

The refrigerant enters the evaporator (refrigeration chamber) at f at lower temperature T_3 and almost at atmospheric pressure p_2, absorbs the heat. The refrigerant leaves at g at which the temperature raises to T_4, but pressure remains at p_2 in vapour phase. Thus, the refrigeration effect is realized in the evaporator. The vapour refrigerant at T_4 and p_2 enters the compressor at h. The cycle continues. The process can be represented on a *p-v* diagram as shown in Figure 5.2.

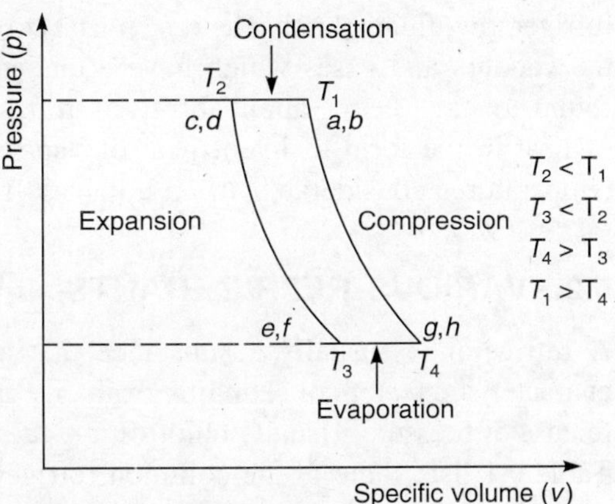

Figure 5.2 *p-v* diagram for the refrigerant in the refrigeration cycle.

5.1.1 Joule–Thomson Effect

Joule–Thomson effect was an important contribution for the invention of refrigeration. This is the outcome of a few classical experiments performed by J.P. Joule, using a very crude experimental setup during 1850's. The experimental setup consisted of two chambers immersed in a pool of water and interconnected with a valve. One chamber was filled with a gas under high pressure. Later the gas was expanded into second chamber by opening the valve. This experiment did not show any change in the temperature nor in the volume. One of the major reasons for non-observance of any conceivable change was due to the inaccuracy of the crude experimental setup. The study showed that neither the internal energy nor the temperature of the system changed inspite of the fact that some changes were brought about in the system.

The significant contribution of this experimental study was its finding that the internal energy of a gas was a function of temperature alone (Jordan and Priester 1971).

Later the experimental setup was modified in collaboration with Sir William Thomson (in the year 1852) into a two thermally insulated chambers with a porous plug. Both the chambers were provided with a provision for measuring temperature and pressure. A real gas was entered into one chamber and it was throttled into the second chamber through the plug. A constant pressure differential was maintained to allow for the throttling of the gas into the next chamber. The temperatures of the gas were noted in both the chambers. It was found that there was a reduction in temperature of the gas under adiabatic throttling. This study and subsequent experiments have conclusively proved that *during adiabatic throttling of a real gas, there was a change in temperature with drop in pressure.* This effect is known as **Joule–Thomson effect**.

We can also define a coefficient popularly known as **Joule–Thomson coefficient** as follows:

$$\left(\frac{\partial T}{\partial p}\right)_H = \mu \tag{5.1}$$

where μ is the Joule–Thomson coefficient. T, p and H stand for temperature, pressure and enthalpy respectively.

The Joule–Thomson effect has an important application in refrigeration. The observation was used for liquefaction of gases. The lowering of temperature of any real gas due to sudden and adiabatic expansion has stemmed the idea for achieving the refrigeration effect. Subsequent research was concentrated on finding out different gases and vapours which can be used as refrigerating fluids (known as *refrigerant*) to achieve better lowering in temperatures. Particularly, the vapours and gases which have a tendency to change the phase by change in pressure, are found to have better applicability as refrigerating gases (refrigerants) because the absorption of heat in the form of latent heat of vaporization during phase change would help reduce the temperatures considerably in an adiabatic system.

5.2 VARIOUS REFRIGERANTS

A refrigerant is usually a substance that absorbs heat on expansion, and it should have the characteristic feature of changing from one phase to other (liquid to vapour phase) on expansion or release of pressure. Usually chlorofluorocarbons, ammonia are some of the common refrigerants. Table 5.1 lists some of the common refrigerants traditionally used. In addition to them, of late, a good number of new refrigerants is being identified which do not contain fluorocarbons in view of the recent concern for environment. For example *R*-134a is one new refrigerant that shows considerable promise in refrigeration.

Table 5.1 Some of the Common Refrigerants

Refrigerant no.	*Chemical name*	*Chemical formula*
R-11	Trichlorofluoromethane or freon-11	CCl_3F
R-12	Dichlorodifluoromethane or freon-12	CCl_2F_2
R-22	Chlorodifluoromethane or freon-22	$CHClF_2$
R-30	Methylene chloride or Dichloromethane	CH_2Cl_2

(*contd.*)

Table 5.1 Some of the Common Refrigerants (*contd.*)

Refrigerant no.	*Chemical name*	*Chemical formula*
R-40	Methyl chloride or Chloromethane	CH_3Cl
R-114	1,2-dichloro-1,1,2,2-tetrafluoroethane	$C_2Cl_2F_4$
R-134a	1,1,1,2-tetrafluoroethane	CF_3CH_2F
R-500(Azeotrope)	73.8% CCl_2F_2 (R-12) + 26.2% CH_3CHF_2 (R-152a)	
R-502(Azeotrope)	48.8% $CHClF_2$ (R-22) + 51.2% C_2ClF_5 (R-115)	
R-717	Ammonia	NH_3
R-744	Carbon dioxide	CO_2

The basic desirable characteristics of a good refrigerant are listed in Table 5.2. The performance characteristics of most of the refrigerants are compared at the evaporating temperature of –15 °C and condensing temperature of 30 °C.

Table 5.2 Desirable Characteristics of a Good Refrigerant

It should be non-toxic.
It should be non-poisonous.
It should be non-explosive.
It should be non-inflammable.
It should be non-corrosive.
It should be a stable gas (chemically stable).
Leakage of it should be easily detectable.
It should not cause damage to environment when leaked.
It should be a low boiling liquid.
It should have high latent heat of vaporization.
The cost of it should be low.
The freezing temperature of the refrigerant should be below the evaporator temperature.
The critical temperature of the refrigerant should be sufficiently high.

5.2.1 Classification of Refrigerants

The refrigerants are classified into three groups by the National Refrigeration Code of USA based on their safety, toxicity, and inflammable nature.

Table 5.3 shows the three groups with their safety limits by classifying them into six classes with an indication that Class 1 is most toxic and Class 6 is the least.

Table 5.3 Classification of Refrigerants

Group	*Refrigerant code*	*Name*	*Class*	*Remarks*
I	R-744	Carbon dioxide	5	Safest of the refrigerants
	R-12	Dichlorodifluoromethane	6	
	R-21	Dichloromonofluoromethane	6	
	R-114	Dichlorotetrafluoroethane	6	
	R-30	Methylene chloride	4	
	R-11	Trichlorofluoromethane	6	
	R-22	Dichlorofluoromethane	5	
	R-113	Trichlorotrifluoroethane	4	

(*contd.*)

Table 5.3 Classification of Refrigerants (*contd.*)

Group	*Refrigerant code*	*Name*	*Class*	*Remarks*
	R-500	Azeotropic mixture	6	
	R-502	Azeotropic mixture	6	
	R-40	Methyl chloride	4	
II	R-717	Ammonia	2	Toxic and somewhat inflammable
	R-113	Dichloroethylene	4	
	R-160	Ethyl chloride	4	
	R-40	Methyl chloride	4	
	R-611	Methyl formate	3	
	R-764	Sulphur dioxide	1	
III	R-600	Butane	5	Inflammable refrigerants
	R-170	Ethane	5	
	R-601	Isobutane	5	
	R-200	Propane	5	

5.3 REFRIGERATION LOAD

Refrigeration load or cooling load is the quantity of the heat to be removed from the cooling chamber or from the materials kept in the chamber or both. This cooling load or cooling capacity is represented customarily in terms of tons of ice which would absorb heat on melting. Thus, refrigeration effect is measured in terms of tons of refrigeration. To understand the concept clearly, let us define what is meant by one ton of refrigeration. One ton of refrigeration is defined as *the quantity of heat equivalent required or absorbed by one ton of ice to melt at* 0 °C *in* 24 hours. In other words, it is nothing but the latent heat of fusion of ice at 0 °C. The latent heat of fusion of ice at 0 °C is 303.8 kJ/kg.

∴ Heat absorbed for melting one ton of ice at 0 °C = 303.8×10^3 kJ.

This quantity of heat is to be removed in 24 hours.

$$\therefore \quad \text{Heat removal rate} = \frac{303.8 \times 10^3}{24 \times 60 \times 60} = 3.516 \text{ kJ/s} = 3.516 \text{ kW}$$

Thus, a refrigeration system which has the capacity to absorb heat at the rate of 3.516 kW is termed as **one ton of refrigeration.**

The heat to be removed from a system is calculated based on the following:

- heat released by respiration of the produce (fruits, vegetables, etc. respire and release heat even after harvesting)
- sensible heat of the produce to bring down the temperature to the desired cooling temperature.
- radiation heat losses by the walls, ceiling, floor, etc.
- heat gained by the doors and other metallic accessories.
- heat generated by lights, personnel moving, etc.

All the above factors are to be considered while deciding the refrigeration load. Appendix 2 lists heats of respiration of some fruits and vegetables.

PROBLEM 5.1 One ton of mango fruits at 35 °C is to be cooled to 4 °C in 8 hours. The radiation and other losses are estimated to be 10 per cent of the refrigeration load. Find the tonnage of refrigeration and horsepower of the motor to be used if the efficiency of the motor is 85 per cent. For want of data let us assume the specific heat of mango is equal to that of water.

Solution Heat to be removed = $m\ C_P\ (\Delta T)$

$$= 1000 \times 4.18 \times (35 - 4)$$

$$= 1,29,500 \text{ kJ}$$

Heat lost @ 10 per cent = 1,29,500 × 0.1 = 12,950 kJ
Total heat to be removed = 1,29,500 × 1.1 = 1,42,450 kJ
Time available to remove the heat = 8 hours

$$\text{Heat removal rate} = \frac{1,42,450}{8} = 17,806 \text{ kJ/hr}$$

$$= 4.95 \text{ kJ/s} = 4.95 \text{ kW}$$

One ton of refrigeration is equivalent to 3.51

$$\text{Tonnage of refrigeration} = \frac{4.95}{3.51} = 1.414$$

$$1\text{hp} = 0.745 \text{ kW}$$

$$\therefore \text{ Horsepower} = \frac{4.95}{0.745} = 6.64 \text{ hp}$$

Efficiency of the motor = 85 per cent

$$\therefore \text{ Horsepower of the motor} = \frac{6.64}{0.85} = 7.8 \text{ hp}$$

PROBLEM 5.2 500 kg of fish at 30 °C are kept in a bunker with 100 kg of blocks of ice at 0 °C. It has taken 8 hours for the ice to melt to water at 0 °C. Maintaining the same heat transfer rate, find what will be the equilibrium temperature of fish and water, and after how many hours this would occur? The specific heat of fish may be taken as 5.2 kJ/kg °C.

Solution Latent heat of fusion of ice = 302.4 kJ/kg
∴ Heat released by melting 100 kg of ice = 302.4 × 100 = 30,240 kJ

$$\text{Heat transfer rate} = \frac{30,240}{8} = 3,780 \text{ kJ/hr}$$

Temp. of fish after 8 hr can be found by making heat balance on 500 kg of fish at 30 °C.
Heat absorbed from fish = 500 × 5.2(30 – T) = 30,240 kJ

$$\therefore \qquad T = 30 - \left(\frac{30,240}{500 \times 5.2}\right) = 18.4 \text{ °C.}$$

Let T_e be the equilibrium temperature.

Heat gained by ice water = $100 \times 4.18\ (T_e - 0)$ kJ

$$\text{Heat lost by the fish} = 500 \times 5.2 \times (T_e - 18.4)^{\dagger}$$

$$\therefore \quad 100 \times 4.18\ (T_e) = -\ 500 \times 5.2\ (T_e - 18.4)$$

$$\frac{100 \times 4.18}{500 \times 5.2} = 0.16 = \frac{18.4}{T_e} - 1$$

$$\therefore \quad T_e = \frac{18.4}{(1 + 0.16)} = 15.9\ °\text{C}$$

Heat lost by fish = $-\ 500 \times 5.2 \times (15.9 - 18.4) = 6599$ kJ

$\cong 6600$ kJ

Rate of loss of heat = 3780 kJ/hr

$\therefore$ Time taken by fish to reach an equilibrium temperature of 15.9 °C = $\dfrac{6600}{3780} = 1.75$ hr

= 105 minutes

5.4 REFRIGERATION TYPES

The refrigeration effects hitherto described can be achieved by the following refrigeration types:

5.4.1 Vapour Compression Refrigeration

Vapour compression refrigeration effect is based on compression and expansion of the vapours and possibly the phase change of the refrigerant in the cycle. It was already described by Figures 5.1 and 5.2.

5.4.2 Absorption Refrigeration

In the absorption refrigeration systems, the vapour is pumped from the evaporator into a liquid having high affinity for the refrigerant, and is absorbed into the liquid. Subsequently, by applying heat, refrigerant in the form of vapours can be driven off from the liquid. It is further passed on to the condenser and is liquefied. The arrangement is schematically shown in Figure 5.3. The pressure is reduced at this stage. Later the liquid refrigerant goes to the evaporator, and the cycle continues.

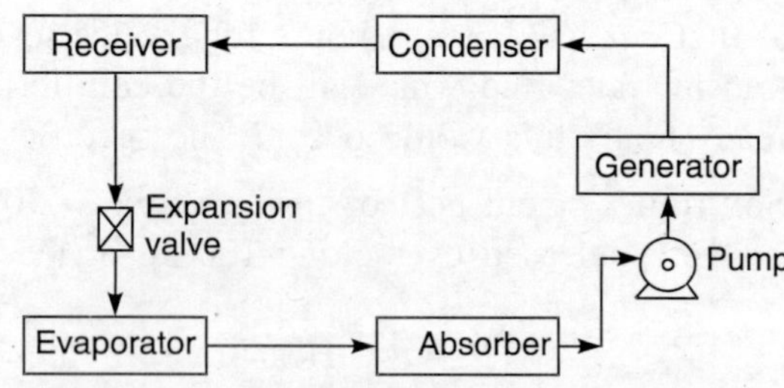

Figure 5.3 Schematic representation of an absorption refrigeration cycle.

A typical absorption system is ammonia–water system. An absorption refrigeration system, thus contains: an evaporator, an absorber, a generator and a condenser.

---

† The negative sign (–) indicates that the heat is lost from the fish, i.e., the system is bereft of that much of heat.

In addition to these main equipments, the system also consists of an expansion valve, pressure reducing valve, and a pump. The refrigeration cycle is characterized by the absence of a compressor, and it (compressor) is replaced by absorber, generator, pump and a reducing valve.

The ammonia vapours coming from the evaporator are absorbed into the water in the absorber. Since the ammonia in the form of vapours is absorbed, the pressure of the system is considerably reduced in the absorber. Thus, more ammonia vapours can be drawn from the evaporator and absorbed in the absorber. The temperature in the absorber is increased because of the release of heat of solution and heat of condensation. The absorber is cooled by employing some cooling arrangement. The strong solution from the absorber is pumped into the generator, where it is slightly heated to drive off the ammonia vapours into the condenser, where it is liquefied. The liquid ammonia is collected in the receiver from where it is flown into the evaporator through the expansion valve.

Comparison of the two systems: The absorption refrigeration has certain advantages over the vapour compression system. The comparison between the two systems is as follows:

- Moving parts are the least in absorption refrigeration systems. It will help reduce the wear and tear on the equipment.
- Absorption systems are usually designed to use steam at low pressures. Hence, any waste/exhaust steam coming from any other equipment can also be used.
- The absorption refrigeration systems can operate at the same capacity at reduced (lower) evaporator pressures also, whereas it is not possible in vapour compression refrigeration systems.
- The absorption units can operate at any load capacities by controlling the quantity of flow of aqua.
- Any leakage of refrigerant from the vapour compression is detrimental to the environment, and need to be attended to immediately; whereas the liquid refrigerant leaving the vapour in the absorption refrigeration systems has no bad effects.
- High capacity refrigeration units (above 1000 tons, etc.) can be made out of the absorption systems only. In case of low capacity refrigeration systems (like domestic units) the absorption units occupy relatively more space as compared to compression systems.
- Absorption systems require less space for higher capacities, and incorporation of automation is easier as compared to compression units.

5.5 APPLICATIONS OF REFRIGERATION IN FOOD PROCESSING

As has been mentioned, refrigeration techniques can be used for preservation of food which are susceptible to spoilage by microbial degradation or by enzymatic action, since the lower temperatures considerably reduce the rate of biochemical reaction of degradation. Therefore, those food produce or food products which are susceptible to microbial spoilage, viz., fruits, vegetables, meat, fish, chicken meat, milk and milk products made out of them, and other processed food products like pasty products, pasta foods, bakery and confectionary items and chocolates, etc. can be preserved under cold storage conditions by applying refrigeration methods.

In addition to preservation, even processing can also be done under cold conditions. Milk based sweets, ice creams, etc. are processed using refrigeration techniques.

Refrigeration and cold storage facilities are also extensively used in food processing for transportation and storage of the fresh produce from time of harvesting until it reaches the consumer. In this respect, the refrigerated transportation of materials from a small distance (a few kilometres) to long distances which extend to thousands of kilometres beyond the trans-national borders is a common practice. Chilled and frozen food industry is one of the largest applications of refrigeration, the units of which are run by diesel or oil engines using liquid fuels. Let us see some of the typical applications.

5.5.1 Fruit and Vegetable Processing

A large variety of fruits and vegetables is stored in fresh form after preliminary cleaning in cold storage. The cold storage only will extend the storage life of the fruits and vegetables from few days to few months. It does not affect taste, texture and nutritive value of the produce. Table 5.4 lists temperature conditions for some typical fruits and vegetables. The storage temperatures are very important because too low temperature would cause *chilling injury* to the fruits.

Table 5.4 Temperature Conditions for the Cold Storage of Some Fruits and Vegetables†

Commodity	*Storage temperature* °C	*Storage period*
Apples	–1–4.5 (90)	3–8 months
Apricots	–0.5–0.0 (90–95)	1–2 weeks
Bananas (green)	11.5–14.5 (85–90)	10–20 days
Beans	4.5–7.5 (90–95)	7–10 days
Beets	0	1–3 months
Black berries	–0.5–0.0 (95)	3 days
Cabbage	0 (90–95)	3–4 days
Carrots	0 (80–90)	4–5 months
Cauliflower	0 (90–95)	2–4 weeks
Cherries	–1 to –0.5 (90–95)	2–3 weeks
Cucumber	7–10 (45–50)	10–14 days
Grapes	10–15 (85–90)	4–6 months
Green leafy vegetables	0 (90–95)	10–14 days
Lemons	0–10 (80–90)	6–8 weeks
Lemon (green)	11–15	1–4 months
Mangoes	12–13 (85–90)	2–3 weeks
Mushrooms	0 (90)	3–4 days
Onions	0 (65–70)	1–8 months
Oranges	0–9 (85–90)	3–12 weeks
Papayas	7 (85–90)	1–3 weeks
Pears	–1 to +1	1–6 months
Pineapples (unripe)	10–13 (85–90)	3–4 weeks
Pineapples (ripe)	7 (85–90)	2–4 weeks
Potatoes	10–13 (90)	–
Pumpkins	10–13 (70–75)	2–3 months
Radish	0 (90–95)	3–4 weeks
Strawberries	0	7–10 days
Tomatoes (unripe)	13–17 (85–90)	4–7 days
Tomatoes (ripe)	7–10 (85–90)	1–3 days
Water melon	2–4.5 (85–90)	5–15 days

†Values in the parenthesis are per cent relative humidity.

Some of the fresh fruits are transported under cold storage conditions with and without controlled atmospheric conditions. Fruits, after harvesting also respire carbon dioxide. By increasing the carbon dioxide content or by reducing the oxygen content or both in the container, the shelf-life can be considerably extended under refrigerated conditions. This technique is called **Modified Atmosphere Packaging (MAP)** or **Controlled Atmospheric Packaging (CAP)**. This has got both pre-harvesting and post-harvesting protocol. The pre-harvesting protocol mostly deals with the spraying of pesticides and irrigation practices to be followed; whereas the post-harvest protocol includes preclearing, desapping (if desired), washing and cleaning, precooling, and cold storage.

Central Food Technological Reasearch Institute (CFTRI) at Mysore has standardized the protocol for export of fresh mangoes and grapes in reefer container by sea, by which the shelf-life of the fruits is extendable to 4–5 weeks.

Another typical example is the freezing of fruits and vegetables. The food material is frozen and maintained under frozen conditions at –18 °C. Another very interesting application is *Individual Quick Freezing* (IQF) in which the material is individually frozen by a blast of cold air (known as **blast freezing**).

The fruit pulps of some fruits are also preserved in cold storage under frozen conditions. The pulp is made from the fruits during the season and frozen. The frozen pulp is used throughout the year. This process is particularly useful for preserving the pulp of custard apple which cannot be preserved by heat treatment. Interestingly, the pulp is used in making of ice creams as one of the flavouring ingredients in addition to offering the body to the ice creams. Hence, the frozen pulp augments the cold processing during making of ice cream.

5.5.2 Fish Processing

Fish processing is one of the largest applications of cold storage facilities. Even at domestic scale, the small-scale vendors use salt-ice mixture to extend the shelf-life of fish by few hours from the time of catching the fish until they are sold in the retail market. Fish are stored under cold conditions for a period of few days to few months. The temperature conditions of storage are indicated in Table 5.5. Prawns are transported in the form of blocks in the frozen condition. Initially the fish or prawns are precleaned and frozen with the moisture present in them using a cold blast of air at –18 to –40 °C. The frozen produce is made in the form of blocks, and the blocks as a whole are exported under refrigerated conditions at –18 °C.

The IQF technique is also much used in processing of fish and prawns for export purposes. Freeze drying is one of the important applications where refrigerated freezing is applied for the material to freeze at low temperature and low pressure below the *triple point* of water. Subsequently, the temperature is risen slowly keeping the pressure constant. The water in the food material sublimes[†]. Later on, the frozen material need not be kept under cold storage. Since the freeze drying takes place by sublimation, the cellular structure of the food material is not distorted, and hence, the material does not get contracted. Further details on freeze drying will be discussed in Chapter 17.

[†]Sublimation is the process in which the material evaporates from solid state (ice) to vapour state directly without passing through the liquid state.

Table 5.5 Temperature Conditions for Cold Storage of Some Animal Products[†]

Commodity	*Storage temperature °C*	*Storage period*
Fish		
Fresh	0.6 to 1.7	5–15 days
Frozen	–23 to –17.38	6–12 months
Brine salted	4.4 to 10	10–12 months
Sheatfish		
Fresh	–1.1 to –3.6	3–7 days
Frozen	–17 to –30	3–8 months
Meat		
Bacon frozen	–23 to –18	4–6 months
Beef fresh	–23 to –18	9–12 months
Beef frozen	–23 to –18	8–10 months
Lamb fresh	0 to 1	5–12 days
Lamb frozen	–23 to –18	4–6 months
Meat fresh	0 to 1	5–10 days
Meat frozen	–23 to –18	8–10 months
Poultry		
Fresh	0	1 week
Frozen	–30 to –18	8–12 months

[†]Relative humidity is 85–95 per cent.

5.5.3 Meat and Poultry Processing

The carcass is stored under refrigerated conditions. The temperature conditions of storage are shown in Table 5.5. Generally the carcass is chilled, after slaughtering the animal and cleaning, by using cold 20 per cent brine solution by spraying. The chilled brine immediately absorbs heat from the carcass. Buffalo meat is exported from India in the refrigerated form. No processing is done to the meat. Just the frozen meat in the form of blocks is exported. The freezing temperatures required are of the order of –29 °C for storage life upto 12 months or more for beef, lamb or pork. The meat quality deteriorates if the temperature exceeds –9 °C.

The poultry meat is also preserved both by chilling and freezing. Freezing is done by air blast at –29 °C to –40 °C. Later the frozen meat is kept with ice flakes. Refrigeration and air-conditioning effects are also used to maintain the poultry sheds usually at about 24 °C. Higher temperatures are detrimental for the birds, and lower the egg laying capacity of the poultry.

5.5.4 Dairy and Dairy Products

Milk preservation by chilling to 2–5 °C is a very common application of refrigeration in dairy. The chilled milk is kept in the storage until it goes to the consumers. Plate heat exchangers are used for chilling of milk. Separation of cream from milk can best be done at 4 °C by keeping the milk at this temperature. Butter also can be stored at 0–4 °C for short period storage. For storing beyond six months, the butter is stored in frozen condition. Some of the milk-based products like *Kheer, rasmalai*, etc. are kept in chilled conditions only until they are served to the consumers.

Ice cream making is one operation where refrigeration is used as a processing step. The milk cream with all ingredients is whipped with a certain amount of mix of air under cold conditions. The temperatures used are anywhere between – 2.5 °C to – 55 °C. The ice cream is manufactured under refrigerated conditions only.

5.5.5 Bakery and Confectionary

In the preparation of bakery products, of course, we do not use very low temperatures. While making the dough for the bread, temperatures of the order of 24–26 °C are most ideal for good leavening to take place. During the process of fermentation, for 3–4 hours, the temperature of the dough increases which can be arrested by maintaining chilled environment.

The chocolates are preserved in cold conditions. Otherwise they melt. The cocoa butter in chocolates starts melting from 30 °C onwards. Hence, the temperatures are kept much below 30 °C. Ideally maintaining 15 °C is best for chocolates to preserve without melting and without deterioration of quality.

5.5.6 Beverages

Most of the carbonated beverages are stored and served in chilled conditions. The solubility of carbon dioxide increases considerably with lowering of temperature, and hence, quenches the thirst. Usually the temperatures maintained are refrigerator conditions (4 °C).

Carbonated alcoholic beverages like beers need to be maintained at refrigerator temperatures only (2–4 °C) to increase the solubility of carbon dioxide.

Symbols

H: enthalpy (kJ/kg)
p: pressure (N/m^2)
T: temperature (°C or K)
v: molar volume (m^3/k mole)

Greek Symbol

μ: Joule–Thomson coefficient

REVIEW QUESTIONS

5.1 What is meant by refrigeration? How does a refrigeration cycle work?

5.2 How does refrigeration help preserve foods?

5.3 What is meant by Joule–Thomson effect?

5.4 Draw p-v diagram for a refrigeration cycle.

5.5 What is meant by a refrigerant? What are the characteristic features of a good refrigerant?

5.6 How do you classify various refrigerants?

5.7 What is meant by one ton of refrigeration?

5.8 What are the two different refrigeration systems? Compare them.

5.9 Describe the absorption refrigeration cycle with a neat diagram.

5.10 What are various applications of refrigeration in food processing?

5.11 Describe various applications of refrigeration in

(i) fruits and vegetable processing industry,
(ii) meat and poultry industry,
(iii) dairy industry, and
(iv) beverage industry.

REFERENCES

Jordan, R.C. and Priester, G.B. (1971), *Refrigeration and Air Conditioning*, 2nd ed., Prentice Hall, Englewood Cliffs, pp. 41–42.

Singh, R.P. and Heldman, D.R. (2001), *Introduction to Food Engineering*, 3rd ed., Academic Press, California, p. 367.

CHAPTER

6

Humidity and Humidification

Dehydration and refrigeration (which we generally call as thermal processing operation) play a very vital role in food processing. Hence, it is essential to understand the relationships and behaviour of air–water vapour systems under various process conditions, viz., temperature, etc. Such a study is known as **humidification operations**. It is the extent of water vapour already present in the air which determines the ability of air to pick up the moisture from the surrounding food material by *convective mass transfer*.

Humidity is a measure of extent of water present in air in general, whereas psychrometry is a study of determination of thermodynamic properties of gas-vapour mixtures. Since we are concerned more with the properties of air–water vapour system in food processing, humidity plays a significant role in food processing operations; especially in the design of equipment for the following:

- refrigeration and air conditioning for storage of fresh produce,
- dryers for drying of cereals and grains, and
- cooling towers.

6.1 HUMIDITY AND OTHER RELATED TERMS

Dry air is a mixture of oxygen and nitrogen as its major components in the molar ratio of 21:79. For gases and vapours, the molar composition is equal to volumetric composition. However, in reality in addition to oxygen and nitrogen, air also contains some other gases as well. The volumetric composition of air is shown in Table 6.1. The molecular weight of the air is 29 (precisely 28.9645), and its universal gas constant is:

$$R = 287.045 \text{ J/kg°C}$$
$$= 287.045 \text{ m}^3\text{Pa/kg°C}$$

The humid air contains traces of moisture in the form of water vapour, as water prevails everywhere, and hence, whenever dry air comes in contact with the water, it picks up moisture

Table 6.1 Percentage Volumetric Composition of Air

Oxygen	20.95
Nitrogen	78.08
Argon	0.934
Carbon dioxide	0.0314
Neon	0.00182
Helium	0.000524
Other gases remaining methane, SO_2, H_2, Kr and Xe	Remaining

depending upon the temperature. The maximum amount of water vapour that can go into the air is determined by the saturated vapour pressure of the water at that temperature if both water and air are brought into intimate contact. If the contact is not intimate and if the mixture does not reach equilibrium, only a fraction of the maximum possible moisture can go into air. Such an air will still have ability to pick up some more moisture from the surrounding material until it (the air) is fully saturated with moisture. Thus far we have made some qualitative statements which need quantification. They can be described in the following sections.

6.1.1 Humidity

Humidity is defined as the quantity (mass) of moisture carried by a unit mass of dry air.

$$H = \frac{\text{kg of moisture}}{\text{kg of moisture-free air}} \tag{6.1}$$

The above expression can also be written in terms of the partial pressure. Since the ratio of partial pressures is equal to the ratio of mole fractions for gases and vapours, we can write,

$$H = \frac{M_w p_w}{M_A p_A} = \frac{M_w (p_w)}{M_A (1 - p_w)} \tag{6.2}$$

where p_w is partial pressure of water in atmospheres, p_A is partial pressure of air in atmospheres, M_w is molecular weight of water = 18, and M_A is molecular weight of air = 29.

Thus,

$$H = \frac{18}{29} \frac{p_w}{(1 - p_w)} = 0.62 \frac{p_w}{(1 - p_w)} \tag{6.3}$$

Obviously, humidity is a dimensionless term and it does not change with temperature or other process parameters. Generally at low temperatures upto room temperature (30 °C), p_w is a very small quantity compared to 1.0, and hence, the denominator in Eq. (6.3) may be taken as 1.0. Thus, Eq. (6.3) can be approximated to

$$H = 0.62\ p_w \tag{6.4}$$

6.1.2 Relative Humidity

Relative Humidity (RH) is the ratio of partial pressure of vapour to the vapour pressure of the liquid at the gas temperature. Generally, the relative humidity is expressed in percentages. Hence,

$$H_R = \frac{p_w}{P_w} \times 100 \tag{6.5}$$

where P_w is the saturated vapour pressure of water at any particular temperature. Obviously, relative humidity indicates to what extent the air is saturated with water; and it also indicates to what extent still the air can absorb moisture from the surroundings. 100 per cent Relative humidity means that the air is fully saturated with moisture, and hence, it can no more absorb any moisture; and 0 per cent relative humidity means that the air is dry. In the above equation P_w changes with temperature. As temperature increases P_w also increases; whereas p_w remains the same. This throws light on the point that as temperature of a given air is increased, its relative humidity decreases. And thus, the air can be made to absorb some more moisture from the surroundings, which it could not have done, had it (air) been at the same temperature.

6.1.3 Saturation Humidity

Saturated air is the one which is fully in equilibrium with the moisture. Hence, the partial pressure of vapour is equal to the vapour pressure of the liquid at that temperature. Hence, Eq. (6.2) can be written for saturation humidity as follows:

$$H_s = \frac{18}{29}\frac{P_w}{(1-P_w)} \tag{6.6}$$

where H_s is the saturation humidity.

6.1.4 Percentage Humidity

The ratio of humidity (H) to the saturation humidity (H_s) is defined as the percentage humidity ($H_{\%}$)

Thus,

$$H_{\%} = \frac{H}{H_s} \times 100 = \frac{p_w/(1-p_w)}{P_w/(1-P_w)} \times 100 \tag{6.7}$$

$$= H_R\left(\frac{1-P_w}{1-p_w}\right) \tag{6.8}$$

Since p_w is always less than P_w except at 100 per cent relative humidity, the percentage humidity is always less than the relative humidity. At 100 per cent relative humidity both H_R and $H_{\%}$ are equal.

$$H_{\%} = H_R \text{ at 100 per cent } RH \tag{6.9}$$

6.1.5 Humid Heat

Humid heat is the specific heat of the wet air (humid air). Hence, it is defined as the quantity of heat required to raise the temperature of one kg of dry air along with the associated water vapour by one degree centigrade.

Thus,

$$C_{P,\text{air}} = C_{PA} + C_{PW}H \tag{6.10}$$

where $C_{P,\text{air}}$ is humid heat of the wet air, kJ/kg°C; C_{PA} is specific heat of dry air kJ/kg°C; C_{PW} is specific heat of water vapour kJ/kg°C.

6.1.6 Humid Volume

It is also known as the specific volume which is the volume occupied by one kg of dry air along with the associated moisture vapour at a given humidity and temperature. The specific volume is calculated using ideal gas law for the volumes of 1 kg of dry air and associated moisture vapour. Thus, the specific volume is the summation of

(i) volume occupied by 1 kg of dry air at temperature T,

(ii) volume occupied by the water vapour associated with the dry air at temperature T and humidity H.

This is calculated by noting that 1 kg mole of the gases or vapours occupy 22.4 m^3 volume at 0 °C. Ideal gas is applied to calculate the volume at any other temperature T.

$$\text{Volume occupied by 1 kg of dry air at temperature } T = \frac{22.4}{29}\frac{(T + 273)}{273} \tag{6.11}$$

in which 29 is the molecular weight of air, and T is in degrees centigrade.
Moisture (vapour) associated with 1 kg of dry air when the humidity is H

$$= H \text{ kg} \tag{6.12}$$

$$\text{Volume occupied by the above water vapour} = \frac{22.4}{18}\frac{(T + 273)}{273}H \tag{6.13}$$

in which 18 is the molecular weight of water.
∴ The humid volume, v_H in m^3/kg is given by

$$v_H = \frac{22.4}{29}\frac{(T + 273)}{273} + \frac{22.4}{18}\frac{(T + 273)}{273}H \tag{6.14}$$

Noting down that $\frac{22.4}{273} = 0.082$, Eq. (6.14) can be written on rearrangement as:

$$v_H = (0.082\ T + 22.4)\left(\frac{1}{29} + \frac{H}{18}\right) \tag{6.15}$$

T is in degrees centigrade.

6.1.7 Dew Point

Dew point is the temperature to which the humid air at a given humidity is to be cooled so that it gets saturated with water vapour. At a given temperature and humidity, moist air contains some moisture depending upon the humidity. Suppose, if this air is cooled, as the temperature decreases the relative humidity increases, and a stage comes where the partial pressure of moisture in the air becomes equal to the vapour pressure of the moisture at that temperature. In a way we can say that the air gets fully saturated with the moisture. Any further decrease in temperature, results in the condensation of moisture from the air. Thus, dew point is defined as the temperature at which mist or dew starts forming. Because dew starts forming on the clean and shiny surfaces, it is known as **dew point**.

6.1.8 Enthalpy of Humid Air

This is also known as the total enthalpy of the air. It is a combination of three terms:

(i) Enthalpy of dry air at temperature T over a reference temperature, T_0 which is equal to C_{PA} $(T - T_0)$.

(ii) Enthalpy of moisture vapour at temperature T over the reference temperature, T_0 which is equal to C_{PW} $(T - T_0)$ H.

(iii) Latent heat of vaporization of liquid water at the reference temperature T_0 which is equal to $\lambda_W H$.

The enthalpy of humid air at a temperature of T is equal to

$$h_y = C_{PA}\,(T - T_0) + HC_{PW}\,(T - T_0) + H\lambda_W \tag{6.16}$$

$$= (T - T_0)\,(C_{PA} + HC_{PW}) + H\lambda_W$$

From Eq. (6.10) we can write

$$h_y = C_{P,\text{air}}\,(T - T_0) + H\lambda_W \tag{6.17}$$

6.1.9 Dry Bulb Temperature

The temperature of humid air (along with the humidity) as measured by a thermometer is known as the **dry bulb temperature**. It is the temperature of the air-water vapour mixture system.

6.1.10 Wet Bulb Temperature

The temperature of the air if it is fully saturated with moisture is known as the wet bulb temperature. Thermometer bulb is moisted or fully wetted with a wick dipped in a pool of water, and the temperature recorded by such a thermometer bulb is known as the **wet bulb temperature.** The wet and dry bulb thermometer is used to measure the wet bulb and dry bulb temperatures, and is shown in Figure 6.1.

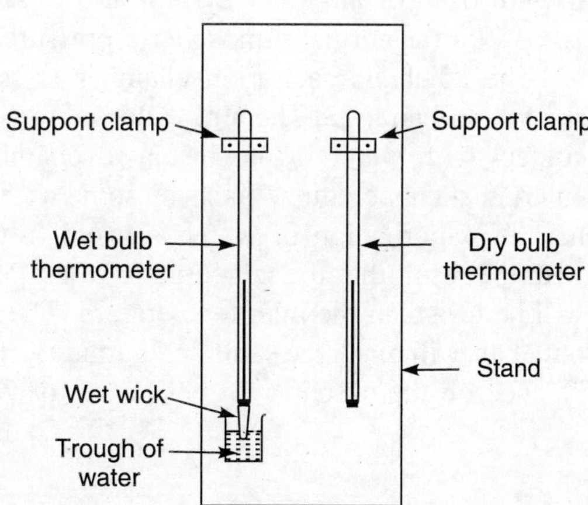

Figure 6.1 Wet and dry bulb thermometer.

The wet and dry bulb temperatures are used to measure the Relative Humidity (RH) of air and absolute humidity (H) of the air by using *psychrometric chart*. The wet and dry bulb thermometer is therefore also known as the **wet and dry bulb hygrometer**, since the *hygrometer* is the instrument used for measuring the humidity. The manufacturers of wet and dry bulb hygrometers often supply a simplified chart to measure the RH based on the wet and dry bulb temperatures. The tabular data on percentage relative humidity versus the wet bulb temperature depression are shown in Appendix 3.

6.1.11 Adiabatic Saturation of Air

Adiabatic saturation of air is followed in many process industries to cool the air as well as to saturate the air with moisture. The greatest application of it is in convective drying of foods where a breeze of air passes through a wet solid during which the air picks up the moisture from the solid under adiabatic conditions. The maximum amount of water that can go into the air is limited by the saturation of air by moisture.

The phase equilibrium diagram for air-water system can be drawn between the saturated mole fraction of water at different temperatures which can be calculated from the saturated quantity of water in the air. This is given by saturated humidity. The equilibrium saturated molar concentration of water in the air-water mixture in the vapour phase (y_s) is given by

$$y_s = \frac{\text{No. of moles of moisture present in vapour form in humid air}}{\text{Total number of moles of water vapour and dry air}} \tag{6.18}$$

Equation (6.18) can also be expressed in terms of saturated humidity as follows:

$$y_s = \frac{\left(\dfrac{H_s}{M_w}\right)}{\left(\dfrac{H_s}{M_w} + \dfrac{1}{29}\right)} \tag{6.19}$$

The data on equilibrium mole fraction of water vapour (given by Eq. 6.18) at different temperatures can be shown in the form of phase-equilibrium diagram (Figure 6.2) for air-water system at a total pressure of 101.3 kPa (at normal atmospheric pressure).

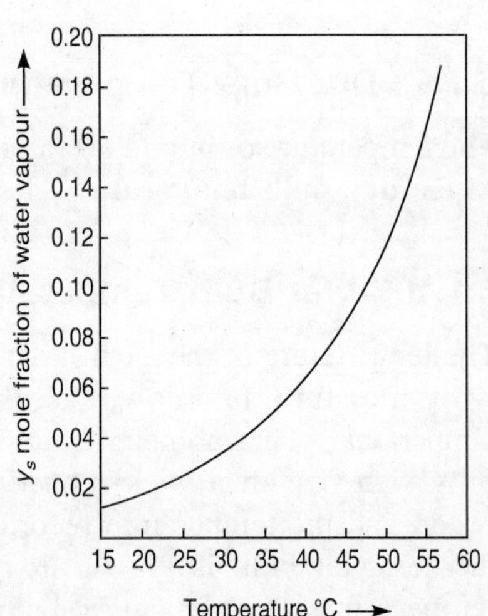

Figure 6.2 Phase equilibrium for air-water vapour system at 101.3 kPa.

The adiabatic saturation chamber is schematically shown in Figure 6.3. The air is allowed to have intimate contact with water in the insulated chamber. The air entering at temperature T picks up some water. The latent heat of vaporization of water is taken from the water with the result that the temperature of outgoing air (T_A) will be less than the inlet temperature. The humidity of outlet air will reach the saturated humidity. The enthalpy balance on the system will enable us to write

$$C_{p,\text{air}}\ (T - T_A) = \lambda_w\ (H_s - H) \tag{6.20}$$

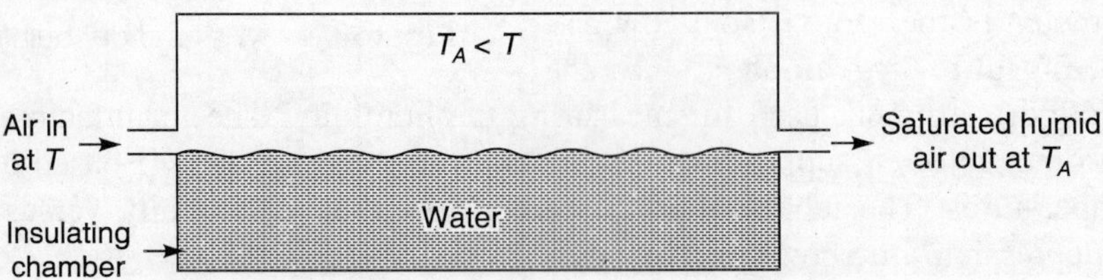

Figure 6.3 Adiabatic saturation of air.

In the above equation, the LHS term indicates heat lost by 1 kg of entering air which results in decrease of temperature from T to T_A. The RHS term indicates the latent heat of vaporization taken up by air to increase the humidity of air from H to H_s. The process is called as **adiabatic saturation** since no heat is supplied or removed from the chamber.

PROBLEM 6.1 The relative humidity of air at 30 °C is 70 per cent. Find its absolute humidity.

Solution

$$H_R = \frac{p_w}{P_w} \times 100 \quad \text{[from Eq. (6.5)]}$$

The P_w for water at 30 °C is 4.24 kN/m^2 (from Appendix 1A)

$$\therefore \quad 70 = \frac{p_w}{4.24} \times 100 \quad \text{[from Eq. (6.5)]}$$

Hence,
$$p_w = 4.24 \times 0.7 = 2.96 \text{ kN/m}^2$$

$$= \frac{2.968}{101.3} = 0.0293 \text{ atm}$$

From Eq. (6.3),
$$H = 0.62 \frac{p_w}{1 - p_w}$$

where p_w is in atmospheres

$$\therefore \quad H = 0.62 \times \frac{0.0293}{(1 - 0.0293)} = 0.0176$$

The humidity of air is 0.0176 kg of moisture/kg of dry air.

PROBLEM 6.2 Find the saturated humidity of the air (in Problem 6.1) at 30 °C.

Solution From Appendix 1, the P_w = 4.2 kN/m^2

$$= \frac{4.24}{101.3} = 0.0418 \text{ atm.}$$

From Eq. (6.6),
$$H_s = \frac{18}{29} \frac{P_w}{1 - P_w} = 0.62 \times \frac{0.0418}{(1 - 0.0418)}$$

$$= 0.027 \text{ kg of moisture/kg of dry air.}$$

PROBLEM 6.3 Find the humid volume of the air in Problem 6.1.

Solution The humidity of air = 0.0176 kg/kg

Temperature = 30 °C

From Eq. (6.15), we can find v_H

$$v_H = (0.082\ T + 22.4)\left(\frac{1}{29} + \frac{H}{18}\right)$$

$$= [(0.082 \times 30) + 22.4]\ [0.034 + (0.0555 \times 0.0176)]$$

$$= 0.882 \text{ m}^3\text{/kg}$$

PROBLEM 6.4 Find the specific volume of air in Problem 6.1 neglecting the presence of moisture, and compare the value with that obtained in Problem 6.3.

Solution The value of *R*, the universal gas constant for air is 287.045 m^3 Pa/kg°C

30 °C = 303 K

$$v_H = \frac{R \times (273 + 30)}{101.3 \times 10^3} = \frac{287.045 \times 303}{101.3 \times 10^3} = 0.8585 \text{ m}^3/\text{kg}$$

as compared to 0.882 m^3/kg in Problem 6.3.

6.2 PSYCHROMETRIC CHART

Psychrometry is the study of behaviour of air–water vapour mixture. In view of the importance of air–water vapour system in determining the drying characteristics of the air vis-a-vis the wet food material, the data on this system are well documented in the form of graphs. The graph or chart showing the relationships of various terms, viz.,

- humidity of air,
- relative humidity of air,
- adiabatic saturation lines,
- humid heat, and
- wet and dry bulb temperatures, etc.

is known as the **psychrometric chart**. It is shown in Appendix 4. This chart is drawn essentially between temperature on the *x*-axis and humidity on the *y*-axis with different percentage saturations as parameters. For example one curve marked as 100 per cent shows the humidity of saturated air. All other curves drawn below this show different levels of saturation. The slanting lines running downwards show the adiabatic cooling lines represented by Eq. (6.20). Thus, this chart is very useful in calculating the humidity, RH or humid heat, etc. based on temperature. In dehydration of foods, the relationships of moisture and temperature of the air–water systems play an important role. We shall see the application of this chart in section 6.2.1.

6.2.1 Utilization of Psychrometric Chart

The Psychrometric chart essentially shows the relationship of humidity and temperature at different levels of saturation of the air with water vapour. This is explained with reference to Figure 6.4. If we know the temperature of air, say T_1, and Relative Humidity (RH) say 40 per cent, we can mark the temperature on the abscissa at point *P* and draw a vertical line. We see it touches the RH line corresponding to 40 per cent at point *Q*. Now, we draw a horizontal line from point *Q* to go to ordinate and measure the humidity H_1, at point *R*.

Similarly, the chart is also used to measure the RH of air with a knowledge of wet and dry bulb temperatures. Let us say T_2 is wet bulb temperature and T_3 is dry bulb temperatures of a typical humid air. We mark T_2 at point *A* on abscissa and draw a vertical line to touch the 100 per cent RH line at point *B*. We draw a horizontal line from here. We mark the dry bulb temperature T_3 at point *C* on abscissa and draw a vertical line to intersect at point *D* with the horizontal line drawn from point *B*. The RH line corresponding to point *D* indicates the RH

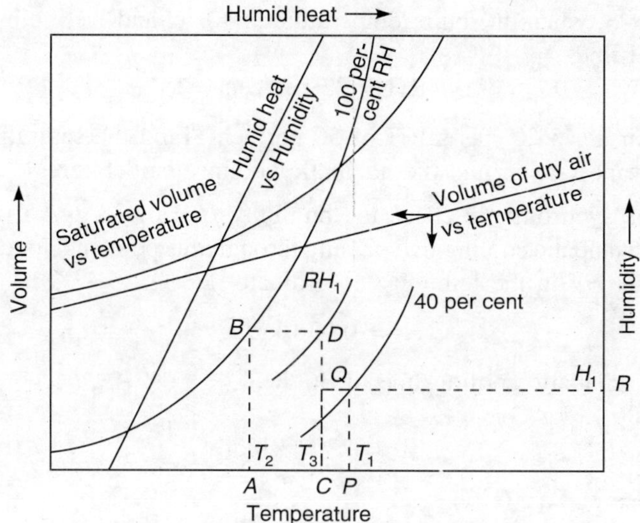

Figure 6.4 A typical psychrometric chart.

of the air. Thus, the psychrometric chart is used for various measurements related to air-water systems. The procedures are explained better with solved Problems 6.5–6.8.

PROBLEM 6.5 The relative humidity of air at 30 °C is 70 per cent. Find its absolute humidity using the psychrometric chart.

Solution As shown in Figure 6.4, we measure the humidity from the psychrometric chart (Appendix 4B) by drawing a vertical line at 30 °C and noting down the point where it touches the 70 per cent RH line. From this line, we draw a horizontal line where it touches the ordinate (humidity) at 0.0185.

$$\text{Humidity} = 0.0185 \text{ kg of water/kg of dry air}$$

PROBLEM 6.6 The wet and dry bulb temperatures of air are 35 °C and 40 °C, respectively. Find the RH of the air by using the psychrometric chart. Compare it with the calculated value.

Solution Note the point where the vertical line drawn at 35 °C touches the 100 per cent RH line. From there we draw a horizontal line until it touches the vertical line drawn at dry bulb temperature of 40 °C. We try to note the RH curve which touches the intersection point. From the psychrometric chart (Appendix 4B) it is slightly above 70 per cent. Hence, it could be about 75 per cent RH. From the psychrometric chart, RH = 75 per cent.

At 100 per cent RH, the partial pressure is equal to the saturated vapour pressure of water vapour

$$P_w = p_w = 0.05622 \text{ bars at } 35\ °\text{C}$$

$$= 5.622 \text{ kPa}$$

At 40 °C,

$$P_w = 0.07375 \text{ bars}$$

$$\cong 7.375 \text{ kPa}$$

From Eq. (6.5),

$$H_R = \frac{p_w}{P_w} \times 100 = \frac{0.05622}{0.07375} \times 100 = 76 \text{ per cent}$$

From Appendix 3, when dry bulb temperature is 40 °C and wet bulb depression is 35 °C (40 °C – 35 °C), we find

$$\text{RH} = 72 \text{ per cent}$$

PROBLEM 6.7 Air at 45 °C has a RH of 50 per cent. Find the saturation humidity, humid heat, latent heat of vaporization and the adiabatic cooling temperature.

Solution From the psychrometric chart, the humidity $H = 0.033$ kg/kg of dry air.

By following the adiabatic cooling line passing through this point and touching the 100 per cent RH curve, we can get H_s on the humidity co-ordinate

$$H_s = 0.036 \text{ kg/kg}$$

This corresponds to adiabatic cooling line of 36 °C.
Similarly $C_{P,\text{air}} = 1.06$ kJ/kg°C

$$\lambda_s = 2392 \text{ kJ/kg}$$

Applying these values in Eq. (6.21),

$$\frac{H_s - H}{T - T_w} = \frac{C_{p,\text{air}}}{\lambda_s}$$

i.e.,

$$\frac{0.036 - 0.033}{(45 - T_w)} = \frac{1.06}{2392}$$

From the above, we get $T_w = 38.2$ °C.

PROBLEM 6.8 Find the humid volume of air at 60 per cent RH and at a temperature of 50 °C.

Solution The humid volume can be calculated by Eq. (6.15)

$$v_H = (0.082\ T + 22.4)\left(\frac{1}{29} + \frac{H}{18}\right)$$

in which T is the temperature in °C

$$\text{RH} = 0.6 = \frac{p_w}{P_w}$$

At 50 °C, $P_w = 0.12335$ bars (Appendix 1B).

Therefore, $$p_w = 0.6 \times 0.12335 = 0.074 \text{ bars}$$

$$H = 0.62\frac{p_w}{1 - p_w} \qquad \text{Eq. (6.3)}$$

$$= 0.62\frac{0.074}{(1 - 0.074)} = 0.0495 \text{ kg/kg}$$

Therefore, $$v_H = (0.082 \times 50 + 22.4)\left(\frac{1}{29} + \frac{0.0495}{18}\right)$$

$$= 0.987 \text{ m}^3\text{/kg}$$

6.3 HUMIDIFIERS AND DEHUMIDIFIERS

Various types of equipments are used in food processing for humidifying and dehumidifying the gas, since both the types of operations are frequently encountered in processing. Dehumidified air is required for dehydration of foods, while humidified air is required for certain germination operations.

Humidifiers are used to increase the humidity of air. Air which is not fully saturated with water vapour is brought into contact with water so that the air gets saturated with water. In this respect, the equipment is almost similar to any gas–liquid contacting equipment. The process is similar to what was describe in section 6.1.11 (Figure 6.3) under adiabatic saturation of air. Humidifiers are generally designed and fabricated by local fabricators. It is almost similar to the design of cooling towers.

A typical humidifier is shown in Figure 6.5 where a jet of water is sprayed through nozzles into a chamber through which air is blown continuously by a blower. The upgoing air comes fully in contact with the water mist and gets saturated. The humid air leaves the humidifier from top, while the unabsorbed water leaves from the bottom. The water gets cooled in the process as the latent heat of vaporization for the water to go into the air stream is taken from the water sprayed, and hence, gets cooled.

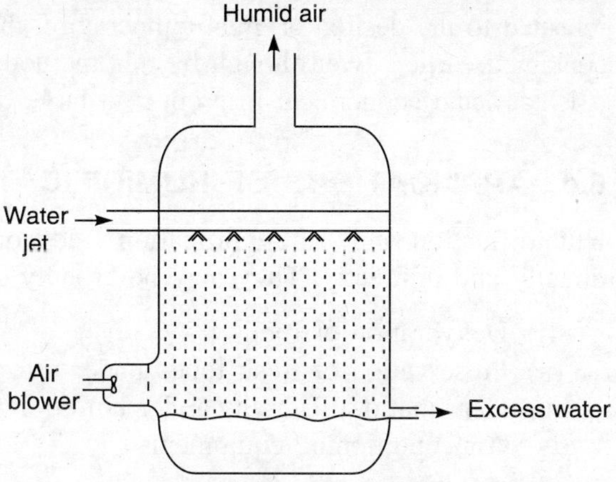

Figure 6.5 Humidifier.

In another typical design, bubble column towers (Figure 6.6) are used for humidification of air. Bubble column tower is a column of water into which air is blown from the bottom through the column of water and gets saturated with water. Humid air leaves from the top. A provision is made to add make-up water into the column. The towers are best and simple in construction except for the pressure drop suffered by the inlet air.

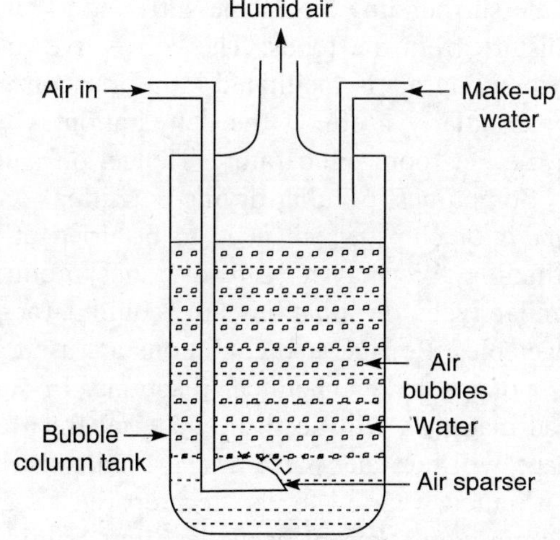

Figure 6.6 Bubble column tower.

Dehumidifiers are used to decrease the humidity in air by removing the moisture from air. The dehumidified air is dry and will be able to take up more moisture from the wet solid food materials during dehydration. Dehumidification can be done by two methods:

(i) Using desiccants to absorb the moisture from the air
(ii) Cooling and reheating the air

Desiccants are generally used when the quantity of dehumidified air required is in small quantities mainly at laboratory or pilot plant scale operations. Air is passed through a loosely packed tower containing a desiccant like anhydrous calcium chloride or anhydrous silica gel. The moisture in the air would be adsorbed by the desiccants. The outgoing air will be fairly dehumidified. However, during the process, the desiccant gets humidified and will be ineffective, and hence, needs to be changed periodically. It is not economical to use the desiccants for large quantities or for continuous use.

Another method of dehumidification is to cool the humid air below the temperature of dew point. Air gets saturated with moisture and excess moisture will condense out. If the air is reheated to the desired original temperature, the air will have moisture much less than what it contained earlier. Even though by this method, we can not get 100 per cent dehumidified air, it is easy and economical to produce a large quantities of air for process operations.

6.4 APPLICATIONS OF HUMIDIFICATION IN FOOD PROCESSING

Humidification and dehumidification find considerable application in food processing both directly and indirectly. They may be broadly described under the following heads:

(i) Dehydration of foods
(ii) Preservation of fresh foods under refrigerated conditions
(iii) Germination of grains under humid conditions
(iv) Air conditioning equipment

As has been mentioned, dehydration is one major application of dehumidification operations. The moisture transmission from the foods depends upon the moisture gradient in the food and in the surrounding air. If the surrounding air is sufficiently dehumidified, it will absorb more moisture from the foods. The convective mass transfer of moisture from the wet food to the surrounding air is facilitated either by concentration gradient or thermal gradience or both. It is particularly useful if the dehydration is to be achieved without raising the temperature of air, i.e., at room temperature. Hence, only concentration gradience alone becomes important.

Sometimes, peculiar drying conditions are required for dehydration of thin wafer foods like *papads* or chips which need to be dried at low temperature using highly humid air to avoid wrinkling. We may have to dry such products under high air flow rates, but with humid air. Similar types of situations are required for preservation of fresh food materials like fruits or vegetables. Refrigerated conditions are used as a processing technique only to delay the process of spoilage by biochemical reactions. In such cases, passage of high humid air through the food materials will keep the material fresh and free from wrinkles. Shriveling in the size and shape will not take place as the food material does not dehydrate.

Malted wearing foods are prepared by germinating the grains to increase the digestibility. Germination takes place under high humid conditions. For this, the wet grains are covered with a wet cloth which dampens the grains all throughout. It may be difficult to dampen the grains in this way in commercial scale operations. The dampening conditions are maintained by maintaining a flow of humid air.

For production of mushrooms, the spawn is cultured under solid state fermentation conditions which need a highly humid environment. The cultivation (processing) sheds are maintained under high humid conditions by blowing air through wet cloths in small scale operations.

In the design of air-conditioning equipment, cooling towers are an integral part. Cold air is blown over the fins through which the compressed refrigerant passes. The temperature raise during the compression stage is taken away by the cooling air which is cooled in the cooling towers. This is an indirect application of humidification operations to food processing.

Symbols

C_p: specific heat
$C_{P,\text{air}}$: humid heat (kJ/kg°C)
H: humidity (kg/kg)
H_R: relative humidity (%)
H_s: saturation humidity
$H_{\%}$: percentage humidity
h: total enthalpy
h_y: enthalpy of humid air
M: molecular weight
P: vapour pressure
p: partial pressure
R: universal gas constant
T: temperature (°C)
y_s: equilibrium mole fraction of water in the humid air

Subscripts

A: air
W: water

Greek Symbols

λ: latent heat of vaporization (kJ/kg)
v_H: humid volume (m^3/kg)

REVIEW QUESTIONS

6.1 Describe humidification process and its importance in food processing.

6.2 What is meant by
(i) partial pressure of a gas?
(ii) vapour pressure of a gas?

6.3 Describe the following terms:
(i) humidity
(ii) relative humidity
(iii) saturation humidity
(iv) percentage humidity
(v) humid heat
(vi) humid volume
(vii) dew point

6.4 Describe a relationship between relative humidity and percentage humidity.

6.5 Describe what is meant by
(i) wet bulb temperature
(ii) dry bulb temperature

6.6 How relative humidity is measured from the wet and dry bulb temperatures?

6.7 Describe adiabatic saturation of air.

6.8 Describe a psychrometric chart.

6.9 How the psychrometric chart is used for measuring various humid properties of air?

6.10 Describe a typical humidification equipment.

6.11 How dehumidification of air is done?

6.12 What are various applications of humidification operations in food processing?

NUMERICAL PROBLEMS

6.1 The wet bulb and dry bulb temperatures of humid air are 35 °C and 45 °C. Find the RH of the air by calculations and also by using the psychometric chart.

[**Ans :** RH (by calculations) = 58.7 per cent
RH (by psychrometric chart) = 55 per cent]

6.2 Air at 60 °C is having a RH of 60 per cent. Find the humidity of air both by calculations and by using the psychometric chart.

[**Ans :** H (by calculations) = 0.084 kg/kg
H (by psychrometric chart) = 0.090 kg/kg]

6.3 Find the humid volume of air at 40 °C and 50 per cent RH. [**Ans:** v_H = 0.9197 m^3/kg
specific volume of dry air from psychrometric chart = 0.885 m^3/kg]

CHAPTER

7

Measurement and Control of Process Parameters

Measurement and control of process parameters like temperature, flow rates, etc. are an important integral part of any food processing operations. Since food materials contain a large number of components that are sensitive to heat and other process parameters, maintaining them in the foods without spoilage even after cooking, which is very vital in food processing. Controlling of the process parameters adds another constraint in foods because they (food materials) contain a large number of sensitive components, viz., carbohydrates, sugars, proteins, minerals, enzymes, vitamins, essential amino acids, salts, and various other micronutrients.

In fact, food is valued for above sensitive and essential components. They are intact in any fresh produce. The moment we cook them to increase the palatability or process them to increase the shelf-life of the produce, the components get affected and deteriorate. Unfortunately, these components are highly varied in nature, and require different processing conditions. For example, the carbohydrates and starches need high temperatures of the order of 100 °C and beyond, for gelatinization and cooking, whereas the vitamins and amino acids are highly sensitive to temperature. Even exposure to light affects the vitamins, whereas high temperatures and pressure cooking are required for inhibiting certain undesirable enzymes like tripsin, etc. Proteins get denatured only at higher temperatures. In this way the conditions vary for processing of various components in the foods. Bereft of these nutrients, food loses its value. Hence, a rigorous control of the process parameters is very vital which can come only from a systematic measurement of them.

7.1 VARIOUS PROCESS PARAMETERS

A large number of process parameters needs to be controlled. They maybe classified as physical parameters, chemical parameters (Figure 7.1). The classification is not exhaustive. There could be many more process parameters; however, we are restricting to a few. Table 7.1 shows various instruments/accessories for measuring and controlling the above process parameters.

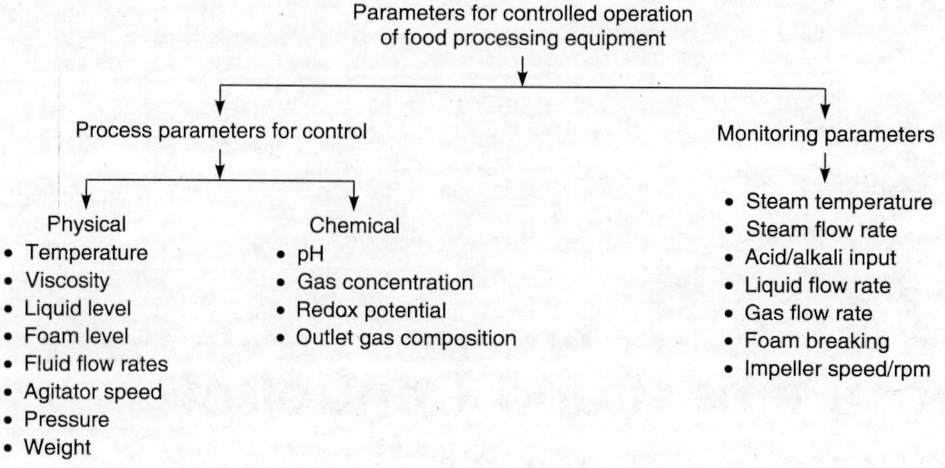

Figure 7.1 Various process parameters to be controlled.

Table 7.1 Various Instruments and Accessories Used to Measure/Control the Process Parameters

Parameter	*Accessory*	*Parameter*	*Accessory*
Temperature	Mercury thermometers (M)	Foam	Foam sensor (M)
	Bimetallic thermometers (M and C)		Adding antifoaming agents (C)
	Pressure bulb thermometers (M)		Mechanical foam breakers (C)
	Thermo couples (M and C)	Feed flow	Orificemeter (M)
	Thermistors (M and C)		Rotameter (M)
pH	pH meter (M)		Venturimeter (M)
	pH electrode (M)		Needle valve (C)
Dissolved oxygen	DO analyzer (M)		Solenoid valve (C)
Pressure	Bourdon pressure gauge (M)	Flow accessories	Valves (C)
	Pressure transducers (M and C)		Bends
Agitator speed	Torsion dynamometer (M)		Elbows
	Tachometer (M)		Pumps (C)
Liquid height	Liquid level sensor (M)	Weight	Load cells (M and C)
	Solenoid valve (C)	Outlet gas analysis	Titrometric methods (M)
			Gas chromatography (M)

M: Measuring C: Controlling

Source: Reprinted from *Introduction to Biochemical Engineering*, Rao, D.G., p. 346, Copyright 2005, with permission from McGraw Hill, New Delhi.

7.1.1 On-line and Off-line Control

If the process parameters are measured and controlled then and there itself while the process is going, we call it as on-line control. For example, we want to measure the temperature in a jam making process. We measure the temperature with a thermometer or thermocouple by inserting it into the broth and measure the temperature. Subsequently, the temperature is compared with the set value. Corrective measures are implemented like reducing or increasing the flow rate of stream, increasing or reducing the wattage input to the heater. It is known as on-line measurement and controlling. Similarly, we could do for measurement of pressure, weight of the system or any other process parameter; compare it with the set value and implement the corrective action immediately.

As contrast to the above, in case of off-line measurement-cum-control, some quantity of material is taken away from the processing vessel to another room/hall and measurements are made. Comparisons are made with the set points, and then corrective measures are implemented. For example, in the measurement of the viscosity of the broth, the material is withdrawn from the processing vessel, and the viscosity of the broth is measured by using some kind of viscometer (Section 9.3). The corrective step depends upon the purpose for which the viscosity is a parameter. If it is for manufacture of some soups or gravies, we may stop the process at the moment when desired consistency is reached.

Of course, the off-line control process has the disadvantage that by the time we measure the process parameter and try to implement the corrective measure the process conditions may change, and warrant a different type of corrective measure. This problem is absent in the case of on-line control as the corrective action is immediate. But the instrumentation in the case of on-line control is elaborative, which may add for the initial cost of the project. Most of the off-line control methods are manual and they do not require elaborative instrumentation.

7.1.2 Critical Parameters

The process parameters can be classified into critical and non-critical parameters. Those parameters which have a direct bearing on the process are known as **critical parameters**. For example, if we consider the hydrogenation of fats, we find that the hydrogen pressure, catalyst concentration, temperature are all critical parameters, and they decide the progress of reaction. The following are some critical parameters:

- Temperature
- Conversion of the reactants
- pH
- Pressure
- Air/gas flow rates
- Agitator speed
- Weight of the dried product

7.1.3 Non-critical Parameters

Non-critical parameters are those parameters of the process which are not directly involved with the progress of processing. Sometimes these parameters could be outcome of some other process parameters. It is imperative that they need to be controlled, but delay in controlling them does not affect drastically. Some of such parameters are:

- Viscosity of the broth
- Liquid level

- Extraction time in batch solvent extraction process
- Foam control
- Final moisture content
- Outlet gas composition

For example in the case of liquid extraction of active components from the nuts or solids, the time of contact does not matter much as long as a minimum amount of time of contact is given. Further contact time does not extract more once the equilibrium is reached. Thus, the time of extraction is a non-critical parameter.

7.2 MEASUREMENT OF VARIOUS PARAMETERS

The different methods and instruments used for measurement of various process parameters are temperature, pressure, weight, pH, and flow rate of fluids.

The instruments (accessories) used for measurement of above process parameters have already been indicated in Table 7.1.

7.2.1 Temperature

Temperature of the contents in a vessel or a pipe often need to be measured, so that overheating does not spoil the food materials while underheating does not cook the foods. Temperatures are measured usually by thermometers of various types as follows:

Mercury-in-glass-thermometer: This is the simplest and most versatile instrument to measure the temperatures of fluids. If the vessel is large, usually a thermowell is incorporated which contains another contacting fluid like mercury or glycerol, into which the thermometer tip is inserted.

Pressure bulb thermometer: This is similar to a pressure gauge connected to a bulb containing a fluid. As the temperature is increased, the fluid expands and makes a pointer attached to it move. The movements of the pointer are initially calibrated to measure the temperature.

Bimetallic thermometer: It consists of two metals or metal alloys which have different coefficients of thermal expansion. The two metal strips are pasted to each other and are bent in shape. One end of the strips is fixed. When the strips are exposed to temperatures they expand to different extents, and hence, the loose end of the bend moves to different extents. The movements are calibrated through a pointer to different temperatures. The concept is shown in Figure 7.2.

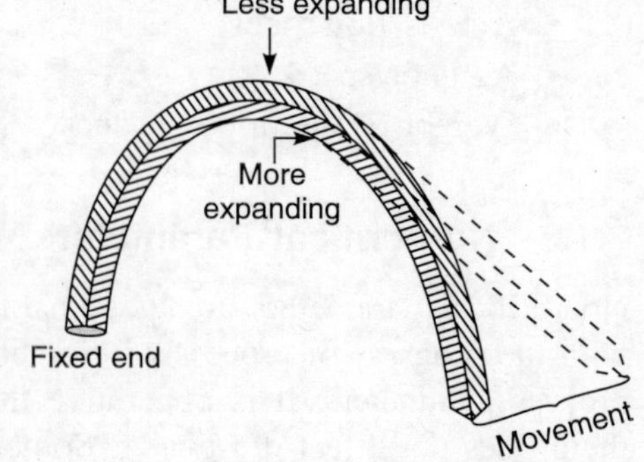

Figure 7.2 Concept of bimetallic thermometer.

Resistance thermometer: The electrical resistance of metals/conductors vary with temperature. This character is made use of in the electrical resistance thermometer. The bulb of the thermometer consists of the resistance material (resistor) which is exposed to heat. The resistance is put in a *Wheatstone bridge* circuit and is balanced. When the temperature changes (or increases) the resistance of the sensing element also changes which would imbalance the Wheatstone bridge. These imbalances are calibrated to the temperature of the system. These thermometers are very accurate, and can measure upto 0.25 °C fluctuations as well with a fair amount of accuracy.

In addition to thermometer, we also use the following instruments.

Thermocouples: When two dissimilar metals are joined as junctions and maintained at different temperatures, certain amount of emf (millivolts) generates and flows through the metals. The quantity of the millivolts produced is proportional to the temperature of one junction (known as **hot junction**) with reference to the other (known as **cold junction**). This classical observation of Seebeck in the year 1821, has enabled measurement of temperature using different types of metals and metal alloys. Some commonly used thermocouples are:

- Iron-constantan
- Chromel-alumel

By keeping the cold junction at reference temperature, and the other junction at a known temperature, the millivolts generated in the thermocouple can be calibrated against the temperature. Subsequently, the calibrated data can be used to measure the temperatures against the millivolts generated in the thermocouple. The thermocouples can be connected to temperature indicator-cum-controller to measure and control the temperatures.

Thermistors: The function of thermistors is similar to electrical resistance thermometers except that the thermistors are semiconductors. Thermistors are made out of pure oxides of iron, nickel and other metals, and have a characteristic feature of a large change in resistance even for a small change of temperatures. The thermistors are incorporated into Wheatstone bridge circuit, and balanced initially. Changes in temperatures will bring in changes in resistance which would destabilize the Wheatstone bridge circuit. These fluctuations are in turn calibrated with the temperatures.

Pyrometers: These are used for measurement of very high temperatures of the order of 1000 °C and above, and are generally used in metallurgy.

7.2.2 Pressure

Measurement of pressure in a vessel is very important both from process point of view as well as from safety point of view. Incidentally food processing does not involve many high pressure operations except a few like hydrogenation of vegetable oils, manufacture of carbonated beverages, etc.

Usually the pressure in a vessel or pipe is measured using a pressure gauge, popularly known as **Bourdon tube** pressure gauge. The Bourdon tube is a C-shaped tube which has an elliptical cross-section. The process fluid under pressure enters from one end. Since the other end of the tube is closed, the C-shaped tube tries to unwind and straightens. The closed end starts moving. These movements are conveyed to a pointer through a gear-and-pinion movement (Figure 7.3). The pointer moves on a scale which is calibrated. Eventhough the movements of the Bourdon tube are small, they are exaggerated through the gear-and-pinion arrangement by the pointer. Thus, the pressure gauge can be used to measure pressures from fraction of an atmosphere (101.3 kPa) to a few atmospheres. As the range increases, the sensitivity decreases.

Pressure control is usually done by means of *pressure transducers* which have a tendency to make or break an electric circuit when the pressure exceeds the set point. This will help activate or deactivate a *solenoid valve* which will release the pressure from the process vessel. Pressure release is also done by using *pressure release valve* or *safety valve* which is made out of a soft metal. This breaks and gives way for the pressure to release. Once it is broken, it has to be replaced by a new one.

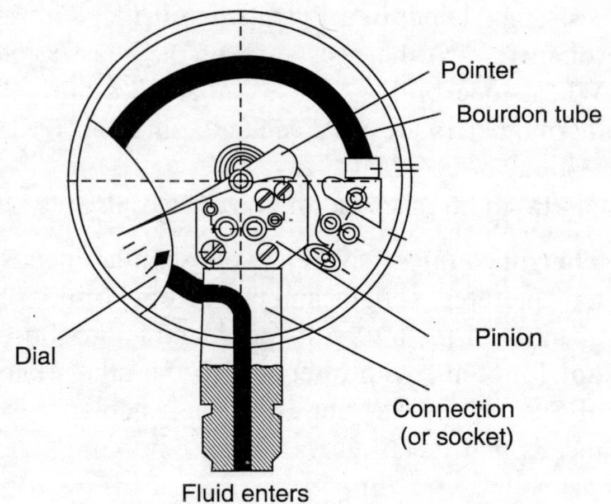

Figure 7.3 Bourdon pressure gauge.

7.2.3 Weight

In process operations, the *load cell* is frequently used to measure and control the weight. The material to be weighed is kept on a platform under which a compression load cell is kept. The load cell is fitted with an electrical resistance strain gauge. When the load increases, the strain on the cell increases, which in turn increases the electrical resistance proportional to the strain. The variations in electrical resistance can always be measured using Wheatstone bridge circuit which gets destabilized. The disturbances (dampenings) in the Wheatstone bridge circuit are precalibrated to the weights. This is how a load cell is used to measure the weight of an object.

The load cell can also be used to control the weight. The moment the electrical resistance crosses the set value, it makes or breaks an electric circuit. In process operations, it may give an alarm indicating that the process vessel has reached the set weight, and hence, the process is to be terminated. If it is used while adding a fluid to vessel, the moment the weight crosses the set point, the load cell deactivates the solenoid value which in turn stops further addition. The controlling methodology depends upon the type of process operations.

7.2.4 pH Measurement and Control

Maintenance of pH is not a routinely desired operation in food processing except in the fermentation processes. The pH of the broth is measured using a pH electrode usually calomel/mercury electrodes. The electrode is connected to the *pH meter* which measures the pH of the solution.

Controlling of pH is done either by adding an acid (sulphuric acid) if the solution is too basic, or by adding an alkali (usually ammonium hydroxide or sodium hydroxide) if the solution is too acidic. The pH meter is connected to pH controller which in turn activates a pinch cock or a solenoid valve to add an acid or alkali as the case may be.

7.2.5 Flow of Fluids

Flow rates of fluids passing through pipes or conduits can be measured by a large number of flow meters which are based on pressure drop or change in the area of cross-section. The following are a few:

- Orificemeter,
- Venturimeter,
- Rotameter, etc.

They are described in Chapter 8 (Section 8.7).

7.3 CONTROLLING METHODS

Precise control of most of the process parameters is a prerequisite for a successful operation in food processing. However the controlling methods are the same like any other process operation; could it be a chemical plant, petrochemical plant, fertilizer plant or a fermenter or a food processing plant. Almost in all these cases, the process parameters are the same; could it be temperature, pressure, weight, pH, flow rate of fluids, liquid level control, foam, agitator speed, and agitator shaft power.

In view of this, we can say that a control loop strategy essentially consists of a process parameter, a measuring/sensing element, a set point, a controlling loop, and a final control element.

For example, in case of making of jam, we heat all the ingredients along with the fruit pulp and sugar until the final brix reaches 68°. But in a controlling strategy, we may not measure the brix. We measure the temperature of the broth. So the sensing element is a temperature sensor. The moment the temperature reaches 105.5 °C, we stop cooking as the temperature controller signals for stoppage of heating and further processing. Here, the measurement of temperature is an indirect indication of the brix. Thus, in this case,

- The process parameter is brix.
- Measuring or sensing element is thermocouple.
- Set point is temperature.
- Controlling loop is stoppage of heating.
- Final control element is the shutting off of the solenoid valve/steam valve to stop supply of steam.

Thus, controlling can be done by activating/deactivating a valve, a pump or some other instrument/accessory. The controlling can be done in the following ways:

7.3.1 Manual Control

In manual control, the process parameter to be controlled is adjusted manually by comparing the measured parameter with the set value. When the measured value reaches the set value, corrective action is implemented. For example, in the control of heating of a vessel with the contents, the temperature is measured using a thermometer or a thermocouple. The measured value is compared with the set value. Then the corrective measure of adjusting the steam valve/adjusting the wattage input to the electrical heating element is implemented manually (Figure 7.4). Thus, the control of heating is achieved. Similarly other process parameters can also be controlled manually. This control system is simplest and easy to implement except that it involves human bias/experience in adjustment of the process parameter. Since no instrumentation is involved in operation, the initial investment costs are low, but the operational costs are high, because of high labour costs. The controlling operation is also unreliable since it depends upon the skill of the operator.

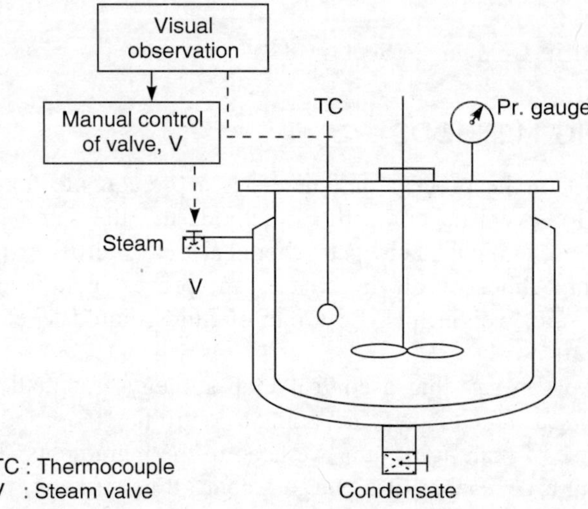

Figure 7.4 Control of temperature by adjusting the steam valve manually.

Source: Reprinted from *Introduction to Biochemical Engineering*, Rao, D.G., p. 114, Copyright 2005, with permission from McGraw Hill, New Delhi.

7.3.2 Automatic Control

In automatic control systems, the controlling of the process parameter is done by an instrument, and hence, human biased error can be avoided. The instrument measuring the process parameter produces a signal or an impulse which in turn will be compared with the set value in the control loop. Based on the comparison, an output signal is generated which implements the controlling strategy. Thus, the whole controlling is done by an input-output signal. We shall revert back to the earlier example of controlling the temperature. The temperature is measured by a thermocouple which generates an output in the form of emf (millivolts). The thermocouple is connected to a temperature controller for which the temperature is set. Based on the comparison of the measured value and set value, the controller activates a solenoid valve to continue or stop the flow of steam. Accordingly, the temperature is maintained. The arrangement is shown in Figure 7.5.

The efficiency of automatic control depends upon how the controlling strategy is implemented. The automatic control is done by any one of the following ways:

On-off controller: If the temperature is maintained by totally cutting off the steam supply or totally allowing the steam supply to pass through in full swing, we find fluctuations in

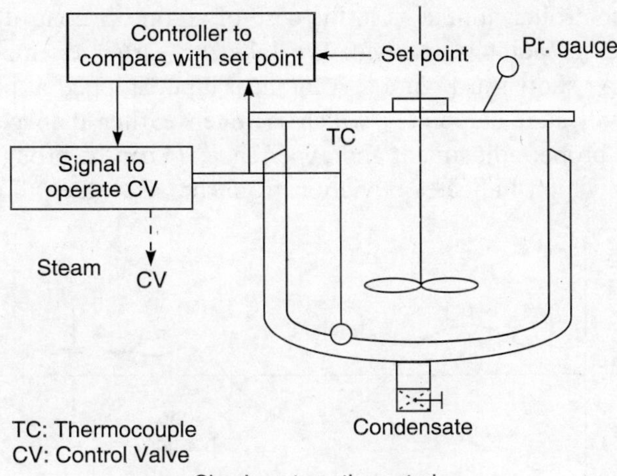

Figure 7.5 Control of temperature by adjusting steam valve with automatic control arrangement.

Source: Reprinted from *Introduction to Biochemical Engineering*, Rao, D.G., p. 114, Copyright 2005, with permission from McGraw-Hill, New Delhi.

temperature from the set point. The fluctuations are mainly due to overshooting or undershooting of the process parameter even after shutting off or putting on of the monitoring parameter. The pattern is shown in Figure 7.6. At point *a* when the temperature has reached the set point, the steam supply is stopped. Still the temperature raises upto point *b*, and subsequently starts falling down upto point *c*. The moment the temperature starts falling down below the set point, the steam supply starts. But the temperature will not pick up instantaneously, it falls upto point *d* and starts raising upto point *e*. Again the cycle of stoppage of steam supply starts. The raise continues upto point *f* and falls down to point *g*. Thus, the cycle continues with a positive and negative deviation of the temperature to an extent of T_0 °C. Even though we set the temperature at T °C, it deviates as $T \pm T_0$. This is the disadvantage with an on-off controller.

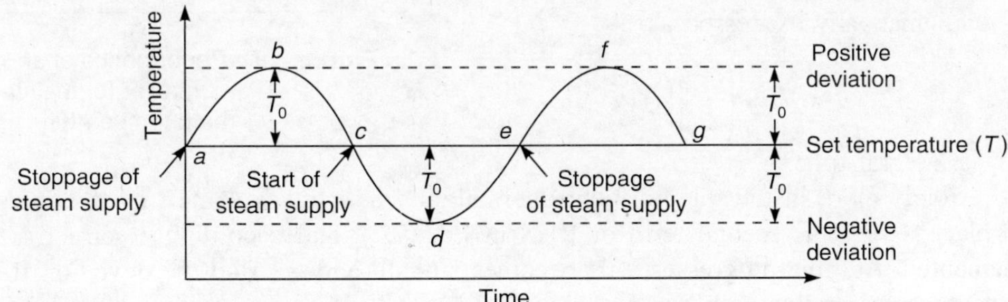

Figure 7.6 Performance of an on-off controller for maintaining the bath temperature.

Proportional controller: We have seen the difficulty in using the on-off controller which causes a positive and negative drift from the set process parameter. Subsequently attempts have been made to overcome this difficulty. This has resulted in emergence of a number of complex controllers. We shall describe them briefly here. One of them is a proportional controller.

In a proportional controller, unlike as in the case of an on-off controller, the on-off process starts much before the set point is reached. We take the earlier example of maintaining the constant temperature. As shown in Figure 7.7, the heat input stopped at point z instead of point a. Similarly heating will start at point y which is much earlier than point c. This will help reduce the drift T_0. By proper adjustment and experience, T_0 can be brought down to minimum. T_0 is known as off-set which indicates deviation from the set value.

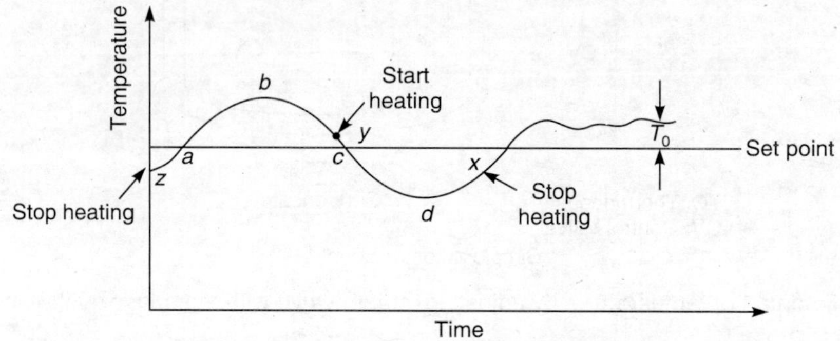

Figure 7.7 Performance of a proportional controller for maintaining the bath temperature.

The performance of a proportional controller can be mathematically expressed as:

$$I = I_0 + k_p\varepsilon \tag{7.1}$$

where I stands for output signal, I_0 stands for controller output signal without error, ε stands for error signal, and k_p stands for controller sensitivity coefficient. Depending upon the error, the corrective action is proportionately adjusted.

Integral controller: It also functions like proportional controller except that the error signal is integrated over the time frame, and the corrective action is implemented to the output signal. The performance is shown in Figure 7.8. The performance of an integral controller can be mathematically expressed as:

$$I = I_0 + \left(\frac{1}{t_i}\right)\int(\varepsilon dt) \tag{7.2}$$

where t_i is integral time.

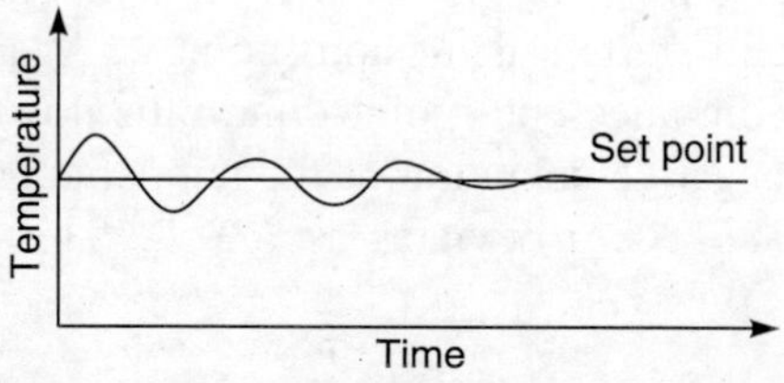

Figure 7.8 Performance of an integral controller for maintaining bath temperature.

The error is more in the integral controller initially as it takes some time for integration to take place as per the second term of RHS in Eq. (7.2), and accordingly corrective action is implemented. As time progresses, this becomes smooth and we find the deviation from the set point decreases and smoothens of.

Derivative controller: The derivative controller functions by sensing the rate of change of error signal and implements the controlling strategy. The performance of a derivative controller can be mathematically expressed as:

$$I = I_0 + t_d \frac{d\varepsilon}{dt} \tag{7.3}$$

where t_d is derivative time rate constant.

The correction for the output signal is based on the rate of change of error. If there is no rate of change of error (still there could be error!), controlling strategy will not work as the second term in RHS of Eq. (7.3) becomes zero; and hence, the error becomes constant.

P-I-D controller: The Proportional-Integral-Derivative (P-I-D) controller takes care of the set-backs in each of the controllers, and reduces the offset to the minimum. It is shown in Figure 7.9. Eventhough the drift (off set) is slightly there in the initial stages (which is much less compared to the integral controller), it vanishes as time progresses. The temperature is well controlled, and is almost equal to the set point.

Thus, the automatic controllers can be made to perform to the desired setting. Table 7.2 shows a comparison between the manual and automatic controllers.

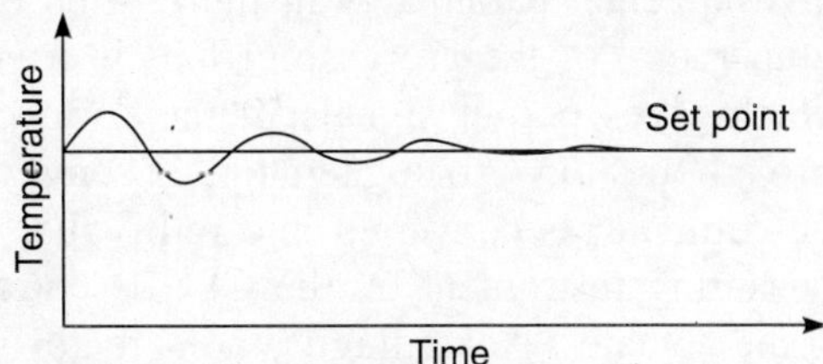

Figure 7.9 Performance of a P-I-D controller for maintaining bath temperature.

Table 7.2 Comparison of the Manual and Automatic Control Systems

Manual control	*Automatic control*
Human biased error is involved	Human biased error is absent
Manual controlling can be improved with experience	Experience does not matter
Initial investments are low	Initial investments are high
Operating costs are high	Operating costs are low
Suitable for small-scale operations	Suitable for large-scale operations
Failure errors are not fatal	Failure of the controller to respond results in fatal accidents
Routine maintenance is not required	Routine maintenance is required to check whether all controllers are functioning properly

7.3.3 Computer Control

Computers have revolutionized the whole of process control operations in any process industry. Whenever a large number of process parameters need to be controlled with precision, especially like those in a petrochemical complex, computer control is the only option left out. However, the use of computers in food processing is slightly less, and it is of recent origin. Another characteristic feature of the computer controlled systems is that the controlling is done on-line, but the controlling process is done from a remote place. Particularly in case of petrochemical complexes in which the explosion hazards are more likely to occur, the computer control room is far away from the process hall, anywhere between 200–500 metres away.

The computers have high memory power, and hence, they can handle a large number of process parameters; and all the process parameters can be controlled by a single computer. The automatic controlling units are required, but they are fed with instructions through the computer with precision. The computer is interphased with the process equipment, particularly the process control units through an Analogue-to-Digital (A/D) converter. Initially the operator issues instructions to the computer. Sensors send input signals to the computer. The logic control unit in the computer compares the sensor inputs with set values, and gives information to the operator. Based on the outcome of the logic control unit (comparing the measured values with the set values), computer gives manipulation—instructions to the actuator. Thus, the actuator manipulates the controlling parameter. Computer also gives communicative messages to the operator through various output devices.

The physical changes in a process vessel are manifested in the form of some quantifiable environmental parameters in the on-line experiments. For example, if we want to control the temperature of the processing fluid in a vessel, the temperature is measured by thermocouple which gives output signals in the form of millivolts, which in turn is fed to the computer through an A/D (Analogue/Digital) converter. The set temperature data which is already fed to the computer is converted into millivolts, and will be compared with the output stimuli of the measuring instrument by the CPU (Central Processing Unit). Instructions will then be issued to the actuator. Accordingly, the actuator will increase or decrease the electrical wattage input or steam valve to maintain the temperature.

Figure 7.10 shows interphasing a computer to the process vessel to activate a pump. Figure 7.11 shows a computer controlled single-screw extruder (Wenger, USA) of apacity 1 ton/hr.

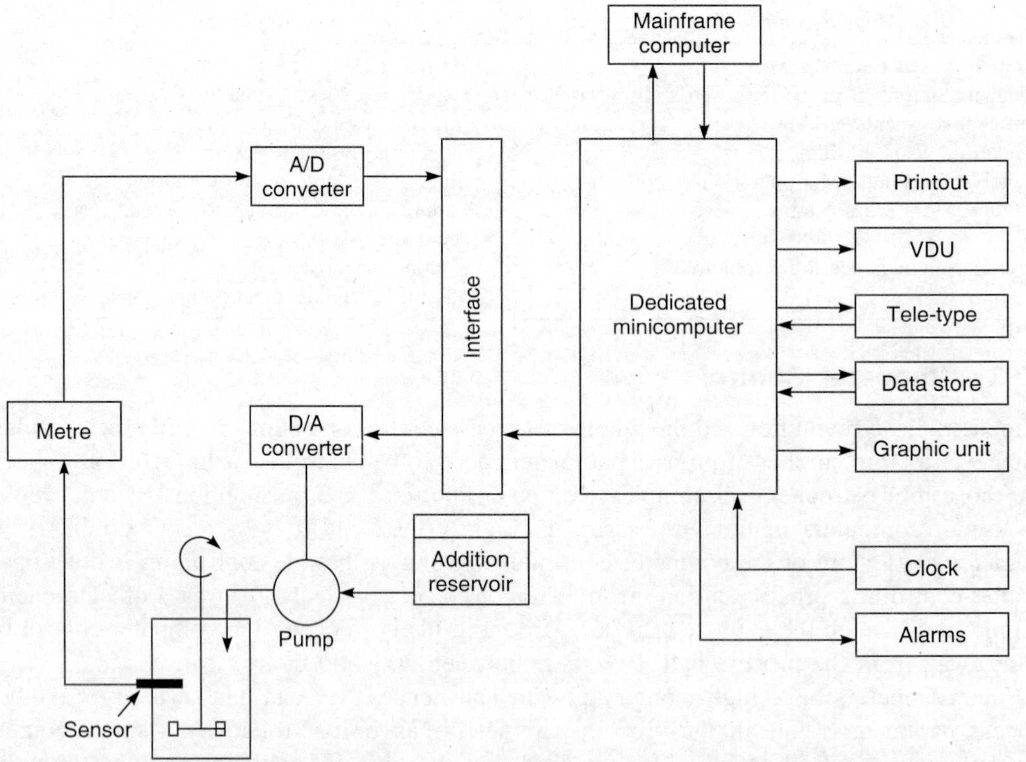

Figure 7.10 Interphasing a computer with a process equipment to control the operation of a pump.

Source: Reprinted from *Principles of Fermentation Technology*, Stanbury, P.F. and Whitaker, A., p. 164, Copyright 1993, with permisson from Elsevier.

Figure 7.11 Computer controlled twin-screw extruder.

Source: Printed with permission from M/s Telangana Foods, Hyderabad.

Symbols

I: output signal of controller

I_0: controller output signal without error

k: sensitivity coefficient

T: set temperature

T_0: drift in temperature

t: time function

Subscripts

D: derivative controller

I: integral controller

P: proportional controller

P-I-D: Proportional-Integral-Derivative controller

Greek Symbol

ε: error signal

REVIEW QUESTIONS

7.1 What are various process parameters that need to be controlled in a food processing plant?

7.2 What is the importance of process control in food processing operations?

7.3 What is meant by on-line process control and off-line process control?

7.4 Differentiate between the critical and non-critical parameters.

7.5 Describe the instruments used for measurement of temperature.

7.6 How do you measure and control the pressure in any process line?

7.7 Describe the functioning of a load cell.

7.8 What are different controlling methods?

7.9 Describe manual controlling of temperature in a process plant.

7.10 Describe what are various automatic controlling methods in a food processing plant.

7.11 Describe an on-off controller.

7.12 Describe a proportional controller.

7.13 Describe an integral controller.

7.14 Describe a derivative controller.

7.15 Describe a P-I-D controller.

7.16 Describe how a computer is interphased in a process plant for process control.

7.17 How do you measure the following process parameters:

(i) pressure,
(ii) weight,
(iii) pH, and
(iv) flow rate.

REFERENCES

Rao, D.G. (2005), *Introduction to Biochemical Engineering*, McGraw-Hill, New Delhi, p. 346.

Stanbury, P.F. and Whitaker, A. (1993), *Principles of Fermentation Technology*, Pergamon Press, Oxford, p. 164.

Part III

Unit Operations

CHAPTER

8

Fluid Mechanics

A fluid is a substance which is amenable to distortion when a force is applied upon it. It takes the shape of the container unlike solids. Thus, vapours, gases and liquids are classified as fluids. A special characteristic feature of fluids is that they flow through pipes, conduits or open channels whenever there is a pressure differential. They flow from high level to low level under gravity. Whenever they are contained in a vessel, they exert a pressure on the walls of the vessel as well as on the bottom of the vessel.

The fluids are differentiated from solids mainly in terms of their flowability. Solids generally do not flow. Of course solids in fine powder form flow through pipes and conduits like fluids, the concept of which is used in combustion of pulverized coal in burners like liquid fuels. They may only slide in conduits whenever there is a potential difference. Hence, fluids have *viscosity* (which is a measure of the resistance to flow) while solids do not have.

The fluids have also got the typical physical property called **density**, which is defined as the mass of the material per unit volume. For some fluids, the density changes with the pressure whereas for others it does not. The former types of fluids are called **compressible fluids**, whereas the latter types of fluids are known as **incompressible fluids**. Mostly the gases and vapours are compressible fluids, whereas the liquids are incompressible fluids. The density of solids is not affected by pressure at all.

Motion of fluids through pipes and conduits when there is a pressure loss or energy loss is a subject matter of *fluid dynamics*, whereas the properties exerted by stagnant fluids are studied as *fluid statics. Fluid mechanics* is a study of both fluid dynamics and fluid statics. However, we restrict ourselves mostly to fluid dynamics in the fluid mechanics, as the motion of fluids through pipes and conduits is more important and relevant to us in food engineering.

8.1 FLUID PRESSURE

Fluid pressure is an important concept. Fluids in static form exert pressure on the walls of the container also; but this pressure is found to be independent of the orientation of any internal surface/wall upon which the pressure acts [McCabe, et al. 1993 (a)]; i.e., the pressure in any direction of the wall will be the same. The pressure changes only with increase/decrease in heights which we generally call as *hydrostatic head.* The pressure at point *a* in a column of fluid shown in Figure 8.1 is p_a. It could be any pressure acting on the column, like atmospheric pressure. Hence, pressure at point *b* includes the pressure at point *a* and gravitational pull of the liquid column above *b* and is given by:

$$p_b = p_a + \rho_l\,(\Delta H) \tag{8.1}$$

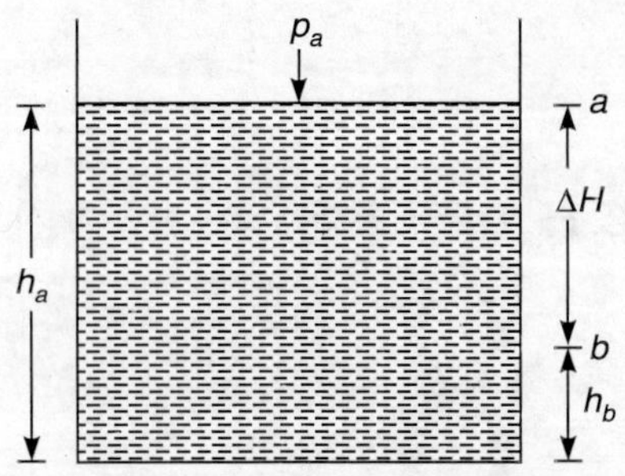

Figure 8.1 Column of fluid in a rectangular container of cross-sectional area *A*.

where ρ_l is the density of the fluid and ΔH is the height difference between the points *a* and *b*. If there are no additional pressures/forces acting at point *a*, then pressure exerted by the fluid is known as the **hydrostatic head**, and is given by:

$$\Delta p = p_b - p_a = \rho_l\,(\Delta H) \tag{8.2}$$

Δp given by Eq. (8.2) is known as the **gauge pressure**, if p_a is the atmospheric pressure. Thus, we define *gauge pressure is the one which is exerted over and above the atmospheric pressure.* It is the pressure measured by the pressure gauge which is obviously above the atmospheric pressure. Thus, if we want to measure absolute pressure, we add atmospheric pressure value to the gauge pressure.

$$p_{abs} = p_g + \text{atm. pressure} \tag{8.3}$$

The hydrostatic head at point *b* is given by Eq. (8.2) and it is considered as the *fundamental equation of fluid pressure.*

PROBLEM 8.1 Find the absolute and gauge pressures exerted by a liquid column of 2 m height at the bottom of the tank. The specific gravity of the liquid is 1.1.

Solution Density of the fluid = ρ_f = 1.1 × 1000 = 1100 kg/m^3

$$\begin{aligned}
p_g &= \rho_f\,(\Delta H) \\
&= 1100 \times 2 = 2200 \text{ kg/m}^2 \\
&= \frac{2200}{10^4} \text{ kg/cm}^2 = 0.22 \text{ kg/cm}^2 \\
&= 0.22 \times 101.325 = 22.3 \text{ kPa}^{\dagger} \\
p_{abs} &= 101 + 22.3 = 123.3 \text{ kPa}
\end{aligned}$$

Pressure and atmospheric pressure are also denoted by various units, viz., psi (pounds per square inch) feet of water head, inches of mercury head, millimeters of mercury head. Thus,

$$\begin{aligned}
1 \text{ atm} &= 14.7 \text{ psi} \\
&= 33.9 \cong 34 \text{ feet of water head}
\end{aligned}$$

---

†kPa stands for kilo Pascal, N/m^2 stands for Newton/square metre, and MN/m^2 stands for mega Newton per square metre. The pressure is denoted by kg/cm^2 or N/m^2 or Pascal.

= 29.6 inches of mercury
= 760 mm of mercury
= 76 cm of mercury
= 1 kg/cm^2
= 101325 Pascals
= 101.3 kPa
= 101325 Newtons/m^2
= 1 × 10^5 Newtons/m^2
= 0.1 MN/m^2

8.2 MANOMETER

It is generally made out of glass in the form of U, and hence, is popularly known as **U-tube manometer**. The two limbs of the U-tube are connected to a pipe through which fluid is passing. Due to the flow, there will be difference in pressure at two limbs of the U-tube. The differential pressures are in turn calibrated to the volumetric flow rates of the fluids passing through the tubes. To increase the pressure differential, an orifice plate is inserted in the flow path; orificemeter will be described later in the chapter. A typical U-tube manometer is shown in Figure 8.2. The U-tube portion between a' and b' is filled with the manometer fluid (density ρ), whereas the columns aa' and bb' are filled with the fluid (density ρ_f) passing through the pipe.

$$\text{Pressure at } a' = p_a + h_a\,\rho_f \tag{8.4}$$
$$\text{Pressure at } b' = p_b + h_b\,\rho_f \tag{8.5}$$
$$\text{Pressure at } a'' = \text{pressure at } b' + \rho(\Delta H) \tag{8.6}$$

Obviously pressures at points a' and a'' are equal. Hence,

$$p_a + h_a\,\rho_f = p_b + h_b\,\rho_f + \rho(\Delta H) \tag{8.7}$$

or
$$(p_a - p_b) = \Delta p = h_b\,\rho_f - h_a\,\rho_f + \rho(\Delta H)$$

h_a can be written as $h_b + \Delta H$.

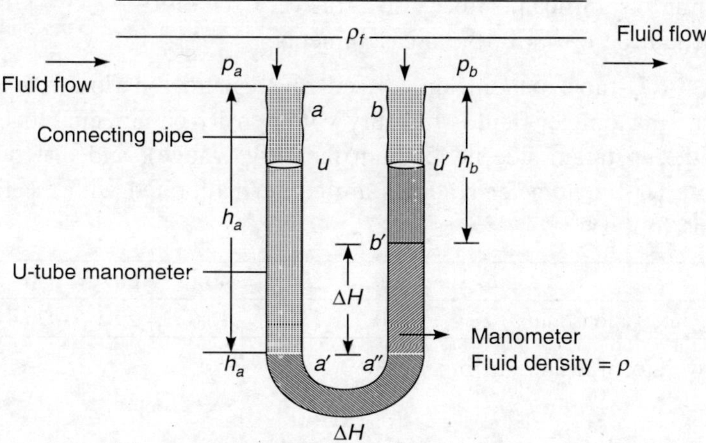

Figure 8.2 U-tube manometer.

Hence, $(\Delta p) = h_b\ \rho_f - h_b\ \rho_f - (\Delta H)\ \rho_f + \rho\ (\Delta H)$

$$\therefore \quad (\Delta p\ ) = (\Delta H)\ (\rho - \rho_f) \tag{8.8}$$

Thus, the pressure drop ($-\Delta p$) between the points a and b can be calculated by measuring the difference in the levels of the manometer fluid in two limbs (ΔH). It may be noted from the above equation that the distance between the ports of the pipe and limbs of the U-tube manometer is immaterial. Hence, the manomter can be at any distance away from the pipe. The following points are important to note while using the U-tube manometer:

(i) The distance between *au* and that between *bu′* are not important.
(ii) The manometer fluid should not be same as that flowing in the pipe.
(iii) The manometer fluid should be immiscible with the fluid in the pipe.
(iv) The density of the manometer fluid should be more than that of the fluid in the pipe, i.e. $\rho > \rho_f$
(v) The manometer fluid should be so chosen that $(\rho - \rho_f)$ is small, so that (ΔH) can be large for a given pressure drop, and hence, can be measured more accurately.
(vi) ΔH can be calibrated with the fluid flow rate so that the U-tube manometer can be used to measure the flow rates of the fluids passing through the pipes by noting down (ΔH).

If the ΔH values are less, an *inclined manometer* (Figure 8.3) can be used so that

$$\Delta H = l \sin \theta \tag{8.9}$$

or

$$\Delta p = (\rho - \rho_f)\ l \sin \theta \tag{8.10}$$

The leg a should have an enlargement as shown in Figure 8.3. This will help when there is a change in level during operation, the fall in the level in the bulged portion is negligible, whereas the increase in the level in the measured column is significant. Otherwise ΔH is very small, and there could be errors in the measurements.

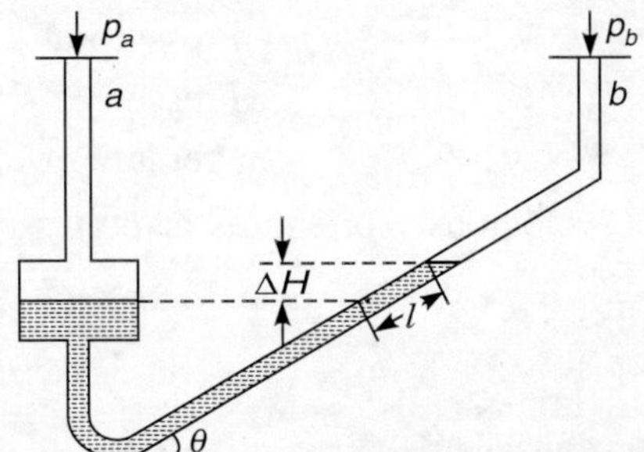

Figure 8.3 Schematic representation of an inclined manometer.

PROBLEM 8.2 A U-tube manometer is used to measure the flow rate of water flowing through a pipe. The manometer fluid is mercury with density being equal to 13.68×10^3 kg/m^3. The pressure on the upstream side is 450 mm mercury (gauge), and that on the downstream side is 0.5 kg/cm^2. The manometer readings are earlier calibrated with the flow rate of water and is tabulated as follows:

ΔH, mm	10	20	40	60	80	100
Water flow rate gallons/min	4.0	7.7	15.0	22.5	26.4	30.5

Calculate the flow rate of water in lpm.

Solution

Pressure on upstream side, $p_a = 450$ mm of Hg

$$= \frac{450}{760} = 0.592 \text{ atm.}$$

Pressure on downstream side, $p_b = 0.5$ kg/cm^2

$$= 0.5 \text{ atm.}$$

$$\Delta p = 0.592 - 0.5 = 0.092 \text{ atm.}$$

$$= 0.092 \times 10^4 \text{ kg/m}^2$$

$$\rho - \rho_f = (13.6 - 1.0) \times 10^3 \text{ kg/m}^3$$

$$\Delta p = (\Delta H)\,(\rho - \rho_f)$$

$$\Delta H = \frac{0.092 \times 10^4}{12.6 \times 10^3} = 0.073 \text{ m}$$

$$= 73 \text{ mm}$$

We can draw a calibration graph with the above tabular data, and can be used for measuring the flow rate corresponding to 73 mm of ΔH. From the calibration chart (not shown), the flow rate = 24.4 gpm = 24.4 × 4.546 = 110.9 lpm.

By interpolation method: Flow rate corresponding ΔH of 73 mm

$$= 26.4 - \left\{\frac{26.4 - 22.5}{(80 - 60)} \times (80 - 73)\right\} = 25 \text{ gpm.}$$

8.3 BERNOULLI'S EQUATION

Bernoulli's equation is related to equating the hydrostatic heads or energy of the fluid at any two different points in a closed conduit/pipe. It is a corollary of energy balance equation or momentum balance equation. The equation can be applied for flow through pipes without friction and flow through pipes with friction. It can even be applied when the flow is assisted by a pump also. Initially we consider flow through the pipe without friction.

8.3.1 Bernoulli's Equation without Friction

When a fluid is passing through the pipe which does not offer resistance for the flow, we can write the energy balance at any two points, and equate them. Let us consider two points at positions a and b in the pipe (Figure 8.4)

The head due to pressure at $a = \dfrac{p_a}{g \cdot \rho_a}$

The head due to pressure at $b = \dfrac{p_b}{g \cdot \rho_b}$

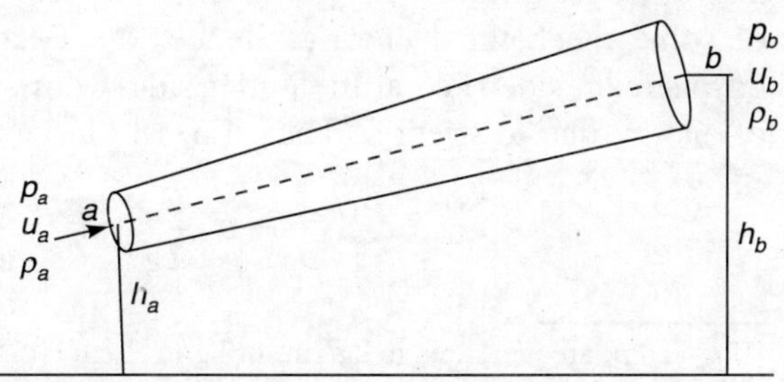

Figure 8.4 Flow of fluid through a pipe of varying cross-section.

The head due to velocity or kinetic energy at $a = \frac{u_a^2}{2g}$

The head due to velocity or kinetic energy at $b = \frac{u_b^2}{2g}$

The potential head at $a = h_a$
The potential head at $b = h_b$
Making energy balance equation,

$$\frac{p_a}{g \cdot \rho_a} + \frac{u_a^2}{2g} + h_a = \frac{p_b}{g \cdot \rho_b} + \frac{u_b^2}{2g} + h_b \qquad (8.11)^{\dagger}$$

Equation (8.11) is Bernoulli's equation for flow of fluids through pipes without considering the friction for flow. Each term in above equation is in the units of metres which is in the form of hydrostatic head. The above equation is very useful in finding the velocity of flow or the pressure at the outlet position based on the information at the inlet. The equation tells us that the pressure energy or kinetic energy or potential energy of the fluid is inter-convertible to others. The higher pressure at inlet of the pipe can only be converted into either higher kinetic energy by increasing u_b or the fluid can go to higher heights by increasing the potential energy.

8.3.2 Bernoulli's Equation with Correction for Kinetic Energy Factor

When a fluid is flowing through a pipe, whatever be the linear velocity (m/s) we are measuring is only the average velocity; different layers of the fluid in the pipe will have different velocities varying from zero (since the pipe is stagnant) to a maximum velocity. Hence, accordingly the kinetic energy term will also be varying. To account for this variability we incorporate a correction factor known as **kinetic energy correction factor**, usually represented by α.

The head due to kinetic energy $= \alpha \frac{\bar{u}^2}{2g}$.

The value of α may be taken as 2.0 for laminar flow and 1.05 for highly turbulent flow [McCabe, et al., 1993 (b)].

8.3.3 Bernoulli's Equation with Correction for Friction in the Pipe

When the fluid is flowing through a pipe, it experiences friction at the walls of the pipe, and also amongst different layers of the fluid. The frictional resistance for the flow results in loss of some mechanical energy. Hence, the Bernoulli's equation must be incorporated on the downstream side (i.e., at the outlet side) with a frictional loss term which accounts for the loss of energy due to friction. Thus, Eq. (8.11) becomes

$$\frac{p_a}{g \cdot \rho_a} + \alpha_a \frac{\bar{u}_a^2}{2g} + h_a = \frac{p_b}{g \cdot \rho_b} + \alpha_b \frac{\bar{u}_b^2}{2g} + h_b + H_f \qquad (8.12)$$

---

†If p_a and p_b are measured in kg/cm² or kg/m² Eq. (8.11) would be:

$$\frac{p_a}{\rho_a} + \frac{u_a^2}{2g} + h_a = \frac{p_b}{\rho_b} + \frac{u_b^2}{2g} + h_b.$$

All the terms in the above equation will have units of metres. H_f represents energy lost due to friction in the overall length of the pipe, unlike other terms which represent value at the designated points *a* and *b*. Different forms of frictions which prevail during flow of fluids through pipes and conduits will be dealt separately in the later part of this chapter.

8.3.4 Pumping Energy Term in Bernoulli's Equation

Additional energy is required for the fluid flowing through the pipes:

(i) to compensate for the energy lost due to friction or
(ii) to increase the kinetic energy of the fluid at the outlet or
(iii) to increase the pressure at the outlet of the fluid or
(iv) to increase the potential head of the fluid.

Such additional energy is provided to the system by incorporating a pump in the flow line. If H_p is the head contributed due to the pumping energy of the pump, it will be added at LHS of the equation, and Eq. (8.12) is written as:

$$\frac{p_a}{g \cdot \rho_a} + \alpha_a \frac{\bar{u}_a^2}{2g} + h_a + H_p = \frac{p_b}{g \cdot \rho_b} + \alpha_b \frac{\bar{u}_b^2}{2g} + h_b + H_f \qquad (8.13)$$

Since all mechanical pumps will have invariably some loss of energy because of friction within the pump, an efficiency factor is also to be incorporated into H_p. Thus,

$$H_p = \eta W_p$$

where W_p is the energy to be supplied to the pump and η is the efficiency factor which is always < 1.0 Eq. (8.13) can be written as:

$$\frac{p_a}{g \cdot \rho_a} + \alpha_a \frac{\bar{u}_a^2}{2g} + h_a + \eta W_p = \frac{p_b}{g \cdot \rho_b} + \alpha_b \frac{\bar{u}_b^2}{2g} + h_b + H_f \qquad (8.14)$$

PROBLEM 8.3 A liquid fruit juice is pumped using a one h.p. pump from a storage tank at the ground level to the processing zone at 10 m high through a 5.08 cm (2 inch) diameter pipe. The efficiency of the pump is 65 per cent. The discharge velocity of juice at the outlet is 260 lpm. The specific gravity of juice may be taken as 1.1. Find the frictional losses in the pipe.

Solution Volumetric flow rate of juice = 26 lpm

$$= \frac{260 \times 10^{-3}}{60} = 4.33 \times 10^{-3} \text{ m}^3/\text{s}$$

Density of fruit juice = 1.1×10^3 kg/m^3
Mass flow rate = $4.33 \times 10^{-3} \times 1100 = 4.763$ kg/s
Area of cross-section of the pipe at the outlet = A_b

$$A_b = \pi\, (2.54 \times 10^{-2})^2 = 2.03 \times 10^{-3} \text{ m}^2$$

$$u_b = \frac{4.33 \times 10^{-3}}{2.03 \times 10^{-3}} = 2.136 \text{ m/s}$$

Since the fruit juice is being pumped from a tank at atmospheric pressure, and is getting discharged into a tank at atmospheric pressure

$$p_a = p_b = 1 \text{ atm}$$

Hence,

$$\frac{p_a}{g \cdot \rho_a} = \frac{p_b}{g \cdot \rho_b}$$

If

$$h_a = 0;\ h_b = 10 \text{ m}$$

Hence, Eq. (8.14) can be written as

$$\eta W_p = \alpha_b \frac{\bar{u}_b^2}{2g} + h_b + H_f$$

In the preceding equation, units of W_p is to be suitably modified.

$$\text{h.p. of motor} = 1 = 746 \text{ watts}$$

$$1 \text{ kg. m/s} = 9.807 \text{ W}$$

$$1 \text{ h.p.} = \frac{746}{9.807} = 76 \text{ kg-m/s}$$

$$\eta \text{ of pump} = 0.65$$

∴ Power supplied by the pump = 0.65 × 76 = 49.4 kg-m/s

Mass flow rate of fluid = 4.763 kg/s

$$\text{Energy component of the motor per unit wt} = \frac{49.4}{4.763} = 10.37 \text{ kg-m/kg}$$

∴

$$\eta W_p = 10.37 \text{ m}$$

Bernoulli's equation:

∴

$$\eta W_p = \alpha_b \frac{\bar{u}_b^2}{2g} + h_b + H_f$$

i.e.,

$$10.37 = \frac{(2.136)^2}{2 \times 9.807} + 10 + H_f$$

∴

$$H_f = 10.37 - 0.273 - 10 = 0.097 \cong 0.1 \text{ m.}$$

We get the energy lost by the friction by multiplying the hydrostatic head with the fluid rate. Therefore, H_f = 0.1 × 4.763 = 0.4763 kg.m/s = 0.4763 × 9.807 = 4.67 watts.

8.4 FLOW THROUGH PIPES AND CONDUITS

Fluids passing through a pipe or flowing over a stagnant surface have an interesting flow characteristics with varying velocity profiles. The fluid flows in the form of different layers one over the other. There will be always a layer of constant thickness which is adhering to the surface and does not move at all, i.e., its velocity is zero. The layer above it will have some velocity. As the layer thickness increases, the velocity also increases slowly and reaches the maximum velocity which is the velocity of the flowing fluid.

8.4.1 Boundary Layer

The situation is schematically represented in Figure 8.5. All the fluid elements having different velocities which are increasing with thickness constitute what we call as the **boundary layer**. The velocity profile is shown in Figure 8.5(b). This kind of boundary layer separation takes place even in case of flow of fluids through circular pipes which are stagnant. This layer is customarily called as **hydrodynamic boundary layer** to distinguish it from the *thermal boundary layer* (which we will study in section 11.2.1). Since the fluid elements in the boundary layer have varying velocities, they will also exhibit varying characteristic features in *transport phenomenon.* The boundary layer thickness diminishes with increase in velocity.

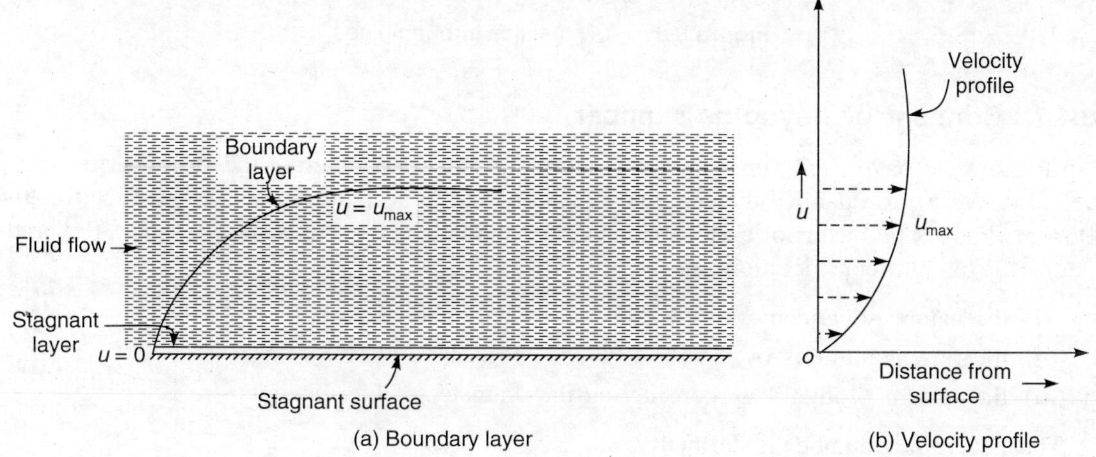

Figure 8.5 Boundary layer separation.

8.4.2 Flow Characterization

Flow of fluids through pipes is an interesting phenomenon. Depending upon the fluid velocity, the flow could be stream-lined or otherwise. In a stream-lined flow, which occurs when the fluid velocity is low, all the layers of fluid move in an orderly manner as shown in Figure 8.6(a). There is no intermixing of one layer with the other. If the velocity of the flow is increased at this stage, the disturbances in the flow start raising up, and chaotic situation starts prevailing [Figure 8.6(b)]. If at this stage, if the velocity is further increased the chaotic flow becomes fully disturbed. There will be complete mixing of all the layers with each other as shown in Figure 8.6(c).

The smooth flow as shown in Figure 8.6(a) is known as **laminar flow**, and the flow shown in Figure 8.6(c) is **turbulent flow.** The intermediate flow is known as **transient flow**.

One of the main reasons for disturbances in the flow is fluid velocity, the other things being

- fluid characteristics, and
- channel characteristics.

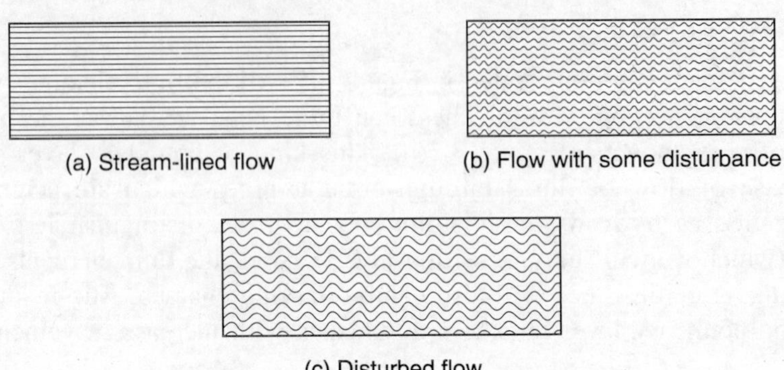

Figure 8.6 Flow of fluids through pipes.

8.4.3 Concept of Reynolds Number

Up to now we have used some qualitative statements to characterize the flow based on the velocity. We try to quantify the same now. One of the important concepts to characterize the flow of fluids is by means of a dimensionless number called **Reynolds number**. The Reynolds number takes into consideration

(i) the nature of fluid by considering its viscosity and density.
(ii) the flow conditions by the fluid velocity (linear velocity).
(iii) the system geometry by considering the diameter of the tube/pipe.

Thus, Reynolds number is defined as:

$$N_{Re} = \frac{\rho d u}{\mu} \tag{8.15}$$

where ρ is the density of the fluid, kg/m^3; d is the diameter of the pipe, m; u is the linear velocity of the fluid in the pipe in terms of m/s; and μ is velocity of the fluid, Pas (kg/m.s).

The linear velocity of the fluid is obtained by dividing the volumetric flow rate with the internal cross-sectional area of the pipe.

The Reynolds number is the *ratio of the kinetic forces to the viscous forces of the fluid flow*. Thus, the numerator in Eq. (8.15) represents the kinetic forces of the fluid mainly because of the fluid velocity. The denominator indicates viscous forces represented by viscosity, i.e., if the viscosity is high, the Reynolds number will be low and the flow will be stabilized. Similarly if the velocity is high, the Reynolds number will be high and the flow will be disturbed. The flow characterization is as follows:

If $N_{Re} < 2100$, the flow is laminar
$N_{Re} > 4000$, the flow is turbulent
N_{Re} is in between 2100 to 4000, the flow is transient (8.16)

Reynolds number is an important concept. It characterizes the mixing pattern which affects heat transfer and mass transfer characteristics. Hence, we will be using this number in transport phenomena very frequently, whether it is momentum transport, mass transport or heat transport.

PROBLEM 8.4 Orange juice is passing through an SS pipe, 1″ diameter, schedule 40. The specific gravity of orange juice is 1.05 and its viscosity is 2.5 cp. The volumetric flow rate of the juice is 60 lpm. Calculate the Reynolds number, and find whether the flow is laminar or turbulent.

Solution Cross-sectional area of the pipe of 1″ diameter

Schedule 40 = 0.006 ft ^{2} (Appendix 5)

$$= 0.006 \times (30)^2 = 5.4 \text{ cm}^2 = 5.4 \times 10^{-4} \text{ m}^2$$

$$\text{Diameter of pipe} = d = 1.049'' = 2.625 \text{ cm}$$

$$= 2.625 \times 10^{-2} \text{ m}$$

$$\text{Volume flow rate} = 60 \text{ lpm} = \frac{60 \times 10^{-3}}{60} = 1 \times 10^{-3} \text{ m}^3/\text{s}$$

$$u = \frac{1 \times 10^{-3}}{5.4 \times 10^{-4}} = 1.85 \text{ m/s}$$

$$\text{Viscosity of juice} = 2.5 \text{ cp} = 2.5 \times 10^{-3} \text{ Pas}$$

$$\text{Density of juice} = 1.05 \times 10^3 \text{ kg/m}^3$$

$$\therefore \quad N_{Re} = \frac{\rho du}{\mu} = \frac{(1.05 \times 10^3) \times (2.625 \times 10^{-2}) \times (1.85)}{2.5 \times 10^{-3}}$$

$$= 2.04 \times 10^4$$

Since, the $N_{Re} > 4000$, the flow is turbulent.

8.5 FRICTION IN PIPES

When the fluid is flowing through pipes, some amount of energy is lost depending upon the smoothness or roughness of the surface of the pipe. Rarely surface of the pipe could be perfectly smooth. The fluid layer just adhering to the walls of the pipe will have the velocity of the surface; since the pipe is stagnant, the layer will have zero velocity. The layers over it will be moving gradually because of the drag by the upper layers which are moving. This results in shear stress on different layers of the fluid. If we visualize an element of fluid layer of radius r_i in a bulk fluid flowing through the pipe (Figure 8.7), it will be under equilibrium of the pressure p acting in the direction of the flow and the shear stress τ_i acting in the opposite direction. By balancing both the forces acting on the element of the fluid, we can write

$$\frac{-dp}{dL} = \frac{2\tau_i}{r_i} \tag{8.17}$$

Similarly by extending the concept to the wall of the pipe, we can write

$$\frac{-dp}{dL} = \frac{2\tau}{r} \tag{8.18}$$

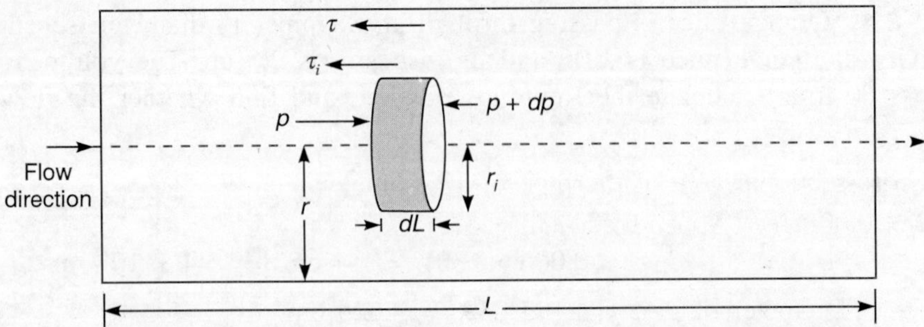

Figure 8.7 Fluid element in a fluid stream under equilibrium.

Therefore,

$$\frac{\tau_i}{r_i} = \frac{\tau}{r}$$

i.e., the shear stress will be increasing proportionate to the radius as we go towards the wall of the pipe. Armed with the above equation in our hands, it will be now possible for us to draw a relationship between the shear and the energy lost due to friction in the pipes. Applying the Bernoulli's equation along with the correction for friction in the pipe (Eq. 8.12) in which the potential head and kinetic head are the same in a horizontal pipe.

$$\frac{\Delta p}{g\rho} = H_f \tag{8.19}$$

The energy is lost due to the friction at the walls of the pipe. Hence, this friction is known as **skin friction**, and accordingly in the above equation H_f is replaced by H_{fs} to mean energy lost due to skin friction. Thus,

$$\frac{\Delta p}{g \cdot \rho} = H_{fs} \tag{8.20}$$

Replacing Δp by Eq. (8.18),

$$H_{fs} = \frac{2\tau\Delta L}{\rho r g} = \frac{4\tau\Delta L}{\rho d g} \tag{8.21}$$

Equation (8.21) shows the relationship between the skin friction and shear at the wall. In Eq. (8.21), τ is in the units of N/m^2. If τ is represented in kg/cm^2 or kg/m^2, the above equation becomes

$$H_{fs} = \frac{4\tau}{\rho d}(\Delta L) \tag{8.21a}$$

8.5.1 Friction Factor

Friction factor is an important parameter particularly in turbulent flow of fluids. It is usually denoted by symbol, f and is *the ratio of the wall shear stress to the kinetic energy head* given by $(\rho u^2/2g)$. Thus,

$$f = \frac{\tau_w}{\rho u^2/2g} = \frac{2g\tau_w}{\rho u^2} \tag{8.22}$$

Here τ_w stands for shear stress at the wall. Replacing τ_w by Eq. (8.18) as $(\Delta p/\Delta L) \times (r/2)$ and r by $(d/2)$ we get

$$f = \frac{(\Delta p)}{\Delta L}\frac{gd}{2\rho u^2}$$

In the preceding equation, u is the average velocity of fluid in m/s which is obtained by dividing the volumetric flow rate of the fluid with the cross-sectional area of the pipe. Sometimes, it is customary to add subscripts to (Δp) in the above equation to indicate that the pressure drop is only due to skin friction and does not include *from friction*, Eq. (8.22) is written as

$$f = \frac{(\Delta p_s)}{\Delta L}\frac{gd}{2\rho u^2} \tag{8.23}$$

If form friction is also included, $(p_a - p_b)$ in the Bernoulli's equation is not just equal to (Δp_s) it could be much more. The relationship between the friction factor f in Eq.(8.23) and the Reynolds number was found to be (Mc Cabe, et al., 193(c)).

$$f = \frac{16}{N_{Re}} \tag{8.24}$$

$$= \frac{16\mu}{\rho du}$$

Combining Eqs. (8.23) and (8.24) we get

$$\frac{(\Delta p_s)}{\Delta L} = \frac{32\mu u}{gd^2} \tag{8.25}$$

$$= \frac{32\mu u}{d^2} \quad \text{if } \Delta p_s \text{ is in N/m}^2$$

Equation (8.25) is the popular *Hagen–Poiseuille equation* which gives a relationship for the pressure drop per unit length of the pipe $(\Delta p/\Delta L)$ in terms of the fluid property (μ), diameter of the pipe (d) and the flow condition represented by average velocity (u).

PROBLEM 8.5 For Problem 8.4, find the pressure drop per unit length of the pipe using Hagen–Poiseuille equation:

Solution From Eq. (8.25)

$$\frac{\Delta p}{\Delta L} = \frac{32\mu u}{gd^2} = \frac{32 \times 2.5 \times 10^{-3} \times 1.85}{9.81 \times (2.625 \times 10^{-2})^2}$$

$$= 21.9 \frac{\text{kg/m}^2}{\text{m}} \qquad (\because \quad 1 \text{ kg} = 9.81 \text{ newton})$$

Therefore, $\dfrac{\Delta p}{\Delta L} = 21.9 \times 9.81 = 214.8 \text{ N/m}^2$ per metre

8.5.2 Friction Factor Chart

Depending upon the roughness of the pipe's inner surface, the friction factor also varies. However, its effect (the effect of roughness) is insignificant in the laminar zone. The friction factor chart drawn between f and N_{Re} is a very useful document for design purposes to evaluate the friction factor and accordingly the pressure drop. It is shown in Figure 8.8 [after Mc Cabe, et al., 1993(d)].

Figure 8.8 is applicable only for Newtonian fluids[†]. For non-Newtonian fluids, the charts are slightly different and was dealt by Dodge and Metzner (1959).

PROBLEM 8.6 From Problem 8.4, find the friction factor of the pipe for the flow conditions.

Solution The Reynolds number is 2.04×10^4

From Figure 8.8, the friction factor for Reynolds number for a smooth pipe is 0.0063.

8.6 PIPELINE FITTINGS

Fluid flow lines consist of pipes and, or tubes through which the fluids flow. For this purpose, the pipes are made of standard dimensions. The notation of pipes and tubes is used interchangeably, even though some subtle difference is there. Pipes generally are those which have large diameters and higher wall thickness. The diameter range at least starts from ¼ inch (6 mm) and goes upto 30 inch or so. Since the wall thicknesses are high, they have a definite length, usually 10 ft (3.048 m) or 20 ft (6.096 m). Tubes usually come in large lengths in the form of bundles and diameters less than 0.25 inch (6 mm). Pipes are described in the form of a *schedule number* along with the diameter. For example, when we say 2 inch schedule 40 SS pipe, we mean it is a pipe of the following description (Appendix 5):

Nominal pipe size	= 2 inch
Outside diameter	= 2.375 inch
Wall thickness	= 0.154 inch
Inside diameter	= 2.067 inch
Cross-sectional area of metal	= 1.075 inch^2
Inside sectional area	= 0.0273 ft^2
Circumference outside	= 0.622 ft
Circumference inside	= 0.541 ft
Pipe weight	= 3.65 lb/ft length

Schedule 40 means above data. Dimensions for standard sizes for steal pipes are given in Appendix 5 which are based on ASME (American Society of Mechanical Engineers) standards.

Since the pipes and tubes come in definite sizes, they need to be joined for making up to a definite length. They need to be bent to fit into a definite setup. The flow rate of the fluids also need to be controlled in any process operation. All these fabrication methodologies require a large number of fittings. Table 8.1 illustrates a few of them. They are briefly described here:

[†]Newtonian fluids are those which maintain a proportionality relationship between the shear stress shear strain. Fluids which do not maintain this relationship are known as **non-Newtonian fluids**, and will be discussed in Chapter 9.

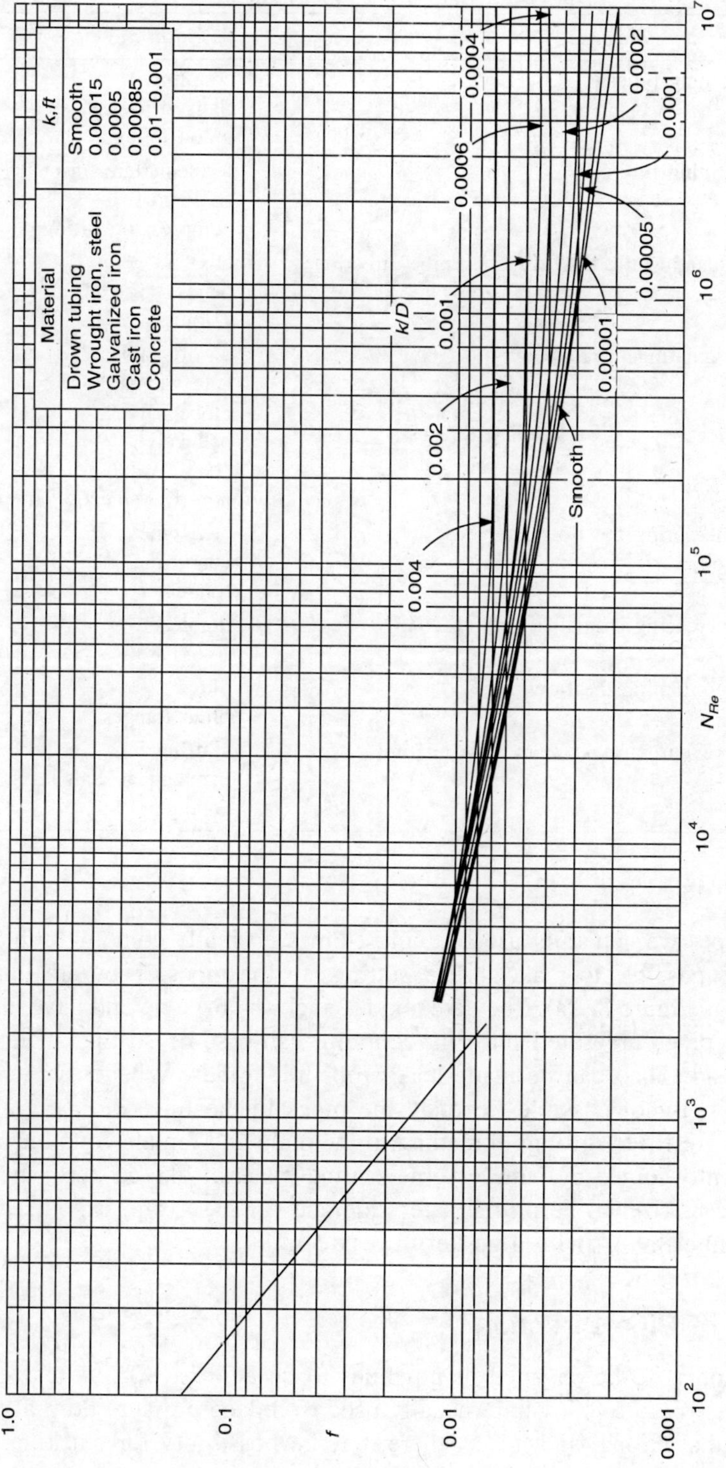

Figure 8.8 Friction factor chart.

Table 8.1 Pipeline Fittings

Purpose	*Fittings*
For joining pipes	couplings flanged joints union joints
For joining tubes	ferules (flare fittings) soldered fittings compression fitting
For changing the flow direction/pipe direction	elbows bends *T*-joints
For regulating the flow	on-off valves gate valves globe valves ball valves needle valves non-return/check valves
For measuring the flow rate	orificemeter venturimeter rotameter
For increasing a small length (1″–6″)	nipples *T*-joints
For terminating the flow	plugs blind flanges
For preventing leaks at moving/rotating parts	stuffing box mechanical seals O-rings

8.6.1 Pipe Joints

For joining two pipes we generally use couplings. Pipes normally come with threaded endings. If it is not so it is possible to make thread cuttings to the pipes. They are joined simply by using pipe couplings Figure 8.9(a). They are about 2 inch (5 cm) long and have inside threading. The two threaded pipes are fitted into the coupling using some sealing material like teflon tape or cotton thread. They can be made leak proof for liquids. Whenever there are bends in the pipelines and it is not possible to rotate the pipes in the line, we use union joints. The pipes to be joined are fitted through threading to the male and female component of the union. They are subsequently joined by putting some sealing material like gasket in between them. It is shown in Figure 8.9(b). To be on the safer side and for gases and vapours, we usually use flanged joints [Figure 8.9(c)] with a gasket in between.

8.6.2 Change of Pipe Direction

If the flow direction is to be changed or pipelines to be bent by 90°, we use elbow joints or bends. Elbow joints [Figure 8.9(d)] are usually used to reduce pressure drop at the bend. They are not generally that strong, and hence, there may be possibility for cracking. In such cases

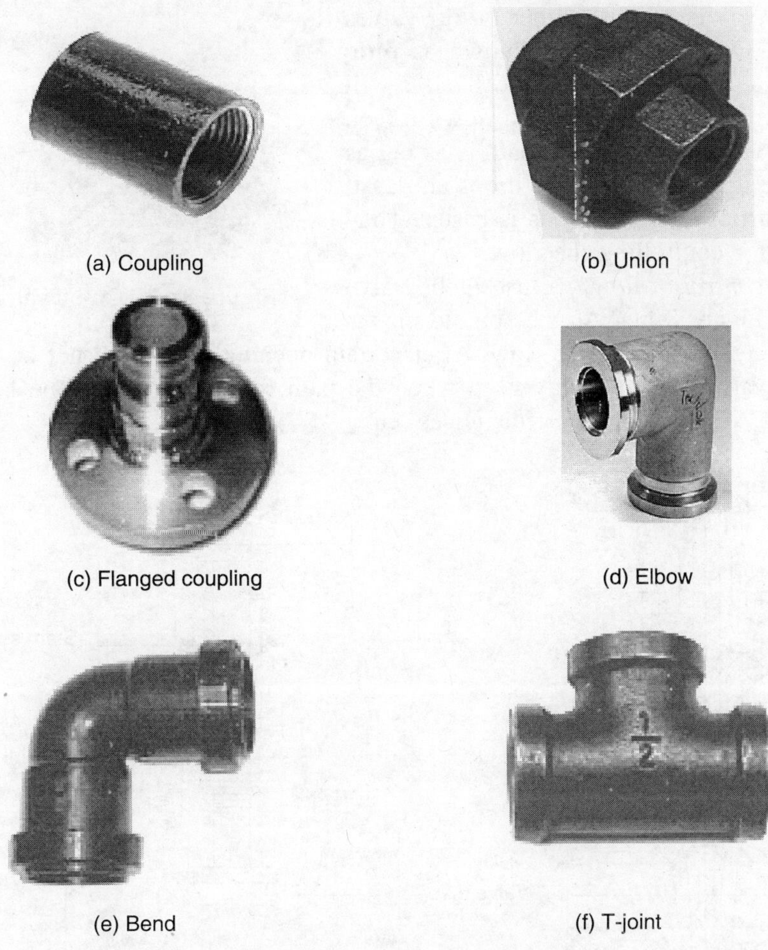

(a) Coupling

(b) Union

(c) Flanged coupling

(d) Elbow

(e) Bend

(f) T-joint

Figure. 8.9 Various fittings used in pipe lines.

we use bends [Figure 8.9(e)]. The pressure drops are more with bends as compared to elbows. T-joints [Figure 8.9(f)] are used whenever we want to take an additional connection or join some gadget (like valves/taps) to a running line. Generally T-joints come with inside threading.

8.6.3 Flow Regulation

For regulating the flow, we use different types of valves depending upon the purpose. Flow regulation may be of the following types:

(i) On-off control
(ii) Rough control
(iii) Fine control
(iv) Non-return control

For *on-off control*, we use either on-off valves or ball valves. On-off valves are known as **plug cocks**. A simple quarter turn of the handle will totally close the flow/or allow the full flow. In fact, the size of the valve (Figure 8.10) path is as big as that of the pipe itself. The pressure drops are least, and flow of slurries or suspensions is easier. They are not meant for controlling the flow.

Figure 8.10 On-off valve or ball valve.

For *rough control of flow*, we use globe valve or gate valve (Figure 8.11). As the spindle moves upwards, they create path for the flow. After certain opening, there will not be any regulation of flow. Gate valves are generally used when the path for flow needs to be larger for using for flow of fine suspensions like fruit juices, etc.

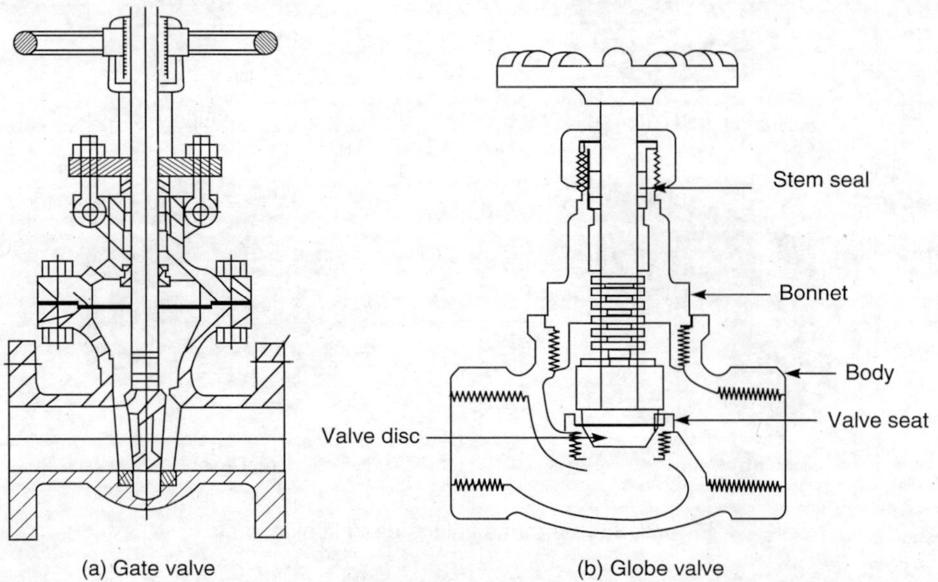

Figure 8.11

For *fine control of flow*, we need to go only for needle valve. The spindle is in the form of a fine needle (Figure 8.12). As the spindle moves, flow regulates. Very precise control is possible with these types of valves. When used with slurries and suspensions the needle path may be blocked and further flow is not possible. Hence, needle valves are used for flow regulation of gases/vapours. Generally needle valves are not used with liquids.

For a *unidirectional flow* non-return valves are used. They are also known as foot valves because mostly the non-return valves are used in the suction line of centrifugal pumps[†] to avoid *priming*. Since the valve is located at the foot of the suction line, it is popularly known as **foot valve**.

[†]More discussion on centrifugal pumps will be given in Section 8.9.

They allow the flow of fluids in one direction only. They do not allow the fluids to flow through in the opposite direction even if the pressure is high in the opposite direction. This will facilitate process fluids not to enter into the originating zone. They do not possess any handle to operate. They are just connected in the line. The flow of fluid in the flowing direction allows the ball to lift or gate to open (Figure 8.13). The moment the flow is stopped, the ball falls into the position or the gate closes the path by means of spring action.

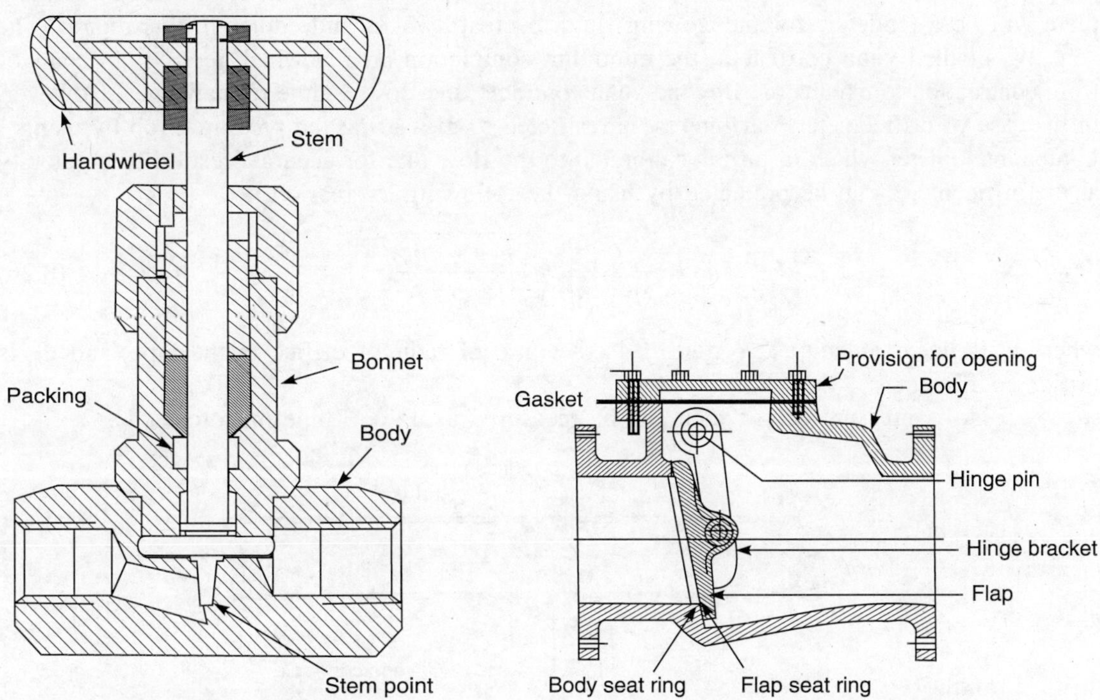

Figure 8.12 Needle valve.

Figure 8.13 Unidirectional flow valve.

8.7 FLOW MEASURING INSTRUMENTS

Measurement and regulation of flow of fluids are necessary in most of the process industries including chemical industries and food processing industries. Measurement is the primary operation, and regulation follows later on. Fluid flow rates, flowing through the pipes and conduits, can be measured by,

(i) measuring the pressure drop using an orificemeter or venturimeter or a pitot tube.

(ii) changing the buoyancy forces on a float using rotameter which is also known as **area flow meter**.

(iii) measuring some related parameter like the resistance of a wire as in hot wire anemometer.

There are various other types of meters also like turbine meter, etc. Of all these, the orifice-meter, venturimeter and rotameter are very important for food engineers, and hence, will be discussed in some detail here.

8.7.1 Orificemeter

An orificemeter as shown in Figure 8.14 consists of a carefully machined orifice plate with a hole, the diameter of which is less than that of the pipe. Since the diameter of the orifice plate is lower, it causes obstruction for free flow of the fluids. This results in considerable pressure drop. The flow at point *a* is normal. When the fluid passes through the orifice, the fluid layers jump over the orifice and create a wake zone at *W*. Here certain amount of vacuum forms, and there will be a tendency for the flowing fluid to create wakes and eddies in this region. The zone *W* is called **vena contracta**, the minimum contraction area at which jet of fluid changes from contraction to expansion. Beyond vena contracta, the flow regime reestablishes. However, in the case of orificemeter, our interest is restricted to measuring the pressure drop by using a U-tube manometer which in turn is calibrated to the flow rate for accurate results. The velocity at the orificemeter can be predicted by using the following expression:

$$u_o = \frac{C_o}{\sqrt{1-\beta^4}} \sqrt{\frac{2g(p_a - p_b)}{\rho}} \tag{8.26}$$

where u_o is linear velocity at orifice, β is the ratio of radii of orifice to the pipe, and C_o is orifice coefficient.

$(p_a - p_b)$ is the pressure drop which can be measured using a U-tube manometer.

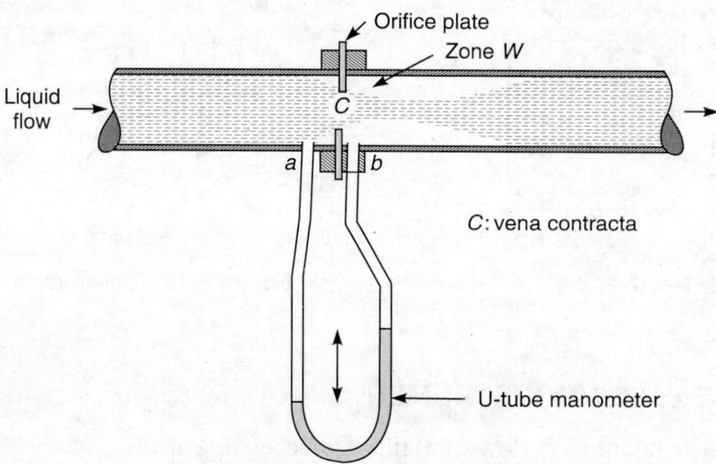

Figure 8.14 Orificemeter.

One of the reasons how errors can creep in calculation of u_o using Eq. (8.26) is prediction of C_o. If N_{Reo} (Reynolds number at orifice) is greater than 30,000, C_o may be taken as constant at 0.61 and is independent of β.

Orificemeter is simple in construction and is very easy to use. The orifice plate can be changed to get different β values to suit to getting adequate Δp for any flow range. A major complication with orificemeter is that because of the sharpness of the orifice, the jet in the downstream flow side may separate out and continue to exist, without mixing with other fluid stream which is adjacent to the metallic walls of the pipe. This disadvantage can be overcome in the venturimeter.

8.7.2 Venturimeter

Venturimeter is also similar to orificemeter except that the vena contracta and beyond in orificemeter is made up of the metallic portion, and it runs as a single metallic piece (Figure 8.15). Whatever we have seen as a wake zone in orificemeter, is absent in the venturimeter. Hence, the fluid going as jet in the downstream side is absent in the venturimeter. It is more or less like a correction to the orificemeter. Connection features and other things are same as that of orificemeter.

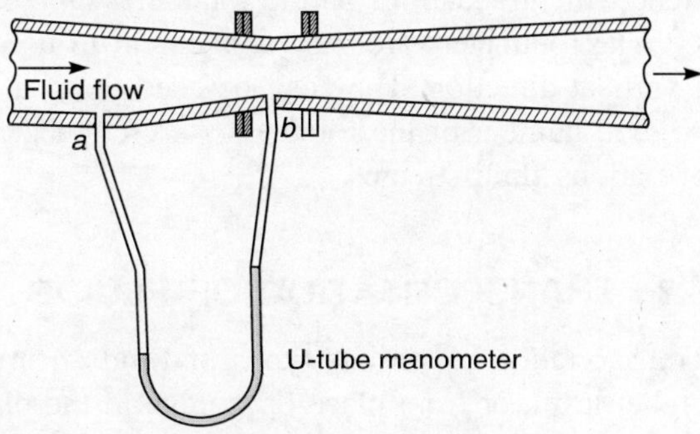

Figure 8.15 Venturimeter.

In the venturimeter, we do not have the liberty to change the throat diameter, as we could change the orifice plate in orificemeter. The velocity at the venturi throat (u_v) is given by,

$$u_v = \frac{C_v}{\sqrt{1-\beta^4}}\sqrt{\frac{2g(p_a - p_b)}{\rho}} \tag{8.27}$$

where C_v is the venturi coefficient.

For a well-designed venturimeter, C_v is of the order of 0.98 – 0.99.

8.7.3 Rotameter

Rotameter is considered as an *area flow meter*. In case of orificemeter and venturimeter, we have a throat of constant area through which the fluid flows. The fluid suffers a drop in pressure, which is proportional to the flow rate. Hence, we take this as an advantage to measure the flow rates. In the case of rotameter, we allow the area to change with the flow rate, keeping pressure almost constant. Rotameter is made out of a variable diameter glass tube in which a bob floats (Figure 8.16). As the flow rate increases, the bob is pushed upwards. The bob is in equilibrium between the gravitational pull acting downwards and the buoyancy push acting upwards. All other things being equal, the buoyancy force is proportional to the linear velocity of the fluid. As the volumetric flow rate of the fluid increases, the bob is also pushed up. As the bob goes up, the cross-sectional area of the pipe increases, and hence, the linear velocity does not change with the result that the buoyancy forces do not change. Thus, the bob is kept in equilibrium in the new position.

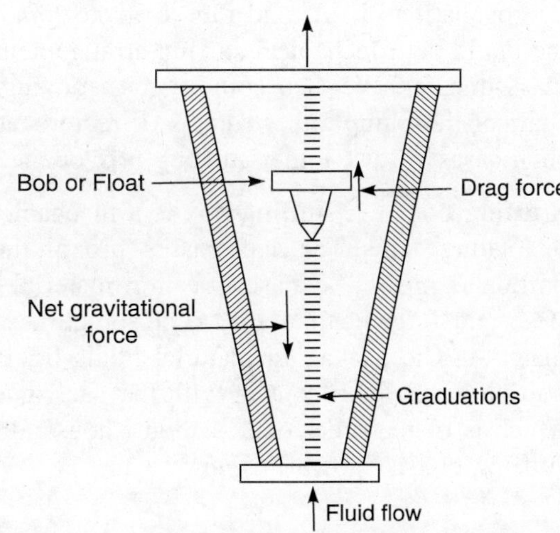

Figure 8.16 Rotameter.

In other words, the bob position changes proportional to the volumetric flow rate of the fluid. Hence, the graduations on the rotameter tube can be used to calibrate the flow rate of the fluid.

The rotameters are very convenient to use. They need to be connected in the pipe lines in a vertical direction. They can be used up to a system pressure of 5 atmospheres (0.5 MN/m^2). A good number of the rotameters are connected in the process lines to measure the flow rates of various fluid streams.

8.8 TRANSPORTATION OF FLUIDS

Transportation of various types of fluids from one place to other, or from a lower level to a higher level, or from place of storage to the place of processing, is an important unit operation in any food processing industry. It could be transportation of fruit juices/pulps, transportation of milk in a dairy industry, transportation of various organic solvents in any process industry, transportation of oils and fats in a solvent extraction unit or transportation of water, steam and, or furnace oil in any utility section of any process industry. The fluids are pumped to increase the potential energy, or kinetic energy of the fluids by using various types of mechanical pumps. The fluid quantities are regulated through valves, or measured by means of various measuring instruments, they may pass through bends, elbows and T-joints in pipelines which may be smooth or rough. The process vessels may be fitted with mechanical seals or stuffing box wherever movable rotating shafts are involved.

Thus, transportation of fluids is a multifarious activity involving a large number of accessories, fittings, flow measuring instruments, etc. Some of them we have already studied in this chapter. We study regarding various types of pumps in Section 8.9. We also come across moving/rotating parts under higher pressure or under normal pressure in process operations. For example, during hydrogenation of oils and fats to make *Vanaspati*, we pass hydrogen gas under pressure into the liquid oil. Mechanical stirring arrangement is also required to ensure good mixing and good mass transfer. We also come across moving/rotational parts in the operation of pumps. The shaft of the pump either rotates or reciprocates in a cylindrical chamber. To prevent leaks in such cases we use either stuffing box or mechanical seals.

Stuffing box: A stuffing box is a mechanical gadget used to prevent leaks with any moving or rotating shafts. The shaft passes through the stuffing box with a small stepping. The stepping portion is pressed against a sealing material which is usually a malleable metal alloy of lead (Rao and Raghunathan 1985), or sometimes even an asbestos rope is used as the packaging material. The packaging material holds tightly against the rotating shaft. The rotating shafts' portion which is in contact with the packaging materials will have a very fine smooth surface which is of the order of 2–4 thou (thousandth of an inch, i.e., 2"/1000 to 4"/1000.) A typical stuffing box is shown in Figure 8.17.

8.8.1 Recommended Practice for a Fluid Line

The fluid pipelines should be straight as far as possible to reduce pressure drop during fluid flow. Wherever it is essential to turn the flow direction, only right angle bends are recommended.

While laying the fluid lines, technical considerations (like pressure drop, etc.) should be given priority over aesthetic sense of beauty. The following are some of the recommended practices:

(i) The valves and other accessories should be approachable, and should be kept in vertical positions as far as possible.

(ii) Pressure gauges and flow measuring instruments should be mounted in such a way that they are easily noticeable.

(iii) The direction of flow of fluids should be indicated in the pipelines.

(iv) Standard notation of colouring should be painted on the pipes containing different fluids. For example, chlorine line should be painted in yellow colour, hydrogen line should be red, etc.

(v) The pipes should be joined with couplings or union joints, with gaskets so that they can be dismantled easily in case of any blockage or leakage.

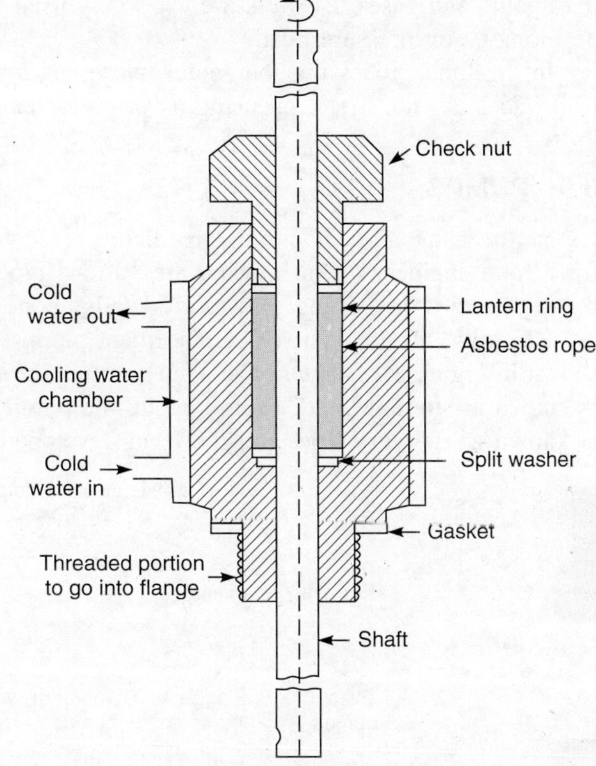

Figure 8.17 Stuffing box.

(vi) The fluid lines should preferably be drawn from the top to bottom, and should never be from bottom to top if it is not assisted by a mechanical pump. This may create a stagnant liquid column in the pipelines, when the fluid flow is not there. The liquids may corrode the pipelines, or deposit the salts on the pipe surfaces.

(vii) The steam lines should have drain pipes in between to remove the condensate if any.

8.8.2 Fluid Transportation Methods

For transportation of fluids, we use various types of gadgets mainly to increase the,

- potential energy of the fluid.
- kinetic energy of the fluid.
- pressure energy of the fluid.

This is achieved by using pumps, fans, blowers, and compressors.

Pumps are generally for positive displacement of liquids for transportation against gravity and friction, or for increasing the kinetic energy or pressure energy of the fluids. Fans, blowers and compressors are generally used for gases or vapours. Fans and blowers do not generate any pressure energy, whereas compressors are used exclusively to develop the pressure energy of vapours and gases. Blowers are especially used to transport large volumes of gases without developing any pressure head.

In the food processing, our main interest is in transportation of various liquids like milk, fruit juices, solvents, etc. by using a variety of pumps.

8.9 PUMPS

As has been mentioned, pumps are mainly used for transportation of liquids. It is sufficient for a food engineer to know what are various types of pumps and what are their operational features, rather than going into more details about the design features of the pumps, etc.

Basically, all the positive displacement pumps can be classified into various categories as shown in Figure 8.18. They have a suction line through which the liquid is drawn, and the head is known as **suction head**. From beyond the pump and up to the point of delivery, the section is known as **delivery line**, and the head developed is known as **delivery head** (Figure 8.19).

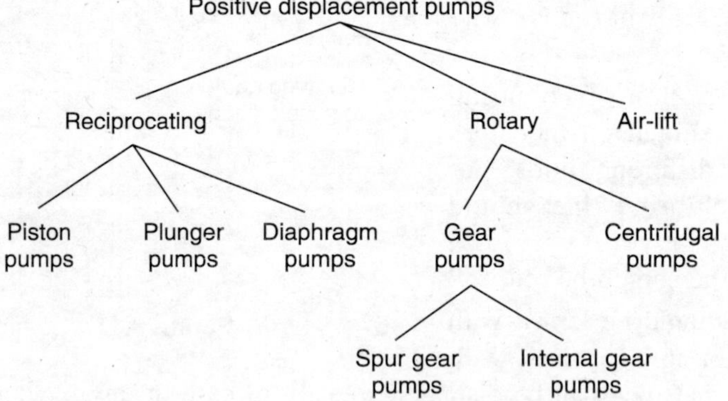

Figure 8.18 Various types of positive displacement pumps.

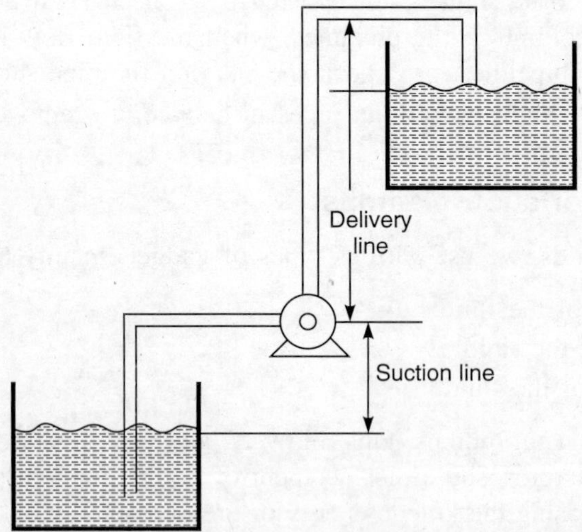

Figure 8.19 Suction and delivery lines of a positive displacement pump.

8.9.1 Reciprocating Pumps

Reciprocating pumps are generally used for transportation of liquids to increase pressure energy or kinetic energy. As shown in Figure 8.18, they include piston pumps, plunger pumps and diaphragm pumps.

Piston pump: Piston pump (shown in Figure 8.20) consists of a cylindrical chamber (*C*) in which a tightly fitted piston (*P*) moves in a reciprocating direction. Mechanical seals are used to prevent leaks at position *a*. The piston rod is attached to a reciprocating slide which in turn is connected to a rotating wheel in an accentric position by a pinion arrangement. As the wheel rotates, the piston moves in the cylindrical chamber. Figure 8.20 shows a double acting pump. As the piston moves backwards the valves at position b_1 and b_3 open while the valves at position b_2 and b_4 close. Hence suction is created in the front portion of the cylinder which allows fluid to enter into *F*, and allows the fluid in the rear position *R* to go out through valve b_3 (which is now open). When the piston moves forward, valves b_2 and b_4 open while valves b_1 and b_3 close. Thus, the fluid which has earlier entered into front portion *F* will escape through b_4 into delivery line while valve b_2 allows fluid into rear portion at *R*. Thus, there is a continuous suction and delivery of the fluids. They can disharge fluids up to 50 atm. pressures.

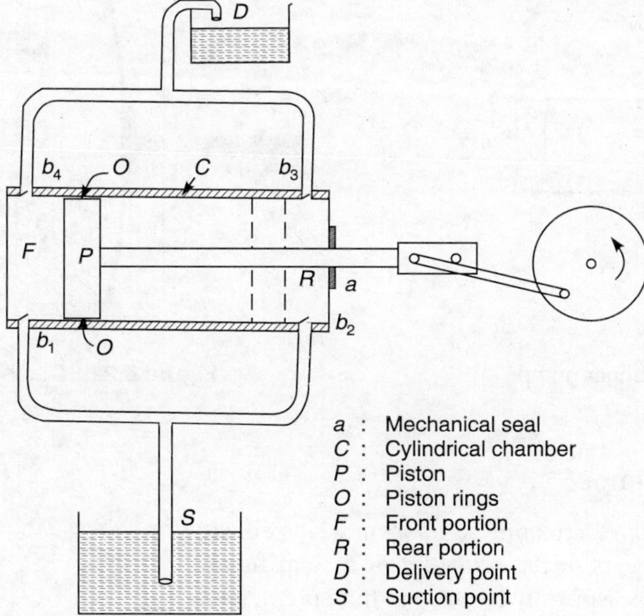

Figure 8.20 Piston pump.

In the single acting pumps, valves b_2 and b_3 will not be there. Hence, fluid enters when the piston moves backwards, and the fluid leaves when piston moves forward. Thus, it gives only alternate discharge of fluid in a sinusoidal manner. Hence, single acting pumps are not normally preferred.

Plunger pumps: The plunger pumps are used to develop very high pressures up to 1500 atm. or more. The construction features and working mechanism are similar to a single acting piston pump, with a plunger in place of piston (Figure 8.21). The plunger is driven by a motor. It has a thick walled cylindrical chamber in which a tightly fitting plunger reciprocates. It can be only single acting pump. The discharge quantities are generally low, but against high pressures.

Diaphragm pumps: A flexible diaphragm (*A*) is the reciprocating component in the diaphragm pump (Figure 8.22). In fact, the diaphragm does not reciprocate, it only vibrates. Hence, the outputs of these pumps are always low with the upper limit being approximately 500 lpm. Since there are no reciprocating parts in the pump, there will not be any leakage problems. Hence, there is no necessity to especially use any mechanical seals, or packaging materials to prevent leaks. They can pump against pressures up to 100 atm. Instead of a piston or a plunger moving in a cylindrical chamber as in case of reciprocating pumps, the diaphragm just vibrates, and sucks the fluid in one direction of vibration and delivers in the other direction of vibration. The diaphragms are made out of some flexible materials like plastic, rubber or some soft metal or metal alloys, which are all essentially non-corrosive in nature. Hence, these pumps can be used for fluids which are toxic or corrosive in nature. This is the special characteristic feature of these pumps.

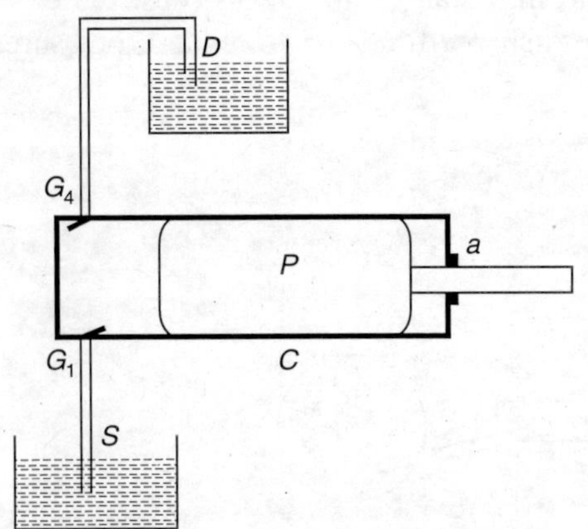

Figure 8.21 Plunger pump.

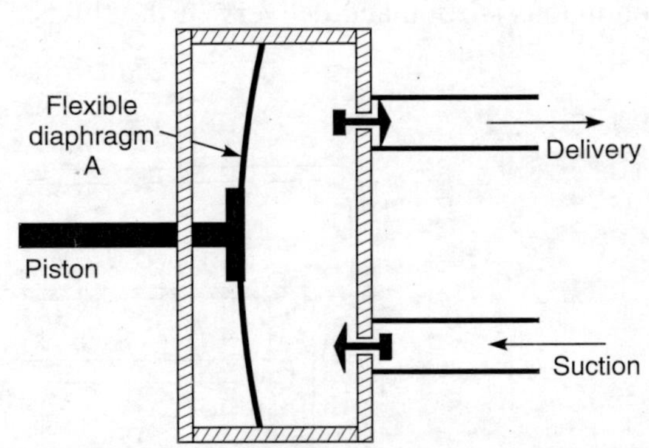

Figure 8.22 Diaphragm pump.

8.9.2 Rotary Pumps

In rotary pumps, we have a rotating component which either rotates the toothed gears or the propeller of a centrifugal pump. The gear pumps generally create very high pressures, whereas the centrifugal pumps are used mainly to transport liquids, particularly the viscous fluids.

Spur gear pump: It contains two closely rotating toothed gears in opposite direction. When the gears are rotating, they create some vacuum which sucks the liquids from the suction line (*S*). The liquid is delivered through the delivery line (*D*). A schematic representation is shown in Figure 8.23.

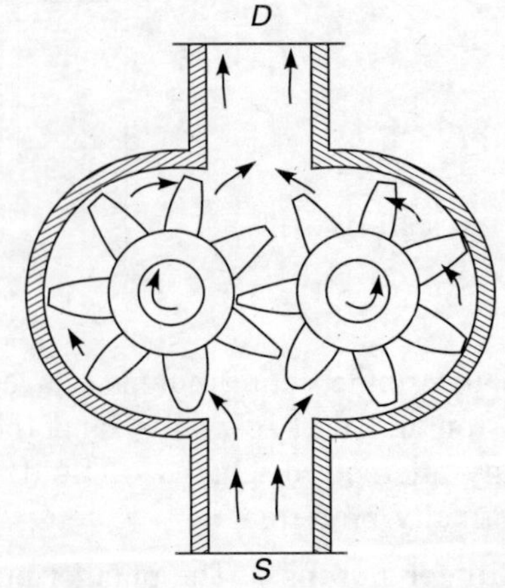

Figure 8.23 Spur gear rotary pump.

Internal gear pump: It consists of two gears, a pinion gear (*E*) and a ring gear (*R*) as shown in Figure 8.24. The ring gear is coaxial with the casing, but the pinion is mounted accentrically with respect to the casing. Thus, in the RHS of the casing we find some empty space between the gears. The empty space is filled with a stationary crescent. The pinion gear is driven by an external motor. It also rotates the ring gear in the same direction. The ring gear is not separately driven by a motor. When the gears are rotating, they suck the liquid from the suction line (*S*). The liquid passes through the space between the gears, and finally is delivered through the dischare line (*D*) as shown in Figure 8.24.

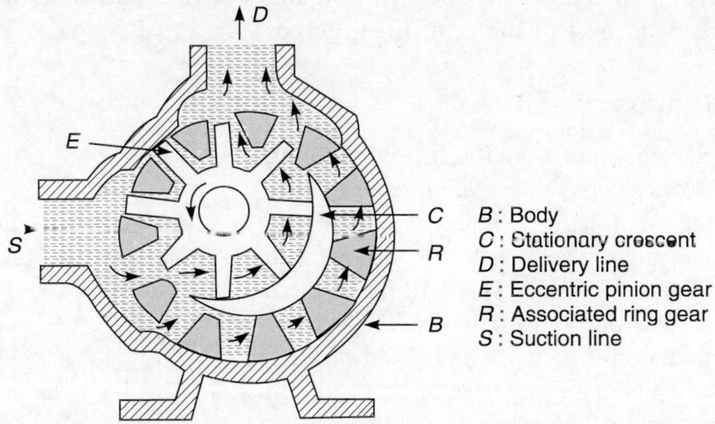

Figure 8.24 Schematic diagram of an internal gear pump.

Centrifugal pumps: In the centrifugal pumps, the mechanical energy of the liquid is increased by centrifugal force imparted to the liquid by the vanes. Hence, the design of the vanes is very important for proper functioning of the pump. The centrifugal pumps are very versatile, simple and compact in appearance. They are very easy to operate, and hence is used in most of the process industries and in domestic lines. They are essentially meant for lifting the fluids, rather than imparting pressure energy to the fluids.

It consists of an impeller *I* (Figure 8.25) with vanes fixed on it. The impeller is rotated by an external motor. The fluid enters the suction line, which is along the axis, and the delivery line is perpendicular to the suction line. When the impeller rotates, the vanes also rotate. The

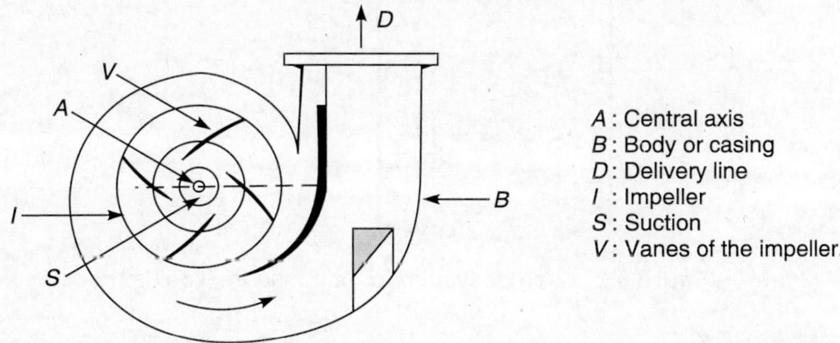

Figure 8.25 Centrifugal pump.

vanes are filled with the fluid, and thus, they impart centrifugal force to the fluid. This is mainly the reason why *priming*[†] is essential for the centrifugal pumps. The liquid rises in the delivery line creating space for fresh liquid to enter into the suction line. Thus, the pump pumps the liquids through pipes, and lifts liquids to greater heights.

Since the impeller and vanes come in direct contact with the liquids, centrifugal pumps generally cannot be used for corrosive and toxic fluids, and also at higher temperatures. The foot valve at the bottom of the suction line is very important. If it fails or leaks, the whole liquid in the suction line drains, and air enters into the suction line. When we use the pump next time, we have to fill the suction line with liquid, which is known as **priming**. This is a recurring nuisance with most of the centrifugal pumps.

8.9.3 Air-lift Pumps

Air-lift pumps are generally used for lifting/pumping water or liquids from deep wells. Air under pressure is pumped into the suction line, which is submerged in the liquid. Thus, air pressure lifts the liquids and creates suction in the suction line. This will make room for fresh liquid to enter into the suction line. Thus, the liquid rises in the suction line and is delivered through the delivery line. A schematic representation of air-lift pump is shown in Figure 8.26.

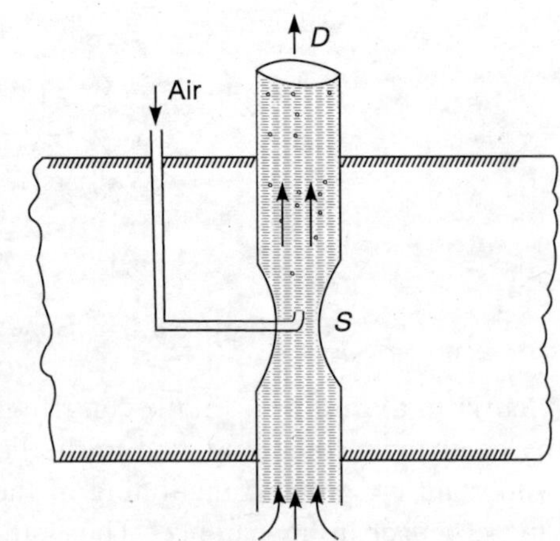

Figure 8.26 Air-lift pump.

Air-lift systems are also used for pumping liquids from tanks by applying positive air pressure on the surface of the liquid. This will make the liquid raise through the discharge tubes, which is submerged in the fluid. A schematic representation is shown in Figure 8.27. By controlling the flow rate with a valve, we can get continuous regulated supply of feed liquid from the tank.

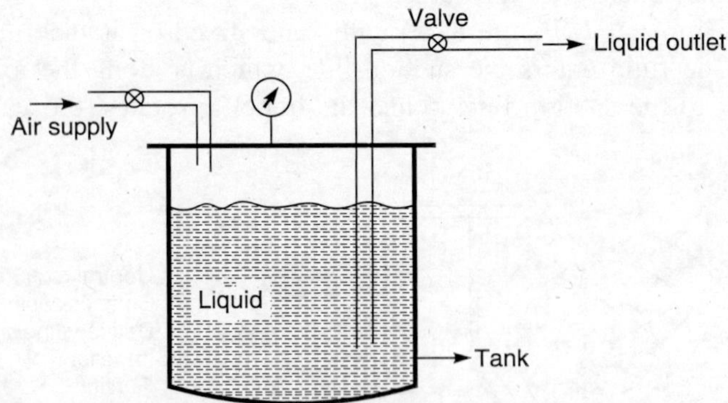

Figure 8.27 Air-lift system for pumping the liquids.

[†]Filling of the suction line and the vanes of the impeller with the fluid is known as **priming**.

Table 8.2 summarizes various characteristic features of different pumps and their applications.

Table 8.2 Characteristic Features and Applications of Various Pumps

Pump	*Characteristic feature/application*
Piston pump	It is a reciprocating pump. Simple in construction. Can be double acting. Develops pressure upto 50 atm. or more. The piston comes in contact with the fluid. There are moving and rotating parts. They contain check valves in the discharge line.
Plunger pump	It is a reciprocating pump. Discharge rates are moderate. Single acting. Develops pressure upto 1500 atm. or more. The plunger comes in contact with the fluid. There are moving and rotating parts. They contain check valves in the discharge line.
Diaphragm pump	It is a reciprocating pump. Discharge rates are very low, can be used with toxic and, or corrosive fluids. Develops pressures upto 100 atm. There are no moving or rotating parts. Hence, the problem of leakages is eliminated.
Spur gear pump	It is a rotating pump. The clearance between the teeth is very small. They do not need a check valve in the discharge line. The rotating teeth come in contact with the fluid. They develop pressure upto 200 atm. Operate well with moderately viscous fluids like lubricating oils, etc.
Internal gear pump	It is a rotating pump. The suction and the discharge lines are at right angles to each other. The gear teeth and the crescent come in contact with each other.
Centrifugal pump	It is a rotating pump. Very simple in construction, easy to operate. It is a most versatile pump, generally used very commonly in all process operations for transportation of fluids to different heights to increase the potential energy. They do not develop very high pressures. The rotating vanes come in contact with the fluid. Check valve is necessary. Priming is required without which the pump does not operate. Priming is indeed a nuisance with centrifugal pumps.
Air-lift pump	Air or gas is used to impart energy to the liquid to raise. There are no rotating or reciprocating parts involved. No leakage problems at all. Used for lifting water from an artesian well, or any liquid from a tank.

Symbols

A: cross-sectional area (m^2)
C_v: venturi coefficient
C_o: orifice coefficient
d: diameter of the pipe
f: friction factor [Eq. (8.22)]
g: acceleration due to gravity (m/s^2)
g_c: gravitational factor = 9.81
h: height (m)
ΔH: height diference (m)
H_f: head lost due to friction (m)
H_p: head due to pump (m)
ΔL: an elemental length
l: length in inclined manometer fluid
N_{Re}: Reynolds number
p: pressure (N/m^2)
Δp_s: pressure drop due to skin friction
$\bar{u}$: average velocity
u: linear velocity (m/s)
u_o: velocity at orifice (m/s)
u_v: velocity at venturi (m/s)
W_p: energy to be supplied by the pump

Subscripts

a: position a
abs: absolute
b: position b
f: fluid
g: gauge
l: liquid
w: wall

Greek Letters

ρ: density (kg/m^3)

α: kinetic energy correction factor

μ: viscosity of the fluid

β: ratio of orifice diameter to the diameter of pipe

η: efficiency of pump

θ: angle of inclined manometer

REVIEW QUESTIONS

8.1 Differentiate between the fluids and solids.

8.2 What are the values for atmospheric pressure in different units?

8.3 Describe the working of a U-tube manometer.

8.4 What is an inclined manometer?

8.5 What is Bernoulli's equation, and how it is useful in fluid mechanics?

8.6 Describe Bernoulli's equation with friction in the pipe lines and a pump attached in the fluid line. Indicate the units of each term in SI units.

8.7 What are various flow regimes in a pipe, and how do you characterize the flow in a pipe?

8.8 Derive Hagen–Poiseuille equation.

8.9 What are different types of pipe joints? Explain them with neat diagrams.

8.10 Distinguish between an elbow and bend.

8.11 What are various pipeline fittings, and describe them in a tabular form.

8.12 How do you control/regulate the fluid flow in a food process industry?

8.13 Describe various types of valves with special features of each in a tabular form.

8.14 What are various flow measuring instruments and describe them with neat diagrams?

8.15 Describe the working of an orificemeter with a neat diagram.

8.16 Describe the working of a venturimeter with a neat diagram.

8.17 Describe the working of a rotameter with a neat diagram.

8.18 Write a note on transportation of fluids.

8.19 Describe what are recommended practices for laying a fluid line.

8.20 What are various functions of using a pump in a fluid line?

8.21 Classify various types of pumps, and describe their special characteristic features in a tabular form. Describe any one of them with a neat diagram.

8.22 Describe reciprocating pumps with neat diagrams.

8.23 Describe rotary pumps with neat diagrams.

8.24 Describe air-lift pumps with neat diagrams.

8.25 Write short notes on the following:

(i) density,
(ii) fluid pressure,
(iii) Reynolds number,
(iv) skin friction factor,
(v) non-return valves,
(vi) vena contracta,
(vii) stuffing box, and
(viii) priming in a centrifugal pump.

NUMERICAL PROBLEMS

8.1 The atmospheric pressure is 1 kg/cm^2. Convert into N/m^2. (**Ans:** 10^5 N/m^2)

8.2 A U-tube manometer is used to measure the flow rate of a fruit juice (specific gravity = 1.05) flowing through a pipe. The pressures on the upstream and downstream sides of the flow are respectively, 500 mm of mercury and 0.55 atm. Calculate the level difference in mm in the U-tube manometer containing a fluid whose density is 12.5 g/cm^3. (**Ans:** 96 mm)

8.3 Milk is being pumped through a pipe whose suction pipe diameter is 7.5 cm from a tank to a processing vessel at 2 kg/cm^2 pressure through a pipe whose diameter is 5 cm. The processing vessel is at a height of 5 m above the ground level. The flow rate of milk is 200 lpm. The specific gravity of milk is 0.95. The milk is also heated in the delivery line with the result, its density has fallen to 900 kg/m^3. What should be the h.p. of the motor considering that the pipes are smooth and offer no resistance for the flow. The pump efficiency may be taken as 75 per cent. (**Ans:** 1.55 h.p.)

8.4 For the above problem, what is the Reynolds number of milk in the delivery line, considering that the viscosity of milk is equal to that of water. (**Ans:** 7.65×10^4)

8.5 Orange juice (specific gravity = 1.05) is passing through a 1.5″ pipe @ 5 lpm, the viscosity of the juice is 2.5 cp. Find the friction factor. (**Ans:** 0.0135)

8.6 For the data of above problem, calculate the pressure drop per unit length of the pipe.

(**Hint:** Use Hagen–Poiseuille equation), $\left(\textbf{Ans:} \, 3.51 \dfrac{\text{N/m}^2}{\text{m}} \right)$

REFERENCES

Dodge, D.W. and Metzner, A.B. (1959), Turbulent flow of non-Newtonian systems, *AIChEJ*, **5**, pp. 189–204.

McCabe, W.L., Smith, J.C. and Harriott, P. (1993), *Unit Operations of Chemical Engineering*, 5th ed., McGraw-Hill, Inc., New York, (a) p. 28, (b) p. 75, (c) p. 89, (d) p. 99.

Rao, D.G. and Raghunathan, T.S. (1985), Design of a stuffing box for a laboratory scale autoclave, *Chemical Engg. World*, **20**(7), pp. 47–49.

CHAPTER

9

Rheology of Foods

Rheology of foods plays an important role in food processing. Rheology deals with the deformation taking place in liquid/pasty foods when pressure (stress) is applied on the food materials. It could be flow of fruit juices or milk through pipes or conduits, or flow of viscous fluids like corn syrup in confectionary, or blood through veins; deformations taking place in the doughs in bakery industry; movement of semi-solid foods in chocolate making or ice cream industry, or *khoa* making in dairy industry; movement of thick fluids as in oleoresin industry; or mixing of suspensions of fermentation broths as in fermented foods processing.

When a stress is applied to any fluid including food materials, there develops a strain in the material which causes the deformation of the material. The stress, which has the units of pressure and is the force acting on unit area (N/m^2 or Pascal) is known as **shear stress** and is usually represented by τ. The strain developed in the body is known as **shear strain** and has the units of s^{-1}. It is usually represented by γ or du/dy, which is the velocity gradient along the distance or thickness. *Rheology* is study of relationship of shear stress and shear strain, and constitutes the subject matter of this chapter.

9.1 VISCOSITY

Viscosity is defined as the resistance to the flow of fluids. If a fluid is held in between two parallel plates, the fluid may be conceived to be consisting of a series of parallel layers as a pack of cards (Figure 9.1). The fluid is initially in stagnant position and stable. Let us apply a force F axially along the upper plate CD. The plate CD moves forward. Let the plate CD move with velocity u. As the plate CD moves with velocity u, the layer of fluid first adjacent to plate CD and adhering to it will also move with velocity u. The next layer which is below it will also be dragged by the upper layer of fluid, and hence, it moves; but it does not move with velocity u. It moves with velocity slightly less than u. The layer below it will be moving with a velocity still less than the earlier upper layer. The trend continues upto the bottom most layer which is stagnant, as it is adhering to the bottom plate AB which is stationary. Thus, the velocity of different layers vary between u and zero; and the velocity profile along the thickness y, can be

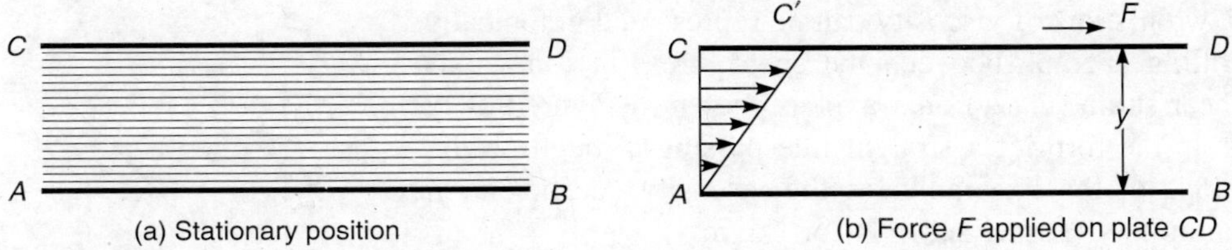

(a) Stationary position (b) Force F applied on plate CD

Figure 9.1 Velocity gradient of fluid between two plates because of viscosity.

represented by AC'. If the thickness (or distance) y is measured from the moving plate, the velocity profile can be drawn as in Figure 9.2. The velocity profile (or gradient) occurred because of the *resistance* for flow that is there in between the layers. If there is no resistance for the flow, probably all the layers would have been moving with the same velocity, u; and there would not have been any velocity gradient at all along the thickness. This resistance for the flow is known as **viscosity**.

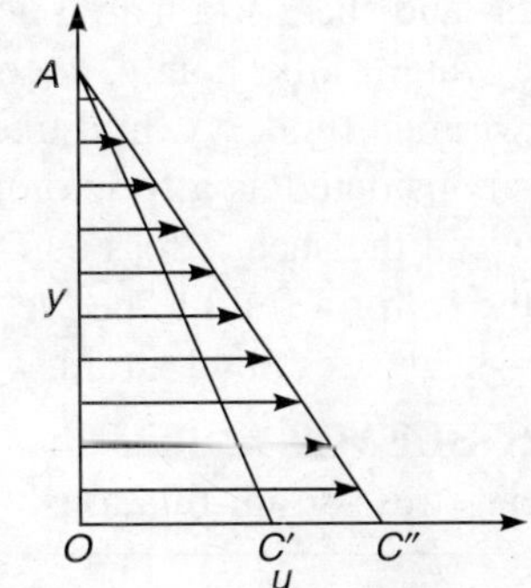

Figure 9.2 Velocity profile from the moving plate (AC' for force F, AC'' for force F_1, and $F_1 > F$).

If the force F_1 applied on the plate CD is more (i.e., $F_1 > F$) it would move the layers faster, creating a different velocity profile AC'' (Figure 9.2). Thus, more is F, more is the velocity gradient $\Delta u/\Delta y$ or du/dy. In other words, we can say that du/dy is proportional to F, or F is proportional to du/dy. Now we can write

$$F \propto \frac{du}{dy} \tag{9.1}$$

If the plate CD has an area of A, F/A is known as the **pressure acting** on the plate CD. Since the pressure is acting axially along the plate, we call the pressure as stress or more systematically *shear stress*, τ. Since A is constant, we can still write

$$\tau \propto \left(\frac{du}{dy}\right) \tag{9.2}$$

The velocity gradient, du/dy is also known as the strain created in the body of the fluid. This strain is known as **shear strain** or **shear rate** and is customarily represented by γ. In the above equation, if proportionality is removed, we can write

$$\tau = \text{constant}\left(-\frac{du}{dy}\right)$$

Or more systematically

$$\tau = \mu\left(-\frac{du}{dy}\right) \tag{9.3}$$

where μ is the constant of proportionality and is known as **coefficient of viscosity**; the negative sign in the parenthesis indicates that as distance y increases, the velocity u decreases (as shown in Figure 9.2). Eq. (9.3) which shows a proportionality relationship between τ and du/dy is known as **Newton's law of viscosity**.

Newton's law of viscosity can be represented graphically by plotting a graph between the shear stress (τ) on y-axis and shear strain (du/dy) on x-axis as shown in Figure 9.3. It results in the form of a straight line passing through origin. The slope of the line indicates the viscosity coefficient, μ with units Pa s (Pascal seconds) or (kg/m^{2}s).

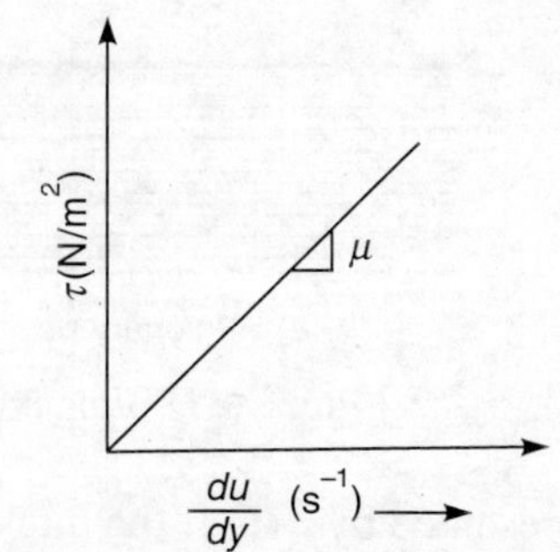

Figure 9.3 Shear stress-shear strain relationship for Newtonian fluids.

All the fluids that obey Newton's law of viscosity [Eq. (9.3)], and show a linear relationship between the shear stress and shear strain are known as **Newtonian fluids**. Fluids like water, milk, honey, coconut oil and glycerol, etc. are all Newtonian fluids. A fluid like honey may be viscous, but still it is considered as a Newtonian fluid because the relationship between the shear stress and shear strain is linear. Hence, the criterian for a fluid to be Newtonian is not its viscosity, but the relationship it exhibits between the shear stress and strain which should be linear.

PROBLEM 9.1 Find the viscosity of a solution of water-honey exhibiting the following shear stress-strain relationship:

τ, Pa	0.215	0.432	0.650	1.1	1.485	2.1	2.62	3.12
γ, s^{-1}	10	20	30	50	70	100	120	140

Solution The shear stress-strain relationship is drawn in Figure 9.4. The slope of the line is measured.

$$\text{Slope} = \frac{\Delta y}{\Delta x} = \frac{3.1}{140} = 0.022 \text{ Pa s}$$

The viscosity of the solution = 0.022 Pa s

9.1.1 Effect of Temperature on Viscosity

Many food materials are handled at different temperatures. Foods may be handled for packaging, transportation or for consumption. In all these cases, the flowability of foods plays a major role, and flowability is inversely proportional to viscosity. The effect of temperature on viscosity is also known to be inversely proportional to liquids; i.e., as temperature increases, viscosity decreases which in turn reduces the resistance to flow, and hence, helps increase the flowability. The effect of temperature on viscosity can be represented by Arrhenius type relationship (Rao 1986).

$$\mu = \mu_0 e^{E/RT} \tag{9.4}$$

where μ_0 is Arrhenius factor, E is activation energy in J/g mole, T is temperature in degrees Kelvin, and R is universal gas constant = 8.23 J/g mole K.

Equation (9.4) can be linearized by taking logarithms on both sides.

$$\ln(\mu) = \ln(\mu_0) + \left(\frac{E}{R}\right)\left(\frac{1}{T}\right) \tag{9.5}$$

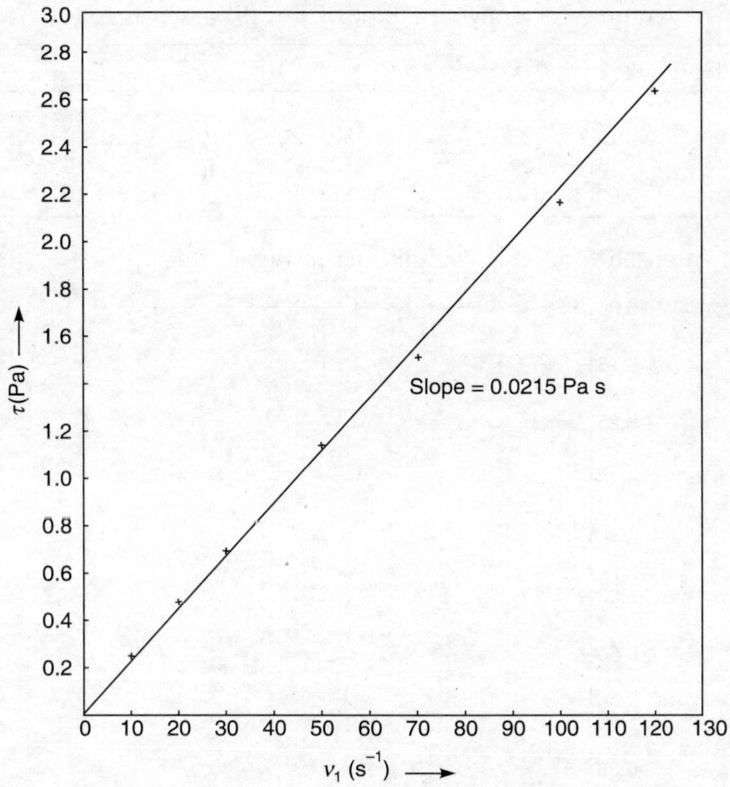

Figure 9.4 Shear stress vs. shear strain relationship for Problem 10.1.

To evaluate the effect of temperature on the viscosity, we need to find the viscosity at different temperatures and use the data to plot Eq. (9.5). Taking $\ln(\mu)$ on the y-axis and $(1/T)$ on the x-axis as shown in Figure 9.5, the data yields a straight line. The slope of the line is equal to (E/R) and the intercept is equal to $\ln(\mu_0)$, from which μ_0 can be found.

$$\text{Slope} = \frac{E}{R}$$

$$\text{Intercept} = \ln(\mu_0)$$

The above procedure is illustrated with a numerical example.

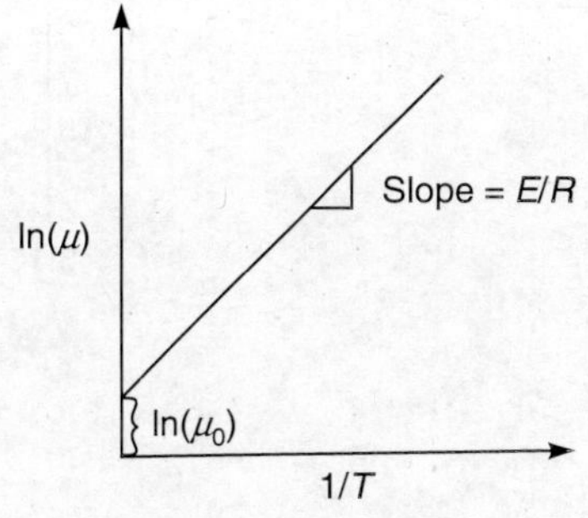

Figure 9.5 Effect of temperature on viscosity.

PROBLEM 9.2 The temperature vs. viscosity data for a dilute sugar solution are given below. Establish a relationship based on Arrhenius type of equation.

Temperature °C	25	30	35	40
$\mu \times 10^3$, Pa s	2.0	1.47	1.07	0.8

Calculate the viscosity at 32 °C.

Solution The data required for drawing a plot of Eq. (9.5) are as follows:

Temperature °C	*Viscosity* μ, Pa s	*Temperature* T (K)	$1/T$	$\ln(\mu)$
25	2.0×10^{-3}	298	3.36×10^{-3}	–6.21
30	1.47×10^{-3}	303	3.3×10^{-3}	–6.52
35	1.07×10^{-3}	308	3.25×10^{-3}	–6.84
40	0.8×10^{-3}	313	3.19×10^{-3}	–7.13

The data are drawn in the form of a straight line in Figure 9.6.

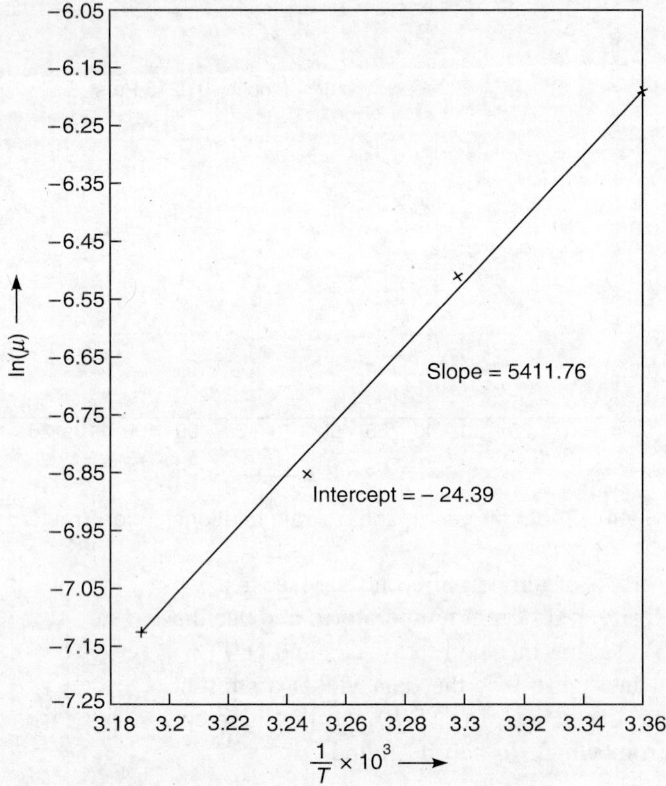

Figure 9.6 (1/T) vs. ln(μ) data for Problem 9.2.

$$\text{Slope} = \frac{-6.21-(-7.13)}{3.36\times10^{-3}-3.19\times10^{-3}} = 5411.76$$

$$\text{Intercept} = -24.39 = \ln(\mu_0)$$

$$\therefore \qquad \mu_0 = e^{-24.39} = 2.54 \times 10^{-11}$$

To calculate viscosity at 32 °C:

$$T = 273 + 32 = 305 \text{ K}$$

$$\mu_{32\,^\circ\text{C}} = 2.54 \times 10^{-11} \times [e^{(5411.76/305)}]$$

$$= 2.54 \times 10^{-11} \times e^{17.74} = 2.54 \times 10^{-11} \times 5.08 \times 10^{+7}$$

$$= 1.29 \times 10^{-3} \text{ Pa s}$$

9.2 NON-NEWTONIAN FLUIDS

The concept of viscosity has emerged from the linear relationship of shear stress and shear strain which is known as **Newton's law of viscosity** [Eq. (9.3)], and the fluids obeying this law will all have constant viscosity for any shear rate, and are known as **Newtonian fluids**. All the fluids, which do not obey the Newton's law of viscosity, and hence, exhibit varying viscosity with shear rate, are known as **non-Newtonian fluids**. For such fluids, the plot of τ vs. γ is not a straight line passing through the origin (as was shown in Figure 9.3). The plots are curves as shown in Figure 9.7

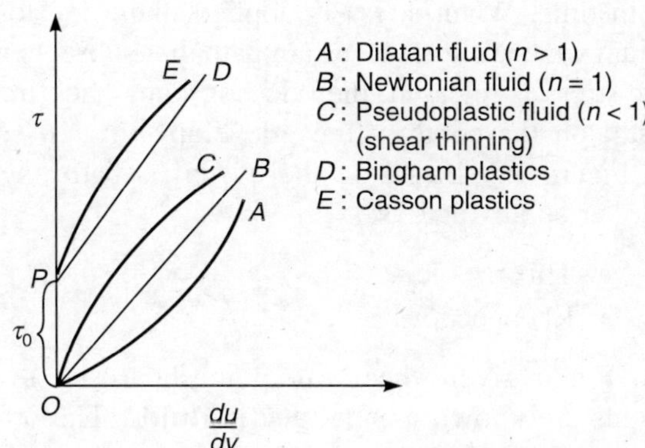

Figure 9.7 Shear stress vs. shear strain relationships for various non-Newtonian fluids.

The curve *OA* is concave in shape and represents a class of non-Newtonian fluids known as **dilatant fluids**. Fluids such as solution of corn flour, sugar and starch are some common examples of dilatant fluids. Similarly, the curve *OC* which is convex shaped represents **pseudoplastic fluids**. Most of the biological fluids, starch suspensions, cellulose acetate solutions are some common examples of pseudoplastic fluids.

There are some other types of fluids which behave like Newtonian fluids showing a linear relationship between shear stress and shear rate, but they are displaced from the origin. They start flowing only after exceeding certain initial stress (known as **threshold stress**) and are represented by straight line *PD*. Such types of fluids are known as **Bingham plastics**. Tooth paste contained in ą tube is a very common example of Bingham plastics. Unless we give an initial stress, it will not start flowing. Later on for the tooth paste, the shear rate is proportional to the shear stress. Some plastic melts, cooking fats, soap slurries are examples of Bingham plastics. The Newton's law of viscosity represented by Eq. (9.3) is modified for Bingham plastics as:

$$\tau - \tau_0 = \mu\left(-\frac{du}{dy}\right) \tag{9.6}$$

where τ_0 is threshold shear stress.

There is yet another type of fluid represented by curve *PE*, which is displaced from origin by τ_0, and shows convex shaped behaviour like pseudoplastic fluids. Such types of fluids are known as **casson plastics**, and are represented by equation

$$\tau^{1/2} - \tau_0^{\ 1/2} = K\gamma^{1/2} \tag{9.7}$$

in which *K* is known as the consistency index. Blood is a very common example of casson plastics. Tomato sauce, orange juice also show casson plastic behaviour. (Atkinson and Mavitona 1991).

So far we have discussed a group of non-Newtonian fluids which have a stress-strain relationship that does not depend on time. They are also known as time-independent non-Newtonian fluids. There are certain other types of non-Newtonian fluids whose stress-strain relationship varies with time. When stress is applied, they develop a shear strain which follows a certain path, but subsequently when the stress is released, they do not come back in the same path; on the contrary they develop *hysterisis* as shown in Figure 9.8. Based on the hysterisis path, they can be further subdivided as:

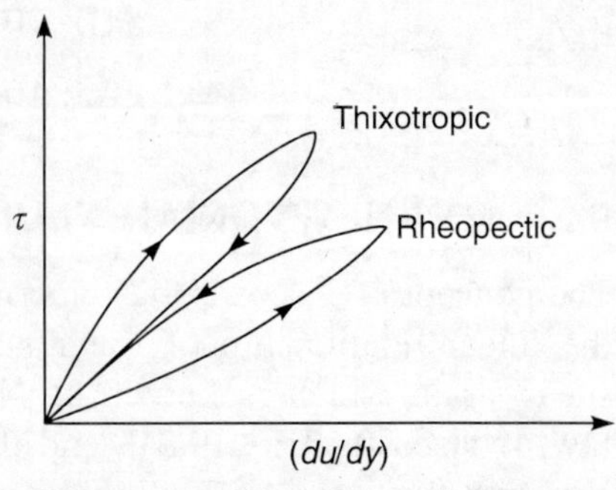

Figure 9.8 Time-dependent non-Newtonian fluids.

- Thixotropic
- Rheopectic

For a given shear rate, the shear stress may increase in the reverse path, and such fluids are known as **rheopectic fluids**. The other variety of fluids is that the shear rate may decrease in the reverse path. They are known as **thixotropic fluids**. Some high polymeric melts, food materials such as condensed milk, egg yolk were reported to exhibit thixotropic behaviour. Bentonite clay solutions and suspensions of gypsum are some classical examples of rheopectic fluids (Skelland 1967). In food processing, rarely we come across rheopectic foods (Rao 1986).

9.2.1 Two-parameter Model

Most of the food materials and suspensions fall under the category of non-Newtonian fluids and are represented by either pseudoplastic nature or dilatant nature (shown in Figure 9.7). The rheological data (τ vs. γ relationship) can be conveniently represented by a two-parameter model which is popularly known as **power law fluid model**.

$$\tau = K\gamma^n \tag{9.8}$$

where n is the flow behaviour index and K is the consistency index.

Equation (9.8) is also known as **Ostwald-de Waele model**.

Equation (9.8) represents behaviour of Newtonian, pseudoplastic and dilatant fluids, based on the value of n.

$n > 1$ for dilatant fluids, $n < 1$ for pseudoplastic fluids, and $n = 1$ for Newtonian fluids, and $K = \mu$.

Obviously for non-Newtonian fluids, the viscosity changes with shear stress, and hence, cannot have a single value for viscosity. Here we define a term known as **apparent viscosity** (as contrast to absolute viscosity) and is represented by μ_a.

$$\mu_a = \frac{\tau}{\gamma} = K\gamma^{n-1} \tag{9.9}$$

For pseudoplastic fluids, $n < 1$. Hence $n - 1$ becomes negative; as γ increases μ_a decreases, and liquid becomes thinner. These kind of fluids are known as **shear thinning fluids**. Concentrated fruit juices, melted chocolate, dairy cream, fruit and vegetable purees, and some gum solutions are typical shear thinning fluids. For dilatant fluids, $n > 1$; hence, $n - 1$ is positive. As γ increases μ_a also will be increasing. This kind of fluids is known as **shear thickening fluids**. We come across shear thickening fluids rarely in food processing. Cooked starch solutions, honeys obtained from different sources are a few foods that exhibit shear thickening behaviour (Rao 1986).

9.2.2 Evaluation of Rheological Parameters

We use the power-law fluid model and evaluate the flow behaviour index and consistency index.

$$\tau = K\gamma^n$$

By taking logarithms on both sides, the power law fluid model can be linearized

$$\ln(\tau) = \ln(K) + n \ln(\gamma) \tag{9.10}$$

For this purpose, we apply a certain amount of shear stress, and note down the shear rate, or vice-versa. Having obtained the stress vs. shear rate data, we draw a plot on log-log graph as shown in Figure 9.9. The data result in the form of a straight line.

$$\text{Slope} = n$$

$$\text{Intercept} = \ln K$$

Once K and n values are known, we can evaluate the shear rate developed by applying any known shear stress. We shall discuss in the next section how to evaluate τ and γ by using various types of viscometers.

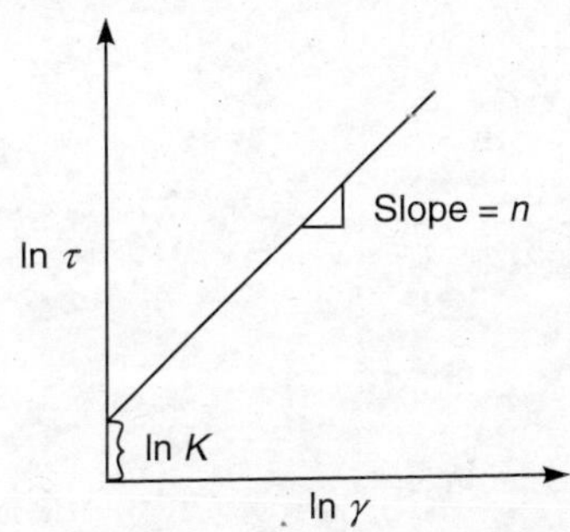

Figure 9.9 Evaluation of rheological parameters.

PROBLEM 9.3 In the process of making *khoa,* milk is concentrated. At a certain stage, milk thickness and solids start forming. The slurry milk starts behaving like a non-Newtonian fluid. Data are collected on shear stress vs. shear rate and are presented here.

γ (s^{-1})	30	60	120	180
τ (N/m^2)	186	330	589	780

Calculate K and n.

Solution We shall apply Eq. (9.10) to calculate K and n. We prepare the following table for drawing a graph similar to Figure 9.9:

γ (s^{-1})	τ (N/m^2)	ln γ	ln τ
30	186	3.4	5.23
60	330	4.1	5.8
120	589	4.8	6.38
180	780	5.2	6.66

The data are plotted in Figure 9.10.

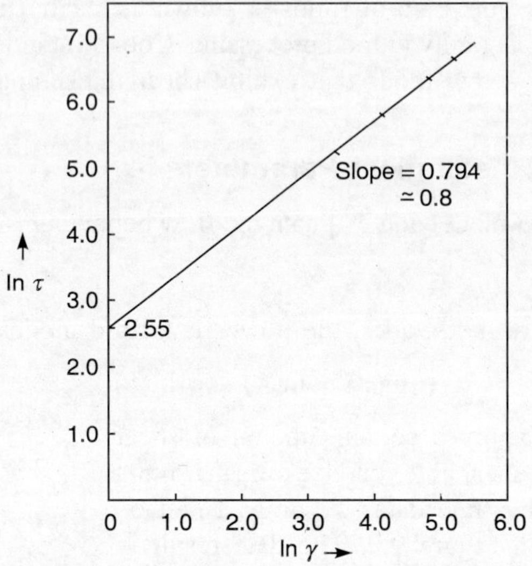

Figure 9.10 Solution of Problem 9.3.

$$\text{Slope} = \frac{(6.66 - 5.23)}{(5.2 - 3.4)} = \frac{1.43}{1.8} = 0.794 = 0.8 \text{ (approx.)}$$

$$\text{Intercept} = \ln K = 2.55$$

Therefore, $K = 12.8 \text{ Pa s}^n$

PROBLEM 9.4 Walde, et al. (1997) reported the rheological data for a solution of dried gum karaya as follows:

γ (s^{-1})	4.95	6.36	9.02	16.4	23.33
τ (Pa)	9.63	10.8	12.8	15.89	18.4

Calculate K and n.

Solution The data required for drawing a plot of Eq. (9.10) are calculated as follows:

γ (s^{-1})	τ (Pa)	ln γ	ln γ
4.95	9.63	1.6	2.265
6.36	10.8	1.85	2.38
9.02	12.8	2.2	2.55
16.4	15.89	2.8	2.77
23.3	18.4	3.15	2.91

The data are plotted on Figure 9.11.

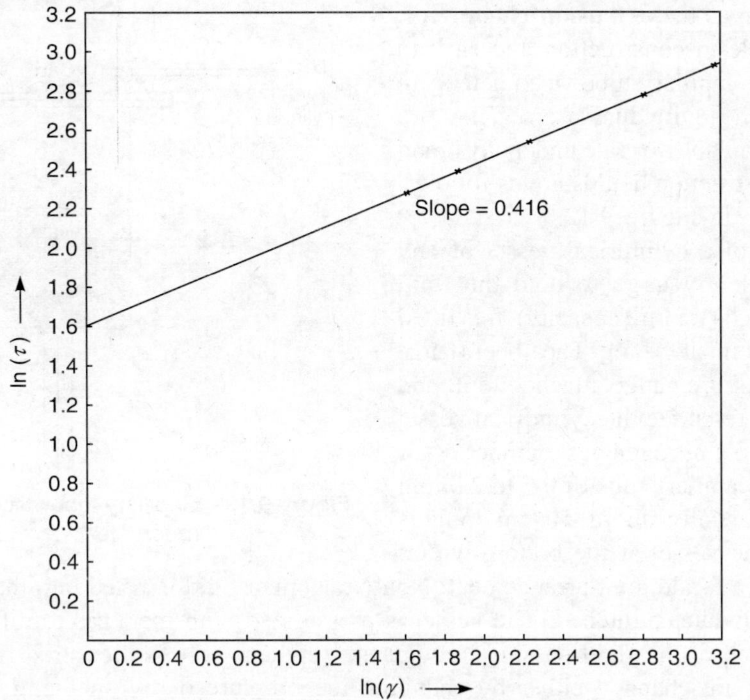

Figure 9.11 Data for Problem 9.4.

$$\text{Slope} = \frac{(2.91 - 2.265)}{(3.15 - 2.91)} = 0.416$$

$$\text{Intercept} = 1.6 = \ln K$$

Therefore,
$$K = 4.95 \text{ Pa s}^n.$$

9.3 MEASUREMENT OF VISCOSITY

Viscosity or the rheological data of a fluid are measured by using an instrument called **viscometer**. In a viscometer we apply different shear stresses. We measure the deformity that takes place in the fluid, in the form of shear rate or shear strain. In a way it is nothing but evaluation of flow consistency index (K) and flow behaviour index (n). There are various types of viscometers available for evaluating the rheological data. Each viscometer is characteristic in the sense that it has some parameter which represents or corresponds to shear stress. Similarly the deformation or shear rate also occurs in the form of some other parameter. By measuring that parameter, we calculate the shear rate. Once, the shear stress and shear rate data are available, we can easily calculate K and n by using Eq. (9.10).

The following are the most versatile viscometers even though a large number of them are described in some standard textbooks [For example, Skelland (1967)].

9.3.1 Capillary Tube Viscometer

It is also known as the **extrusion rheometer**. It is very simple in construction and easy to operate. In fact, capillary tube viscometers are fabricated by the individual users. They are generally not available for sale under any brand name. The construction details are as follows, and it is shown in Figure 9.12.

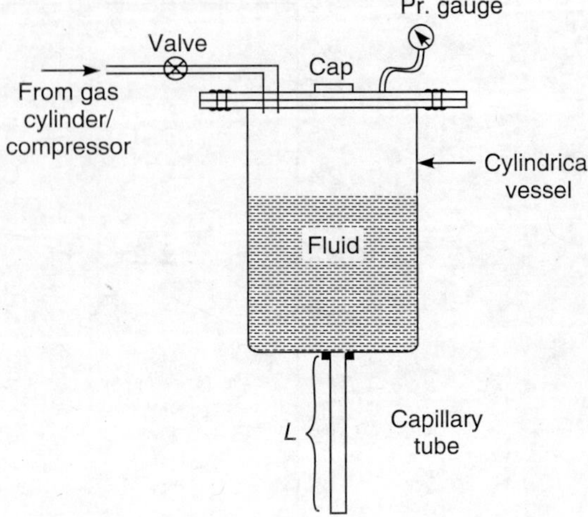

Figure 9.12 Capillary tube viscometer/extrusion rheometer.

It consists of a cylindrical vessel of any capacity sufficiently large to hold the fluid (usually about 1.0–1.5 litre capacity). It is fitted with a flanged joint at the top. The upper flange has a cap, a pressure gauge attached to it, and a provision to pressurize the cylindrical vessel by connecting to a nitrogen gas cylinder or air compressor. A capillary tube of 0.5 to 2.0 mm diameter and usually about 50 cm long is connected to the vessel at the bottom with a bolt and nut or a ferule joint arrangement. Non-Newtonian fluid is filled into the vessel through the cap which is later tightened. The vessel is pressurized with inert gas to different pressures and the flow rate of the fluid passing through the capillary tube is measured from the bottom. The flow rates are changed either by changing the pressure or by changing the diameter of the capillary or by changing the length of the capillary or by changing all of them. Data can be collected at various temperatures also by circulating hot fluid around the cylindrical vessel (the arrangement is not shown in the Figure 9.12). In some arrangements, the capillary tube is kept inside the cylindrical vessel which will make the fluid passing through the capillary is also kept very much at the same temperature as the fluid. The arrangement is shown in Figure 9.13 along with the provision to circulate hot water to maintain different temperatures.

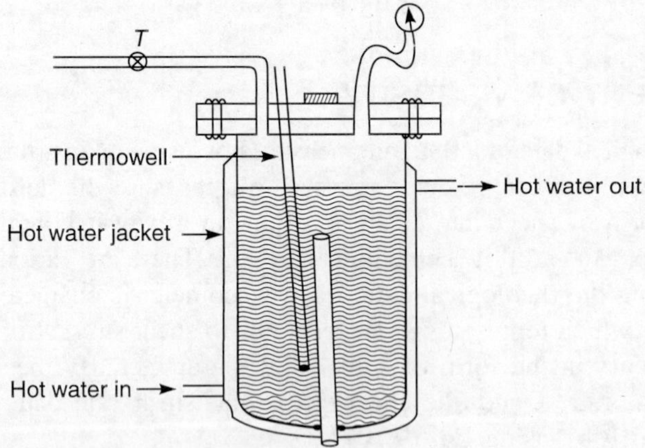

Figure 9.13 Capillary viscometer with capillary tube inside the vessel and provision to maintain different temperatures.

For collecting the rheological data, we need to measure the shear stress and shear rate. Robinowitsch–Mooney equation (Skelland 1967) gives expression for calculating shear stress and shear rate in terms of some measurable parameters from the viscometer. They are

$$\tau = \frac{d(\Delta p)}{4L} \tag{9.11}$$

and

$$\frac{du}{dy} = \gamma = \frac{8u}{d} \tag{9.12}$$

where (Δp) is the pressure differential applied on the fluid and measured by the pressure gauge, d and L are dia and length of the capillary tube, u is linear velocity of the fluid in m/s.

If v is the volumetric flow rate of the fluid (in m^3/s) through the capillary tube of diameter d (in m) we have

$$u = \frac{(v)}{(\pi d^2)/4} \tag{9.13}$$

After applying the Robinowitsch-Mooney equation, the power law fluid model (Eq. 9.8) can be written as:

$$\tau = \frac{d(\Delta p)}{4L} = K\left(\frac{8u}{d}\right)^n \tag{9.14}$$

As we did earlier for linearizing the equations we can rewrite Eq. (9.14) by taking logarithms on both sides as follows:

$$\ln\left(\frac{d\,\Delta p}{4L}\right) = \ln(K) + n\ln\left(\frac{8u}{d}\right) \tag{9.15}$$

The above equation is used to find K and n by collecting data on u by varying (Δp) or d or L (or by varying all), and drawing a plot of $\ln[d(\Delta p)/(4L)]$ vs. $\ln(8u/d)$. By measuring slope and intercept, we get

$$\text{Slope} = n$$

$$\text{Intercept} = \ln K$$

$$\text{Hence } K = \exp\text{ (intercept)}$$

This is explained with the following numerical problem:

PROBLEM 9.5 An extrusion rheometer is used to collect the data to measure the power law fluid model coefficients. The instrument is fitted with a capillary tube of 1 mm diameter and 0.5 m length. The following are the data:

Δp (N/m^2)	0.5×10^5	0.6×10^5	0.7×10^5	0.8×10^5
$v \times 10^{10}$ (m^3/s)	1.132	1.386	1.644	1.91

Calculate K and n.

Solution

$$\tau = \frac{d(\Delta p)}{4L} = \frac{0.001(\Delta p)}{4 \times 0.5} = 5 \times 10^{-4}\ (\Delta p)\ \text{N/m}^2$$

$$\gamma = \frac{8u}{d} = \frac{8v}{d[(\pi/4)d^2]} = \frac{32v}{\pi d^3} = 1.0185 \times 10^{10}\ (v)$$

v (m³/s)	Δp (N/m²)	τ (N/m²)	γ s⁻¹	ln τ	ln γ
1.132×10^{-10}	0.5×10^5	25	1.153	3.22	0.142
1.386×10^{-10}	0.6×10^5	30	1.411	3.4	0.345
1.644×10^{-10}	0.7×10^5	35	1.675	3.55	0.516
1.91×10^{-10}	0.8×10^5	40	1.943	3.7	0.664

The above tabular data are drawn in Figure 9.14 using the last two columns.

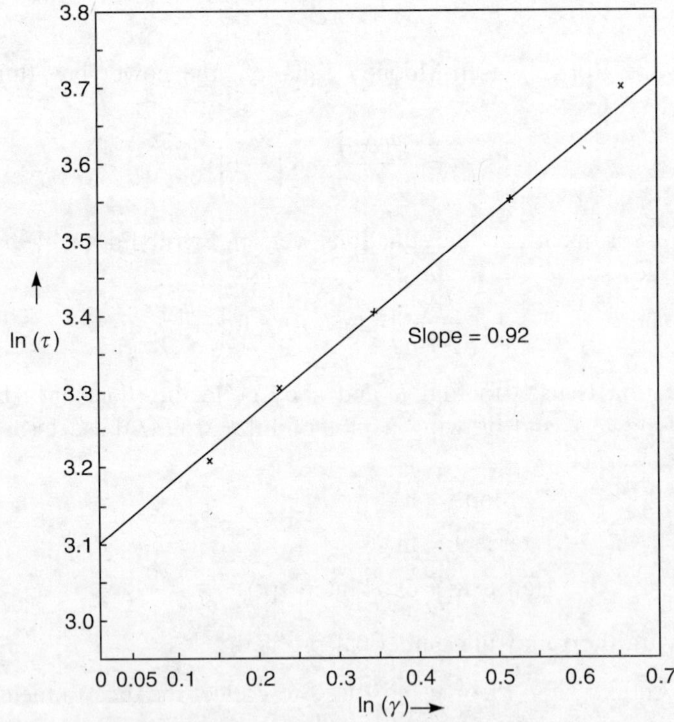

Figure 9.14 ln(γ) vs. ln(τ) data for Problem 9.5.

$$\text{Slope} = \frac{(3.7 - 3.07)}{(0.66 - 0)} = \frac{0.63}{0.685} = 0.92 = n$$

$$\text{Intercept} = 3.1$$

$$\therefore \quad K = e^{3.1} = 22.2$$

Power law fluid model can be written as:

$$\tau = 22.2(\gamma)^{0.92}$$

9.3.2 Plate and Cone Viscometer

It consists of a horizontal stationary plate on which rotates a vertical cone with its tip touching the plate. The cone is in inverted position as shown in Figure 9.15, and touching the flat plate at point P, and has a diameter D. In fact, the cone also looks like a flat plate as the angle of the cone is very small which is generally of the order of 3° to the horizontal plate. The fluid is kept in between the plate and the cone as shown in Figure 9.15. The cone rotates very slowly on the central axis by means of a small motor (which is not shown in the figure). The movement of the cone generates shear strain in the fluid. In other words, the fluid offers resistance for the motion of the cone. The angular velocity and the torque (t) of the cone are measured. Since the movement of the cone is very low, it is reasonable to assume that streamlined flow conditions exist forming concentric circles about the central axis of rotation of the cone. Since the angle (θ) of the cone is very small, the error that creeps in because of the above assumption is very small. The temperature of the fluid can be controlled and maintained by circulating constant temperature fluid through the stationary plate.

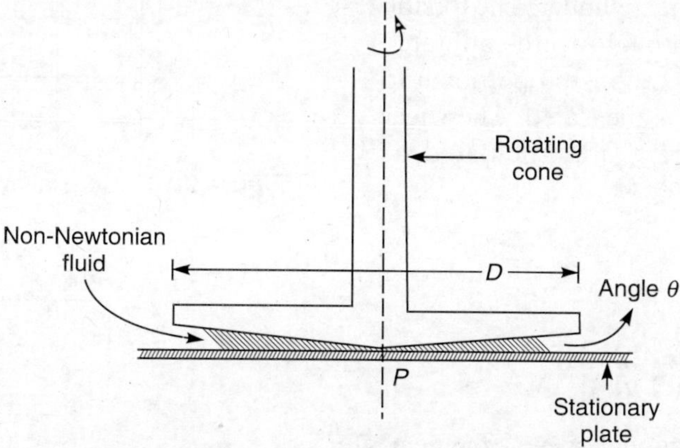

Figure 9.15 Plate and cone viscometer.

The plate and cone viscometer is very simple to operate. There are a few limitations of the viscometer. Corrections need to be made for the edge-effects, turbulence generated in the fluid, and temperature effects. Except for the few drawbacks, it is very easy to use and operate.

The shear stress and shear rate are given by (Skelland 1967):

$$\frac{du}{dy} = \gamma = \frac{2\pi N}{\tan(\theta)} \cong \frac{2\pi N}{\theta} \tag{9.16}$$

when θ is very small. N is the rotational speed of the cone in rps.

$$\tau = \frac{12t}{\pi D^3} \tag{9.17}$$

where D is the diameter of the cone, and t is the torque.

9.3.3 Coaxial Cylinder Viscometer

Coaxial cylinder viscometers are very popular and are very convenient to use, mainly because the commercial models are available in the market with computer support and software. One popular model is *Brookfield viscometer.* It consists of a stationary outer cylindrical cup into which an inner cylinder is suspended with a torsion wire (Figure 9.16). The top end of the wire is supported and connected to a geared motor to rotate slowly at different speeds (not shown in the figure). The outer cylindrical cup and the inner cylinder are closely placed, the clearance between them being very small. The inner cylinder does not touch the bottom surface of the outer cylindrical cup. The fluid fills in between the two coaxial cylinders and the bottom space. In fact, the inner cylinder is suspended in the fluid, and it rotates very slowly. Because of viscous nature of the fluid and depending upon the magnitude of the viscosity, it develops torque in the torsion wire of the inner cylinder. The torsion wire is connected through a pre-calibrated torsion spring. The torque is proportional to the shear stress, and is measured. The shear stress and shear strain are calculated as follows (Skeland 196):

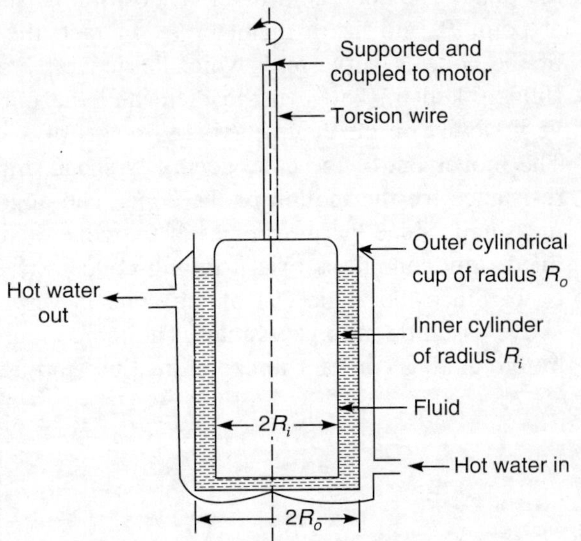

Figure 9.16 Coaxial cylinder viscometer.

$$\tau = \frac{t}{2\pi R_i^2 L} \tag{9.18}$$

$$\frac{du}{dy} = \gamma = \frac{4\pi N}{[1-(R_i/R_o)]} C_R \tag{9.19}$$

where N is the rotational speed of inner cylinder,

C_R is a constant to be evaluated.

C_R values are tabulated in the published literature for different (R_i/R_o) and other parameters (Skelland 1967). Different rotational speeds (N) give different values for γ, which in turn generate different τ values.

Thus, τ and γ data can be generated using Eq. (9.18) and Eq. (9.19). Most of the commercial equipment (like Brookfield viscometer shown in Figure 9.17) are provided with coefficients to calculate τ and γ. The modern versions of the Brookfield viscometer have not only got facility to give directly values for τ and γ, they even calculate K and n along with the graph

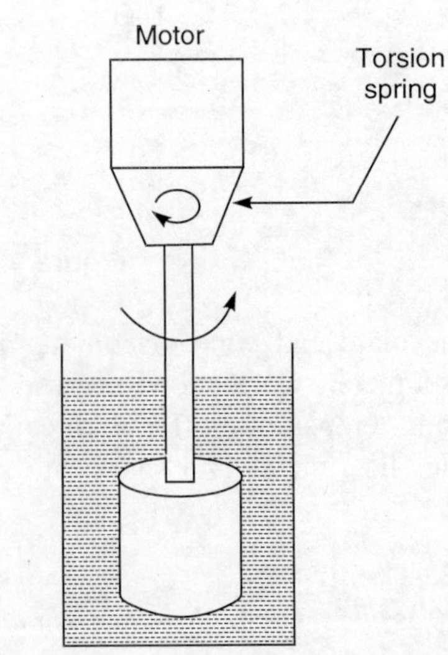

Figure 9.17 Schematic diagram of a Brookfield-type viscometer.

as shown in Figure 9.7 on a computer screen supported by suitable software. The instrument is very robust and easy to operate, and hence, finds wider application in the industry.

There will be a provision also to vary the temperature of the fluid by circulating hot water from a constant temperature bath through the outer jacket of the outer cylindrical cup. This will enable us to collect data at different temperatures. In some instruments, the outer cylindrical cup rotates while the inner cylinder is stationary. In either case, the torque is generated in the torsion wire.

The only drawback with this viscometer is that sometimes, the particles in the fluid suspensions may settle down causing some errors in the evaluations. If the suspended particles are big and uneven, they may even block the annular space at certain positions and give very erratic results. In spite of these drawbacks, and in spite of the fact that it is less accurate as compared to the capillary tube viscometer, still it finds wider application in the industries because of its robust nature in performance.

Symbols

A: area (m^2)
C_R: coefficient of the viscometer [Eq. (9.19)]
D: diameter of the cone in Eq. (9.17)
d: diameter of the capillary tube (m)
E: activation energy (kJ/kg mole)
F: force applied (Newton)
K: consistency index
L: length of the capillary tube (m)
N: rotational speed (rps)
n: flow behaviour index
Δp: pressure differential
R: Universal gas content
R_o: radius of the outer cylindrical cup (m)
R_i: radius of the inner cylinder (m)
T: temperature in degrees Kelvin
t: torque
u: linear velocity of the fluid (m/s)
v: volumetric flow rate (m^3/s)
y: distance or thickness (m)

Greek Letters

θ: angle of the cone with the ground
μ: viscosity (Pa s)
μ_0: Arrhenius factor for viscosity
μ_a: apparent viscosity (Pa s)
γ: shear rate (s^{-1})
τ: shear stress (N/m^2)
τ_0: threshold stress (N/m^2)

REVIEW QUESTIONS

9.1 What is meant by rheology? What are its applications in food processing?

9.2 Define viscosity.

9.3 What is Newton's law of viscosity?

9.4 What is meant by Newtonian fluids?

9.5 What is the effect of temperature on viscosity?

9.6 What is meant by non-Newtonian fluids?

9.7 What are different non-Newtonian fluids? Explain with suitable examples.

9.8 What do you mean by thixotropic and rheopectic fluids?

9.9 What is power law fluid model? How do you evaluate the model coefficients?

9.10 Describe a capillary tube viscometer with a neat diagram and explain how it is used for evaluating the rheological data. What are its advantages and drawbacks?

9.11 Describe a plate and cone viscometer with a neat diagram and explain how it is used for evaluating the rheological data. What are its advantages and drawbacks?

9.12 Describe a coaxial cylinder viscometer with a neat diagram and explain how it is used for evaluating the rheological data. What are its advantages and drawbacks?

NUMERICAL PROBLEMS

9.1 A concentrated orange juice had the following rheological data. Calculate K and n.

γ, s^{-1}	20	40	60	80	100
τ, N/m^2	90	150	200	246	290

(**Ans:** K = 10.5, n = 0.72)

9.2 The data from a capillary tube viscometer is presented here. Find the power law fluid model indices.

Diameter of capillary d, mm	*Length of capillary L*, cm	*Volumetric flow rate*, ml/min	Δp, N/m^2
2.0	50	5	0.75×10^5
1.5	60	5	2.0×10^5
2.0	60	5	0.9×10^5
1.5	50	5	1.67×10^5

(**Ans:** K = 4.58, n = 0.6)

REFERENCES

Atkinson, B. and Mavitona, F. (1991), *Biochemical Engineering and Biotechnology Handbook*, Macmillan, Basingstoke.

Rao, D.G. (2005), *Introduction to Biochemical Engineering*, Tata McGraw-Hill, New Delhi, Ch. 13.

Rao, M.A. (1986), *Rheological properties of fluid foods*, in Engineering Properties of Foods, (Eds.) Rao, M.A. and Rizvi, S.S.H., Marcel Dekker, Inc., New York, pp. 1–47.

Skelland, A.H.P. (1967), *Non-Newtonian Flow and Heat Transfer*, John Wiley & Sons, Inc., New York, pp. 27–49.

Walde, S.G., Balaswamy, K., Shivaswamy, R., Chakkaravarthi, A. and Rao, D.G. (1997), "Microwave drying and grinding characteristics of gum karaya (*Sterculia urens*)", *J. Food Engg.*, **31**, pp. 305–313.

CHAPTER

10

Heat Transfer by Conduction

Heat transfer plays a major role in food processing. Food products are processed for preservation by thermal processing. We call it as thermal processing because it could be heating or cooling. In either case, the heat transfer takes place, but only the direction of transfer changes. Heating of foods to inactivate the microorganisms or enzymes spoiling the foods is a very popular approach which was known from the days of Nicolas Appert in the year 1795, who could first successfully produce biologically stable foods (free from being fermented or putrified) by canning process. From then onwards, thermal processing started. Original concept of Appert was that the spoilage of foods was mainly due to the air present at the time of processing, and hence, efforts were concentrated more to see that the foods were canned by hermetically sealing them to prevent air from entering. Subsequently the concept changed radically with the findings of Louis Pasteur who for the first time proposed that the spoilage of foods was mainly due not to air present, but the microorganisms present in air which caused deterioration of foods. Hence, inactivation of microorganisms by thermal processing (heating) could be an alternative for the prevention of food spoilage.

When the concept of canning was proposed, availability of energy was not a big constraint. On the contrary availability of energy (heat) was taken for granted. Of late, when the energy became a scarce source, thoughts were diverted towards efficient transfer of heat to make effective utilization of it. Thus, heat transfer became an important area in all fields of engineering including food engineering.

Heat is known to transfer from a source at higher temperature to a body (sink) at a lower temperature. It is like any other transfer process for which rate of transfer is proportional to the driving force; i.e.,

$$\text{Rate process} \propto \text{driving force}$$

$$\propto (\text{resistance})^{-1} \tag{10.1}$$

$$\therefore \quad \text{Rate process} = \frac{\text{Constant (Driving force)}}{\text{Resistance}} \tag{10.2}$$

The driving force in case of heat transfer is temperature difference. Heat transfer also depends upon whether the materials are in contact or otherwise. Based on these things, heat transfer is conceived to take place by any one of the following three fundamental modes: conduction, convection, and radiation.

10.1 CONDUCTION

If we put a metallic rod in a flame of fire, initially the other end of the rod is at room temperature, whereas the one in contact with the flame is hot. Slowly as time progresses, the temperature of the rod goes on increasing, and the cold end also gets heated up. Here there is no physical movement of the rod or the flame, but there is transfer of heat from the hot end to the cold end. The transfer of heat mainly takes place by the movement of molecules in the metallic rod. In other words, we can say that the heat is conducted due to the *invisible movement* of the molecules in the rod which we call the process as **conduction**. Thus conduction process of heat transfer takes place in metallic rods, or solids which have a *tendency* to transfer heat. Solids can be classified into two categories: those which have a tendency to transfer heat are known as **conductors**; those which have a tendency not to conduct heat are known as **insulators** or **non-conductors** of heat.

Let us see the process of heat transfer through a slab of thickness, B, shown in Figure 10.1. The temperature on the hot side is T_1 and that on the cold side is T_2. Initially the temperature profile is as shown where the temperature on the hot side is almost T_1 and that on the cold side is T_2. As time progresses let us say at time t_1, the temperature profile is as shown. As time increases, the temperature profile also changes, and will become linear as shown at time t, which is known as the steady state temperature profile. Further, the temperature profile does not change. The temperature profiles at times t_1, t_2, t_3, etc. are all known as transient heat transfer profiles or *unsteady state heat transfer* profiles, whereas that at time t is known as temperature profile for *steady state heat transfer*.

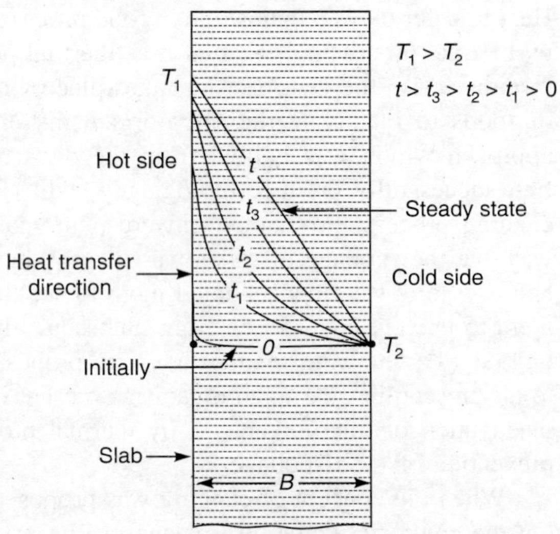

Figure 10.1 Heat transfer through a slab thickness B.

10.1.1 Fourier Law

The rate of heat transfer (or simply heat transfer rate, represented by q) per unit area is directly proportional to the temperature difference (ΔT) and inversely proportional to the thickness of the slab (B). It is represented mathematically as:

$$\frac{q}{A} \propto (\Delta T) \quad \text{and}$$

$$\propto \left(\frac{1}{B}\right)$$

or combining the above expressions, we can write

$$\frac{q}{A} \propto \frac{(\Delta T)}{B} \tag{10.3}$$

or

$$\frac{q}{A} = \text{Constant } \frac{(\Delta T)}{B}$$

$$= k\frac{(\Delta T)}{B} = \frac{\Delta T}{(B/k)} \tag{10.4}$$

in which k is the thermal conductivity of material in kJ/sm°C or kW/m°C, and resistance for heat transfer

$$R = \left(\frac{B}{k}\right) \tag{10.5}$$

In differential form, the above equation can be written is:

$$\frac{dq}{dA} = \frac{k(-dT)}{dx} \tag{10.6}$$

x is measured in the direction of heat transfer. Therefore as x increases, T decreases; and hence, dT/dx is preceded by a negative sign in Eq. (10.6).

Equation (10.6) is known as **Fourier law of conduction**, and is similar to Newtons law of viscosity [Eq. (9.3)] for momentum transfer. Equation (10.6) also gives an expression for **thermal conductivity**, which is an indication of how good a material is for conducting heat. Hence, we choose materials with high thermal conductivity to transfer heat; and we prefer materials with low conductivity to be used as insulators, where we do not want heat to be lost by transfer. Metals are highly conducting in nature, and hence, have higher thermal conductivities. For example metals like aluminum, copper and silver have higher thermal conductivities at 40 °C as 2300, 2100 and 1100 W/m°C, respectively as contrast to insulating materials like asbestos which has 0.17 W/m°C at 38 °C and cotton has 0.06 W/m°C. Thermal conductivity data of some metals and non-metals are given in Appendix 6.

Most of the food materials contain high quantities of water, and hence, the thermal conductivity of most of the foods is equal to or very close to that of water which is 0.628 W/m°C. It is interesting to note that the thermal conductivity of ice is almost four times that of water (2.6 W/m°C); and hence, naturally frozen foods will have higher thermal conductivity than the normal foods. Therefore, they lose heat very quickly when exposed to outside environment.

The thermal data of some food materials is given in Appendix 7.

PROBLEM 10.1 Find the rate of heat loss through a stainless steel slab 10 cm thick which is maintained at 100 °C on hot side and 30 °C on the cold side. The thermal conductivity of steel is 16.37 W/m°C.

Solution It is a straight case of application of Eq. (10.4) which states

$$\frac{q}{A} = k\frac{(\Delta T)}{B}$$

where

$$\Delta T = 100 - 30 = 70\ °C$$
$$B = 10\ cm = 0.1\ m$$
$$k = 16.37\ W/m°C$$

Therefor, $$\frac{q}{A} = 16.37 \times \frac{70}{0.1} = 11{,}459\ J/m^2s$$
$$= 11.459\ kW/m^2$$

10.2 HEAT CONDUCTANCE IN SERIES

Some insulating materials for which the thermal conductivity is very low like cotton, glass wool or silk wool are very good at low temperatures, but they cannot withstand high temperatures. Materials like brick or asbestos have reasonably low thermal conductivities and can withstand very high temperatures also. That is why we find most of the furnace walls are made up of composite walls where a brick wall is constructed along the furnace wall, and subsequently we use glass wool or cotton insulation over it. Thus, it is very common to use composite walls in furnaces or refrigeration units to reduce heat losses as much as possible. Hence, it is necessary to know what would be the total conductance of heat through the composite walls or what would be the total resistance offered by the composite materials for heat transfer.

The heat conductors (slabs) in series are shown schematically in Figure 10.2 for three slabs A, B and C of thickness B_A, B_B, B_C, and having thermal conductivities k_A, k_B and k_C. The temperature differences between each slab is ΔT_A, ΔT_B and ΔT_C, whereas the overall temperature difference is ΔT.

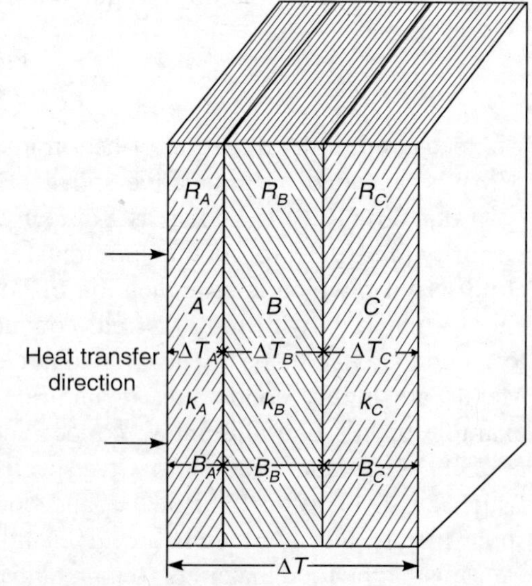

Figure 10.2 Heat conductors in series.

Therefore, $$\Delta T = \Delta T_A + \Delta T_B + \Delta T_C \tag{10.7}$$

The surface area of each slab is equal to A. Let the heat transfer rate through each slab be q_A, q_B and q_C, and the over all heat transferred is equal to q.

By Eq. (10.4), we can write

$$\Delta T = \left(\frac{q}{A}\right)\left(\frac{B}{k}\right) = \left(\frac{q}{A}\right)(R) \tag{10.8}$$

Similarly for slab A, $$\Delta T_A = \left(\frac{q_A}{A}\right)\left(\frac{B_A}{k_A}\right) = \left(\frac{q_A}{A}\right)(R_A) \tag{10.9}$$

$$\Delta T_B = \left(\frac{q_B}{A}\right)\left(\frac{B_B}{k_B}\right) = \left(\frac{q_B}{A}\right)(R_B) \tag{10.10}$$

and
$$\Delta T_C = \left(\frac{q_C}{A}\right)\left(\frac{B_C}{k_C}\right) = \left(\frac{q_C}{A}\right)(R_C) \tag{10.11}$$

Whatever be the amount of heat that is passing through slab A will have to pass through slab B, and will also pass through slab C, since there is no accumulation of heat in between the slabs. Hence,

$$q_A = q_B = q_C = q \tag{10.12}$$

Now, we replace q_A, q_B and q_C with q in Eqs. (10.9) to (10.11).

$\therefore$
$$\Delta T_A = \left(\frac{q}{A}\right)(R_A)$$

$$\Delta T_D = \left(\frac{q}{A}\right)(R_B)$$

and
$$\Delta T_C = \left(\frac{q}{A}\right)(R_C)$$

Now, we can write from Eq. (10.7)

$$\Delta T = \left(\frac{q}{A}\right)(R_A) + \left(\frac{q}{A}\right)(R_B) + \left(\frac{q}{A}\right)(R_C) \tag{10.13}$$

$$= \frac{q}{A}\ (R_A + R_B + R_C)$$

And ΔT is also equal to (q/A) (R) by Eq. (10.8)

Therefore,
$$\left(\frac{q}{A}\right)(R) = \left(\frac{q}{A}\right)(R_A + R_B + R_C) \tag{10.14}$$

and
$$R = R_A + R_B + R_C \tag{10.5}$$

or
$$\left(\frac{1}{R}\right) = \frac{1}{(R_A + R_B + R_C)}$$

This means that the overall resistance is equal to the sum of individual resistances.

i.e,
$$\frac{B}{k} = \left(\frac{B_A}{k_A}\right) + \left(\frac{B_B}{k_B}\right) + \left(\frac{B_C}{k_C}\right) \tag{10.16}$$

or in other words, since

$$\text{Resistance} = \frac{1}{\text{Conductance}}.$$

we can write

$$(\text{Overall conductance})^{-1} = [(\text{Conductance of } A)^{-1} + (\text{Conductance of } B)^{-1} + (\text{Conductance of } C)^{-1}]^{-1} \tag{10.17}$$

PROBLEM 10.2 A furnace is made out of a steel wall followed by a brick wall followed by a sil-o-cel brick wall. The inside temperature of the furnace is 400 °C and the outside temp is 40 °C. Find the rate of heat loss per m^2 and temperatures at the junction of the steel and brick wall and at the junction of the brick and sil-o-cel brick walls. Properties of the materials are as follows:

Material	*Thickness* (cm)	*Thermal Conductivity* (W/m°C)
Steel (*A*)	1.2	16.3
Brick (*B*)	20	0.72
Sil-o-cel brick (*C*)	12	0.14

Solution Let us assume the cross-sectional area be equal to 1 m^2. We find out the resistance of each material

$$R_A = \frac{B_A}{k_A} = \frac{1.2\times10^{-2}}{16.3} = 7.36\times10^{-4}\ \text{m}^2/\text{W}$$

$$R_B = \frac{B_B}{k_B} = \frac{0.2}{0.72} = 0.28\ \text{m}^2/\text{W}$$

$$R_C = \frac{B_C}{k_C} = \frac{0.12}{0.14} = 0.857\ \text{m}^2/\text{W}$$

Therefore, $$R = R_A + R_B + R_C = 1.138\ \text{m}^2/\text{W}$$

$$\Delta T = 400 - 40 = 360\ °\text{C}$$

$$\frac{q}{A} = \frac{\Delta T}{R} = \frac{360}{1.138} = 316.4\ \text{W/m}^2$$

The rate of heat loss per unit area = 316.4 W/m^2.

To find temperature at the junction of steel and brick walls (T_{AB})

$$\frac{q}{A} = \frac{\Delta T_A}{R_A}$$

i.e., $$316.4 = \frac{\Delta T_A}{7.36\times10^{-4}} = \frac{400 - T_{AB}}{7.36\times10^{-4}}$$

Therefore, $$400 - T_{AB} = 316.4 \times 7.36 \times 10^{-4} = 0.23$$

and $$T_{AB} = 400 - 0.23 \cong 400\ °\text{C}$$

Thus, we do not find much difference in temperature inside the furnace and at the junction of the steel and brick walls.

To find the temperature at the junction of brick and sil-o-cel brick walls (T_{BC}).

T_{BC} can be calculated either by using R_B or R_C.

Let us do it with R_C

$$\frac{q}{A} = \frac{\Delta T_C}{R_C}$$

i.e.,
$$316.4 = \frac{\Delta T_C}{0.857} = \frac{T_{BC} - 40}{0.857}$$

Therefore,
$$T_{BC} = (316.4 \times 0.857) + 40 = 311\ °C$$

Let us do it with R_B:

$$\frac{q}{A} = \frac{\Delta T_B}{R_B} = \frac{400 - T_{AC}}{0.28}$$

$$316.4 \times (0.28) = 400 - T_{BC}$$

Therefore,
$$T_{BC} = 400 - (316.4 \times 0.28) = 311\ °C$$

10.3 HEAT FLUX THROUGH CYLINDRICAL PIPES

So far we have seen heat conductance through slabs which have a definite heat transfer area because their cross-sections are in the form of squares or rectangle, and hence, easy to calculate. In many process operations, we come across the circular pipes of definite thickness through which the heat transfer takes place. In such cases whether the heat transfer area is to be based on inner diameter of the pipe or outer diameter of the pipe; and obviously both are not same. If outer radius is r_o, inner radius is r_i, length of the pipe is L, and thickness of the pipe wall is B, which is also equal to $(r_o - r_i)$, as shown in Figure 10.3.

$$r_o = r_i + B$$

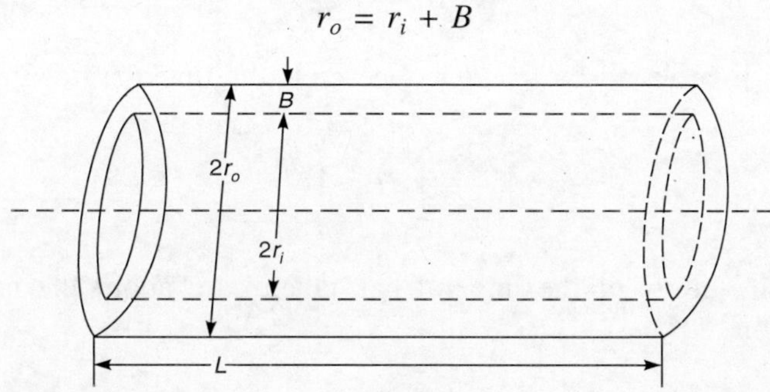

Figure 10.3 A circular pipe with thickness *B* for heat transfer.

Hence, area of the outside surface of pipe $= 2\pi r_o L$

$$= 2\pi(r_i + B)L$$

Area of inside surface of pipe is $2\pi r_i L$.

One possible approach to get the heat transfer area is to take average of both areas as follows:

$$A_{Av} = \frac{(2\pi r_o L + 2\pi r_i L)}{2}$$

$$= \frac{2\pi L(r_o + r_i)}{2} \tag{10.18}$$

A systematic and rigorous way of doing could be as follows:

Let us consider a small and thin circular cross-section of the pipe, at a radius of r and having a thickness of dr, as shown in Figure 10.4. Since it is a thin cylinder, its heat transfer area could be $2\pi rL$, where L is the length of the pipe. Heat transfer by conduction is given by Eq. (10.6).

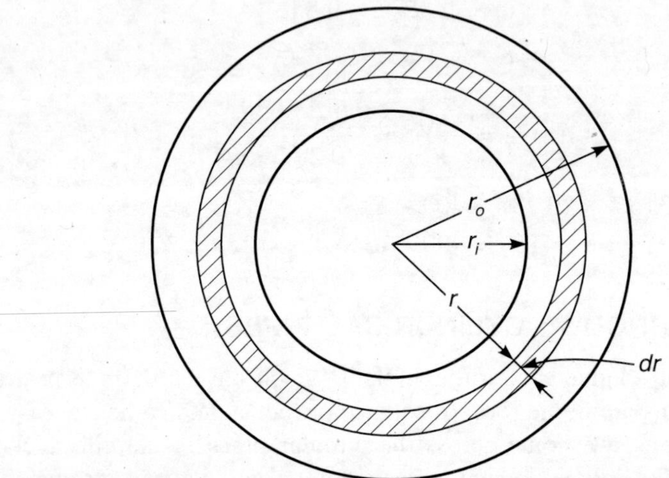

Figure 10.4 Flow of heat through a cylindrical pipe.

$$\frac{q}{2\pi rL} = -\frac{kdT}{dr} \tag{10.19}$$

The term (dT/dx) is replaced by (dT/dr) since dr is the thickness of the hypothetical pipe. On rearrangement

$$\frac{2\pi kL}{q}\int_{T_o}^{T_i} dT = \int_{r_i}^{r_o}\frac{dr}{r} \tag{10.20}$$

The change in the limits of the integral has taken care about the negative sign in Eq. (10.19). On integration

$$\frac{2\pi kL}{q}(T_i - T_o) = \ln\left(\frac{r_o}{r_i}\right) \tag{10.21}$$

or

$$q = \frac{2\pi kL(T_i - T_o)}{\ln(r_o/r_i)} \tag{10.22}$$

If we multiply and divide the above equation by $(r_o - r_i)$, we get

$$q = \frac{2\pi kL(r_o - r_i)}{\ln(r_o/r_i)} \times \frac{(T_i - T_o)}{(r_o - r_i)} \tag{10.23}$$

The above equation can be written as

$$q = (2\pi r_L L)(k)\frac{\Delta T}{B} = A_L(k)\frac{(\Delta T)}{B} \tag{10.24}$$

where
$$r_L = \frac{(r_o - r_i)}{\ln(r_o/r_i)} \tag{10.25}$$

and
$$A_L = 2\pi r_L L \tag{10.26}$$

in which r_L is called as the **log mean radius** and A_L is known as **log mean area**.

PROBLEM 10.3 Find the heat transfer rate per one metre length of circular copper pipe of 10 cm id and 12.5 cm od. The inside temperature of the pipe is 100 °C and the outside temperature of the pipe is 40 °C. Also calculate the percentage error involved by using average radius in place of log mean radius. The thermal conductivity of the metal is 850 W/m°C.

Solution Given data

$$r_i = 5 \text{ cm} = 5 \times 10^{-2} \text{ m}$$

$$r_o = 6.25 \text{ cm} = 6.25 \times 10^{-2} \text{ m}$$

$$L = 1 \text{ m}$$

$$T_i = 100\ °\text{C}$$

$$T_o = 40\ °\text{C}$$

$$B = (r_o - r_i) = 1.25 \times 10^{-2}$$

$$r_L = \frac{(r_o - r_i)}{\ln(r_o/r_i)} = \frac{(6.25 - 5)\times 10^{-2}}{\ln(6.25/5)} = 0.056 \text{ m}$$

Therefore,
$$A_L = 2\pi r_L L = 2 \times \pi(0.056) \times 1 = 0.352 \text{ m}^2$$

$$q = (A_L)\ k\frac{\Delta T}{B} = 0.352 \times 850 \times \frac{(100 - 40)}{1.25 \times 10^{-2}} = 1496 \times 10^3 \text{ W}$$

$$= 1496 \text{ kW}$$

q based on average radius:

$$\text{Average radius} = \frac{(5 + 6.25)}{2} \times 10^{-2} = 5.625 \times 10^{-2} \text{ m}$$

$$\text{Average area} = 2\pi r L = 2 \times \pi \times 5.625 \times 10^{-2} \times 1$$

$$= 0.3534 \text{ m}^2$$

$$q = kA\frac{\Delta T}{B} = 0.3534 \times 850 \times \frac{60}{1.25 \times 10^{-2}} = 1442 \times 10^3 \text{ W}$$

$$= 1442 \text{ kW}$$

Percentage decrease in q by using average radius $= \dfrac{1496 - 1442}{1496} \times 100 = 3.6$ per cent

10.4 UNSTEADY STATE HEAT CONDUCTION

Upto now we have seen the steady state heat conduction process through solids which have a constant temperatures on the heating side and cooling side. Since sufficient time has been given, steady state conditions prevail and make the temperature gradiance in the slab to be a straight line as shown in Figure 10.1. But, in real situations, many times steady state conditions may not prevail. This kind of heat transfer by conduction is known as **unsteady state heat transfer**. Let us consider a situation where a solid slab is being heated from both sides (Figure 10.5). It could be a slice of bread in a toaster, or it could be slab of frozen meat kept outside or kept in an oven where it is heated by radiation, or a slab of *bajji* or *wada* or *papad* being deep fat fried in an oil medium. In all these cases, heat transfer within the solid slabs is by conduction.

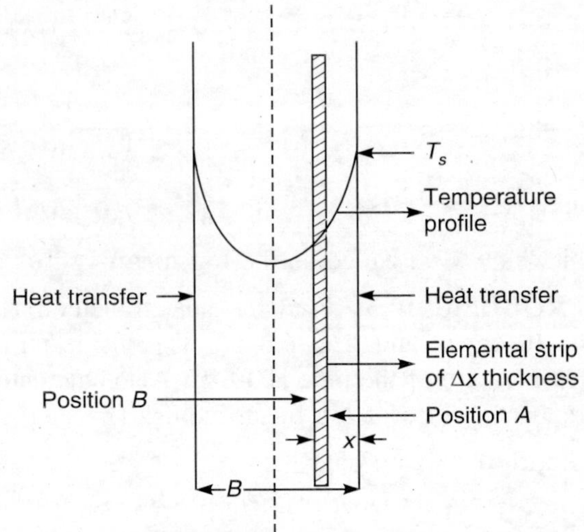

Figure 10.5 Schematic representation of unsteady state heat conduction in solid slab.

In the unsteady process, the temperature within the slab varies both with time t, and position in the slab, i.e., distance x. Let us visualize a small elemental piece of thickness dx at a distance x from the heating surface as shown in Figure 10.5. Let us say the temperature profile at the entrance of the strip is $\partial T/\partial x$. We use partial derivatives here because T varies with t and x. Hence, heat entering the strip in a time interval of dt is given by $-kA(\partial T/\partial x)dt$ as given by Eq. (10.6), i.e.,

$$\text{Heat entering strip at position } A = -kA\left(\frac{\partial T}{\partial x}\right)dt \tag{10.27}$$

The variation in temperature profile in the strip is given by

$$\frac{\partial}{\partial x}\frac{(\partial t)}{\partial x}dx$$

Hence, the temperature profile at position B is:

$$\frac{\partial T}{\partial x}+\frac{\partial}{\partial x}\left(\frac{\partial T}{\partial x}\right)dx \tag{10.28}$$

Heat leaving the strip at position B in time interval, dt

$$-kA\left(\frac{\partial T}{\partial x}+\frac{\partial}{\partial x}\left(\frac{\partial T}{\partial x}\right)dx\right)dt \tag{10.29}$$

Therefore, heat accumulation in the strip is the difference of Eq. (10.27) and Eq. (10.29) and is equal to

$$-kA\frac{\partial T}{\partial x}dt + kA\left[\frac{\partial T}{\partial x} + \frac{\partial}{\partial x}\left(\frac{\partial T}{\partial x}\right)dx\right]du \tag{10.30}$$

$$= kA\frac{\partial^2 T}{\partial x^2}(dx)(dt)$$

Heat accumulation in the strip = (Mass of the strip) × (Specific heat) × (Temperature variation with time) × (Incremental time)

$$= \rho(Adx)C_p\left(\frac{\partial T}{\partial t}\right)dt \tag{10.31}$$

where C_p is specific heat of the material.

$$kA\frac{\partial^2 T}{\partial x^2}(dx)(dt) = \rho C_p(Adx)\frac{\partial T}{\partial t}(dt) \tag{10.32}$$

Dividing both sides by $\rho AC_p dxdt$, we get

$$\frac{k}{\rho C_p}\frac{\partial^2 T}{\partial x^2} = \left(\frac{\partial T}{\partial t}\right) \tag{10.33}$$

Here we define a term known as **thermal diffusivity** (α) which has the units of m^2/s as follows:

$$\alpha = \frac{k}{\rho C_p} \tag{10.34}$$

Thermal diffusivity data for some foods is given by Appendix 9, and the specific heat data in Appendix 8.

Combining Eq. (10.33) and Eq. (10.34), we get

$$\frac{\partial T}{\partial t} = \alpha\frac{\partial^2 T}{\partial x^2} \tag{10.35}$$

Equation (10.35) is the expression for unsteady state heat conduction through solid materials. The equation is applicable both for heating and cooling of solids.

Integration of Eq. (10.35) gives an expression (McCabe, et al., 1993) for the temperature of the solid slab heated from both sides with a constant surface temperature of T_s and having a thickness of $2s$,

$$\frac{T_s - T_{av}}{T_s - T_o} = \frac{8}{\pi^2}\left[\exp\left\{-\left(\frac{\pi}{2}\right)^2 N_{Fo}\right\} + \frac{1}{3^2}\exp\left\{-\left(\frac{3\pi}{2}\right)^2 N_{Fo}\right\} + \frac{1}{5^2}\exp\left\{-\left(\frac{5\pi}{2}\right)^2 N_{Fo}\right\} + \ldots\right] \tag{10.36}$$

where T_s is constant surface temperature, T_{av} is average temperature of slab at a time interval of t_T, T_o is initial temperature of the slab.

$$N_{Fo} = \frac{\alpha t_T}{s^2} \tag{10.37}$$

here N_{Fo} is a dimensionless number and is known as **Fourier number**, t_T is total time period, s is half thickness of the slab and is equal to $B/2$.

Half slab thickness is used since the heating is from both sides. If the heating is only from one side, as in the case of baking of *chapathi* or *dosa* on a hot pan then s is replaced by full thickness $2s$ (B).

Similar type of expression can also be derived for infinitely long cylinder of radius, $r^\dagger$, and is given by (McAdam 1954).

$$\frac{T_s - T_{av}}{T_s - T_o} = 0.692 \exp(-5.78\, N_{Fo}) + 0.131 \exp(-30.5\, N_{Fo}) + 0.0534 \exp(-74.8\, N_{Fo}) + \cdots \tag{10.38}$$

in which $N_{Fo} = \alpha t_T / r^2$, where r is the radius of the cylinder.
Similarly for a sphere of radius r, the above expression becomes (Carslaw and Jaega 1959).

$$\frac{T_s - T_{av}}{T_s - T_o} = 0.608 \exp(-9.87\, N_{Fo}) + 0.152 \exp(-39.5\, N_{Fo}) + 0.0676 \exp(-88.8\, N_{Fo}) + \cdots \tag{10.39}$$

In all the preceding three expressions, if $N_{Fo} > 0.1$, only the first term of the series is significant, and rest of the terms may be neglected. By truncating to the first term, the above expressions can be rearranged to give an expression for total time t_T for achieving the desired average temperature T_{av} of the body (initially at a temperature of T_o) when the surface temperature is T_s.

For an infinite slab,

$$t_T = \frac{1}{\alpha}\left(\frac{2s}{\pi}\right)^2 \ln\left(\frac{8(T_s - T_o)}{\pi^2 (T_s - T_{av})}\right) \tag{10.40}$$

For an infinitely long cylinder,

$$t_T = \frac{r^2}{5.78\,(\alpha)} \ln\left(\frac{0.692(T_s - T_o)}{(T_s - T_{av})}\right) \tag{10.41}$$

For a sphere,

$$t_T = \frac{r^2}{9.87\,(\alpha)} \ln\left(\frac{0.608(T_s - T_o)}{(T_s - T_{av})}\right) \tag{10.42}$$

The term $\frac{T_s - T_{av}}{T_s - T_o}$ in Eq. (10.36), Eq. (10.38) and Eq. (10.39) can be drawn against N_{Fo}. The plots are almost straight lines except at very low Fourier numbers. The plots are shown in Figure 10.6 (McCabe, et al., 1993).

---

†Where L is much large compared to radius r.

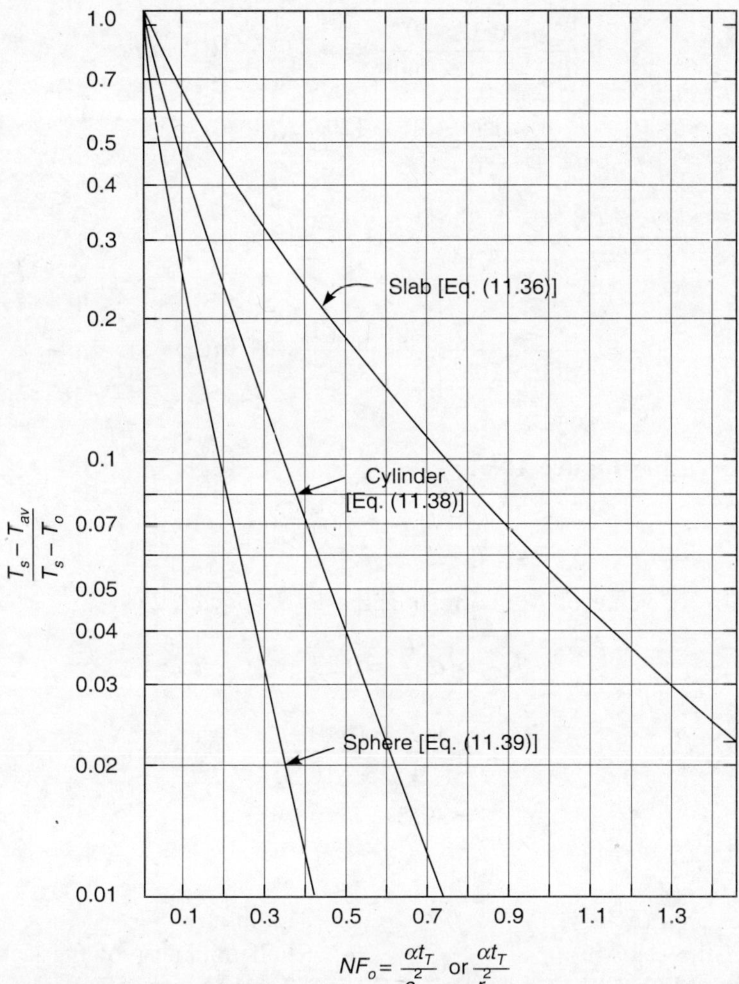

Figure 10.6 Average temperature vs. the Fourier number plots for a large slab, infinite cylinder and sphere.

Source: Reprinted from *Unit Operation of Chemical Engineering*, McCabe, et al., p. 302, Copyright 1993 with permission from McGraw-Hill, New York.

PROBLEM 10.4 Cooked and mashed potato is used to make cutlets of 12 mm thick by heating on a hot pan which is at 150 °C. The initial temperature of the unbaked cutlet is 30 °C. The thermal conductivity of the potato mash is 0.37 W/m°C, the specific heat is 3.8 kJ/kg°C, and the density is 990 kg/m^3. Calculate how much time it takes for the average temperature of the cutlet to reach 100 °C. The dimensions of the cutlet are such that it may be assumed to be an infinite slab.

Solution It is a straight case of application of Eq. (10.40) except that we use 2s in place of s as the heating is only from one side. Hence, s = 12 mm

$$t_T = \frac{1}{\alpha}\left(\frac{2s}{\pi}\right)^2 \ln\left(\frac{8(T_s - T_o)}{\pi^2 (T_s - T_{av})}\right)$$

$$\alpha = \frac{k}{\rho C_p} = \frac{0.37\times10^{-3}}{3.8\times990} = 9.835 \times 10^{-8}\ \text{m}^2/\text{s}$$

$$\frac{T_s - T_o}{T_s - T_{av}} = \frac{150-30}{150-100} = \frac{120}{50} = 2.4$$

$$\frac{8(T_s - T_o)}{\pi^2 (T_s - T_{av})} = \frac{8}{(3.14)^2}\times 2.4 = 1.945$$

Therefore,
$$t_T = \frac{1}{9.835\times10^{-8}}\left[\left(2\times\frac{12\times10^{-3}}{\pi}\right)^2 \ln(1.945)\right] = 394.76s$$

$$\cong 395s$$

Alternate way by using Figure 10.6:

$$\frac{T_s - T_{av}}{T_s - T_o} = \frac{1}{2.4} = 0.417$$

From Figure 10.6, for value of N_{Fo} corresponding to 0.417 is 0.27

$$N_{Fo} = \alpha\frac{t_T}{s^2} = 0.27$$

Therefore,
$$t_T = \frac{0.27\times(0.012)^2}{9.835\times10^{-8}} = 395.32s$$

$$= 395s$$

Symbols

A: heat transfer area (m^2)
B: thickness of the slab (m)
C_p: specific heat (kJ/kg°C)
k: thermal conductivity (J/sm°C)
L: length (m)
N_{Fo}: Fourier number
q: heat transfer rate (kJ/s or kW)
R: Resistance for heat transfer (m^2 °C/W)
r: radius (m)
s: half thickness of the slab (m)
T: Temperature (°C)
ΔT: temperature difference (°C)
T_{av}: average temperature at time t_T
t: time (s or h)
t_T: total time of heat transfer (s or h)
x: distance measured in x-direction (m)

Subscripts

A: slab A
A_v: average
B: slab B
C: slab C
i: inner
L: logarithmic
o: outer or initial
s: surface
T: total

Greek Symbols

α: thermal diffusivity (m^2/s)
ρ: density (kg/m^3)

REVIEW QUESTIONS

10.1 What are the applications of heat conduction in food processing?

10.2 Describe the conductive method of heat transfer.

10.3 What is Fourier's law? What is its importance in food processing?

10.4 Derive an expression for heat conductance through slabs in series.

10.5 Derive an expression for heat flux through a cylindrical pipe.

10.6 What is meant by log mean radius?

10.7 Derive an expression for unsteady state heat conduction in a slab.

10.8 What is meant by Fourier number?

10.9 Write expressions for total time, during unsteady state heat conduction, to attain a certain temperature (T) in
  (i) long slabs,
  (ii) long cylinders, and
  (iii) spheres.

10.10 What is meant by thermal diffusivity and what are its units?

NUMERICAL PROBLEMS

10.1 The maximum tolerable heat loss through a furnace wall is 1.3 kW/m^2. A brick wall is constructed next to the furnace wall to insulate the heat loss. The temperature on either side of the wall is 200 °C and 40 °C. What should be the thickness of the brick wall if the thermal conductivity of the brick is 0.72 W/m°C. (**Ans:** B = 0.2 m)

10.2 A cold storage room is maintained at –18 °C. The room is made out of 12 mm thick SS wall (k = 16.3 W/m°C). The metallic wall is followed by a 40 cm thick teflon layer (k = 0.94 W/m°C) which is supported by a 45 cm thick brick wall (k = 0.7 W/m°C). Calculate the heat transfer per m^2 if the outside temperature is 30 °C.
(**Ans:** Q/A = 44.89 W/m^2)

10.3 Find the heat transfer area for a cylindrical pipe with 7.5 cm id and 10 cm od and 1 m long. (**Ans:** A_L = 0.273 m^2)

10.4 A block of meat 2.5 cm thick taken out from cold storage freezing room has an initial temperature of –18 °C and is kept outside at room temperature at 28 °C. After one hour, what is the average temperature of meat block? The thermal conductivity of meat block is 0.065 W/m °C; density is 990 kg/m^3 and the specific heat is 3.5 kJ/kg °C. Assume the meat slab to be an infinite slab. (**Ans:** T_{av} = 15.12 °C)

REFERENCES

Carslaw, H.S. and Jaeger, J.C. (1959), *Conduction of Heat in Solids*, Oxford University Press, Fair Lawn, N.J., p. 234.

McAdams, W.H. (1954), *Heat Transmission*, 3rd ed., McGraw-Hill, New York, p. 232.

McCabe, W.L., Smith, J.C. and Harriott, P. (1993), *Unit Operations of Chemical Engineering*, 5th ed., McGraw-Hill, New York, p. 301.

CHAPTER

11

Heat Transfer by Convection

We have seen in the earlier chapter how the heat flows through solids without any physical movement of particles or molecules or layers of the solid medium from a body at a higher temperature to a body at a lower temperature which is known as **conduction**. In this chapter, we will study the process of heat transfer in fluids (liquids, vapours and gases) from a metallic surface or a solid surface at higher temperature to the fluids at lower temperature and are in contact with it. Here we see that the fluid layers in contact with the hot surface virtually move upwards because of density differences caused by temperature differences. This kind of heat transfer in fluids due to physical movement of layers or droplets of the fluids is known as **convection**. The most common example is heating of water in a vessel or in a beaker.

If we start heating water in a vessel, the water gets heated up. Slowly some bubbles start forming at the bottom of the vessel if the surface temperature is high. (The temperature difference is of the order of 50 °C or more). These bubbles consist of water vapour. They raise to the top because of density difference. This is an example of convection, and the type of boiling (heating) is known as **nucleate boiling**. If the temperature is not very high, we still find some movement of layers of water slowly raising to the top. The upper layers which are at a relatively less temperature, and hence, are dense will go to the bottom. Thus, there will be a continuous exchange of layers of less density from the bottom going to the top, and the layers of higher density going to the bottom (Figure 11.1) until the temperature of all layers becomes the same which is almost equal to the temperature of the surface. Later there will not be any transfer of heat.

If the temperature of the surface is very high (the temperature difference is of the order of 100 °C), the bubbles at the bottom coalesce and form into a thin quiescent film of vapour (Figure 11.2). The heat transfer has to take place through the film. This kind of boiling is known as **film boiling**. We observe this kind of phenomenon only when the heat transfer is taking place to the fluids with the phase change. The point at which transition occurs between the nucleate boiling and film boiling is known as **Leidenfrost point** and is shown in Figure 11.3.

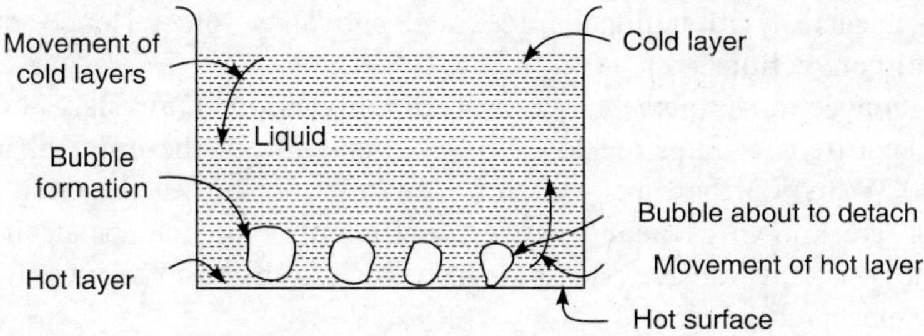

Figure 11.1 Convective heating process.

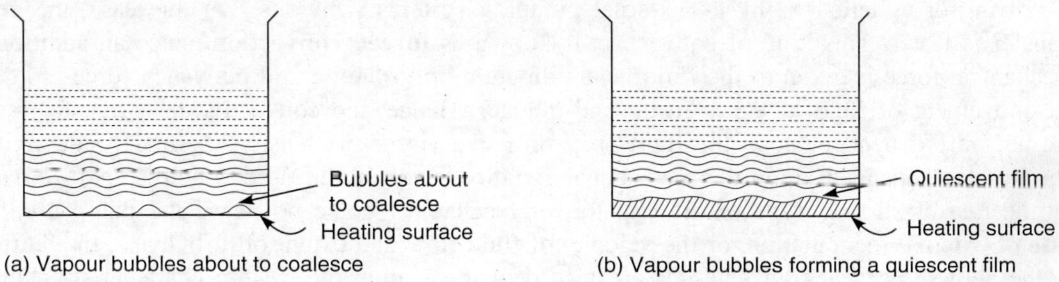

Figure 11.2 Concept of film boiling.

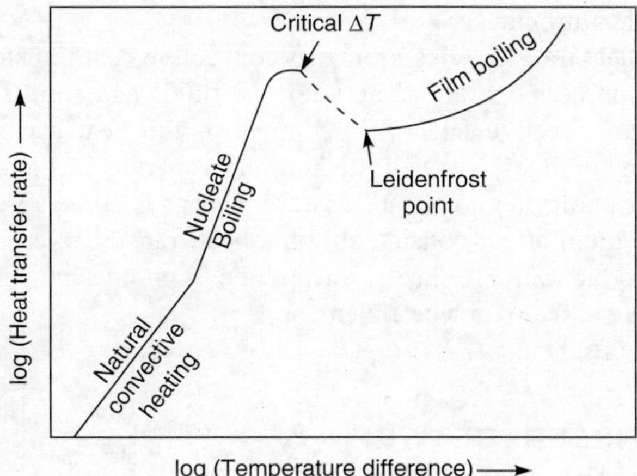

Figure 11.3 Heat transfer process by convection with phase-change.

11.1 NATURAL CONVECTION AND FORCED CONVECTION

The type of heat transfer process which is shown in Figure 11.1 where heat transfer takes place mainly due to the movement of hot layers and cold layers because of density differences which occur due to temperature differences is known as **natural convection**. This is in contrast to any assisted movement of different layers. The movement of different molecules or layers of the fluids is by means of gravitational forces and buoyancy force. Hence, the convection is called as **natural convections**.

The natural convection is also known as **free convection**. Obviously, in case of natural convection, the heat transfer rates are low, and are generally of the order of 60 W/(m^2°C) for air, and 60–3000 W/(m^2°C) for water systems. The heat transfer rates mentioned above in the units of W/m^2°C are known as **heat transfer coefficients**, and will be dealt in detail in the next section. The low heat transfer rates have necessitated the need to increase them by using some additional mechanical force.

Let us imagine a situation where the heating process is assisted by means of a mechanical stirring given to the liquid. The movement of different layers is faster. It could be by way of stirring or agitation or by an external pumping systems. Obviously in this case, the heat transfer is faster. This kind of convection is known as **forced convection**, since an additional mechanical force is given to the fluid layers, in addition to the natural buoyancy forces so that the movement of fluid layers is faster and quicker. Hence, the convective heat transfer rates are not only dependent upon the fluid properties like viscosity, density, thermal conductivity, but are also dependent upon the mechanical agitation given to the fluid which is characterized by the rate of stirring, type of the agitator, or mechanical horse power of the pump used in case of external recirculation, or the velocity of fluid flow and extent of turbulence. In addition to the above, the heat transfer rates are also dependent upon the system characteristics, viz., diameter of the conduit, smoothness of the surface of the conduit, thermal conductivity of the metal, presence or absence of baffles in the vessel which break the formation of vortex and help in creation of better turbulence.

Indeed convection heat transfer (also known as convective heat transfer) is the major method of heat transfer in any process operations between the fluids and solid surfaces, which contain the fluids. Hence, the convective heat transfer rates or the heat transfer coefficients which determine the heat transfer rates are well studied and reported in engineering literature. The influence of the process parameters on heat transfer rates has become a subject of major interest, since any variations in them affect considerably the heat transfer rates which is not so in case of heat transfer by conduction. The heat transfer rate is more characterized by a parameter known as **convective heat transfer coefficient** or simply heat transfer coefficient.

11.2 HEAT TRANSFER BETWEEN A METALLIC WALL AND THE FLUID

Whenever the heat is transferring from a metallic wall to a fluid adjacent to it (it could be in the reverse direction also), it is presumed that there exists a fluid film of constant thickness (Figure 11.4) in which actually lies the resistance to heat transfer. This concept was originally proposed by Prandtl (Heldman and Lund 1992). The thickness of the film may vary depending upon the agitation or movement of the fluid. More is the agitation, thinner is the

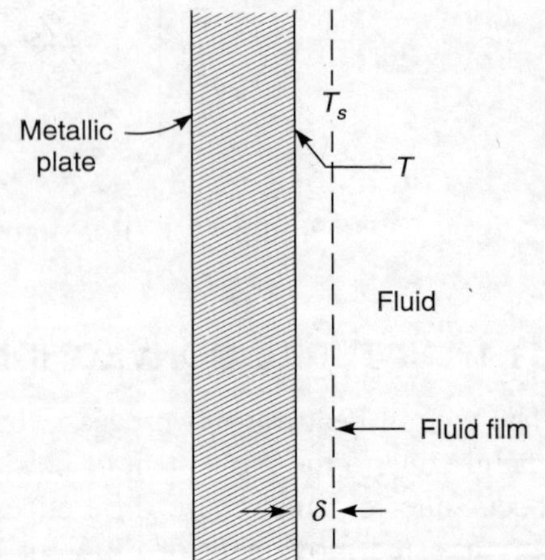

Figure 11.4 Convective heat transfer.

film; but it cannot be totally done away with. Within the film, the temperature varies from the wall temperature (T_s) to the average temperature of the fluid.

11.2.1 Thermal Boundary Layer

The film within which the temperature varies is known as the **Thermal Boundary Layer** (TBL) as contrast to the Hydrodynamic Boundary Layer (HBL) which we studied in Chapter 8 and is shown in Figure 11.5. The relative thickness of both the boundary layers are dependent upon a dimensionless number known as **Prandtl number (N_{Pr})** and is defined as:

$$N_{Pr} = \frac{C_p \mu}{k} \tag{11.1}$$

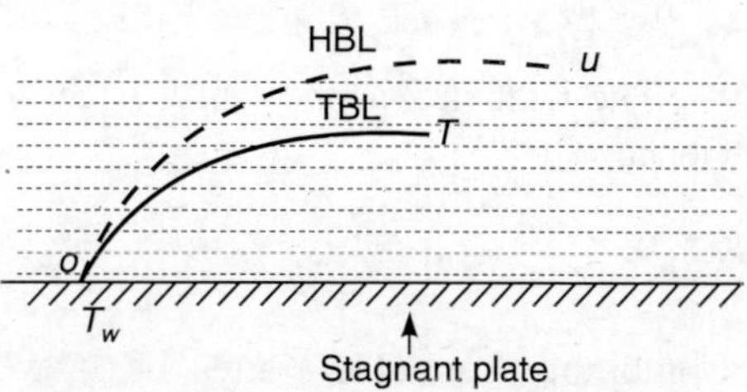

Figure 11.5 Boundary layer concept (TBL: thermal boundary layer, HBL: hydrodynamic boundary layer, $T > T_w$).

HBL is thicker than the TBL if $N_{Pr} > 1.0$ which is mostly the case with most of the liquids. HBL is equal to TBL if $N_{Pr} \cong 1.0$ which is mostly true for most of the gases and vapours. HBL is thinner than TBL if $N_{Pr} < 1.0$ which happens rarely and is true for molten metals, etc.

The thermal boundary layer concept is very important in understanding the process of heat transfer in most of the process operations. This prohibits us from using very high wall temperatures in heating liquid foods because the temperature of the liquid food in the stagnant film adhering to the wall is equal to the temperature of the wall, and hence, may deteriorate or char the food. This is precisely the reason why the milk heated in a *SS* vessel on a gas stove invariably results in charring the milk at the edges. This is also precisely the reason why we always prefer to use steam heated jacketed *SS* vessels where we can have a better control on temperature rather than using direct flame stoves like gas stoves, etc.

11.3 HEAT TRANSFER COEFFICIENT

As originally proposed by Prandtl, and as has already been mentioned, the resistance to heat transfer lies in the fluid film adjacent to the hot plate/surface. The heat transfer rate may be represented by Fourier law through the stagnant fluid film as follow:

$$\frac{q}{A} = \frac{k}{\delta}(T_s - T) \tag{11.2}$$

where δ is the film thickness and k is thermal conductivity. Since it is difficult to measure δ, we combine (k/δ) into a term called as the **convective heat transfer coefficient** and is abbreviated as h. Equation (11.2) can be written as:

$$q = hA(T_s - T) \tag{11.3}$$

or in differential form, it can be written as:

$$\frac{dq}{dA} = h\,(T_s - T) \tag{11.4}$$

in which h has the units of kJ/(sm^2°C) and A is the area in m^2. T is the average temperature of the fluid and T_s is surface temperature. The reciprocal of the heat transfer coefficient is known as the **resistance**.

$$R = \frac{1}{h} \tag{11.5}$$

The heat flux at the wall/surface is given by the conductive heat transfer through the wall. Hence,

$$\frac{dq}{dA} = -k\left(\frac{dT}{dy}\right)_w \tag{11.6}$$

Combining Eqs. (11.4) and (11.6), we write

$$h = -k\frac{(dT/dy)_w}{(T_s - T)} \tag{11.7}$$

In the above equation, it may be noted that h is always positive. Since y is measured in the direction of heat transfer, dT/dy is negative.

If Eq. (11.7) is multiplied on both sides by d/k, we get

$$\frac{hd}{k} = \frac{-d(dT/dy)_w}{(T_s - T)} \tag{11.8}$$

In Eq. (11.8), hd/k is dimensionless, and is known as **Nusselt number** represented by N_{Nu}. Nusselt number is one important dimensionless number, and is very useful in convective heat transfer. As we see later, most of the equations for heat transfer are based on Nusselt number.[†]

The heat transfer coefficient h is dependent upon a number of fluid properties, and system parameters.

Fluid properties which affect h are density, viscosity, thermal conductivity, specific heat, etc. and system properties which affect the h are diameter of pipe or length of pipe, velocity of the fluid, roughness of the surface walls, etc. Thus,

$$h = f(\rho, d, u, \mu, C_p, k) \tag{11.9}$$

Combining Eqs. (11.2) and (11.3) we have

$$h = \frac{k}{\delta} \tag{11.10}$$

or from Fourier law [Eq. (11.6)]

$$h = \frac{k}{x}$$

where x is the thickness.

From Eq. (11.10), we can write Nusselt number

$$N_{Nu} = \frac{hd}{k} = \frac{d}{\delta} \quad \text{or} \quad \frac{d}{x} \tag{11.11}$$

Thus, *Nusselt number is the ratio of the pipe diameter to the thickness of the laminar layer.*

[†]Other dimensionless numbers and predictive equations for h will be discussed in Section 11.8.

PROBLEM 11.1 A fruit juice is passing through a hot pipe of 7.5 cm diameter and 3.3 m long @ 60 lpm and the wall temperature is 120 °C. The juice is entering at 25 °C and is leaving at 65 °C. Find the heat transfer coefficient and Nusselt number if the fluid properties are assumed to be those of water.

Solution The fluid properties are:

$$\rho = 1000 \text{ kg/m}^3$$

$$k = 0.6 \text{ W/m°C} = 0.6 \times 10^{-3} \text{ kW/m°C.}$$

$$C_p = 4.2 \text{ kJ/kg°C}$$

$$\mu = 1 \times 10^{-3} \text{ Pa s}$$

Heat transfer rate for the fluid $= q = \dot{m}C_p(\Delta T)$

$$\dot{m} = \text{Volume flow rate} \times \text{density}$$

$$= \frac{60}{60} \times 10^{-3} \times 1000 = 1 \text{ kg/s}$$

$$q = 1 \times 4.2 \times (65 - 25) = 168 \text{ kJ/s}$$

$$q = hA(\Delta T)$$

Now, we find the heat transfer area based on the inside diameter only

$$A = \pi dL = \pi \times 7.5 \times 10^{-2} \times 3.3 = 0.778 \text{ m}^2.$$

The temperature difference ΔT is measured on the basis of average temperature difference of the fluid† and the wall temperature. It is shown in Figure 11.6.

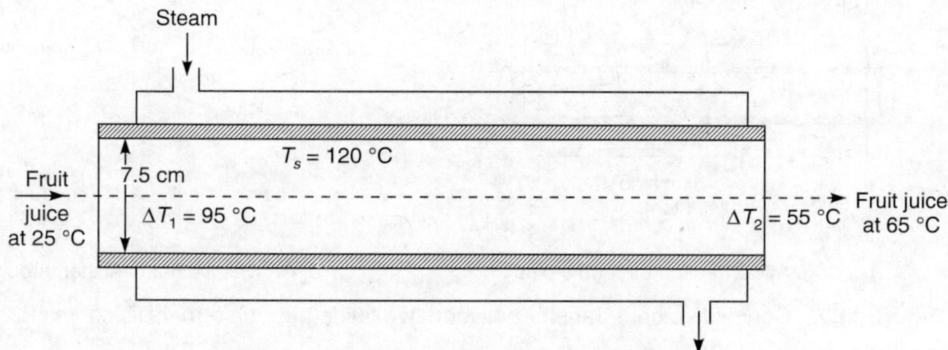

Figure 11.6 Flow conditions for Problem 10.1.

$$\Delta T_1 = 120 - 25 = 95 \text{ °C}$$

$$\Delta T_2 = 120 - 65 = 55 \text{ C}$$

†In such cases in place of average temperature difference, the log mean temperature difference is used and will be discussed in Section 11.6.

$$\therefore \quad \Delta T = \frac{\Delta T_1 + \Delta T_2}{2} = \frac{95 + 55}{2} = 75\ °C$$

$$\therefore \quad 168 = h \times 0.778 \times 75$$

$$h = \frac{168}{0.778 \times 75} = 2.879\ kJ/sm^2\ °C$$

$$N_{Nu} = \frac{hd}{k} = \frac{2.879 \times 7.5 \times 10^{-2}}{0.6 \times 10^{-3}}$$

$$= 360$$

Note that all units should be consistent. Hence, k is multiplied by 10^{-3} to make it in kJ/s m°C in place of the original units in J/s m°C.

11.4 OVERALL HEAT TRANSFER COEFFICIENT

When the heat transfers or exchanges between two fluids (one of them is a cold fluid and the other is a hot fluid) through a solid metallic wall, the following three heat transfer processes occur, and is shown in Figure 11.7, if the heating is taking place when the cold fluid is passing through the tube, and the hot fluid is outside the tube (known as shell side).

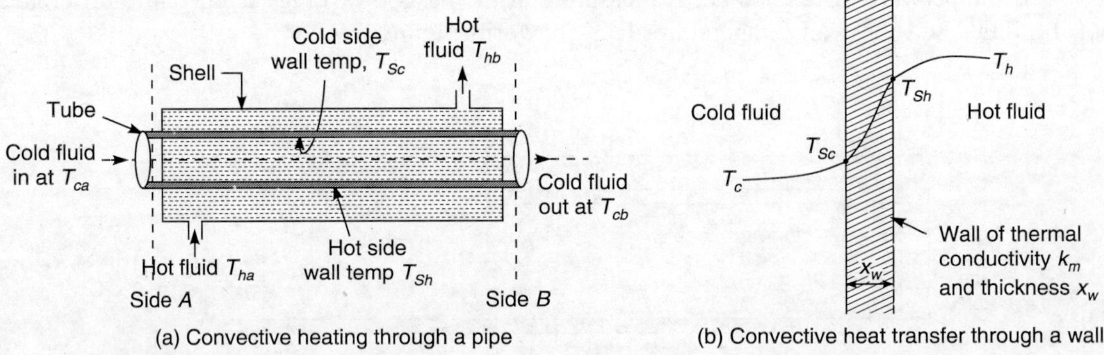

Figure 11.7 Convective heat transfer between two fluids through a metallic conductor.

(i) Heat transfer between the wall and the hot fluid in the shell side by convection which is designated as outside heat transfer process and the heat transfer coefficient is represented by h_o.

(ii) Heat transfer through the metallic wall by conduction.

(iii) Heat transfer between the hot wall and the cold fluid in the tube side by convection which is designated as inside heat transfer process, and the heat transfer coefficient is represented by h_i.

Inside heat transfer rate $= \dfrac{dq}{dA_i} = h_i\,(T_{Sc} - T_c)$ (11.12)

Outside heat transfer rate $= \dfrac{dq}{dA_o} = h_o\,(T_h - T_{Sh})$ (11.13)

Heat transfer rate through the wall $= \dfrac{dq}{dA_L} = \dfrac{k_m}{x_w}\,(T_{Sh} - T_{Sc})$ (11.14)

Now, the overall temperature difference is $(T_h - T_c)$

Therefore, $$(T_h - T_c) = (T_h - T_{Sh}) + (T_{Sh} - T_{Sc}) + (T_{Sc} - T_c) = \Delta T \qquad (11.15)$$

and $$(T_h - T_c) = \Delta T = dq\left(\frac{1}{dA_i(h_i)} + \frac{x_w}{k_m}\frac{1}{dA_L} + \frac{1}{dA_o(h_o)}\right) \qquad (11.16)$$

Mulitplying and dividing the RHS by dA_o, we get

$$(T_h - T_c) = \frac{dq}{dA_o}\left(\frac{dA_o}{dA_i}\frac{1}{(h_i)} + \frac{x_w}{k_m}\frac{dA_o}{dA_L} + \frac{1}{(h_o)}\right) \qquad (11.17)$$

Since it is not always possible to measure the wall temperatures (T_{Sc} and T_{Sh}), we define only one heat transfer coefficient, known as **overall heat transfer coefficient** based on the temperatures of the hot side and cold side, viz., T_h and T_c. We can now write

$$q = UA(\Delta T) \qquad (11.18)$$

where U is the overall heat transfer coefficient. In differential form, we can write

$$\frac{dq}{dA} = U(\Delta T) \qquad (11.19)$$

The overall heat transfer coefficient can now be defined based on either inside conditions or outside conditions and accordingly called as:

- Inside overall heat transfer coefficient (U_i) or
- Outside overall heat transfer coefficient (U_o).

Accordingly we can write Eq. (11.19) as:

$$\frac{dq}{dA_i} = U_i\,(\Delta T) \qquad (11.20)$$

or $$\frac{dq}{dA_o} = U_o\,(\Delta T) = U_o(T_h - T_c) \qquad (11.21)$$

In Eq. (11.17) $$\frac{dA_o}{dA_i} = \frac{d_o}{d_i} \text{ and } \frac{dA_o}{dA_L} = \frac{d_o}{d_L}$$

Now, Eq. (11.17) can be written as:

$$\frac{dq}{dA_o} = \frac{T_h - T_c}{\dfrac{d_o}{d_i}\dfrac{1}{h_i} + \dfrac{x_w}{k_m}\dfrac{d_o}{d_L} + \dfrac{1}{h_o}} \qquad (11.22)$$

Combining Eqs. (11.21) and (11.22),

$$U_o = \frac{1}{\dfrac{d_o}{d_i}\dfrac{1}{h_i} + \dfrac{x_w}{k_m}\dfrac{d_o}{d_L} + \dfrac{1}{h_o}} \tag{11.23}$$

Or in other words

$$\frac{1}{U_o} = \frac{d_o}{d_i}\frac{1}{h_i} + \frac{x_w}{k_m}\frac{d_o}{d_L} + \frac{1}{h_o} \tag{11.24}$$

And similarly

$$\frac{1}{U_i} = \frac{1}{h_i} + \frac{x_w}{k_m}\frac{d_i}{d_L} + \frac{d_i}{d_o}\frac{1}{h_o} \tag{11.25}$$

Or we can also simply write for a slab

$$\frac{1}{U} = \frac{1}{h_i} + \frac{x_w}{k_m} + \frac{1}{h_o} \tag{11.26}$$

Since $(1/U)$ is the overall resistance (R) for heat transfer, $(1/h_i)$ is the resistance for heat transfer based on the internal conditions (R_i), and $(1/h_o)$ is the resistance for heat transfer based on outside conditions (R_o).

$$R = R_i + R_m + R_o \tag{11.27}$$

where R is the resistance for heat transfer, and R_m is the resistance for heat transfer by the metallic wall. The above equation is similar to the summation of the individual resistances being equal to the overall resistance which we derived in Chapter 10 (Section 10.2).

PROBLEM 11.2 100 lit of dilute sugar syrup is heated in a 80 cm diameter, hemispherical jacketed *SS* vessel with a little agitation using steam at 2 atm. pressure. The wall thickness of the vessel is 3 mm. The sugar syrup, whose physico-chemical properties may be assumed to be those of water, is heated from 28 °C to 75 °C. What is the heat transfer rate? What is the quantity of steam required, if the inside heat transfer coefficient is 2 kW/m^2°C, and the outside coefficient is 1.5 kW/m^2°C.

Solution

Approach:

(i) Initially we calculate what is the quantity of heat required to raise the temperature of 100 lit dilute sugar syrup from 28 °C to 75 °C. From this, we can find out the quantity of steam required.

(ii) We calculate the overall heat transfer coefficient using Eq. (11.26).

(iii) Then we calculate the heat transfer rate.

The arrangement is shown in Figure 11.8.

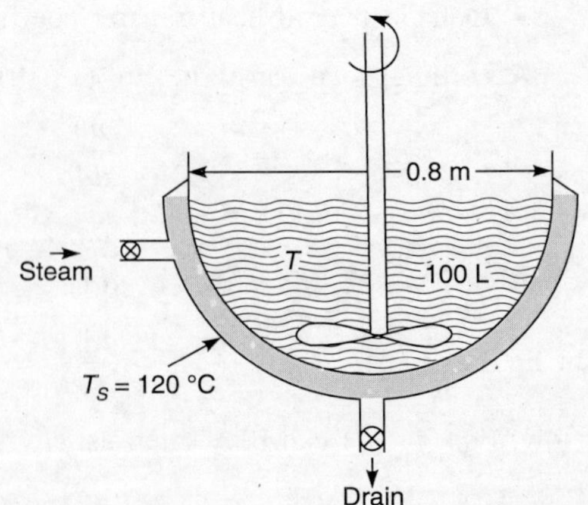

Figure 11.8 Heating sugar syrup in a steam jacketed vessel (Problem 11.2).

Physico-chemical and thermal properties of sugar syrup are:

$$\rho = 1000 \text{ kg/m}^3$$

$$\mu = 1 \times 10^{-3} \text{ Pa s}$$

$$C_p = 4.2 \text{ kJ/kg°C}$$

$$k = 0.628 \text{ W/m°C (average at 40 °C)}$$

$$= 0.628 \times 10^{-3} \text{ kW/m°C}$$

Thermal conductivity of steel = 43 W/m°C = 43×10^{-3} kW/m°C (Appendix 6).
Thickness of wall = 3 mm = 3×10^{-3} m
ΔT = Temperature raise of sugar syrup = 75 °C – 28 °C = 47 °C
Quantity of heat required to raise the temperature of 100 lit sugar syrup from 28 to 75 °C = $mC_p\,(\Delta T)$

$$Q = (100 \times 10^{-3} \times 1000) \times (4.2) \times (75 - 28) = 19{,}740 \text{ kJ}$$

Using Eq (11.26)

$$\frac{1}{U} = \frac{1}{2} + \frac{3\times10^{-3}}{43\times10^{-3}} + \frac{1}{1.5} = 0.5 + 0.069 + 0.667$$

$$= 1.2367 \text{ m}^2\text{ °C/kW}$$

Therefore, $$U = 0.8085 \text{ kW/m}^2\text{ °C}$$

$$q = UA\,(\Delta T)$$

In the preceding equation,

$$A = 2\pi r^2 = 2 \times \pi \times (0.4)^2 = 1 \text{ m}^2$$

where r is radius of the vessel = 40 cm

To find the value for q, we need to know what is ΔT. The steam temperature outside the vessel corresponding to 2 atm. pressure is 120 °C (from Steam Tables in Appendix 1B). The inside temperature is averaged out† to be

$$\frac{28 + 75}{2} = 51.5$$

Therefore, $$\Delta T\ 120 - 51.5 = 68.5 \text{ °C}$$

and $$q = 0.8085 \times 1 \times 68.5 = 55.38 \text{ kJ/s.}$$

The quantity of steam required = $\dfrac{Q}{\lambda_s}$

where λ_s is latent heat of vapourization of steam and is noted from steam tables (Appendix 1).

$$\lambda_s = 2706.3 - 503.71 = 2202.6 \text{ kJ/kg.}$$

Therefore, quantity of steam required = $\dfrac{19740}{2202.6}$ = 8.96 kg ≅ 9 kg.

†Some errors must have crept into the calculations. The problem is to be attempted more as an unsteady state heat transfer process. The equations we used are all applicable for steady state heat transfer.

11.5 UNSTEADY STATE HEAT TRANSFER

Upto now we have only considered steady state heat transfer; but in a case like that in Problem 11.2, where the material is heated from an initial temperature to a final temperature, in which case the heat transfer process is going in an unsteady manner, and hence, ΔT goes on changing as time progresses. In this case, we follow the approach of heat transfer for a spherical body, for which the rate of heat transfer for the sphere is approximated to be 5 k/r_m [McCabe, et al., 1993 (a)]. That is, the whole fluid is presumed to be contained in a metallic spherical body of radius r_m, and is being heated. Hence, the volume of the fluid contained in the spherical vessel is $(4/3)\pi r_m^3$ and the heating surface area is $4\pi r_m^2$.

We can write heat balance for unsteady state conditions in which the rate of temperature change is dT/dt and the temperature change is ΔT

$$\Delta T = (T_s - T)$$

where T_s is the constant surface temperature and T is the temperature of the fluid at time t.

$$\rho C_p \left(\frac{4\pi r_m^3}{3}\right)\frac{dT}{dt} = h_i A \Delta T \tag{11.28}$$

in which

$$h_i = \frac{5k}{r_m} \tag{11.29}$$

and

$$A = 4\pi r_m^2$$

Therefore,

$$\rho C_p \frac{(4\pi r_m^3)}{3}\frac{dT}{dt} = \frac{5k}{r_m}(4\pi r_m^2)(T_s - T)$$

On rearrangement

$$\frac{dT}{T_s - T} = \frac{15}{r_m^2}\left(\frac{k}{\rho C_p}\right)dt \tag{11.30}$$

On integration between the limits of initial temperature T_0 to the final temperature T at time t,

$$\ln\left(\frac{T_s - T_0}{T_s - T}\right) = \frac{15}{r_m^2}\alpha t \tag{11.31}$$

Equation (11.31) seems to be over simplification, and does not account for any forced convection conditions prevailing in the heating system. Irrespective of the forced convection conditions prevailing in the vessel, we get the same value for t for a given T_s, T_0 and T, since the alpha (α) for the fluid does not change. This needs a relook into the situation and the validity of Eq. (11.31).

Alternatively, we can write Eq. (11.28) as follows (Singh and Heldman 1993)

$$\rho C_p \frac{dT}{dt} = h_i A (T_s - T) = q \tag{11.32}$$

where V is the volume of the fluid, A is the heat transfer area and h_i is the inside heat transfer coefficient.

The preceding equation is applicable only where there is no internal resistance to heat transfer or negligible internal resistance. This kind of a situation occurs with most of the foods with high thermal conductivity or if the liquid foods are well stirred so that there will not be any temperature gradients within the liquid foods.

Equation (11.32) can be integrated with the boundary conditions that

$$T = T_0 \text{ when } t = 0$$

$$T = T \text{ when } t = t$$

$$\ln \frac{(T_s - T_0)}{T_s - T} = \frac{h_i A t}{\rho V C_p} \tag{11.33}$$

Equation (11.33) can be applied for heating or cooling of liquid foods with negligible internal resistance to heat transfer. The above equation was used by Al-Rawahi and Rao (2016) to find unsteady state h of petroleum crude with different viscosities, and reported that h varied as $\mu^{0.3257}$ which is almost equivalent to that given by Chilton-modified Dittus-Boelter equation (Eq 11.66) for agitated vessels.

PROBLEM 11.3 For the Problem 11.2, find the time taken for the to reach the final temperature, and also compare the time taken by Problem 11.2.

Solution It is a direct case of Eq. (11.33)

$$T_s = 120 \text{ °C}$$

$$T_0 = 28 \text{ °C}$$

$$T = 75 \text{ °C}$$

$$k = 0.628 \times 10^{-3} \text{ kW/m°C} = 0.628 \times 10^{-3} \text{ kJ/sm°C}$$

$$\rho = 1000 \text{ kg/m}^3$$

$$C_p = 4.2 \text{ kJ/kg°C}$$

Therefore,

$$\alpha = \frac{k}{\rho C_p} = \frac{0.628 \times 10^{-3}}{1000 \times 4.2} \cong 1.5 \times 10^{-7} \text{ m}^2/\text{s}$$

$$\frac{T_s - T_0}{T_s - T} = \frac{120 - 28}{120 - 75} = 2.04$$

$$\ln\left(\frac{T_s - T_0}{T_s - T}\right) = \ln (2.04) = 0.715$$

We apply Eq (11.33)

$$\ln\left(\frac{T_s - T_0}{T_s - T}\right) = \frac{h_i A t}{\rho V C_p}$$

where

$$V = 100 \text{ litres} = 0.1 \text{ m}^3$$

$$A = 1 \text{ m}^2$$

$$h_i = 2 \text{ kW/m}^2 \,°\text{C}$$

$$0.7151 = \frac{2 \times 1}{1000 \times 0.1 \times 4.2} \times t$$

Therefore,
$$t = \frac{0.7151 \times 1000 \times 0.1 \times 4.2}{2} = 150 \text{ s} = 2.5 \text{ minutes}$$

Now, we will see how the results will be by using Eq. (11.31).

We have to calculate r_m assuming all 100 litres of the fluid is contained in a spherical container

$$100 \times 10^{-3} = \frac{4}{3}\pi r_m^3$$

Therefore,
$$r_m = \left(\frac{0.1 \times 3}{4\pi}\right)^{1/3} = 0.2879 \text{ m}$$

Now, we apply Eq. (11.31)
$$\ln \frac{T_s - T_0}{T_s - T} = \frac{15\alpha t}{r_m^2}$$

Therefore
$$0.7151 = \frac{15}{(0.2879)^2} \times 1.5 \times 10^{-7} \times t$$

and
$$t = \frac{0.7151 \times (0.2879)^2}{15 \times 1.5 \times 10^{-7}} = 26{,}350 \text{ s}$$

$$= 7.3 \text{ hours.}$$

This looks to be ridiculous. Let us see what is the value of $\alpha t/r_m^2$ whether it is greater than 0.1 for which Eq. (11.31) applies.

$$\frac{\alpha t}{r_m^2} = \frac{1.5 \times 10^{-7} \times 26{,}350}{(0.2879)^2} = 0.047$$

The value is much less than 0.1. Hence, the result is presumably erratic.

By assuming average temperature for the fluid and heat transfer by steady state:

$$Q = 19{,}740 \text{ kJ}$$

$$q = 55.38 \text{ J/s}$$

$$\therefore \quad t = \frac{19740}{55.38} = 356.4 \text{ s}$$

$$\cong 6 \text{ min.}$$

11.6 COCURRENT AND COUNTER CURRENT FLOW

When the fluid is exchanging heat with another fluid in a typical shell and tube heat exchanger (as shown in Figure 11.9), there are two possible ways of flow of the fluids, viz.,

- Cocurrent, and
- Counter current

If the direction of fluid flow in the shell side and tube side is the same as shown in Figure 11.9(a) it is known as **cocurrent flow** conditions. The temperature profiles along the length of the heat exchanger for both the flow conditions are also shown in the Figure 11.9.

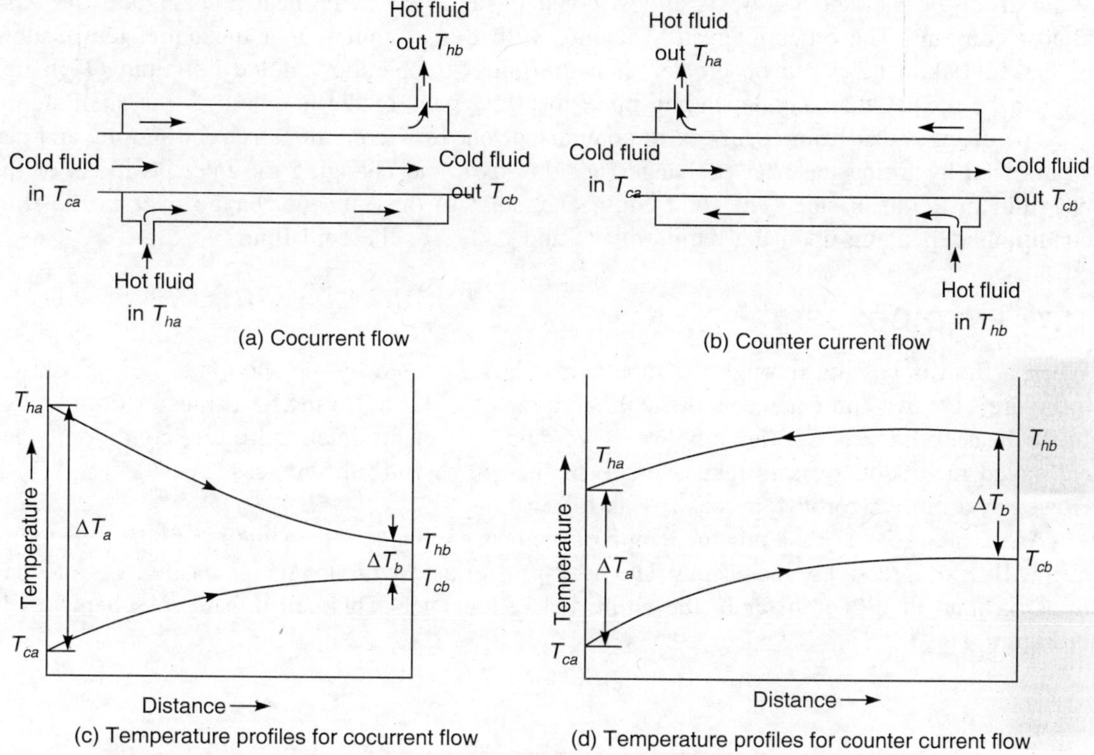

Figure 11.9 Cocurrent and counter-current flow systems with temperature profiles.

11.6.1 Temperature Profiles for Cocurrent Flow Conditions

As can be seen in the figure, ΔT_a and ΔT_b are the temperature differences at the inlet (position *a*) and outlet (position *b*) conditions. They are quite different; i.e., the value of ΔT_a is much larger compared to ΔT_b. Thus, the fluid entering at position *a* will have initially a very high temperature difference which gradually diminishes as we proceed along the length of the heat exchanger. This is both advantageous and disadvantageous depending upon the type of the material. The heat transfer rates are high initially and will gradually come down. So this kind of flow conditions are suitable for such materials which are highly viscous such as corn syrup, etc. and need high temperature differences initially to set the motion. But high ΔT_s are detrimental while heating certain fermentation fluids such as microbial suspensions which are subjected to high thermal shocks initially due to high ΔT. Another characteristic feature of the system is that the outlet temperature of the cold fluid (T_{cb}) can never be more than the outlet temperature of the hot fluid (T_{hb}); i.e., $T_{cb} \ngtr T_{hb}$ however much is the flow rate of the hot fluid in the shell side.

11.6.2 Temperature Profiles for Counter Current Flow Conditions

Unlike in the earlier case, ΔT_a and ΔT_b are not much different; and hence, the fluid being heated in the tube side is always subjected to almost constant ΔT which helps the cold fluid not to suffer from thermal shocks. ΔT is almost constant, and hence, the heat transfer rates are also almost constant. The other interesting feature with this system is that the outlet temperature of the cold fluid (T_{cb}) can be greater than the outlet temperature of the hot fluid (T_{ha}); i.e., T_{cb} can be greater than T_{ha} depending upon the flow rates of the hot fluid in the shell side.

The change over from cocurrent to counter-current flow conditions is very simple, and can be affected by tilting the heat exchanger upside down and changing the inlet positions of the hot fluid in the shell side. The food engineer resorts to this kind of change over in the flow conditions depending upon the requirements and nature of the cold fluid.

11.7 LMTD CONCEPT

When a fluid is passing through the tube and is being heated by the shell side fluid as shown in Figure 11.9, we come across a difficult situations to take ΔT in the heat transfer Eq. (11.18). In the heat exchanger, ΔT varies between ΔT_1 and ΔT_2, at the inlet and outlet conditions. One envisaged methodology is to take average of the temperature differences, i.e., $(\Delta T_1 + \Delta T_2)/2$. However, a more rigorous approach is as follows.

As we are progressing along the length of the heat exchanger, the quantity of heat transferred also will be progressively increasing. The heat transfer area will also be increasing, we may call q_T as the rate of heat transfer in the entire heat exchanger. This can be shown schematically in Figure 11.10.

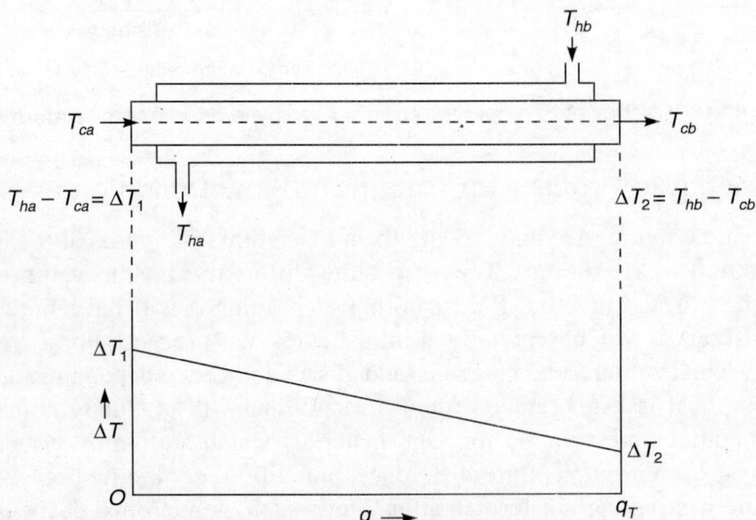

Figure 11.10 ΔT vs q in a counter-current heat exchanger.

Using the overall heat transfer coefficient, we can write

$$\frac{dq}{dA} = U(\Delta T) = U(T_h - T_c) \tag{11.34}$$

where T_h is the average temperature of the hot fluid and T_c is the average temperature of the cold fluid.

Integration of Eq. (11.34) gives the heat transfer rate in the entire area of the heat exchanger. The integration is possible with the following assumptions:

- The overall heat transfer coefficient is constant in entire range of temperatures.
- The physico-chemical properties of the fluids do not change considerably with the temperature.
- Heat lost to the environment from the shell is negligible.
- Flow through the pipe is steady.

The slope of the line in the Figure 11.10 is given by

$$\frac{d(\Delta T)}{dq} = \frac{\Delta T_2 - \Delta T_1}{q_T} \tag{11.35}$$

where q_T is heat transfer rate in the entire heat exchanger.

We replace dq in Eq. (11.35) from Eq. (11.34)

$$\frac{d(\Delta T)}{U dA\,(\Delta T)} = \frac{\Delta T_2 - \Delta T_1}{q_T} \tag{11.36}$$

$$\int_{\Delta T_1}^{\Delta T_2} \frac{d(\Delta T)}{\Delta T} = \frac{U(\Delta T_2 - \Delta T_1)}{q_T} \int_0^{A_T} dA$$

$$\ln\left(\frac{\Delta T_2}{\Delta T_1}\right) = \frac{U A_T\,(\Delta T_2 - \Delta T_1)}{q_T} \tag{11.37}$$

On rearrangement

$$q_T = U A_T \left(\frac{\Delta T_2 - \Delta T_1}{\ln(\Delta T_2/\Delta T_1)}\right) \tag{11.38}$$

We may also write above equation in the standard notation as follows:

$$q_T = U A_T\,(\Delta T_L) \tag{11.39}$$

where

$$\Delta T_L = \frac{\Delta T_2 - \Delta T_1}{\ln(\Delta T_2/\Delta T_1)} \tag{11.40}$$

ΔT_L is known as the log mean temperature difference, and is popularly known as **LMTD**.

Application of preceding equation has the following constraints:

(i) It may give erratic results if q is not changing linearly with ΔT as is drawn in Figure 11.10.

(ii) U should not vary drastically with ΔT.

PROBLEM 11.4 A dilute pineapple juice is heated in a double pipe heat exchanger from 28 °C to 75 °C by heat exchanging with hot water flowing in shell in counter-current direction and is cooled from 95 °C to 85 °C. Calculate the log mean temperature difference, and compare with average temperature difference.

Solution It is similar to the one shown in Figure 11.10.

$$T_{ca} = 28 \text{ °C}$$

$$T_{cb} = 75 \text{ °C}$$

$$T_{hb} = 95 \text{ °C}$$

$$T_{ha} = 85 \text{ °C}$$

Therefore,

$$\Delta T_1 = T_{ha} - T_{ca} = 85 - 28 = 57 \text{ °C}$$

$$\Delta T_2 = T_{hb} - T_{cb} = 95 - 75 = 20 \text{ °C}$$

Therfore,

$$\text{LMTD} = \frac{\Delta T_1 - \Delta T_2}{\ln(\Delta T_1/\Delta T_2)} = \frac{57 - 20}{\ln(57/20)} = \frac{37}{\ln(2.85)}$$

$$= 35.33 \text{ °C}.$$

$$\text{Average } \Delta T = \frac{\Delta T_1 + \Delta T_2}{2} = \frac{57 + 20}{2} = 38.5 \text{ °C}$$

11.8 CORRELATIONS FOR HEAT TRANSFER COEFFICIENT

Heat transfer coefficients are to be generally measured experimentally. However, some correlations are available which are all based on the physico-chemical and thermal properties of the fluid, and the flow conditions and system geometry. Most of these parameters are characterized by means of various dimensionless numbers. Use of dimensionless numbers for predictive purposes is a very common practice in engineering design, and hence, a good number of them is available in chemical engineering literature.

We have already come across three such numbers so far viz.,

$$\text{Reynolds number, } N_{Re} = \frac{\rho d u}{\mu} \tag{8.15}$$

$$\text{Nusselt number, } N_{Nu} = \frac{hd}{k} \tag{11.11}$$

$$\text{Prandtl number, } N_{Pr} = \frac{C_p \mu}{k} \tag{11.1}$$

Few more numbers are defined here:

Grashof number (N_{Gr}) is a dimensionless number which relates a large number of parameters viz., a characteristic dimension of the system like diameter of the pipe (d), the acceleration due to gravity (g), the coefficient of thermal expansion (β), viscosity and density of the fluid, and the temperature difference (ΔT)

$$N_{Gr} = \frac{d^3 g\beta\rho^2(\Delta T)}{\mu^2} \tag{11.41}$$

The number is a ratio of the force of gravity to the buoyant forces resulting due to ΔT of the fluid, and is associated with free convection when the fluid movement is mainly due to gravitational pull and buoyancy forces, resulting due to temperature difference. *L*/*d* ratio also plays a significant role if the flow is in laminar conditions.

Pecklet number (N_{Pe}) is a product of the Reynolds number and Prandtl number.

$$N_{Pe} = (N_{Re})(N_{Pr}) = \frac{\rho du}{\mu}\frac{C_p\mu}{k}$$

$$= \frac{\rho duC_p}{k} \tag{11.42}$$

Stanton number (N_{St}) is associated with N_{Nu}, N_{Pr} and N_{Re} as follows:

$$N_{St} = \frac{N_{Nu}}{N_{Pr} \times N_{Re}} = \frac{N_{Nu}}{N_{Pe}} = \frac{h}{C_pG} \tag{11.43}$$

where G is known as **mass flux** and is equal to ρu.

Rayleigh number (N_{Ra}) is a product of Grashof number and Prandtl number

$$N_{Ra} = N_{Gr} \times N_{Pr} \tag{11.44}$$

Graetz number (N_{Gz}) is applicable for heat flow to fluids from a hot surface under laminar conditions and is defined as:

$$N_{Gz} = \frac{\dot{m}C_p}{kL} = \frac{\pi}{4}[N_{Re} \times N_{Pr}(d/L)] \tag{11.45}$$

where is mass flow rate in kg/s and L is the length of the pipe or plate. This is also similar to N_{Pe}. We may also rite

$$N_{Gz} = \frac{\pi}{4}\frac{d}{L}(N_{Pe}) \tag{11.46}$$

The Graetz number is also related to Fourier number [defined by Eq. (11.37)] as follows:

$$N_{Gz} = \frac{\pi}{N_{Fo}} \qquad (11.47$$

where $$N_{Fo} = \frac{\alpha t_T}{r_m^2} \tag{11.37}$$

and $$t_T = \frac{L}{u} \tag{11.48}$$

After having defined a number of dimensionless numbers, we shall see how these are useful for calculating the heat transfer coefficients. By and large most of these correlations are written in the form of expressions relating Nusselt number to Reynolds number and Prandtl number. If the viscosity of the fluid changes significantly with temperature as compared to the bulk temperature of the fluid, we also add an additional term, called (μ/μ_w) where μ_w is the viscosity of the fluid at the wall temperature. In the light of the above discussion, we may wite,

$$N_{Nu} = f\left[N_{Re}, N_{Pr}, N_{Gr}\left(\frac{L}{D}\right), \left(\frac{\mu}{\mu_w}\right)\right] \tag{11.49}$$

In case of free convection, N_{Gr} is important and L/D is also associated with the correlations particularly in case of laminar flow. Generally we may write,

$$N_{Nu} = \psi\,(N_{Re})^a\,(N_{Pr})^b\left(\frac{d}{L}\right)^c\left(\frac{\mu}{\mu_w}\right)^{0.14} \tag{11.50}$$

where

ψ is a constant and is of the order of 0.023 in case of forced convection.

a is a coefficient and varies between 0.6 to 0.8 and is generally of the order of 2/3.

b is a coefficient which is generally 1/3.

$(\mu/\mu_w)^{0.14}$ is represented by ϕ_v.

$$\phi_v = \left(\frac{\mu}{\mu_w}\right)^{0.14} \tag{11.51}$$

The last term in RHS, $(\mu/\mu_w)^{0.14}$ is generally taken as 1.0, since in food processing we generally do not tend to use high wall temperatures, and hence, variation of μ_w as compared to μ is not very significant.

11.8.1 Equations for Free (Natural) Convection

There are some correlations available for calculating heat transfer coefficients for water and air in free convection mode (McAdams 1954).

For horizontal cylinder,

$$h = 1.3196\times\left(\frac{\Delta T}{d}\right)^{0.25} \quad \text{for air} \tag{11.52}$$

$$h = 291.1\times\left(\frac{\Delta T}{d}\right)^{0.25} \quad \text{for water} \tag{11.53}$$

The above equations are applicable both for heating and cooling. Similarly for a vertical cylinder or vertical plate (diameter is used for cylinders, L is used for plates)

$$h = 1.3683\times\left(\frac{\Delta T}{d}\right)^{0.25} \quad \text{for air} \tag{11.54}$$

$$= 127.1\times\left(\frac{\Delta T}{d}\right)^{0.25} \quad \text{for water} \tag{11.55}$$

11.8.2 Equations for Forced Convection

The forced convection process may be looked upon into two regimes; viz., heat transfer in laminar flow conditions, heat transfer in turbulent conditions. Most of the reports were made on turbulent conditions.

Laminar flow conditions: At low *Graetz numbers*, the Nusselt number approaches a limiting value of 3.66, and it is very difficult to calculate accurate values for heat transfer coefficients [McCabe, et al., 1993 (b)]. If the Graetz numbers are more than 20, we can write with a reasonable amount of accuracy

$$N_{Nu} \cong 2.0\ (N_{Gz})^{1/3} \tag{11.56}$$

The above equation is applicable for both heating and cooling, and is sometimes associated with viscosity change term as follows:

$$N_{Nu} = 2.0\ (N_{Gz})^{1/3}\ (\phi_v) \tag{11.57}$$

In some textbooks, the coefficient 2.0 is replaced by more accurate 1.75 and the equation is written as (Toledo 1997).

$$N_{Nu} = 1.75\ (N_{Gz})^{1/3}\ (\phi_v) \tag{11.58}$$

The above equation is modified for **non-Newtonian fluids** as follows by replacing the viscosities with apparent viscosities:

$$N_{Nu} = 2.0\left(\frac{3n+1}{4n}\right)^{1/3} (N_{Gz})^{1/3} \left(\frac{\mu_a}{\mu_{a,w}}\right)^{0.14} \tag{11.59}$$

where n is flow consistency index of power – law fluid model in Eq. (9.8), and

$$\mu_a = K\ \gamma^{n-1} \tag{9.9}$$

in which K is the flow behaviour index of the power law fluid model [Eq. (9.8)] for non-Newtonian fluids.

Turbulent flow conditions: *Dittus–Boelter equation* adequately represents the turbulent heat transfer during flow through pipes,

$$N_{Nu} = 0.023\ (N_{Re})^{0.8}\ (N_{Pr})^{1/3} \tag{11.60}$$

Equation (11.60) is modified by adding the terms for viscosity correction, and is known as *Sieder–Tate* equation.

$$N_{Nu} = 0.023\ (N_{Re})^{0.8}\ (N_{Pr})^{1/3}\ \phi_v \tag{11.61}$$

The above Dittus–Boelter equation is also represented as follows:

$$N_{Nu} = 0.023\ (N_{Re})^{0.8}\ (N_{Pr})^{0.4} \tag{11.62}$$

However, Eq. (11.62) is found to be applicable for the following conditions both for heating and cooling, for flow of fluids in tubes without phase change,

$$\left.\begin{array}{l} 10^4 \le N_{Re} \le 1.2\times10^5 \\ 0.7 \le N_{Pr} \le 120 \quad \text{and} \\ L/D \ge 60 \end{array}\right\} \tag{11.63}$$

In **stirred vessels** agitated by an impeller, the heat transfer coefficient is very much influenced by the degree of agitation, the effect of which is represented by $N_{Rei.}$ where N_{Rei} is the impeller Reynolds number and is given by

$$N_{Rei} = \frac{N_i d_i^2 \rho}{\mu} \tag{11.64}$$

$$N_{Nu} = 0.87\,(N_{Rei})^{0.62}\,(N_{Pr})^{1/3}\,\phi_v \tag{11.65}$$

Equation (11.65) is applicable for agitated vessels for heat transfer by a *helical coil* in the vessel.

When the heat transfer takes place in an **agitated jacketed vessel**, the above equation is modified (Chilton, et al., 1944) as follows:

$$N_{Nu} = 0.36\,(N_{Rei})^{0.67}\,(N_{Pr})^{0.33}\,\phi_v \tag{11.66}$$

where N_{Nu} is based on the stirred vessel diameter (d_t) and is defined as:

$$N_{Nu} = \frac{h_j d_t}{k} \tag{11.67}$$

in which h_j is the jacketed vessel heat transfer coefficient.

11.8.3 Applications of the Design Equations

The preceding equations described so far in Section 11.8.1 and Section 11.8.2 are useful for calculating the heat transfer coefficients based on the system parameters, fluid properties and flow conditions. But, how do they help us?

The whole design of heat exchanges revolves around finding out the heat transfer area to meet a desired requirement of heat load. The heat load is calculated based on the fluid properties (viz., ρ, C_p, k and μ) and process requirements (viz., $\dot{m}$ and ΔT) using the energy balance equation (Section 3.4). This gives an idea of how much heat is required in how much time, and hence, what should be the heat transfer rate, q. Once we know q and h (by Sections 11.8.1 or 11.8.2), we can calculate the heat transfer area A by using the equation

$$q = hA(\Delta T) \tag{11.3}$$

The procedure for evaluation of h is described in the following numerical problems:

PROBLEM 11.5 In a fruit juice making unit, sugar syrup is heated in a stirred vessel of diameter 2 m with a helical coil inside the vessel. The diameter of the impeller is 0.67 m and rotates at 60 rpm. The viscosity of the syrup is 3×10^{-3} Pa s; the other properties are similar to those of water. Neglecting the effect of temperature on the viscosity of the syrup, calculate the heat transfer coefficient.

Solution The following are the given data:

$$d_t = 2 \text{ m}$$

$d_i = 0.67$ m
$N_i = 60$ rpm $= 1$ rps $= 1\ s^{-1}$
$\mu = 3 \times 10^{-3}$ Pa s $= 3\times10^{-3}$ kg/ms
$k = 0.628$ W/m°C $= 0.628 \times 10^{-3}$ kJ/ms°C
$\rho = 1000$ kg/m^3
$C_p = 4.2$ kJ/kg°C

Approach: We calculate N_{Rei} and N_{Pr}. We shall use Eq. (11.65) to calculate N_{Nu}. Then we calculate h by Eq. (11.8).

By Eq. (11.64),we have

$$N_{Rei} = \frac{N_i d_i^2 \rho}{\mu} = \frac{1\times(0.67)^2\times1000}{3\times10^{-3}} = 1.5 \times 10^5$$

$$N_{Pr} = \frac{C_p\mu}{k} = \frac{4.2\times3\times10^{-3}}{0.628\times10^{-3}} = 20$$

$$N_{Nu} = 0.87\ (N_{Re})^{0.62}\ (N_{Pr})^{1/3} = 0.87\ (1.5 \times 10^5)^{0.62}\ (20)^{1/3}$$

$$= 0.87 \times 1618.7 \times 2.717 = 3826.4$$

$$N_{Nu} = \frac{hd_t}{k} = 3826.4$$

Therefore,
$$h = \frac{3826.4\times0.628\times10^{-3}}{2} = 1.2\ \text{kW/m}^2{}^\circ\text{C}$$

PROBLEM 11.6 In Problem 11.5, if the vessel is jacketed and the heating takes place by steam in the jacket, find the heat transfer coefficient.

Solution

Approach: It is a straight case of application of Eq. (11.66) which is applicable for stirred jacketed vessel.

$$N_{Rei} = 1.5 \times 10^5$$

$$(N_{Pr})^{1/3} = (20)^{1/3} = 2.717$$

$$N_{Nu} = 0.36\ (N_{Rei})^{0.67}\ (N_{Pr})^{1/3}$$

$$= 0.36 \times (1.5 \times 10^5)^{0.67} \times 2.717 = 2873$$

$$= \frac{hd_t}{k}$$

Therefore,
$$h = \frac{N_{Nu}\times k}{d_t} = \frac{2873\times0.628\times10^{-3}}{2} = 0.9\ \text{kW/m}^2{}^\circ\text{C}$$

From the above two examples, it is clear that the heat transfer rates are higher in case of helical coils submerged directly in the fluid of the vessel as compared to jacketed vessels where heating takes place only all along the wall from outside.

PROBLEM 11.7 The helical coil heated stirring tank of Problem 11.5 is used to maintain the sugar syrup solution at 35 °C by passing hot water through the helical coil. The inlet and outlet temperatures of the heating water is 50 and 38 °C. The heat load required for the tank is 500 kW. The heat transfer coefficient for the heating water is 10 kW/m^2°C. The helical coil is made up of *SS* which has thermal conductivity of 60 W/m°C. What length of the cooling coil is required if the outside diameter of the *SS* pipe is 5 cm and the pipe thickness is 3 mm.

Solution

$$h_i = 1.2 \text{ kW/m}^2\text{°C}$$
$$k_m = 60 \times 10^{-3} \text{ kW/m°C}$$
$$B = \text{pipe thickness} = 3 \times 10^{-3} \text{ m}$$
$$q = 500 \text{ kW}$$

Approach:

(i) Initially we calculate overall heat transfer coefficient (U) by applying Eq. (11.26).
(ii) We apply Eq. (11.18) to calculate A by using average temperature difference.
(iii) Next we use A to calculate the length of the pipe by noting $A = 2\pi rL$.

By Eq. (11.26), we have

$$\frac{1}{U} = \frac{1}{1.2} + \frac{3 \times 10^{-3}}{60 \times 10^{-3}} + \frac{1}{10}$$

$$= 0.983 \frac{\text{m}^2\text{°C}}{\text{kW}}$$

Therefore, $$U = 1.017 \text{ kW/m}^2\text{°C}$$

By Eq. (11.3), we have

$$500 = 1.017 \times A \times (\Delta T)$$

$$\Delta T = \frac{(50 - 35) + (38 - 35)}{2} = 9 \text{ °C}$$

Therefore, $$A = \frac{500}{1.017 \times 9} = 54.63 \text{ m}^2$$

$$A = \pi dL = \pi \times 5 \times 10^{-2} \times L$$

∴ $$L = \frac{54.63}{\pi \times 5 \times 10^{-2}} = 347.76 \text{ m}$$

The length of the coil is too long to be able to manage with a reasonable extent of convenience, and the cost of the tube will be prohibitively high.

11.9 HEAT TRANSFER BY CONDENSING VAPOURS

Most of the thermal processing (transfer of heat) in food industries is done by condensation of steam inside the jacket of the steam heated jacketed kettles. The main advantage of this is that a constant temperature is maintained throughout the surface of the vessel, and the steam temperature can be varied by varying the pressure of the steam coming from the boiler. Another very common example in food processing is the use of condensers for condensing the vapours in the steam by recirculating cold water outside (shell) of the condenser. The most common example is condensation of water vapours from the steam in preparation of distilled water. Condensation of alcohol and/rectified spirit is a very common process operation. In all distillation processes†, the distillate is condensed to separate the low boiling liquid from the high boiling liquid mixture. In the desolventization of the extract in the extraction processes (will be studied in detail in Chapter 19), the solvent vapours are condensed to recover and reuse the solvent (we come across this situation more in the solvent extraction of spices for the preparation of spice oleoresins).

In this way, we come across condensation of vapours in the condensers for recovery of volatiles, or condensation of steam in the steam heated jacketed vessels for heating the contents. In either case, it is the transfer of heat by condensation. Upto now we have seen from process point of view. Condensation of vapours, per se, is very effective for heat transfer because during condensation, the vapours release the *latent heat of vaporization* which is usually high. Hence, better heat transfer can take place, or in other words, higher heat transfer coefficients can be realized.

When the vapours are condensing, two types of condensation is possible to occur, viz., dropwise condensation, and film-type condensation. The type of condensation depends upon the nature and cleanliness of the surface.

11.9.1 Dropwise Condensation

When a vapour condenses in the form of fine drops on the clean surface of a heat exchanger, the drops coalesce and fall in the form of rivulets. It is a common experience to see drops forming in the form of mist on pitcher containing cold water or cold fluid because of the condensation of moisture vapour present in the outside air.

In case of dropwise condensation, most of the heat transfer area is in direct contact with the hot vapours, and hence, the heat transfer coefficients are higher, which are of the order of 115 kW/m^2°C [McCabe, et al., 1993(c)]. The basic criteria for the dropwise condensation is the cleanliness of the surface and cleanliness of the vapours. In case of mixed vapours, more so with non-condensable vapours, dropwise condensation is hardly possible. If the cooling surface is not wetted by the liquid, dropwise condensation is impossible. If the surface is smooth, the drops do not stick, and hence, they fall in the form of rivulets creating facility for more drops to form, which in turn helps dropwise condensation. To induce and sustain dropwise condensation, sometimes some drop promoters are used. Oleic acid is generally effective on most of the surfaces. Mercaptans are used on copper alloys.

In spite of the above advantage of dropwise condensation, it is generally not realizable in practice in view of the associated problems in maintaining and sustaining the drops. Another important factor in case of condensation of steam by dropwise condensation is that we do not realize much advantage of higher heat transfer coefficients because its contribution to overall heat transfer coefficient becomes less significant when compared to high heat transfer resistance in the inside of the tubes. Hence, in general, we realize mostly the film-type condensation only.

†Distillation will be studied in more detail as a unit operation in Chapter 18.

11.9.2 Film-type Condensation

When the drops condense on a surface, they coalesce to form into a thin film on the heat transfer surface; and the whole condensing liquid falls in the form of a film all along the heat transfer area. This film offers resistance for the heat transfer to take place. The basic studies on film-type condensation were reported by Nusselt [McCabe, et al., 1993 (c)]. Film-type condensation is based on the assumption that the condensing vapour and liquid are in thermodynamic equilibrium, and hence, the resistance to heat transfer wholly lies in the condensing liquid film, which will be falling under gravity from the heat transfer surface. Some assumptions in case of the film-type condensation are:

(i) The velocity of liquid layer at the surface of the wall is zero as the wall is stagnant.
(ii) The velocity of the falling film is not affected by the velocity of the vapours.
(iii) The wall temperature and the vapour temperatures are not constant.
(iv) The condensate temperature is same as the temperature of condensation.
(v) All the physico-chemical and thermal properties of the condensing liquid are taken at the mean film temperature.

The film thickness will be increasing as we move from top to bottom of the tube, whereas the heat transfer will be decreasing as shown in Figure 11.11. Most of the film forms on the upper part of the tube, and the increase in thickness as we descend is very less. Since the stagnant film forms as a barrier between the surface and the condensing vapours, the heat transfer is assumed to take place through the film solely by conduction according to the Nusselt theory; and hence we can write

$$h_x = \frac{k}{\delta} \tag{11.10}$$

where δ is the local film thickness. We notice from Eq. (11.10) that the local heat transfer coefficient is inversely proportional to the film thickness. The same is depicted in Figure 11.11. k is thermal conductivity of the liquid at the average temperature of the liquid generally referred to as T_f and is known as **reference temperature**. It is given by [McCabe, et al., 1993 (c)]

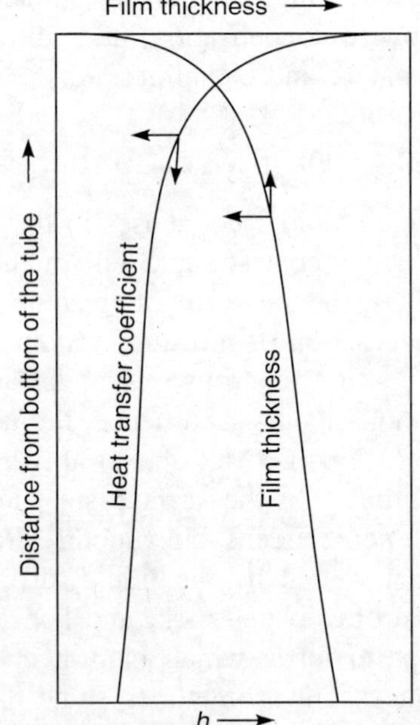

Figure 11.11 Variation of film thickness and h along the height of the tube in a falling film condenser.

$$T_f = T_v \frac{-3\Delta T_0}{4} \tag{11.68}$$

and
$$\Delta T_0 = (T_v - T_s) \tag{11.69}$$

where,

T_f is reference temperature,
ΔT_0 is average temperature difference,

T_v is temperature of condensing vapours, and
T_s is wall temperature on the outside surface of the tube.

For a *vertical tube,* the average heat transfer coefficient is given by

$$h = 1.13\left(\frac{k^3\rho^2 g\lambda}{\Delta T_0 L\mu}\right)^{1/4} \tag{11.70}^{\dagger}$$

The coefficient 1.13 is applicable in FPS system as well as in SI units also.

A similar type of equation for *horizontal tube* is:

$$h = 0.725\left(\frac{k^3\rho^2 g\lambda}{\Delta T_0 d_o\mu}\right)^{1/4} \tag{11.71}$$

in which d_o is the outside diameter of the tube. Generally vapours are condensed outside the tubes[‡], and hence, d_o is used in place of L in Eq. (11.70).

11.9.3 Design of a Condenser

In most of the desolventization processes for the extraction of oleoresins, the solvent is removed from the active principle by distillation. The distillate contains the vapours which are condensed in a condenser to recover the solvent which is reused for extraction process. The following numerical example describes the design features of a condenser:

PROBLEM 11.8 A shell and a vertical tube condenser with a bundle of 10 tubes each of 1/2″ schedule 40 *SS* tubes of 5′ length is used for condensing acetone vapours from a desolventization unit. Cooling water is flowing inside the tubes with an average temperature of 40 °C. The water side heat transfer coefficient is 1.7 kW/m^2°C.

(a) Find the heat transfer coefficient of the condensing acetone.

(b) What would be the h for a horizontal condenser of similar configuration? There is no fouling factor and the tube-wall heat transfer resistance is negligible.

Solution

Approach:

(i) The physico-chemical and thermal properties of the vapour are noted from any standard reference handbook like *Perry's Chemical Engineers' Handbook* (Perry and Green, 1984).

(ii) These properties are to be evaluated at the reference temperature, T_f to be calculated by Eq. (11.68). For applying Eq. (11.68) we require the wall temperature T_s which must be evaluated from h. Hence, initially we assume T_s and follow trial and error method.

The data for acetone as available from *Perry's Chemical Engineers' Handbook* are taken here.

Boiling temperature = 56.5 °C

[†]The original coefficient of 0.943 is increased by 20 per cent and is made 1.13 to account for any variations in flow conditions from laminar to turbulent.

[‡]If the vapours contain uncondensable gases also (like air), then the mixture of vapours and condensable gases are condensed inside the tubes so that the uncondensable gases are continuously swept away from the heat transfer area by the incoming steam (McCabe, et al., 1993).

$$\text{Density } (\rho) = 792 \text{ kg/m}^3$$

$$k \text{ at } 30\text{ °C} = 0.176 \text{ W/m°C}$$

$$k \text{ at } 75\text{ °C} = 0.164 \text{ W/m°C}$$

Hence, k at average temperature (52.5 °C) = 0.17 W/m°C

$$C_p = 2.25 \text{ kJ/kg°C}$$

$$\lambda = 516.2 \text{ kJ/kg}$$

$$\mu = 0.256 \times 10^{-3} \text{ Ns/m}^2 = 0.256 \times 10^{-3} \text{ kg/ms}$$

(a) **Vertical tubes:** The condensing temperature is 56.5 °C (T_v). The wall temperature could be somewhat less than this, but it should be more than the cooling water temperature, 40 °C. Hence, we may choose as first guess T_s to be 48 °C.

Therefore,

$$T_f = T_v - \frac{3}{4}(T_v - T_s)$$

$$= 56.5 - 0.75\ (56.5 - 48)$$

$$= 50.125 \text{ °C}$$

$$\Delta T_0 = 56.5 - 50.125 = 6.375$$

Since we do not have any source to collect the physico-chemical and thermal properties of acetone at different temperatures, we presume that they do not change considerably with temperature, and hence, the data are valid at 50 °C also.

$$\text{Tube length} = 5' = 5 \times 0.3 = 1.5 \text{ m}$$

By Eq (11.70),

$$h = 1.13\left(\frac{k^3\rho^2 g\lambda}{\Delta T_0 L\mu}\right)^{1/4}$$

$$= 1.13\left(\frac{(0.17\times10^{-3})^3\times(792)^2\times9.81\times516.2}{(6.375)\times(1.5)\times(0.256\times10^{-3})}\right)^{1/4}$$

$$= 1.13\ (4.78)^{1/4} = 1.48 \times 1.13 = 1.67 \text{ kW/m}^2\text{°C}$$

This is outside heat transfer coefficient (h_o).

Hence,

$$h_o = 1.67 \text{ kW/m}^2\text{°C}$$

and

$$h_i = 1.7 \text{ kW/m}^2\text{°C (given)}$$

Now, we calculate ΔT_i by

$$\Delta T_i = \frac{\dfrac{1}{h_i}}{\left(\dfrac{1}{h_i}\right)+\left(\dfrac{d_i}{d_o}\dfrac{1}{h_o}\right)}\Delta T$$

d_i and d_o for scheduled 40–0.5″ *SS* pipe are:

$$d_i = 0.622'' = 0.622 \times 2.54 \times 10^{-2}$$

$$= 0.0158 \text{ m}$$

$$d_o = [0.622 + (2 \times 0.109)] \times 2.54 \times 10^{-2} = 0.021 \text{ m}$$

Therefore,
$$\Delta T_i = \frac{\dfrac{1}{1.7}}{\dfrac{1}{1.7} + \left(\dfrac{0.0158}{0.0213} \times \dfrac{1}{1.67}\right)}(56.5 - 40)$$

$$= 7.32 \text{ °C}$$

Therefore, the wall temperature, $T_s = 40 + 7.326 = 47.3$ °C. This is sufficiently close to 48 °C which we have assumed. Hence, further trials are not necessary. We shall try to see whether the flow is laminar. The perimeter of the 1/2″ schedule 40-*SS* pipe is 0.220′.

Hence,
$$\text{Area } (A) = \text{Length} \times \text{Perimeter}$$

$$= (5 \times 0.220) = 1.1 \text{ ft}^2 = 1.1 \times (0.3)^2$$

$$= 0.1 \text{ m}^2$$

$$q = h_o A\,(T_v - T_s) = 1.8 \times 0.1 \times (56.5 - 47.3)$$

$$= 1.656 \text{ kW}$$

Mass flow rate of vapours,
$$\dot{m} = q/\lambda = \frac{1.656}{516.2}$$

$$= 3.2 \times 10^{-3} \text{ kg/s}$$

The mass flow rate per unit length of the periphery of the tube = m_v

$$\frac{\dot{m}}{\pi d_o} = \frac{3.2 \times 10^{-3}}{\pi \times 0.0213} = 0.0479 \text{ kg/s m}$$

The value of ($4\,m_v/\mu$) is a measure of Reynolds number to find whether the flow is laminar or turbulent [McCabe, et al., 1993(c)].

Therefore,
$$\frac{4m_v}{\mu} = \frac{4 \times 0.0479}{0.256 \times 10^{-3}} = 748$$

Since it is much less than 2100, laminar flow conditions prevail.

(b) **For horizontal tubes:** Now, we assume a slightly higher wall temperature as the heat transfer is better in case of horizontal tubes. We assume T_s to be 52 °C. Here the number of tubes also matter.

$$N_t = \text{Number of tubes} = 10$$

$$\Delta T_0 = 56.5 - 52 = 4.5 \text{ °C}$$

We do not resort to calculation of T_f as anyway we do not have data for finding the physico-chemical and thermal parameters at different temperatures. Otherwise systematically speaking, we should calculate T_f and then note the value of parameters of the fluid at T_f.

$$T_f = 56.5 - 0.75\,(56.5 - 52) = 53.125\ ^\circ\text{C}.$$

We calculate h_o by using Eq. (11.71)

$$h_o = 0.725\left(\frac{k^3\rho^2 g\lambda}{N_t\,(\Delta T_0)\,(d_o)\,(\mu)}\right)^{1/4}$$

$$= 0.725\left(\frac{(0.17\times10^{-3})^3\,(792)^2\times9.81\times516.2}{10\times(4.5)\times(0.0213)\,(0.256\times10^{-3})}\right)^{1/4}$$

$$= 0.725\,(63.6)^{1/4} = 2.047\ \text{kW/m}^2\,^\circ\text{C}.$$

Now, we calculate ΔT_i as we did earlier.

$$\Delta T_i = \frac{\dfrac{1}{1.7}}{\dfrac{1}{1.7}+\dfrac{0.0158}{0.0213}\times\dfrac{1}{2.047}}(56.5-40)$$

$$= 10.2\ ^\circ\text{C}$$

Therefore, the wall temperature (T_s) = 40 + 10.2 = 50.2 °C. It is close to our assumed value of 52 °C.

We note that the heat transfer coefficient in case of horizontal tubes (2.047 kW/m^2°C) is greater than that (1.67 kW/m^2°C) for vertical tubes.

Symbols

A: heat transfer area (m^2)
A_L: log mean area
A_T: total heat transfer area of the heat exchanger
a: coefficient in Eq. (11.50)
b: coefficient in Eq. (11.50)
C_p: specific heat (kJ/kg°C)
c: coefficient in Eq. (11.50)
d: diameter
d_i: impeller diameter (m)
d_t: tank diameter (m)
d_L: log mean diameter
G: max flux (kg/m^{2}s)
g: acceleration due to gravity (m/s^2)
h: heat transfer coefficient [kJ/(m^{2}s°C)]
h_j: jacketed vessel heat transfer coefficient [kJ/(m^{2}s°C)]
K: low behaviour index in power law fluid model [Eq. (9.8)]
k: thermal conductivity [W/mK or kJ/ms°C]
L: length (m)
$\dot{m}$: mass flow rate (kg/s)
m_v: mass flow rate per unit length of periphery (kg/sm)
N_{Gr}: Grashof number
N_{GZ}: Graetz number defined by Eq. (11.46)
N_i: impeller speed (rps)
N_{Nu}: Nusselt number
N_{Pe}: Pecklet number defined by Eq. (11.42)
N_{Pr}: Prandtl number
N_{Ra}: Rayleigh number defined by Eq. (11.44)
N_{Re}: Reynolds number
N_{Rei}: impeller Reynolds number defined by Eq. (11.64)

N_{St}: Stanton number defined by Eq. (11.43)
n: flow constituency index of non-Newtonian fluid in power law fluid model [Eq. (19.8)]
Q: quantity of heat (kJ)
q: heat transfer rate (kJ/s or kW)
R: resistance for heat transfer (m^2°C/kW)
r_m: radius of the metallic vessel (m)
T: temperature (°C or K)
T_c: average temperature of the cold fluid
T_f: reference temperature (°C)
T_h: average temperature of the hot fluid
T_v: Temperature of condensing vapours
t: time (s or h)
t_T: total time taken as defined by Eq. (11.48)
ΔT: temperature difference
U: overall heat transfer coefficient (kW/m^2°C)
u: velocity (m/s)
V: volume (m^3)
x: thickness or distance measured in x-direction (m)
x_w: wall thickness (m)
y: distance measured along the heat transfer direction

Subscripts

a: inlet position
b: outlet position
c: cold fluid
f: fluid
h: hot fluid
i: inside
L: log mean
m: metal surface
o: outside
s: surface
w: wall
x: x-direction
0: initial

Greek Letters

γ: shear strain (s^{-1})
δ: thickness of the film (m)
μ: viscosity of the fluid (kg/ms)
μ_a: apparent viscosity of non-Newtonian fluid given Eq. (9.9)
λ_s: latent heat of vaporization of steam (kJ/kg)
β: coefficient of thermal expansion
ρ: density of the fluid (kg/m^3)
α: thermal diffusivity (m^2/s)
ψ: coefficient in Eq. (11.50)
φ_v: a parameter represented by Eq. (11.51)
λ: latent heat of vaporization (kJ/kg)

REVIEW QUESTIONS

11.1 Write a note on convective heat transfer.

11.2 What are meant by natural convection and forced convection?

11.3 What is meant by thermal boundary layer? Compare it with the hydrodynamic boundary layer.

11.4 What is meant by heat transfer coefficient?

11.5 What is meant by overall heat transfer coefficient? Write an expression for the same incorporating individual heat transfer coefficients.

11.6 Write a note on the following:

(i) Nusselt number,
(ii) Prandtl number,
(iii) Pecklet number,
(iv) Stanton number,
(v) Graetz number,
(vi) Grashof number, and
(vii) Rayleigh number.

11.7 Write a note on the unsteady state convective heat transfer process.

11.8 Derive an expression for the overall heat transfer coefficient during unsteady state convective heat transfer.

11.9 Explain the cocurrent and counter-current flow patterns, and draw the temperature profiles.

11.10 Describe the concept of LMTD.

11.11 What are various equations for calculating heat transfer coefficients?

11.12 How do you go about for design of an equipment for convective heat transfer process?

11.13 Describe the process of heat transfer by condensing vapours.

11.14 Write a note on dropwise condensation.

11.15 Write a note on film-type condensation.

11.16 Describe systematically how do you design a condenser.

NUMERICAL PROBLEMS

11.1 Milk is heated in a plate heat exchanger of 0.5 m^2 plate area, and is maintained at 110 °C using steam at a flow rate of 0.3 kg/s. Milk at a mass flow rate of 2 kg/s is entering at 5 °C. Find at what temperature the milk would leave the heat exchanger and calculate the heat transfer coefficient.
(**Hint:** Note the C_p of milk is taken as that of water and latent heat of steam from Appendix 1B) (**Ans:** T_2 = 84.5 °C, h = 19 kW/m^2°C)

11.2 A sugar solution is heated in a double pipe heat exchanger from initial temperature of 32 °C to 82 °C. The steam is passed in the outer shell and is condensing at 100 °C. Find the LMTD in cocurrent and counter-current flow conditions and compare with the average temperature differences. What are your observations with reference to the flow conditions.
(**Ans:** LMTD is same in either flow, LMTD = 37.62 °C ΔT_{Av} = 43 °C)

11.3 A mechanically stirred jacketed vessel is used to warm water by utilizing the steam condensate coming at 100 °C from another processing unit. The tank diameter is 5 m and the impeller diameter is 1.7 m. The impeller rotates at a speed of 75 rpm. Find the heat transfer coefficient. (**Ans:** h = 2.26 kW/m^2°C)

REFERENCES

Al-Rawahi, M.S.S. and Rao, D.G. (2016), Studies on unsteady state heat transfer to petroleum crude oil, *Chemical Technology*, 11(3), pp. 114–117.

Chilton, T.H., Drew, T.B. and Jabans, R.H. (1944), Heat transfer coefficient in agitated vessels, *Ind. Engg. Chem.*, **36**, pp. 510–516.

Heldman, D.R. and Lund, D.B. (1992), *Handbook of Food Engineering*, Marcel–Dekker, Inc., New York, p. 257.

McAdams, W.H. (1954), *Heat Transmission*, 3rd ed., McGraw-Hill, New York.

McCabe, W.L., Smith, J.C. and Harriott, P. (1993), *Unit Operation of Chemical Engineering*, 5th ed., McGraw-Hill, New York (a) p. 327, (b) pp. 338–339, (c) pp. 374–381.

Perry, R.H. and Green, D.W. (1994), *Perry's Chemical Engineer's Handbook*, 6th ed., McGraw-Hill, New York.

Singh, R.P. and Heldman, D.R. (1993), *Introduction to Food Engineering*, 2nd ed., Academic Press, Inc., San Diego, p. 191.

Toledo, R.T. (1997), *Fundamentals of Food Process Engineering*, 2nd ed., CBS Publishers and Distributors, New Delhi, pp. 268–273.

CHAPTER

12

Heat Transfer Equipment

As has already been mentioned, thermal processing plays a prominent role in food industries for processing and preservation of food materials. The thermal processing could be either by way of heating or cooling; in either case it is transfer of heat, the only difference is in the direction of heat transfer. In food processing industries, heat energy is transferred by different methods and modes. Various methods of heating sources are:

- Electric resistance heating
- Passing hot water or thermic fluids
- Condensation of steam generated in boilers, by various methods
- Passing hot exhaust gases

Various methods of heat transfer are:

- Conduction from hot or cold metallic plates
- Convective heat transfer
- Conduction-convection heat transfer
- Radiative heat transfer as in ovens and furnaces
- Infrared heating
- Microwave or dielectric heating

Various types of heat transfer equipment used in any processing industry including food processing industries at pilot-plant scale and in small scale units operating in batch mode (generally) are shown in Figure 12.1 (Doran 1995). The most commonly used set-up is a jacketed vessel [Figures 12.1(a) and (b)] which consists of a vessel with a jacket outside through which steam or hot fluid is passed. Heat transfer takes place through the wall by conduction, and to the liquid by convection. It is versatile and simple to use except that it has no flexibility to increase the heat transfer area. The only way to increase the heat transfer rates is by increasing ΔT. Figure 12.1(c) shows vessel wound outside by a heating coil through which the hot fluid flows

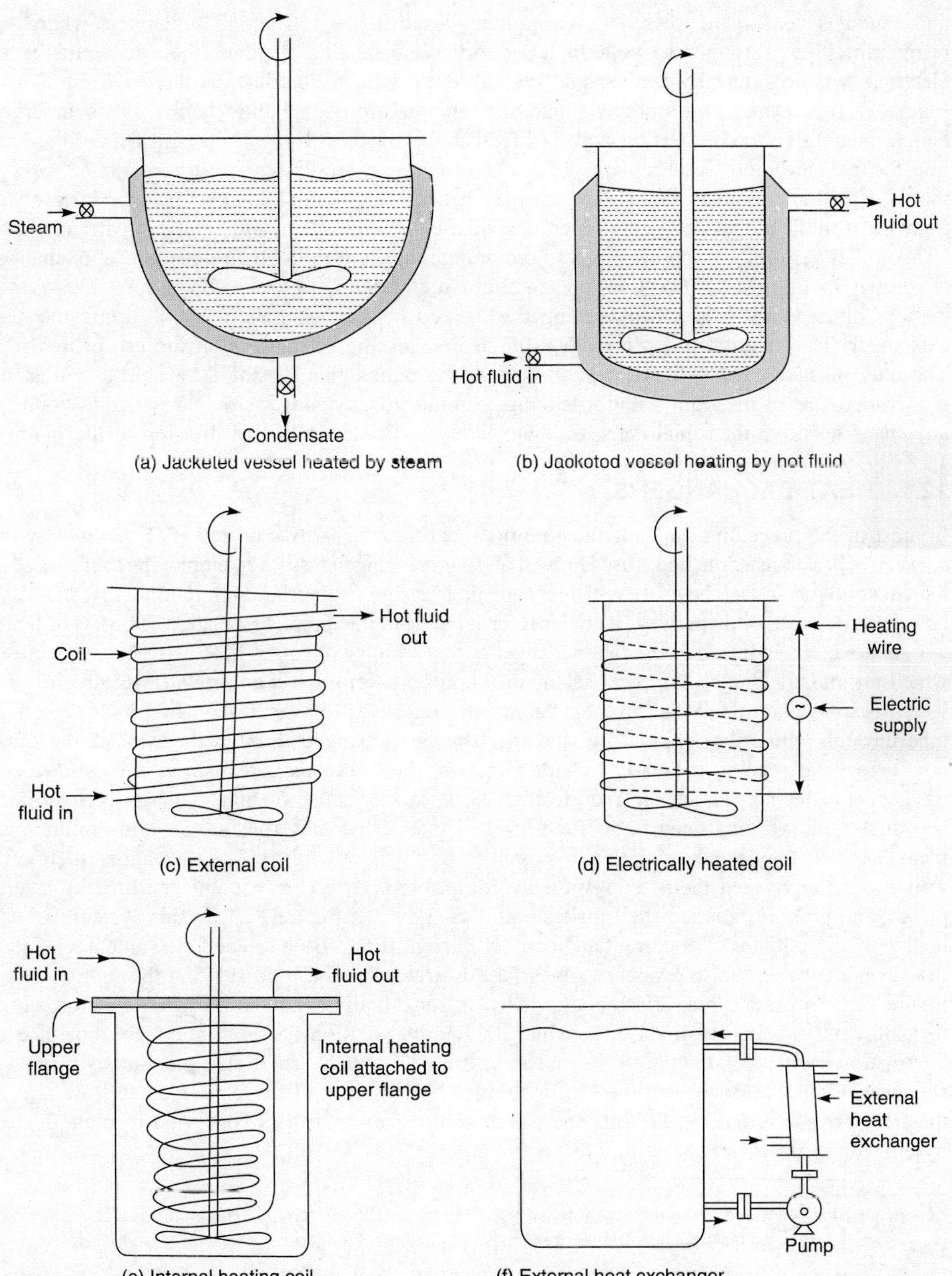

Figure 12.1 Heat transfer systems for food processing.

and conducts heat to the vessel. It is frequently used in the refrigerated chambers where the refrigerant flows through the coil. In laboratory scale heating systems, the coil could be an electrical winding (usually a nichrome coil) through which electricity is passed as shown in Figure 12.1(d). Depending on the wattage, the temperature rises. Figure 12.1(e) is a submerged helical heating coil. The heating coil is directly in contact with the fluid, and hence, the heat transfer rates are high (see Problems 11.5 and 11.6). It is, no doubt, a best arrangement except that it is difficult to clean the coil which may become a potential source for bacterial growth. Sometimes, if the alignment is not proper, the coil may pose problem with the stirring arrangement.

Another very interesting system is to recirculate the fluid through an outside heat exchanger as shown in Figure 12.1(f). It is an excellent method in the sense that it provides wider degrees of freedom in terms of varying the desired heat transfer areas by changing the heat exchanger. The external pump provides the desired mixing effect without the use of a stirrer. The only disadvantage with the system is with the pump; that it should be able to withstand the temperature of the fluid without leaking. Another disadvantage is that if the fluid contains any shear-sensitive microbial cells, they are likely to be destroyed by abrasion in the pump.

12.1 HEAT EXCHANGERS

In most of the preceding systems, the common heat transfer is by convection. Thus, mostly the convective heat transfer method is used in most of the processing industries by employing a unit called as **heat exchanger**. In fact the heat exchangers and their design features are so important that standards have been devised for them by the Tubular Exchanger Manufacturers Association (TEMA) of USA.

As the name indicates, the heat exchanger is one which exchanges heat between two sources which are mostly fluids. A typical schematic representation of it was shown in Figure 12.7(a) which is a standard shell-and-tube exchanger. It consists of an outer shell in which there is a tube through which the fluids flow, and the heat is exchanged through the wall of the inner tube. We have seen in the earlier chapter that the heat transfer rate increases by increasing the heat transfer area. So if we require high heat transfer rates, or high quantity of heat is to be transferred, we may need to go for very long tubes that may run into several hundreds of metres, which may not be practically possible. Long tubes will tend to sag under their own weight. One envisaged methodology for avoiding this is to go for bending the tube at several places, and provide separate shell for each unit as shown in Figure 12.2. In this system, we can realize the advantages of cocurrent and counter-current flow systems. The long (inner) tube adds the advantage where long residence times[†] are desirable for the fluid flowing through the tube. Figure 12.3 shows a typical flash cooker used for enzymatic conversion of liquid starch slurry to maltodextrin using α-amylase, in which the starch solution passes through the hot pipe at 95 °C along with the enzyme. Steam passes through the outer shell also continuously (not as shown in Figure 12.3) to heat the starch solution. The length of the pipe and the flow rate of the fluid are such, that by the time the starch solution exits, it gets converted to maltodextrin by enzymatic hydrolysis.

[†]Residence time is defined as the time spent by any element of the fluid in the tube, and long residence times are particularly advantageous where heat transfer proceeds with chemical reaction.

$$\text{Residence time} = \frac{\text{Volume of vessel or tube}}{\text{Volumetric flow rate of the fluid}}$$

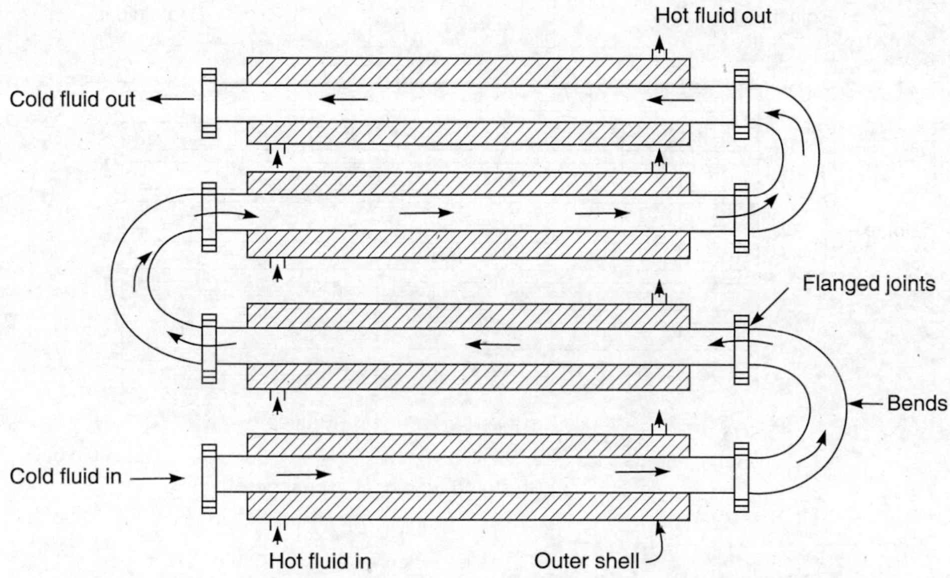

Figure 12.2 Long double pipe heat exchanger joined at the ends with bends.

Figure 12.3 Flash cooker for enzymatic conversion of liquid starch slurry to maltodextrin using amylase. (Reproduced with permission from M/s. Sukhjit Starch Industries, Nizamabad, Telangana).

However, since this type of arrangement as shown in Figure 12.2 is cumbersome and occupies more place, we look for an alternative way for arrangement of the pipes in the shell.

12.1.1 1-1 Pass Shell and Tube Heat Exchanger

A bundle of tubes is made and the tubes are attached to circular plates at either end. The bundle is enclosed in an outer shell as shown in Figure 12.4(a). The pipes are as per ASTM (American Society for Testing and Materials) standards. The pipes arrangement is also specified, viz., triangular pitch or square pitch. A triangular pitch arrangement is shown in Figure 12.4(b). The triangular pitch is good for heat transfer, but it has the disadvantage while cleaning as there will not be a through movement. The tubes are bundled and enclosed inside a shell which is made out of a rolled sheet. Generally the shell is made out of MS as the food material does not come in contact with the outer shell.

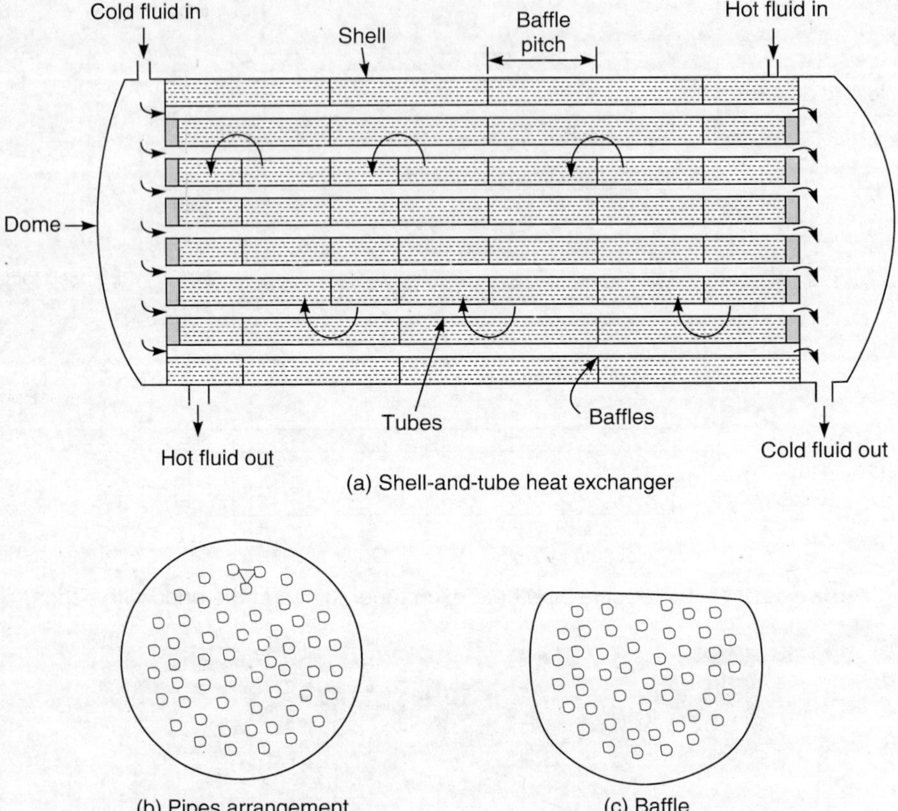

(a) Shell-and-tube heat exchanger

(b) Pipes arrangement

(c) Baffle

Figure 12.4 Single pass 1-1 counter-current heat exchanger.

Baffles (shown in Figure 12.4(c) are provided inside the shell. They serve two purposes, viz.,

(i) They deflect the shell side fluid so that channeling does not take place.

(ii) They provide support for the tubes so that they do not sag.

The distance between the baffles is known as **baffle pitch**, and it is in between the 1/5th diameter to one diameter of the shell.

In this heat exchanger, the flow of the shell-side fluid and that of the tube-side fluid are in counter-current direction. The tube-side fluid has only one pass, and the shell-side fluid has also one pass; i.e., they have a single path for flow. Hence, it is known as **Single pass 1-1 heat exchanger**.

We shall discuss the alternate arrangement in the following sections.

12.1.2 1-2 Heat Exchanger

The 1-2 heat exchanger has the advantage of both cocurrent flow and counter-current flow. Hence, it is known as **Parallel-counter flow heat exchanger**. It is called as 1-2 heat exchanger because the shell-side fluid has one way of direction of flow, whereas the tube-side flow has both cocurrent and counter-current (two way of direction) flow. A schematic representation is shown in Figure 12.5. Since the total fluid passes through half the number of tubes in any one direction, the heat exchanger will have high fluid flow rates through the tubes which will help turbulence, but increase pressure drop through the tubes. Except for higher pressure drops, it is preferred over the 1-1 heat exchanger.

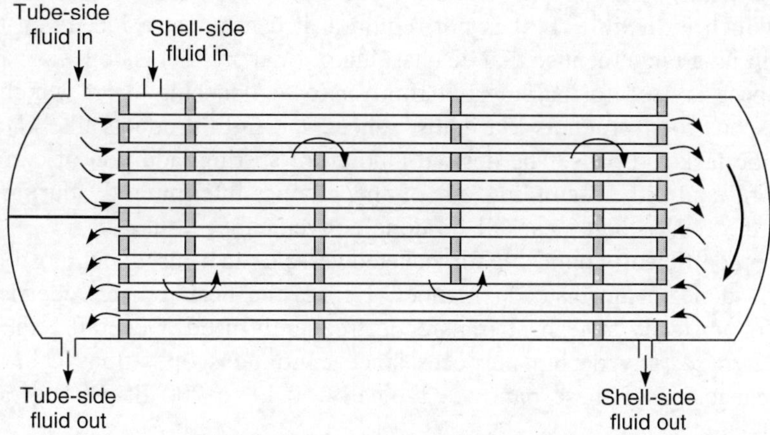

Figure 12.5 1-2 heat exchanger.

12.1.3 2-4 Heat Exchanger

In this heat exchanger we have two paths for the shell-side fluid and four passes for the tube-side flow (Figure 12.6) by virtue of which we realize both cocurrent and counter-current flow conditions. Since the tube-side fluid has four passes, the fluid will have velocity almost four times that in a similar single pass heat exchanger. The tube-side heat transfer coefficients are almost $4^{0.8} = 3.03$ times compared to 1-1 heat exchanger (McCabe, et al., 1993). Mechanical design features like flanges, baffles, bolts and nuts are all taken care by the equipment manufactures.

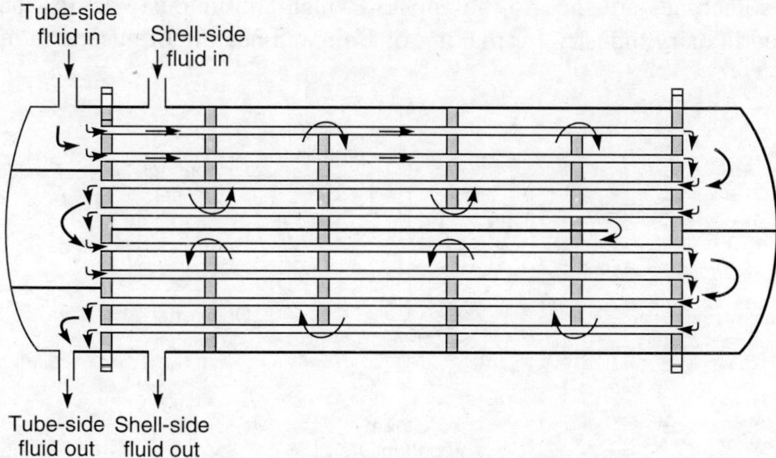

Figure 12.6 2-4 heat exchanger.

12.2 JACKETED PANS

Jacketed pans are frequently used in food processing industries because of their simplicity, ease of operation and versatility in use. They are also known as jacketed vessels. They consist of a hemispherical vessel usually made out of *SS* to which an annular jacket is provided [shown in Figure 12.1(a)]. They may or may not have stirring arrangement; usually they are supplied with a stirrer which is flexible. If it is not required, it can be set aside. Even the vessels are supplied with tilting arrangement so that discharging of final product is both easy and convenient. The jacketed space is used for passing hot fluid or for condensing steam which releases heat. Most of the viscous food products like jams, sauces and milk products like *khoa* are made in small scale in the jacketed pans. The desired amount of stirring, addition of certain ingradients during processing and either sampling out of the product intermittently during processing or discharging of the final product can all be done conveniently.

Heat transfer coefficients applicable for jacketed vessels can be used; but mostly the processing in these pans is done in an unsteady manner. Hence, the heat transfer coefficient equations [Eq. (11.33)] for unsteady state heat transfer is frequently used. Generally, the jacketed pans are not considered for any design purposes since continuous operations cannot be taken up in them. The capacities of these pans are typically 50 lit to 200 lit. Rarely 500 lit capacity jacketed vessels are made.

12.3 SCRAPED SURFACE HEAT EXCHANGERS

Some of the food products are highly viscous. Another peculiarity with most of the liquid foods is that there will be change in viscosity as the fluids get concentrated during processing by removal of moisture; and they (liquid foods) change their rheological behaviour from Newtonian to non-Newtonian fluids. This behaviour of foods necessitates the use of a special type of heat exchanger which continuously scraps the surface. This continuous scraping helps renew the liquid food layers adhering to the heat transfer surfaces. The scraping is reported to improve the heat transfer coefficient by 5 times in case of non-Newtonian foods (Uhl and Gray 1966).

In agitated vessels containing anchor type impeller (Figure 12.7), a scraper in the form of a teflon strip is sometimes attached to the anchor, which continuously scraps the surface. This is specially used in dairy industry for making of *khoa* and heating of ghee, and in confectionary

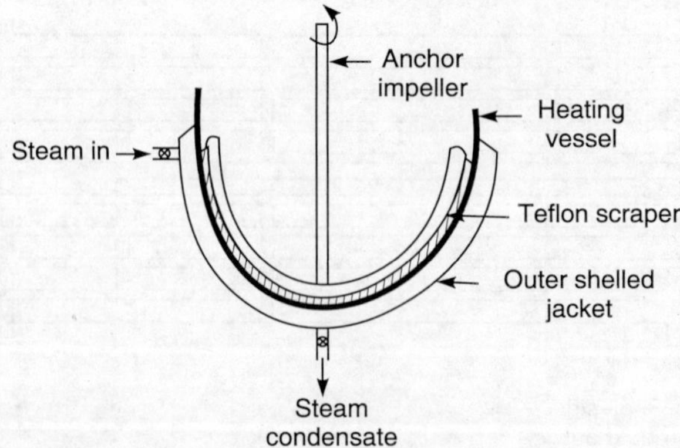

Figure 12.7 Anchor-type agitated vessel with scraper.

industry to heat the liquid glucose with other ingredients like colours, flavours, etc. These vessels can be operated only in batch manner. Some more discussion on typical scrapers was given by Uhl and Gray (1966).

Scraped surface heat exchangers are made for continuous operation. They are a double pipe heat exchanger with a provision for entry of steam or cooling water in the jacket. The inner tube is fairly large (typically 100 to 300 mm in diameter). A scraper is attached centrally to the tube which has one or more blades, and rotates slowly by means of a motor. The viscous fluid flows in the tube, and is continuously agitated and scraped from the walls by the blades as shown in Figure 12.8. A provision is made to clean the inside of the tube and also the scraper by lowering or sliding the rotor with blades. They are generally designed for heavily fouling products. They can handle fluids containing even 75 per cent solids which are more required in the case of crystallization. Discussion on scraped surface evaporators, which are almost similar to scraped surface heat exchangers, will be given in Section 15.5.7.

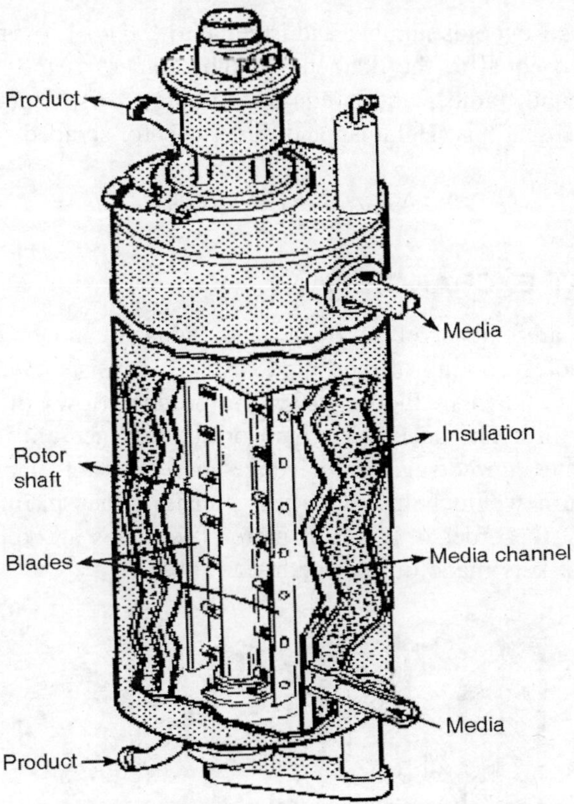

Figure 12.8 Scraped surface heat exchanger.

The heat transfer coefficients (h_j) for a scraped surface heat exchanger can be derived from first principles (McCabe, et al., 1993) which is as follows:

$$h_j = 2\ (k\rho C_p nB/\pi)^{1/2} \tag{12.1}$$

where k, ρ and C_p, respectively are the thermal conductivity, density and specific heat of the fluid, n is the agitator speed in revolutions per second, and B is the number of blades carried by the shaft. Equation (12.1) seems to be an oversimplification of the system, and mainly lacks the viscosity term which is very vital in case of scraped surface heat exchangers. However, one point which is brought to light very clearly by the above equation is that the heat transfer coefficient (and hence, the rate of heat transfer) is influenced by two factors.

(i) Heat conduction by the thin layer at the heating surface which is dependent upon the properties of the fluid.
(ii) The rate at which the thin layer is scraped and then mixed with the bulk product.

Hence, we present here an empirical equation reported by Skelland (1958).

$$\left(\frac{h_j d_i}{k}\right) = 4.9\ (N_{Re})^{0.57}\ (N_{Pr})^{0.47} \left(\frac{d_i n}{u}\right)^{0.17} \left(\frac{d_i}{L}\right)^{0.37} \tag{12.2}$$

In Eq. (12.2) N_{Re} is Reynolds number and is equal to $(\rho d_i u/\mu)$ in which d_i is inside diameter of the tube and is also equal to the diameter of the scraper, as the scraper is closely fitting to the tube. N_{Pr} is Prandtl number and is equal to $(C_p\mu/k)$. L is length of the heat exchanger.

Another empirical equations (Heldman and Lud 1992) for scraped surface heat exchanger is:

$$N_{Nu} = \left(\frac{2}{\pi}\right) (N_{Re} N_{Pr} n)^{0.5} \tag{12.3}$$

12.4 PLATE HEAT EXCHANGERS

Plate heat exchangers are extensively used in dairy industry, and food and beverage industry for quick heating or quick cooling of the liquid foods. It consists of a series of *SS* plates held together in a frame as shown in Figure 12.9. The product flows in between the two plates in the form of a thin film. The heating medium or cooling medium is passed in the alternate plate spacing. The plates are corrugated to induce some amount of mixing and turbulence in the fluids which in turn would help better heat transfer. The spacing between the plates is very small which is of the order of 1/32″ or more. The plates are kept in position, and made leak-proof by using gaskets made out of synthetic rubber.

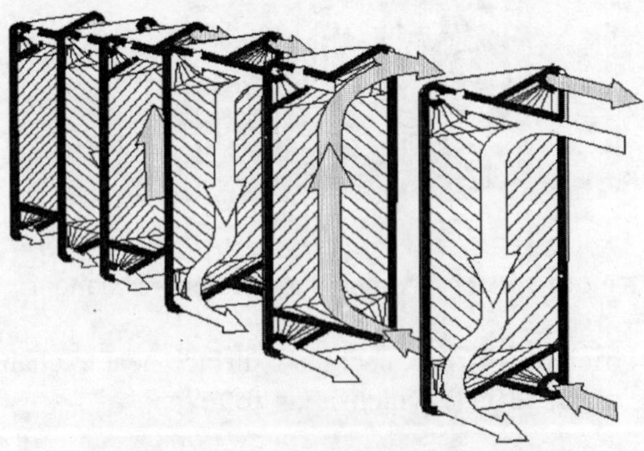

Figure 12.9 Plate heat exchanger.

Heat exchanges through the metallic plates. Since the spacing between the plates is very small, the fluid flows in the form of a thin film, and gets heated up quickly. Hence, this heat exchanger is more useful for sterilization of heat sensitive foods like milk. It is extensively used for Ultra High Temperature (UHT) treatment, or High Temperature and Short Time (HTST) treatment. The only drawback with this heat exchanger is that it can be used only for foods with low viscosity. Fluids with viscosities less than 5 Pa s can be operated. Temperatures can go upto 200 °C.

The heat transfer coefficients are typically 4.5 to 7 kW/m^2°C. The heat exchanger capacities also vary depending on the number of plates and area of each plate, and is of the order of 5000 to 20,000 kg/h. The individual plate areas vary between 0.02 m^2 to 2 m^2.

Since the rubber gaskets are used, the operating pressures need to be restricted to 700 kPa (7 atm pr) and hence, the flow rates can be 1.5–2 m/s. Since the spacing between the plates is usually very small, it cannot be used for slurries and suspensions. Also the entire equipment needs to be sterilized before every operation since the chances of contamination are more because of the small spacing available between the plates. The contamination of the residual proteinaceous materials is known as **fouling.**

In spite of these limitations, plate heat exchangers are used for ultra high temperature operations, serialization of foods and HTST processing because of the simplicity of equipment in terms of cost economics and the floor space occupied by them. Thermal efficiency more than 90 per cent is realized in plate heat exchangers. Capacity of operation can be quickly altered in the processing place by increasing or decreasing the number of plates.

Plate evaporator which is similar to plate heat exchanger will be discussed in Section 15.5.6.

Symbols

B: no. of blades on the scraper
C_p: specific heat of liquid (kJ/kg°C)
d_i: inside diameter of the tube (m)
h_j: jacketed vessel heat transfer coefficient (kW/m^2°C)
k: thermal conductivity of fluid (W/m°C)
L: length of the tube (m)
N_{Re}: Reynolds number
N_{Pr}: Prandtl number
N_{Nu}: Nusselt number
n: agitator speed (s^{-1})
u: linear velocity of fluid (m/s)

Greek Symbols

μ: viscosity (Pa s)
ρ: density (kg/m^3)

REVIEW QUESTIONS

12.1 What are various types of heat transfer equipments you come across for operating in small-scale level? What are relative advantages of each?

12.2 Describe a typical shell-and-tube heat exchanger with a neat diagram.

12.3 Describe a counter-current flow 1-1 heat exchanger.

12.4 Describe a 1-2 heat exchanger with a neat diagram. What are the advantages of it over 1-1 heat exchanger?

12.5 Describe a 2-4 heat exchanger with a neat diagram.

12.6 Describe a scraped surface heat exchanger. What are its special features?

12.7 Describe a plate heat exchanger. Explain where we get better advantage of using this heat exchanger.

REFERENCES

Doran, P.M. (1995), *Bioprocess Engineering Principles*, Academic Press, London, pp. 164–167.

Heldman, D.R. and Lund, D.B. (1992), *Handbook of Food Engineering*, Marcel–Dekker, Inc., New York, pp. 266–269.

McCabe, W.L., Smith, J.C. and Harriott, P. (1993), *Unit Operations of Chemical Engineering*, 5th ed., McGraw-Hill, New York, Chapter 15.

Skelland, A.H.P. (1958), Correlations of scraped film heat transfer in the votator, *Chem. Engg. Sci*, **7**, pp. 166–175.

Uhl, V.W. and Gray, J.B. (1966), *Mixing*, Vol. 1, Academic Press, New York, pp. 298–303.

CHAPTER

13

Heat Transfer by Radiation

Of the various heat transfer mechanisms we have seen, radiative heat transfer mechanism is quite different. It does not involve any movement of molecules either visibly or invisibly. Radiation heat waves are a form of electromagnetic waves, and pass through space at the speed of light. They do not require any medium to pass through. Virtually all bodies which are all above 0 K will emit radiations, which is known as **thermal radiation**. There are various other ways by which a body emits radiations, viz., bombardment with electrons, or treatment with an electric discharge, etc. But in the present chapter we restrict ourselves only to the thermal radiation studies.

Before we try to understand the basics of radiation, let us see where do we come across such a mechanism of heat transfer in food processing. In a baking oven, we find the food material is kept in an environment which is hot by virtue of the fact that the walls that constitute the oven are hot. But there is no medium between the material and hot walls; still heat transfers from the walls to the body, and heat the body. Similarly in a chilling chamber, the walls of the chamber are very cold. When a material is kept in the chamber, it gets cooled because of the radiation transfer of heat from the body to the walls of the cooling chamber. Infact, the transfer of heat from sun to the earth through space is by radiation only. There is no medium in between to transfer the heat. Similarly a furnace wall, a boiler wall, or a steam line which are hot will be loosing heat by radiation. Thus, radiation is an important mode of heat transfer particularly in food processing.

One important aspect of radiation is that, since heat transfers in the form of electromagnetic waves without the need for any medium, the heat absorbed by a body by radiation depends upon the nature of the receiving body. For example when solar radiations are reaching the earth, the temperatures of different bodies will be different depending upon the nature of the body; a black body will receive more heat and will be more hot as compared to another material which is white in colour, and so will be receiving (absorbing) less heat. It will be reflecting more heat and hence, will attain lower temperatures. The nature of the body in terms of its ability to absorb heat, reflect heat or transmit heat will decide to what extent the body will be heated by radiations.

13.1 BASIC PRINCIPLES OF RADIATION

When a ray of radiant heat of any wavelength falls on a surface, it undergoes the following three processes as shown in Figure 13.1:

- Some amount of heat will be absorbed by the body.
- Some amount of heat will be transmitted by the body.
- Rest of the amount of heat will be reflected by the body.

The fraction of the heat radiation received and absorbed by a body is known as **absorbivity (α)** of the body; and the fraction of heat that is absorbed and transmitted by the body is known as **transmissivity (τ)** of the body; whereas the fraction of heat radiation reflected by the body without absorbing is known as **reflectivity (γ)** of the body.

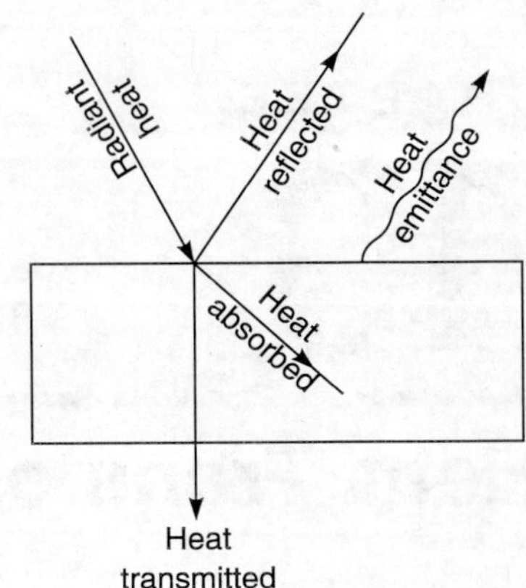

Figure 13.1 Radiant heat dissipation by a body.

Therefore, $$\alpha + \gamma + \tau = 1.0 \tag{13.1}$$

From the above discussion, it is clear that *radiation as such is not heat, and if it is absorbed and converted into heat it is no more a radiation* (McCabe, et al., 1993).

From Eq. (13.1), it is clear that the maximum value of α, the absorbivity, is 1.0; and such a body which absorbs all the radiant heating falling on it is known as **black body**. All other bodies will have absorbivity less than 1.0 and are known as gray bodies. In reality, there is no body which can be called as black body. Some materials like carbon soot, some grades of carbon black attain almost near to black body with α approaching 1.0.

13.1.1 Emissivity

The emission of radiations by a body is independent of the emission of radiation by other bodies in the sight. That is why if two bodies at two different temperatures are kept closeby the hotter body loses heat by radiation faster than it receives radiation heat from the colder body, and hence, the hot body gets cooler. In the same manner, the cold body receives radiation at a rate faster than it loses heat by emission, and hence, it gets heated up. Thus, both the bodies will attain equilibrium temperature.

A black body will have maximum emission power as compared to any other body (McCabe, et al., 1993). So all other bodies (gray bodies) will have emission powers less than that of the black body. Hence the **emissivity** (ε) of any body is defined as the ratio of emissive power (W) of the body to that of the black body (W_b)

Therefore, $$\varepsilon = \frac{W}{W_b} \tag{13.2}$$

where W and W_b are in the units of W/m^2.

The emissive power is equivalent to the heat flux from a body by radiation. The emissivities of various bodies are tabulated in Table 13.1.

Table 13.1 Emissivity of Various Surfaces

Material	Wavelength and average temperatures				
	9.3 μm 38 °C	5.4 μm 260 °C	3.6 μm 540 °C	1.8 μm 1370 °C	0.6 μm *Solar*
Metals					
Aluminium					
Polished	0.04	0.05	0.08	0.19	~ 0.3
Oxidized	0.11	0.12	0.18		
24-ST weathered	0.4	0.32	0.27		
Surface roofing	0.22				
Anodized (at 1000 °F)	0.94	0.42	0.60	0.34	
Brass					
Polished	0.10	0.10			
Oxidized	0.61				
Chromium					
Polished	0.08	0.17	0.26	0.40	0.49
Copper					
Polished	0.04	0.05	0.18	0.17	
Oxidized	0.87	0.83	0.77		
Iron					
Polished	0.06	0.08	0.13	0.25	0.45
Cast, oxidized	0.63	0.66	0.76		
Galvanized, new	0.23	–	–	0.42	0.66
Galvanized, dirty	0.28	–	–	0.90	0.89
Steel plate, rough	0.94	0.97	0.98		
Oxide	0.96	–	0.85	–	0.74
Magnesium	0.07	0.13	0.18	0.24	0.30
Silver					
Polished	0.01	0.02	0.03	–	0.11
Stainless steel					
18–8, polished	0.15	0.18	0.22		
18–8, weathered	0.85	0.85	0.85		
Steel tube					
Oxidized	–	0.80			
Tungsten filament	0.03	–	–	~ 0.18	0.36[a]
Zinc					
Polished	0.02	0.03	0.04	0.06	0.46
Galvanized sheet	~ 0.25				
Building and insulating materials					
Asphalt	0.93	–	0.9	–	0.93
Brick					
Red	0.93	–	–	–	0.7
Fire clay	0.9	–	~ 0.7	~ 0.75	
Silica	0.9	–	~ 0.75	0.84	
Magnesite refractory	0.9	–	–	~ 0.4	
Enamel, white	0.9				
Paper, white	0.95	–	0.82	0.25	0.28
Plaster	0.91				
Roofing board	0.93				

(contd.)

Table 13.1 Emissivity of Various Surfaces (*contd.*)

Material	*Wavelength and average temperatures* 9.3 μm 38 °C	5.4 μm 260 °C	3.6 μm 540 °C	1.8 μm 1370 °C	0.6 μm *Solar*
Enameled steel, white	–	–	–	0.65	0.47
Paints					
Aluminized lacquer	0.65	0.65			
Lacquer, black	0.96	0.98			
Lampblack paint	0.96	0.97	–	0.97	0.97
Red paint	0.96	–	–	–	0.74
Yellow paint	0.95	–	0.5	–	0.30
Oil paints (all colours)	~ 0.94	~ 0.9			
White (ZnO)	0.95	–	0.91	–	0.18
Miscellaneous					
Ice	~ 0.97[b]				
Water	~ 0.96				
Carbon, T-carbon, 0.9% ash	0.82	0.80	0.79		
Wood	~ 0.93				
Glass	0.90	–	–	–	(Low)

[a]At 3315 °C
[b]At 0 °C

Source: Adapted from Kreith, F. (1976), *Principles of Heat Transfer*, 3rd ed., pp. 236–237.

13.1.2 Kirchhoff's Law

When two bodies are in temperature equilibrium and are brought together, ratios of their emissive power to absorbivity are equal and dependent upon the temperature. This is the statement of **Kirchhoff's law**, and can be mathematically represented as:

$$\frac{W_1}{\alpha_1} = \frac{W_2}{\alpha_2} \tag{13.3}$$

where W_1 and W_2 stand for emissive power (or radiating power) of substances 1 and 2. If one of the bodies is a black body (say body 2). Eq. (13.3) becomes

$$\frac{W_1}{\alpha_1} = W_b \tag{13.4}$$

since

$$\alpha_2 = 1$$

or

$$\alpha_1 = \frac{W_1}{W_b} \tag{13.5}$$

This is same as Eq. (13.2), and hence,

$$\alpha_1 = \varepsilon_1 \tag{13.6}$$

Thus, when a body is in temperature equilibrium with its surroundings, the absorbivity of the body is equal to its emissivity. This is an alternate statement of Kirchhoff's law and finds wider application. The use of Kirchhoff's law can better be explained with a day-to-day experience why we prefer to coat the roof surfaces with white paints rather than black paint for minimum heat absorbivity (or gain) during summer seasons (Singh and Heldman 1993).

From Table 13.1, we have the emissivity data for white paint and black paint at short wavelengths (solar radiations) and large wavelengths (at normal temperature).

Black paint, ε = 0.97 (solar)

= 0.96 (longer wavelength)

White paint, ε = 0.18 (solar)

= 0.95 (longer wavelength)

Assuming that there is no transmissivity, i.e., $\tau = 0$ reflectivity of the white paint,

$$\gamma_{\text{white paint}} = 1 - 0.18 \text{ for solar wavelength}$$

$$\gamma_{\text{white paint}} = 0.82$$

Similarly, $\gamma_{\text{black paint}} = 1 - 0.97 = 0.03$

Similarly for longer wavelengths

$$\gamma_{\text{white paint}} = 1 - 0.95 = 0.05$$

$$\gamma_{\text{black paint}} = 1 - 0.96 = 0.04$$

Thus, the total reflectivity by solar radiation and by longer wavelength radiations put together for black paint is 0.03 + 0.04 = 0.07. Whereas that for white paint is 0.82 + 0.05 = 0.87.

Or in other words the total absorbivity for white paint by all radiations is 1 – 0.87 = 0.13, whereas that for black paint is 1 – 0.07 = 0.93; i.e., 93 per cent of the radiating heat is absorbed by the black paint, where only 13 per cent of the radiating heat is absorbed by the white paint. Hence, the room remains relatively cooler if its roof is painted with white paint as contrast to black paint.

13.1.3 Stefan–Boltzmann Law

The radiating heat from the black body is given by a law known as **Stefan–Boltzmann law** which states that the emissive radiations or radiating heat flux is proportional to the 4th power of temperature on absolute scale (degrees Kelvin). Thus,

$$W_b = \left(\frac{q}{A}\right) \propto T^4 \tag{13.7}$$

or

$$W_b = \left(\frac{q}{A}\right) = \sigma T^4 \tag{13.8}$$

where q is the rate of heat transfer in watts, and A is the surface area in m^2. σ is known as **Stefan Boltzmann constant** and is equal to 5.669×10^{-8} W/m^{2}K^4. In fact the above equation was originally derived by Stefan using empirical approaches based on the radiating energy from black bodies, but subsequently Boltzmann provided rationale for the above equation based on the laws of thermodynamics and electromagnetism. Thus, Eq. (13.8) is popularly known as **Stefan–Boltzmann law**. Equation (13.8) can also be applied for other bodies giving emissive radiations as follows by including the emissivity term:

$$W = \left(\frac{q}{A}\right) = \sigma \varepsilon T^4 \tag{13.9}$$

PROBLEM 13.1 In a process plant, steam line is drawn from the boiler to the process vessels. The steam line is covered with glass wool and is covered with polished aluminium sheets. The outside temperature of the pipeline is 38 °C. Calculate the heat energy emitted per unit area per unit time.

Solution Given data

Emissivity (ε) for polished aluminium at 38 °C = 0.04 (from Table 13.1).

$$\text{Temperature} = 38 + 273 = 311 \text{ K}$$

$$\text{Area} = 1 \text{ m}^2$$

$$\text{Time} = 1 \text{ s}$$

Approach: We use Eq. (13.9)

$$\frac{q}{A} = \sigma \varepsilon T^4 = 5.669 \times 10^{-8} \times 0.04 \times (311)^4$$

$$= 21.2 \text{ W/m}^2$$

The rate of heat energy emitted from the steam pipe = 21.2 J/sm^2

13.2 RADIATION BETWEEN TWO BODIES

As has been mentioned earlier, if two bodies at temperatures T_1 and T_2 with their emissivities ε_1 and ε_2, are enclosed in a container, heat exchanges between the two bodies depending upon:

- the orientation of surfaces (whether parallel to each other or otherwise),
- their emissivities, and
- type of surfaces (whether the surfaces are clean, smooth, convex-shaped or concave-shaped).

If the two surfaces are parallel and plane

$$\left(\frac{q}{A}\right) = C\sigma (T_1^4 - T_2^4) \tag{13.10}$$

where C is a constant and

$$C = \frac{1}{\varepsilon_1} + \frac{1}{\varepsilon_2} - 1 \tag{13.11}$$

If the surfaces are not parallel, the angle of vision becomes important. Distance between the two surfaces also would matter.

Considering above factors

$$q = \sigma AF\,(T_1^4 - T_2^4) \tag{13.12}$$

in which A is arbitrarily chosen area of any one surface. F is a dimensionless geometric factor to account for the

- geometries of two surfaces,
- their spatial relationship, and
- surface area (A) chosen for consideration, etc.

Since consideration of all above parameters is complex and are not normally required by food engineers, we prefer to restrict ourselves to contemplate on Eq. (13.10), and the importance of heat transfer to a small body from its surroundings. Further discussion on the matter [Eq. (13.12)] can be found in any standard textbooks (McCabe, et al., 1993).

13.2.1 Radiations from Surroundings

Many process applications in food engineering are based on the radiative heat required by a substance from the surroundings, as in the case of a loaf of bread or biscuit placed in a hot baking oven. In all these cases, the substance is a small body compared to the surroundings radiating heat. The emissivity of the body is considered in Eq. (13.10) and can be written as

$$q = A\sigma\varepsilon\,(T_1^4 - T_2^4) \tag{13.13}$$

in which T_1 is temperature in degrees Kelvin of the surroundings radiating heat and T_2 is the temperature in degrees Kelvin of the body receiving heat. We try to write the radiative heat transfer process similar to Eq. (11.18).

$$q = h_r\,A\,(\Delta T) = h_r\,A\,(T_1 - T_2) \tag{13.14}$$

where h_r is the radiative heat transfer coeffcient. In Eq. (13.14), it does not matter whether T_1 and T_2 are in °C or K; as the ΔT remains same.

Equating Eqs. (13.13) and (13.14)

$$q = h_r\,A\,(T_1 - T_2) = A\sigma\varepsilon\,(T_1^4 - T_2^4) \tag{13.15}$$

$$T_1^4 - T_2^4 = (T_1^2 + T_2^2)\,(T_1^2 - T_2^2) = (T_1^2 + T_2^2)\,(T_1 - T_2)\,(T_1 + T_2) \tag{13.16}$$

Therefore,

$$h_r = \sigma\varepsilon\,(T_1^2 + T_2^2)\,(T_1 + T_2) \tag{13.17}$$

If

$$T_{Av} = \frac{T_1 + T_2}{2}$$

and if $\Delta T << T_1$ or T_2, we may write with a reasonable amount of accuracy (Earle 1969).

$$h_r = \varepsilon\sigma 4\,(T_{Av})^3 \tag{13.18}$$

$$= (0.227)\,\varepsilon\left(\frac{T_{Av}}{100}\right)^3 \tag{13.19}$$

PROBLEM 13.2 A piece of meat carcase is kept in a deep freezer maintained at –18 °C. Calculate the radiative heat transfer if the meat carcase is at 25 °C and has an average area of 0.045 m^2. The emissivity of carcase may be taken as 0.82.

Solution

$$T_1 = 25 + 273 = 298$$

$$T_2 = -18 + 273 = 255$$

$$\sigma = 5.669 \times 10^{-8}\ \text{W/m}^2\text{K}^4$$

$$\varepsilon = 0.82$$

$$A = 0.045\ \text{m}^2$$

$$T_1^4 = 7.88 \times 10^9$$

$$T_2^4 = 4.22 \times 10^9$$

Therefore, by Eq. (13.13),

$$q = 5.669 \times 10^{-8} \times 0.82 \times 0.045(7.88 - 4.22) \times 10^9 = 7.64\ \text{W}$$

Let us see how much different q will be if we use Eqs. (13.19) and (13.14),

$$T_{Av} = \frac{298 + 255}{2} = 276.5\ \text{K}$$

Therefore,

$$h_r = 0.27 \times 0.82 \left(\frac{276.5}{100}\right)^3 = 3.93\ \text{W/m}^2\text{K}$$

and

$$q = 3.93 \times 0.045\ (298 - 255) = 7.61$$

The values of q are reasonably comparable.

PROBLEM 13.3 A loaf of bread is passing through a baking oven, the walls of which are maintained at a constant temperature of 220 °C. The bread has an area of 0.09 m^2 and is at 100 °C. The emissivity of bread may be taken as 0.52. In addition to radiation heat, there is convective heat also by air at 220 °C. Calculate the heat transfer rate.

Solution

Approach: Here we consider both radiative heat and convective heat, (i) the radiative heat is calculated by Eq. (13.13) or (13.14), (ii) the conductive heat is calculated by Eqs. (11.52) and (11.18).

Given:

$$T_1 = 220\ °\text{C} = 493\ \text{K}$$

$$T_2 = 100\ °\text{C} = 373\ \text{K}$$

$$\Delta T = 120\ °\text{C}$$

$$A = 0.09\ \text{m}^2$$

$$\varepsilon = 0.52$$

$$\sigma = 5.669 \times 10^{-8}$$

$$T_1^4 = 5.9 \times 10^{10}$$

$$T_2^2 = 1.94 \times 10^{10}$$

Heat transfer by radiation: Eq. (13.13) gives

$$q = 0.09 \times 5.669 \times 10^{-8} \times 0.52\ (5.9 - 1.94) \times 10^{10}$$

$$= 105 \text{ W}$$

Let us calculate h_r by Eq. (13.19)

$$T_{Av} = \frac{493 + 373}{2} = 433$$

$$h_r = 0.27 \times 0.52 \left(\frac{T_{Av}}{100}\right)^3 = 9.58 \text{ W/m}^2\text{K}$$

Therefore, $$q = 9.58 \times 0.09\ (493 - 373) = 103.5 \text{ W}$$

Since in this case ΔT is not very small as compared to T_1 or T_2, we better use the value of q calculated by Eq. (13.13) only

$$q_{\text{radiation}} = 105 \text{ W}$$

Heat transfer by natural convection: We consider the loaf of bread to be a cube and try to find out the size of one side by equating the surface area given by $6x^2$ to be 0.09, where x is the length of one side

$$6x^2 = 0.09$$

Therefore, $$x = 0.1225 \text{ m}$$

We calculate h_c by Eq. (11.52)

$$h_c = 1.3196 \times \left(\frac{\Delta T}{x}\right)^{0.25} \tag{11.52}$$

$$= 1.3196 \times \left(\frac{120}{0.1225}\right)^{0.25}$$

Therefore,

$$h_c = (1.3196) \times 5.59$$

$$= 7.376 \text{ W/m}^2\text{K}$$

Therefore, $$q = 7.376 \times 0.09\ (120) = 79.67 \text{ W}$$

$$q_{\text{convection}} = 79.67 \text{ W}$$

$$q_T = q_{\text{radiation}} + q_{\text{convection}} = 105 + 79.67$$

$$= 184.67 \cong 185 \text{ W}$$

PROBLEM 13.4 In a canning process, 100 cans per minute are to be heated from 60 °C to 100 °C, and demoistured of the adhering moisture to the cans. Each can contains on average 0.3 ml of water adhering to it. The cans have an emissivity of 0.7 and an average area of each can is 0.047 m², and its weight is 100 g. The specific heat of can material is 0.25 kJ/kg°C. Calculate the temperature of the oven considering the radiative heat transfer only.

Solution

Approach:

(i) The approach for this problem is almost similar to earlier problems using Eq. (13.13) except that we first find heat requirement load.

(ii) Heat load is determined by calculating the heat required for raising the temperature of 100 cans per minute with adhering water from 60 °C to 100 °C and later latent heat vaporization of water (0.3 ml) at 100 °C.

Assumption: For all practical purposes we assume the oven to be a radiating oven, and all the area of the can is totally surrounded by radiating heat.

Given: Noted from literature:

$$\rho_{water} = 1000 \text{ kg/m}^3$$

$$C_p \text{ of water} = 4.18 \text{ kJ/kg°C}$$

$$\lambda_s \text{ at } 100 \text{ °C} = 2671.1 - 419.04 = 2257 \text{ kJ/kg}$$

$$A = 0.0385 \times 100 = 3.85 \text{ m}^2$$

$$\sigma = 5.669 \times 10^{-8} \text{ W/m}^2\text{K}^4$$

$$T_r = 100 + 273 = 373 \text{ K}$$

Heat load for cans in one minute:

$$Q_{cans} = mC_p\,(\Delta T) = 100 \times 100 \times 10^{-3} \times 0.25\,(100 - 60)$$

$$= 100 \text{ kJ/minute} \qquad (1)$$

$$Q_{water} = mC_p\,(\Delta T\,) = 100 \times 0.3 \times 10^{-6} \times 1000 \times 4.18\,(100 - 60)$$

$$= 5.016 \text{ kJ/min} \qquad (2)$$

$$Q_{steam} = m\lambda_s = 100 \times 0.3 \times 10^{-6} \times 1000 \times 2257$$

$$= 67.71 \text{ kJ/min} \qquad (3)$$

Total heat requirement Q per minute = (1) + (2) + (3)

$$= 172.68 \text{ kJ}$$

Therefore,
$$q = \frac{172.68}{60} = 2.878 \text{ kJ/s} = 2.878 \times 10^3 \text{ W}$$

Now, we apply Eq. (13.13)

$$q = A\sigma\varepsilon\,(T_1^4 - T_2^4)$$

$$= 0.047 \times (100/60) \times 0.7 \times 5.669 \times 10^{-8} \times (T_1^4 - 373^4)$$

$$= 3.108 \times 10^{-9}\,(T_1^4 - 1.96 \times 10^{10})$$

Therefore,
$$T_1^4 = 9.49 \times 10^{11}$$

$$T_1 = 987 \text{ K} = 714 \text{ °C}$$

Note: In fact, there will be some convective heat transfer also by natural convection as we have seen in Problem 13.3. If that is also considered, the temperature could be slightly less.

13.3 APPLICATION OF RADIATION HEAT TRANSFER IN FOOD PROCESSING

As has already been mentioned, radiative heat transfer plays a major role in food processing both for heating and cooling: heating in baking ovens, and cooling in deep freezers. In addition to the above, radiation dryers are also very popular.

13.3.1 Heating in Baking Ovens

Baking ovens are invariably heated by radiation heaters. The baking oven walls are maintained at high temperatures almost of the order of 250–300 °C in which the fermented dough is kept. It (the dough) puffs due to liberation of carbon dioxide, and makes loaves of breads.

Similarly biscuits and pasta products are also heated in radiation heaters. After making the formulation, the dough is stretched in the form of sheets, and cut into the shape of the biscuit. Later the biscuits are baked in the baking oven to cook the dough to form crispy biscuits. The temperature control in the baking oven is an important parameter which attributes the crispiness and taste to the biscuits.

Radiation dryers are used for drying the grains or other solid materials so that their moisture is considerably brought down for safe storage or for further processing. Especially grains before going for grinding are invariably dried to bring down the moisture content for easy grinding in continuous-radiation dryers in which the heat is provided by IR (infrared) rays to the continuously moving trays containing the grains or solids.

Cans or bottles before being used for filling-in operation are sterilized by radiation heaters to avoid moisture contamination. For puffing of cereals or pulses in puffing cookers, dry heat is provided by radiation heaters to the moistened grains in continuous operations.

13.3.2 Cooling in Deep Freezers

Deep freezers are used for cooling or chilling of food materials. It could be for the purpose of processing or preservation. The deep freezers are cooled by refrigeration systems from outside. Ultimately we get a deep freezing chamber with cold walls.

Most of the horticultural produce (fruits and vegetables) or marine produce (fish, prawns, shrimp or crabs) are cooled for preservation purposes[†] in cold chambers either for further use or for further processing. Ice creams and a variety of dairy products are kept in cold storage until they are sold to the customers. The cold chamber walls are maintained at low temperatures which absorb heat from food materials by radiation.

Symbols

A: area (m^2)
C: constant defined by Eq. (13.11)
F: dimensionless factor defined by Eq. (13.12)
h_r: radiative heat transfer coefficient (W/m^2K)
h_c: convective heat transfer coefficient (W/m^2K)
q: heat transfer rate (W)
T: temperature in degrees Kelvin
ΔT: temperature difference
W: emissive power (W/m^2)
W_b: emissive power of black body (W/m^2)
x: length of any one side of a cube (m)

[†]Of course, in Individual Quick Freezing (IQF) operations, the materials are cooled by a cold blast of air, in which case the heat transfer is by convection and conduction and not by radiation.

Subscripts

1, 2: bodies 1 and 2

Greek Symbols

α: absorbivity
γ: reflectivity
τ: transmissivity
ε: emissivity
σ: Stefan–Boltzmann constant

REVIEW QUESTIONS

13.1 Describe the importance of radiation heat transfer.

13.2 Explain the terms
(i) absorbivity
(ii) reflectivity
(iii) transmissivity
(iv) emissivity

13.3 What is a black body?

13.4 What is Kirchhoff's law and what is its significance?

13.5 What is Stefan–Boltzmann law?

13.6 Describe the process of radiative heat transfer between two bodies.

13.7 Describe the radiative heat transfer from surroundings.

13.8 Describe various applications of radiation heat transfer in food processing.

NUMERICAL PROBLEMS

13.1 Calculate the rate of heat emitted per unit area from a polished copper surface, the temperature of which is 40 °C. (Hint: Note emissivity from Table 13.1)
(**Ans:** q/A = 21.77 W/m^2)

13.2 Meat is roasted in a tandoori oven at a temperature of 200 °C. The meat piece has an area of 0.05 m^2 and emissivity of 0.82, and is at a temperature of 85 °C. Calculate the heat transfer rate considering both convective and radiating heat. (**Ans:** q_T = 123.4 W)

REFERENCES

Earle, R.L. (1969), *Unit Operations in Food Processing*, Pergamon Press, Oxford, pp. 88–92.

Kreith, F. (1976), *Principles of Heat Transfer*, 3rd ed., Intex Educational Publishers, New York, pp. 236–237.

McCabe, W.L., Smith, J.C. and Harriott, P. (1993), *Unit Operations of Chemical Engineering*, 5th ed., McGraw-Hill, New York, pp. 397–416.

Singh, R.P. and Heldman, D.R. (1993), *Introduction to Food Engineering*, 2nd ed., Academic Press, New York, pp. 183–188.

CHAPTER

14

Microwave Heating

Microwaves are electromagnetic waves of certain frequencies as contrast to radio frequencies. The Federal Communications Commission (USA) has allocated certain frequencies for microwaves, so that they do not interfere with the radio frequencies. The purpose for which these microwaves are used has been categorized as Industrial, Scientific and Medical (ISM) frequencies. These are shown in Table 14.1.

Table 14.1 Microwave Frequency (in MHz†) Assigned for ISM Use

	Frequency	*Wavelength*	*Penetrating depth for highly absorbing material*	*Countries permitted*
Radio	13.56 ± 6.68 KHz			
	27.12 ± 160 KHz	~10 m		
	40.68 ± 20 KHz			
Microwave	896.0			UK
	915.0 ± 25‡	33.0 cm	5 cm	North and South America
	2450.0 ± 50‡	12.2 cm	1–2 cm	All over the world
	5800.0 ± 75			
	24225.0 ± 125			

†MHz stands for mega Hertz, the units for microwave frequencies.
‡Microwave frequencies assigned for microwave ovens.

How the microwaves are used for heating purposes is a very interesting phenomenon. Water consists of dipoles. A dipole is a configuration which has two charges, opposite in sign and equal in magnitude, and are held at a fixed distance apart. Thus, the negatively charged oxygen atom and positively charged hydrogen atoms constitute a configuration, which we call as dipole. When the water molecule is exposed to a highly fluctuating electric field, the dipoles change their orientation. The electric field applied is of very high frequency and makes the water molecules reorient themselves (or rotate) some millions of times every second (Figure 14.1). This would

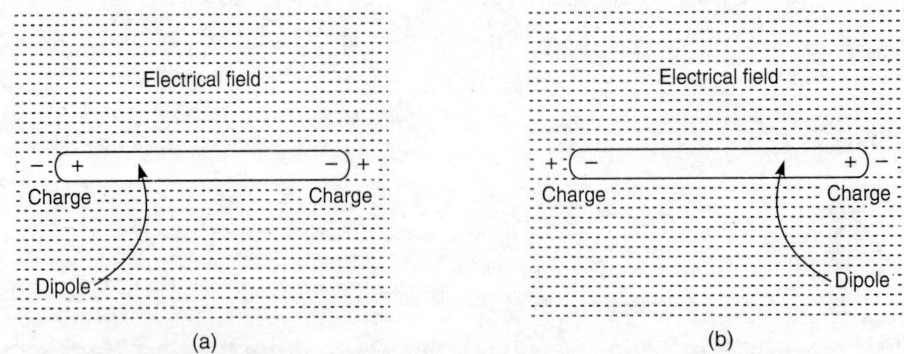

Figure 14.1 Orientation of dipoles in a changing electric field.

generate heat due to *molecular friction*. At molecular level, the heat generation is due to disruption of weak hydrogen bonds which is caused due to rotation of dipoles of free water molecules and due to electrophoretic migration of ions in dissolved form (free salts) in a rapidly changing polarity of electric field which is of the order of 21.45 billion times per sec. This has nothing to do with heating sources or container of the food material. The container does not get heated at all. In fact, we have to use such containers, which do not interfere with the microwaves. The heat is generated within the food materials because of the moisture content in the food. Obviously, if the material is totally dry and does not contain any water molecule-dipoles, microwaves do not generate any heat in such materials, no matter how long the material is exposed to microwaves. For example, if a dry paper or paper plate is put in microwave oven, it never gets heated up. On the contrary, if a wet paper is put, it will get heated up because the dipoles of water molecules become active, and generate heat by molecular friction. Food materials generally contain, in addition to water, the following as well:

- Salts
- Proteins
- Carbohydrates
- Lipids
- Ash content which is due to the silicious material present in the foods

Among these, only water, salts and ash interact with the microwaves to generate heat.

14.1 THEORY OF MICROWAVE HEATING

When a food material is placed in a microwave path, some amount of microwave energy is absorbed by the dipole molecules of the food and generate heat. The amount of energy absorbed by the foods to generate heat is considered to have been *lost* by the microwaves, and is determined by the *loss factor* of the food material. The loss factor (represented by ε'') is also known as **dielectric loss** or **loss tangent**, and it represents the material's ability to absorb the waves. Thus, it is a desirable quality of the food material.

The governing energy equation for microwave heating is given by (Datta 1990)

$$\frac{\partial T}{\partial t} = \alpha \nabla^2 T + \frac{Q}{\rho C_P} \tag{14.1}$$

Equation (14.1) represents temperature (T) at time (t) in which ρ, α and C_p are density, thermal diffusivity and specific heat of food material and Q is the heat generated per unit volume of material. In other words, Q represents electromagnetic energy that has converted into heat energy in the microwave oven. If E is the electric field intensity (v/m), then Q is given by:

$$Q = 2\pi f \varepsilon_0 \varepsilon'' E^2 \tag{14.2}$$

where ε_0 is the dielectric constant, f is the microwave frequency (2450 MHz for most microwave ovens or 950 MHz for some industrial ovens). The unit of f in the above is in Hz.

Equation (14.2) is derived from Maxwell's equations which govern propagation of electromagnetic waves in a dielectric medium. The application of above equation is not straight forward, since it is not possible to find the intensity of electric field (E) inside a food material. It depends upon various factors, viz., dielectric constant of food, loss factor, the placement of the food material inside the oven, the design of the oven, orientation of microwaves in the cavity, and the packaging material covering the food.

Reasonably we can assume an exponential variation of the electric field intensity from the boundary of the food material (x)

$$E = E_0 e^{-\alpha_e x} \tag{14.3}$$

where α_e is a coefficient related to the dielectric properties of the food, and E_0 is the electric field at the surface of the food material. We may also introduce here a new term called *penetration depth* which is defined as the distance inside the food material at which the heat generation drops to 37 per cent (e^{-1}) as compared to that on the surface (Datta, 1990). Thus

$$D_p = \frac{1}{2\alpha_e} = \frac{\lambda_0}{2\pi(2\varepsilon^1)^{1/2}} \left(\left\{ 1 + \left(\frac{\varepsilon''}{\varepsilon^1} \right)^2 \right\}^{-1/2} - 1 \right)^{-1/2} \tag{14.4}$$

where ε^1 is the relative dielectric constant and λ_0 is free space wavelength.

Since, it is difficult to measure or predict Q in Eq. (14.2), we use an indirect method of evaluating it by measuring the temperature profiles with time and modifying Eq. (14.1) by neglecting the first term in RHS, assuming that there are no temperature gradients axially. Thus,

$$Q = \rho C_p \frac{\partial T}{\partial t} \tag{14.5}$$

Combining Eqs. (14.5) and (14.2) we can calculate the intensity of electrical energy.

14.2 MICROWAVE PROPERTIES OF FOODS

There are two important properties of foods which relate to the generation of heat by microwaves. They are:

dielectric property (ε_0 or ε')
loss factor (ε'')

We also define a term called **loss tangent**

$$\tan \delta = \frac{\varepsilon''}{\varepsilon_0} \tag{14.6}$$

The dielectric properties are also depending upon the chemical composition and physical structure. Presence of salts or colloidal solids have a significant effect on the relative dielectric factors (Mudgett 1982). Temperature also affects the dielectric constant and loss factor. The values for potato are shown in Table 14.2 (Mudgett 1986).

Table 14.2 Microwave Properties of Raw Potatoes at Two Different Microwave Frequencies

Temperature	*Dielectric constant*		*Loss factor*		*Loss tangent*	
(°C)	915 MHz	2450 MHz	915 MHz	2450 MHz	915 MHz	2450 MHz
0	71	67	16	21	0.23	0.30
20	65	64	19	15	0.29	0.23
40	60	59	24	13	0.41	0.23
60	54	54	31	14	0.58	0.27
80	49	49	35	17	0.80	0.34
100	45	45	48	18	1.06	0.41

Source: Reprinted with permission from, Microwave properties and heating characteristics of foods, Mudgett, R.E., *Food Technology*, **40**(6), pp. 86–87, Copyright 1986 with the permission from Institute of Food Technologists, USA.

The dielectric properties of some semisolid foods are shown in Table 14.3 (Mudgett 1982) and those of some commercial oils and fats in Table 14.4. A compilation of dielectric properties various food materials is given in Appendix 10 (Ayappa, et al., 1991).

Table 14.3 Dielectric Measurements of Semisolid Food Products at Various Frequencies and Temperatures

Product	*Frequency* (MHz)	*Temperature* (°C)	ε_0	ε''
Beef	915	25	55	22
		65	51	33
	2450	25	52	17
		65	48	18
Pork	915	25	54	23
		65	52	34
	2450	25	51	17
		65	49	20
Ham	2800	20	54	28
		60	64	41
Potato	900	20	68	20
		60	59	26
	3000	25	66	19
		65	64	17
Carrot	2800	20	72	18
		60	61	15

Source: Reprinted with permission from Electrical properties of foods in microwave processing, Mudgett, R.E., *Food Technology*, **36**(2), p. 111, Copyright 1982 with the permission from Institute of Food Technologists, USA.

Table 14.4 Dielectric Data on 11 Commercial Fats and Oils

Sample	*Description*	*Dielectric properties*	300 MHz			1000 MHz			3000 MHz		
			25 °C	49 °C	82 °C	25 °C	49 °C	82 °C	25 °C	49 °C	82 °C
Soybean salad oil	Refined, bleached and deodorized	ε_0	2.853	2.879	2.862	2.612	2.705	2.715	2.506	2.590	2.594
		tan δ	0.0559	0.0480	0.0322	0.0544	0.0645	0.0517	0.0551	0.0648	0.0617
		ε''	0.159	0.138	0.092	0.158	0.174	0.140	0.138	0.168	0.160
Corn oil	Refined, bleached and deodorized	ε_0	2.829	2.868	2.861	2.638	2.703	2.713	2.526	2.567	2.587
		tan δ	0.0615	0.0468	0.0359	0.0664	0.0643	0.0538	0.0566	0.0645	0.0630
		ε''	0.174	0.134	0.103	0.175	0.174	0.146	0.143	0.166	0.163
Cottonseed cooking oil	Refined, bleached and deodorized	ε_0	2.825	2.859	2.834	2.629	2.669	2.673	2.515	2.536	2.554
		tan δ	0.0606	0.0463	0.0363	0.0660	0.0641	0.0547	0.0568	0.0651	0.0627
		ε''	0.171	0.132	0.103	0.174	0.171	0.146	0.143	0.165	0.160
Lard	Deodorized	ε_0	2.718	2.779	2.770	2.584	2.651	2.656	2.486	2.527	2.541
		tan δ	0.0564	0.0493	0.0394	0.0612	0.0600	0.0516	0.0509	0.0608	0.0583
		ε''	0.153	0.137	0.109	0.158	0.159	0.137	0.127	0.154	0.148
Tallow	Deodorized tallow used for deep-fat frying	ε_0	2.603	2.772	2.765	2.531	2.568	2.610	2.430	2.454	2.492
		tan δ	0.0485	0.0509	0.0381	0.0582	0.0567	0.0512	0.0487	0.0582	0.0576
		ε''	0.126	0.141	0.105	0.147	0.146	0.134	0.118	0.143	0.144
Tex 425	Blend of meat and vegetable fats. Especially suitable for deep-fat frying	ε_0	2.681	2.723	2.765	2.596	2.650	2.662	2.491	2.529	2.546
		tan δ	0.0543	0.0525	0.0402	0.0578	0.0575	0.0519	0.0497	0.0583	0.0573
		ε''	0.146	0.143	0.111	0.150	0.152	0.138	0.124	0.147	0.146
Kremit	Hydrogenated all-vegetable shortening. The standard all-purpose vegetable shortening	ε_0	2.683	2.777	2.772	2.530	2.654	2.665	2.420	2.534	2.550
		tan δ	0.0524	0.0505	0.0371	0.0582	0.0576	0.0514	0.0482	0.0578	0.0571
		ε''	0.141	0.140	0.103	0.147	0.153	0.137	0.117	0.146	0.146
Specification	Partially hydrogenated cottonseed oil. Designed for deep-fat frying	ε_0	2.755	2.804	2.793	2.622	2.620	2.628	2.497	2.499	2.515
		tan δ	0.0573	0.0497	0.0391	0.0625	0.0588	0.0522	0.0533	0.0600	0.0587
		ε''	0.157	0.141	0.109	0.164	0.154	0.137	0.133	0.150	0.148
Kremax	Selectively hydrogenated all-vegetable shortening. Designed especially for deep-fat frying.	ε_0	2.693	2.765	2.760	2.550	2.649	2.657	2.466	2.521	2.546
		tan δ	0.0523	0.0512	0.0392	0.0616	0.0557	0.0507	0.0511	0.0568	0.0570
		ε''	0.141	0.142	0.108	0.157	0.148	0.135	0.126	0.143	0.145
Conventionally rendered bacon fat		ε_0	2.753	2.799	2.767	2.615	2.655	2.637	2.498	2.539	2.526
		tan δ	0.0622	0.0533	0.0356	0.0625	0.0606	0.0547	0.0532	0.0597	0.0588
		ε''	0.172	0.149	0.099	0.163	0.161	0.144	0.133	0.152	0.148
Microwave rendered bacon fat		ε_0	2.742	2.796	2.772	2.601	2.655	2.660	2.487	2.536	2.546
		tan δ	0.0576	0.0460	0.0354	0.0624	0.0607	0.0549	0.0507	0.0599	0.0590
		ε''	0.158	0.129	0.098	0.162	0.161	0.143	0.126	0.152	0.150

Source: Reprinted with permission from Dielectric properties of potatoes and potato chips, Pace, W.E., Westphal, W.B., Goldblith, S.A., and Van Dyke, D., *Journal of Food Science*, **33**, pp. 31–33, Copyright 1968, with the permission from Institute of Food Technologists, USA.

14.3 COMPARISON OF MICROWAVE AND CONVENTIONAL HEATING

As has been mentioned earlier, in microwave heating systems, heat is not applied to the food material; on the contrary, heat is made to generate within the body of the food materials by molecular friction caused due to rotation of water dipoles by application of high frequency electric field. And as such, the generation of heat in the material varies depending upon the water content in the material. If a material with uneven moisture distribution is subjected to microwaves, heat generates only in those areas where moisture content is high, and heat does not generate in other areas where there is no moisture at all. Thus, it selectively dries the material. This advantage we do not realize in conventional heating/drying systems where the heat transfer is by conduction or convection or radiation.

For pasteurization or sterilization: Microwave heating methods are preferred to the conventional heating because the former is fast and generates heat uniformly in the material. This helps in a better degree of sterilization. Particularly for solid and semisolid foods, the heat transfer by conventional means depends upon the slow thermal diffusion process, whereas with microwave heating, we realize the benefits of High Temperature Short Time (HTST) processing. Figure 14.2 illustrates the advantage of microwave heating over conventional heating with typical time–temperature histories (Datta and Liu 1992). The figure shows that the temperatures in microwave oven reached 115 °C in less than 3 min. whereas it took more than 65 min. to reach

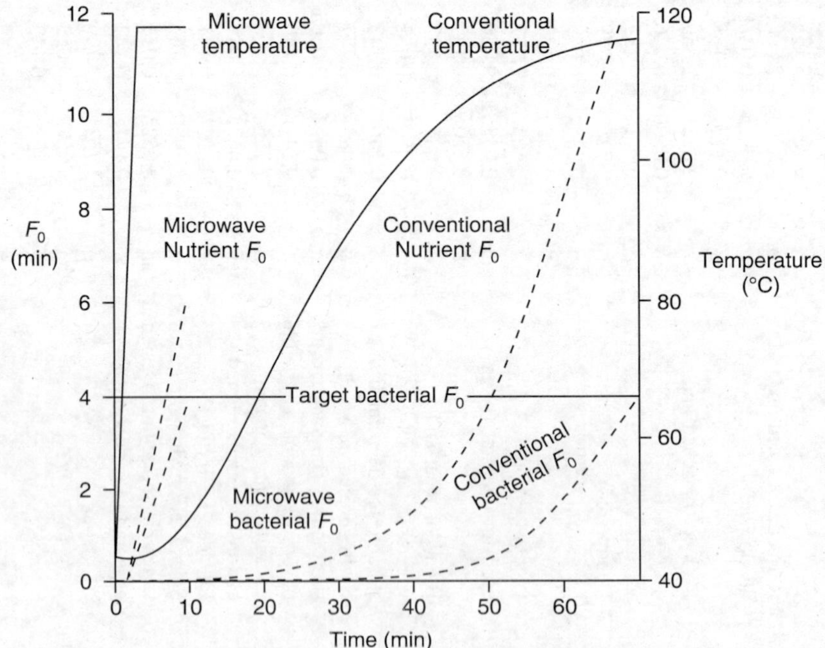

Figure 14.2 Quality parameters for microwave and conventional heating systems—Data on accumulated lethality (F_0) vs. time.

Source: Reprinted with permission from Optimization of quality in microwave heating, Datta, A.K. and Hu, W., *Food Technology*, **46**(3), p. 54, Copyright 1992 with the permission from Institute of Food Technologists, USA.

the same temperature in a conventional heating method. Datta and Liu (1992) have also shown the beneficial effects of microwaves over conventional heating, by comparing a parameter (F_0) with the cumulative volume fraction of the food material (Figure 14.3). The parameter F_0 signifies the microbiological safety of the process based on the time–temperature history at the coldest point. The figure compares the thermal time distribution for the food material to reach approximately the same minimum temperature. The microwave heating shows a smaller spread in distribution in the entire volume of the food material, which obviously means more uniform heating. One important conclusion from the figure is that the microwave heated sample has much lower average thermal time (which is of the order of few seconds) as compared to conventional heating. This will help in much reduced loss in nutrients, etc.

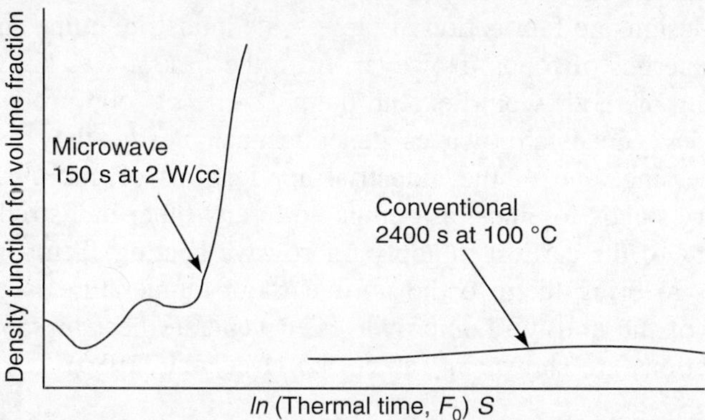

Figure 14.3 Thermal time distribution for microwave and conventional heating systems.

Source: Datta, A.K. and Liu, J. (1992), Thermal time distribution for microwave and conventional heating of food, *Trans. Inst. Chem. Engineers*, **70**(C), pp. 83–90.

14.3.1 Benefits of using Microwave Heating

Here we summarize some of the advantages of using microwave energy for heating/cooking/drying as compared to the conventional heating systems.

- The processing time is very little, i.e., heating is faster. For example, defrosting of meat blocks takes very long times, which is of the order of 3–4 days or so, whereas the same thing can be done within five minutes of time with microwaves. This also helps in saving of labour, and hence, labour costs.
- The running costs of microwave heating are less, because the energy conversion efficiency in microwaves is of the order of 85 per cent, i.e., 85 per cent of electrical energy supplied goes to the food material for heating.
- The size of the equipment is small because of quicker processing efficiency.
- The power control is instantaneous, i.e., the moment the power supply is switched off, heating stops unlike in conventional heating systems.
- The quality of the product is better because of uniform heating. Even though it is debatable about the nutritional safety of foods, by and large, the product appearance and colour are better.
- The rejections are less as the product quality is better.

14.4 APPLICATIONS IN FOOD PROCESSING

Microwaves find innumerable applications in food processing because of their obvious advantages over conventional heating. They are used in varied areas such as melting of vegetable fats and lard, and softening of confectionary products. Puffing and cooking of certain grains to make snack foods can be best done by microwaves. Tempering of the defrosted (frozen) foods is a major application area of microwave heating. Sterilization and pasteurization of meat products and pasta products can also be done by microwave heating. Pasteurization of ready meal packs and bread can be effectively and conveniently done by microwave heating.

Even though microwaves were first introduced as early as in mid-1940s, they were first taken up more by domestic users than by the industry. One of the major reasons for it could be lack of expertise for design and fabrication of large-scale industrial units, and lack of knowledge of the microwave properties of foods. However, from the time it was known in the year 1895, that thermal processing of foods would extend the shelf-life of food products, food technology has taken major strides. Since microwaves generate heat *in situ,* they have become obvious choice for food processing. Out of the industrial applications of microwave heating systems, probably food industry stands foremost as compared to any other industrial sectors. Potato chip making industry is one of the earliest to apply microwave heating. Tempering (thawing) of the frozen meat products to bring them to the normal room temperature is one of the important applications because of the ability of microwaves to generate heat *in situ* in the foods. Some of the typical applications are shown in Table 14.5 (Decareau 1986).

Table 14.5 Applications of Microwaves in Food Processing

Process	*Products*
Tempering	Meat, fish, poultry
Cooking	Bacon, meat patties, sausage, potatoes, sardine, chicken
Drying	Pasta, onions, rice cakes, egg yolk, snack foods, seaweed
Vacuum drying	Orange juice, grains, seeds
Freeze drying	Meat, vegetables, fruits
Pasteurization	Bread, yogurt
Sterilization	Pouch-packed foods
Baking	Bread, donuts
Roasting	Nuts, cocoa beans, coffee beans
Blanching	Corn, potatoes, fruit
Rendering	Lard, tallow

Source: Reprinted with permission from, Microwave food processing equipment throughout the world, Decareau, R.V., *Food Technology*, **40**(6), p. 99, Copyright 1986, with the permission from Institute of Food Technologists, USA.

Drying of food products: Since microwave heating is generated more in places where moisture is high, microwave heating is effective for drying of solids with uneven moisture distribution. Drying takes place only at the evaporating temperature of water, and hence, is effective for food products. Potato chips industry has been using the microwaves for drying of fried chips.

Microwave drying is relatively more effective with those materials like gums which are susceptible to case-hardening. Walde, et al. (1997) reported the effectiveness of microwave drying for gum karaya *(Sterculia urens*). Microwaves are also used for drying of grains like wheat (Walde, et al., 2002) and maize (Velu, et al., 2005) before milling them. However, in the former process structural changes in gluten protein were noticed which made the wheat flour unsuitable for making bakery products. Microwave drying of garlic cloves and grits were reported by Chakkaravarthi and Rao (2002) and found that microwave heating/drying was very suitable for drying of garlic cloves before grinding them to make garlic powder. The energy required for grinding was reported to be less than that in cabinet tray drying based on the Bond's work index values. However, subsequently Prabhakar Rao, et al. (2007) reported that microwave drying had affected the volatile oil composition of essential oil of garlic, which is responsible for most of the medicinal and therapeutic characteristics of garlic.

Vacuum dehydration: Microwaves can also be effectively utilized for vacuum dehydration system. Under vacuum, the boiling point of water decreases and vacuum drying takes place at lower temperatures. For concentration of citrus juices, microwave-vacuum dehydration was first used in France (Decareau 1985). This technique can also be used for vacuum dehydration of grapes for production of grape puffs (McKinney, et al., 1983).

Sterilization and pasteurization: The mechanism of microwave sterilization of foods is not yet fully understood. It could be

- The effect of heat generated that may sterilize the food products by destroying microorganisms.
- The microwaves may be directly acting on the solid microorganisms to destroy them which could be a non-thermal process.

Most of the theoretical studies are in support of the former.

Irrespective of the method how they pasteurize the foods, microwaves are proved to be effective for pasteurization and sterilization. Some commercial establishments are operating to produce 13 million ready to eat meal packs per day by sterilizing the packed foods. Yet another industrial application of microwave heating is in the pasteurization of raw milk.

Puffing of foods: Microwave heating is ideal for making puffed food products because of very high heat transfer rates which are more than the rates of moisture loss. This technique is used in preparation of most of the puffed snack foods from cereals/pulse grains like rice, jowar or Bengal gram. This avoids the usage of sand roasting, etc. which is likely to contaminate the product.

Thawing-tempering of frozen foods: This could be one of the potential areas for application of microwaves. Their ability to generate heat inside the food materials (and hence, the heat transfer is from inside the food material to the surface) will considerably bring down the thawing time to an hour or so which otherwise would take a few days. For example, barrels of frozen fruits take almost a week's time to thaw under normal conditions. Mostly, the processing industries rely upon large quantities of frozen foods which are frozen for preservation after harvest during season to use them throughout the year. In the conventional heating where heat transfer is only by conduction from the surface of the food material to inside will not be very effective. Overheating will harden the surface after removing the surface moisture, and hence, heat transfer to inside the surface of the material is not very effective. Continuous thawing of frozen fish is commercially practiced in fish processing industry.

Precooking: Heat-and-serve food items can better be heated by microwave heating than by conventional heating method. Even though it is a domestic application, it is one of the niche areas where microwave heating can alone be effective.

Caramelization of sugars: Microwave heating technique can also be used as a process tool for caramelization of sugars to develop a toffee-like flavour. Havisha *et al.* (2016) has used the technique to make date (*Phoenix dectylifera*) toffees by caramelizing the sugars in the date pulp carefully under certain ideal conditions so that charring does not take place. The process has resulted in making date toffees with good shelf life and nutrients intact.

14.5 MICROWAVE HEATING EQUIPMENT

Microwave heating systems essentially consist of a microwave generator, known as *magnetron* and a *wave guide* to direct the generated microwaves to a position where the food material is placed in the *oven cavity*. The food material is cooked in the oven cavity by the heat generated by microwaves. The quantity of heat generation depends upon

(i) the dielectric property of food, and
(ii) the loss factor.

Thus, microwave heating systems must be developed carefully on a selective basis. Due consideration should be given to mutual interaction between

- the equipment parameters which are related to the processing equipment, and
- the product characteristics which are based on the dielectric properties of food, and loss factor which correlate the electric energy converted to microwave heating energy.

They can be briefly described as follows:

- Equipment parameters, viz., magnetron, wave guide, and oven cavity.
- Product parameters, viz., dielectric properties, thermal properties, and loss factor.

The heating process is governed by:

- internal heat generated by absorption of microwave energy.
- heat transfer by conduction, convection and evaporating of moisture from the surface of the food product.

14.5.1 Magnetron

Thus, as far as the design of equipment is concerned, magnetron is considered as the heart of microwave oven, and is an essential component. It is a cylindrical diode and generates microwaves. It consists of a *central cathode* and *outer anode* (Figure 14.4). The cathode is a metal cylinder and is coated with an electron emitting material which emits (boils off) the electrons on heating. The cathode is centrally located in the magnetron. A ring of resonant cavities around the magnetron forms the anode. It is maintained at large positive potential relative to cathode. An electrostatic field is set up between the cathode and anode. When high voltage is applied, the electrons give up energy to form rapidly oscillating microwave energy. A strong electromagnetic field is created next to the anode and cathode. The orientation of the magnetic field is at right angles to the axis of electrostatic field. This field makes and bends the electrons to circle around (instead of getting diverted to the anode) in the cavity between the anode and cathode in a *high-energy swarm* (Sharma, et al., 1999).

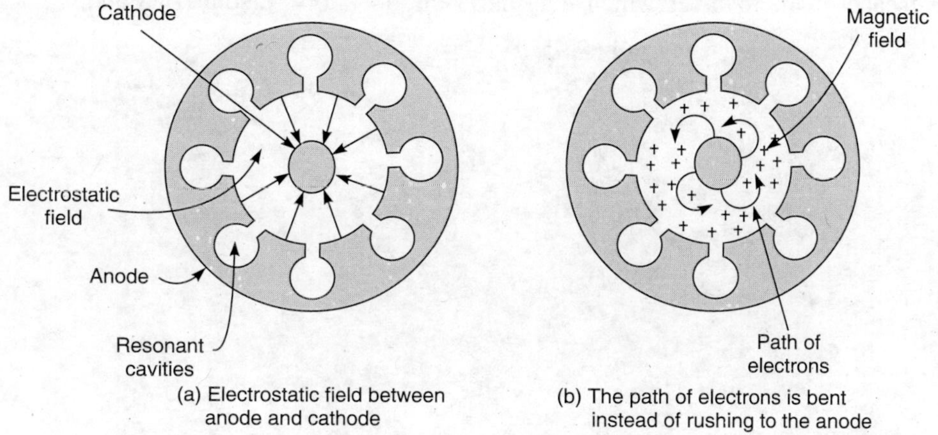

Figure 14.4 Magnetron.

Source: Reprinted from, *Food Process Engineering: Theory and Laboratory Experiments*, Sharma, S.K., Mulvaney, S.J. and Rizvi, S.S.H., p. 251, Copyright (1999), with permission from John Wiley and Sons, Inc.

The wave-guide directs the energy internally to transfer it (energy) to the heating chamber. A rotating fan (or antenna) is used in batch heating oven, or food material itself is rotated on a round table so that all parts of the food material are exposed to the microwaves. This reduces the *shadowing effect*.

In continuous heating ovens, where the food material passes on a conveyor-belt in a tunnel oven, the antenna design is so made that all parts of the food material is exposed to the microwaves. Industrial microwave ovens are operated in the energy range of 30–120 kW. Further details on the microwave ovens are available in any standard textbooks (Copson 1975) on microwave heating. A schematic representation (Figure 14.5) of a general continuous microwave heating system was given by Decareau (1985) for finish drying of potato chips.

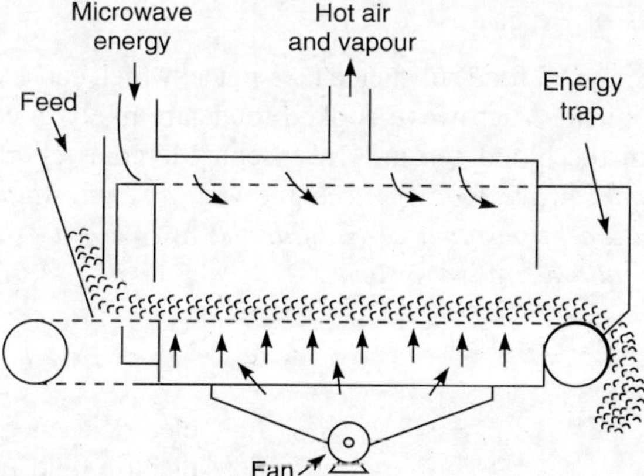

Figure 14.5 Schematic representation of a continuous microwave finish dryer for potato chips.

Source: Reprinted from, *Microwaves in Food Processing Industry*, Decareau, R.V. Dehydration, p. 95, Copyright (1985), with permission from Elsevier.

While fabricating the microwave ovens, care should be taken to see that the heating chambers and tunnels (in case of continuous systems, Figure 14.6) are properly sealed to prevent the leakage/escape of microwaves which may harm the operators/personnel around.

Figure 14.6 A 300 kW MW oven with conveyor for continuous heating and drying applications.

14.6 HAZARDS OF MICROWAVE HEATING

Even though so much has been said about the advantages of microwave heating and its utilization in food processing, there is another strong school of thought which *believes* that microwaves are very bad for health, and hence, should be banned for heating of food materials meant for human consumption. Dr. Joseph Mercola (2005)[†] in his article on "Hidden hazards of microwave cooking" categorizes the hazards into following three major categories:

I. Cancer-causing effects
II. Nutritive destruction of foods
III. Biological effects of exposure

Some alterations in elemental food substance take place, which cause disorders in the digestive system by unstable catabolism. Microwave cooked foods are likely to cause a higher percentage of cancerous cells within the blood serum. Dr. Joseph Mercola reports *In a statistically high percentage of persons, microwave foods caused stomach and intestinal cancerous growths, as well as general degradation of peripherial cellular tissues, with a gradual break down of the function of the digestive and excretive systems.*

Symbols

C_p: specific heat
D_p: penetration depth
E: electrical energy intensity (v/m)
E_0: electric field at surface of food material
F_0: microbiological safety parameter (Figure 14.3)

[†]Dr. Joseph Mercola (http://www.mercola.com/article/microwave/hazards.com)

f: microwave frequency (Hz)
Q: heat generated per unit where (joules/m^3)
T: temperature
t: time
x: distance in food material (m)

Greek Letters

α: thermal diffusivity (m^2/s)
α_e: coefficient in Eq. (14.3)
ε_0: dielectric constant
ε': relative dielectric constant
ε'': loss factor
ρ: density
λ: free space wavelength in Eq. (14.4)
λ_0: free space wavelength
δ: tan δ is loss tangent defined by Eq. (14.6)

REVIEW QUESTIONS

14.1 What are microwaves, and what are their frequency ranges?

14.2 How microwaves generate heat in a food material?

14.3 What is meant by loss factor?

14.4 Describe an equation that relates microwave heat generated to the applied electric field.

14.5 Write the governing energy equation for microwave heating.

14.6 What is meant by penetration depth?

14.7 Compare the microwave heating with conventional heating citing some examples.

14.8 What are the benefits of using microwave heating systems in food processing?

14.9 Describe various applications of microwave heating in food processing.

14.10 How microwaves are effective for pasteurization or sterilization of food materials?

14.11 Describe the functioning of magnatron in microwave heating system.

14.12 Describe a continuous microwave heating system with a neat diagram.

14.13 Describe briefly the hazards of microwave heating when applied in food processing.

REFERENCES

Ayappa, K.G., Davis, H.T., Crapiste, G., Davis, E.A. and Gordon, J. (1991), Microwave heating: An evaluation of power formulations, *Chem. Engg. Sci.*, **46**(4), pp. 1005–1016.

Chakkaravarthi, A. and Rao, D.G. (2002), Microwave drying and grinding characteristics of garlic (*Allium sativum*), *Indian Chemical Engineer*, **44**(3), pp. 180–182.

Copson, D.A. (1975), *Microwave Heating*, AVI, West Port, pp. 262–265.

Datta, A.K. (1990), Heat and mass transfer in the microwave processing of food, *Chem. Engg. Progress*, June, pp. 47–53.

Datta, A.K. and Hu, W. (1992), Quality optimization of dielectric heating process, *Food Technol.*, **46**(3), pp. 53–56.

Datta, A.K. and Liu, J. (1992), Thermal time distribution for microwave and conventional heating of food, *Trans. Inst. Chem. Engineers*, **70**(C), pp. 83–90.

Decareau, R.V. (1985), *Microwaves in the Food Processing Industry*, Academic Press, Inc., Orlando.

Decareau, R.V. (1986), Microwave food processing equipment throughout the world, *Food Technol.*, **40**(6), pp. 99–105.

Gerling, J.E. (1986), Microwaves in the food industry promise and reality, *Food Technol.*, **40**(6), pp. 82–83.

Havisha, G., Lavanya, E., Asifullah, A.M., Meyyappan, N. and Rao, D.G. (2016), A process for making fruit toffees by microwave heating, *Indian Patent No.* 350323.

McKinney, H.F., Wear, F.C., Sandy, H.L., Petrucci, V.E. and Clary, C.D. (1983), Process for making hollow dried grapes, *US Pat.*, 4418, 083 (Nov. 23).

Mudgett, R.E. (1982), Electrical properties of foods in microwave processing, *Food Technol.*, **36**(2), pp. 109–126.

Mudgett, R.E. (1986), Microwave properties and heating characteristic of foods, *Food Technol.*, **40**(6), pp. 84–93.

Pace, W.E., Westphal, W.B., Goldblith, S.A. and Vau Dyke, D. (1968), Dielectric properties of potato and potato chips, *J. Food Sci.*, **33**, pp. 31–33.

Prabhakar Rao, P.G., Nagender, A., Jagan Mohan Rao, L. and Rao, D.G. (2007), Studies on the effects of microwave drying and cabinet tray drying on the chemical composition of volatile oils of garlic powders, *European Food Research and Technology*, **224**, pp. 791–795.

Schiffmann, R.F. (1986), Food product development for microwave processing, *Food Technol.*, **40**(6), pp. 94–98.

Sharma, S.K., Mulvaney, S.T. and Rizvii, S.S.H. (1999), *Food Process Engg: Theory and Lab. Experiments*, Willey Interscience, New York, p. 251.

Velu, V., Nagender, A., Prabhakar Rao, P.G. and Rao, D.G. (2005), Dry milling characteristics of microwave dried maize grains (Zea Mays. L), *J. Food Engg.*, **74**, pp. 30–36.

Walde, S.G., Balaswamy, K., Shivaswamy, R., Chakkaravarthi, A. and Rao, D.G. (1997), Microwave drying and grinding characteristics of gum karaya (*Sterculia urens*), *J. Food Engg.*, **31**, pp. 305–313.

Walde, S.G., Balaswamy, K., Velu, V. and Rao, D.G. (2002), Microwave drying and grinding characteristic of wheat (*Triticum aestivum*), *J. Food Engg.*, **55**, pp. 271–276.

CHAPTER

15

Evaporation

Evaporation is a frequently used unit operation in food processing; sometimes, it is used as a processing tool for concentration, whereas at other times, it is used as a preprocessing step for concentration of liquids before they are subjected to other processing operations like dehydration or crystallization. Essentially it is classified as a concentration step, and hence, it is sometimes known as **concentration**. In fact, the equipment used for evaporation of solvent to concentrate the liquor is known as **concentrator**.

Now, let us see what is evaporation. It is a step used for removal of solvent from a liquor (solution) by application of heat by virtue of which the solvent is evaporated into vapour by phase-change which is later condensed. In that respect it is different from other concentration operations like membrane separation processes in which the phase change of the solvent does not take place, and hence, is not discussed here. If the solids are not in a state of solution, but are suspended in the solvent, the solids are removed by physical separation methods like filtrations or centrifugation, and it will be discussed separately in respective chapters (Chapter 20 and 21). The situation is different if the solute is in a dissolved form as a solution. In majority of operations, the concentration of the liquor is desired, and hence, the solvent after evaporation is not desired as in the case of concentration of fruit juices to reduce their bulk for storage and transportation. In some instances like water purification, water admixed with minerals and solvents is evaporated. And the vapours are condensed as pure water, and the concentrated liquor is rejected. The condensed water is used for human consumption, or for process operations like feed water, etc. It is known as **water distillation** because the vapours (distillate) are condensed just as what we do in case of distillation. However, it is different from distillation. In evaporation even if the vapours are a mixture more than one component no efforts are made to condense the vapours. Moreover, distillation is an operation in which both mass transfer and heat transfer are considered; and hence, distillation is considered as a process of *simultaneous heat and mass transfer*[†].

Evaporation is also different from drying since in the letter, the final product is a solid, whereas in evaporation, the final product is a concentrated solution or a slurry of solids.

[†]Further details on distillation will be dealt in Chapter 18.

Drying also involves simultaneous heat and mass transfer like distillation. It is also different from crystallization in which the interest is in concentrating a solution by removal of solvent (usually water) until a stage comes where the solids (solute) start crystallizing out as the relative solubility is affected by concentrating the solution.

Thus, evaporation is unique for concentrating the dissolved solute in the solution by vaporization. Hence, evaporation is considered purely as a heat transfer operation; and thus, is always dealt in any textbook on unit operations as a part of heat transfer subject.

15.1 SINGLE EFFECT EVAPORATOR

The whole of evaporation process can be fragmented into two categories: various components, and operational aspects.

The evaporator operates on the principles of energy balance and material balance.

15.1.1 Various Components of a Single Effect Evaporator

For easy comprehension of the principles, let us start with a single effect evaporator† shown in Figure 15.1. It essentially consists of:

(i) a heating source,
(ii) a vapour chamber, and
(iii) a condenser

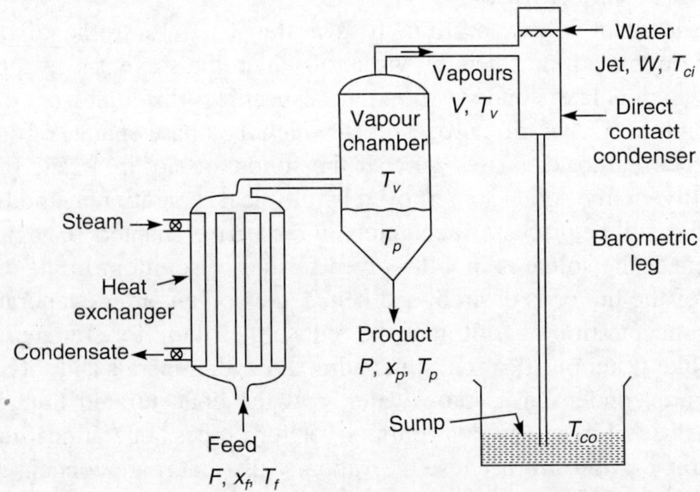

Figure 15.1 Schematic representation of a single effect evaporator.

The feed (which is usually a dilute solution) is fed into heat exchanger where the solution is heated beyond the boiling point of the solvent. The heat exchanger is a typical shell and tube heat exchanger where the solution passes through the tubes, and steam passes through

†This is as contrast to a multiple effect evaporators which will be dealt later.

the shell side. Various types of heat exchanger tubes are in vogue, like short tubes; long tubes; horizontal tubes, etc.

The solvent evaporates in the heat exchanger. The vapour liquid mixture enters the vapour chamber and gets separated. The vapour, being light, goes upwards and enters a condenser. The concentrated liquor leaves from the bottom of the vapour chamber. If the concentration is not adequate, a part of the liquor is recycled and added to the feed (which is not shown in Figure 15.1 deliberately to avoid confusion during material balance).

The vapours that enter the condenser need to be condensed. Generally in case of evaporation, the condensate is not the desired product. Hence, these vapours are directly mixed with water, since the vapours are usually water vapours. In this case, the condensation is very fast because of the direct contact of the vapours with liquid. The barometric head is to be properly maintained so that the condensate, along with the cooling waters, falls into the sump. In many cases, the liquid in the sump is cooled and recycled to the shower to dispense the liquid in the form of fine droplets in the condenser.

If the condensate is the desired product and cannot be mixed with the solvent/water, we need to use a separate shell and a tube type of condenser in which the vapours are condensed by indirect contact with cold liquid. Relatively it is difficult to condense the vapours, more so if the vapour is a mixture of condensable and non-condensable gases/vapours. Hence, we prefer to use a direct contact condenser in which the condensed vapours are mixed with the cooling fluid.

15.1.2 Operational Features

The operational features of an evaporator need a special mention because certain operational difficulties arise in case of evaporation due to changes in liquid characteristics during concentration. They are as follows:

Boiling Point Rise (BPR): The BPR of the solution during concentration is an uninevitable nuisense with any evaporation process. As the solids' concentration increases, the boiling point of the solution also increases due to colligative properties (the solute admixed with the solvent will elevate the boiling point of the solution or depress the freezing point of the solution). Hence, the temperature difference (ΔT) of the heating medium in the shell side of the heat exchanger and the fluid in the pipes of the heat exchanger, will decrease. This results in reduced heat transfer rates, since

$$q = UA\,(\Delta T) \tag{11.18}$$

If ΔT decreases, q also decreases, and thus, affects the heat transfer rate.

Change (increase) in viscosity: This is caused due to the removal of the solvent. The increased viscosity affects the flowability of the fluid in the pipes and in the vapour chamber. It also reduces the heat transfer rates. Sometimes the increase in viscosity is so high, as in the preparation of condensed milk, etc., it virtually becomes impossible to pass the concentrated liquor.

Rheological characteristics: They also change as the solids' concentration builds up. The fluid behaviour changes from Newtonian to non-Newtonian characteristics which affects both the heat transfer and momentum transfer.

Foaming: Most of the food materials contain proteinaceous materials which are susceptible to foaming. The foaming affects the flowability of the fluids, chokes the pipes and tubes, and hinders the separation of vapour and liquid. Hence, methods need to be devised to reduce the foaming. One of the ways of doing it is to add some kind of silicon materials or other materials which affect the surface characteristics and reduce the foaming. Addition of a foreign substance is not always desirable in food processing.

Scale formation: Some materials are susceptible to formation of scale at high temperatures during flow through pipes and conduits. The scale formation affects the heat transfer coefficients, and hence, is not desired. If the scale formation is severe, the process needs to be stopped once in a while, and the tubes need to be cleaned either mechanically or by using some solvents or acids. This affects the production schedule.

Sensitivity of material to temperatures: Most of the food materials are sensitive to heat, and hence, cannot be heated over longer periods or subjected to high temperatures. But they should be exposed to high temperatures to realize higher heat transfer rates. Most of the food materials, because of their thermal sensitive nature, cannot be exposed to high temperatures to remove the solvent by evaporation. For example, during concentration of fruit juices, the aroma and flavour components are lost. Hence, evaporation is generally carried out under vacuum which reduces the boiling point of the solvent. Of course, applying vacuum has other advantages also!

15.2 PRINCIPLES OF EVAPORATION

The basic principles of evaporation involve mass balance and energy balance. It is also necessary to be able to estimate the boiling point rise of the thick liquor as concentration proceeds.

15.2.1 Mass Balance

Mass balance is made on the mass flow rate of the inlet and outlet fluids which yields,

$$F = P + V \tag{15.1}$$

where F is the feed rate, kg/s; P is thick liquor (product) flow rate in the product stream, kg/s; and V is the vapour flow rate to the condenser, kg/s.

A mass balance on the total solids yields,

$$F(X_f) = P(X_p) \tag{15.2}$$

15.2.2 Heat Balance

Similarly, *energy balance* is also made on the fluid streams which consists of:

(i) enthalpy required to raise the temperature of feed from T_f to T_p.

(ii) latent heat of vaporization to vaporize the solvent vapours (V kg/s).

(iii) heat to be removed from the vapours in the condenser to condense the vapours, and also enthalpy to be removed from the condensed liquid from the boiling point to the desired temperature level.

Heat exchanger load: Heat to be supplied in the heat exchanger to the feed is:

$$F \times C_{pf} \times (T_p - T_f) + V \times \lambda_v \tag{15.3}$$

where T_f is initial temperature of feed, T_p is the temperature of the product and λ_v is latent heat of vaporization of the vapours. This heat is to be supplied by condensing the steam in the steam chest, and a part of its enthalpy. However, generally we neglect the enthalpy component of the condensed steam.

Heat supplied by condensing steam $= \dot{m}_s \lambda_s$ (15.4)

Therefore, $$\dot{m}_s \lambda_s = F C_{pf} (T_p - T_f) + V \lambda_v \tag{15.5}$$

where $\dot{m}_s$ is mass flow rate of the steam in kg/s, and λ_s is the latent heat of vaporization associated with the steam in kJ/kg.

Condenser load: The heat carried by the vapours is to be removed in the condenser by the condenser liquid.

Heat contained in the vapours at temperature $T_v = V\lambda_v$ (15.6)

Enthalpy of the condensing liquid $= VC_{pv} (T_v - T_{co})$ (15.7)

Heat to be supplied by the condenser liquid is:

$$WC_{Pc} (T_{ci} - T_{co}) = V\lambda_v + VC_{Pv} (T_v - T_{co}) \tag{15.8}$$

where,

W is the condenser liquid (usually water) flow rate in kg/s,

C_{Pc} is the specific heat of condenser liquid,

T_{ci} is condenser liquid inlet temperature,

T_{co} is condenser liquid outlet temperature,

C_{Pv} is specific heat of vapours in kJ/kg°C, and

T_v is temperature of the vapour.

15.2.3 Boiling Point Rise (BPR)

As has been mentioned, boiling point elevation is a common phenomenon in evaporation operations. There are some empirical expressions available for calculating boiling point rise (Toledo, 1997).

$$\Delta T_b = 0.51m \tag{15.9}$$

where ΔT_b is BPR in °C which is above the boiling point of pure water at the given pressure, m is the molality of the solution. The above equation holds good for most food products which contain organic compounds as the soluble solids in the water solvents.

In some cases, the boiling point rise is not only due to colligative properties, it could be also due to increase in pressure at the bottom of the product liquor which is due to the hydrostatic head of the column and is given by

$$p = \rho g h \tag{15.10}$$

where

p is the pressure exerted by the column of liquid in P_a;
ρ is the density of the liquor, kg/m^3; and
h is the hydrostatic column height of the liquor m.

In many instances, the contribution of p is negligible.

One of the important rules that helps us find out the boiling point elevation is *Duhring's rule* which states that *the ratio of the temperatures at which two solutions exert the same vapour pressure is constant.* The Duhring's rule can also be stated in a more straight forward manner as *the boiling point of a given solution is a linear function of the boiling point of pure water at the same pressure.* Figure 15.2 shows Duhring's plots for sodium chloride solutions of different concentrations. In the literature, we may find Duhring's plots for different aqueous solutions of inorganic compounds and salts.

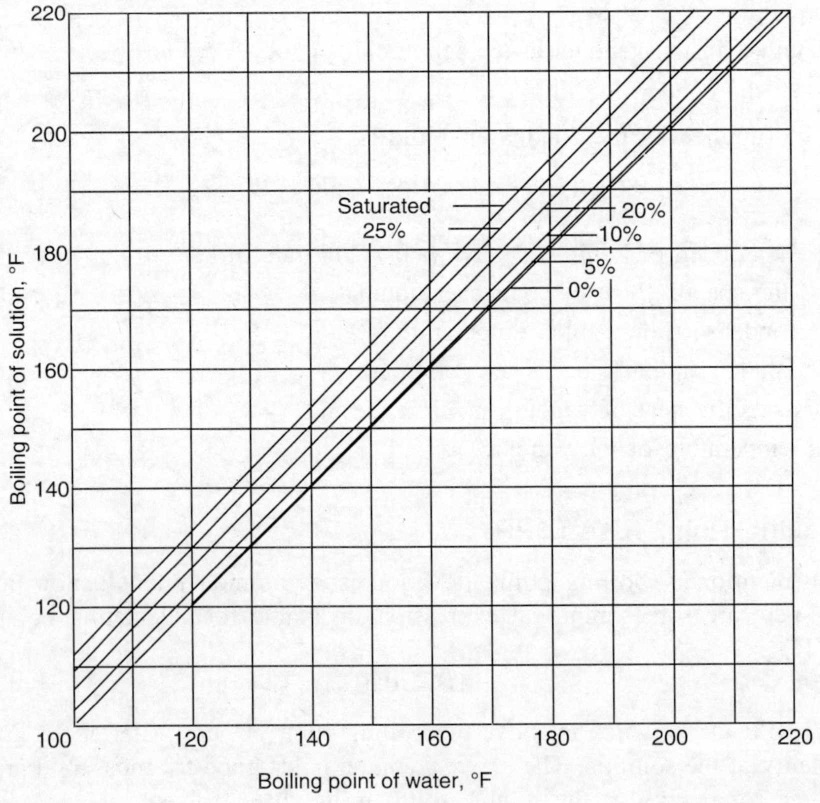

Figure 15.2 Duhring's plot showing boiling point elevation for sodium chloride solutions.

PROBLEM 15.1 Find the boiling point and BPR of a 20 per cent sodium chloride solution under a pressure of 70 kPa using the Figure 15.2, compare the value with that calculated by Eq. (15.9).

Solution From the steam tables, the boiling point of pure water at 70 kPa is 90 °C. From Figure 15.2, the boiling point of 20 per cent, NaCl solution corresponding to 90 °C is 96 °C.

i.e., the boiling point of 20 per cent NaCl solution under 70 kPa pressure 96 °C.

Therefore, $$\text{BPR} = 96 - 90 = 6\ °\text{C}$$

$$\Delta T_b = 6\ °\text{C}$$

Let us calculate ΔT_b by Eq. (15.9)

$$\Delta T_b = 0.51m$$

where m is molality, i.e., moles of solute in 1000 gm of water. Molecular weight of NaCl is 58.5.

Therefore, $$m = \frac{20/58.5}{80/1000} = 4.27$$

and $$\Delta T_b = 0.51 \times m = 0.51 \times 4.27 = 2.18\ °\text{C}$$

The value is considerably different.

Note: Equation (15.9) is applicable if the solute is an organic compound like sucrose, etc. Hence, the difference in the calculated value.

PROBLEM 15.2 A solution with an initial solid concentration of 10 per cent is being concentrated in a single effect evaporator to a final solid concentration of 40 per cent under a vacuum of 40 kPa. The boiling point rise of the solution is negligible. Steam at a pressure of 101 kPa (gauge) is used to concentrate the liquor. The feed is entering at a temperature of 35 °C and the temperature of the final liquor is 86 °C corresponding to a vacuum of 40 kPa prevailing in the evaporator. If the feed rate is 1000 kg/h, find the quantity of steam required and the heat transfer area of the evaporator. The specific heat of the feed which is reasonably constant is 5 kJ/kg°C, and the overall heat transfer coefficient is 1.9 kW/m^2°C.

Solution Material balance (Refer to Figure 15.1)

$$F = 1000\ \text{kg/h}$$

$$X_f = 10\ \text{per cent} = 0.1$$

$$X_p = 40\ \text{per cent} = 0.4$$

Basis: 1 hour

Applying Eqs. (15.1) and (15.2)

$$1000 \times 0.1 = P \times 0.4$$

and $$1000 = P + V$$

Therefore, $$P = \frac{1000 \times 0.1}{0.4} = 250\ \text{kg}$$

and $$V = 1000 - 250 = 750\ \text{kg}$$

Water to be removed from the solution for concentrating the liquor from 10 per cent to 40 per cent is 750 kg/hr.

Heat balance: Heat to be supplied to the feed (Q)

= Heat to be supplied to rise the temperature of feed from 35 °C to 86 °C

+ latent heat of vaporization at 86 °C for vaporizing V kg of water from the feed.

$= mC_p\,(\Delta T) + V\lambda_v$

$= 1000\ (5)\ (86 - 35) + 750 \times \lambda_v$

λ_v is noted from steam tables at 86 °C = (2653.6 – 359.93)

$= 2293.67 \cong 2294$ kJ/kg

$$Q = (1000 \times 5 \times 51) + (750 \times 2294) = 1.975 \times 10^6 \text{ kJ}$$

Steam to be supplied = $m_s \times 2201.6$

Heat transferred by steam = $m_s\lambda_s$ kJ

λ_s at 101 kPa (gauge) = (2706.3 – 504.7) = 2201.6 kJ/kg

(at 2 Atm. absolute pr. from Appendix 1B)

$$Q = 1.975 \times 10^6 = m_s \times 2201.6$$

Therefore, $m_s = 897$ kg

$$\text{Steam economy} = \frac{750}{897} = 0.836 \text{ kg of water/kg of steam.}$$

Heat transfer area:

$$q = \text{Heat transfer rate} = \frac{Q}{3600} = \frac{1.975\times10^6}{3600} = 548.6 \text{ kJ/s} = 548.6 \text{ kW}$$

$$q = UA(\Delta T)$$

where

U is the overall heat transfer coefficient which is equal to 1.9 kW/m^2°C;

A is the heat transfer area, m^2;

ΔT is the temperature difference (T_s – 86); and

T_s is steam temperature corresponding 101 kPa (guage) = 120.23 °C

Therefore, $548.6 = 1.9 \times A \times (120.23 - 86)$

$$A = \frac{548.6}{1.9\times34.23} = 8.44 \text{ m}^2$$

Note: The above problem is a oversimplification. When the solution is concentrated upto 40 per cent, there will be definitely BPR. This will affect the (ΔT) in heat transfer rate, and hence, will slightly increase A.

PROBLEM 15.3 A sugar solution is concentrated in a single effect evaporator from an initial concentration of 10 per cent to a final concentration of 35 per cent solids. The height of the sugar solution in the evaporator is 0.2 m. The vacuum in the evaporator is 30 kPa. Feed enters the evaporator at 50 °C and the density of feed is 990 kg/m^3. Steam is supplied to the steam chest at the rate of 900 kg/hr at a temperature of 120 °C. The specific heat of the feed is 4.2 kJ/kg°C. What would be the feed rate?

Solution
Approach:

(i) Since the feed rate is not known, we start with heat balance.
(ii) Initially we try to find out the BPR to establish ΔT.

The molecular weight of sucrose is 180.

$$\text{Molality of the solution} = \frac{\text{Moles of solute}}{\text{1000 gm of solvent}} = \frac{35/180}{65/1000} = 3.0$$

$$= m$$

By Eq. (15.9)

$$\Delta T_b = 0.51 \times 3.0 = 1.53$$

We find out the total pressure on the liquor = Pressure exerted by the vapours + Hydrostatic head of the solution [which is given by Eq. (15.10)].

$$p = \rho gh = 990 \times 9.81 \times 0.2 = 1942.4 \text{ Pa} = 1.94 \text{ kPa}$$

Pressure in the system exerted by vapours = 101.3 – 30 = 71.3 kPa
Therefore, total pressure on the liquor = 1.94 + 71.3 = 73.24 kPa = 0.723 bars.
Boiling point corresponding to 73.24 kPa is noted from stream tables for the value of 0.7 bars = 90 °C
(Otherwise exact value is to be noted by interpolation method)

Therefore, liquor temp = 90 + 1.53 = 91.53 °C

The steam temperature in the steam chest = 120 °C

Therefore, $\Delta T = 120 - 91.53 = 28.47$ °C

Now we shall make heat balance:

$$\lambda_s \text{ at } 120\ °\text{C} = 2706.3 - 504.7 = 2201.6 \text{ kJ/kg}$$

Heat supplied in one hour = 2201.6 × 900 = 1.98 × 10^6 kJ/hr.
Heat required by feed at inflow rate of F kg/hr and the vapour leaving @ V kg/hr is

$$(F \times 4.2 \times 28.47) + V\,(2660.1 - 376.8) = 119.6(F) + 2283.3(V)$$

$$1.98 \times 10^6 = 119.6(F) + 2283.3(V) \qquad (1)$$

From the mass balance, we have

$$F = P + V$$

$$F(X_f) = P(X_p)$$

i.e.,

$$0.1(F) = 0.35(P)$$

or

$$P = \frac{0.1}{0.35}(F) = 0.286(F)$$

$$F = 0.286(F) + V$$

or

$$0.714F = V$$

Substituting the value of V in Eq. (1), we get

$$1.98 \times 10^6 = 119.6(F) + 2283.3\ (0.714)(F) = 1750(F)$$

Therefore,
$$F = \frac{1.98\times10^6}{1750} = 1131.3 \cong 1131 \text{ kg/h}$$

$$V = 0.714 \times 1131 = 808 \text{ kg/h}$$

$$\text{Steam efficiency} = \frac{808}{900} = 0.9 \text{ kg of water evaporated/kg of steam.}$$

15.3 MULTIPLE EFFECT EVAPORATORS

We have seen upto now the vapour (V kg) generated in the evaporator is going to condenser for condensation, in which case the heat associated with it (which is equal to $V\lambda_s$) is going as a waste into condenser. It would be a wonderful idea if the vapours from the evaporator are passed into the steam chest of a subsequent evaporator so that the heat associated with it could be effectively utilized. Subsequently, the vapour generated in the second evaporator could be passed into third evaporator and so on. This concept is known as **multiple effect evaporation**. This will obviously raise the steam economy considerably, which we measure as the kg (quantity) of water evaporated per kg of steam.

Obviously the vapours from the first evaporator can be effectively used in the steam chest of second evaporator if and only if the temperature of the liquid in the second evaporator is less than the vapour temperature of the first evaporator. This is possible only if the pressure in the second evaporator is less than that in the first evaporator. Generally we employ slight vacuum in the first evaporator to bring down the boiling point of the liquid so that the heat sensitive materials are not destroyed during evaporation. So it becomes obligatory for us to employ still more vacuum in the second evaporator and in subsequent evaporators. The reduced pressure in the subsequent effect evaporators also facilitate the transfer of liquid into subsequent sections by vacuum without employing external pumps in between the evaporators. The concept of multiple effect evaporators is schematically shown in Figure 15.3.

The temperature in each effect is T_1, T_2, and T_3, whereas the pressure in each is p_1, p_2 and p_3. Obviously $T_1 > T_2 > T_3$, and hence, $p_1 > p_2 > p_3$. Since $p_1 > p_2$, the liquid in evaporator 1 flows into evaporator 2 by suction and so on. Generally, the heat transfer area of each effect evaporator is taken as equal; i.e., the construction is such that all evaporators have equal surface area. This will economise the fabrication costs of evaporators.

Therefore,
$$A_1 = A_2 = A_3 \qquad (15.11)$$

Neglecting the enthalpy required to raise the temperature of the feed, almost all the latent heat of the steam in evaporator 1 is used to heat liquid in 1. The same liquor goes into the second evaporator, and is vaporized by the vapours coming from evaporator 1. Same is the case for evaporator 3, and so on. Hence,

$$q_1 = q_2 = q_3 = q \qquad (15.12)$$

i.e., the heat transfer rate in each evaporator is the same.

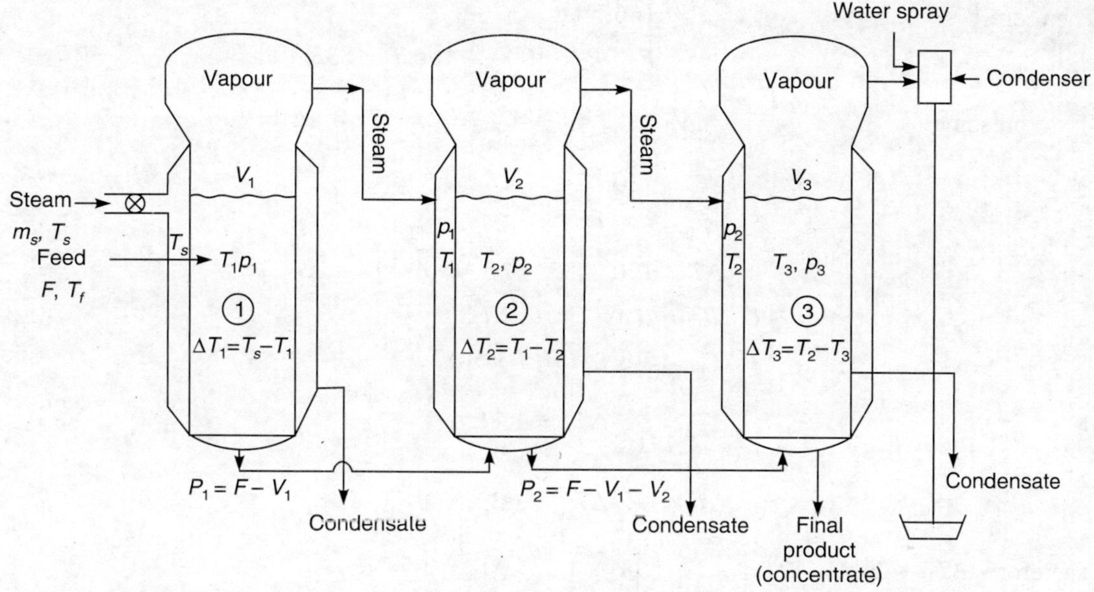

Figure 15.3 Schematic representation of a triple effect evaporation (Forward feed).

$$q_1 = U_1A_1\ (\Delta T_1) = U_1A_1\ (T_s - T_1)$$

$$q_2 = U_2A_2\ (\Delta T_2) = U_2A_2\ (T_1 - T_2)$$

$$q_3 = U_3A_3\ (\Delta T_3) = U_3A_3\ (T_2 - T_3)$$

Applying Eqs. (15.11) and (15.12)

$$U_1(\Delta T_1) = U_2(\Delta T_2) = U_3(T_3) = \frac{q}{A} \tag{15.13}$$

and

$$\Delta T = \Delta T_1 + \Delta T_2 + \Delta T_3 = (T_s - T_3) \tag{15.14}$$

The design of a multiple effect evaporator is explained with Problem 15.4:

PROBLEM 15.4 A triple effect evaporator is used to concentrate a fruit juice at the rate of 1000 kg/h flow rate with 5 per cent solids to 30 per cent solids. The temperature of the steam in the first effect evaporator is 120 °C and the boiling point in the last affect in 55 °C. The heat transfer coefficient of each evaporator is 2, 1.8 and 1.5 kW/m^2°C. Assuming that there is no boiling point elevation and the specific heat of feed is 4.5 kJ/kg°C in the entire range of temperature, find the steam consumption, ΔT in each evaporator and the heat transfer area of any one evaporator, since all evaporators are identical.

Solution Basis: 1 hour

$$\text{Mass balance } F = P + V$$

$$F(X_f) = P(X_p)$$

i.e.,

$$1000 \times 0.05 = P \times 0.3$$

Therefore, $P = 166.7$ kg

$$V = F - P = 1000 - 166.7 = 833.3 \text{ kg}$$

If we presume $V_1 = V_2 = V_3 = \frac{V}{3} = \frac{833.3}{3} = 277.8$ kg

$$T_s = 120\ °C$$

$$T_3 = 55\ °C$$

$$\Delta T = 120 - 55 = 65\ °C = \Delta T_1 + \Delta T_2 + \Delta T_3$$

$$U_1\Delta T_1 = U_2\Delta T_2 = U_3\Delta T_3$$

$$2(\Delta T_1) = 1.8\ (\Delta T_2) = 1.5(\Delta T_3)$$

$$\Delta T_2 = \frac{2}{1.8}\Delta T_1 = 1.11\ \Delta T_1$$

$$\Delta T_3 = \frac{2}{1.5}\Delta T_1 = 1.33\ \Delta T_1$$

Therefore, $\Delta T_1 + \Delta T_2 + \Delta T_3 = (1 + 1.11 + 1.33)\ \Delta T_1 = 65\ °C$

$$\Delta T_1 = 18.87\ °C$$

$$\Delta T_2 = 1.11 \times 18.89 = 20.95\ °C$$

$$\Delta T_3 = 65 - 18.87 - 2.95 = 25.18\ °C$$

$$T_1 = T_s - \Delta T_1 = 120 - 18.87 = 101.13\ °C$$

$$T_2 = T_1 - \Delta T_2 = 101.13 - 20.95 = 80.18\ °C$$

$$T_3 = T_2 - \Delta T_3 = 80.18 - 25.18 = 55\ °C$$

$$q = mC_p\ (\Delta T) = 1000 \times 4.5 \times 65 = 2.92 \times 10^5 \text{ kJ/h}$$

$$= 81.25 \text{ kW} = U_1A_1\Delta T_1$$

$$= 2.0 \times A_1 \times 18.87$$

Therefore $A_1 = \frac{81.25}{2 \times 18.87} = 2.153 \text{ m}^2$

Let us see whether A_2 and A_3 are equal to A_1

$$A_2 = \frac{81{:}25}{1.8 \times 20.95} = 2.154 \text{ m}^2$$

$$A_3 = \frac{81.25}{1.5 \times 25.18} = 2.15 \text{ m}^2$$

We also find A by finding total heat transfer coefficient

$$\frac{1}{U} = \frac{1}{U_1} + \frac{1}{U_2} + \frac{1}{U_3} = \frac{1}{2.0} + \frac{1}{1.8} + \frac{1}{1.5} = 1.72$$

Therefore, $$U = \frac{1}{1.72} = 0.58 \text{ kW/m}^2\text{°C}$$

$$q = UA\Delta T = 0.58 \times A \times 65$$

Therefore, $$A = \frac{81.25}{0.58 \times 65} = 2.152 \text{ m}^2$$

p_1 = Pressure in first evaporator corresponding to a temperature of 101.13 °C

$$= 10.33 + \frac{120.8 - 101.33}{(105 - 100)} \times (101.13 - 100)$$

= 105.73 kPa

Similarly, p_2 = Pressure in the second evaporator corresponding to a temperature of 80.18 °C
= 47.36 kPa
= 101.3 – 47.36 = 53.94 kPa vacuum

p_3 = Pressure in the third evaporator corresponding to a temperature of 55 °C
= 15.74 kPa
= 101.3 – 15.74 = 85.56 kPa vacuum

15.3.1 Types of Feeding Methods

What we have seen in Figure 15.3 is a **forward feed** arrangement in a multiple effect evaporation systems, in which the feed enters into the first effect evaporator, flows to successive evaporators and leaves as a thick concentrated liquor from the last evaporator. The arrangement is shown in Figure 15.4(a). It requires a pump to feed the feed into the first evaporator, and a pump to remove the concentrated liquor from the last evaporator after concentration. The flow of liquor into subsequent evaporators is by means of vacuum prevailing in the subsequent evaporators, i.e., the flow is in the direction of decreasing pressure, and hence, does not require any pumping system. This type of feeding is simple and easy to operate. Only draw back with this system is that we will be withdrawing the concentrated liquor, may be even sometimes with crystallized salts under low temperature which prevails in the last evaporator.

Another type of feeding mechanism which is opposite to the forward feed is **backward feed** arrangement, in which the feed enters from the last evaporator and leaves through the first evaporator as concentrated liquor after concentration [Figure 15.4(b)]. In this system of feeding, the concentrated liquor will be flowing opposite to the direction in which the pressure is decreasing. Hence, we need to pump the liquor from one evaporator to the other, whereas in forward feed evaporation we take advantage of the reducing pressures as we go forward. However, one of the best advantages with the backward feed is that the liquor which is getting concentrated will be evaporating and will be discharged from the first evaporator in which the temperature is high. High temperature helps easy concentration and discharge. The lean feed can also be easily evaporated in the last effect.

After having seen some of the advantages and disadvantages of both types of feeding arrangements, a third arrangement which is a combination of both forward feed and backward

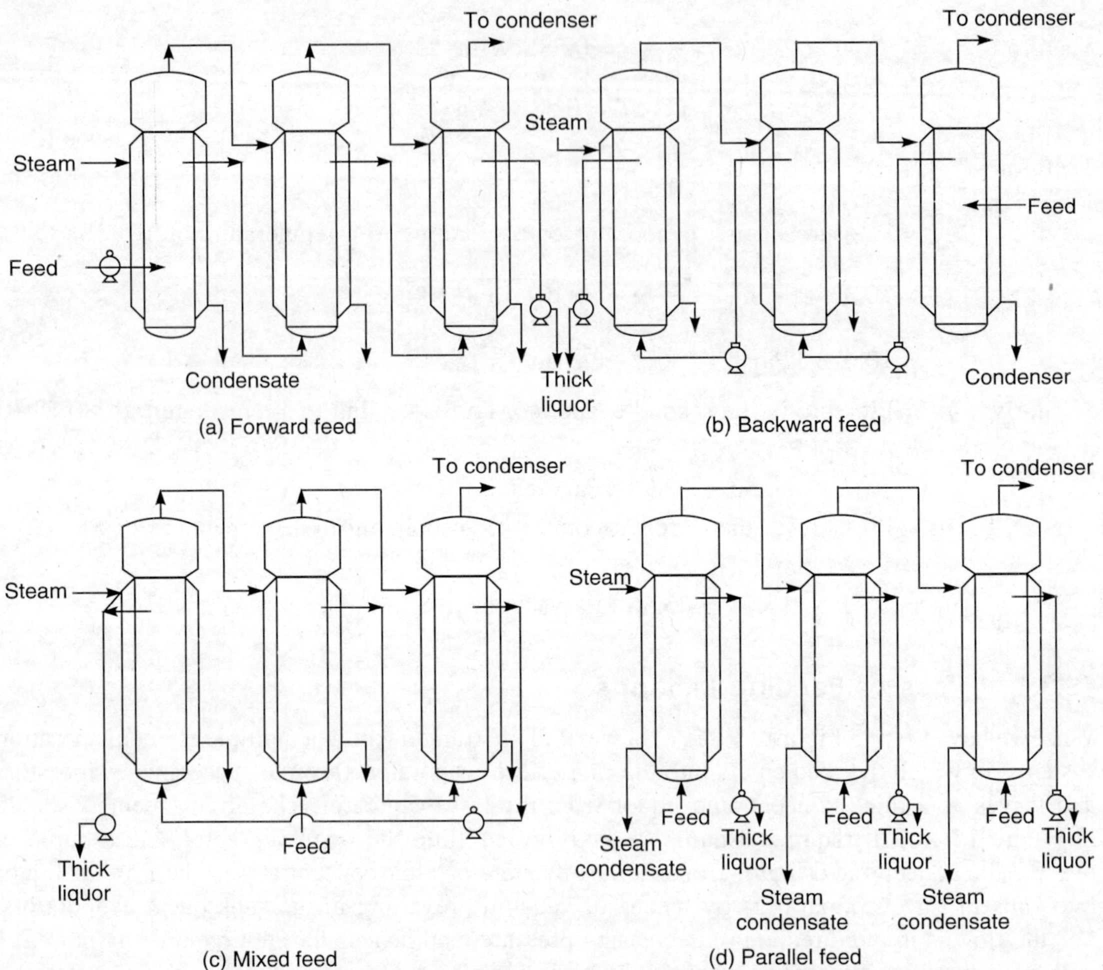

Figure 15.4 Feeding arrangement in multiple effect evaporator.

feed is contemplated; and this is known as **mixed feed** arrangement [Figure 15.4(c)]. In this method, the feed enters somewhere in between and the liquor moves forward without the help of any pumps. Thus, we realize the advantages of a forward feed partially. Even feeding can be done without a pump because vacuum prevails in the evaporator into which feeding is done. The concentration of the final liquor is done in the first evaporator in which the temperature is high; the advantage which we get in backward feed. Thus, mixed feed arrangement is a combination of both forward and backward feeding methods.

Another method of feeding is **parallel feed** [Figure 15.4(d)], in which feeding is done into every evaporator, and the concentrated liquor is also discharged from every evaporator. The vapour from one evaporator will alone go into the heating chest of the subsequent evaporator to be used as the heating source. In this method, we realize high capacities and high steam efficiencies. This kind of feeding arrangement is used when the formation of crystals is fast in the concentrated liquor, and hence, causes difficulty in transferring the concentrated liquor from one evaporator to the other.

A brief discussion on the various methods of feeding is well presented in the textbook of Chemical Engineering by Coulson, et al. (1991) with numerical examples.

15.4 VAPOUR COMPRESSION EVAPORATION SYSTEMS

We have seen how the vapour coming out of a single effect evaporator is condensed either directly or indirectly, and the heat associated with it is wasted instead of being economically utilized. We have also seen that such heat associated with the vapours can be used in the steam chest of the subsequent evaporators in a multiple effect evaporator and realize higher steam economy; but the pressures in the subsequent evaporators are to be in the reducing order, necessitating the use of vacuum to evaporate the final concentrated liquor at low temperatures. We have to think of a means by which the heat associated with the vapours is to be properly utilized without having the drawbacks of the multiple effect evaporators. One of the envisaged methodologies for this would be to replenish the heat loss in the vapours by other means. It can be done either by,

(i) recompressing the vapours using a compressor to the desired pressure and temperature, and hence, it can be reused as the heating steam.

or

(ii) adding high pressure and high temperature steam from another source to the vapours by using *steam-jet ejectors* and bring the vapours to the desired level of temperature and pressure to be used in the steam chest of the evaporator as a heating source.

Another method of replenishing the vapour is *evaporation at low temperatures using a heat pump cycle*. However, this concept will not be considered in this book.

15.4.1 Vapour Recompression System

In a vapour recompression system, the vapours coming out of the evaporator are passed through a mechanical compressor (rotary compressors) in which the vapours (steam) are compressed, and hence, the temperature and pressure of the vapours increase which are then recirculated into the steam chest of the evaporator (Figure 15.5). Any small amount of the additional steam required is added from outside.

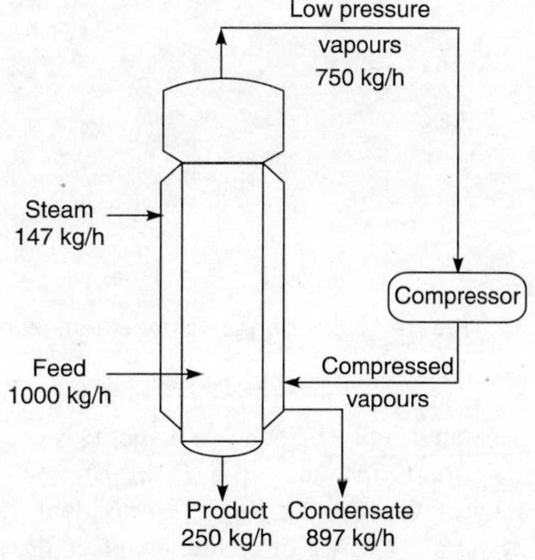

Figure 15.5 Vapour recompression system.

Let us consider Problem 15.2, in which 1000 kg/h of feed is concentrated from 10 per cent solids to 40 per cent solids by using 897 kg/h of steam at 101 kPa (gauge) and 120.23 °C. In this the product generated is 250 kg/h and the vapours generated are 750 kg/h at a temperature of 86 °C (corresponding to 40 kPa vacuum). These vapours are recompressed in a centrifugal compressor to the same pressure and temperature of the feed steam, i.e., 120.23 °C and 101 kPa (gauge). From the steam source, after first one hour is over, we will be adding only 897 – 750 = 147 ≅ 150 kg/h of steam to replenish the deficit in the steam requirement. After the first one hour is over, we will be feeding only 150 kg/h of steam to vapourize 750 kg/h of water from the feed, which pushes the steam economy to 750/150 = 5.0 as compared to 0.836.

In spite of the attractive steam economy the system has an inherent drawback of using a compressor to compress the steam at high temperature, and hence, is susceptible to wear and tear. Generally vapour recompression systems are used in small and medium size units.

15.4.2 Steam-jet Ejector System

This is also known as the **thermal recompression** system as contrast to the mechanical recompression systems we have used in Section 15.4.1. In this system, the vapours coming out of the evaporator are mixed with a high pressure and high temperature steam in a steam–jet ejector (Figure 15.6). In this all the quantity of vapours coming from the evaporator cannot be used. Only a part of it is recycled. The remaining vapours are sent to condenser. Generally the high pressure steam is at 800–1000 kPa (8–10 atm. pressure) and is at a temperature of 170–180 °C. The ratio of the high pressure steam required to the quantity of water to be evaporated is about 0.5.

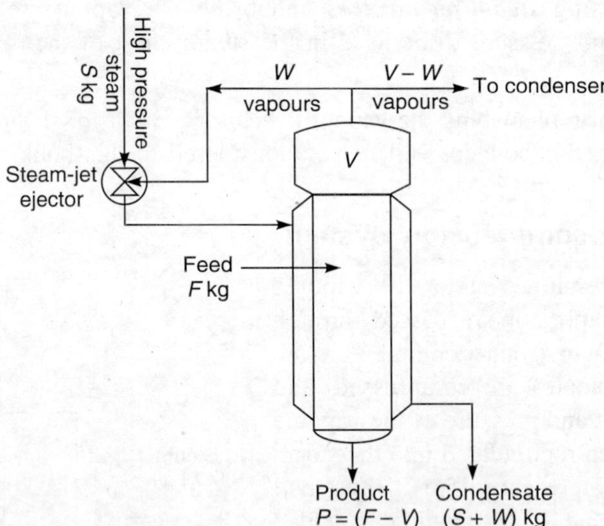

Figure 15.6 High pressure steam-jet ejector system for thermal recompression of vapours.

Coulson, et al. (1991) explained this with an example in which out of the 1.75 kg of vapours generated, only 0.75 kg of vapours were mixed with 1 kg of high pressure steam at 928 kPa at ~176 °C to make up 1.75 kg of low pressure steam at 170 kPa at 115 °C. The remaining 1 kg of vapours are let out to condenser. By this method we are using only 1 kg of steam to evaporate 1.75 kg of water. It makes the steam economy to be 1.75 which is normally realized in a double effect evaporator.

If high pressure steam is available from another source, steam-jet ejector system is a better one as no moving parts are involved (like compressors, etc.), and hence, the wear and tear of the machinery is the least. The jet-ejectors made out of corrosion resistant materials will last long.

15.5 EVAPORATION EQUIPMENT

A number of evaporators is used in the food industry; some of them are specific while others are general. Sometimes more than one type of evaporator can be used for same purpose. In such cases, the choice is based on economic considerations. The various factors that need to be considered in case of choice of an evaporator are all the factors that affect the rate of evaporation, and are summarized in Table 15.1. For meeting these requirements, a number of evaporators are available which can be classified based on the length of tubes, flow of the liquid film, forced circulation of the liquid inside the tubes. The classification is shown in Figure 15.7 and will be described here.

Table 15.1 Factors that Affect the Rate of Evaporation

Boiling point elevation
Rate at which heat can be transferred to the liquid
Thermal sensitivity of the food material
Loss of volatiles
Foaming nature of the food material
Build-up of non-condensable gases in the evaporator
The pressure at which evaporation takes place
Viscosity and change of viscosity of the liquid as evaporation takes place
Changes in rheological properties of foods with evaporation/concentration
Physical and chemical properties of foods
Fouling of heat transfer surface
Frequency of cleaning and sanitation of the evaporators

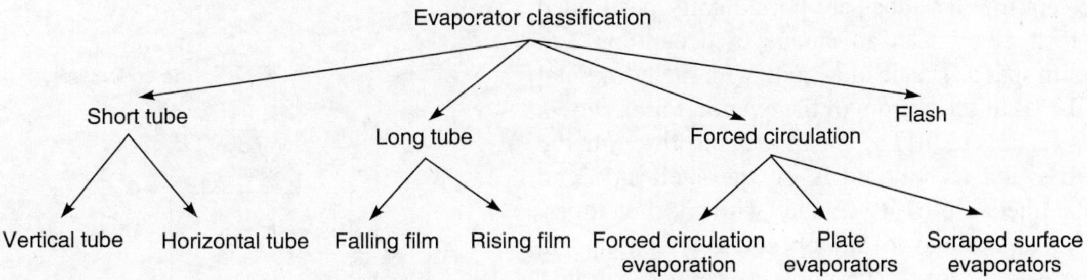

Figure 15.7 Classification of evaporators.

15.5.1 Short Tube Vertical Evaporator

It is also known as the *Calendria* evaporator, and is one of the oldest systems of evaporators in use because of some of the obvious advantages, such as:

- simple in construction,
- high heat transfer rates at high ΔT,

- ease of cleaning, and
- inexpensive.

It consists of a steam chest of a bundle of short vertical tubes through which the steam passes. The steam chest is placed in a big shell which contains the solution to be concentrated as shown in Figure 15.8. The tubes are generally 1.2 m long and L/D ratio is between 20 to 40 (Coulson, et al., 1991).

The heating zone is a bundle of tubes which is fitted inside the evaporator. The feed enters from the top, and gets heated up in the calendria zone. The concentrated liquor is discharged from the bottom. The steam condensate also discharges from the steam chest. The upper portion is the vapour separator which helps separate the vapours, and send them to condenser for condensation.

Another design of a vertical tube evaporator is basket type evaporator in which the liquid flows through the tubes, and the steam passes through the shell side. The main advantage of this design is that the heating element can be easily dismantable from the rest of the evaporator which gives flexibility for its ease of clearing.

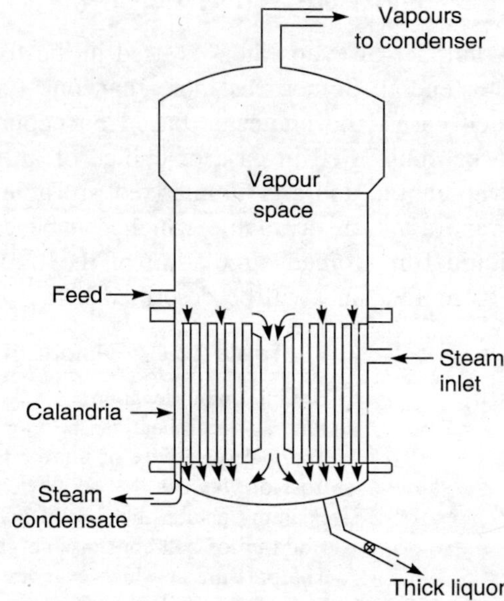

Figure 15.8 Short tube (calandria) evaporator.

15.5.2 Horizontal Tube Evaporator

It is almost similar to the vertical tube evaporator except that the tubes are horizontally positioned which gives it an advantage of occupying less room space. The heating zone with the horizontal tubes is at the bottom of the evaporator as shown in Figure 15.9. The steam passes through the tubes, the feed liquid is on the shell side and boils because of its contact with the hot tubes. The vapours go to the top and get separated in the vapour zone. The thick concentrated liquor discharges from the bottom. Since the feed liquid flows outside the tubes, the tubes hinder the agitation in the liquid resulting lower overall heat transfer coefficients.

One of the advantages of this system is that the horizontal tube heat exchanger can be easily dismantled from the evaporator portion which provides flexibility for cleaning of the heat exchanger.

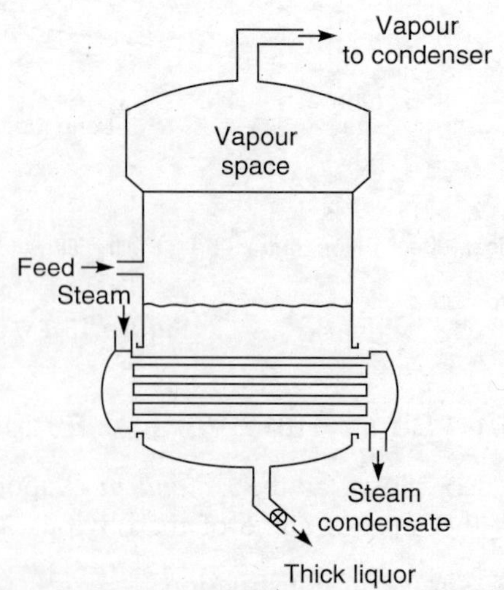

Figure 15.9 Horizontal tube natural circulation evaporator.

15.5.3 Long Tube Rising Film Evaporator

In the long tube vertical rising film evaporator, the feed enters at the bottom of a bundle of long vertical tubes which is enclosed in a shell, heated by steam (Figure 15.10). In some evaporators, the feed is initially heated separately in a heat exchanger and then enters the evaporator tubes which are also heated (Figure 15.11). The heated vapours along with the liquid rise through long vertical tubes about 10–15 m long. As the vapours rise fast, the liquid also moves fast. Hence, high convective heat transfer coefficients are realized. The residence times

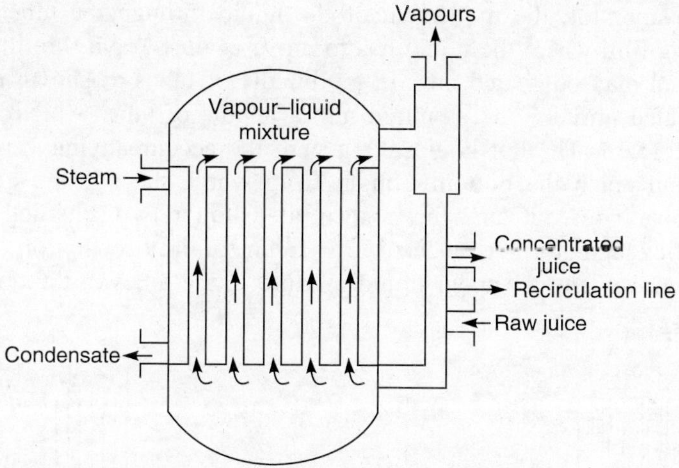

Figure 15.10 Rising film evaporator.

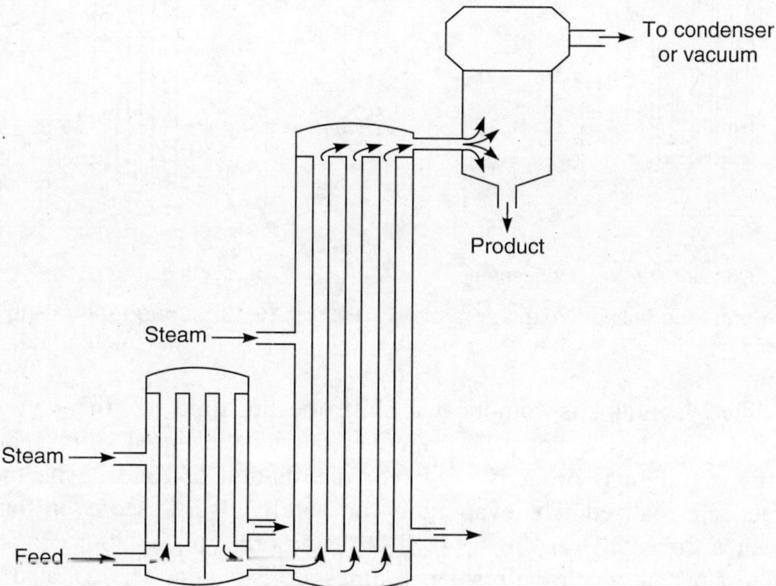

Figure 15.11 Long tube rising film evaporator with feed heated initially.

of the liquids are also low; but the evaporator is suitable only for low viscosity fluids, as the liquid film has to go up against the gravity. Since the residence times are low, the evaporator is used for heat-sensitive foods. The ΔT should be at least 14 °C. Low temperature differences do not allow quick formation of vapours, and hence, liquid movement will be slow. Thus, the operation of evaporator is very sensitive to operating conditions.

15.5.4 Long Tube Falling Film Evaporator

In a falling film evaporator, the movement of the liquid through the tubes is by gravity, in contrast to the rising film where the liquid has to move against gravity. In this evaporator also, either the liquid feed may be heated only in the bundle of tubes as shown in Figure 15.12 or the feed can be heated initially and admitted into a bundle of tubes which are also heated by steam as in Figure 15.13. The hot liquid along with the accompanying vapours moves down along the tubes, and reach the bottom trough, from where the mixture goes into a vapour separator. The vapour goes out, and the concentrated liquor is discharged from the bottom. Generally, the vapour separator is connected to vacuum for easy separation of vapours, as well as to facilitate the movement of vapour-liquid mixture in the tubes in the downward direction.

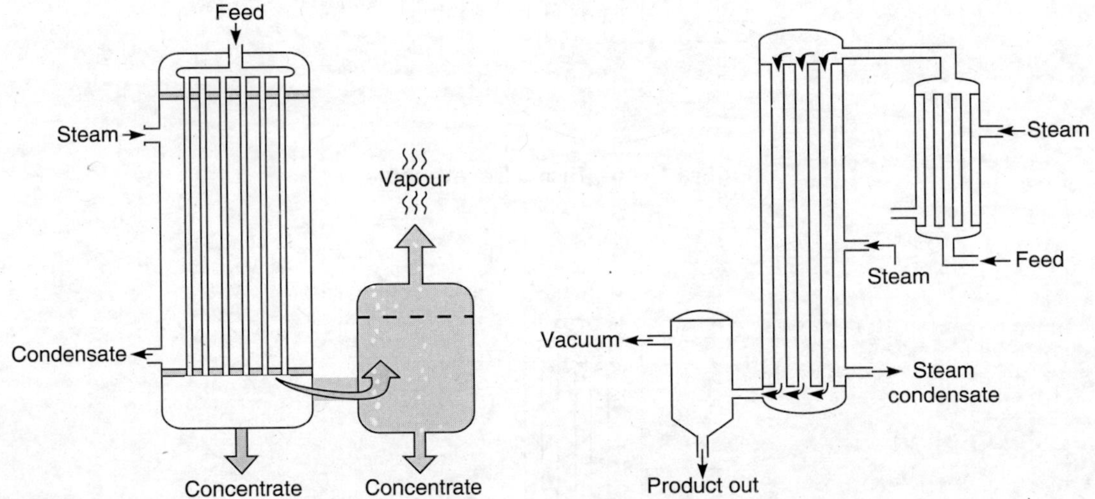

Figure 15.12 Long tube falling film evaporator.

Figure 15.13 Long tube falling film evaporator with the feed heated initially.

Design of this evaporator is complicated since the hot fluid has to move downwards. If the entire wall of the tubes is not properly covered with the liquid film, heat transfer rates become low. If the design is properly made for distribution of fluid in the tubes, very high heat transfer rates are realized. The evaporator can be effectively used even for viscous fluids also. The residence times are very low which is of the order of 20–30 seconds as contrast to rising film evaporators in which the residence times are 3–4 min. The ΔT also need not be as high as that in rising film evaporators.

15.5.5 Forced Circulation Evaporator

In a forced circulation evaporator, an external pump is used to recirculate the feed through an external heat exchanger (also known as Calandria), to heat and concentrate the liquid feed. It vaporizes in the heat exchanger and goes to the vapour separator as shown in Figure 15.14. Vapours go to condenser. The thick liquor is discharged. Certain amount of hydrostatic head (ΔH) is maintained above the heat exchanger to prevent the liquid boiling in the tube itself. Vacuum or low pressure is maintained in the vapour-liquid separator, so that the hot fluid flashes, and the vapour is separated.

The external pump for recirculation would help maintain high liquid flow rates which is of the order of 2–5 m/s through the tubes which is considerably higher as compared to the natural circulation evaporators. The forced circulation helps achieve high heat transfer rates (heat transfer coefficient varying between 3–6 kW/m^2°C), and hence, the size of the equipment is smaller, which results in low capital costs. However, use of an external pump adds for operating costs, which may off-set the gain in capital costs. Forced circulation systems are required for concentrating viscous fluids like tomato paste or in the crystallization step as in sugar industry.

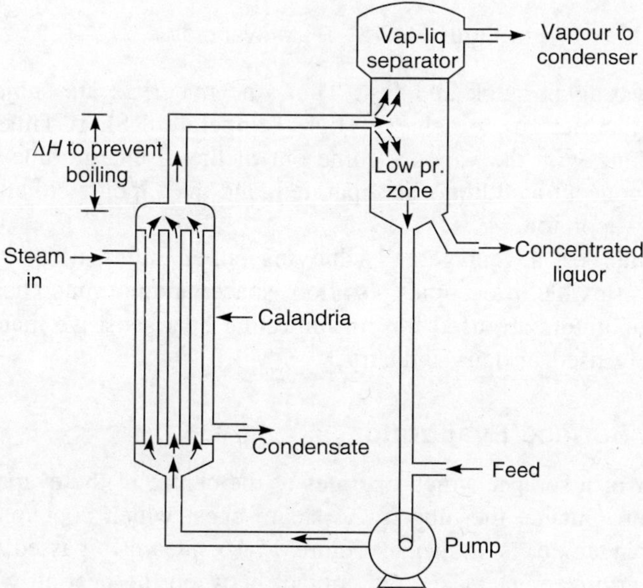

Figure 15.14 Forced circulation evaporator.

15.5.6 Plate Evaporators

Plate evaporators are almost similar to plate heat exchangers (Section 12.4). They consist of a series of plates which have an arrangement for the steam to flow through pipes in one plate, and the liquid flows through the product plate. Thus, each unit (compartment) consists of a steam plate and product plate, and are joined by placing a gasket in between. Figure 15.15 shows a plate evaporator arrangement of APV Inc., company falling/rising film plate evaporator (who are the world leading manufacturers of evaporators). Since the material flows in the form

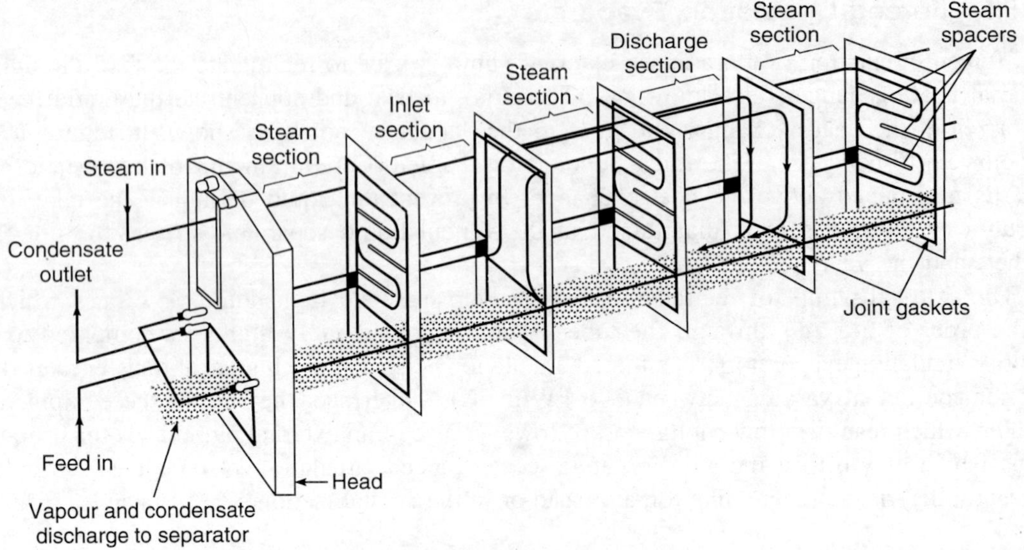

Figure 15.15 Plate evaporator.

of thin films, the heating is quick and fast. Thus, the materials are subjected to heating for a short time which facilitate us to achieve High Temperature Short Time (HTST) operation. The heated liquid along with the vapours come out of the discharge tube into a vapour-liquid separator where the concentrated liquor is separated, and the vapours either go for condensation or to a next effect evaporator.

It has some of the best advantages of achieving either falling film or rising film or both. The operation of it is flexible. It occupies less floor space; and not much head space is required. Mostly the plate evaporators are used for concentrating heat sensitive materials like milk, and hence, are extensively used in dairy industry.

15.5.7 Scraped Surface Evaporator

It consists of a rotor or a scraper which operates at the centre of the evaporator, and is rotated by an external motor. Outside the tube is the steam chest which contains the heating source, viz., steam. The arrangement is shown in Figure 15.16. Steam is passed through the heating, zone. The scraping arrangement makes the contact between the heating wall and the liquid in the form of a thin layer which will be renewing quickly. The arrangement is similar to scraped surface heat exchanger (Section 12.3) except that the liquid is allowed to evaporate in the case of scraped surface evaporators. The heated liquid along with the vapours will be moving downwards, and be discharged at the bottom in a vapour-liquid separator. The thick concentrated liquor is collected at the bottom, and the vapours are taken to a condenser for condensation.

The scraped surface evaporators are best suited for high viscous fluids (like corn syrup or tomato puree) which cannot be concentrated in any other type of evaporator. Since the material is not let to be in contact with the hot surface, the evaporator is also best suited for heat sensitive materials like tomato puree. High temperature differentials (ΔT) can be maintained in this evaporator, and hence, can be used in single effect evaporators with higher thermal efficiency or steam economy.

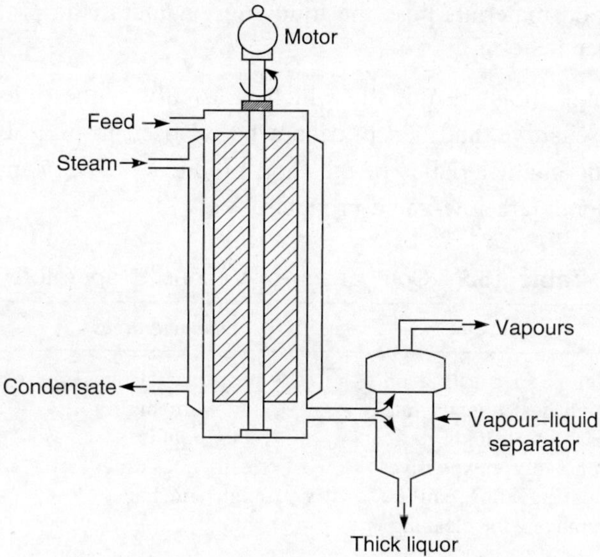

Figure 15.16 Scraped surface evaporator.

15.5.8 Flash Evaporator

Indeed flash evaporator is not an equipment, but an operation. In flash evaporation, the liquid is not allowed to boil in the tubes, but is only superheated to a slightly higher pressure. The high temperature-high pressure liquid is suddenly flashed into a low pressure (vacuum) zone. The liquid vaporizes in the vapour zone under low pressure and the concentrated liquid flows down. In fact, the flash evaporation is used in the *evaporator-crystallizer* systems. It is particularly used in the desalination operations. Baker (1963) reported an excellent work on flash evaporation.

Flash evaporator is also used for recovery of aroma from fruit juices, by flashing the hot fruit juice into a packed bed or a perforated plate column maintained at very low pressure. The feed juice is preheated to a temperature of 50–65 °C, and is flashed into a packed bed maintained at 3.45 kPa. The evaporative cooling reduces the temperature of the liquid. The vapours, as they go up in the packed bed get richer in the aroma vapours, which are condensed in a surface condenser using refrigerated cooling system.

15.5.9 Choosing the Correct Evaporator

Table 15.2 tabulates the relative advantages and disadvantages of various evaporators with their possible applications. Various criteria that need to be considered for selection of a correct type of evaporator is based on the following because of heat sensitive nature of food materials and their desirable organoleptic characteristic (Moore and Hesler 1963).

- Organoleptic characteristics of the food product
- Product quality
- Product characteristics
- Operating economy
- Initial investment

Particularly, the food materials pose the following characteristic problems which normally we do not find in other materials:

- They contain high moisture initially which is usually of the order of > 80 per cent.
- They are heat sensitive, and hence, cannot be exposed to high temperatures.
- The organoleptic characteristics of the food products are very important.
- Normally the foods are low-value products.

Table 15.2 Comparison of Various Evaporators

Evaporator	*Advantages*	*Disadvantages*	*Applications/Remarks*
Vertical short tube evaporator	• simple in construction • high heat transfer rates • ease of cleaning • relatively inexpensive • heating unit can be easily removed for cleaning	• heat transfer rates are low with highly viscous liquids. • residence times are high and hence, low output.	• Mostly used in sugar and salt industry. OHTCs[†] are of the order of 0.57–2.8
Horizontal short tube evaporator	• relatively inexpensive • less head room space is required • can be operated both batch and continuously	• liquid circulation is poor • cannot be used with high viscosity liquids • OHTCs are lower	
Long tube falling film evaporator	• for heat sensitive foods • high H T rates • short residence times (5–30 s) • can be used for viscous liquid • ΔT can be low	• design is complicated as the liquid should flow in a thin flow	• used in dairy industry and for concentration of fruit juices. OHTCs are of the order of 2–3
Long tube rising film evaporator	• short residence times (10–60 s) • can be used for heat sensitive liquids • high OHTCs are possible (2.25 to 6.0)	• ΔT should be high • HT rates are low for low ΔT • very sensitive to operational changes • cannot be used for high viscosity liquid	• used in dairy industry and for concentration of fruit juices. OHTC < 0.3 for high viscous solutions
Forced circulation evaporator	• through flow rates are high • smaller in size and hence, capital cost is less • can be used for viscous liquids. High OHTCs are possible (3–6 kW/m^2°C)	• external pumping costs are extra • operational costs are high	• useful for meat extracts • processing of tomato products • used for salt and other crystallizing solutions
Plate evaporator	• low floor space • low head space • flexibility • easy accessibility • HTST operation is possible	• high viscosity liquids cannot be processed	• concentration of corn syrup • used in dairy industry • OHTCs are of the order of 2–3

(contd.)

Table 15.2 Comparison of Various Evaporators (*contd.*)

Evaporator	*Advantages*	*Disadvantages*	*Applications/Remarks*
	• useful for high heat sensitive materials • contact times are 2–30 s		
Scraped surface evaporator	• useful for high viscous liquids • for heat sensitive liquids • useful for liquid with tendency to foul • contact times are 20–30 s • high heat transfer rates	• cost of construction is high • high operating costs	• gelatine solutions • tomato puree • OHTCs are of the order of 2–3 for low viscosity liquids and 1.7 for high viscosity liquids
Flash evaporator		• scale formation is more	• used for aroma recovery • used in desalination operations

†OHTCs (Overall Heat Transfer Coefficients) are in kW/m^2°C.

The major problem we face in evaporation is that always a certain portion of the food material remains in contact with transfer area even in a single pass evaporator also. This causes spoilage of some heat sensitive food materials. Hence, efforts should be made always to avoid recirculatory type of evaporators; a true single pass evaporator should be used as far as possible.

All things being equal, a Rising-falling Film Concentrator (RFC) or single pass down flow is preferred in view of the cost economics as contrast to an agitated or wiped film evaporator (Moore and Hesler 1963).

15.6 APPLICATIONS OF EVAPORATION IN FOOD PROCESSING

As has been mentioned earlier in the chapter, evaporation finds extensive application in food processing. It could be for concentration of fruit juices for preservation and transportation, or concentration of sugar juice for crystallization, or evaporating liquid milk to dry further into milk powder, or concentration of milk to make a condensed milk for preservation, or it could be flash evaporation of fruit juices for recovery of aroma. Thus, the applications are varied. Table 15.3 highlights some of the applications of various types of evaporators. Broadly the applications may be classified as follows:

Fruits and vegetables processing: In fruit and vegetable processing, usually the fruit juice is concentrated to bring down the bulk for easy and economical transportation. The pulp of the fruits is concentrated for preservation purposes. For example, tomato pulp is concentrated to make double strength or triple strength puree.

The aroma from fruit juices like orange or pine apple is recovered by flash evaporation, and will be added after concentrating the pulp. Even in the manufacture of jam, the fruit pulp is evaporated with sugar in the open pans until the required brix (68°) is achieved. For making the pineapple bar (candy) the first juice is concentrated before dehydrating in a cabinet tray dryer.

Table 15.3 Comparision of OHTCs (Cover all Heat Transfer Coefficients) in Some Evaporators

Evaporator	*Residence times (approx.)*	*OHTC* (kW/m^2°C)	
		Low viscosity	*High viscosity*
Open or vacuum pan	30 min to several hours	0.5–1.0	< 0.5
Vertical short tube	–	0.57–2.8	–
Rising film	10–60 s	2.25–6.0	< 0.3
Falling film	5–30 s	2–3	–
Plate (three stage)	2–30 s	2–3	–
Scraped surface	20–30 s	2–3	1.7

Source: Adapted from Earle, R.L. (1983), *Unit Operations in Food Processing*, 2nd ed., Pergamon Press, Oxford, pp. 105–115.

Dairy industry: Milk is concentrated in open pans to make *Khoa.* Milk is also reasonably concentrated before spray drying it into powder. It would make the drying process economical. Concentrated milk is preserved as it is, and is known as **condensed milk**.

Sugar and salt industry: Sugar industry is one which extensively uses evaporation. The sugar juice extracted by crushing the sugar cane is evaporated and concentrated carefully to super saturation level so that the nucleation of crystallization sets in. Similarly in salt processing also, the sea water is evaporated in open fields so that salt crystals start forming, which are further refined.

Improving the aroma: In processing of cocoa beans, the extract of the cocoa is concentrated so that the unpleasant volatiles evaporate out and give a good flavour to the cocoa extract. Similarly, the concentration of milk also improves the flavour of the milk and imparts the characteristic aroma of milk.

Desolventization: During the extraction of oleoresins from the spices by using organic solvents, the extract contains both the active principles of the spice (oleoresin) and the organic solvent. In further processing, the solvent is evaporated from the extract to recover the oleoresin. The solvent vapours are condensed for reuse. Desolventization is also known as distillation because the solvent vapours are recovered and condensed for reuse, whereas normally in evaporation, we do not recover the condensate.

Symbols

A: heat transfer area (m^2)
C_p: specific heat (kJ/kg°C)
F: feed flow rate (kg/s)
g: acceleration due to gravity (9.81 m/s^2)
h: column height (m)
m: mass
m: molality in Eq. (15.9)
$\dot{m}$: mass flow rate (kg/s)
P: product flow rate (kg/s)
p: pressure (Pa)
Q: quantity of heat (kJ)
q: heat transfer rate (kW)
ΔT: temperature rise/temperature difference (°C)
ΔT_b: boiling point rise
T: temperature (°C)
U: heat transfer coefficient (W/m^2°C)
V: vapour flow rate (kg/s)
W: condenser liquid flow rate (kg/s)
X: mass fraction of solids

Subscripts

c: condenser liquid
f: feed
i: inlet
o: outlet
p: product
s: steam
v: vapour phase
1, 2, 3: no effect of evaporator

Greek Symbols

λ: latent heat of vaporization (kJ/kg)
ρ: density (kg/m^3)

REVIEW QUESTIONS

15.1 How evaporation is different from distillation and drying?

15.2 What are various components of an evaporator?

15.3 What is meant by BPR and how does it affect the evaporation process?

15.4 What are various operational features that you consider as important in evaporation?

15.5 How do heat and mass balance equations help in the design of an evaporator?

15.6 What is Duhring's rule?

15.7 What is meant by a multiple effect evaporator?

15.8 What is meant by steam economy in evaporation?

15.9 What are various feeding methods in a multiple effect evaporators? What are the relative advantages of each method?

15.10 What is meant by vapour compression evaporation? How does it help steam economy?

15.11 What are different methods of vapour recompression?

15.12 How do you classify various evaporation equipment?

15.13 Describe a short tube evaporator system with a neat diagram.

15.14 Describe a horizontal tube evaporator system with a neat diagram.

15.15 Describe a falling film evaporator system with a neat diagram.

15.16 Describe a rising film evaporator system with a neat diagram.

15.17 Describe a forced circulation evaporator system with a neat diagram.

15.18 Describe a plate evaporator system with a neat diagram.

15.19 Describe a scraped surface evaporator system with a neat diagram.

15.20 What is meant by flash evaporation?

15.21 Describe in a tabular form the advantages and disadvantages of various evaporation systems.

15.22 What are various applications of evaporation in food processing?

NUMERICAL PROBLEMS

15.1 A single effect evaporator is used to concentrate a solution containing 5 per cent solids at the flow rate of 1000 kg/h and initial temperature of 40 °C. The vacuum in the evaporator is 30 kPa. Steam is supplied into the steam chest at a gauge pressure of 30 kPa. The average density and specific heat of the feed may be taken as 990 kg/m^3 and 4.5 kJ/kg°C. The boiling point rise may be neglected. Find the steam economy. If the overall inside heat transfer coefficient is 1.8 kW/m^2°C, find the evaporator area required. Draw a diagram to show the flow conditions.

(**Ans:** P = 125 kg, V = 875 kg, T_s = 106.67, T_b = 90 °C, λ_s = 2226.3 kJ/kg, λ_v = 2283.3 kJ/kg, m_s = 931 kg/h and A = 19.2 m^2)

15.2 In a triple effect evaporator, a fruit juice containing 10 per cent solids is being concentrated to 35 per cent solids. The flow rate of the fruit juice is 1500 kg/h. Steam is supplied to the first effect evaporator at 101.3 kPa (gauge), and the vacuum in the last evaporator is 30 kPa. Neglecting the sensible heat effects and BPR, calculate the steam requirement and heat transfer area if the heat transfer coefficients of first, second and third effect evaporators are respectively 1.8, 1.6 and 1.4 kW/m^2°C.

(**Ans:** m_s = 367 kg/h , A_1 = 14.17 m^2, A = 42.52 m^2, Steam economy = 2.92)

REFERENCES

Baker, R.A. (1963), The flash evaporator, *Chem. Engg. Progress*, **59**(6), pp. 80–83.

Coulson, J.M., Richardson, J.F., Backhurst, J.R. and Harker, J.H. (1991), *Chemical Engineering*, Vol. 2., 4th ed., Pergamon Press, Oxford, pp. 623–664.

Earle, R.L. (1983), *Unit Operations in Food Processing*, 2nd ed., Pergamon Press, Oxford, pp. 105–115.

Moore, J.G. and Hesler, W.E. (1963), Evaporation of heat sensitive materials, *Chem. Engg. Progress*, **59**(2), pp. 87–92.

Toledo, R.T., (1997), *Fundamentals of Food Process Engineering*, 2nd ed., CBS Pub. and Distributors, New Delhi, pp. 437–454.

CHAPTER

16

Diffusion and Mass Transfer

Mass transfer is the term which is generally used in chemical engineering and food engineering terminology to mean exclusively mass of a substance transferred from one phase to the other phase by means of concentration difference of the substance in the two phases aided or unaided by turbulence. For example, in the preparation of carbonated beverages, the carbon dioxide gas in the gas phase when brought into contact with liquid water in the liquid phase, the carbon dioxide gas diffuses into the water through a boundary layer that is perceived to have been existing between the gas-liquid phases (carbon dioxide-water phases). The transfer of the gas continues until the concentration of carbon dioxide in the water reaches the equilibrium state, what we call as equilibrium solubility. Thus, the gaseous carbon dioxide transfers into the liquid phase.

This explanation obviously means that if we transfer a material from one place to other, for example, we transfer one ton of fruits from ripening shed to the processing hall, the operation *cannot be considered as mass transfer.* It is only termed as *material handling* (which we shall discuss in Chapter 24). Similarly if we transfer 100 gallons of water from a water sump to an overhead tank by using a pump, it is only considered as material handling as contrast to mass transfer.

Sometimes, we may specifically use the word as mass transport which is a synonym of mass transfer. From the earlier discussion, it is clear that if concentration difference is the driving force, we call it as mass transfer. We recall our earlier expression for transfer rate as:

$$\text{Transfer rate} \propto \text{driving force}$$
$$\propto \text{interfacial area or cross-sectional area.}$$

We may now write,

$$N'_A = \text{constant } (A)\ (\Delta C) \tag{16.1}$$

where N'_A is moles or mass of component A transported per unit time (kg/s), ΔC is the concentration difference in moles or mass per unit volume, and is the driving force for mass transfer, A is the area of contact.

Generally the constant is known as the **mass transfer coefficient**.

Mass transfer in food processing plays a major role because most of the transfer processes (unit operations) for processing or separation or mixing of components in a multi-phase systems do occur in food processing. As contrast to chemical engineering studies where multi-phase mass transport is handled more for chemical reaction systems, what we call as classical *mass transfer with chemical reaction,* in food engineering these studies are restricted to unit operations. Thus, mass transfer in food engineering may be studied for the following systems:

- diffusional/convective mass transfer
- mass transfer with phase change
- liquid-liquid extractions

The liquid-liquid extraction studies will be separately taken up in Chapter 19 and dehydration in Chapter 17. Hence, suffice it is in this chapter to concentrate on the basic principles of mass transfer by diffusion and convection (both natural and forced), for which the concentration difference (ΔC) is the driving force.

The preceding systems involve all the following varieties of phases, viz.,

- gas-liquid
- liquid-liquid/fluid-fluid
- gas-solid
- liquid-solid
- gas-liquid-solid phases

16.1 PRINCIPLES AND THEORY OF DIFFUSION

Diffusion is defined as the transportation of an individual component through a mixture under the influence of physical stimulus known as the **concentration difference**. In most of the cases in multi-phase systems, it means crossing the barriers caused by phase differences, and the transportation continues until the concentration of the individual component known as **solute** reaches equilibrium concentration in both the phases. We call it *molecular diffusion* if the transport takes place solely by concentration differences only; the solute moving from a region (or phase) of higher concentration to the region (or phase) of lower concentration. If two miscible liquids like water and alcohol are kept in two chambers separated by a barrier or screen, no mixing takes place. If we remove the barrier, alcohol starts moving from chamber A into chamber B, and similarly water moves from chamber B to chamber A (Figure 16.1).

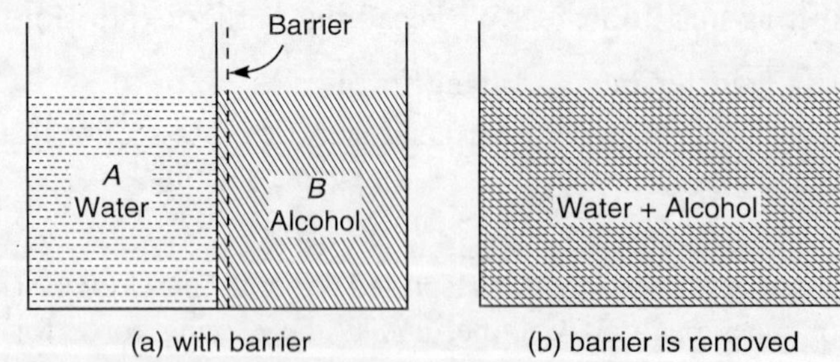

Figure 16.1 Diffusion of alcohol and water into each other.

This is not a case of multiphase diffusion, as no separate phases are involved in this case. If no external agitation is given, probably it would take pretty long time (a few days) to reach equilibrium. This movement is only by diffusion. However, if the two fluids are mixed with an agitator, they would reach the equilibrium concentration quickly. The word equilibrium is more relevant in case of two phase systems like air-water in which oxygen from air is partially miscible with water. Equilibrium concentration is not strictly applicable to water-alcohol system since both are freely and fully miscible.

Now, let us see the situation where the two fluids are not fully miscible like oxygen-water systems. Oxygen from air diffuses into water by molecular diffusion. It is schematically represented by Figure 16.2.

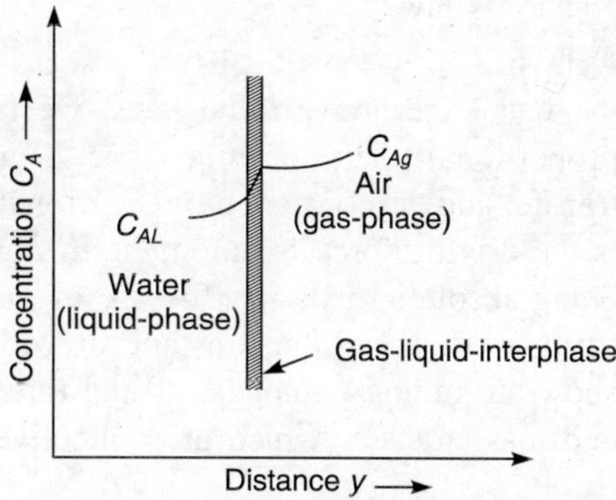

Figure 16.2 Diffusion of oxygen from air into water through gas-liquid-interphase.

It is conceived that there exists a gas-liquid interphase (film) between the two phases. Oxygen has certain concentration in air (usually 21 per cent). From this phase oxygen diffuses through the film existing in between the gas and liquid phases, and enters into liquid phase, in which it has an equilibrium solubility of 10 ppm at normal temperature and pressure. Now, at this stage, if the oxygen in the water is depleted by any means, then some more oxygen enters into water phase. Thus, continuous diffusion of oxygen into water continues. This is known as **continuous molecular diffusion**. As the diffusion takes place, there will be some depletion oxygen in the air, which will equilibrate. Similarly, if the oxygen is depleted from water, it will also equilibrate. By agitating the fluids either in the gas phase or in liquid phase, we can reduce the concentration differences in the respective phases, but not in the film. Similarly, we also cannot do away with the film. The agitation can help reduce the film thickness, but cannot totally do away with the film. Higher is the agitation, less thick is the film.

The mass flux (J_A) is the rate of transfer of mass (or moles) per unit area and is obtained by dividing the mass (moles) flow rate by the cross-sectional area.

$$J_A = \frac{N'_A}{A} \tag{16.2}$$

where J_A is in kg/m^{2}s and N'_A is the mass flow rate in kg/s. The mass flux is proportional to the concentration gradience.

$$J_A \propto -\frac{dC_A}{dy} \tag{16.3}$$

or

$$J_A = -D_{AB}\frac{(dC_A)}{dy} \tag{16.4}$$

where D_{AB} is the diffusivity of component A into component B, y is the distance measured from lower concentration to higher concentration as shown in Figure 16.2. Thus, the direction of y is opposite to the direction of mass flow. We use negative sign in Eq. (16.3) to indicate that the direction of mass transfer is opposite to the direction of the concentration gradient; and it is always from higher concentration to lower concentration zone. Equation (16.4) is known as the classical **Fick's law of diffusion**. This is similar to Fourier's law of heat conduction in heat transfer, and is also similar to Newton's law of viscosity for momentum transfer.

Fourier law
$$\frac{q}{A} = -k\left(\frac{dT}{dy}\right) \tag{10.6}$$

Newton's law
$$\tau = -\mu\left(\frac{du}{dy}\right) \tag{9.3}$$

In all the above three laws, we find that rate of transfer per unit area (flux or stress) is proportional to the driving force. Temperature gradiance is the driving force in case of heat transfer and the constant of proportionality is thermal conductivity, whereas the velocity gradiance is the driving force for momentum transfer in fluid mechanics with constant of proportionality being absolute viscosity. In case of mass flux, the concentration gradiance is the driving force for mass flux, and the constant of proportionality is **diffusivity coefficient**. Thus, the three laws show an **analogy** amongst all the three transport processes, viz., fluid mechanics, heat transfer and mass transfer which are collectively known as **transport phenomena**.

16.2 ROLE OF DIFFUSION IN MASS TRANSFER

From the preceding discussion, it is clear that mass transfer takes place even if there is no mixing or agitation which we call as being due to molecular diffusion. It was also clear that mass transfer due to diffusion takes place even if there is vigorous agitation also because of the existence of the phase boundary in the form of a stagnant film[†] which continues to exist irrespective of agitation, but its thickness only varies with the degree of agitation. The diffusion could be in one direction or in both directions.

In case of drying of solids, the moisture in the form of vapours goes into the surrounding stream of air, in which case it is only in one direction in which moisture vapours diffuse. Similarly in case of extraction, the solute alone diffuses from one phase into the other until the equilibrium reaches. In case of gas absorption, for example, carbonation of beverages, carbon dioxide alone diffuses into the liquid beverage and gets absorbed. For dissolving a solid in a liquid as in case of dissolving crystal sugar in water to make a sugar syrup, diffusion of sugar takes place from solid phase to liquid phase purely by diffusion alone where there is no mixing or agitation at all. Contrary to this, in case of distillation, when the vapours are going up in the distillation column, they will be getting richer in the more volatile component present in the down coming liquid phase by diffusion from the liquid phase to vapour phase; the liquid which is coming down will be richer in the less volatile component present in the vapour phase by its diffusion from the vapour phase into the liquid phase. This is known as **counter-current diffusion**.

16.3 CONVECTIVE MASS TRANSFER

So far, we have restricted our attention more to diffusion, and more specifically to molecular diffusion where the mixing of the phases by an external agitation was not considered. In fact,

---

[†]We shall discuss it in more detail in Section 16.4.

mixing plays a major role in mass transfer when physically the layers of the fluid are in virtual motion, and hence, mass transfer takes place by *turbulence* or by the movement of eddies from a region of higher concentration to one at lower concentration. This kind of mass transfer we call as **convective mass transfer**†. Mass transfer by convection takes place in addition to that by diffusion, and obviously the convective mass transfer is much faster depending upon either the degree of external agitation or the movement of layers due to differences in densities. Hence, the convective mass transfer is dependent upon the following parameters:

(i) Extent or degree of agitation
(ii) Type of agitator
(iii) Method of dispersion or sparging of one component into the other
(iv) Flow rates of fluids in case of continuous operations
(v) Temperature differences which cause thermal diffusion

We shall try to characterize the mass transfer process by considering these parameters. Let us recall Eq. (16.1) according to which the mass transfer is proportional to both the area of contact, and concentration difference.

Therefore, mass transfer rate = constant (A) (ΔC). The constant is known as the **mass transfer coefficient**.

Generally, the convective mass transfer takes place when one phase is dispersed in another phase whether it is:

- Liquid-liquid systems,
- Gas-liquid systems or
- Solid-liquid systems.

and is schematically shown in Figure 16.3. The one that is in larger quantities is called as **continuous phase**, and the one that is in small quantities and is dispersed into the continuous phase is known as **dispersed phase**. The size of the bubbles or droplets of dispersed phase varies, depending upon the degree agitation. Hence, their interfacial area available in a given volume is more relevant. The mass transfer rate also varies accordingly, and hence, we define a new term for mass transfer as N_A which is the rate of mass transfer per unit volume. We also define here a new term for area as **interfacial area** which is the area of the dispersed phase in a unit volume of the vessel (which is volume occupied by the dispersed phase and continuous phase).

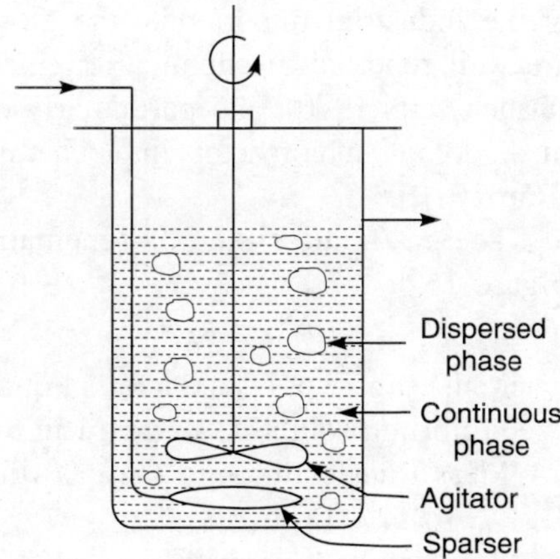

Figure 16.3 Schematic representation of convective mass transfer.

$$a = \frac{\text{Area of the disperesed phase}}{\text{Total volume}} \tag{16.5}$$

†This is similar to convective heat transfer wherein the heat transfer takes place virtually by movement of different layers of the fluids either naturally by density differences or by forced circulation.

$$= \frac{m^2}{m^3} = m^{-1}$$

where a is the interfacial area and has units of (m^{-1}).
Now, we write

$$N_A = k_L(a)(\Delta C) \tag{16.6}$$

where

k_L is the mass transfer coefficient, and has units of m/s,
N_A is mass transfer rate per unit volume, and has units of kg/(m^3s), and
k_La is known as mass transfer rate and has units of s^{-1}. Eq. (16.6) can also be written as:

$$N_A = \frac{(\Delta C)}{R} \tag{16.7}$$

where R is the resistance for mass transfer, and

$$R = \frac{1}{(k_La)} \tag{16.8}$$

16.3.1 Mass Transfer in Gas-liquid Systems

Gas-liquid systems are frequently encountered in chemical engineering, biochemical engineering and food engineering systems more than the mass transfer through other phases. It could be aerated water industries, or carbonation of beverages with CO_2 or hydrogenation of fats with hydrogen gas. In view of the industrial importance, the gas-liquid systems are well reported in chemical engineering literature (Danckwerts 1970), and particularly oxygen transfer methodology in bioreactors in biochemical engineering (Rao 2010).

The gas-liquid system is schematically shown in Figure 16.4.

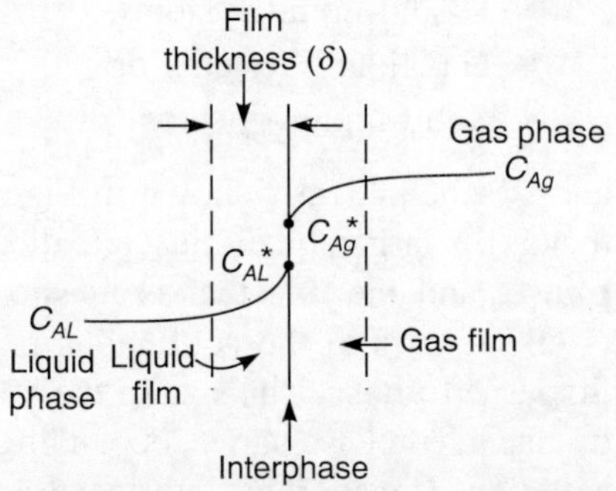

Film
thickness (δ)
Gas phase
C_{Ag}
C_{Ag}^*
C_{AL}^*
C_{AL}
Liquid phase
Liquid film
Gas film
Interphase

Figure 16.4 Schematic representation of gas-liquid system with film forming on either side.

$$N_{AL} = k_La(C_{AL}^* - C_{AL}) \tag{16.9}$$

where the subscript L stands for a liquid side, and C_{AL}^* is equilibrium solubility of gas A in the liquid phase.

Mass transfer through the gas film is given by

$$N_{AG} = k_g(a)\ (C_{Ag} - C_{Ag}^*) \tag{16.10}$$

where k_g is the mass transfer coefficient in the gas phase.

C_{AL}^* and C_{Ag}^* are equilibrium concentrations of solute A in the liquid phase and gas phase, respectively. They are not equal, but

$$\frac{C_{AL}^*}{C_{Ag}^*} = \text{constant} = \alpha \tag{16.11}$$

where α is known as the **distribution coefficient**.

We shall concentrate more on studying k_L or k_La and try to derive equations for evaluating (k_La) both by predictive equations and by experimental means. Before we go ahead, let us see the process of gas transfer from a bubble into bulk liquid.

Figure 16.5 shows a schematic representation of a gas-liquid system. Various steps involved in the transport of gas from a gas bubble (which may or may not be a mixture of several gases) into bulk liquid are:

Step 1: Transportation of the gaseous component from the bulk gas to gas-liquid interphase; relatively the transfer is fast and does not offer much resistance for mass transfer.

Step 2: The gaseous component crosses the gas-liquid interphase and enters into the liquid film. In this step also the resistance is negligible.

Step 3: Diffusion of the gaseous component through the quiescent (stagnant) liquid film into the bulk liquid. The resistance for mass transfer lies in the film, as the quiescent liquid film offers resistance for diffusion of gas through it; and hence, is the rate limiting step.

Step 4: Transportation of the gaseous component into the relatively well mixed bulk liquid. Since the bulk liquid is well mixed, the mass transfer is quick, and hence, it does not offer any resistance for the mass transfer.

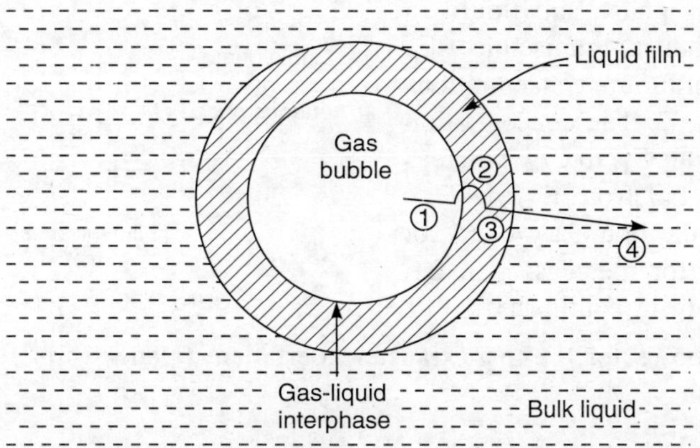

Figure 16.5 Schematic representation of gas transfer from gas bubble into bulk liquid.

Thus, out of the four steps involved, step 3 is the one which offers resistance for the mass transfer to take place. In case if the gaseous component has to go to the active site of a solid catalyst (as in the case of hydrogenation of oils and fats by hydrogen on nickel catalyst), all these four steps repeat amongst which the step involving diffusion of gas through the relatively stagnant liquid film surrounding the solid catalyst particle would be the rate limiting one.

16.4 THEORIES OF MASS TRANSFER

Interphase mass transfer involving the diffusion of the solute has always attracted the attention of various research workers in view of its industrial importance. Various theories have been proposed from time to time for the purpose of evaluating the mass transfer coefficient or more precisely to find a relationship between the diffusivity coefficient and mass transfer coefficient. We shall take them up one by one.

16.4.1 Film Theory

Of the various theories, one of the oldest and most classical theory is **film theory** which was first proposed by Whitman (1923) originally based on the concept of diffusional-layer of Nernst who proposed the idea as early as in 1904. The film theory proposed the idea of existence of a constant thickness film between two phases at the interface (Figure 16.6). Generally the concentration gradients do not exist within the phases as they can be equilibrated by agitation in the liquid side. As has been mentioned, agitation in the liquid phase can only reduce the thickness of the film, but cannot totally do away with it, and the mass transfer takes place solely by diffusion in the film. Hence, the diffusional resistance for the mass transfer solely lies in the stagnant film. In Figure 16.6, the two phases are immiscible, and hence, a phase boundary exists between them. The two different phases could be gas and liquid, or liquid in aqueous phase and liquid in organic phase, and both are immiscible. C_{A1}^* and C_{A2}^* indicate equilibrium concentrations in phase 1 and phase 2. The concentration in phase 1 will fall from C_{A1} to C_{A1}^*, whereas that in phase 2 will fall from C_{A2}^* to C_{A2}. It is presumed that the boundary does not offer any resistance for the mass transfer, and C_{A1}^* and C_{A2}^* are in equilibrium.

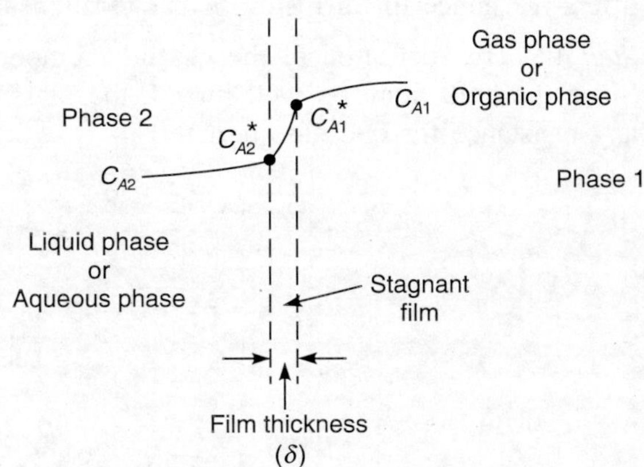

Figure 16.6 Concept of film theory.

Film theory proposes for the mass transfer coefficient (Danekwerts 1970) that

$$k_L = \frac{D_{AB}}{\delta} \tag{16.12}$$

where δ is the film thickness. It indicates the hydrodynamic characteristics of the system and is a function of system geometry, extent of agitation and fluid properties.

Film theory originally proposed the concept of probable existence of a thin, constant thickness film between two phases. The transportation of the solute into solvent is only by diffusion through the film which is the rate limiting step. The main drawback of the theory is that it conceives the discontinuity between two phases and are separated by a constant thickness stagnant film. Equation (16.12) proposes that k_L is directly proportional to the diffusivity coefficient which was later experimentally found to be not true.

16.4.2 Penetration Theory

Penetration theory model for mass transfer is based on unsteady state mass transfer. It was originally proposed by Higbie (1935), and is based on the concept that every element of the fluid surface is exposed to the gas for the same amount of the time θ, before it is replaced by the bulk liquid. However, during the process of exposure, the film behaves like a stagnant fluid and infinitely deep, and absorbs the gas with a constant concentration at the surface. Since it is an unsteady state process, the gas (solute) composition changes both with distance and time in the film. The rate of change of solute concentration $\partial C_A/\partial\theta$ is equal to the rate of gas absorption, and can be represented by *Fick's second law of diffusion* in one direction along x-axis. Thus,

$$\frac{\partial C_A}{\partial \theta} = \frac{D_{AB}\,\partial^2 C_A}{\partial x^2} \tag{16.13}$$

Equation (16.13) is applicable with the following boundary conditions,

when
$$\left.\begin{aligned} \theta = 0, x > 0, C_A &= C_{Ao} \\ \theta > 0, x = 0, C_A &= C_A^* \\ \theta > 0, x = \infty, C_A &= C_A \end{aligned}\right\} \tag{16.14}$$

For the above boundary conditions, Eq. (16.13) can be solved which gives an expression for mass flux as follows (McCae, et al., 1993):

$$N_A^1 = 2\left(\frac{D_{AB}}{\pi\theta}\right)^{1/2}(C_A^* - C_A) = k_L\,(C_A^* - C_A) \tag{16.15}$$

Therefore,
$$k_L = 2\left(\frac{D_{AB}}{\pi\theta}\right)^{1/2} = \left(\frac{4D_{AB}}{\pi\theta}\right)^{1/2} = 1.13\left(\frac{D_{AB}}{\theta}\right)^{1/2} \tag{16.16}$$

The depth of penetration in the above model is assumed to be infinitely long and is defined as that depth at which the concentration change is 1 per cent of the final value, ad is equal to $3.6\sqrt{(D_{AB}\theta)}$ (Danckwarts 1951). In the above equation, in case of gas absorption, θ is also defined as the time taken by a single gas bubble of diameter d_B to travel a distance equal to its diameter in the bulk liquid. Thus,

$$\theta = \frac{d_b}{u_b} \tag{16.17}$$

where u_B is the linear velocity of the bubble in m/s. Equation (16.16) predicts that mass transfer coefficient, k_L, is proportional to $D_{AB}^{1/2}$ unlike as in the case of film theory where it is proportional $D_{AB.}$ There is some evidence that mass transfer coefficient varies only as 1/2 power of diffusivity coefficient in packed beds and stirred vessels (Danckwerts 1970). Thus, Higbie's model scores over film theory in this respect. But still, the film theory is always credited for conceiving the concept of formation of film between two phases.

Only serious drawback of the penetration theory model is that it conceives the same time of exposure for all the elements of surface, which seems to be unrealistic.

16.4.3 Surface Renewal Theory

The surface renewal model was originally proposed by Danckwerts (1951), which is a further extension of Higbie's penetration theory model. It also conceives the concept of unsteady state diffusion process (transient diffusion), but refines the Higbie's model by introducing the concept that the film is randomly renewed by a fresh batch of liquid from the bulk liquid at every fixed interval of time. Thus, the model defines

$$k_L = \sqrt{(D_{AB}\, s)} \tag{16.18}$$

where s is the fractional rate of surface renewal and has the units of s^{-1}. Therefore, $1/s$ indicates how many times the surface renews in unit time. Equation (16.18) also shows that the k_L is varying as 1/2 power of D_{AB} (i.e., $k_L \alpha \sqrt{D_{AB}}$) which was experimentally shown to be realistic. During the penetration process, the gas (solute) diffuses through the film, hence, the solute concentration is maximum at the beginning of the film, and it will reach minimum value by the time it reaches the other end of the film. In the mean time, randomly the film is replaced and merges with the bulk liquid, and accordingly increases the solute concentration in the bulk. When a fresh film is formed it will have the composition of the solute which is equal to that prevailing in the bulk liquid (fluid) before the film is formed. Thus, the mass transfer continues to take place through the film which is renewing randomly at fixed intervals.

The surface renewal model seems to be better applicable to situations of mass transfer with chemical reaction of pseudo–first order reaction systems. Rao (1990) applied the same successfully to model oxidation of cyclohexane.

16.5 DETERMINATION OF MASS TRANSFER RATES

Let us consider Eq. (16.9) which gives a relationship for mass transfer.

$$N_{AL} = k_L a\ (C_{AL}^* - C_{AL}) \tag{16.9}$$

where $k_L a$ is termed as mass transfer rate and has unit of s^{-1}. By multiplying $k_L a$ with ΔC we get mass transfer in terms of kg/(m^3.s). Hence, to measure N_{AL} we require the values of ($k_L a$) and (ΔC). Generally we club k_L and a together and call it as ($k_L a$) since, very interestingly in case of fluid-fluid or gas-liquid mass transfer, the factors which help increase k_L are all detrimental for a and vice versa. For example, if bubble size is big, k_L is high, but a is small, and vice versa. Hence, measurement techniques have been devised for ($k_L a$) as a whole, even though there is a number of theoretical equations available for calculation of k_L and a separately.

16.5.1 Measurement of $k_L a$

Efforts are made considerably to develop methods for measurement of $k_L a$ of gas-liquid systems and are summarized in Danckwerts' book (Danckwerts 1970). A review on the same has been made (Rao 2010). Essentially there are three methods which are important, they are discussed as follows:

Static method: This method is also known as **oxygen balancing method** since it is based on oxygen balance to the system by measuring oxygen entering the system and leaving the system. It is based on Eq. (16.9)

$$N_A = (k_L a)\ (C_{AL}^* - C_{AL})$$

where N_A is the rate of gas absorption in unit volume and is given by

$$N_A = \frac{(v_g C_{Ag})_{\text{in}} - (v_g C_{Ag})_{\text{out}}}{V} \tag{16.19}$$

where v_g is the volumetric flow rate of the gas A, C_{Ag} is the concentration of A in the gas phase. V is the total volume of the liquid. The subscripts 'in' and 'out' indicate inlet and outlet flow conditions.

Combining Eqs. (16.19) and (16.9), we get

$$N_A = \frac{(v_g C_{Ag})_{\text{in}} - (v_g C_{Ag})_{\text{out}}}{V} = (k_L a)(C_{AL}^* - C_{AL}) \tag{16.20}$$

Since it is difficult to measure $(C_{Ag})_{\text{in}}$ and $(C_{Ag})_{\text{out}}$ we replace them by applying ideal gas law as follows:

$$C_{Ag} = \frac{p_A}{RT}$$

where T is temperature in degrees Kelvin and R is universal gas constant and

$$R = 8.314 \text{ kJ/(kg mol.K)}$$

Therefor, Eq. (16.20) becomes

$$\left(\frac{\dfrac{(v_g p_{Ag})_{\text{in}}}{T} - \dfrac{(v_g p_{Ag})_{\text{out}}}{T}}{R} \right) = (k_L a)\ (C_{AL}^* - C_{AL}) \tag{16.21}$$

By measurement of partial pressure of A in inlet and outlet gas streams, and by measurement of C_{AL} in the liquid, we can calculate $(k_L a)$. Obviously Eq. (16.21) can be applied if and only if we can devise effective methods for measurement of p_{Ag} and C_{AL}. Any error in these measurements may lead to error in $(k_L a)$.

Dynamic gassing out method: The dynamic gassing out method or simply known as **dynamic method** is an unsteady state process of gas absorption. The liquid bulk is made bereft of the gaseous component either by reaction or by sparging out another gas. Suppose if we want to strip off oxygen from the aqueous medium, we sparge nitrogen, and bring the concentration to zero which is shown by point C in Figure 16.7. Upto point B, the gas concentration in the liquid bulk is constant. At point B, we stop supplying gas, and sparge another gas to strip off A from the bulk liquid. The concentration of A falls to point C. At point C, again we start admitting gas A; slowly the concentration will be building up and goes to an average concentration value $\overline{C}_{AL}$ at point R.

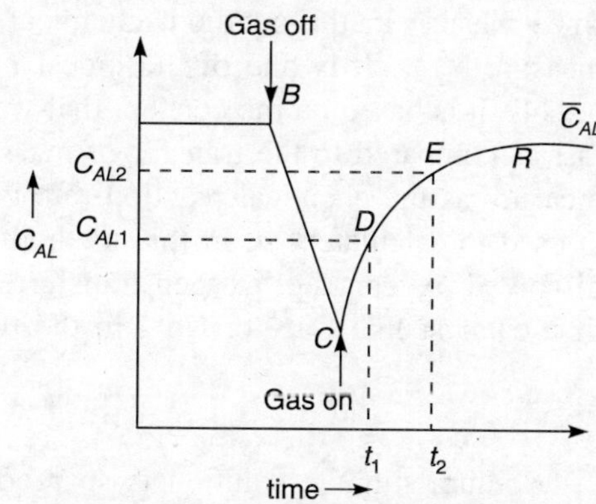

Figure 16.7 Measurement of $k_L a$ by dynamic method.

Now, we choose any two points like *D* and *E* in the *CR* zone and note down the time periods t_1 and t_2 and concentrations C_{AL1} and $C_{AL2.}$ The rate of change of concentration in the *CR* zone is given by

$$\frac{dC_{AL}}{dt} = k_La\ (\bar{C}_{AL} - C_{AL}) \tag{16.22}$$

Assuming k_La to be constant in the *CR* zone, we can integrate Eq. (16.22) to get

$$k_La = \left(\frac{\ln\left(\dfrac{\bar{C}_{AL} - C_{AL1}}{\bar{C}_{AL} - C_{AL2}}\right)}{(t_2 - t_1)} \right) \tag{16.23}$$

From Eq. (16.23), we can calculate k_La by experimentally measuring C_{AL1}, C_{AL2}, t_1, t_2 and, $\bar{C}_{AL}$. This is based on measurement at two points *D* and *E*. To improve the accuracy of measurement, we can measure C_{AL} at a number of points in *CR* zone. We can now daw a graph between ln $[(\bar{C}_{AL} - C_{AL1})/(\bar{C}_{AL} - C_{AL2})]$ and $(t_2 - t_1)$ as shown in Figure 16.8. Care should be taken while noting $(t_2 - t_1)$ since t_1 is initial time corresponding to C_{AL1}. The units of C_{AL} do not matter much in calculation since the term in Eq. (16.23) on the RHS ln $[(\bar{C}_{AL} - C_{AL1})/(\bar{C}_{AL} - C_{AL2})]$ is the ratio of concentrations. It could be in the form of kg/m^3 or ppm or partial pressures. The method is explained with a worked example.

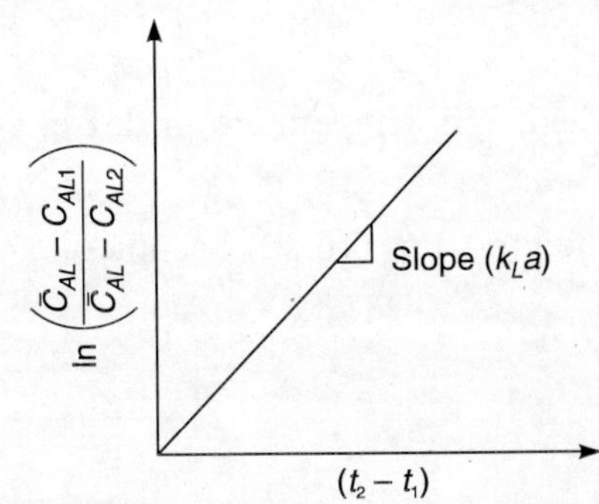

Figure 16.8 Plot for measurement of k_La by dynamic method.

The above is an excellent experimental method to measure k_La in any operating vessel, except that it needs a very efficient instrument to measure the changes in gas compositions very quickly which is of the order of a few seconds. Some of the other limitations associated with this method are fairly dealt and discussed by Van't Reit (1979).

Sulphite oxidation method: Sulphite oxidation method is also known as the chemical method since we use in this method, chemical reaction between the sodium sulphite and oxygen to measure (k_La). It is one of the oldest methods developed as early as 1944 by Cooper, et al., (1944). It is based on the concept that we choose a chemical reaction system which is extremely fast as compared to the transfer of mass for the chemical reaction to proceed. Hence, what we measure as the chemical reaction rate is actually the mass transfer rate only. This information is used to calculate k_La. In this method sodium sulphite is oxidized to sodium sulphate by the dissolved oxygen (i.e. oxygen transferred from gas phase into the bulk liquid) in the presence of a catalyst like Cu^{2+} or Co^{2+} in the form of copper acetate or cobalt acetate.

$$SO_3^{-2} + \frac{1}{2}O_2 \xrightarrow{Co^{2+}\ \text{cat}} SO_4^{-2} \tag{16.24}$$

Sodium sulphite solution is sparged with oxygen for the preceding reaction to take place in a known amount of time. Since the reaction rates are extremely fast, the reaction proceeds the moment oxygen is available in the dissolved form in the sodium sulphite solution. We measure the initial and final concentrations of sodium sulphite, and the time of sparging of gas. From the amount of sodium sulphite reacted, we can know the oxygen transferred in the known amount of time, which gives the oxygen transfer rate. Cooper, et al. (1944) sparged oxygen for 4–20 min. Since the reaction is very fast, the concentration of oxygen in the liquid phase is virtually zero. Hence, we write Eq. (16.9) as follows:

$$N_A = k_L a C_{AL}^*$$

And N_A is equal to r_A, the rate of chemical reaction of A.

Therefore,

$$r_A = N_A = (k_L a)\ C_{AL}^* \tag{16.25}$$

The method has the best advantage that it is one of the simplest methods, and allows for measurement of $k_L a$ in the working units, which gives us an indication of what would be the probable mass transfer rates to help in design of mass transfer equipment. The major drawback of the system is that it is heavily dependent upon the reaction conditions, viz.,

(i) Concentration of the catalyst
(ii) Type of the catalyst
(iii) Initial concentration of the sulphite, etc.

A further detailed discussion on the method can be found in Danckwerts' book (Danckwerts 1970).

PROBLEM 16.1 Dynamic gassing out method is used to measure the $k_L a$ in a vessel. The oxygen gas is stripped from the liquid medium, and then readmitted. The concentration of oxygen in the liquid at different time periods in the build-up zone is measured as follows:

Time, s	0	5	10	15	20	25
O_2 Conc., ppm.	0.5	1.7	2.7	3.5	4.2	4.7

The average concentration was noted to be 7 ppm. Calculate $k_L a$.

Solution Table 16.1 shows data required for plotting Eq. (16.23) in Figure 16.9. Slope of the line plotted in Figure 16.9 is 0.042.
Hence, $k_L a = 0.042\ s^{-1}$.

Table 16.1 Data Required for Plotting Eq. (16.23)

Time, s	*Concentration*, ppm	$(t_2 - t_1)$, s	$\dfrac{\bar{C}_{AL} - C_{AL1}}{\bar{C}_{AL} - C_{AL2}}$	$\ln\left(\dfrac{\bar{C}_{AL} - C_{AL1}}{\bar{C}_{AL} - C_{AL2}}\right)$
0	0.5			
5	1.7	5	1.226	0.204
10	2.7	10	1.512	0.413
15	3.5	15	1.857	0.619
20	4.2	20	2.321	0.842
25	4.7	25	2.826	1.04

Slope of the line = 1.05/25 = 0.042 s^{-1}

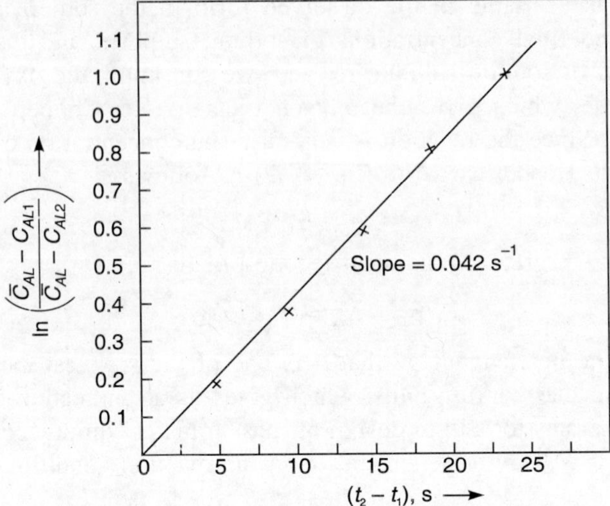

Figure 16.9 Plot of Eq. (16.23) for Problem 16.1.

16.5.2 Correlations for k_L

A number of correlations is available for calculation of k_L which is due to the good efforts made in the last half a century particularly in the stirred reactor systems. To make them widely applicable the correlations are made in the form of non-dimensional numbers. They are described here. The mass transfer coefficient is related by Sherwood number which in turn is related to Reynolds' number and Schmidt number.

Sherwood number (N_{Sh}): Sherwood number is defined as:

$$N_{Sh} = \frac{\text{Total mass transferred}}{\text{Mass transferred due to molecular diffusion}}$$

$$= \frac{k_L d}{D_{AB}} \tag{16.26}$$

Schmidt number (N_{Sc}): It is defined as:

$$N_{Sc} = \frac{\text{Molecular diffusion of momentum}}{\text{Molecular diffusion of mass}}$$

$$= \frac{\mu}{\rho D_{AB}} \tag{16.27}$$

Grash of number (N_{Gr}) This is defined as:

$$N_{Gr} = \frac{d_B^3 \rho_L g(\rho_L - \rho_g)}{\mu_L^2} = \frac{d_B^3 \rho_L g(\Delta\rho)}{\mu_L^2} \tag{16.28}$$

where d_B is the diameter of the bubble in m. It is applicable for mass transfer of the solute from bubbles or drops in the form of dispersions into the bulk liquid. If it is a gas-liquid system ρ_L and ρ_g stand for density of liquid phase and gas phase and μ_L is the viscosity of the liquid.

We also define one more number, known as Rayleigh number (N_{Ra}) which is a product of Schmidt number and Grashof number

$$N_{Ra} = N_{Sc} \times N_{Gr} \tag{16.29}$$

$$= \frac{d_B^3 g(\rho_L - \rho_g)}{\mu_L D_{AB}}$$

The Sherwood number relates mass transfer coefficient to various properties of the fluid and the dimensions of the system[†]. In general, Sherwood number correlates for forced convection as:

$$N_{Sh} = f(N_{Re}, N_{Sc}) \tag{16.30}$$

A number of empirical correlations is available in the literature, but all of them are based on the following inherent assumptions:

- The physical and thermal properties do not change appreciably during processing.
- Absolutely no chemical or biochemical reactions do take place in the fluids.
- Diffusion is purely by molecular diffusion; and there are no thermal or pressure diffusions.
- No viscous dissipation.

The liquid side mass transfer coefficient (k_L) for gas-liquid or fluid-fluid systems with bubble diameters less than 2.5 mm is correlated by

$$N_{Sh} = 2.0 + 0.31 \, (N_{Gr})^{1/3} \, (N_{Sc})^{1/3}$$

$$= 2.0 + 0.31 \, (N_{Ra})^{1/3} \tag{16.31}$$

For bubbles larger than 2.5 mm

$$N_{Sh} = 0.42 \, (N_{Gr})^{1/3} \, (N_{Sc})^{1/3} \tag{16.32}$$

For laminar flow of fluids over a *flat plate* for which $N_{Re} < 5 \times 10^5$ and $N_{Sc} \geq 0.6$, we can write if the flow is over entire surface of the phase,

$$N_{Sh} = \frac{k_L \times L}{D_{AB}} = 0.664 \, (N_{Re,\,L})^{1/2} \, (N_{Sc})^{1/3} \tag{16.33}$$

In which $N_{Re,\,L}$ is the Reynolds' number for the plate for which the characteristic dimension is the length of the plate (L) facing the flow, and hence,

$$N_{Re,\,L} = \frac{\rho L u}{\mu} = \frac{Lu}{\gamma} \tag{16.34}$$

where γ is known as **kinematic viscosity** and is the ratio of absolute viscosity and density, and has units of m^2/s.

Thus,

$$\gamma = \frac{\mu}{\rho} \tag{16.35}$$

[†]Sherwood number is similar to Nusselt number in convective heat transfer given by Eq. (11.11).

Similarly, for turbulent flow over a flat plate, when $N_{Re} > 5 \times 10^5$ and $0.6 < N_{Sc} < 3000$

$$N_{Sh} = \frac{k_L \times L}{D_{AB}} = 0.0296\,(N_{Re,\,L})^{4/5}\,(N_{Sc})^{1/3} \tag{16.36}$$

For flow through pipes: For flow when $N_{Re} < 10{,}000$,

$$N_{Sh} = \frac{k_L d}{D_{AB}} = 1.86\left[\frac{(N_{Re}\,N_{Sc})}{(L/D)}\right]^{1/3} \tag{16.37}$$

For turbulent flow conditions when $N_{Re} > 10{,}000$

$$N_{Sh} = 0.023\,N_{Re}^{0.8} N_{Sc}^{1/3} \tag{16.38†}$$

Flow over spherical bodies: When there is a relative motion between a spherical body and the surrounding fluid, as in the case of a packed beds where fluid is passing through stagnant spherical particles or as in the case where a solid sphere is freely falling in a stagnant liquid the mass transfer can be correlated by

$$N_{Sh} = 2.0 + (0.4\,N_{Re}^{1/2} + 0.06\,N_{Re}^{2/3})\,N_{Sc}^{0.4} \tag{16.39}$$

For a *freely falling liquid drops,* the above equation modifies to

$$N_{Sh} = 2.0 + 0.6\,N_{Re}^{1/2}\,(N_{Sc})^{1/3} \tag{16.40}$$

For a low viscosity drop falling through a viscous liquid

$$N_{Sh} = 1.13\,N_{Re}^{1/2}\,(N_{Sc})^{1/2} \tag{16.41}$$

PROBLEM 16.2 In a typical tutifruity making process, the raw papaya cubes are immersed in salt water in a square shaped vessel of 40 cm side. The room temperature is 30 °C. Air flow rate over the vessel is 0.5 m/s, and relative humidity of air is 55 per cent. Calculate the rate of water evaporation from the vessel when the water level is upto the brim of the vessel. Other desired physico-chemical data may be noted down from the literature. For want of data and for other practical considerations, the salt solution may be assumed to behave like water.

Solution First of all, we shall note down whatever the data are given, so that we can know what we need to collect from the literature.

Room temperature = 30 °C
Dimension of the vessel = 40 cm = 0.4 m
Air flow rate = 0.5 m/s
RH of air = 55 per cent

We may have to apply Eq. (16.33) or (16.36) depending upon the Reynolds' number range. Since the flow rate is very small, the flow seems to be laminar, and hence, Eq. (16.33) is likely to apply. For this, we need to find Reynolds' number and Schmidt number by Eqs. (16.34) and (16.27).

$$N_{Re,\,L} = \frac{\rho L u}{\mu} = \frac{L u}{\gamma}$$

[†]Note the similarity of Eqs. (16.38) and (11.62) for convective mass transfer.

and $$N_{Sc} = \frac{\mu}{\rho D_{AB}} = \frac{\gamma}{D_{AB}}$$

We have to note down data from literature for μ, ρ and D_{AB} from the book of Chandrasekharan and Venkateswarlu (1974). Actually we have to get the data at 30 °C, but we will note down data at 25 °C which is available in the book.

$$D_{AB} = 25.6 \text{ mm}^2\text{/s} = 2.56 \times 10^{-5} \text{ m}^2\text{/s}$$

$$\rho_{air} = 1.197 \text{ kg/m}^3$$

$$\mu_{air} = 18.46 \ \mu\text{Ns/m}^2$$

$$= 18.46 \times 10^{-6} \text{ Pa s} = 1.846 \times 10^{-5} \text{ kg/ms}$$

$$\gamma_{air} = 1.568 \times 10^{-5} \text{ m}^2\text{/s}$$

Therefore, $$N_{Re, L} = \frac{0.4 \times 0.5}{1.568 \times 10^{-5}} = 1.2755 \times 10^4$$

and $$N_{Sc} = \frac{1.568 \times 10^{-5}}{2.56 \times 10^{-5}} = 0.6125$$

Since $N_{Re, L}$ is $< 5 \times 10^5$, Eq. (16.33) is applicable.

$$N_{Sh} = \frac{k_L L}{D_{AB}} = 0.664\ (1.2755 \times 10^4)^{1/2}\ (0.6125)^{1/3} = 63.7$$

$$\frac{k_L \times 0.4}{2.56 \times 10^{-5}} = 63.7$$

Therefore, $$k_L = 4.07 \times 10^{-3} \text{ m/s}$$

Now, using k_L we shall try to find the rate of moisture transfer or rate of loss of moisture into air by using Eq. (16.1)

$$N_A' = k_L A\ (\Delta C)$$

ΔC is the concentration difference of the moisture vapour in the tank and that in the air. The moisture vapour present in air is given by relative humidity of air which is 55 per cent of the actual saturated concentration. We note down the saturated concentration from the steam tables for water vapour at 30 °C.

$$\Delta C = (C_{AS} - C_{A,\text{ air}})$$

where C_{AS} is saturated concentration

$$C_{AS} = \frac{1}{\upsilon_{w,g}} = \frac{1}{35.93} = 0.0278 \text{ kg/m}^3$$

where $\upsilon_{w,g}$ is the specific volume of the water vapour in m^3/kg at the desired temperature.
$C_{A,\text{ air}}$ at 55 per cent RH = $0.55 \times 0.0278 = 0.0153$ kg/m^3

$$N_A' = 4.07 \times 10^{-3} \times (0.4 \times 0.4)\ (0.0278 - 0.0153)$$
$$= 8.135 \times 10^{-6} \text{ kg/s} \cong 0.03 \text{ kg/h}$$

16.6 APPLICATIONS OF MASS TRANSFER IN FOOD PROCESSING

The applications of mass transfer or diffusional mass transfer in food processing are many, even though they are not comparable to those in chemical engineering. We shall briefly see where we come across these systems in food processing.

(i) Carbonation of beverages is a classical example of diffusional mass transfer in gas-liquid systems in which carbon dioxide is dissolved in water or some fruit juices to make the carbonated beverages for RTS (Ready To Serve) purpose; or it could be transfer of oxygen in fermentation broths.

(ii) Extraction of active principles from oleoresins using an organic solvent is an example of liquid-liquid extraction systems.

(iii) Extraction (or leaching) of oleoresins or spice oils from the spices is an example of solid-liquid leaching; making a decoction of coffee beans using hot water is also an example of leaching.

(iv) Removal of moisture from porous solids like food grains for drying of grains is an example of diffusional mass transfer of moisture vapours into surrounding air stream; the whole study of dehydration is a classical example of simultaneous heat and mass transfer. It could be drying of mango bar in which the moisture from the mango pulp is transferred into the hot air in a cabinet tray dryer; or it could be removal of moisture from the cooked baby foods or weaning foods in a drum dryer; or it could be removal of moisture from milk to make a spray dried milk powder in a spray dryer.

(v) Dissolving of sugar or salt into water is a solid-liquid system where the dissolution of the solid is restricted by the equilibrium solubility of the solid in the liquid.

(vi) Hydrogenation of oils and fats by hydrogen gas using solid raney nickel catalyst is a classical example for gas-liquid-solid systems operating in slurry reactors where the dissolved hydrogen gas saturates the unsaturated bonding of the fatty acids on the active sites of raney nickel solid catalyst.

(vii) Distillation of fermented molasis broth to separate alcohol from the rest of the water mixture; in the distillation column, the vapours as they go up get richer in more volatile alcohol vapours giving up the less volatile water to the down coming liquid stream; this is a classical example of simultaneous heat and mass transfer operation like drying.

(viii) In the crystallization of sugars from the concentrated sugar solution after evaporation, the solid sugars crystallize by transfer from the sugar solution to the solid form.

16.7 DIFFUSION OF GASES/VAPOURS THROUGH SOLID FILMS

One of the important applications of diffusion in food packaging is to predict how much the gases/vapours diffuse through the polymeric films which are used as packaging materials. Various polymeric materials used for food packaging are:

- PE (Polyethylene)
  - High Density Polyethylene (HDPE)
  - Low Density Polyethylene (LDPE)

- PP (Polypropylene)
- PVC (Polyvinyl Chloride)
- PVA (Polyvinyl Acetate)
- PETP (Polyethylene Terephthalate)
- Polyvinyledene chloride (Saran)
- Nylon-6
- Nylon-11

The gases and vapours that are generally considered for diffusion are oxygen, air, carbon dioxide, nitrogen, water vapour, various volatile materials (flavours) of foods like spices, etc. When a food material is packed in an air tight container, apparently we feel that the food material is safe; and no flavours are lost from it nor any entry of outside water vapour or air into the food material. But in reality, it is not the case. Food material is not all that safe because of the diffusion. However, in packaging terminology we use the word **permeability** or migration in place of diffusion. Permeability coefficient (P) has a slightly different definition, it is *defined as the product of diffusivity and solubility*, and it has the units of

$$\frac{\text{(Rate of gas/Vapour diffusing) (Film thickness)}}{\text{(Area) (Pressure differential across the film)}}$$

$$= \left(\frac{\text{kg}}{\text{m}^2\text{s}}\right)\left(\frac{\text{m}}{\text{Pa}}\right)$$

An expression for permeability of a gas through a polymeric film can be derived from the Fick's law of diffusion [Eq.(16.4)].

$$J_A = -D_{AB}\frac{dC_A}{dx}$$

The permeability process can be represented by Figure 16.10, which is similar to Figure 16.2, except that there is no equilibrium solubility what we had in the case of gas-liquid interphase. C_{A1} and C_{A2} are concentrations of the gas on either side of the film. Now, Eq. (16.4) can be written by removing the derivatives [integrating Eq. (16.4)] as follows:

$$J_A = D_A\frac{(C_{A1} - C_{A2})}{x} \tag{16.42}$$

In the above equation, we deliberately write D_A in place of D_{AB} since it represents only permeation of gas A. The concentration terms in Eq. (16.42) are replaced by the partial pressure and solubility terms using *Henry's law*.

$$C_A = Sp \tag{16.43}$$

where S is the solubility and p is the partial pressure of the gas.

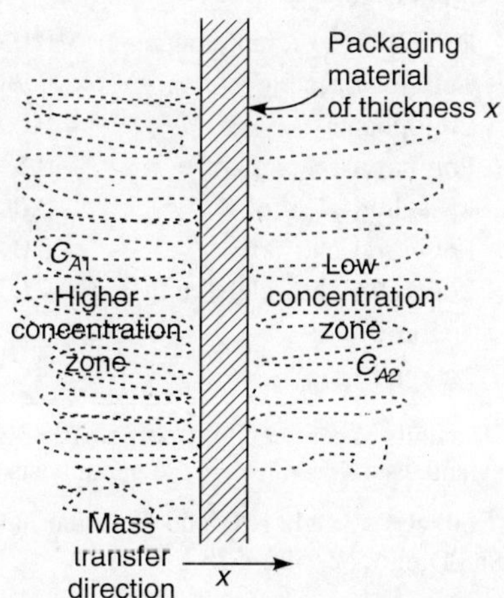

Figure 16.10 Permeability of a gas through a packaging film.

Eq (16.42) can be written as:

$$J_A = \frac{m_A}{At} = D_A S \frac{(p_{A1} - p_{A2})}{x} \tag{16.44}$$

where m_A is mass of gas transferred in t seconds though an area of A m^2.

Therefore,

$$\frac{m_A}{t} = D_A SA \frac{(p_{A1} - p_{A2})}{x} \tag{16.45}$$

In Eq. (16.45) $D_A S$ is known as **permeability coefficient *P*.**

$$P = D_A S \tag{16.46}$$

i.e.,

$$J_A = \frac{(p_{A1} - p_{A2})}{x} \tag{16.47}$$

The permeability coefficients for common gases like oxygen, nitrogen, carbon dioxide, hydrogen (also known as **fixed gases**[†]) and moisture through a number of polymeric materials are given in Table 16.2 mostly at 25 °C. The permeability changes with temperature and follows Arrhenius type of relationship.

$$P = P_0\, e^{-Ep/RT} \tag{16.48}$$

where

E_p is the activation energy for permeation

T is temperature in degrees Kelvin.

Table 16.2 Permeability Coefficient of some Gases/Vapours through Various Polymeric Films*

Polymeric film	*Permeability coefficient*[‡] × 10^{+17}				
	O_2	CO_2	N_2	H_2O	H_2
Polyethylene(density: 0.914)	2.88	17.325	0.848	50.67	–
Polyethylene(density: 0.964)	0.403	2.32	0.125	6.76	–
Polypropylene	2.3 (30°)	12.65 (30°)	0.385 (30°)	28.7	2.46 (20°)
Polyethylene teraphthalate	0.035	0.23	0.0057	73.19	–
Cellulose acetate	0.78 (30°)	31.21 (30°)	0.245 (30°)	3096.7	–
Cellophane	0.0021	0.0065	0.0028	1070	
Polyvinyl acetate	0.5 (30°)	–	–	–	–
Polyvinyl alcohol	0.0089	0.0165	0.0009 (14°)		
Polyvinyl chloride	0.0453	0.216	0.01	154.8	0.102
Polyvinyl chloride (Saran)	0.0053 (30°)	0.041 (30°)	0.00082 (30°)	0.28	–
Nylon-6	0.038 (30°)	0.121 (20°)	0.0083 (30°)	99.7	–
Nylon-11	–	1.375 (40°)	–	–	–

[‡]The units of permeability coefficient = (kgm)/(m^2sPa). The data are reported at 25 °C, unless otherwise mentioned in the paranthesis. The values in the paranthesis correspond to temperature °C.

*__Source:__ Singh, R.P. and Heldman, D.R. (2001), "Introduction to Food Engineering.", 3rd ed., Academic Press, San Diego, pp. 522–523.

[†]Fixed gases are those which have low boiling point.

PROBLEM 16.3 Dry grapes (raisins) are stored in a polyethylene pouch of 0.05 mm thickness. The EMC (Equilibrium Moisture Content) of the grapes is 7 per cent which corresponds to RH of 43 per cent. The pouch size is 15 cm × 15 cm. If the RH of the air is 90 per cent, find moisture transmission rate at 25 °C.

Solution Given: $x = 0.05 \text{ mm} = 5 \times 10^{-5} \text{ m}$

$A = 2 \times 15 \times 15 \times 10^{-4} = 0.045 \text{ m}^2$

RH of air = 90 per cent

ERH inside the pouch = 43 per cent

The vapour pressure of water vapour at 25 °C is noted from the steam tables

$p_s = 0.03166 \text{ bars} = 3.207 \text{ kPa}$

At 90 per cent RH, $p_{A1} = 0.9 \times 3.207 = 2.886 \text{ kPa}$

At 43 per cent RH, $p_{A2} = 0.43 \times 3.207 = 1.38 \text{ kPa}$

The permeability is noted from Table 16.2.

$$P = 50.67 \times 10^{-17} \frac{\text{kgm}}{\text{m}^2\text{sPa}} = 50.67 \times 10^{-14} \frac{\text{kgm}}{\text{m}^2\text{skPa}}$$

Now, we apply Eq. (16.45), making sure that unit of all terms are consistent.

$$\frac{m_A}{t} = \frac{PA[(p_{A1}) - (p_{A2})]}{x} = 50.67 \times 10^{-14} \times 0.045 \times (2.886 - 1.38)/(5 \times 10^{-5})$$

$$= 6.86 \times 10^{-10} \text{ kg/s} = 5.93 \times 10^{-5} \text{ g/day}$$

$$\cong 59 \ \mu\text{g/day}$$

The diffusional studies help us find out the moisture transportation rate through the barrier films. In Problem 16.3, instead of using LDPE, had we used HDPE, the Moisture Transmission Rate (MTR) would have come down by 50.67/6.76 = 7.5 times, and the MTR would have been $(6.86 \times 10^{-10})/7.5 = 9.15 \times 10^{-13}$ kg/s or 7.8 μg/day. Thus, these studies would help us predict the weight gain by the material over a period of time and possibility of spoilage of the material with time due to migration.

Symbols

A: cross-sectional area (m^2)
a: interfacial area (m^{-1})
C: concentration
ΔC: concentration difference (kg/m^3)
C_A^*: equilibrium solubility of A
$\overline{C}_{AL}$: average concentration of A in the liquid phase
C_{A0}: initial concentration of A (kg/m^3)
D_A: diffusivity of A through the polymeric film (m^2/s)
D_{AB}: diffusivity of A into B (m^2/s)
d_b: diameter of the gas bubble (m)
E_P: activation energy for permeation (kJ/mol)
g: acceleration due to gravity (m/s^2)
J_A: mass flux of component A (kg/m^2s)
k_g: mass transfer coefficient in gas phase
k_L: mass transfer coefficient (m/s)
L: Length/breadth (m)
m: mass transferred (kg)
N_A': mass of component A transported, (kg/s)
N_A^1: mass flux by Eq. (16.15) (kg/m^2s)
N_A: mass transfer rate per unit volume (kg/m^3s)
N_{Gr}: Grashof number defined by Eq. (16.28)

N_{Ra}: Rayleigh number defined by Eq. (16.29)
N_{Re}: Reynolds number
$N_{Re,L}$: plate Reynolds number defined by Eq. (16.34)
N_{Sc}: Shmidt number defined by Eq. (16.27)
N_{Sh}: Sherwood number defined by Eq. (16.26)
P: permeability coefficient
P_0: coefficient defined by Eq. (16.48)
p: partial pressure (MN/m^2 or Pa)
p_s: saturated vapour pressure (kPa)
q: heat transfer rate (W)
R: universal gas constant (= 8.314 kJ/kg mole°C)
R: resistance for mass transfer (s)
r_A: chemical reaction rate of A
S: solubility [mol or kg/(m^3 Pa)]
s: surface renewal time, (s^{-1})
T: temperature (K)
t: time (s)
u: linear velocity (m/s)
u_b: bubble velocity (m/s)
V: volume (m^3)
v: volumetric flow rate (m^3/s)
x: distance in x-direction (m)
x: film thickness in Figure 16.10
y: distance (m)

Subscripts

A: component A
B: component B
g: gas phase
L: liquid phase
o: initial conditions

Superscript

*: equilibrium value

Greek Symbols

α: distribution coefficient
δ: film thickness (m)
θ: time of exposure (s)
μ: viscosity (kg/ms)
γ: kinematic viscosity (μ/ρ), given by Eq. (16.35)
$\upsilon_{w,g}$: specific volume of water vapour (m^3/kg)
τ: shear stress

REVIEW QUESTIONS

16.1 Write a note on diffusional mass transfer.

16.2 What is meant by diffusion?

16.3 What is Fick's law of diffusion?

16.4 Write a note on convective mass transfer.

16.5 What is meant by mass transfer coefficient?

16.6 Write the analogy between momentum, heat and mass transfer.

16.7 Describe the process of mass transfer in gas-liquid systems.

16.8 What is meant by distribution coefficient?

16.9 What are various theories of mass transfer? Explain them briefly.

16.10 Explain the film theory of mass transfer. What are its merits and demerits?

16.11 Explain the penetration theory of mass transfer. What are its merits and demerits?

16.12 Explain the surface renewal theory of mass transfer. What are its merits and demerits?

16.13 What are different methods of measuring mass transfer rates (k_La)?

16.14 How do you measure k_La by static method? What are its merits and demerits?

16.15 How do you measure k_La by dynamic method? What are its merits and demerits?

16.16 How do you measure k_La by sulphite oxidation method? What are its merits and demerits?

16.17 Explain the following dimensionless numbers:

(i) Sherwood number
(ii) Schmidt number
(iii) Grashof number
(iv) Rayleigh number

16.18 What are different predictive equations for calculating mass transfer coefficient?

16.19 What are various applications of mass transfer in food processing?

16.20 Write a note on diffusion of gases/vapours through solid films. What are its applications in food technology?

NUMERICAL PROBLEMS

16.1 In a typical gas absorption equipment, carbon dioxide is dissolved in water. The diffusivity of carbon dioxide into water is 1.8×10^{-9} m^2/s. Calculate the mass transfer coefficient using film theory if the stagnant liquid film thickness is about 3 microns.

(**Ans:** $k_L = 6.0 \times 10^{-4}$ m/s)

16.2 A spherical ball of jaggery of 7.5 cm dia is placed in a stream of water flowing at the rate of 0.1 m/s. The temperature is reasonably constant at 28 °C, and the diffusivity of jaggery into water may be approximately taken as 4.5×10^{-6} cm^2/s. Calculate the mass transfer coefficient. [**Hint:** Use Eq. (16.40)]

(**Ans:** $k_L = 4.07 \times 10^{-6}$ m/s)

REFERENCES

Chandrasekharan, K.D. and Venketeswarlu D. (1974), *SI Units in Chemical Engineering and Technology*, Chemical Engineering Development Centre, IIT-Madras, Chennai.

Cooper, C.M., Fernstram, G.A. and Miller, S.A. (1944), Performance of agitated gas-liquid contactors, *I & EC.*, **36**(6), pp. 504–509.

Danckwerts, P.V. (1951), Significance of liquid-film coefficient in gas absorption, *I & EC.*, **43**, pp. 1460–1467.

Danckwerts, P.V. (1970), *Gas-liquid Reactions*, McGraw-Hill, New York.

Higbie, R. (1935), The rate of absorption of pure gas into a still liquid during short periods of exposure, *Trans. A I Ch E*, **31**, p. 365.

McCabe, W.L., Smith, J.C. and Harrvott, P. (1993), *Unit Operations of Chem. Engg.*, 5th ed., McGraw-Hill, New York, pp. 647–676.

Rao, D.G. (1990), A semi-empirical model for oxidation of cyclohexane, *I & EC.*, Process Design & Develop., **29**, pp. 696–699.

Rao, D.G. (2010), *Introduction to Biochemical Engineering*, 2nd ed., Tata McGraw-Hill, New Delhi.

Singh, R.P. and Heldman, D.R. (2001), *Introduction to Food Engineering*, 3rd ed., Academic Press, San Diego, pp. 522–523.

Van't Reit, K. (1979), Review of measuring methods and results in non-viscous gas-liquid mass transfer in stirred vessels, *I & EC., Process Design & Develop.*, **18**(31), pp. 357–364.

Whitman, W.G. (1923), Preliminary experimental confirmation of two film theory of gas absorption, *Chem. Met. Engg.*, **29**, pp. 146–148.

CHAPTER 17

Dehydration

Dehydration is one of the most important unit operations in food processing. Probably food industry is one process industry, in which dehydration is used maximum as compared to any other process industries; and hence, is well studied in food engineering as a unit operation, and in food technology as a processing tool. Thus, in food processing dehydration serves dual purposes:

(i) As a unit operation for removal of last traces of water.

(ii) As a technological tool for production of dehydrated foods.

We shall look into these details subsequently.

In dehydration, we notice that moisture in the liquid form evaporates, and diffuses into the surrounding air in the form of vapours by application of heat. More is the heat applied, more is the loss of moisture. Thus, it involves both heat transfer and mass transfer. Hence, it is categorized as a unit operation involving *simultaneous heat and mass transfer*.

It is different from evaporation since evaporation is a heat transfer process, in which no diffusional mass transfer occurs, whereas in dehydration, we encounter both heat transfer and mass transfer. In dehydration, the end product is invariably a dry solid; yet it is different from crystallization, in which the solids are crystallized out of a solution by using solubility difference; and drying is usually used to dry the crystallized solids to get in a pure form.

Even though it is a simultaneous heat and mass transfer operation, it is different from distillation in which the separation of one liquid from the other is achieved by using the differences in volatilities of both the liquids. Distillation is restricted to separation of one liquid from the other, and probably both the liquids are valuable, and hence, are desired. Thus, dehydration (or drying) is unique.

Even though we use the term drying for removal of solvents or water from the solids, sometimes even the liquids are dried by removing the last traces of moisture contained in them by various methods. But we do not consider here such operations.

Dehydration is also used as a technological tool in food processing for preservation of foods over a period of time. By removal of moisture, the *water activity* in solids is reduced, and hence, microorganisms which cause spoilage of foods cannot survive in the absence of water. The activity of microorganisms or enzymes is drastically affected if the water activity is low. This principle is made use of in food technology as a powerful tool for post harvest preservation of foods. It could be fruits, vegetables, spices, condiments, cereals, pulses, meat, fish, or some of the convenience foods and ready to eat foods; all are preserved by simple application of dehydration technique.

Based on the characteristics of the materials, all the food materials can be conveniently classified for the purpose of dehydration into those,

(i) which are available in the form of liquid solutions and gels; viz., fruit juices, milk, gelatinized products and some solutions having dissolved solutes come under this category.

(ii) which have capillary porous structure and rigid; viz., cereals, pulses, etc.

(iii) which are colloidal in nature, and have capillary porous structure, such as meats, fruits, vegetables and tissues.

In processing of all the above varieties of products, mostly our main concern is preservation. Now, let us see what are various parameters that determine the drying characteristics of foods. Some of them are:

(i) Structure of food materials

(ii) Equilibrium of moisture with the surroundings

(iii) Temperature and humidity of the surroundings

(iv) Transport phenomena of moisture migration

(v) Type of drying method used

(vi) Type of dryer used

An understanding of some of the above, and their interdependence of one over the other, coupled with the interaction of simultaneous heat and mass, make dehydration a process based on sound scientific principles, which otherwise could have been reduced to an art.

17.1 VARIOUS DRYING METHODS

Drying takes place by a number of methods. They are all essentially classified into the following six methods:

(i) *Convective drying* or contact drying is the one in which the material is in direct contact with the drying air; the moisture from the solid material evaporates directly into the stream of surrounding air. Thus, drying is dependent upon the temperature, flow rate and conditions of air.

(ii) *Conductive drying* or indirect contact drying is the one in which solid material receives heat from a hot surface, i.e., the drying material is dried by indirect contact with the heat source.

(iii) *Vacuum drying* is the process in which the wet solids are dried under vacuum. There may not be any convective heat transfer at all. Removal of moisture is facilitated by application of vacuum, and drying takes place at a much lower temperatures.

(iv) In *freeze drying* the material is initially cooled and high vacuum is applied; this transfers the moisture from solid phase (ice) to vapour phase by sublimation without passing through the liquid phase.

(v) In *radiation drying* the heat is supplied to the material by means of radiation. It is neither by direct contact with the hot air (heating source) nor by indirect contact with the heating source.

(vi) *Microwave heating-cum-drying* is an altogether a different heating method in which heat is generated in the food material by application of electromagnetic waves which in turn would generate heat by molecular friction, and the quantity of heat generated is dependent upon the moisture content of the food material.

We shall discuss various types of dryers based on all the first four methods in the *drying equipment section.*

17.2 DRYING PRINCIPLES

Drying of food material takes place by exposure of the wet material to a stream of air. Depending upon the relative humidity of air, the moisture diffusion into air varies. So also is the temperature of air, since drying is a simultaneous heat and mass transfer operation. However, more important is the relative humidity of air. The moisture content in the food material equilibrates with the surrounding air depending upon the nature of the material; the moisture content which equilibrates with the surrounding air is known as the **Equilibrium Moisture Content (EMC)** for the corresponding Relative Humidity (RH) of air. Such relationship is known as **EMC-RH data** for the food material. It is obvious that all materials do not have same EMC-RH data. Depending upon the porous nature of the material, the vapour pressure exerted by the moisture in the food material equilibrates with the moisture in the surrounding air which depends upon its relative humidity. Figure 17.1 shows such curves for some typical food materials.

Now, we shall define certain terms related to moisture contained by the materials.

Moisture content: It is a fractional term and is defined as the ratio of moisture content present anytime to the total weight of the wet solid (which is a summation of moisture content and weight of the dry solids). Thus,

$$x = \frac{\text{kg of moisture present}}{\text{kg of the wet material}}$$

$$= \frac{\text{kg of moisture present}}{(\text{kg of moisture} + \text{kg of dry solids})} \tag{17.1}$$

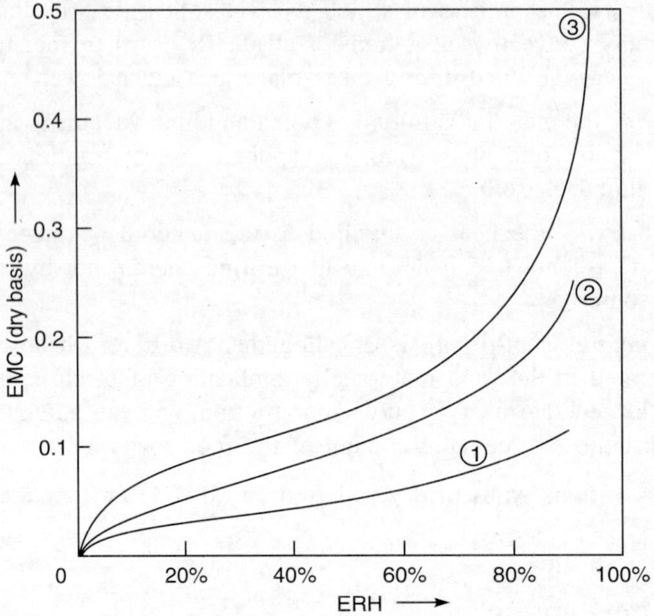

Figure 17.1 EMC-RH data at 20 °C for ① cellulose ② Egg albumin ③ potatoes.

Sometimes we deliberately use the subscript w to write it as x_w which indicates that the moisture content is based on the wet solids, and hence, is known as **moisture content on wet basis**.

Moisture content on dry basis: It is defined as the ratio of the moisture content present at any time to the weight of dry solids. Thus,

$$X = \frac{\text{Weight of moisture present at any time}}{\text{Weight of dry solids}} \tag{17.2}$$

$$X = \frac{x}{1-x} \tag{17.3}$$

or

$$x = \frac{X}{1+X} \tag{17.4}$$

Generally we use in food engineering the moisture content based on dry weight basis. It has the advantage as can be seen from Eq. (17.2) that the denominator remains constant as we compare on the basis of the weight of dry solids which obviously remains constant. So it becomes easy for us for making comparison during the process of drying, whereas if we use Eq. (17.1) the denominator will be changing as the drying continues.

Another interesting feature of Eq. (17.2) is that it need not be a fraction of one, we use the word *ratio*. For example, most of the fruits and vegetables contain moisture content of the order of 90 per cent or so; i.e., every kg of wet material contains 0.9 kg of water and 0.1 kg of dry solids. Hence, in such a case,

$$X = \frac{0.9}{0.1} = 9.0$$

or if we speak in terms of percentages, it is 900 per cent.

whereas $x = \frac{0.9}{1.0} = 0.9$ or in terms of percentages, it is 90 per cent.

Equilibrium moisture content (X^*): It is defined as the moisture content of the material which is in equilibrium with the relative humidity of the surrounding air; i.e., the vapour pressure of the moisture in the food is equal to the partial pressure of moisture in the humid air. It is generally represented with a superscript *. It is better explained with an EMC-RH curve (Figure 17.2). The moisture content at *P* corresponding to the relative humidity *E*, is called the **equilibrium moisture content** of the food material which is in equilibrium with the relative humidity at *E*. Obviously we can only remove the moisture in excess of X^*. Once if the moisture content reaches the equilibrium value, the drying rate falls to zero, and no more moisture can be removed with that air.

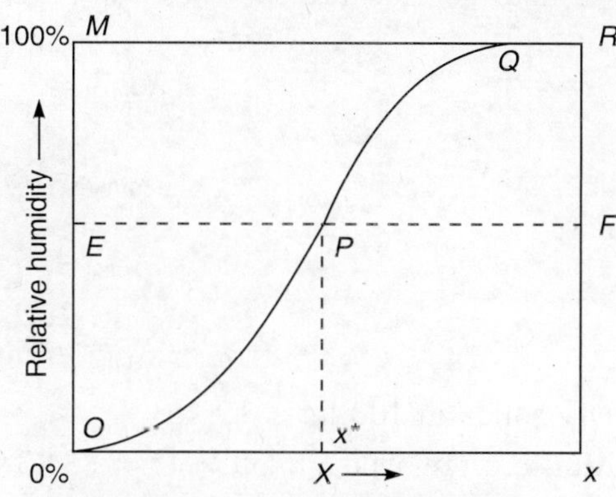

Figure 17.2 A typical EMC-RH curve.

Free moisture content: Free moisture content is the one which is in excess of the equilibrium moisture content, and hence, can be removed. *PF* represents the free moisture content in Figure 17.2.

Unbound moisture content: It is the moisture content in the food material which exerts an equilibrium vapour pressure equal to the saturated vapour pressure of water at that temperature. In simple words, we can say that it is the moisture content available in unbound form, and hence, can easily be removed. *QR* represents the unbound moisture content in Figure 17.2.

Bound moisture: Bound moisture is all the moisture present excluding (in excess of) the unbound moisture. This includes the equilibrium moisture also; and hence, all of it cannot be removed. *MQ* represents the bound moisture in Figure 17.2.

PROBLEM 17.1 Onions with 85 per cent moisture content on wet basis is dried to a final moisture content of 10 per cent. If we dry 100 kg of onions, what is the final weight of the product, and how much water is removed?

Solution Initial weight of onions = 100 kg

Moisture content on wet basis = 85 per cent

Water present in onions = 85 kg

Dry weight of the solids = 100 – 85 = 15 kg

Final moisture content = 10 per cent

Moisture present in the product is 10 kg, and the dry weight of the solids is 90 kg.

Therefore, 15 kg of dry weight of solids contain (15/90) × 10 = 1.66 kg of water

Final weight of product = 15 + 1.66 = 16.66 kg

Water removed = 85 – 1.66 = 83.33 kg

PROBLEM 17.2 Repeat the Problem 17.1 by converting the moisture content into dry weight basis.

Solution

$$x_0 = 0.85$$

we find the moisture content on dry weight basis by using Eq. (17.3)

$$X_{\text{initial}} = \frac{0.85}{1 - 0.85} = 5.66$$

$$x_{\text{final}} = 0.1$$

$$X_{\text{final}} = \frac{0.1}{1 - 0.1} = 0.11$$

$$X_{\text{initial}} - X_{\text{final}} = 5.66 - 0.11 = 5.55 \text{ kg of water/kg of dry solids}$$

Dry solids in 100 kg = 15 kg
Moisture removed = 5.55 × 15 = 83.23 kg
Final weight of the product = 100 – 83.23 = 16.76 kg.

17.2.1 EMC-RH Data

EMC-RH data play an important role in dehydration since it is only the moisture in excess of the EMC that can be removed with the existing air. Figure 17.1 shows data for some typical foods. It is also customary practice to fit the data to some of the imperical expressions so that EMC at any other conditions can be evaluated. Toledo (1997) gave a number of such expressions. Rao, et al. (1992) used the Henderson's equation to model the EMC-RH data of toria (*Brassica campestris*) seeds. The toria seeds of known weights were kept in desiccators containing sulphuric acid of different normalities, (Table 17.1) at room temperature, and final constant weight of the seeds was noted when the seeds equilibrated with the surrounding RH of the air in the desiccator. The EMC was noted based on the final weight of the sample and weight of the dry solids. The EMC was correlated with the corresponding RH, and the data were shown in Figure 17.3. Alternatively different RHs in the desiccators can also be maintained by using saturated salt solutions as shown in Table 17.2.

Table 17.1 Maintaining Specific RH at 25 °C Using H_2SO_4 of Different Normalities

RH(%)	*Specific gravity of* H_2SO_4 *solution*	*Normality of* H_2SO_4 *solution*
0	1.835	22.0
5	1.608	19.5
10	1.552	18.0
15	1.512	16.8
20	1.478	15.8
25	1.450	14.8
30	1.425	13.9

(contd.)

Table 17.1 Maintaining Specific RH at 25 °C Using H_2SO_4 of Different Normalities (*contd.*)

RH(%)	*Specific gravity of H_2SO_4 solution*	*Normality of H_2SO_4 solution*
35	1.40	13.9
40	1.377	13.1
45	1.355	12.3
50	1.335	11.5
55	1.314	10.8
60	1.293	10.0
65	1.271	9.2
70	1.248	8.3
75	1.224	7.4
80	1.197	6.3
85	1.165	5.2
90	1.127	3.9
95	1.076	2.4
100	1.00	0.0

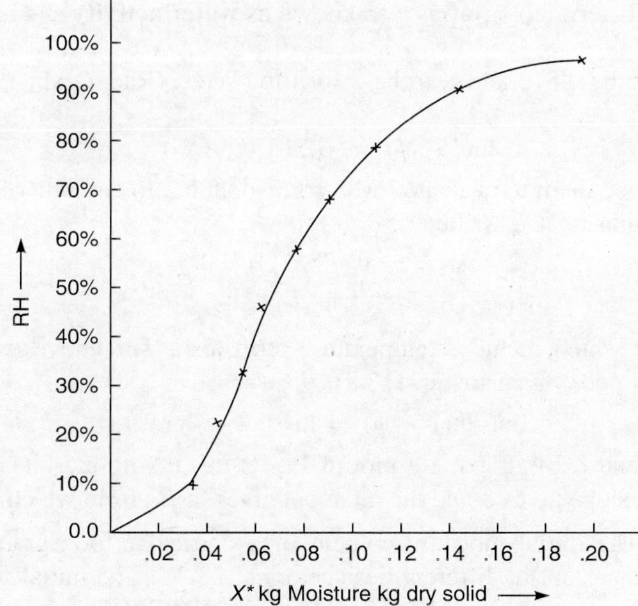

Figure 17.3 EMC-RH data for toria seeds at 30 °C.

Source: This article was published in *J. Food Engg.*, Vol. 17, Rao, D.G., Sridhar B.S. and Nanjundaiah, G., Drying of toria, (*Brassica campestris*, Var. toria) seeds: Part 1, Diffusivity characteristics, p. 53, Copyright Elsevier (1992).

Table 17.2 Maintaining RH Using Saturated Salt Solutions in Water at 20 °C

Salt	*Chemical formula*	*RH, ψ*
Lithium chloride	$LiCl$	0.126
Potassium acetate	CH_3COOK	0.20
Calcium chloride	$CaCl_2$	0.323
Potassium carbonate	K_2CO_3	0.45
Sodium hydrogen sulphate	$NaHSO_4$	0.52
Sodium nitrite	$NaNO_2$	0.66
Sodium chlorate	$NaClO_3$	0.75
Ammonium chloride	NH_4Cl	0.792
Potassium bromide	KBr	0.84
Potassium chromate	K_2CrO_4	0.88
Sodium sulphate heptahydrate	$Na_2SO_4{\cdot}7H_2O$	0.95
Lead nitrate	$Pb(NO_3)_2$	0.98

The data were fitted to Henderson equation

$$(1 - \psi) = \exp(-\beta(T)(X^*)^{\alpha}) \tag{17.5}$$

where ψ is the RH on decimal basis, and T is temperature in degrees Kelvin, β is constant, and α is coefficient.

In food technology terminology ψ is also known as **water activity** and is usually represented by a_w.

β was reported to be 0.513 and α to be 2 for toria seeds (Rao, et al., 1992), and Eq. (17.5) can be written as:

$$\ln(1 - \psi) = -0.513\ (T)\ (X^*)^2 \tag{17.6}$$

Equation (17.6) enabled them to evaluate EMC corresponding to any other RH and temperature.

Henderson equation is also written as:

$$(1 - \psi) = \exp(-\beta'(X^*)^{\alpha}) \tag{17.7}$$

$$-\ln(1 - \psi) = \beta'\ (X^*)^{\alpha} \tag{17.8}$$

where β' is constant which includes temperature term also. The above equation is linearized by taking logarithms once again to Eq. (17.8),

$$\ln(-\ln(1 - \psi)) = \ln \beta' + \alpha \ln(X^*) \tag{17.9}$$

We can find β' and α by drawing a plot of Eq. (17.9) taking $\ln(-\ln(1 - \psi))$ on y-axis and $\ln(X^*)$ on x-axis. The slope gives α and the intercept gives $\ln(\beta')$ from which β' can be calculated.

PROBLEM 17.3 The initial moisture content of a particular food grain is 27 per cent on wet basis. The grains are put in different desiccators at 28 °C. The initial weights and the final equilibrated weights are shown in Table 17.3. Find the EMC-RH data and evaluate constants of Henderson equation.

Table 17.3 Data on Initial Weight and Final Equilibrated Weight of the Sample for Problem 17.3

Initial weight (g)	10.4	10.7	10.2	10.5	10.6	10.8	10.25	10.3	10.32
Final weight (g)	7.759	8.062	7.753	8.048	8.187	8.418	8.065	8.203	8.362
RH *percentage*	10	20	30	40	50	60	70	80	90

Solution From the initial moisture content (27 per cent) on wet basis, we find what is the dry weight of the solids. The difference between the final weight of the sample and the dry weight gives us the moisture present finally, based on which we can find equilibrium moisture content (X^*) on dry weight basis.

$$\text{Dry weight of solids} = \text{Initial weight } (1 - \text{Moisture content})$$

$$= \text{Initial weight } (1 - x)$$

where $x = 27$ per cent $= 0.27$

For first data point,

$$\text{dry weight of solids} = 10.4^* \ (1 - 0.27) = 7.592 \text{ g}$$

Final equilibrated moisture constant = 7.759 – 7.592 = 0.167 g

$$X^* = \frac{0.167}{7.592} = 0.022$$

Similarly, we calculate for all other data points and the results are given in Table 17.4

Table 17.4 Data for Problem 17.3

Initial weight (g)	*Dry weight of solids* (g)	*Final weight* (g)	X^*	$\ln(X^*)$	RH(%)	ψ (*decimal*)	$-\ln(1-\psi)$	$\ln(-\ln(1-\psi))$
10.4	7.592	7.759	0.022	–3.86	10	0.1	0.1054	–2.25
10.7	7.811	8.062	0.0314	–3.46	20	0.2	0.2231	–1.5
10.2	7.446	7.753	0.0413	–3.187	30	0.3	0.3567	–1.03
10.5	7.665	8.048	0.05	–3.0	40	0.4	0.5108	–0.67
10.6	7.738	8.187	0.058	–2.85	50	0.5	0.6931	–0.37
10.8	7.884	8.418	0.0678	–2.69	60	0.6	0.9163	–0.087
10.25	7.482	8.065	0.078	–2.55	70	0.7	1.204	0.186
10.3	7.519	8.203	0.091	–2.4	80	0.8	1.609	0.47
10.32	7.534	8.362	0.110	–2.21	90	0.9	2.3025	0.83

We write here Eq. (17.9)

$$\ln(-\ln(1 - \psi)) = \ln \beta' + \alpha \ln(X^*)$$

Now, we draw the plot of $\ln(-\ln(1 - \psi))$ on y-axis and $\ln(X^*)$ on the x-axis and is shown in Figure 17.4.

$$\text{Slope} = 1.86$$

$$\text{Intercept} = 4.925$$

Therefore, $\beta' = e^{4.925} = 137.7$

$$\text{Temperature} = 28\ ^\circ\text{C} = 273 + 28 = 301 \text{ K}$$

Therefore, $\beta = \dfrac{137.7}{301} = 0.457$

The Henderson equation is:

$$-\ln(1 - \psi) = 0.457\ (T) \times (X^*)^{1.86}$$

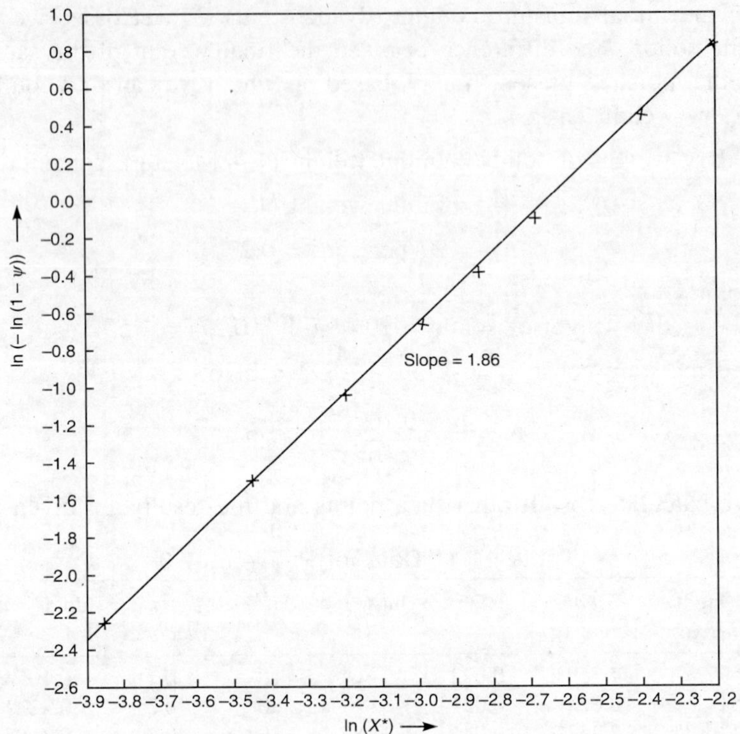

Figure 17.4 Data for Problem 17.3.

17.2.2 Drying Curve

When the wet solids are exposed to a stream of air, depending upon the relative humidity of air and temperature, the moisture from the solids evaporates into the air. As time progresses, the moisture content in the solids will be reducing. We can draw a plot between the moisture content and time which results in a typical plot as shown in Figure 17.5 by Rao, et al. (1992) for drying of toria seeds (*Brassica campestris*). Initially when the moisture content is high, the rate of loss of moisture is also fast. The rate of loss of moisture is defined as the drying rate and is mathematically represented as ($-dX/dt$). It is obtained by drawing tangents to the drying curve at different points, and the tangents represent the drying rate at that particular time or at that particular moisture content.

Sometimes, the drying curve data are represented in the form of a polynomial equation of the following type:

$$X = a_0 + a_1 t + a_2 t^2 + a_3 t^3 + \cdots \tag{17.10}$$

and the above expression is differentiated to obtain the drying rate (dX/dt)

$$-\frac{dX}{dt} = a_1 + 2a_2 t + 3a_3 t^2 + \cdots \tag{17.11}$$

The **drying curve** is drawn between the drying rate and moisture content as shown in Figure 17.6 for the data of Figure 17.5. In Figure 17.6, we find distinctly two regions, namely one in which the drying rate is constant and the other in which the drying rate is falling down.

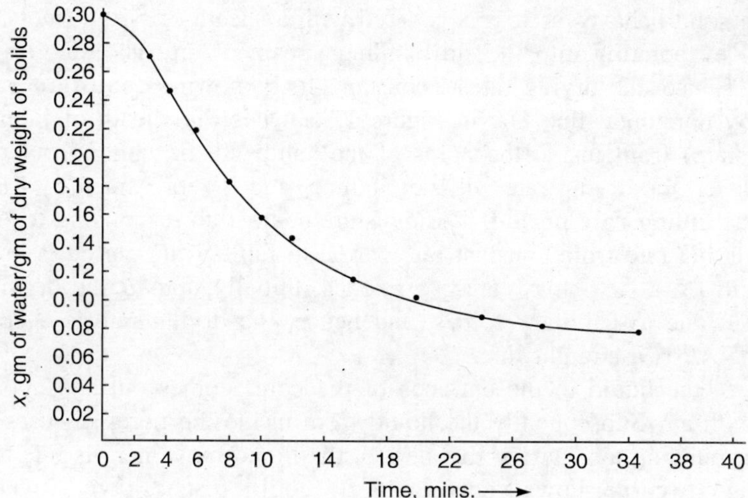

Figure 17.5 Drying curve for toria seeds at 50 °C.

Source: Reprinted from, Drying of toria (*Brassica campestris*, var. toria) seeds: Part 1, Diffusivity characteristics, Rao, D.G., Sridhar, B.S. and Nanjundaiah, G., *J. Food Engg.*, **17**, p. 54, Copyright (1992), with permission from Elsevier.

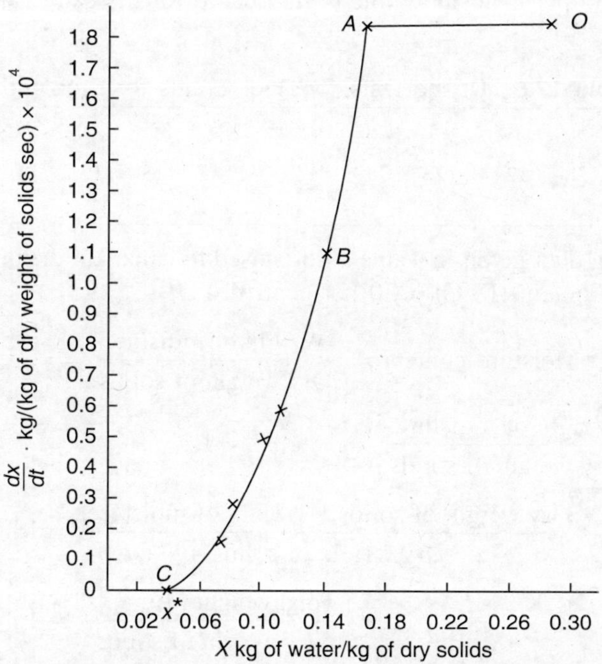

Figure 17.6 Drying rate curve for toria seeds at 50 °C.

Source: This article was published in *J. Food Engg.*, Vol. 17, Rao, D.G. Sridhar B.S. and Nanjundaiah G., Drying of toria (*Brassica campestris*, var. toria) seeds: Part 1, Diffusivity characteristics, p. 55, Copyright Elsevier (1992).

Initially, the solid behaves as if it is covered with a sheath of liquid (moisture) which will be continuously evaporating into the surrounding stream of air. The air gets fully saturated with the liquid. Hence, the drying rate is constant. It is known as **constant rate period**, and is represented by horizontal line OA in Figure 17.6. Once the sheath of liquid is exhausted, the liquid (moisture) from inside the pores of the solid will be coming out due to capillary forces; but however, the drying rate will start falling, and is represented by *ABC*. The portion *ABC* is known as **falling rate period**. The moisture content corresponding to *A* at which there is a change in drying rate from constant rate period to falling rate period is known as **critical moisture content** (X_C). *ABC* may have two zones. Initially upto *B* the drying is due to the liquid oozing out due to capillary forces, and hence, the drying rate is a constantly falling drying rate. Line *AB* is a straight line.

After point *B*, the liquid in the entrance of the pores comes out and gets exhausted, the place is filled with air. Subsequently the liquid from inside the pores diffuses through the air film into the surrounding air. During this period, the liquid movement is due to diffusion, and hence, the drying rate curve shows a curved plot *BC* until it reaches point *C* where the moisture content reaches the equilibrium moisture content, and hence, there will be no further transport of liquid into the air stream; with the result the drying rate becomes zero at point *C*. The entire plot *OABC* is known as **drying rate curve**, and is an important concept in drying process.

PROBLEM 17.4 The data for drying of wet ragi grains is shown in Table 17.5. The IMC on dry basis (db) is 30 per cent. Draw the drying curve for the same, and find out the critical moisture content.

Table 17.5 Drying Data for Ragi Grains for Problem 17.4

Time (min)	0	5	10	15	20	25	30	35	40	45	50	55	60
Weight of grains (g)	130	128	125	122	119	116	114	113	112.2	111.6	111.2	110.8	110.6

Solution The drying data given in Table 17.5 is used to calculate the dry weight of the solids.

Initial Moisture Content (IMC) = 30 per cent (db)

$$\text{Moisture content} = \frac{\text{Weight of moisture}}{\text{Dry weight of solids}} = 0.3$$

$$1 + \frac{\text{Weight of moisture}}{\text{Dry weight of solids}} = 1 + 0.3 = 1.3$$

i.e.,

$$\frac{\text{Dry weight of solids} + \text{Weight of moisture}}{\text{Dry weight of solids}} = 1.3$$

i.e.,

$$\frac{\text{Total weight of grains}}{\text{Dry weight of solids}} = 1.3$$

Therefore,

$$\text{Dry weight of solids} = \frac{\text{Total weight of grains}}{1.3} = \frac{130}{1.3} = 100 \text{ g}$$

$$X \text{ at 5 min time} = \frac{128 - 100}{100} = 0.28$$

Thus, we can find out X at various time periods, and are tabulated in Table 17.6. X vs. t data are shown in Figure 17.7. From the graph, we notice that constant rate period prevails upto 24 min.

Table 17.6 Drying Rate Data for Ragi Grains for Problem 17.4

Time (min)	*Weight of grains* (g)	X	$-(dX/dt)$
0	130	0.3	–
5	128	0.28	6.316×10^{-3}
10	125	0.25	6.316×10^{-3}
15	122	0.22	6.316×10^{-3}
20	119	0.19	6.316×10^{-3}
25	116	0.16	4.25×10^{-3}
30	114	0.14	2.73×10^{-3}
35	113	0.13	1.94×10^{-3}
40	112.2	0.122	–
45	111.6	0.116	1.06×10^{-3}
50	111.2	0.112	
55	110.8	0.108	7.0×10^{-4}
60	110.6	0.106	5.0×10^{-4}

Therefore,

$$\frac{dX}{dt} = \frac{0.28 - 0.16}{(5 - 24)} = -\,6.316 \times 10^{-3}$$

$$-\frac{dX}{dt} = 6.316 \times 10^{-3} \text{ kg of moisture/[(kg of dry solids)(min)]}$$

$$= 6.316 \times 10^{-3} \text{ min}^{-1}$$

At the time period of 25 min,

we draw a tangent at this point an find out $-\dfrac{dX}{dt}$

$$-\frac{dX}{dt} = \frac{0.2 - 0.115}{35 - 15} = 4.25 \times 10^{-3} \text{ min}^{-1}$$

Similarly at 30 min,

$$-\frac{dX}{dt} = \frac{0.182 - 0.1}{45 - 15} = 2.73 \times 10^{-3} \text{ min}^{-1}$$

Thus, the rate data are calculated at every time interval by drawing the tangent at that point and measuring the slope. The data are tabulated in Table 17.6, and plot is drawn in Figure 17.8.

17.2.3 Drying under Constant Drying Conditions

The drying rate curve shown in Figure 17.6 is applicable for constant drying conditions; viz., the temperature of air, flow rate of air, RH of air are all constant. If there are changes in the air, a different type of drying rate curve results (occurs). Since most of the industrial operations are carried out under a set pattern of conditions, the drying conditions may be assumed to be

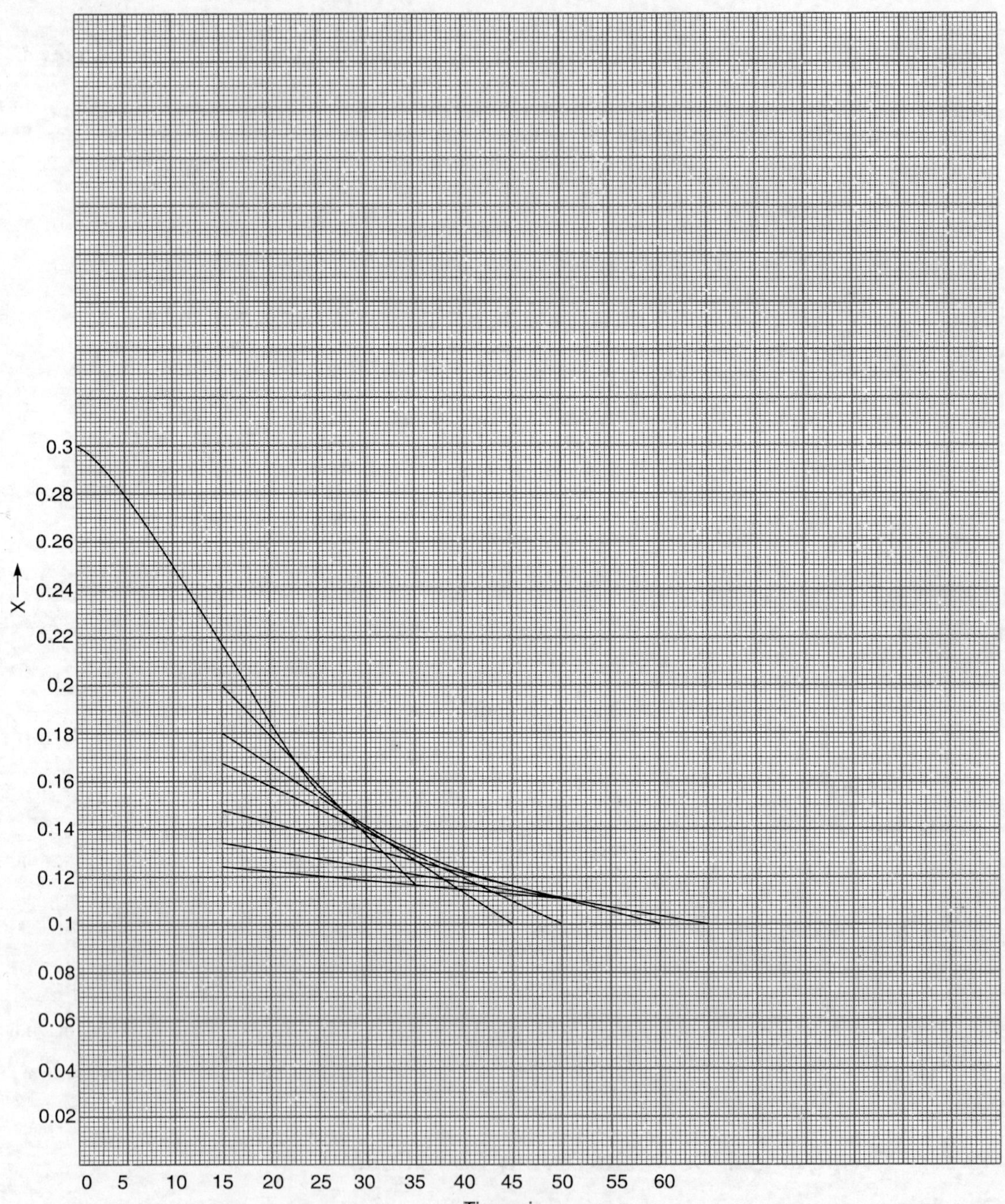

Figure 17.7 *X* vs. *t* data for Problem 17.4.

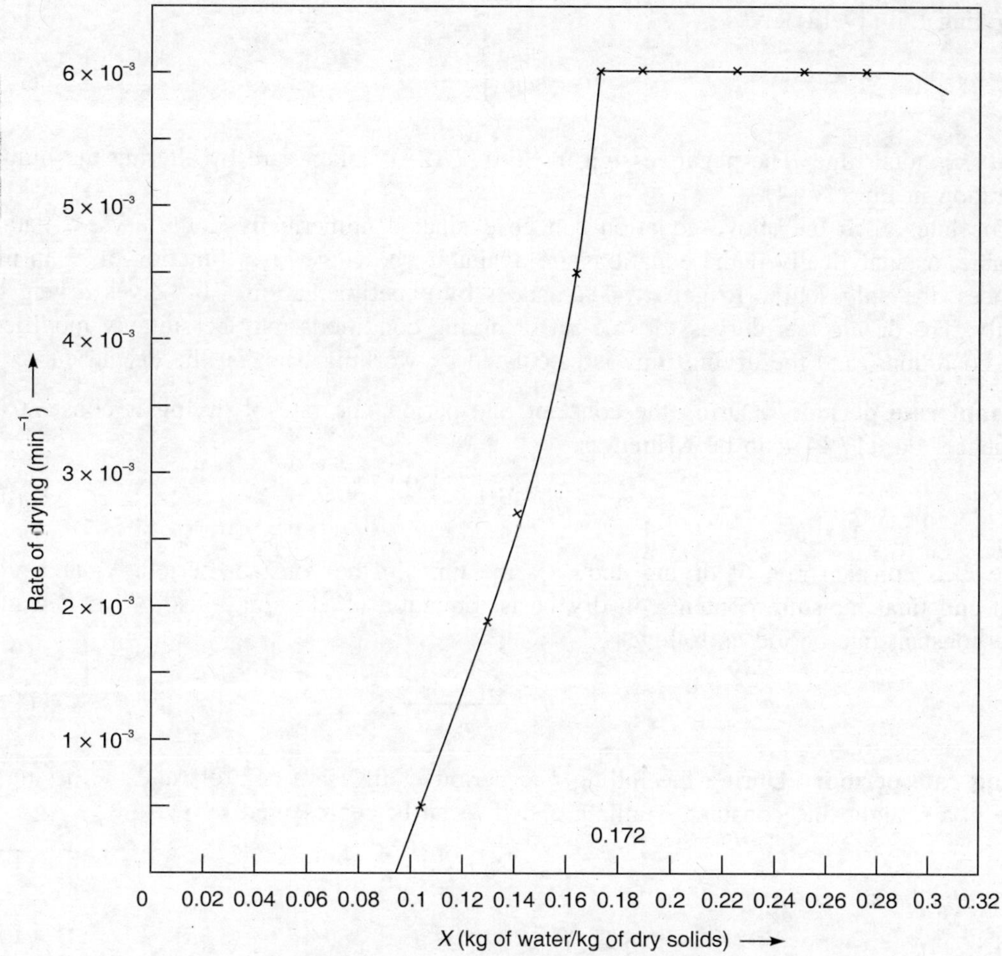

Figure 17.8 Drying curve for Problem 17.4.

reasonably constant, and hence, the following discussion holds good for finding out the total drying time to bring down the moisture content between two limits.

$$\text{Rate of drying } r = -m_s \frac{dX}{dt} \tag{17.12}$$

where r is rate of drying $= \dfrac{\text{kg of water evaporated}}{\text{time}}$, and m_s is weight of dry solids.

Alternately, the drying rate can also be defined on the basis of unit area of cross-section of drying as follows:

$$R = \frac{\text{kg of water evaporated}}{\text{(Unit area) (Unit time)}} \tag{17.13}$$

Integrating Eq. (17.12) gives

$$t_T = m_s \int_{x_2}^{x_1} \frac{dX}{r} \tag{17.14}$$

where t_T is total time. The negative sign in Eq. (17.12) is taken care by altering the limits of integration in Eq. (17.14).

The integral in the above equation can be evaluated numerically if the r vs. X data are available, or analytically if the equations are available showing r as a function of X. In many instances, the only source to get r vs. X data is by experimentation. This gives drying time directly. The drying rate curves for one set of drying conditions may be suitably modified to other conditions, and the drying time is calculated by working back on the drying curve.

Constant rate period: During the constant rate period, the rate of drying is constant at r_c and hence, Eq. (17.14) can be written as:

$$t_c = m_s \frac{(X_1 - X_2)}{r_c} \tag{17.15}$$

where r_c is constant rate of drying and t_c is the time in the constant period. X_1 and X_2 are initial and final moisture contents on dry basis. Equation (17.15) may also be written in the entire constant rate period as follows:

$$t_c = m_s \frac{(X_1 - X_c)}{r_c} \tag{17.16}$$

Falling rate period: During the falling rate period if diffusion controls and the drying rate curve is a straight line constantly falling, and if r can be represented by

$$r = aX + b \tag{17.17}$$

where a and b are constants, then

$$dr = a(X) \tag{17.18}$$

and

$$t_f = \frac{m_s}{a} \ln\left(\frac{r_1}{r_2}\right) \tag{17.19}$$

where r_1 and r_2 are drying rates at two positions. The constant a is given by

$$a = \frac{r_c - r_2}{X_c - X_2} \tag{17.20}$$

Combining Eqs. (17.19) and (17.20),

$$t_f = m_s \frac{(X_c - X_2)}{(r_c - r_2)} \ln\left(\frac{r_1}{r_2}\right) \tag{17.21}$$

Therefor,

$$t_T = t_c + t_f = m_s \left[\frac{(X_1 - X_c)}{r_c} + \frac{(X_c - X_2)}{(r_c - r_2)} \ln\left(\frac{r_1}{r_2}\right)\right] \tag{17.22}$$

If the entire falling rate period is a simple straight line, we get

$$t_T = \frac{m_s}{r_c}\left[(X_1 - X_c) + X_c \ln\left(\frac{X_c}{X_2}\right)\right] \tag{17.23}$$

PROBLEM 17.5 From the drying rate curve of Problem 17.4, calculate the total drying time to bring down the moisture content from 0.3 to 0.106 for 1.3 kg of raw material.

Solution Since we have only experimental rate data and no relationship to correlate rate with moisture content, we adopt the technique of numerical integration to evaluate t_T by Eq. (17.14). Hence, we follow the graphical integration procedure. The integral is calculated by the area of the graph between the limits $X_1 = 0.3$ to $X_2 = 0.106$ drawn between X and $1/r$. The data are as shown in Table 17.7. Plot is drawn between X vs. $\left(\frac{1}{r}\right)$ in Figure 17.9.

Table 17.7 Data for Problem 17.5

X	$r = -(dX/dt)$ min^{-1}	$(1/r)$ min
0.3	6.316×10^{-3}	158.33
0.28	6.316×10^{-3}	158.33
0.25	6.316×10^{-3}	158.33
0.22	6.316×10^{-3}	158.33
0.19	6.316×10^{-3}	158.33
0.172	6.316×10^{-3}	158.33
0.16	4.25×10^{-3}	235.3
0.14	2.73×10^{-3}	366.3
0.13	1.94×10^{-3}	515.5
0.122	–	–
0.116	1.06×10^{-3}	943.4
0.112	–	–
0.108	7.0×10^{-4}	1428.6
0.106	5.0×10^{-4}	2000.0

$$\text{Area under the graph} = 59.5835\ \frac{\text{kg of moisture}}{\text{kg of dry solids}} \times \frac{\text{min}}{\text{kg moisture evaporated}}$$

i.e.,

$$\int_{0.106}^{0.3} \frac{dx}{r} = 59.5835\ \frac{\text{min}}{\text{kg dry solids}}$$

$$\text{IMC} = \frac{0.3\ \text{kg of moisture}}{1\ \text{kg of dry solids}}$$

Weight of raw material = 1.3 kg

1.3 kg of raw material contains 1 kg of dry solids

m_s = Weight of dry solids = 1.0 kg

$$t_T = m_s \int_{0.106}^{0.3} \frac{dx}{r} = 1.0 \times 59.5835 = 59.6\ \text{min} \approx 1\ \text{hr}$$

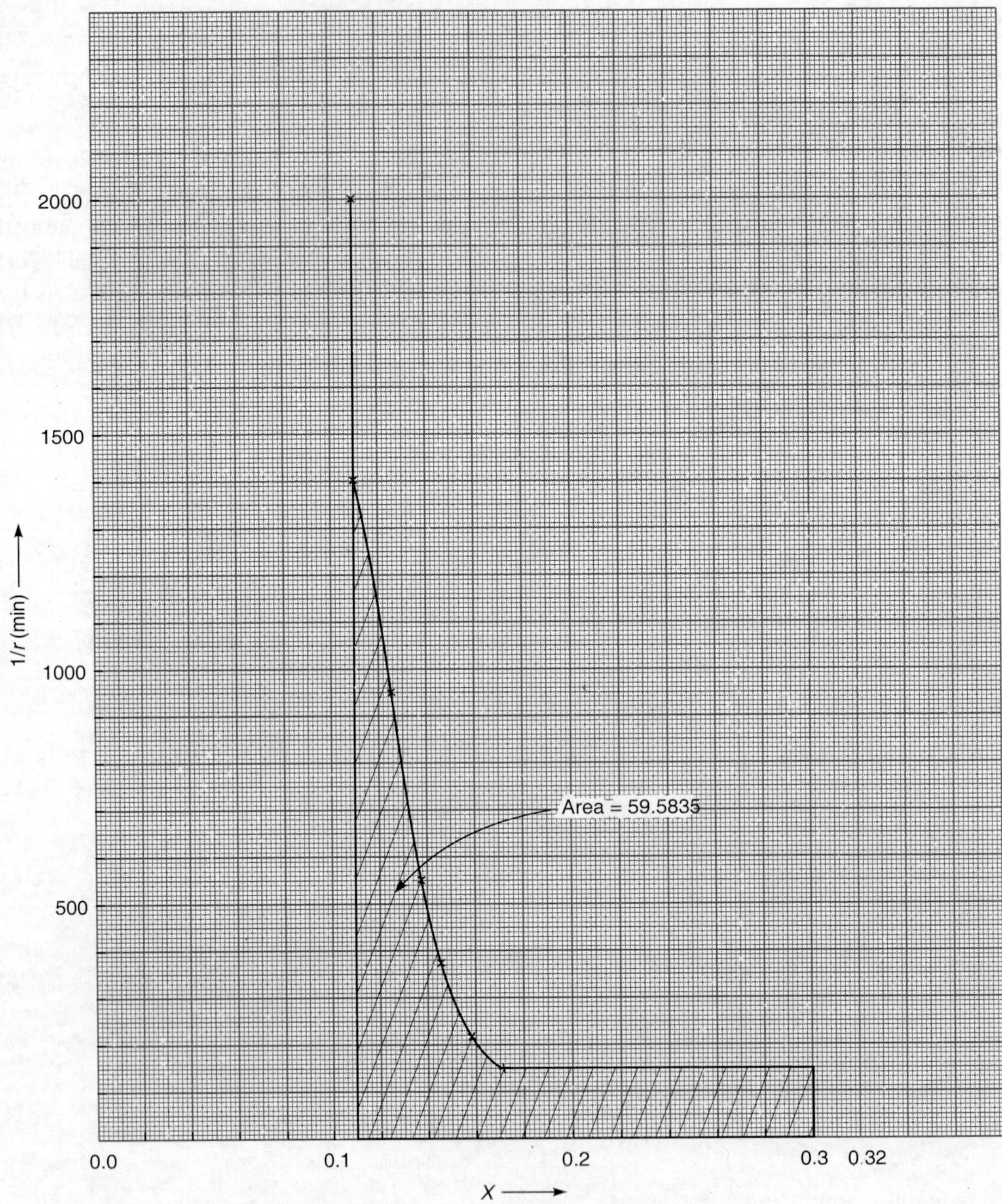

Figure 17.9 Data of X vs. (1/r) for Problem 17.5.

17.2.4 Heat Transfer in Drying

Heat transfer plays a major role in drying process. More is the quantity of heat transferred, more is the moisture evaporated from the solids, and hence, more is the drying rate. The heat transfer may take place by all three modes, viz., conduction, convection or radiation or by microwaves. A relationship between the rate of mass evaporation and quantity of heat to be supplied can be established by applying energy balance approach as follows. If $\dot{m}_v$ is the rate of moisture evaporation, the quantity of heat required to evaporate is equal to the mass of the moisture multiplied by the latent heat of vaporization. If we take the above in terms of rate terms,

$$q = \dot{m}_v \lambda_T \tag{17.24}$$

The heat transfer rate (q) is given by:

$$q = h_s A (T_a - T_s) \tag{17.25}$$

Therefore,

$$\dot{m}_v = \frac{h_s A (T_a - T_s)}{\lambda_T} \tag{17.26}$$

where h_s is heat transfer coefficient in kJ/m^2°Cs, T_a is temperature of drying air, and T_s is temperature at the interface of the drying solids and air, and hence, is usually equal to the wet bulb temperature of the air.

For cross flow conditions (air flowing parallel to the surface of the solids), h_s can be reasonably estimated by (McCabe, et al., 1993):

$$h_s = \frac{8.8 G^{0.8}}{d_e^{0.2}} \tag{17.27}$$

where G is mass flux of air in kg/m^{2}s, d_e is equivalent diameter of air flow duct in m, and h_s is in the units of W/m^2°C.

In the above equation, care should be exercised for appropriate units to suit to the numerical constant. The limitations of the above equation is that it is applicable in the Reynolds number range 2600–22,000 and the air temperature is approximately 95 °C.

Similarly, for cross flow conditions (air flowing perpendicular to the solids) Eq. (17.27) modifies to,

$$h_s = 24.2 G^{0.37} \tag{17.28}$$

the range of air velocities for the above equation is 0.9–4.5 m/s.

The moisture evaporation rate ($\dot{m}_v$) in Eq. (17.26) is also given by,

$$\dot{m}_v = k_g A (p_s - p_a) \tag{17.29}$$

where k_g is the gas phase mass transfer coefficient, p_s is the partial pressure of water vapour at the solid surface, and p_a is the partial pressure of water vapour in the air.

In terms of relative humidity Eq. (17.29) can be written as:

$$\dot{m}_v = k_g' A (H_s - H_a) \tag{17.30}$$

where H_s is the humidity of air at saturation, H_a is the humidity of drying air.

and $$k_g' = 1.8\ k_g \tag{17.31}$$

By combining Eqs. (17.26) and (17.29)

$$k_g = \frac{h_s\,(T_a - T_s)}{\lambda_T\,(p_s - p_a)} \tag{17.32}$$

or $$h_s = \frac{k_g\,\lambda_T\,(p_s - p_a)}{(T_a - T_s)} \tag{17.33}$$

The measurement of gas phase mass transfer coefficient is not that easy, and hence, is usually predicted by using a dimensionless number called **Lewis number** (N_{Le}) where

$$N_{Le} = \frac{h_s}{k_g' C_p} \tag{17.34}$$

For air-water system,

$$N_{Le} = 1.0$$

$$k_g' = \frac{h_s}{C_p} \tag{17.35}$$

where C_p is the specific heat of air, which is almost equal to 1 kJ/kg°C, and hence, in SI units

$$k_g' = h_s \tag{17.36}$$

PROBLEM 17.6 A particular food material is being dried in a cabinet tray dryer from an Initial Moisture Content (IMC) of 70 per cent to a Final Moisture Content (FMC) of 10 per cent on wet basis. Air enters at 10 per cent RH at 65 °C and leaves at 35 °C at 75 per cent RH. Find the quantity of air required to dry 1 kg of the raw material, if the product temperature is 30 °C.

Solution

Approach: We shall first try to find out what is the amount of moisture to be removed. Next we find out the moisture pick-up by the air per kg of dry air. Then we try to find out what is the quantity of air required to dry 1 kg of raw material.

Step 1: Moisture to be removed from 1 kg of raw material.

IMC = 70 per cent (wb)

Initial moisture = 0.7 kg

Dry weight of solids = 0.3 kg

FMC = 10 per cent (wb)

Final moisture = 0.1 kg

Final dry weight of solids = 0.9 kg

Hence, moisture associated with 0.3 kg of dry solids in the final product

$$= 0.1\times\frac{0.3}{0.9} = 0.033 \text{ kg}$$

Moisture to be removed from the produce = 0.7 – 0.033 = 0.667 kg.

Step 2: Moisture pick-up by kg of dry air.

Entering air:

RH = 10 per cent

Temperature = 65 °C

Humidity = 0.02 kg of moisture/kg of dry air from the psychrometry chart (Appendix 4).

$$H = 0.62 \frac{p_w}{(1-P_w)} \quad (6.3)$$

where

p_w is partial pressure of water vapour.

P_w is the saturated vapour pressure at 65 °C and is noted from Steam Tables (Appendix 1B).

$$P = 0.2501 \text{ bars}$$

$$\text{RH} = \frac{p_w}{P_w}\times 100 \quad (6.5)$$

i.e.,

$$0.1 = \frac{p_w}{0.2501}$$

$$p_w = 0.02501 \text{ bars}$$

$$H = 0.62 \frac{0.02501}{(1-0.02501)} = 0.016 \text{ kg moisture/kg of dry air}$$

Leaving air:

RH = 75 per cent

Temperature = 35 °C

Humidity = 0.027 kg of moisture/kg of dry air

(Psychrometric chart)

we calculate by using Eqs. (6.3) and (6.5)

$$P_w \text{ at } 35\ ^\circ\text{C} = 0.05622 \text{ bars}$$

$$p_w = 0.75 \times 0.05622 = 0.04216 \text{ bars}$$

$$H = 0.062 \frac{0.04216}{(1-0.04216)} = 0.0273 \text{ kg of moisture/kg of dry air}$$

Moisture gained by kg of dry air = 0.0273 – 0.016

= 0.0114 kg of moisture/kg of dry air.

Step 3: Dry air required to remove 0.667 kg of moisture

$$= \frac{0.667}{0.0114} = 58.53 \text{ kg}$$

1 kg of entering air contains (1 – 0.016) kg of dry air.

$$\text{Therefore, air required for drying 1 kg of produce} = \frac{58.53}{(1-0.016)} = 59.5 \text{ kg}$$

PROBLEM 17.7 For the drying process shown in Problem 17.6, find the quantity of heat required, and the thermal efficiency of the drying process.

Solution

Approach: First we find the quantity of heat required to vaporize 0.667 kg of moisture. Next we find the enthalpies of entering air and leaving air using the humid heat. Then we calculate thermal efficiency based on the heat absorbed by water to vaporize and the heat lost by the drying air.

Step 1: Heat required to vaporize the 0.667 kg of moisture = $m\lambda_T$

$$= 0.667\ (2556.4 - 125.66)$$

$$= 1621.3 \text{ kJ}$$

λ_T is noted from Steam Tables at 30 °C = 2556.4 – 125.66

$$= 2430.74 \text{ kJ/kg}$$

Step 2:

Entering air:

$H = 0.016$

Temperature = 65 °C

C_p of air at 65 °C (338 K) = 1.0082 kJ/kg°C[†]

C_p of water vapour at 65 °C = 1.925 kJ/kg°C

$$\text{Humid heat} = (C_p)_{\text{air}} + (C_p)_{\text{water}} \times H \tag{6.10}$$

$$= 1.0082 + (1.925 \times 0.016) = 1.034 \text{ kJ/kg°C}$$

Leaving air:

$H = 0.0273$

Temperature = 35 °C

C_p of air at 35 °C (308 K) = 1.005 kJ/kg°C

C_p of water vapour = 1.88 kJ/kg°C

$$\text{Humid heat} = 1.005 + (1.88 \times 0.0273)$$

$$= 1.056 \text{ kJ/kg°C}$$

$$(C_p)_{\text{air, average}} = \frac{1.034 + 1.056}{2} = 1.045 \text{ kJ/kg°C}$$

[†]Noted from Chandrasekhar, K.D. and Venkateshwarlu, D. (1974), *SI Units in Chemical Engineering and Technology*, Chemical Engineering Education Development Centre, IIT Madras, p. 145.

Air used = 59.5 kg

$$\Delta T = 65 - 35 = 30\ °C$$

Therefore, heat lost by air = 59.5 × 1.045 × 30 = 1865.6 kJ

Step 3:

$$\text{Thermal efficiency} = \frac{\text{Heat gained by solids to vaporize moisture}}{\text{Heat lost by drying air}}$$

$$= \frac{1621.3}{1865.6} = 0.869$$

Thermal efficiency = 86.9 per cent

17.2.5 Diffusion in Drying of Solid Foods

As has been mentioned, the moisture migration in the falling rate zone of solid foods is essentially due to diffusion. When the foods are dried slowly, moisture transfer is by diffusion from solid foods into the surrounding air. Since drying is an unsteady state process, we can apply Fick's second law for transient movement [Eq. (16.13)]

$$\frac{\partial C}{\partial \theta} = D_v \frac{\partial^2 C}{\partial x^2} \tag{16.13}$$

which is applicable for long slabs. The same on modification to spherical particles can be written as:

$$\frac{\partial C}{\partial t} = D_v \left(\frac{\partial^2 C}{\partial r^2} + \frac{2}{r}\frac{\partial C}{\partial r} \right) \tag{17.37}$$

with certain boundary conditions (Rao, et al., 1992).

In Eq. (17.37), we replace the concentration terms with moisture content terms, and on integration we get

$$\frac{X - X^*}{X_0 - X^*} = 0.608 \exp\left(-9.87 \frac{D_v t}{r^2} \right) + 0.152 \exp\left(-39.5 \frac{D_v t}{r^2} \right) + \cdots \tag{17.38}$$

where

X^* is equilibrium moisture content

X_0 is IMC

t is the time taken to bring down the moisture content from X_0 to X in the falling rate zone.

In most of the cases, the terms on the RHS other than the first term become insignificant, and hence, Eq. (17.38) can be written as:

$$\frac{X - X^*}{X_0 - X^*} - 0.608 \exp\left(- 9.87 \frac{D_v t}{r^2} \right) \tag{17.39}$$

We experimentally measure time (t) to bring down the moisture content from X_0 to X in the falling rate period, and calculate D_v by applying Eq. (17.39).

PROBLEM 17.8 Studies on drying of toria seeds were reported by Rao, et al. (1992) as follows at 50 °C: Time taken to bring down moisture content from 0.156 to 0.076 was 24 min while the EMC was 0.035. The average radius of the seeds is 0.8 mm. Calculate the diffusivity coefficient at 50 °C.

Solution It is a straight case of application of Eq. (17.39)

$$\frac{X - X^*}{X_0 - X^*} = \frac{0.076 - 0.035}{0.156 - 0.035} = 0.3388$$

$$= 0.608 \exp\left(-9.87 \frac{D_\upsilon t}{r^2}\right)$$

or

$$\frac{0.3388}{0.608} = 0.5573 = \exp\left(-9.87 \frac{D_\upsilon t}{r^2}\right)$$

$$\ln(0.5573) = -\,9.87\left(\frac{D_\upsilon t}{r^2}\right)$$

Therefore,

$$\frac{D_\upsilon t}{r^2} = -\frac{0.5846}{-9.87} = 0.0592$$

$$D_\upsilon = 0.0592 \frac{(0.8 \times 10^{-3})^2}{(24 \times 60)} = 2.63 \times 10^{-11} \text{ m}^2\text{/s}$$

Effect of temperature on diffusivity: The effect of temperature on diffusivity can be represented by Arrhenius type of equation as follows:

$$D_\upsilon = D_0 e^{-E/RT} \tag{17.40}$$

where D_0 is a constant, E is the activation energy in J/mole, T is temperature in degrees Kelvin, and R is universal gas constant = 8.3144 J/mole K.

To evaluate the constants D_0 and (E/R), we measure D_υ at different temperatures and then we can find out D_0 and (E/R) by linearizing Eq. (17.40) and plotting on a graph.

By taking logarithms on both sides of Eq. (17.40)

$$\ln(D_\upsilon) = \ln(D_0) - \left(\frac{E}{R}\right)\left(\frac{1}{T}\right) \tag{17.41}$$

We draw a plot taking $\ln(D_\upsilon)$ on y-axis and $(1/T)$ on x-axis, we measure the slope and intercept

$$\left.\begin{aligned} \text{Slope} &= -\left(\frac{E}{R}\right) \\ \text{and intercept} &= \ln(D_0) \end{aligned}\right\} \tag{17.42}$$

from which

$$D_0 = \exp(\text{intercept}) \tag{17.43}$$

The procedure is explained with a numerical example.

PROBLEM 17.9 The diffusivity data of toria seeds were reported by Rao, et al. (1992) at different temperatures (Table 17.8). Establish a relationship between diffusivity coefficient and temperature.

Table 17.8 Diffusivity vs. Temperature Data for Toria Seeds for Problem 17.9

Temperature °C	50	60	70	80
Diffusivity m^2/s	2.85×10^{-11}	5.55×10^{-11}	9.98×10^{-11}	2.03×10^{-10}

Solution It is a straight case of application of Eq. (17.41). The data on $1/T$ vs. $\ln(D_\upsilon)$ are presented in Table 17.9, and plotted on Figure 17.10.

Table 17.9 Data for Problem 17.9

T (*Temperature*) K	D_υ (m^2/s)	$1/T$	$\ln D_\upsilon$
323	2.85×10^{-11}	3.096×10^{-3}	– 24.28
333	5.55×10^{-11}	3.0×10^{-3}	– 23.61
343	9.98×10^{-11}	2.915×10^{-3}	– 23.03
353	2.03×10^{-10}	2.833×10^{-3}	– 22.29

$$\text{Slope} = \frac{-24.29 + 22.2}{(3.1 - 2.812) \times 10^{-3}} = -7257$$

$$\text{Intercept} = \ln D_0 = \ln(D_\upsilon) - (-7257)\left(\frac{1}{T}\right)$$

$$= -24.29 + (7257 \times 3.1 \times 10^{-3}) = -1.79$$

Therefore, $D_0 = 0.1664$

Hence, the diffusivity equation is:

$$D_\upsilon = 0.1664\, e^{-(7257/T)}$$

17.3 MODELLING OF A DRYING BED

Most of the grains like cereals, pulses, oil seeds are dried to safe moisture limits before storing. They may be dried in drying yards, bins or tray dryers. Most common method of storage is in the form of deep beds in bins. Modelling such deep beds in simulated conditions is often required to get a better understanding of the drying process, and to assess the influence of various process parameters on the drying conditions. The deep bed of drying grains may be considered to be consisting of a series of thin layers. The **thin layer models** of drying beds are extensively reported in literature in view of their applications. One such study was reported by Rangroo and Rao (1992) on the drying of toria seeds. The studies of course can be extended to model the drying process of any bed of grains. The studies were reported to assess the influence of various operating parameters, viz.,

- air flow rate,
- bed height,

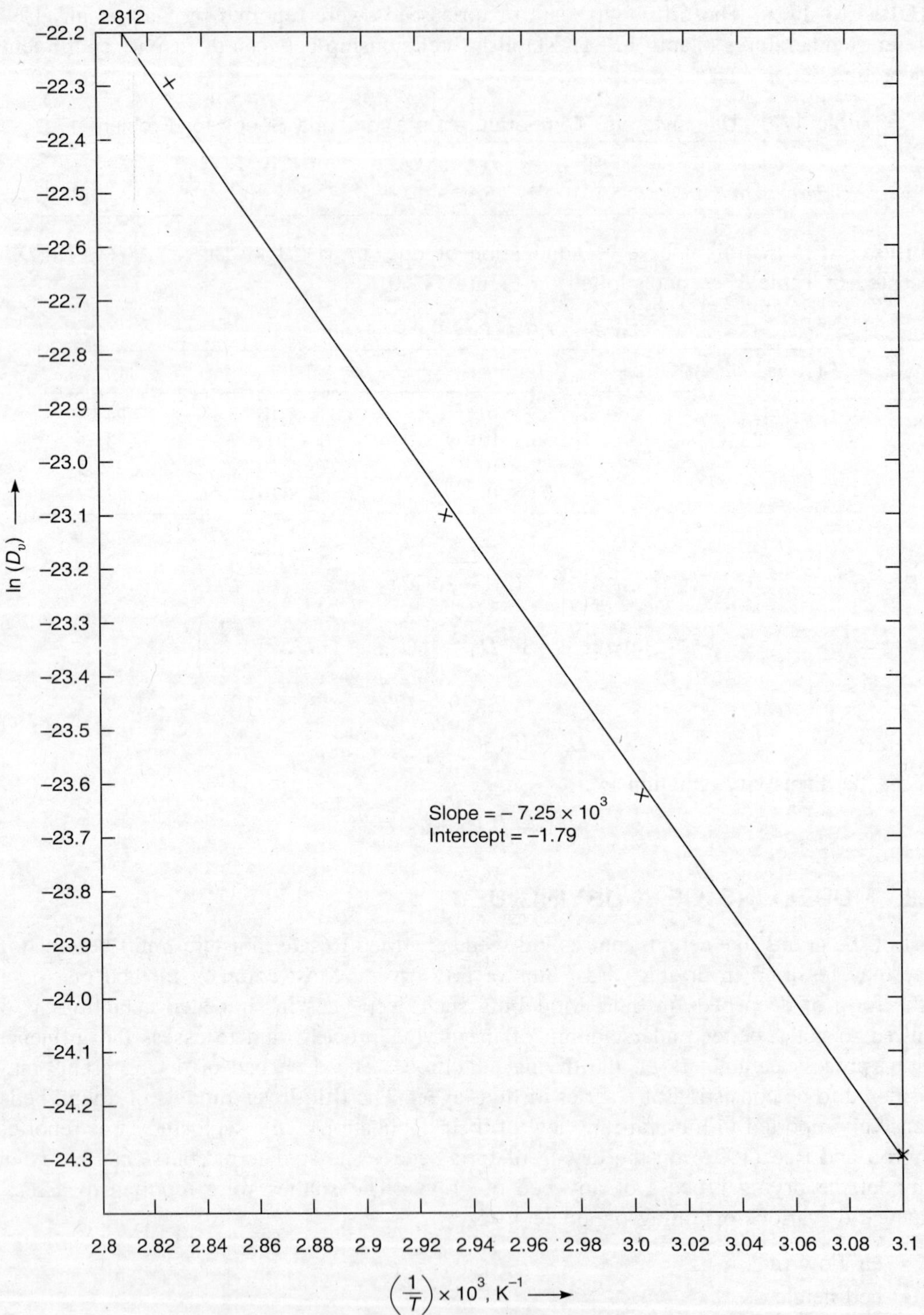

Figure 17.10 (1/*T*) vs. ln (D_v) data for Problem 17.9.

- temperature of air,
- relative humidity of air, and
- initial moisture content of grains;

on the drying time or final moisture content.

The model predicted the time taken to reach a desired moisture content. Data were collected for a set of 27 experiments using different operating parameters and measuring the final moisture for a given time. The data were modelled using Page's equation (Page 1949) which gives a relationship between the Moisture Ratio (MR) and time. The moisture ratio is defined as the ratio of moisture removed to the total removable moisture.

$$\frac{m_2 - m_e}{m_0 - m_e} = \frac{\text{Moisture removed}}{\text{Total removable moisture}} = \text{MR} = \exp(-kt^N) \tag{17.44}$$

where m_2 is the final moisture content in kg, m_e is equilibrium moisture content in kg, m_0 is initial moisture content, t is time taken in s, and k and N are model constants to be evaluated.

The LHS of Eq. (17.44) can also be written in the form of moisture content on decimal basis, i.e., X.

$$\frac{X_2 - X^*}{X_0 - X^*} = \text{MR} \tag{17.45}$$

However, we proceed with Eq. (17.44) as was originally proposed. The experimental data on drying were collected to find X_2 and t using a through flow (air flow perpendicular to the bed of grains) dryer with seeds of different initial moisture contents x on wet basis (Table 17.10). m_e at different conditions were calculated using the EMC-RH data modelled by Henderson equation [Eq. (17.5)].

Table 17.10 *k* and *N* Values for Various Operating Conditions

Run no.	m_0 (g)	T (°C)	V_1 (ms^{-1})	$V^\dagger$(s^{-1})	x_0	ψ	ln (k)	N
1.	39.5	70.0	0.43	29.32	0.148	0.067	–4.299	0.6642
2.	38.9	70.0	0.43	29.77	0.252	0.069	–3.476	0.5904
3.	34.0	70.0	0.43	34.07	0.308	0.078	–2.682	0.4221
4.	39.5	70.0	0.65	44.32	0.165	0.085	–3.075	0.5180
5.	33.0	70.0	0.65	53.05	0.199	0.082	–2.826	0.4952
6.	39.9	70.0	0.65	43.88	0.250	0.067	–1.834	0.3407
7.	39.4	70.0	0.84	57.43	0.096	0.099	–4.186	0.6260
8.	39.4	70.0	0.84	57.43	0.172	0.084	–2.827	0.4550
9.	38.9	70.0	0.84	58.16	0.253	0.072	–1.280	0.3105
10.	39.0	85.0	0.43	29.70	0.149	0.037	–6.410	0.9500
11.	38.9	85.0	0.43	29.77	0.190	0.038	–4.220	0.6670
12.	39.0	85.0	0.43	29.70	0.247	0.038	–4.094	0.6850
13.	39.0	85.0	0.65	44.89	0.112	0.037	–4.698	0.7363
14.	39.5	85.0	0.65	44.32	0.163	0.048	–3.402	0.5225
15.	39.5	85.0	0.65	44.32	0.301	0.044	–2.566	0.4278
16.	39.5	85.0	0.84	57.28	0.167	0.043	–3.617	0.5959
17.	39.0	85.0	0.84	58.01	0.230	0.039	–2.832	0.5486
18.	33.6	85.0	0.84	67.34	0.324	0.038	–2.632	0.4628

(contd.)

Table 17.10 *k* and *N* Values for Various Operating Conditions (*contd.*)

Run no.	m_0 (g)	T (°C)	V_1 (ms^{-1})	$V^{\dagger}$(s^{-1})	x_0	ψ	ln(k)	N
19.	40.5	100.0	0.43	28.60	0.174	0.025	–4.505	0.7422
20.	38.3	100.0	0.43	30.24	0.348	0.025	–3.637	0.5277
21.	33.7	100.0	0.65	51.95	0.085	0.023	–5.098	0.9021
22.	39.0	100.0	0.65	44.89	0.118	0.023	–4.179	0.7000
23.	40.0	100.0	0.65	43.77	0.165	0.026	–3.068	0.5820
24.	31.5	100.0	0.65	55.58	0.258	0.027	–2.073	0.4800
25.	39.5	100.0	0.84	57.28	0.094	0.027	–5.242	0.8640
26.	39.0	100.0	0.84	58.01	0.112	0.021	–3.764	0.7080
27.	36.5	100.0	0.84	61.99	0.262	0.025	–1.781	0.4575

$^{\dagger}V = V_1$/bed *ht*

Source: This article was published in *J. Food Engg*., Vol. 17, Rangroo, S. and Rao. D.G., Drying of toria, (*Brassica campestris, Var. toria*) seeds: Part 2, Drying conditions, p. 63, Copyright Elsevier (1992).

Equation (17.44) can be written as follows by taking logarithms on both sides:

$$\ln(\text{MR}) = -\,kt^N \tag{17.46}$$

Once again taking logarithms to above equation

$$\ln(-\ln\,(\text{MR}) = \ln(k) + N\,\ln(t) \tag{17.47}$$

By plotting ln(–ln(MR)) on y-axis vs. ln(t) on x-axis, values of ln(k) and N were noted from the intercept and slope of the plot. The values ln(k) and N were presented in Table 17.10.

ln(k) and N were related to the operating parameters by the following equations:

$$\ln(k) = f_1\,(T,\ V,\ x_0,\ \psi) \tag{17.48}$$

$$N = f_2\,(T,\ V,\ x_0,\ \psi) \tag{17.49}$$

Various possible functions of f_1 and f_2 are tabulated in Table 17.11. The coefficients A_0, A_1, A_2, ..., A_{12} and B_0, B_1, ..., B_{12} for all the equations were all evaluated by *multiple regression analysis*. All the coefficients are tabulated in Table 17.12. The 13-parameter equation [Eq. (17.55)] was found to be the best. With this equation, we can predict the constants ln(k) and N, which in turn will help us find the moisture content m_2 by using Eq. (17.44).

Table 17.11 Various Functions of f_1 and f_2

f_1	R for f_1	f_2	R for f_2	Equation number
$A_0 + A_1T + A_2V + A_3x_0 + A_4\psi$	0.8752	$B_0 + B_1T + B_2V + B_3x_0 + B_4\psi$	0.9007	(17.50)
$A_0 + A_1T^2 + A_2V^2 + A_3x_0^2 + A_4\psi^2$	0.8398	$B_0 + B_1T^2 + B_2V^2 + B_3x_0^2 + B_4\psi^2$	0.8695	(17.51)
$A_0 + A_1\sqrt{T} + A_2\sqrt{V} + A_3\sqrt{x_0} + A_4\sqrt{\psi}$	0.8903	$B_0 + B_1\sqrt{T} + B_2\sqrt{V} + B_3\sqrt{x_0} + B_4\sqrt{\psi}$	0.9412	(17.52)
$A_0 + A_1\ln(T) + A_2\ln(V) + A_3\ln(x_0) + A_4\ln(\psi)$	0.9002	$B_0 + B_1\ln(T) + B_2\ln(V) + B_3\ln(x_0) + B_4\ln(\psi)$	0.9198	(17.53)

(*contd.*)

Table 17.11 Various Functions of f_1 and f_2 (*contd.*)

f_1	*R for* f_1	f_2	*R for* f_2	*Equation number*
$A_0 + A_1T + A_2 \ln(T) + A_3V + A_4 \ln(V) + A_5x_0 + A_6 \ln(x_0) + A_7\psi + A_8 \ln(\psi)$	0.9614	$B_0 + B_1T + B_2 \ln(T) + B_3V + B_4 \ln(V) + B_5x_0 + B_6 \ln(x_0) + B_7\psi + B_8 \ln(\psi)$	0.9561	(17.54)
$A_0 + A_1T + A_2\sqrt{T} + A_3 \ln(T) + A_4V + A_5\sqrt{V} + A_6 \ln(V) + A_7x_0 + A_8\sqrt{x_0} + A_9 \ln(x_0) + A_{10}\psi + A_{11}\sqrt{\psi} + A_{12} \ln(\psi)$	0.9269	$B_0 + B_1T + B_2\sqrt{T} + B_3 \ln(T) + B_4V + B_5\sqrt{V} + B_6 \ln(V) + B_7x_0 + B_8\sqrt{x_0} + B_9 \ln(x_0) + B_{10}\psi + B_{11}\sqrt{\psi} + B_{12} \ln(\psi)$	0.9600	(17.55)
$A_0 + A_1T + A_2T^2 + A_3 \ln(T) + A_4V + A_5V^2 + A_6 \ln(V) + A_7x_0 + A_8x_0^2 + A_9 \ln(x_0) + A_{10}\psi + A_{11}\psi^2 + A_{12} \ln(\psi)$	0.9615	$B_0 + B_1T + B_2T^2 + B_3 \ln(T) + B_4V + B_5V^2 + B_6 \ln(V) + B_7x_0 + B_8x_0^2 + B_9 \ln(x_0) + B_{10}\psi + B_{11}\psi^2 + B_{12} \ln(\psi)$	0.9583	(17.56)

Source: This article was published in *J. Food Engg.*, Vol. 17, Rangroo, S. and Rao, D.G., Drying of toria, (*Brassica campestris, Var. toria*) seeds: Part 2, Drying conditions, p. 64, Copyright Elsevier (1992).

Table 17.12 *A* and *B* Coefficients Obtained Using Various Equations

A Coeff.	*Eq.* (17.52)	*Eq.* (17.53)	*Eq.* (17.54)	*Eq.* (17.55)	*B Coeff.*	*Eq.* (17.52)	*Eq.* (17.53)	*Eq.* (17.54)	*Eq.* (17.55)
A_0	–17.25	–6.286	150.85	–54.563	B_0	2.789	1.616	–9.689	–11.54
A_1	0.444	–0.1967	0.5687	0.5846	B_1	–0.0788	–0.3002	–0.0456	–0.0372
A_2	0.6181	2.243	–50.94	–8.4703	B_2	–0.0611	–0.2243	3.917	0.2493
A_3	9.804	2.111	–0.2089	–12.243	B_3	–1.395	–0.2952	0.0299	1.9028
A_4	10.50	0.401	11.418	–0.425	B_4	–2.170	–0.2084	–1.528	0.0839
A_5			–16.308	6.0359	B_5			1.0922	–1.489
A_6			5.4355	0.9054	B_6			–0.5366	0.993
A_7			–2.687	–86.50	B_7			–0.6573	–19.01
A_8			–0.6164	121.152	B_8			–0.0370	34.42
A_9				–7.266	B_9				–4.126
A_{10}				–174.65	B_{10}				38.236
A_{11}				171.79	B_{11}				–34.967
A_{12}				–10.535	B_{12}				1.811

Source: This article was published in *J. Food Engg.*, Vol. 17, Rangroo, S. and Rao, D.G., Drying of toria (*Brassica campestris,* Var. toria) seeds: Part 2, drying conditions, p. 64, Copyright Elsevier (1992).

By feeding the above simulated model to a *computer controlled drying system*, we can calculate the time required to reach the desired final moisture content m_2, and instruct the computer to stop the drying process once m_2 is reached. Above kind of model studies made for toria can be done for any grain in a similar way.

17.4 DRYING EQUIPMENT

A number of dryers are available for industrial operations (Table 2.1). Their main purpose is to reduce the heat loss, and to recirculate a part of the air to save energy. Some dryers are specially suitable for certain type of materials. This has necessitated the design and fabrication of various types of dryers, which are classified as shown in Table 17.13. Some of them are discussed as follows:

Table 17.13 Classification of Various Types of Dryers

Parameter	*Batch*	*Continuous*
Storage dryer	Bin dryer	–
Fixed tray dryers	Cabinet tray dryer	Tunnel dryer
Drying under vacuum	Vacuum shelf dryer	Continuous belt dryer
	Agitated pan dryer	
Moving bed dryer	Turbo dryer	Pneumatic conveyor bed dryer
	Vibro-fluidized bed dryer	Screw conveyor dryer
		Belt conveyor dryer
		LSU† dryer
Roller dryer	–	Drum dryer
		Rotary dryer
Product in suspension	Fluidized bed dryer	Fluidized bed dryer
		Pneumatic dryer
		Ring dryer
		Spray dryer
Freeze dryer	Batch freeze dryer	Continuous freeze dryer

†LSU dryer: Lusiana State University dryer

Bin dryer: Bin dryers are also known as **deep bed dryers**. They are generally cylindrical or rectangular containers with a mesh at the base (Figure 17.11). They are used to store the grains after they are already dried in some other drying equipment. In fact the bin dryers are also known as **silos**. The dried grains are stored in bins with a provision for very slow flow rate of air (~0.5 m^3/sm^2 of bin area). The bins are usually of size to carry a few tons of grains. The drying air is mainly used to prevent the material from picking up moisture from atmosphere. Here the drying takes place mostly in the falling rate zone. There will be a large gradiance of moisture content as we go from bottom to top. Hence, mostly discharging is done from the bottom. Fresh material is loaded from the top either manually or by means of bucket elevators.

Thus, bin dryers are eventually a bulk storage facility for grains.

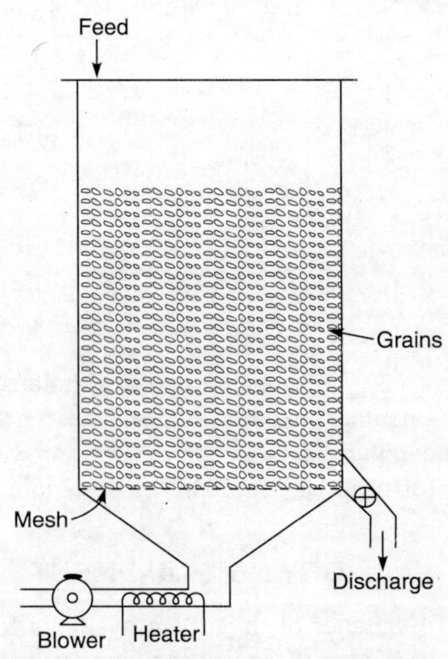

Figure 17.11 Bin dryer.

Cabinet tray dryer: Cabinet tray dryer is one of the most versatile and most frequently used dryer in food processing. It is simple in construction, most flexible to use, easy to operate with very low initial investment. The product quality is not very good, and hence, is generally used for drying of low cost foods. It takes longer times for drying, probably of the order of few hours. It is frequently used for removing initial moisture before shifting the product to some other dryer.

A schematic diagram of the dryer is shown in Figure 17.12, which consists of a heating chamber, heated by steam or electrical coils or by flue gases by combustion of agro waste or by any other heating source. The drying air is drawn into the chambers by a blower, and the air is heated by contact with the heating elements. It consists of a number of trays, usually multiples of 12, and are made out of stainless steel or aluminium. They may be perforated or plain and are usually of the size 40 cm × 100 cm × 2.5 cm. The food material to be dried is filled into these trays. For 12 and 24 tray dryer, the trays are stacked directly into the heating chamber. If the number is more (viz., 48, 96 and 192), the trays are stacked into a trolley which is later pushed into the heating chamber. The hot convective air currents either pass on the food materials (cross flow) or pass through the food material (through flow), and dry the food material. Usually a part of the air is recirculated, and the rest is drawn afresh through the air inlet, which usually consists of a filter screen to filter the air before entering into the dryer.

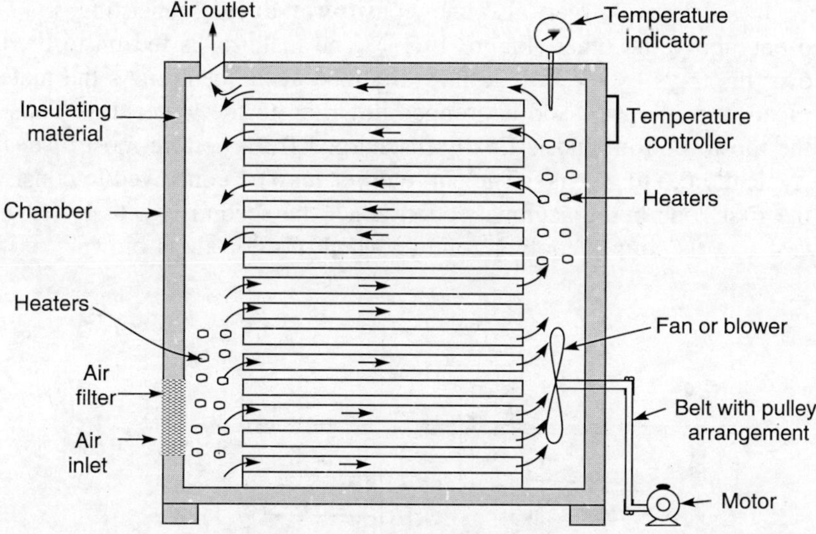

Figure 17.12 Cabinet tray dyer.

The cabinet tray dryers are usually operated in a batch manner. The temperature controller-cum-indicator keeps the heating under control to maintain the temperature of the dryer reasonably constant. Usually the temperatures are of the order of 50–75 °C. The drying is stopped intermittently, and the material is mixed or shuffled to avoid over heating locally.

Even though the temperature controller is installed, the temperature is seldom constant in all the corners of the dryer. Local over heating, particularly, at the edges is a very common phenomenon in the cabinet tray dryers which make them unfit for heat sensitive materials. Since the drying times are longer, heat sensitive volatiles may be lost during drying. Browning of the food material due to long contact with hot air may be possible. Throughputs are low because of long drying times and batch operation.

Most of the above drawbacks are overcompensated in view of the low cost of the dryer and its operation.

Tunnel dryer: The tunnel dryer is only an extension of cabinet tray dryer in which the drying operation is continuous. The trays are loaded into trolleys which move slowly through the tunnel through which hot air passes. The trays with food material are loaded into the trolley at one end, and are discharged at the other end. When a trolley comes out of the tunnel, it is disengaged from the train, and the material is withdrawn. At the same time, another trolley with trays is filled with the material and is connected to the train of trolleys at the inlet position. This makes the drying process continuous; since the wet food material is fed continuously at one end, and the dried product is discharged continuously at the other end. The movement of the trolleys in the tunnel is so adjusted that the residence time is adequate to remove the moisture from the food material. In fact the trolleys move very slowly. In the operation, the tunnel dryer is almost similar to a biscuit making oven except that the heating in a tunnel dryer is by convection by a current of hot air flowing either counter-currently or co-currently, whereas the baking oven is heated by radiation heat.

Grapes are usually dried in tunnel dryers to make raisins.

Turbo dryer: It is a vertical dryer (also known as **tower dryer**) consisting of a series circular trays mounted one above the other (Figure 17.13). The material is fed on to the top tray. Hot air is blown over the trays by the turbine fans arranged at the centre. As the material is partly dried, it is scraped from the tray, and is dropped down on to the lower tray. Thus, the material moves from the top to bottom and is finally discharged from bottom-most discharge tray. Air velocities are typically 0.6 to 2 4 m/s. The entire dryer may be conceived to consist of typically three zones; the first zone being acclimatization zone, the second is actually drying zone, and the third (lowest one) is simply cooling zone in which no hot air is blown.

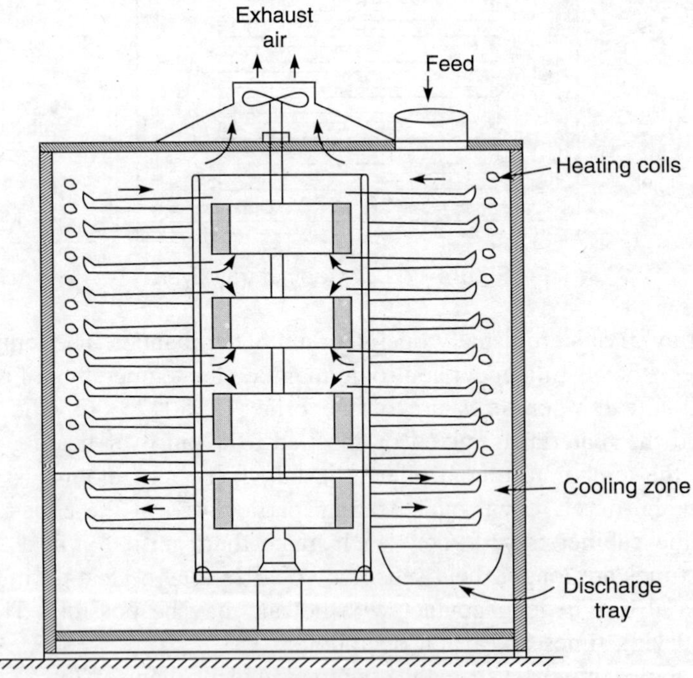

Figure 17.13 Turbo dryer.

In a way it may be considered as a continuous tray dryer since the material is fed continuously on the upper tray and is discharged continuously from the bottom tray. The type of problem of over heating and under heating particularly at the edges which is experienced in case of tray dryers is not faced in the turbo dryers since the air is blown intermittently on to each tray by the turbo fans. This is a particular advantage with turbo dryers.

Vacuum shelf dryer: Vacuum shelf dryer is almost akin to a tray dryer, in which trays are fixed. In this dryer the heat transfer is by means of conduction. It consists of a chamber in which there is a number of fixed trays fitted to the cabinet. The plates are heated by means of steam. The chamber is connected to a vacuum system, and the vacuum is created in the entire chamber. Hence, the doors used for the chamber should be leak-proof. A schematic representation is shown in Figure 17.14.

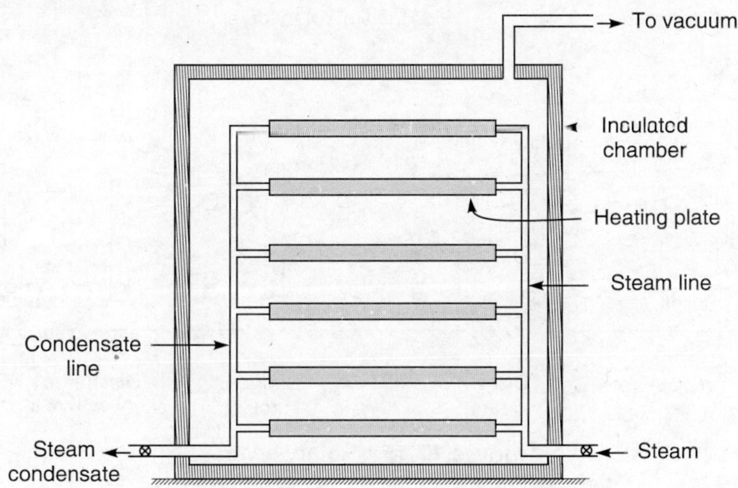

Figure 17.14 Vacuum shelf dryer.

Since vacuum is applied to the system, moisture evaporation takes place at a much lower temperature. Hence, the dryer is particularly used to dry food materials which are heat sensitive like vitamins and proteinaceous materials. The heating of food material on the chamber plate is by means of conduction. In some systems even radiation heat is applied from the dryer walls. Generally convective heat transfer will not be there as there is no flow/blowing of air. This is an excellent method for drying of food materials at much lower temperatures and also for making powders from slurries, which would not form powder, but only form cakes in a cabinet tray dryer. For making fruit juice powders from some fruit pulps, for making crude pappain from papaya latex, for making crystalline annatto dye (*Bixa orellana*) by solvent extraction of the seeds and for many such drying applications the vacuum shelf dryer is particularly useful.

Rotary dryer: Rotary dryers are generally used for drying wet granules/grains which do not have tendency to stick. It is a continuous dryer, and consists of rotating horizontal cylindrical drum which is slightly inclined to the outlet. Wet material is fed in the upper end, and the dry material is discharged from the other end. The cylindrical drum contains lifting flights as shown in Figure 17.15. As the drum rotates, the flights lift the material to the top from where it tumbles down. Hot air flows from the bottom end upwards, whereas the material flows counter-currently from the upper side to bottom side (Figure 17.15). Thus, the material is heated by direct contact with hot air by convective heating. In some arrangements the hot air passes through the hollow spacing provided outside the drum, and then the hot air also passes through the drum. With this arrangement the drum is also heated, and provides conductive heat to the solids when they are in contact with the cylindrical drum, convective heat is also provided wen the solids are tumbling down.†

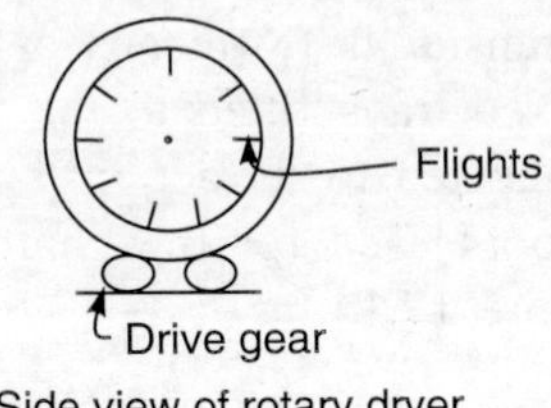

Side view of rotary dryer

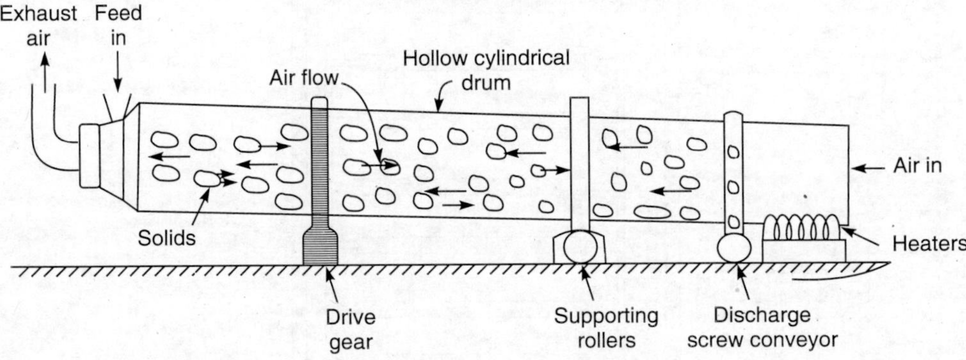

Figure 17.15 Rotary dryer.

Usually the solids are discharged at the outlet through a screw conveyor. The dryer drum rotates on drum-supporting rolls. Hot air can be blown from the discharge end; in this case the whole drum is under slight positive pressure. In some arrangements, the hot air is sucked from the other end (upper end), here the whole dryer is maintained under slight vacuum. This arrangement will help material not to dust, and hence, is more useful to those materials which have a tendency to dust.

The air temperatures are usually of the order of 120–175 °C and the air velocities range from 2000–25,000 kg/m^{2}h for coarse particles. The heat transfer rates are calculated by using the following empirical dimensional equation (McCabe, et al., 1993).‡

$$q = 4.372 \times 10^{-4}\ dLG^{0.67}\ (\Delta T) \qquad (17.50)$$

where d is diameter of the dryer in m, L is length of the dryer in m, q is heat transfer rate in W/m^2°C, and ΔT is the log mean of wet bulb depressions at inlet and outlet conditions of dryer.

---

†The outside heating arrangement is not shown in Figure 17.15.

‡The original equation in FPS units is:

$$q = \frac{0.5\,G^{0.67}}{D} V\Delta T$$

where G is in lb/ft^2 h, V is in ft^3, D is in ft, ΔT in °F and q is in Btu/ft^2.h°F.

The rotary dryers are generally used for drying salt, sugar or any granular solids which do not stick. If the product should be clear and should not be exposed to very high temperature and at the same time it needs to be dried fast, then the rotary dryers are the best choice.

Fluidized bed dryer: Fluidized bed dryers are the best and fastest drying systems for drying wet granular solids which do not tend to stick or agglomerate. The granules should not be very fine and should be resistant to abrasion. Hence, mostly the food grains or seeds are preferably dried in fluidized bed dryers.

A typical fluidized bed dryer, shown in Figure 17.16, consists of a vertical cylinder vessel with a perforated bottom. The dryer can be operated both continuously or in a batch manner depending upon whether the feed is fed continuously or batch wise. The dryer is supplied with hot air from the bottom. If the flow rate of the air is low, the solids are in a packed condition like a packed bed as shown in Figure 17.16(a). As the flow rate of the air is increased, the buoyancy forces pushing the granules will increase. Accordingly pressure drop across the bed also increases. Beyond a certain flow rate of the air, the buoyancy forces are more than the gravitational force acting on the granules (solids) downwards. From this stage onwards, the solids will raise, and will be in a state of suspension. The point at which the fluidization starts is known as **incipient fluidization**. At this stage, the bed does not really fluidize, but the solids are loosened, and they will not be in contact with each other. Hence, at this stage, the pressure drop slightly falls. If we draw a plot [as shown in Figure 17.16(c)] between pressure

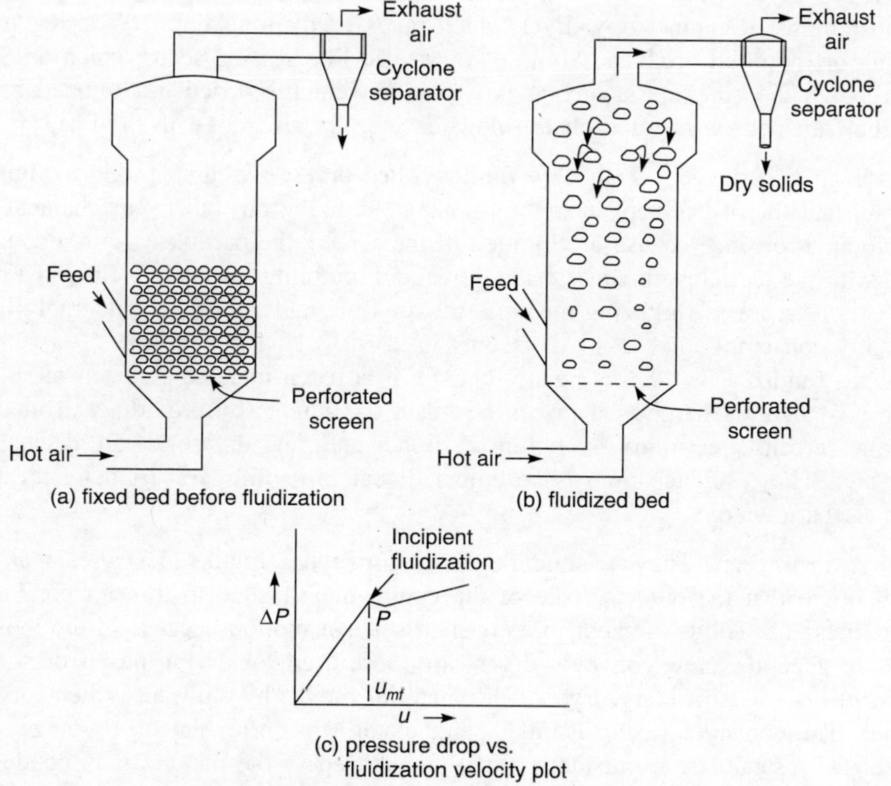

Figure 17.16 Fluidized bed dryer.

drop (Δp) and linear velocity (u), initially as u increases, Δp increases until point P at which fluidization just starts, and Δp falls slightly. The velocity of air corresponding to point P is known as **minimum fluidization velocity**. Beyond this point, any increase in u will increase Δp, and all the solids start rising up. Now, the solid bed is no more a packed bed. The solids move, jump and tumble like fluid, and the bed is in a state of fluidization. Hence, the bed is known as the **fluidized bed**. Since the entire surface area of the solid is exposed to hot air, the drying is fast.

The exhaust air after drying the solids will leave the dryer through a cyclone separator to avoid carrying of any solids or dust. As the material is dried, its bulk density will come down; hence, the material rises to a greater height. Taps can be provided at this level to collect the dry material. Generally, the fluidized bed dryers are operated under entrainment conditions where the solids are also carried along with the exhaust air and are separated in the cyclone separator. The residence time of the material in the dryer is very little which is hardly of the order 30 to 120 s for removing surface moisture, and 15–30 min for removing moisture by internal diffusion also.

Except for consumption of huge quantities of air for fluidizing the solids, fluidized bed dryers are the best for drying non-sticky granular solids. Hence, mostly food grains like wheat are dried in fluidized bed dryers, before they are sent for milling where moisture needs to be low, and the grain should be hard to grind well. Vegetables like mushrooms are dried conveniently in fluidized bed dryers since they contain high surface moisture, and hence, they dry quickly. Drying mushrooms in cabinet tray dryer which takes pretty long time for drying may result in browning of the dried products. Some of oil seeds, like sesame seeds which are soaked in water to remove the outer husk, are quickly and conveniently dried in fluidized bed dryers. Some of the fruit pellets, salt and cheese powder, etc. are also dried in fluidized bed dryers.

Vibro-fluidized bed dryer: The vibro-fluidized bed dryers are almost akin to fluidized bed dryer except that the bed is kept in motion not by fluidizing air, but by mechanical vibration of the perforated screen. Because of vibration of the screen, the particles are in a constant state of motion with the particles moving upside down and tumbling just as in a fluidized bed dryer. Since air is not used as a fluidizing medium, the air flow rates can be considerably lower, and hence, energy consumption is less. Air is only used for drying the solids.

The vibro fluidized bed dryers mostly operate in a batch manner. A batch of solids is fed and dried, and then discharged. They can be made continuous by providing an inclination to the vibrating screen. Wet solids are fed at the upper end, and dry solids are discharged from the other end. The residence time is so adjusted that the solids are dried by the time they reach the discharge end.

Screw conveyor dryer: They are similar to screw conveyors (Figure 24.7) with an arrangement to pass steam so that the outer surface of the conveyor is heated to transfer conductive heat to the materials. The solids are kept in a constant state of motion because of the movement of the screw. In general, screw conveyor dryers are more used for drying pastes or sticky solids which cannot be dried in rotary dryers. The material moves by slow movement of the screw (2–30 rpm). The conveyor casing is filled with material to an extent of 10–60 per cent full, and is heated by steam from outside. Heat is transferred to the materials by conduction. Air is not at all used. The water evaporates from the solids/pastes. Thus, they are indirect type dryers, and hence, the heat transfer coefficients are low, probably of the order of 15–60 W/m^2°C. Sometimes even these dryers are operated under vacuum, and are extensively used for removing the solvent from oil seed cakes after solvent extraction. The purpose is not for recovery of solvent, but to make the cake free of solvent for further use. Hence, these dryers are sometimes called as **desolventizers**.

The schematic representation of the dryer is shown in Figure 17.17. For all practical purposes, the dryer looks like a screw conveyor except that the dryer has got an outside jacketed casing which is used for passing steam.

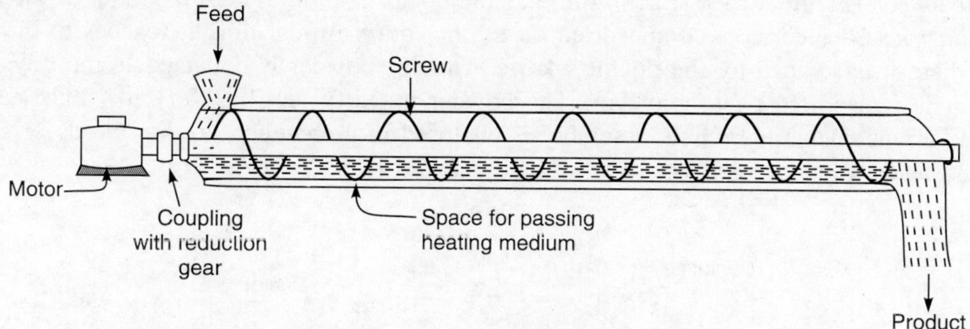

Figure 17.17 Schematic representation of a screw conveyor dryer.

Pneumatic conveyor dryer: The dryer consists of a chute through which the hot air is blown cocurrent with the material (Figure 17.18). In this dryer, the solid material is transported, and also at the same time dried considerably. Mainly it is used to convey the grains to an elevated place for grinding when they also need to be dried so that the grinding can be efficient. Normally the wheat mills are tall, and the grains are fed at the top in between the rollers, from where the ground material flows downward by gravity. For transporting the grains to the upper portion of the mill, pneumatic conveying is used by passing hot air. This dryer can be used only for materials which are non-sticky solids like grains. It is also used for drying flaky materials like desiccated coconut. In such cases they are not used for transportation, but the wet material passes through a circular path and comes back to the original position, and hence, such dryers are also known as **ring dryers** (not shown in Figure 17.18).

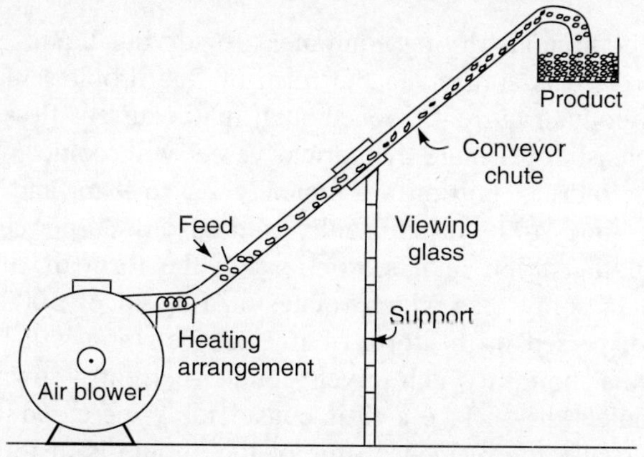

Figure 17.18 Pneumatic conveyor dryer.

Drum dryer: Drum drying is a unique drying process in which both cooking and drying take place. Gelatinization of starch also takes place during drying. Hence, it is used with uncooked/semi-cooked slurries or pastes to make fine and free flowing powders with very low bulk densities.

The drum dryer consists of one or two counter-rotating drums which are heated from inside with steam. The drums are usually made of *SS* and are hollow inside with the diameters ranging from 0.6 to 3.0 m and 0.6 to 4 m long. The drums rotate very slowly in opposite directions revolving at 1–10 rpm. The two drums have a close clearance in between, and the feed slurry is fed in between (Figure 17.19). Each drum has the doctor's knife along the length of the drum. The slurry gets heated, cooked and dried on the hot drum by the time it reaches to the other edge where it is scraped by the doctor's knife. The dry powder is collected in the trays from where it is transported by a conveyor. The powder is fluffy and has very low bulk density. In some arrangements the whole assembly is enclosed in a chamber, so that the dryer can be operated under vacuum also (not shown in Figure 17.19).

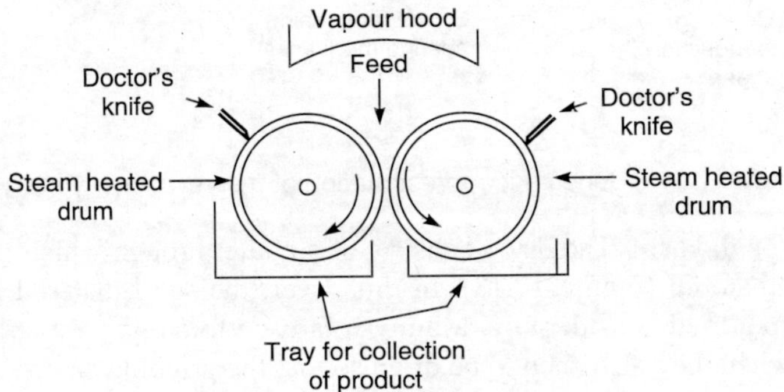

Figure 17.19 Double drum dryer.

The drum dryer is generally used for making baby foods and weaning foods from slurries of cereals, pulses and oil cakes taken in powder form. The slurry gets cooked, the starch gets gelatinized and the material dries into a free flowing, light weight, fluffy or flaky material which can be easily reconstituted with water. Even ready-to-use *sambar* powders can be made by drying the cooked *sambar* slurry.

Spray dryer: It is a continuous drying equipment for drying liquid solutions or slurries, in which the liquid to be dried is sprayed in the form of fine droplets into a current of hot air. Usually the air flow is counter-current, even though the cocurrent flow is also possible.

The spray dryer consists of a huge cylindrical vessel with conical bottom (Figure 17.20). The diameters of the cylindrical portion are typically 2.5 to 9 m, and the height is 2.5 m or more. The liquid or the slurry to be dried usually contains 20–30 per cent solids. It is pumped into the top portion of the dryer, and is atomized in the form of fine droplets by passing through an atomizer or a spray disc which rotates at a speed of 5000 to 10,000 rpm. Thus, the liquid material is dispersed in the form of fine droplets and will be moving downwards. Hot air is blown from the bottom which travels counter-currently upwards. Since the liquid is in the form of fine droplets the surface area is considerably increased, and hence, the rates of drying are very fast. Usually the residence time of the bubble is of the order of 3–6 seconds.

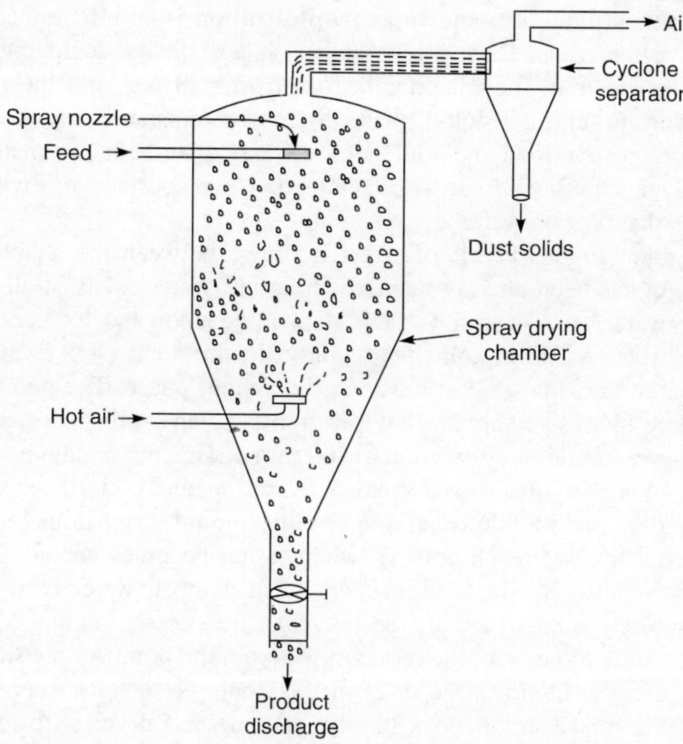

Figure 17.20 Schematic representation of spray dryer.

The contacting air temperature is 220–250 °C. Even though the air temperature is so high, the droplet temperature seldom goes beyond the boiling point of water. The dry product is discharged from the bottom of the dryer. The trajectile of the falling droplet is so manipulated that it never touches the walls of the dryer which are hot, and are at the temperature of the air. If the droplet touches the walls of the dryer, it gets charred. The droplet when it comes in contact with the hot air quickly dries off, and the solid particles that form will be highly porous, hollow, hygroscopic and will have a very low bulk density. These special characteristic features of the dried powder cannot be achieved in any other type of dryer. The low bulk density and high hygroscopic nature of the dried products enable them to be reconstituted quickly when water is added to make the slurries or solutions.

The exhaust hot air leaves through a cyclone separator to separate out accompanying solids; whereas most of the dry powder is discharged from the bottom. Since these dryers are usually very huge and the air temperature is considerably high, the exhaust air carries a good amount of heat, and hence, the thermal efficiency is always poor. If the atomizer tip or the nozzle nose is blocked, the whole operation comes to stand still, and it takes a very long time to restart.

Spray dryers are used in food processing for drying of fruit juices to make easily reconstitutable fruit juice powders. Invariably milk is always made into milk powder by spray drying only. Hence, the name *spray dried milk powder*. Similarly, coconut milk can also be made into powder by spray drying which finds extensive applications in food industry and in cosmetics. Egg powder is made by spray drying of egg yolk. Instant coffee powders are made by spray drying of coffee decoction.

Freeze dryer: Freeze drying, also known as **lyophilization** is an important drying method for highly heat sensitive materials. Drying takes place at a very low temperature and very high vacuum. In fact, the water in the material is first frozen to ice, and the ice is sublimed to vapour without going through the liquid phase. Since the moisture removal is by sublimation, the cellular structure of the food material does not get disturbed, the material does not get contracted, the colour and flavour are well retained. How does it happen? We shall explain this with the phase diagram of water.

The phase diagram (Figure 17.21) of water is drawn between the temperature and vapour pressure of water. As has been already mentioned earlier, water exists in all three phases, viz., solid (ice), liquid (water) and vapour (steam) depending upon the temperature and pressure. In the zone *POQ*, water exists in solid phase only. In the zone *QOR*, water exists in liquid phase only, whereas in the zone *POR* it exists in the vapour phase. The point *O* at which water exists in all the three phases is known as **triple point** which corresponds to a temperature of 0.0098 °C at a pressure of 4.8 mm of mercury (760 mm of Hg corresponding to one atmosphere, 101.3 kPa). So, if we are at point *B*, water exists as ice. At this point if we slowly increase the temperature keeping the pressure constant, ice becomes liquid water in the zone B_1 and B_2, and if we still continue further, we reach point B' where water becomes vapour. In other dryers, we operate in the B_1B_2 zone to B'. But, suppose we are at point *A*, water is in the form of ice. If we keep the pressure constant (it should be less than that corresponding to triple point), and then increase temperature along AA', the moment we cross the point A_1, ice sublimes into vapour without going through the liquid phase. It is in this zone we operate freeze dryer. Except for this one principle, the freeze dryer looks like a vacuum shelf dryer. The food material in the freeze dryer is initially cooled under vacuum, until the pressure reaches below that of the triple point. Later keeping the pressure (vacuum) constant, the temperature is slowly risen, until we cross the point A_1, ic starts subliming and moisture escapes as vapour.

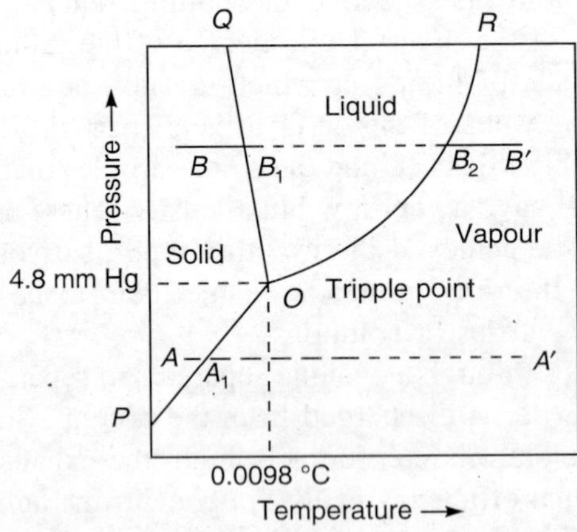

Figure 17.21 Phase diagram of water.

Schematic representation of a typical freeze dryer is shown in Figure 17.22. It consists of a drying chamber with an airtight door and a quick-locking arrangement. The dryer is connected to vacuum system which creates very high vacuum (0.1–1.0 mm of mercury). The plates (trays) are connected to a refrigerated cooling system, so that the heat transfer to the material is by

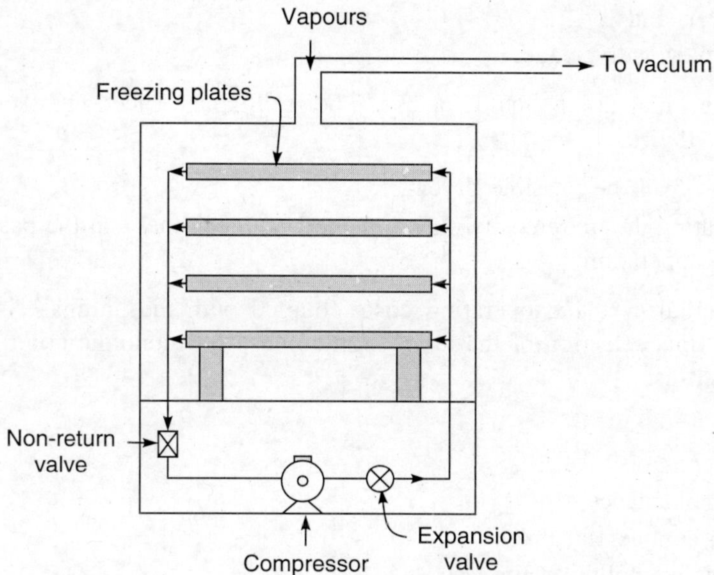

Figure 17.22 Schematic representation of a freeze dryer.

conduction only. Once if point *A* position is reached, the pressure is kept constant, and the temperature of the food material is increased by increasing the temperature of the plates. The material slowly passes through the point A_1 and enters into vapour zone (A_1A'). Once the material is dried, it is discharged by opening the doors. The dryer is now ready for feeding fresh batch.

Except for high initial cost of the dryer because of the vacuum system and refrigeration system and high operating costs, the freeze drying is an excellent drying method. The product quality is excellent. The shape of the material is well retained. Because of low temperature operations, the nutrients are not lost. It is indeed an excellent drying method for highly heat sensitive materials. Freeze dryers are used for drying of blood or vaccines.

Unless the cost of the product is very high, freeze drying method cannot be used. Marine produce like prawns are dried by freeze drying method for preservation. *Lyophilization* method which is frequently used in biotechnological process is indeed freezing only. Sperm samples, most of the enzymes and microorganisms are preserved by lyophilization for subsequent use whenever they are required.

17.4.1 Criteria for Selection of Dryer

So far we have seen a number of drying methods and equipment. For a particular operation, a number of dryers may be useful. For example, for dehydration of a vegetable like carrot or lady's finger (*bhindi*) we may be able to use a

- cabinet tray dryer,
- tunnel dryer,
- vacuum shelf dryer,
- freeze dryer,

- rotary dryer, and
- screw conveyor dryer, etc.

In such a case which dryer should one use? The preliminary selection falls into the following categories (Keey 1972):

(i) Listing of all the possible dryers.
(ii) Eliminating the more costly ones (or uneconomical ones) on the basis of their annual costs of operation.

Thus, the capital cost and operating costs coupled with the purchaser's past experience often decide the final selection of the dryer. Sometimes, the economic criteria is overruled by other factors, such as:

- safety,
- ease of operation,
- ease of maintenance,
- long-term commercial links,
- tolerable noise pollution limits,
- availability of space, and
- availability/unused spare capacity of the existing dryers, etc.

Thus, in a nut shell we can say, cost, convenience, and compatibility will decide the ultimate choice. We have given some clues in a tabular form (Table 17.14) for selection of a dryer for a particular product based on its physio-chemical and thermal properties.

17.5 APPLICATION OF DEHYDRATION IN FOOD PROCESSING

As has been already mentioned, food processing is one of the largest processing operations in which dehydration is extensively used. Some of the applications are:

- Dehydration of fruits and vegetables to extend their shelf-life, and also to use them in powder form in soup mixes, etc.
- Dehydration of fish, shrimp and prawns.
- Dehydration of meat and meat products.
- Preparation of baby food and weaning foods using a drum dryer.
- Drying of cereals and pulses in bins to increase their storage life.
- Drying of processed products like natural crystalline food colours in vacuum shelf dryer.
- Spray drying of milk, or coconut milk in spray dryer.
- Freeze drying of blood, vitamins and some high value food products like prawns, etc.
- Drying of some of the oil seeds or grains in fluidized bed dryers or vibro fluidized bed dryers, etc.

In fact, the information was already provided at the end of description of each specific dryer. In a nutshell, the various applications are given in Table 17.15 taking each dryer into consideration.

Table 17.14 Criteria for Choice of Dryer Based on Material Characteristics

	Bin dryer	*Cabinet tray dryer*	*Tunnel dryer*	*Turbo tray dryer*	*Vacuum shelf dryer*	*Continuous vacuum shelf dryer*	*Fluidized bed dryer*	*Vibro-fluidized bed dryer*	*Screw conveyor dryer*	*Pneumatic conveyor dryer*	*Rotary dryer*	*Drum dryer*	*Foam bed dryer*	*IR dryer*	*Microwave/Dielectric dryer*	*Spray dryer*	*Freeze dryer*
1. Material characteristics																	
(i) liquids																√	√
(ii) solids	√	√	√	√	√	√	√	√	√	√	√				√		√
(iii) small to medium particles		√	√		√	√											
(iv) large particles	√	√	√		√	√	√	√	√	√	√				√		
(v) slurries		√	√	√	√	√						√	√	√	√		√
(vi) pastes		√	√	√	√	√								√	√		√
(vii) free flowing powders		√	√						√		√						
(viii) free flowing granules	√	√	√	√	√	√		√	√	√	√						√
(ix) lumped particles		√	√	√													
(x) continuous sheets												√		√			
(xi) mechanically strong	√	√	√				√			√	√						
2. Initial moisture content																	
(i) high		√	√	√											√	√	√
(ii) moderate to low		√															
(iii) low	√	√	√	√	√	√	√	√	√	√	√			√	√		√
3. Final moisture contents																	
(i) low	√				√	√							√	√		√	√
(ii) moderate		√	√	√	√	√	√	√	√	√	√			√	√		√
4. Drying rates desired																	
(i) low	√	√															√
(ii) medium			√														
(iii) high								√								√	
5. Heat sensitive material					√	√							√			√	√
6. Mode of heat transfer																	
(i) conduction					√	√			√			√					√
(ii) convection		√	√	√			√	√	√	√	√		√		√	√	
(iii) radiation/dielectric														√	√		
7. Mode of operation																	
(i) batch	√	√		√	√								√	√	√		√
(ii) continuous			√			√	√	√	√	√	√	√		√		√	
8. State of material																	
(i) stationary	√	√	√	√	√	√							√	√	√		√
(ii) moving bed							√	√	√	√	√	√				√	
9. Relative motion between air and solids																	
(i) counter-current	–	–	–	–	–	–			√		√					√	
(ii) co-current							√	√		√							
10. Temperature of dying																	
(i) high							√	√	√	√	√	√			–	√	
(ii) low	√	√	√	√	√	√	√	√					√	√			
(iii) Below 0 °C																	√
11. Relative costs (relative to forced air drying)		1				9.3	1.6					1.65					17.8
12. Power consumed in kWh/kg of water removed							3.5			1.8	1.25					2.5	

Table 17.15 Applications of Various Dryers in Food Processing

Name of dryer	*Applications*	*Remarks*
Cabinet tray dryer	Dehydration of fruits and vegetables Dehydration of fruit pulps like mango pulp to make fruit bars Dehydration of grains, cereals and pulses before processing them Preparation of ginger or garlic powders Dehydration of fish or shrimps Dehydration of potato chips Dehydration of green chillies Dehydration of spices like turmeric, pepper, cardamom, ginger Can be used virtually for any product	Can be used for drying of any solids, pastes or slurries Drying times are longer Loss of volatiles is more Very economical to use Initial costs are very low
Vacuum shelf dryer	Can be used wherever a cabinet tray dryer is used Drying of papaya latex to make pappain Drying of annatto extract to make crystalline dye powder	Used for drying of pastes and slurries at lower temperatures
Tunnel dryer	Dehydration of grapes continuously Dehydration of fruits and vegetables	It is a continuous dryer
Bin dryer	For storage of cereals, pulses or grains over a period of time	Used mostly for storage purposes
Fluidized bed dryer	Drying of any non-sticky solids Drying of cereals, pulses or grains Dehydration of red chillies	Used for any non-sticky solids
Rotary dryer	Drying of non-sticky solids Drying of mustard seeds Drying of ragi grains	Used for any small grains which cannot be dried in fluidized beds, bins, etc.
Drum dryer	Preparation of baby foods and weaning foods from cereals and pulses Preparation of dehydrated ready to use *sambar* powder	Can be used where cooking-cum-drying is required
Screw conveyor dryer	Desolventization of oilseed cakes Drying of pastes and sticky solids	Mainly used for desolventization of extracted solids/cakes
Pneumatic conveyor dryer	Dehydration of desiccated coconut powder Drying of any flaky materials Drying and transportation of wheat grains into the grinding mills	Used mainly when conveying and drying are required Used for flaky materials
Spray dryer	Drying of milk to make milk powder Drying of coconut milk to make powder Drying coffee decoction to make instant coffee powder Drying of tea concentrate to make instant tea powder Drying of egg yolk to make egg powder Drying of papaya latex to make papain Drying of molasses to make molasses powder	For drying of any free flowing liquids quickly
Freeze dryer	Drying of prawns Drying of blood Drying of vitamins, etc.	For drying high value, heat sensitive materials

17.6 VARIOUS TYPES OF ADVANCED DRYING TECHNIQUES

Various types of dryers are available in the industry. We have described some of the traditional drying systems with schematic representation of them. In addition to them, as per the requirement of industry, a large number of new drying systems is also available (Kudra and Mujumdar 2002). Some of the basic reasons that warrant one to go for new techniques are:

- Existing drying methods are all not necessarily optimal in terms of energy consumption and product quality.
- Safety and ease of operation.
- Ability and ease of drying process.
- Ability to perform optimally even under large changes in throughputs.
- Impact on environment.
- Noise of operation.

In view of the above Kudra and Mujumdar (2002) enlist a number of novel drying methods, viz.,

(i) impinging steam drying,
(ii) pulsed fluid beds,
(iii) superheated steam drying,
(iv) airless drying,
(v) drying with shock waves,
(vi) sonic drying,
(vii) heat pump drying,
(viii) vacuum jet drying,
(ix) contact sorption drying,
(x) drying in a plasma torch, etc.

But indeed it is questionable how many of them reached commercial success. Some or most of them may be at R and D level. However, some of the hybrid drying technologies are indeed a commercial success. The hybrid drying methods are nothing, but a combination of two or more drying methods, viz.,

(i) microwave-convective drying
(ii) radio frequency-vacuum drying
(iii) spray-fluid bed-vibrated fluid bed drying,
(iv) microwave-vacuum drying, etc.

Symbols

a:	constant in Eq. (17.17)	A:	area
a_0, a_1,...:	coefficients in Eq. (17.10)	A_0, A_1,...:	coefficients in Table 17.11
a_w:	water activity	b:	constant in Eq. (17.17)

$B_0, B_1, \ldots$: coefficients in Table 17.11
C: concentration
C_p: specific heat (kJ/kg°C)
d: diameter (m)
db: dry basis or dry weight basis
d_e: equivalent diameter of the channel (m)
D_0: constant in Eq. (17.40)
D_υ: diffusivity coefficient (m²/s)
E: activation energy (J/mole)
G: mass flux (kg/m²s)
h_s: heat transfer coefficient (kW/m²°C)
H: humidity
H_a: humidity of drying air
H_s: humidity of saturated air
IMC: Initial Moisture Content
k_g: gas phase mass transfer coefficient (kg/s m²Pa)
k_g': gas phase mass transfer coefficient based on humidities $\left(\dfrac{\text{kg}}{\text{sm}^2\ \dfrac{\text{kg of moisture}}{\text{kg of dry solids}}}\right)$
k: constant in Eq. (17.44)
L: length (m)
m: weight of moisture
m_e: equilibrium moisture content
m_s: weight of dry solids (kg)
m_v: weight of moisture evaporated (kg)
$\dot{m}_v$: moisture evaporation rate (kg/s)
MR: Moisture Ratio
N: coefficient in Eq. (17.44)
N_{Le}: Lewis number
p_a: partial pressure of water vapour in the air
p_s: partial pressure of water vapour at solid surface (Pa)
p_w: partial pressure of water vapour
P_w: vapour pressure of water vapour
Δp: pressure drop
q: heat transfer rate (kJ/s or kW)
r: radius in Eq. (17.37)
r: rate of drying (kg/s)
r_c: constant drying rate
R: rate of drying based on unit area (kg/m²s)
R: universal gas constant
t: time (s or h)
t_c: constant rate drying time
t_f: falling rate drying time
t_T: total time (s or h)
T: temperature in K
T_a: air temperature
T_s: temperature at interface or solid surface temperature
ΔT: temperature difference (°C)
u: linear velocity (m/s)
u_{mf}: minimum fluidization velocity
V: airflow rate/bed height in Table 17.10 (s⁻¹)
V_1: air flow rate in Table 17.10 (m/s)
wb: wet basis or wet weight basis
x: moisture content on wet basis
x: distance in x-direction
X: moisture content on dry basis
X_c: critical moisture content
X^*: EMC on dry weight basis

Subscripts

a: air
c: critical
f: falling
s: solid surface
w: water
0,1: initial
2: final

Superscript

*: equilibrium

Greek Symbols

α: coefficient in Eq. (17.5)
β: constant in Eq. (17.5)
β': constant in Eq. (17.7)
λ_T: latent heat of vaporization at temp T_s (kJ/kg)
θ: time (s)
ψ: RH on decimal basis

REVIEW QUESTIONS

17.1 What is meant by dehydration?

17.2 Compare and contrast *dehydration* and *drying*.

17.3 How drying is different from other unit operations like distillation, evaporation, pressing, filtration and centrifugation?

17.4 How do you classify the foods for the purpose of dehydration?

17.5 What are the parameters of foods that determine the drying characteristics?

17.6 What are different types of drying methods?

17.7 Describe briefly the principles of drying.

17.8 Define the following terms:

(i) Moisture content on wet basis
(ii) Moisture content on dry basis
(iii) Equilibrium Moisture Content (EMC)
(iv) Critical moisture content
(v) Free moisture
(vi) Unbound moisture
(vii) Bound moisture

17.9 How do you collect the EMC-RH data for a food material?

17.10 What is the significance of EMC-RH data?

17.11 What is Henderson's equation?

17.12 What is a drying curve?

17.13 How do you evaluate the rate of drying from drying data?

17.14 What is a drying rate curve?

17.15 What is meant by constant rate period and falling rate period? Why do they occur in drying process?

17.16 Write an expression for evaluating heat transfer coefficient under constant drying conditions.

17.17 What is meant by Lewis number? How is it useful in drying process?

17.18 How do you evaluate the diffusion coefficient in drying of solid foods?

17.19 How do you describe the effect of temperature on the diffusivity coefficient?

17.20 What is meant by *thin layer model* in drying of grains?

17.21 How do you model a drying bed?

17.22 What is meant by Page's equation?

17.23 How do you classify the drying equipment?

17.24 Describe with a neat diagram a cabinet tray dryer and its applications in food processing.

17.25 Describe with a neat diagram a turbo dryer and its applications in food processing.

17.26 Describe with a neat diagram a vacuum shelf dryer and its applications in food processing.

17.27 Describe with a neat diagram a tunnel dryer and its applications in food processing.

17.28 Describe with a neat diagram a rotary dryer and its applications in food processing.

17.29 Describe with a neat diagram a fluidized bed dryer and its applications in food processing.

17.30 Describe with a neat diagram a vibro fluidized bed dryer and its applications in food processing.

17.31 Describe with a neat diagram a screw conveyor dryer and its applications in food processing.

17.32 Describe with a neat diagram a pneumatic conveyor dryer and its applications in food processing.

17.33 Describe with a neat diagram a drum dryer and its applications in food processing.

17.34 Describe with a neat diagram a spray dryer and its applications in food processing.

17.35 Describe with a neat diagram a freeze dryer and its applications in food processing.

17.36 What criteria do you follow for selection of a dryer in food processing?

17.37 Why proper selection of a dryer is difficult in food processing?

17.38 What are various applications of dehydration in food processing?

17.39 What are various types of *advanced drying techniques*?

17.40 Why does the need arise to find various advanced drying techniques?

NUMERICAL PROBLEMS

17.1 Cabbage grits containing 80 per cent moisture is dried to 25 per cent in a cabinet tray dryer which has an average moisture removal rate of 10 kg/h. Calculate the time taken to dry 100 kg of cabbage. (**Ans:** 7 hours 20 min)

17.2 During drying of grains, the following data were obtained on the diffusivity coefficient vs. temperature. Establish an expression for diffusion coefficient.

Temperature (°C)	30	35	40	45
$D_v \times 10^{11}$, (m²/s)	2.87	4.16	6.0	8.5

(**Ans:** D_0 = 0.282, E/R = 6971 K)

REFERENCES

Keey, R.B. (1972), *Drying: Principles and Practice*, Pergamon Press, Oxford, pp. 303–322.

Kudra, T. and Mujumdar, A.S. (2002), *Advanced Drying Technologies*, Marcel Dekker, Inc., New York.

McCabe, W.L., Smith, J.C. and Harriott, P. (1993), *Unit Operations of Chemical Engineering*, 5th ed., McGraw-Hill, New York, pp. 767–809.

Page, G.E. (1949), Factors influencing the maximum rates of air drying shelled corn in thin layers, M.S. Thesis, Purdue University, Lafayette, IN.

Rangroo, S. and Rao, D.G. (1992), Drying of toria (*Brassica campestris, var. toria*): Part 2, Drying Conditions, *J. Food Engg.*, **17**, pp. 59–68.

Rao, D.G., Sridhar, B.S. and Nanjundaiah, G. (1992), Drying of toria (*Brassica campestris*) seeds: Part 1, Diffusivity characteristics, *J. Food Engg.*, **17**, pp. 49–58.

Toledo, R.T. (1997), *Fundamentals of Food Process Engineering,* 2nd ed., CBS Publishers and Distributors, New Delhi, pp. 456–506.

CHAPTER

18

Equilibrium-Stage Operations

We have seen in earlier chapters, that if we leave a pure liquid in an open container, the liquid vaporizes and the vapours merge with the surrounding environment. But if the liquid is contained in a closed container, the vapours form, and they exert certain pressure on liquid which we call as vapour pressure. If the liquid is heated, vapour formation is more, and the vapours exert more pressure. If the pressure exerted by vapours is equal to atmospheric pressure, we call that temperature as the boiling point of the liquid. If the liquid is a mixture of two or more components, the liquid mixture boils at a temperature which is between the boiling points of both the liquids. In such cases, the vapour is normally rich in more volatile component (less boiling point), and the liquid mixture is rich in the liquid of high boiling point. Thus, there is an exchange of material between the vapour phase and liquid phase when such vapours and liquids are brought into contact with each other; the former gets richer in more volatile component while the latter gets richer in less volatile component.

We come across equilibrium stage operations whenever there is an interphase mass transfer between two substances. The two phases could be liquid and vapour (or gas) or liquid and liquid (which are immiscible) or a solid and liquid or solid-liquid-gas phases. When such phases are in intimate contact, there will be transfer of mass from one phase to the other. This kind of situation we come across in some unit operations, viz., distillation, drying, gas absorption, liquid-liquid extraction, leaching or crystallization. This can be better explained by taking a specific example like distillation of a liquid mixture containing alcohol-water. If the initial liquid mixture has a composition say 40 per cent alcohol and 60 per cent water; when it is heated the vapours form, but interestingly the vapours will not have the same composition as liquid mixture[†]; the vapour phase is richer in more volatile and the liquid is richer in less volatile (more boiling point).

This concept is used to affect the separation of liquid mixtures. In such a situation, let us imagine that a column is provided in the vapour phase zone (as shown in Figure 18.1); the vapours will be going up. When the vapouring are going up, they will be condensing also,

[†]except in some special cases when at certain composition they form a constant boiling mixture; in such cases, the vapour composition is the same as that of liquid mixture.

and the liquid will be trickling down. But in the process, the down-coming liquid will come in contact with the up-rising vapours. During the process, the vapours give away any high boiling component (water) in them to the liquid stream and the vapours pick up any low boiling component (alcohol) in the liquid stream. This helps better separation to take place. But this is possible only if the liquid and vapours are brought into intimate contact with each other. At any stage, if the liquid and vapour have enough contact, they will exchange completely until a stage comes where there is no further exchange of material. Such a stage is called **equilibrium stage**; and it will be the subject matter of study for this chapter.

So far we have considered situation where the liquid is deliberately boiled to generate vapours and to separate the individual components by condensing the vapours. Another type of situation is that we bring together two components in two different phases, viz., carbon dioxide and water. Here, the idea is not for separating the phases, but for mixing the phases so that the gas component dissolves into the liquid.

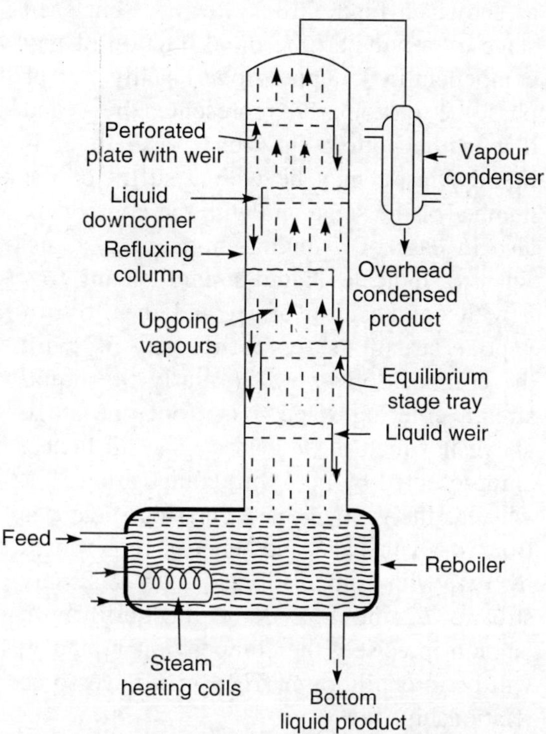

Figure 18.1 A typical distillation column with contacting arrangement and perforated trays.

18.1 CONCEPT OF IDEAL STAGE

As is shown in Figure 18.1, the vapour and liquid come in contact with each other in the sieve plate (perforated tray). If the contact is enough, by the time the streams leave the stage, they reach equilibrium[†]; i.e., any further contact will not be effective for the mass transfer. It is shown in Figure 18.2. In the whole chapter, we follow the notation that the upward flows (overflow) are denoted by V which are normally vapours, and the downside flows (bottom flow) are represented by L. Obviously, the material balance on the stage can be written as:

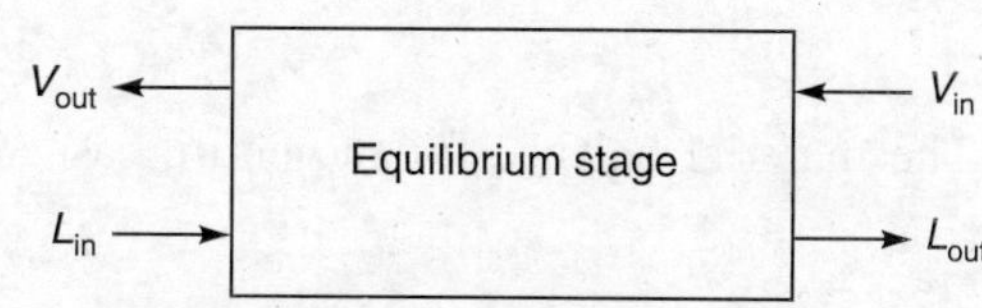

Figure 18.2 Equilibrium stage.

$$L_{in} + V_{in} = L_{out} + V_{out} \tag{18.1}$$

Equation (18.1) holds good irrespective of the fact whether equilibrium is reached or not.

[†]The concentrations of the component in both the phases are not equal; they only reach equilibrium.

If we have a number of stages in a cascade, as shown in Figure 18.3, we represent each stage by a number; the mole fraction of any component in the upper flow (usually vapour phase) is customarily represented by y and that in the bottom flow by x (usually it is liquid phase), may be with a suffix of the number of the stage and/with the component also in case of a multicomponent systems. Thus, L_n indicates liquid stream leaving tray n with composition of x_n, and the flow of vapour stream is represented by V_n with the composition of y_n. Similarly the liquid stream entering stage n is from one stage above it which is $(n - 1)$ stage, and hence, is represented by L_{n-1} with composition x_{n-1}, whereas the vapour phase entering stage n is from $(n + 1)$th stage, and hence, is represented by V_{n+1} with composition y_{n+1}. In this, only streams L_n and V_n will be in equilibrium; or more precisely the composition x_n and y_n will be in equilibrium. They may have some relationship, say

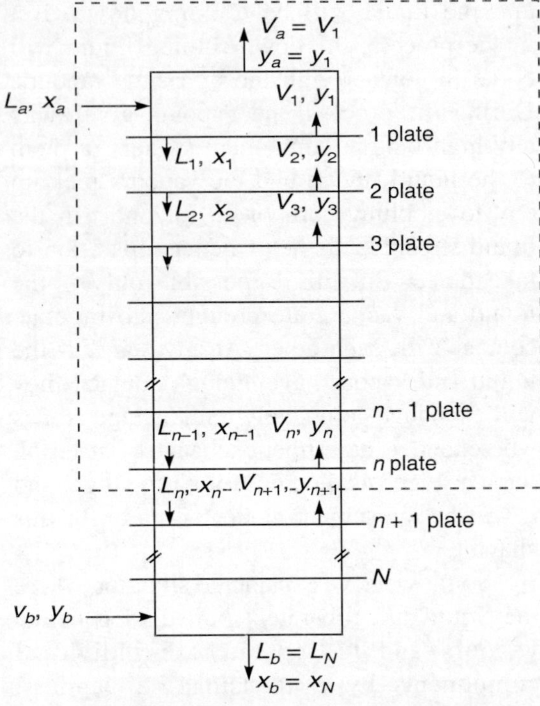

Figure 18.3 Cascade of stages.

$$y_n = f(x_n) \tag{18.2}$$

18.1.1 Material Balance

The material balance on stage n for the stream yields:

$$V_n + L_n = V_{n+1} + L_{n-1} \tag{18.3}$$

whereas for the component A, the material balance is:

$$V_n\, y_n + L_n\, x_n = V_{n+1}\, y_{n+1} + L_{n-1}\, x_{n-1} \tag{18.4}$$

The material balance for the entire cascade is:

$$L_a + V_b = L_b + V_a \tag{18.5}$$

Equation (18.5) holds good for the streams, whereas that for component A is:

$$L_a\, x_a + V_b\, y_b = L_b\, x_b + V_a\, y_a \tag{18.6}$$

$$= L_n\, x_n + V_1\, y_1 \tag{18.7}$$

18.1.2 Operating Line

The operating line is one which gives a relationship between the concentration of the vapour entering a stage (say n) and the concentration of the liquid leaving the stage (stage n), i.e., it basically shows,

$$y_{n+1} = f_1\,(x_n) \tag{18.8}$$

where f_1 is some function. We see now what f_1 is:
From Eq. (18.4), we can write

$$y_{n+1} = x_n\left(\frac{L_n}{V_{n+1}}\right) + \left(\frac{V_n y_n - L_{n-1}x_{n-1}}{V_{n+1}}\right) \tag{18.9}$$

If we consider all the stages above n, we can write

$$y_{n+1} = x_n\left(\frac{L_n}{V_{n+1}}\right) + \left(\frac{V_a y_a - L_a x_a}{V_{n+1}}\right) \tag{18.10}$$

The operating line will help us find the vapour composition y_{n+1} entering stage n based on the composition of the stage n. Once we know y_{n+1} we can know x_{n+1} since they are in equilibrium and can be calculated by Eq. (18.2). If all the flow rates are equal in either phase, i.e., $V_{n+1} = V_n = V_{n-1} = ... = V_1 = V$ and $L_n = L_{n-1} = ... = L_1 = L$.
Equation (18.9) can be simplified and written as:

$$y_{n+1} = x_n\left(\frac{L}{V}\right) + \left(y_n - x_{n-1}\frac{L}{V}\right) \tag{18.11}$$

In such a case, the operating line will become a straight line with slope = L/V and the intercept = $y_n - (L/V)\, x_{n-1}$.

18.1.3 Graphical Procedure to Find Number of Ideal Stages

From the given data and overall material balance, we can get V_a (= V_1), y_a (= y_1), L_a, x_a, V_b, x_b, L_b (= L_n) and x_b (= x_n). From the equilibrium data we can draw the equilibrium curve[†]. The operating line is drawn by Eq. (18.11). The number of ideal stages can be found by step-by-step calculation by drawing horizontal and vertical lines between the operating line and equilibrium curve. Each vertical line joining the equilibrium curve and operating curve represents one ideal state. By this step-by-step procedure, we can calculate graphically the number of ideal stages. The procedure is described with a numerical exercise.

PROBLEM 18.1 After solvent extraction process for oil from oil seeds using n-hexane solvent, the oil cake is stripped with nitrogen to remove the last traces of hexane. To remove the hexane from the nitrogen mixture, a plate column is used with a solvent which absorbs selectively hexane. The initial nitrogen-hexane mixture contains 25 mole per cent hexane and it is desired to absorb 95 per cent of the hexane into the solvent which has an outlet concentration of 10 mole per cent of hexane. The initial solvent is hexane free. Experimentally the equilibrium data was evaluated and found to fit $y^* = 1.8x^*$. Draw the operating line, and find the number of ideal stages using graphical method.

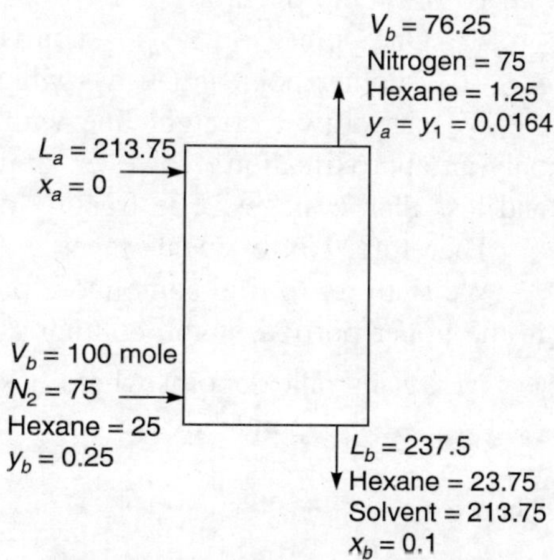

Figure 18.4 Material valance for Problem 18.1.

[†]The equilibrium data for most of the compounds can be found in the published literature. If it is not available, the same can be experimentally evaluated by collecting the V-L-E data for any compound in the mixture.

Solution First we carry out the material balance (Figure 18.4) and find various terms in Eq. (18.11) to draw the operating line.

Basis: 100 moles of entering gas

Therefore, $V_b = 100$ moles

This contains 25 per cent hexane

Therefore, $y_b = 0.25$

The inlet gas mixture contains

25 moles of hexane

75 moles of nitrogen gas

Out of 25 moles of hexane 95 per cent is to be removed; i.e., 25 × 0.95 = 23.75 moles in the bottom flow, whereas 25 – 23.75 = 1.25 mole in the overflow.

The bottom flow can only contain 10 mole per cent of hexane. Hence, bottom flow will be (23.75/0.1) = 237.5 moles which contains 23.75 moles of hexane and 213.75 (237.5 – 23.75) moles of solvent. The data can be consolidated as follows:

Inlet (Moles)		*Outlet (Moles)*	
Bottom flow (V_b):	100	Bottom flow (L_b):	237.5
Hexane = 25		Hexane = 23.75	
Nitrogen = 75		Nitrogen = 213.75	
$y_b = 0.25$		$x_b = 0.1$	
Over flow (L_a):	213.75	Over flow (V_b):	76.25
Hexane = 0		Hexane = 1.25	
Solvent = L_a = 213.75		Nitrogen = 75.0	
$x_a = 0$		$y_a = y_1 = 0.0164$	
Total	**313.75**		**313.75**

To draw the operating line:

One point is (x_a, y_a) which is (0, 0.0164).

Another point is (x_b, y_b) which is (0.1, 0.25).

We can draw a straight line with these two points which is not obviously correct, but is only an approximation. However, drawing this way in Figure 18.5, we get more than 5 stages and less than 6 stages. The fraction of the last stage is 2.2/3.2 = 0.69.

Therefore, No. of ideal stages 5 + (2.2/3.2) = 5.69 ≅ 5.7 (Figure 18.5).

We shall try to find some more points in the operating line by making a balance on hexane in the upper portion of the column.

(i) Let us take a point where hexane = 10 moles = yV

At this point hexane is 10 moles and nitrogen is 75 moles

Therefor, $$y = \frac{10}{10 + 75} = 0.1176$$

Corresponding to this, to find x, we will find what is the quantity of hexane in the liquid stream. Out of 10 moles, 1.25 moles go in the exit stream. Hence, 10 – 1.25 = 8.75 moles are in the liquid stream which also contains 213.75 moles of solvent (L_a).

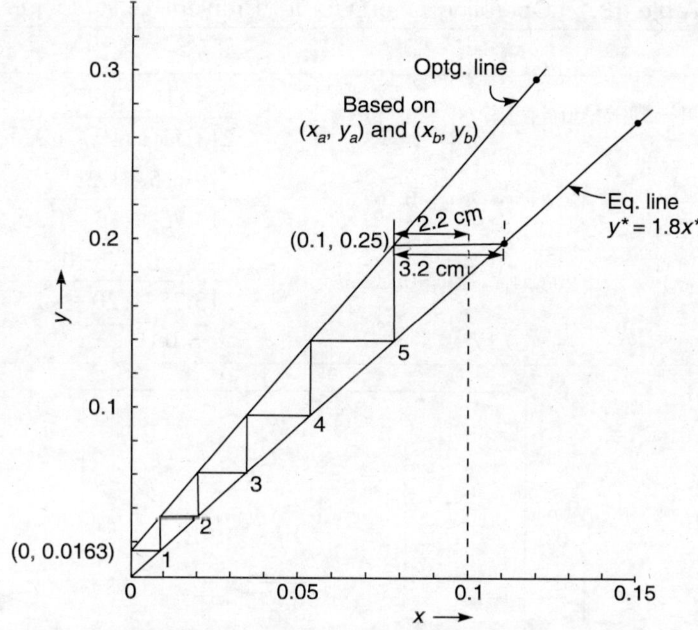

Figure 18.5 Number of ideal stages for Problem 18.1 for a straight optg. line.

Therefore, $$x = \frac{10 - 1.25}{213.75 + (10 - 1.25)} = \frac{8.75}{222.5} = 0.0393$$

(ii) Similarly for yV = 20 moles

$$y = \frac{20}{20 + 75} = 0.2105$$

$$x = \frac{20 - 1.25}{213.75 + (20 - 1.25)} = \frac{18.75}{232.5} = 0.0806$$

Similarly, we find for various yV, the values of y and x are given in Table 18.1. With this data, the operating line is drawn, which is slightly curved and is shown in Figure 18.6. Again following the same procedure, the No. of ideal stages are measured which is found to be more than 4 stages and less than 5 stages.

Table 18.1 Operating Line Data for Problem 18.1

yV (moles)	y	x
1.25	$y_a = 0.0164$	$x_a = 0$
3	3/(3 + 75) = .0385	$\frac{(3 - 1.25)}{213.75 + (3 - 1.25)} = 0.0081$
5	5/(5 + 75) = .0625	$\frac{(5 - 1.25)}{213.75 + (5 - 1.25)} = 0.0172$

(contd.)

Table 18.1 Operating Line Data for Problem 18.1 *(contd.)*

yV (moles)	y	x
10	10/(10 + 75) = 0.1176	$\dfrac{(10-1.25)}{213.75+(10-1.25)} = 0.039$
15	15/(15 + 75) = 0.167	$\dfrac{(15-1.25)}{213.75+(15-1.25)} = 0.06$
20	20/(20 + 75) = 0.21	$\dfrac{(20-1.25)}{213.75+(20-1.25)} = 0.0806$
25	$y_v = 0.25$	$x_b = 0.1$

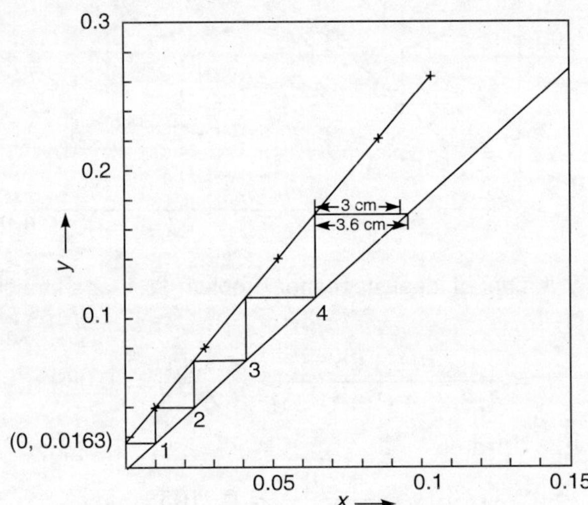

Figure 18.6 No. of ideal stages for Problem 18.1 with curved optg. line.

The fraction of the fifth stage $= \dfrac{3}{3.6} = 0.83$

Therefore, total no. of ideal stages$= 4 + \left(\dfrac{3}{3.6}\right) = 4.83$

Thus, we come across the equilibrium stage operations in most of the separation processes. We study two classical examples where Vapour-Liquid-Equilibrium (VLE) is involved, viz., distillation and gas absorption.

18.2 GAS ABSORPTION

One of the most widely used gas absorption operations in food processing is in carbonation to absorb carbon dioxide in water or fruit juices or synthetic beverages to make carbonated beverages. Gas absorption is also used in hydrogenation of oils and fats by absorbing hydrogen gas in unsaturated liquid oils using a suspended solid catalyst (raney nickel) to make semi-solid fats which are saturated in nature, and find better application in bakery industry and in various other food processing operations in view of their structural advantages.

Coming back to gas absorption as a unit operation, it is imperative that maximum transfer of gas component from the gaseous mixture takes place into the liquid solvent if and only if adequate and intimate contact is brought about between two phases. Initimate contact can be brought about by providing:

- sufficient contact time, and
- adequate contact area.

The contact time beyond the point where the two phases have reached equilibrium will not help in further absorption. However, providing adequate contact area or interfacial area will considerably increase the absorption. For increasing the interfacial area various types of gas absorption equipment are available which will be described briefly here.

18.2.1 Gas Absorption Equipment

As has been mentioned earlier, the whole purpose of an absorption, equipment is to bring an intimate contact between the two phases. Mostly the following types of equipment are used:

(i) packed towers,
(ii) bubble columns, and
(iii) agitated vessels (reactors).

The packed towers are used for physical absorption, whereas the other two are used for absorption with chemical reaction as in the hydrogenation of fats, etc.

Packed towers: It is a very common equipment used to bring intimate contact between the gas phase and liquid phase. A typical equipment is shown in Figure 18.7. It consists of a long cylindrical tower with a provision of liquid to trickle from the top and the gas enters from the bottom. The column consists of a number of packing materials. They may be raschig rings, berl saddles or plastic pal rings; the whole purpose of the packing materials is to provide as much surface area as possible for a given volume of the tower. Some of the packing materials are shown in Figure 18.8. Various characteristics of these dumped packings were given by McCabe, et al. (1993). The packing materials are usually of 6 mm to 75 mm diameter The smaller packings, though they provide more contact area, will cause higher pressure drops for the flow, and hence, are not usually preferred.

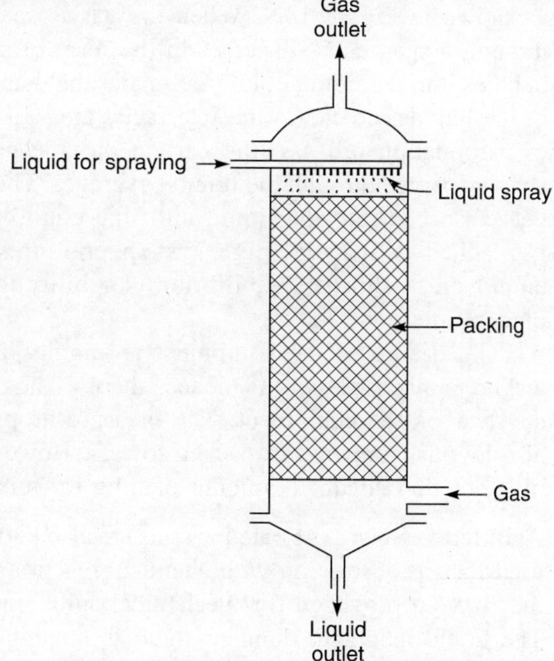

Figure 18.7 A typical packed tower.

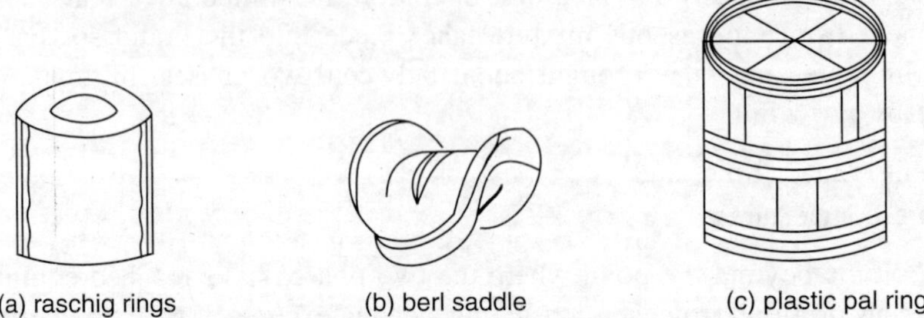

(a) raschig rings (b) berl saddle (c) plastic pal ring

Figure 18.8 Common tower packings.

The packing materials are supported on a screen. The gas to be absorbed is passed from bottom through a gas inlet. Sometimes the gas may be distributed through a gas distributor. The gas moves upwards. The liquid is admitted through liquid inlet from the top through a liquid sprayer. The spraying of liquid is desired to avoid channeling of the liquid when it is flowing downwards. The down coming liquid and up flowing gas will come into intimate contact with each other over the packing materials. The lean gas escapes through the outlet on the top while the concentrated liquid leaves from the bottom.

Bubble columns: The bubble columns are used for both physical absorption as well as for absorption with chemical reaction. In this, the gas phase is bubbled through the liquid column through a sparger as shown in Figure 18.9. When the gas is sparged through a sparger, it disperses in the form of small bubbles into the liquid column. Smaller the diameter of the bubble, greater is the interfacial area. In case of carbonation process, the carbon dioxide gas is simply sparged through the liquid beverages. The gas gets absorbed into the liquid until the equilibrium is reached. Normally, high system pressures are employed to shift the equilibrium solubility to the higher side.

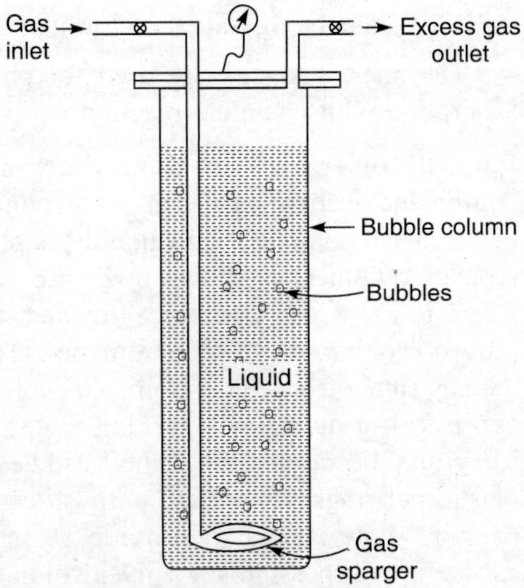

Figure 18.9 Bubble column.

The design of the equipment is simple. There are no moving parts, and hence, there is no wear and tear on the equipment. The pressure drops are also low as compared to packed towers. However, the interfacial area realized is low. The gas may not find adequate amount of time for mixing with the liquid phase.

Agitated vessels: Agitated vessels are also known as **agitated reactors** since they are invariably used for gas absorption with chemical reaction. They are similar to bubble columns except that they have a provision for mechanical agitation, and hence, the columns need not be very tall. The liquid height to diameter ratio is not high. It is almost equal to one. A typical agitated vessel is shown in Figure 16.3 and 22.5. The agitator helps in dispersing the gas into liquid column in the form of small bubbles, and hence, increases the surface area. The smaller bubbles will also have lower rise velocities, and hence, they dwell in the reactor for longer periods of time making the height of the vessel smaller (unlike in bubble columns where the bubble size is big, and hence, the bubble rise-velocity is also high).

Various types of agitators are available in industry. They will be discussed in Chapter 22.

18.3 DISTILLATION

Distillation is essentially a separation process which is used to separate a desired component from a liquid mixture by making use of the differences in the volatilities of different components of the mixture. The difference in volatilities of different components (liquids) is known as **relative volatility**. We shall define relative volatility subsequently. When a liquid mixture is heated in a closed container, as has been mentioned earlier, the mixture boils and the vapours start forming. The vapours will have a different composition compared to the liquid. The vapours are obviously richer in the more volatile component. If such vapours are collected and condensed, the condensed liquid will have a composition different from the original mixture we have taken and is richer in the high volatile component. This concept is essentially made use of in distillation to separate the liquid mixtures. The condensed liquid mixture has to be again distilled to get a still richer high volatile component in the condensed vapours. Thus, by successive distillations, we may be able to separate the high volatile component from the low volatile component. Instead of doing the successive distillations, we attempt to separate high volatile component by using a distillation column (Figure 18.1) in which the liquid and vapour are brought into intimate contact so that they exchange the high volatile and low volatile components from the vapour phase and liquid phase by exchanging heat also. Thus, distillation is a *simultaneous heat and mass transfer operation.*

When the vapours are brought into contact with liquid, both the phases will reach equilibrium with each other if they are conceived to come in contact with each other in an ideal stage. The equilibrium compositions of the vapour and liquid will change only with the temperature of the system. This leads us to a popular concept known as **Vapour-Liquid-Equilibria (VLE)** which is very much desired to draw the equilibrium curve in a distillation process.

18.3.1 Vapour-Liquid-Equilibria

As has been mentioned when a liquid mixture is heated in a closed container, the vapours formed will have a composition which is in equilibrium with the composition of the liquid mixture. A simple expression which relates vapour liquid equilibrium is known as **Raoult's law**, and is represented by

$$p_A = P_A x_A \tag{18.12}$$

where

p_A is the partial pressure of the component A in the vapour phase,
P_A is the vapour pressure of the component A at the temperature, and
x_A is the mole fraction of component A in the liquid mixture.

Usually component A represents the more volatile component in the liquid mixture. Eq. (18.12) which is the statement of Raoult's law is applicable for ideal liquid mixtures, i.e., if the components in the mixture have similar nature like benzene-toluene or methanol-ethanol, etc.

The vapour composition y_A is usually given by

$$y_A = \frac{p_A}{P} \tag{18.13}$$

where P is the total pressure.

Therefore,
$$y_A = \frac{p_A}{P} = \frac{P_A}{P} x_A \tag{18.14}$$

Thus, Eq. (18.14) gives a relationship between y_A and x_A what we call as VLE relationship.

A more generalized expression for VLE can be represented by another expression known as **Henry's law**, and is represented by

$$p_A = H_A x_A \tag{18.15}$$

where H_A is known as Henry's constant of component A. Henry's law is applicable for very dilute solutions.

In certain circumstances, the Henry's constant is equal to the vapour pressure of the pure component.

Raoult's law is more or less true for all substances in their pure state, i.e., as the mole fraction approaches 1.0; whereas the Henry's law is applicable as the concentration approaches zero. In a way we can say that Raoult's law applies for the solvent, whereas the Henry's law applies for the solute. Hence, Henry's law has considerable application in the solubility of gases in liquids as their compositions are usually very small.

If the components are far from ideality, then the VLE data have to be experimentally obtained. A liquid mixture of known composition is taken in a constant boiling still, the vapours are allowed to form. Later they are condensed and refluxed. The vapour condensate and the liquid is analysed for finding y and x at that particular boiling temperature of the mixture. *International critical tables* have tabulated enough data on VLE, but most of it relate to petroleum products. The VLE data for a binary component[†] can be represented on a x–y plot, and is shown in Figure 18.10. But it is very interesting to note that all binary mixtures do not necessarily exhibit equilibrium behaviour as shown in Figure 18.10. For example, binary mixtures like ethanol-water exhibit equilibrium behaviour typically as shown in Figure 18.11. At certain composition corresponding to point P in Figure 18.11, the mixture has same composition both in liquid phase and vapour phase, and they form a constant boiling mixture and the boiling temperature is 78.15 °C which is lower than the normal boiling point of ethanol (78.4 °C) and that of water (100 °C). The boiling point of the mixture corresponding point P is known as minimum boiling point since it is lower than the boiling point of either of the components.

[†]Containing only two components.

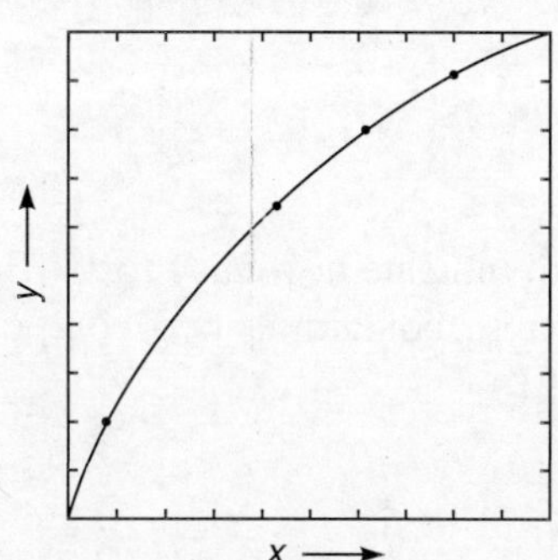

Figure 18.10 A typical VLE curve.

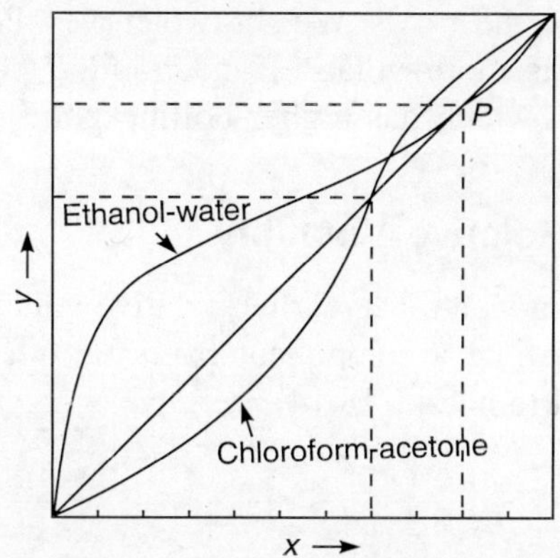

Figure 18.11 Equilibrium diagram for azeotropes.

Once the mixture reaches this composition, any futher heating will result in both the liquid and vapour phase having the same composition, and hence, further separation is not possible by distillation[†]. Point P corresponds to composition on weight basis of 95.6 per cent of alcohol (by weight) and 4.4 per cent of water (by weight). Its molar composition is:

ethanol: 89.43 per cent
water: 10.57 per cent

This mixture of ethanol-water is known as **rectified spirit**.

Similarly, there is another type of mixture, like chloroform and acetone, which also forms a constant boiling point mixture at a molar composition of 65.5 per cent of chloroform and 34.5 per cent of acetone, and this mixture boils at a constant boiling point of 64.5 °C which is more than the normal boiling point of acetone (56.5 °C) and that of chloroform (61.2 °C).

A number of such systems which are of interest in chemical engineering practice are tabulated in Perry's Chemical Engineers' Handbook (Perry and Green 1984).

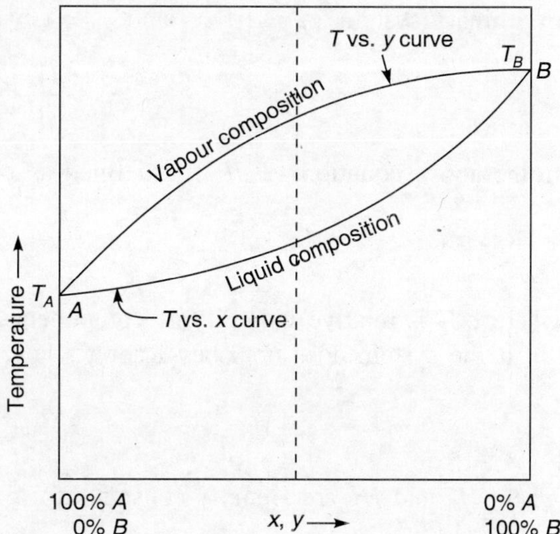

Figure 18.12 Temperature composition diagram for a typical binary system (T_A is BP of A and T_B is BP of B).

Boiling point-concentration diagram: If we take mixtures of different compositions, they boil at different temperatures. At that boiling temperature the vapour composition is different from the liquid composition; but they are in equilibrium. Then we can draw a T–x-y diagram as shown in Figure 18.12 with the temperature on the y-axis, and vapour and liquid compositions

[†]Such mixtures are separated by means of *azeotropic distillation*, by adding a third component like benzene for ethanol-water mixtures.

(y and x) on the x-axis. We get two curves. Point A indicates the boiling point of pure component A (which is more volatile) and point B corresponds to the boiling point of the other pure component which has higher boiling point (less volatile).

18.3.2 Relative Volatility

As has been mentioned earlier, relative volatility is a measure to indicate how easy (or difficult!) it is to separate a component from the other in a binary mixture containing two components A and B. From Eq. (18.14) we have

$$\frac{y_A}{x_A} = \frac{P_A}{P}$$

Similarly,

$$\frac{y_B}{x_B} = \frac{P_B}{P}$$

Therefore,

$$\frac{y_A}{x_A}\frac{x_B}{y_B} = \frac{P_A}{P_B} \tag{18.16}$$

In a binary system, $x_B = 1 - x_A$ and $y_B = (1 - y_A)$, we can write Eq. (18.16) as:

$$\frac{y_A}{x_A}\frac{(1-x_A)}{(1-y_A)} = \frac{P_A}{P_B} \tag{18.17}$$

In the above equation, (P_A/P_B) is defined as **reltive volatility**.

$$\alpha_{AB} = \frac{P_A}{P_B} \tag{18.18}$$

where α_{AB} is relative volatility of component A as compared to B.

If the system does not obey Raoult's law,

$$\alpha_{AB} = \frac{H_A}{H_B} \tag{18.19}$$

where H_A and H_B are Henry's constants for components A and B, respectively.

18.3.3 Batch Distillation

Batch distillation stills are commonly used for small capacity operations of 100–500 litres. It is extensively used in food processing for desolventization of the solvent from an extracted mixture. As the name (batch) indicates that all the contents are fed into the distillation still (Figure 18.13) which is heated by means of steam. The mixture boils, and the vapours start generating. In a simplest system, the vapours are directly taken into the condenser for condensation as shown in Figure 18.13. When the vapours leave the still, they will be in equilibrium with the liquid, but obviously their composition is different and are richer in the more volatile component. Thus, by condensation, we get a distillate which is richer in the more volatile component.

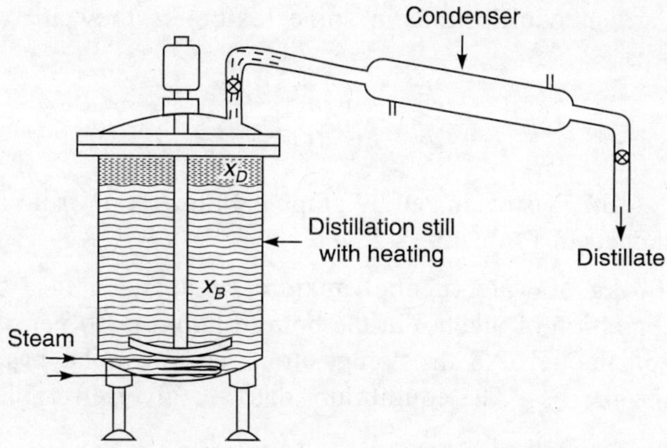

Figure 18.13 Batch distillation unit.

Since the batch distillation is mostly used for desolventization, or for separation of liquids which have wide difference in their boiling points, initially pure component only distills off almost at the boiling point of the more volatile component. As the concentration of the less volatile builds up in the still, both the components start coming out. At that stage, the distillation is stopped.

In a distillation process, we use customarily B for bottom liquid and D for the condensed liquid which is also known as the **distillate**. Similarly, their compositions are represented by mole fractions as x_B and x_D to indicate the mole fraction of bottom liquid and that of the distillate. Let us say a small quantity of the bottom material of dB moles have evaporated from the bottom material of B moles with composition of x_B. This has resulted in reduction of bottom liquid composition by dx_B from original composition of x_B. The quantity dB has evaporated into the vapour phase with composition (mole fraction) of y_D. Incidentally y_D and x_B are in equilibrium.

Volatile component present in bottom stream = Bx_B (18.20)

Let us say a differential quantity $d(Bx_B)$ of it has evaporated into the vapour phase.

Therefore, Qty. of A that has gone into the vapour phase = $dB(y_D)$ (18.21)

$$d(Bx_B) = dB\ (y_D) \tag{18.22}$$

Expanding the differential on the LHS, we have

$$Bdx_B + x_BdB = y_DdB \tag{18.23}$$

$$Bx_B = dB(y_D - x_B) \tag{18.24}$$

$$\frac{dB}{B} = \frac{dx_B}{y_D - x_B} \tag{18.25}$$

If F is the quantity of feed containing x_F fraction of the more volatile component, and integrating Eq. (18.25) between the limits of fluids stream F and B with mole fractions x_F and x_B, we have

$$\ln \frac{F}{B} = \int_{x_B}^{x_F} \frac{dx}{y_D - x_B} \tag{18.26}$$

Since y_D and x_B are in equilibrium, in some textbooks they are written as y^* and x^* Eq. (18.26) is written as:

$$\ln \frac{F}{B} = \int_{x_B}^{x_F} \frac{dx}{(y^* - x^*)} \tag{18.27}$$

The integral in Eq. (18.27) is evaluated by graphical integration from the equilibrium curve. The procedure is explained in Problem 18.2.

PROBLEM 18.2 100 kg of water-alcohol mixture is separated in a batch distillation unit until the left out concentration of alcohol in the bottom liquid is 10 per cent. The feed initially contains 40 per cent of alcohol. All the compositions are on molar basis. Find the moles of alcohol produced in the process. The equilibrium date are given in Table 18.2.

Table 18.2 Equilibrium Data for Ethanol (*A*)-water (*B*) System at 101.3 kPa

x_A* *Mole Fraction in Liquid Phase*	y_A* *Mole Fraction in Vapour Phase*	*Boiling Temperature* °C
0.019	0.17	95.5
0.0721	0.3891	89.0
0.0966	0.4375	86.7
0.1238	0.4704	85.3
0.1661	0.5089	84.1
0.2337	0.5445	82.7
0.2608	0.5580	82.3
0.3273	0.5826	81.5
0.3965	0.6122	80.7
0.5079	0.6564	79.8
0.5198	0.6599	79.7
0.5732	0.6841	79.3
0.6763	0.7385	78.74
0.7472	0.7815	78.41
0.8943	0.8943	78.15

Source: Adapted from *Perry's Chemical Engineers' Handbook* (1994) 6th ed., McGraw-Hill, p. 13.12.

Solution First we shall find out what is *F*.

Feed = 100 kg

Molar composition alcohol = 40 per cent

Mol. weight of feed = (0.4 × 46) + (0.6 × 18) = 29.2

$$F = \frac{100}{29.2} = 3.424 \text{ moles}$$

Alcohol in the feed = 3.424 × 0.4 = 1.37 kg moles

100 kg of feed contains 3.424 kg moles.

It is a straight case of application of Eq. (18.27). We shall evaluate the integral of RHS by graphical integration between limits,

$$x_F = 0.4$$

$$x_B = 0.1$$

The data are presented in Table 18.3 and plotted in Figure 18.14.

Table 18.3 Data for Evaluating Integral of Problem 18.2

x^*	y^*	$y^* - x^*$	$1/(y^* - x^*)$
0.0966	0.4375	0.3409	2.93
0.1238	0.4704	0.3466	2.88
0.1661	0.5089	0.3428	2.92
0.2337	0.5445	0.3108	3.22
0.2608	0.5580	0.2972	3.36
0.3273	0.5826	0.2553	3.92
0.3965	0.6122	0.2157	4.64
0.5079	0.6564	0.1485	6.73

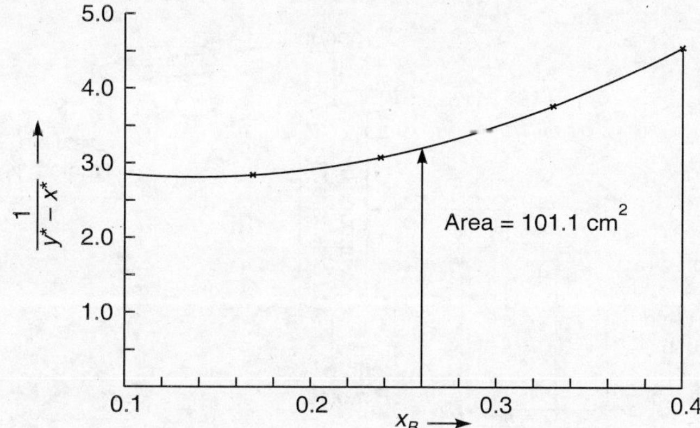

Figure 18.14 Graphical integration for Problem 18.2.

Area under the graph = 101.1 cm^2

Therefore, $$\int_{0.1}^{0.4} \frac{dx_B}{(y^* - x^*)} = 101.1 \times 0.02 \times 0.5 = 1.011$$

By Eq. (18.27)

$$\ln \frac{F}{B} = 1.011$$

Therefore, $$\frac{F}{B} = 2.7483$$

$$B = \frac{3.424}{2.7483} = 1.246 \text{ kg moles}$$

Alcohol in the bottom fluid = 1.246 × 0.1 = 0.1246 moles

Let us do the material balance on ethanol:

Ethanol in the feed = $F \times x_F$ = 3.424 × 0.4 = 1.37 moles

Ethanol in the distillate = 1.37 – 0.1246 = 1.2454 kg moles

Distillate $D = F - B$ = 3.424 – 1.246 = 2.178 moles

18.3.4 Flash Distillation[†]

Let F moles of a binary mixture enter the flash distillation unit which is similar to the one shown in Figure 18.1. We may even simplify the same by incorporating only,

- a heating unit,
- a vapour-liquid separator, and
- a condenser

and is shown in Figure 18.15. Out of the feed of F moles entering the flash distillation chamber, let D moles be evaporating into the vapour phase to come out as distillate, and B moles of it leaving as bottom liquid with composition x_B. The material balance on the fluid yield

$$F = D + B \tag{18.28}$$

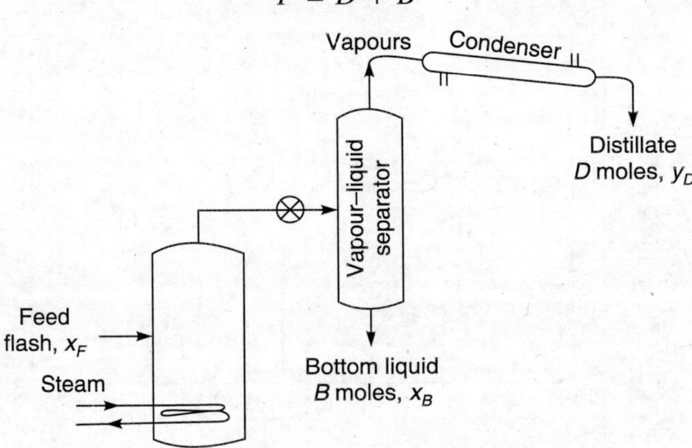

Figure 18.15 Schematic representation of flash distillation set-up.

And material balance on the volatile component is:

$$Fx_F = Dy_D + Bx_B \tag{18.29}$$

By replacing F from Eq. (18.28) into Eq. (18.29), we get

$$(B + D)\, x_F = Dy_D + Bx_B \tag{18.30}$$

from which, we can write on rearrangement,

$$\frac{B}{D} = \frac{y_D - x_F}{x_F - x_B} \tag{18.31}$$

$$y_D = -\frac{Bx_B}{D} + \frac{(B + D)x_F}{D} \tag{18.32}$$

Equation (18.32) can be written by dropping subscripts for y and x as follows:

$$y = -\frac{Bx}{D} + \frac{(B + D)x_F}{D} \tag{18.33}$$

[†]The discussion here is restricted to binary systems only. However, the same methodology applies even for multi-component systems also.

The above is an equation for a straight line with slope = $-(B/D)$; and this line is known as *tie-line*. x and y are obviously the equilibrium values, and hence, they lie (the line touches) on the equilibrium curve.

Initially when $x_B = x_F$ in Eq. (18.33), y_D will be equal to x_F. Hence, one point of the tie line also lies on the diagonal where $y = x_F = x$. Thus, the tie line with slope $-(B/D)$ and passing through the point $y = x = x_F$ gives the values of x_B and y_D when it touches the equilibrium curve.

How the feed F gets distributed between B and D depends upon the enthalpy of the feed, and enthalpies of the liquid and vapours. Hence, D can be increased from a feed of known enthalpy by applying vacuum in the distillation unit (vapour-liquid separator). The utility of above procedure is explained with a numerical example.

PROBLEM 18.3 Alcohol-water mixture in Problem 18.3 is separated in a continuous single stage flash evaporation unit until 60 per cent feed is evaporated. What will be the extent of separation of alcohol in the unit?

Solution

$$F = 3.424 \text{ kg moles}$$

$$x_F = 0.4$$

$$D = 0.6F = 2.05 \text{ kg moles}$$

Therefore,

$$B = F - D = 1.37 \text{ kg moles}$$

$$\frac{B}{D} = \frac{1.37}{2.05} = 0.67$$

Now, we draw the equilibrium curve (Figure 18.20) and diagonal to locate point P (x_F, x_F), i.e., (0.4, 0.4). We draw the tie line from P with slope equal to -0.67, which will touch the equilibrium curve at point Q, from which we can find

$$y_D = 0.46$$

$$x_B = 0.3145$$

Now, we shall do material balance on alcohol to cross check our results.

Alcohol in the feed

$= x_F \times F = 0.4 \times 3.424 = 1.37$ kg moles

Alcohol in the distillate

$= y_D \times D = 0.46 \times 2.05 = 0.943$ kg moles

Alcohol in the bottom liquid

$= x_B \times B = 0.315 \times 1.37 = 0.432$ kg moles

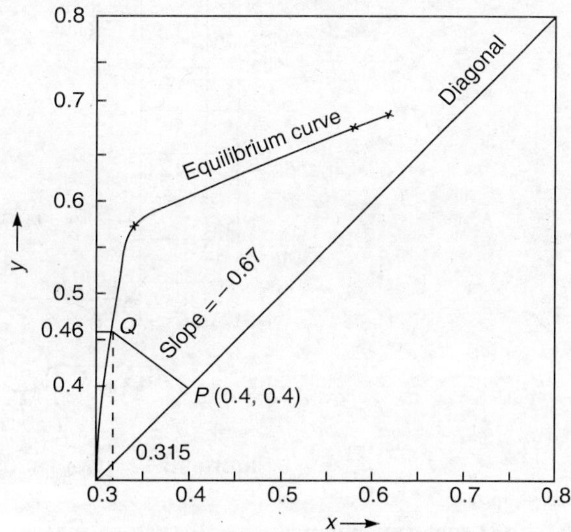

Figure 18.16 Solution of Problem 18.3.

18.3.5 Continuous Distillation with Rectification

Even though flash distillation is also a continuous operation, it is not possible to get the pure bottom liquid or pure distillate, since the vapours also contain some amount of high boiling components. Only successive distillations can alone result in reasonably pure components, or flash distillation can also be used for mixtures of components whose boiling point differences are wide. An alternative for this is to use *rectification.*

In a rectification processing, the vapours and liquid pass in a cascade of stages (Figure 18.3) so that they are brought into intimate contact. In the process of contact, the less volatiles in the vapour stream are exchanged with the more volatiles in the liquid stream. They exchange both mass and heat, and they reach equilibrium. The concept was explained in Section 18.1 with Figure 18.2. The rectification is possible only if there is a cascade of stages or plates. In the rectification process, the feed is admitted at the bottom in the reboiler. Vapours form and go up, and rectification takes place. An improvement over this method of simple rectification is to combine both *rectification and stripping.*

18.3.6 Rectification with Stripping

Instead of admitting the feed at the bottom into the reboiler, the heated feed is admitted somewhere in the middle of the cascade. The liquid feed will be falling down during which time the more volatile in it is stripped, and will be taken up into the rectification section. Thus, all the column (or plates) above the feed plate is considered as rectification section, all the plates below the feed plate including the feed plate are known as **stripping section** (Figure 18.17).

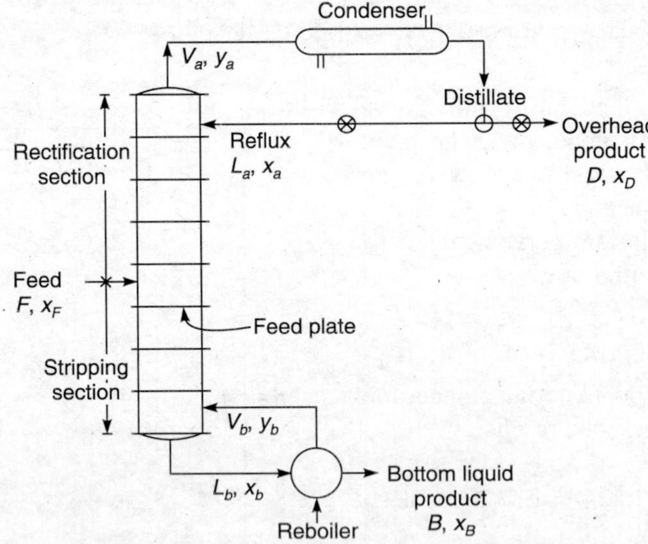

Figure 18.17 Continuous fractionating column.

Overall material balance on the system yields

$$F = B + D$$

Material balance on the volatile component yields

$$F \times x_F = (B \times x_B) + (D \times x_D) \tag{18.34}$$

We use x_D as the mole fraction of the volatile component in the distillate[†]. From Eq. (18.34),

$$\frac{D}{F} = \frac{x_F - x_B}{x_D - x_B} \tag{18.35}$$

Similarly,

$$\frac{B}{F} = \frac{x_D - x_F}{x_D - x_B} \tag{18.36}$$

Refluxing: After the vapours are condensed, a part of it is refluxed into the upper plate (or upper portion of the column). This will facilitate better rectification. But how much fraction of the distillate is refluxed is called **reflux ratio**, and is defined as:

$$R_D = \frac{L}{D} = \frac{V - D}{D} \tag{18.37}$$

where R_D is the reflux ratio and L is the kg moles of distillate refluxed.

The value of R_D varies between minimum and infinity. If the value of R_D is minimum, the number of plates is infinity and if R_D is infinity, the number of plates is minimum. Hence, the working reflux ratio should be more than the minimum. Further details to find $(R_D)_{\min}$ is available in McCabe, et al. (1993).

Operating lines: The operating line equation is derived in the earlier sections as follows:

$$y_{n+1} = x_n \frac{L_n}{V_{n+1}} + \frac{V_a y_a - L_a x_a}{V_{n+1}} \tag{18.10}$$

in which $V_a y_a - L_a x_a$ is equal to Dx_D as shown in Figure 18.17.

$$Dx_D = V_a y_a - L_a x_a \tag{18.38}$$

Replacing Eq. (18.38) into Eq. (18.10)

$$y_{n+1} = x_n \frac{L_n}{V_{n+1}} + \frac{Dx_D}{V_{n+1}} \tag{18.39}$$

In Eq. (18.39) V_{n+1} is replaced by $D + L_n$, since from the Figure 18.3

$$D = V_{n+1} - L_n \tag{18.40}$$

Therefore,

$$y_{n+1} = \frac{L_n}{(L_n + D)} x_n + \frac{Dx_D}{(L_n + D)} \tag{18.41}$$

From Eq. (18.37)

$$R_D + 1 = \frac{L + D}{D} \tag{18.42}$$

$$\frac{R_D}{R_D + 1} = \frac{L}{L + D} \tag{18.43}$$

[†]The distillate composition is represented by x_D and not by y_D, since this is not the equilibrium value.

Replacing Eqs. (18.42) and (18.43) into Eq. (18.41),

$$y_{n+1} = \frac{R_D x_D}{R_D + 1} + \frac{x_D}{R_D + 1} \tag{18.44}$$

$$= x_D$$

The operating line in the rectification section will be touching the diagonal at x_D and will have

$$\text{Intercept} = \frac{x_D}{(R_D + 1)} \tag{18.45}$$

Similarly, in the stripping section, the operating line will be touching the diagonal at x_B and will meet the point of intersection of the feed line and the operating line of rectification section.

Feed line: The feed line is the one which represents the condition of feed. The feed could be

(i) a sub-cooled liquid, i.e., the feed is at temperature less than its boiling point,
(ii) a saturated liquid at its bubble point, i.e., the feed is at its boiling point,
(iii) in a partly vapourized form; i.e., the feed is a mixture of boiling liquid and some vapours,
(iv) a saturated vapour at its due point; i.e., feed is a saturated vapour,
(v) a superheated vapour.

Depending upon the feed conditions, the feed takes away the heat from the up-going vapours to vaporize or to draw its latent heat of vapourization. Hence, location of the feed line is important. The feed conditions are usually represented by a term called q which is defined as:

$$q = \frac{\text{Heat energy required to convert one mole of feed to saturated vapour}}{\text{Molar heat of vaporization of feed}} \tag{18.46}$$

Depending upon whether feed is a cold liquid or superheated vapour, q is given by (McCabe, et al., 1993);

$$q = 1 + \frac{C_{p,L}\,(T_b - T_F)}{\lambda} \tag{18.47}$$

if the feed is a mixture of liquid and vapours. For superheated vapours, Eq (18.47) modifies to

$$q = -\frac{C_{p,v}\,(T_F - T_d)}{\lambda} \tag{18.48}$$

where

$C_{p,L}$ and $C_{p,v}$ stand for specific heat of the liquid feed and that of vapour feed,
λ is the latent heat of vaporization,
T_f is feed temperature,
T_b and T_d stand for bubble point and dew point of feed.

Feed lines for different conditions of feed are shown in Figure 18.18, and are described in Table 18.4. The feed line touches the diagonal line at (x_F, x_F) and will have

$$\text{Slope} = -\frac{q}{1-q} \tag{18.49}$$

(i) If the feed is a sub-cooled liquid, $q > 1$ and the feed line will have a positive slope, and is vertical as shown by line QP_1.

(ii) If the feed is a saturated liquid, $q \cong 1$, and hence, the feed line is a straight vertical line passing through point Q on the diagonal (line QP_2).

(iii) If the feed is a mixture of liquid and vapour, $0 < q < 1.0$, and hence, the feed line will have a negative slope, and bends to the left (line QP_3).

(iv) If the feed is a saturated vapour at its due point, $q = 0$, and hence, the feed is a horizontal line with zero slope (line QP_4).

(v) If the feed is a superheated vapour, $q < 0$, and hence, the feed line bends more towards the left and is shown y line OP_5 in Figure 18.18.

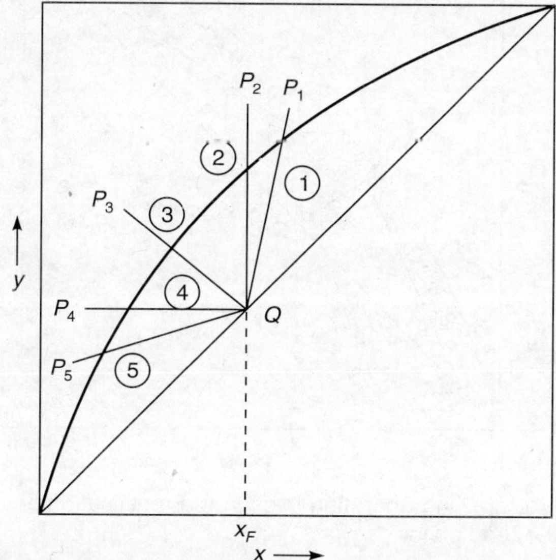

Figure 18.18 Effect of feed conditions on the feed line. ① sub-cooled feed with $q > 1.0$, ② saturated liquid with $q = 1$, ③ mixture of liquid and vapour, $1 < q < 0$, ④ saturated vapour with $q = 0$ and ⑤ superheated vapour with $q < 0$.

Table 18.4 Feed Line Position Depending Upon Feed Conditions

Feed condition	*q*	*Feed line position*
Sub-cooled liquid	> 1.0	↗
Saturated liquid at its bubble point	1.0	↑
Mixture of liquid and vapour	$1 > q > 0$	↖
Saturated vapour at its dew point	0	←
Superheated vapour	< 0	↙

McCabe–Thiele method to find number of ideal stages: Graphical procedure to find the number of ideal stages was proposed by McCabe and Thiele (1925). The step-by-step procedure to find the number of ideal stages is very convenient. The operating line in the rectification section is drawn (Figure 18.19) from point A (x_D, x_D) with an intercept given by Eq. (18.45):

$$\text{Intercept} = \frac{x_D}{R_D + 1}$$

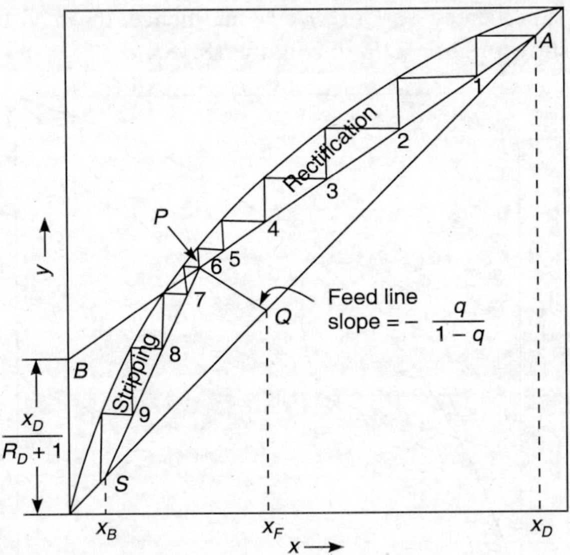

Figure 18.19 Procedure for drawing operating lines in the rectification section and stripping section. Plate 7 is the feed plate.

Subsequently the feed line is drawn with appropriate slope based on the feed conditions and touching the diagonal at Q (x_F, x_F). The feed line will touch the operating line AB at point P. Point S is located on the diagonal at (x_B, x_B). The line joining S and P represents the stripping section and is known as the **operating line** for the stripping section.

Now, the number of ideal stages is found by drawing vertical and horizontal lines between the operating lines and the equilibrium curve. The plate which comes at the point of intersection of the operating lines of rectification section and stripping section is ideal to be the feed plate. The procedure is described in Problem 18.4.

PROBLEM 18.4 A continuous rectification column with stripping is used to distil a 1000 kg mixture of acetic acid and water which contains 40 per cent acetic acid (molar). The feed is a saturated vapour at its boiling point. The reflux ratio used is 2.6. It is desired to purify acetic acid upto 90 per cent concentration (molar) in the bottom product and the top product containing 10 per cent acetic acid (molar). Determine the No. of ideal plates and the location of feed plate. The equilibrium data for acetic acid-water systems is noted from literature and is given in Table 18.5.

Table 18.5 Equilibrium Data for a Water-Acetic Acid System

Temperature °C	*Mol. Fr. of Water in* *Liquid*	*Vapour*
100	1.0	1.0
100.2	0.9578	0.9708
100.5	0.9134	0.9409
100.8	0.8787	0.9186
100.8	0.8556	0.9042
101.5	0.7951	0.8671
102.1	0.7261	0.8239
102.8	0.6750	0.7797
103.5	0.5824	0.7112
104.3	0.5195	0.6580
105.2	0.4498	0.5973
107.8	0.3084	0.4467
110.6	0.1881	0.3063
118.3	0	0

Source: Adapted from *Perry's Chemical Engineers, Handbook*, 6th ed. (1984), pp. 13–14.

Solution Feed contains 40 per cent (molar) acetic acid.

100 kg moles of feed contains 40 kg moles acetic acid + 60 kg moles of water, i.e., 100 kg moles feed weighs (40 × 60 + 60 × 18) kg

$$= 3480 \text{ kg}$$

i.e., one kg mole weighs $\dfrac{3480}{100} = 34.8$ kg

Therefore, molecular weight of feed = 34.8

1000 kg of feed contains = $\dfrac{1000}{34.8} = 28.7$ kg moles

$$F = 28.7 \text{ kg moles}$$

$$x_D = 1.0 - 0.1 = 0.9$$

(since the top product contains 10 mole per cent of acetic acid, and since water is more volatile, we represent mole fractions in terms of the more volatile component.)

Similarly, $$x_B = 1.0 - 0.9 = 0.1$$

and $$x_F = 1.0 - 0.4 = 0.6$$

We draw the feed line and operating lines in Figure 18.20.

We draw feed line *PQ* as a horizontal line, since the feed is a saturated vapour. It will touch the diagonal at *Q*. We mark the point *A* corresponding to $x_D = 0.9$.

By Eq. (18.45), the intercept of the operating line of the rectification is

$$\frac{x_D}{R_D+1} = \frac{0.9}{2.6+1} = 0.25$$

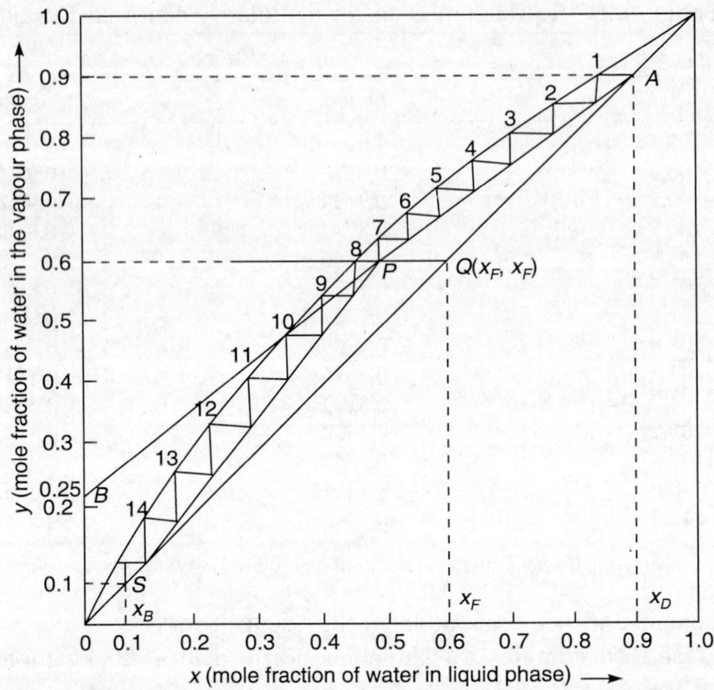

Figure 18.20 McCabe–Thiele method to find number of ideal stages of Problem 19.4.

We mark the intercept at *B*.
By joining *AB*, it intersects the feed line at *P*.
We mark the point *S* on the diagonal corresponding to $x_B = 0.1$.

We join *SP* which makes the operating line in the stripping section. Next we draw a horizontal line from *A* until it touches the equilibrium line. From there we draw a vertical line to the operating line. We follow the step by step procedure and find the number of ideal stages. The procedure is shown in Figure 18.20.

No. of ideal stages is 14. Feed is supplied to the 9th plate.

18.3.7 Steam Distillation

Steam distillation is one of the most convenient methods to distil volatile components (which are heat sensitive, and are insoluble in steam/water) from non-volatile components at a temperature which is much less than the boiling point of the volatile component. This method is frequently used in food processing for removal of some flavours or bad odours from the food products. It is extensively used to distil the volatile oils from the spices like cloves, ginger, garlic or cardamom. In this method, we take advantage of the fact that liquid or liquid mixture boils when the vapour pressure exerted by the components becomes equal to the atmospheric pressure.

In a steam distillation process, the contents (usually a plant material) are taken in a container into which steam is supplied continuously (Figure 18.21). Sometimes even superheated steam

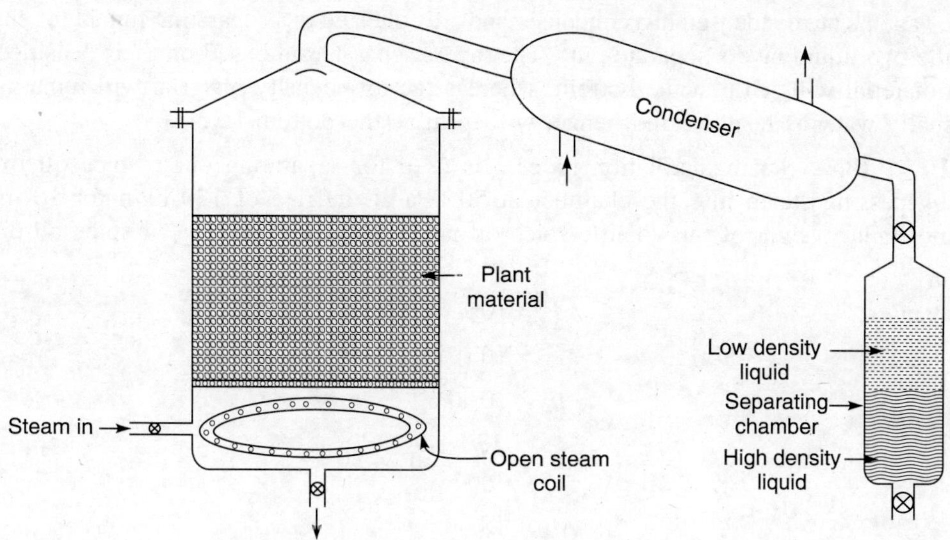

Figure 18.21 Steam distillation set-up.

is supplied. The vapours that form will have the composition proportional to heir partial pressures; i.e.,

$$\frac{y_A}{y_s} = \frac{p_A}{p_s} = \frac{P - p_s}{p_s} \tag{18.50}$$

where y_A and y_s are vapour phase mole fractions of the volatile component A and steam, p_A and p_s are partial pressures of A and steam, and P is the total pressure.

Since the system is binary, the partial pressure of A is equal to $(P - p_s)$. Obvoiusly

$$y_A = \frac{m_A}{M_A} \tag{18.51}$$

and

$$y_s = \frac{m_s}{M_s} \tag{18.52}$$

where m_A stands for mass of the component A and m_s stands for the mass of the steam. Similarly, M_A and M_s are molecular weights of A and steam.

Therefore,

$$\frac{p_A}{p_s} = \frac{(m_A/M_A)}{(m_s/M_s)} \tag{18.53}$$

or if we write in terms of mass fractions,

$$\frac{m_A}{m_s} = \frac{P - p_s}{p_s} \times \frac{M_A}{M_s} \tag{18.54}$$

Equation (18.54) indicates p_A is much less than P_A, the vapour pressure of A. This means that the volatile material is evaporating much below its normal boiling point. Since M_A is much larger than M_s, we get more quantity of volatile oils for a given amount of steam.

The vapours are condensed in a condenser and are collected in a separating funnel (or chamber) where the two liquid layers separate out[†]. The layers separate out based on their densities. High density material will settle at the bottom, whereas the low density material will remain on the top. Usually water is heavier, and hence, will form as the bottom layer.

PROBLEM 18.5 Steam distillation process is used for separating out a spice oil from the spices by passing steam into the chamber at 50 kPa at the rate of 120 kg/h for 10 minutes. If the molecular weight of the volatile spice oil is 165, find the quantity of spice oil distilled.

Solution

Given data:

$$P = 101.3 \text{ kPa}$$

$$M_A = 165$$

$$p_s = 50 \text{ kPa}$$

$$m_s = \frac{120}{60} \times 10 = 20 \text{ kg}$$

$$M_s = 18$$

$$m_A = ?$$

We apply Eq. (18.54)

$$\frac{m_A}{m_s} = \frac{P - p_s}{p_s} \times \frac{M_A}{M_s}$$

i.e.,

$$\frac{m_A}{20} = \frac{101.3 - 50}{50} \times \frac{165}{18} = 9.405$$

Therefore,

$$m_A = 9.405 \times 20 = 188.1 \text{ kg}$$

for 20 kg of steam passed, we get about 188.1 kg of material. The quantity of steam here is only the one which will come into the separating chamber. Initially a good quantity of steam will be utilized for volatilizing the oil and also to bring the plant material (spices) to the temperature prevailing in the distillation unit. We do not account that steam here.

18.3.8 Vacuum Distillation

To carry out distillation at a temperature less than the boiling temperature of the mixture, one another method is to apply vacuum to the distillation column. This will reduce the boiling temperature of the mixture, and hence, distillation can be carried out at a temperature much less than the normal boiling point of the mixture. Vacuum distillation is particularly useful for those components which are heat sensitive, and may degrade if they are heated upto the boiling point of the mixture. As has been mentioned, steam distillation is also one method, but steam distillation can be applied only if the volatile component is not miscible with steam. Even the steam distillation method can also be coupled with vacuum in which case, we call it as *vacuum steam distillation.*

Vacuum can also be applied to the batch distillation systems to carry out separation at lower temperatures. Even vacuum can be applied for continuous distillation systems also, but it is not

[†]It was already mentioned that steam distillation applies only if the component *A* is insoluble in water.

normally done in view of the associated problems in maintaining vacuum to the continuous systems. With modern efficient vacuum producing systems, vacuum distillation finds good application in food processing in view of the heat sensitive nature of most of the food materials.

18.3.9 Applications of Distillation in Food Processing

Distillation is one unit operation which is not extensively used in food processing. The classical applications can be classified as follows:

Brewery industry: In the alcohol manufacture by fermentation of molasses or any fruit juices (like cashew apple juice or banana pulp, etc.), the fermented broth after the fermentation is over, is subjected to distillation to separate out the ethyl alcohol. The distillation step is so important in the brewery industry that the industry itself is called as **Distillary**.

Solvent recovery: Desolventization is not a main operation in any solid-liquid extraction or liquid-liquid extraction units, but it invariably follows the extraction step for recovery of solvent which can be reused. In the solvent extraction of oil seeds, the oil is extracted by bringing in contact the oil seeds with an organic solvent (usually food grade *n*-hexane)[†]. The oil and solvent are separated out from the oil seed cake. The oil is soluble in the solvent. The solvent is distilled from the mixture to have pure edible oil.

Similarly, in the solvent extraction of oleoresins from spices or annatto dye from the annatto seeds, the extract is subjected to distillation to recover the solvent which is subsequently reused.

Steam distillation: Steam distillation of spices, herbs or leaves is a common operation to recover the essential oils (components) from the material. Most of the spices and condiments like cloves, cardamom, pepper, ginger, garlic, cinnamon, curry leaves, etc. are subjected to steam distillation to recover the essential oils from them which have got both culinary and medicinal applications. Most of the aromatic plant materials like citranella or palmorosa are steam distilled to recover the aromatic oils.

Symbols

B: bottom stream or liquid stream
$C_{p,L}$: specific heat of liquid
$C_{p,v}$: specific heat of vapour
D: distillate
F: feed flow rate, (kg moles/s or kg/s)
H: Henry's constant
L: molar liquid phase flow rate (kg moles/s)
M: molecular weight
m: mass (kg)
P: total pressure

[†]The process is discussed in Chapter 19 on extraction.

P_A: vapour pressure of A at the given temperature
p_A: partial pressure of A in the vapour phase
q: a parameter defined by Eq. (18.46)
R_D: reflux ratio
s: steam
T_b: bubble point of feed
T_d: dew point of feed
T_F: feed temperature
t: time
V: molar vapour phase flow rate (kg moles/s)
x: liquid composition
y: vapour composition

Superscript

*: equilibrium values

Subscripts

A: component A, usually more volatile
B: bottom stream or liquid stream
D: distillate
e: equilibrium
F: feed
i: interface
n: stage in the column
s: steam
y: vapour phase, or distillate

Greek Letters

α_{AB}: relative volatility of A wrt B
λ: latent heat of vapourization

REVIEW QUESTIONS

18.1 Describe the concept of equilibrium stage operations.

18.2 What is an ideal stage? Describe the concept with a neat diagram.

18.3 What is the operating line concept in a cascade of ideal stages?

18.4 How do you find the number of ideal stages in equilibrium stage operations?

18.5 Describe the process of gas absorption.

18.6 What are the applications of gas absorption in food processing?

18.7 What are various equipments used for gas absorption?

18.8 Describe the concept of distillation briefly.

18.9 Describe the concept of VLE.

18.10 What is Raoult's law?

18.11 What is Henry's law? How it is useful in distillation?

18.12 Draw the boiling point-concentration diagram for any typical binary mixture.

18.13 What is meant by relative volatility?

18.14 Describe a batch distillation process with a neat diagram.

18.15 Derive a mathematical expression for a batch distillation process to find the quantity of distillate.

18.16 Describe a flash distillation process with a neat diagram.

18.17 Describe a continuous process with rectification.

18.18 What is meant by the stripping section in a rectification unit?

18.19 What is meant by refluxing in a continuous distillation unit?

18.20 What is the significance of reflux ratio in a continuous distillation unit?

18.21 What is the importance of feed conditions in a continuous distillation unit?

18.22 What is meant by feed line in continuous distillation unit? How does it vary with feed conditions?

18.23 Derive an expression for the operating line in a continuous distillation unit.

18.24 Describe the McCabe–Thiele method to find number of ideal stages in a continuous distillation unit.

18.25 Describe the process of steam distillation with a neat diagram.

18.26 Derive the mathematical expression for steam distillation to relate the quantity of distillate to the quantity of steam.

18.27 Describe the process of vacuum distillation.

18.28 What are the various applications of distillation in food processing?

REFERENCES

McCabe, W.L., Smith, J.C. and Harriott, P. (1993), *Unit Operations of Chemical Engineering*, 5th ed., McGraw-Hill, New York, pp. 686–737.

McCabe, W.L. and Thiele, E.W. (1925), Graphical designing of fractionating columns, *Ind. Engg. Chem.*, **17**(6), pp. 605–611.

Perry, R.H. and Green, D. (1984), *Perry's Chemical Engineers' Handbook*, 6th ed., McGraw-Hill Book, New York, pp. 13.57–13.60.

CHAPTER

19

Extraction

Extraction is a typical unit operation which finds extensive applications in chemical engineering and food engineering. Extraction operations can be subdivided into—solid-liquid extraction, and liquid-liquid extraction.

The former is generally called as **leaching** in chemical engineering literature. It is used as a processing tool (technology) for extraction of active principles (solute) from the solids using an organic solvent or water as the extracting liquid source. Generally in food engineering, it could be extraction of vegetable oils from the oil seeds, or extraction of natural food colours, viz., *curcumin* from turmeric or *annatto dye* from annatto seeds (*Bixa orellana*), or extraction of coffee seeds using hot water. If the extraction is done at the boiling point of the solvent, we call it as *decoction*. Hence, coffee extraction is known as **coffee decoction**. Mostly the solvent extraction processes are used for extraction of oleoresins from spices or for extraction of oil from oil seeds. If the oil content in the seeds is high, (like peanuts, cotton seeds, mustard, coconut or gingily seeds), the oil is extracted by mechanical pressing either by using a screw press or traditional *Ghani*. Subsequently, the mechanically pressed solids (known as **oil seed cake**) may be subjected to solvent extraction to extract the residual oil. If the oil content is low in the seeds as in the case of soya bean or rice bran, the oil extraction is done by solvent extraction alone. Mechanical pressing is not at all possible.

Rice bran is even pelletised in the form of pellets before feeding to the extraction to facilitate easy movement of the solvent. If the rice bran is loaded as it is, it would choke the extractor, and does not allow the percolation of solvent through it to extract the oil.

The other extraction operation is liquid-liquid extraction, it is popularly known as **LLE**. It is used more as a *unit operation* for separation of the solute from a mixture of solute and solvent using another solvent in which the solute has got preferential solubility. The two solvents are immiscible and form into two separate phases into which the solute gets distributed depending upon its preferential solubility. This characteristic feature is termed as **partition coefficient** or **distribution coefficient (K)** where

$$K = \frac{C_{A1}}{C_{A2}} \tag{19.1}$$

C_{A1} is concentration of solute A in phase 1 and C_{A2} is that in the phase 2. Phase 1 and phase 2 are immiscible. For example, acetic acid gets distributed between aqueous phase and chloroform phase, and LLE can be used as a tool for separation of acetic acid from water mixture. Similarly, penicillin from the aqueous fermentation broth is extracted into butyl acetate using LLE method which is later recovered by distillation.

However, our interest in food processing is more directed to solid-liquid extraction which is a technological tool for separation of the active principle (solute) from the solid phase. Hence, we restrict our study in this chapter to solid-liquid extraction which is known as **leaching**.

19.1 SOLID-LIQUID EXTRACTION OR LEACHING

The solute present with the solids (known as inerts or inert solids) either in bound or unbound form is extracted using a solvent in which the solute is soluble. If the solute is bound with the solid in its crevices, then the solid needs to be broken down into smaller particles so that the solvent can have good accessibility to the location of the solute to extract it (solute). This is obviously a mass transfer process, and hence, is given by Eqs. (10.1) and (10.2).

$$\text{Rate of transfer} \propto \text{Driving force}$$

$$\propto \text{Interfacial area or area of contact}$$

$$\propto \frac{1}{\text{Resistance}}$$

Hence, we can write

$$\frac{dM}{dt} \propto (C_{AS} - C_A)$$

$$\propto A$$

$$\propto \frac{1}{R}$$

or

$$\frac{dM}{dt} = k_1 A\ (C_{AS} - C_A) \tag{19.2}$$

where M is mass transferred in time t, k_1 is mass transfer coefficient, C_{AS} is concentration of solute (A) at the solid-liquid interphase, C_A is concentration of A at any time t and A is area of contact.

From Eq. (19.2), it is clear that rate of extraction is influenced by,

- concentration difference,
- area of contact (or surface area available),
- solubility of the solute in the solvent,
- viscosity of the solvent,
- diffusivity of the solvent into the solid matrix,
- temperature of extraction, and
- degree of agitation to improve eddy diffusivity.

Most of these criteria can be met by a proper choice of the solvent. Thus, the selection of proper solvent plays a major role. Table 19.1 lists some of the most commonly used solvents in food processing.

Table 19.1 Some Common Solvents Used for Extractions

Solvent	*Remarks*
Hexane (food grade)	Most commonly used organic solvent; highly inflammable
Acetone, Ethyl ether	Preferred if the material is wet
Dichloro ethane	Reduces the risk of explosion hazards
Chloroform	Not used for edible purposes, mostly used for complete extraction of solute for analytical purpose
Methanol, ethanol	–
Ethyl acetate	–

19.1.1 Types of Extraction Processes

Depending upon the need and the quantities to be handled, the extraction processes may be classified as follows:

Single-stage batch extraction: It is a batch extractor in which the solids are brought in contact with the adequate quantity of solvent (Figure 19.1). The contents are kept together (just like in a soaking process) either with or without heating and with or without stirring as the process demands. The solute is extracted into the solvent. It is simple in operation, easy in maintenance and flexible in utilization (i.e., the same vessel can be used for different extraction processes). Generally in batch processing, the throughputs are low, and is restricted to one ton batch.

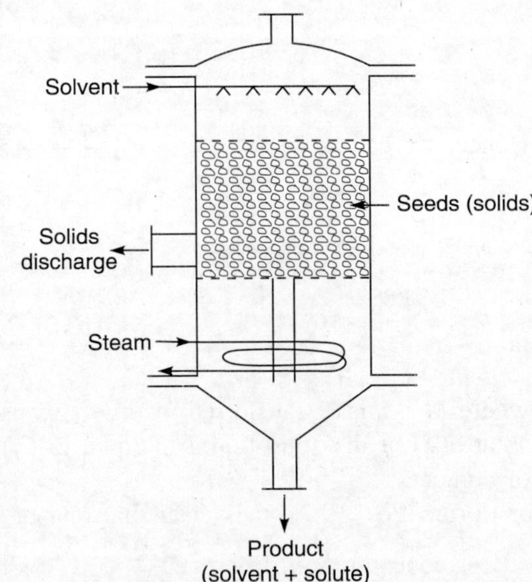

Figure 19.1 Batch extractor.

After the extraction process reaches equilibrium, the solution is decanted. At the time of decanting, the solids are saturated with the solution, and carry the solution with them which contains the solute also. To recover this solute, the contents are washed once again with a fresh batch of solvent to recover the adhering solute. The washings (extraction) continue until the desired extent of solute is extracted. The only disadvantage with this process is that every time the solids retain the adhering solution with them which contains the solute. Thus, the solids retain some amount of solute after every extraction. If we wash more number of times to extract as much of solute as possible, the volume of solvent spent is more, and hence, its recovery by distillation becomes costlier.

For brewing of coffee or tea, batch extractors are used. In the extraction of oleoresins, since the batches are small, generally batch extractors are used. Advantage is also taken that the same batch extractor can be used for extraction of different spices during different seasons.

Multi-stage cross flow extractor: In this extractor also, the solids are contacted every time with a fresh batch of solvent. In that respect, it is similar to batch extractor in which the solids are repeatedly contacted with fresh batches of solvent. Another way of achieving this is similar to **soxhlet extractor**, in which the solids are extracted with a batch of solvent; later the solvent is slowly evaporated from the solution and condensed. The vapours condense and fall into the column containing the solids. The condensing liquid will be in contact with the solids until the liquid fills up to the brim of the solids. Later the solvent with the extracted solute will be siphoned into the bottom vessel in which the liquid is heated for evaporation continuously.

This is an excellent process for recovery of complete solute from the solids, as we do in the laboratory estimations of total fat, etc. The only disadvantage with this process is that the solvent is repeatedly evaporated, and hence, the solute is subjected to continuous heating which may result in its thermal degradation. A lot of energy is consumed in vaporizing and condensing the solvents. Hence, this process is not normally used industrially for separation purposes. It is more used as an analytical tool.

Multi-stage counter-current extraction: In the multi-stage counter-current extraction process, a battery of extractors is used. The solids (seeds) flow in one direction (from right side to left side in Figure 19.2) of the battery while the solvent flows in the opposite direction (from left side to right side). In this system, the fresh solids (seeds) are brought in contact with the solvent which contains a major part of the solute and is about to leave the extraction batteries from the left hand side. Similarly, the solids containing very little amount of the solute before discharge are brought into contact with fresh solvent, and hence, any last traces of the solute adhering to the solids are also extracted almost completely into the solvent. Thus, the solids, even though they carry some amount of the solution before discharge, will be carrying a very little amount of solute with them. In this method, extraction can be almost complete with additional number of extractors in the battery. Addition of more extractors adds to cost of the extraction system. Hence, depending upon the value of the solute, extractors are added. Sometimes it also depends upon the value or purpose for which the spent solids are used. If it is oil seeds, the spent solids

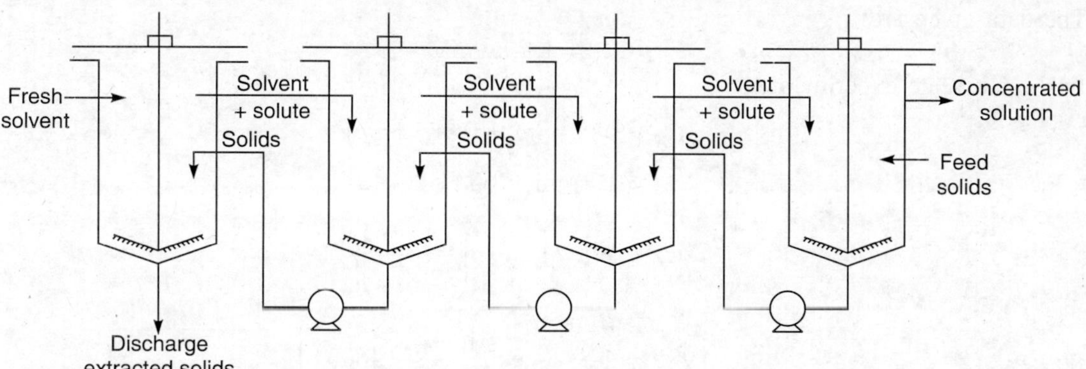

Figure 19.2 Counter-current extraction battery.

should contain as little of oil as possible, not because the oil is costlier, but because the residual oil in the spent solids makes them rancid over a period of time.

19.2 PRINCIPLES OF EXTRACTION

The rate of extraction is given by Eq. (19.2) in which

$$dM = V_v dC_A \tag{19.3}$$

where V_v is the volume

Therefore,

$$\frac{dC_A}{dt} = \frac{k_1}{V_v}(A)(C_{AS} - C_A) \tag{19.4}$$

On rearrangement and integration

$$\ln \frac{C_{AS} - C_{A0}}{C_{AS} - C_A} = t\left(\frac{k_1 A}{V_v}\right) \tag{19.5}$$

where C_{A0} is initial concentration.
If $C_{A0} = 0$, we can write

$$1 - \frac{C_A}{C_{AS}} = \exp\left(\frac{-k_1 At}{V_v}\right) \tag{19.6}$$

or

$$C_A = C_{AS}\left[1 - \exp\left(\frac{-k_1 At}{V_v}\right)\right] \tag{19.7}$$

In Eq. (19.7) (k_1A/V_v) is constant, and is evaluated experimentally, and the same is used in scale-up process. This procedure is described with Problem 19.1.

PROBLEM 19.1 The natural food colour from beet root is to be extracted with water. The saturated concentration of the colour in the water is found to be 1.5 kg/m^3. In a laboratory scale extractor containing about one litre volume, it has taken 10 minutes to extract the colour from beet root to an extent of 985 ppm. Under similar conditions in a commercial plant of 10 m^3 capacity, it is desired to extract 12 kg of the colour into the water. How much time does it take?

Solution It is a direct application of Eq. (19.7)
The data given are:

$$C_{AS} = 1.5 \text{ kg/m}^3$$

In the lab scale experimentation

$$C_A = 985 \text{ ppm} = 0.985 \text{ kg/m}^3$$

$$t = 10 \text{ min} = 600 \text{ s}$$

$$C_A = C_{AS}\left\{1 - \exp\left(-\frac{k_1 At}{V_v}\right)\right\}$$

i.e.,

$$0.985 = 1.5\left[1 - \exp\left\{\left(\frac{-k_1 A}{V_v}\right)600\right\}\right]$$

$$1.5 - 0.985 = 0.515 = \exp\left\{\left(\frac{-k_1 A}{V_v}\right)600\right\}$$

$$\ln (0.515) = -0.663 = -\left(\frac{k_1 A}{V_v}\right)600$$

Therefore,
$$\left(\frac{k_1 A}{V_v}\right) = \frac{0.663}{600} = 1.106 \times 10^{-3} \text{ m}^3/\text{s}$$
$$= 0.001106 \text{ m}^3/\text{s}$$

We use the same value of (k_1A/V_v) in Eq. (19.7) to calculate t, when

$$C_A = \frac{12 \text{ kg}}{10 \text{ m}^3} = 1.2 \text{ kg/m}^3$$

$$1 - \frac{C_A}{C_{AS}} = \exp(-1.106 \times 10^{-3}\ t)$$

i.e.,
$$1 - \frac{1.2}{1.5} = 0.2 = e^{-(0.001106\ t)}$$

$$\ln(0.2) = -1.609 = -1.106 \times 10^{-3}\ t$$

Therefore,
$$t = \frac{1.609}{1.106 \times 10^{-3}} = 1455 \text{ s}$$
$$= 24.25 \text{ min}$$

19.2.1 Counter-current Extraction

The counter-current extraction process is shown schematically in Figure 19.3. We follow the same notation as what we followed for distillation. The upper streams are designated by V and their weight fractions by y, whereas the underflows are represented by L and their weight fraction by x. We make an overall mass balance on he streams:

$$\underset{\text{(Inflow)}}{L_0 + V_{n+1}} = \underset{\text{(Outflow)}}{L_n + V_1} \tag{19.8}$$

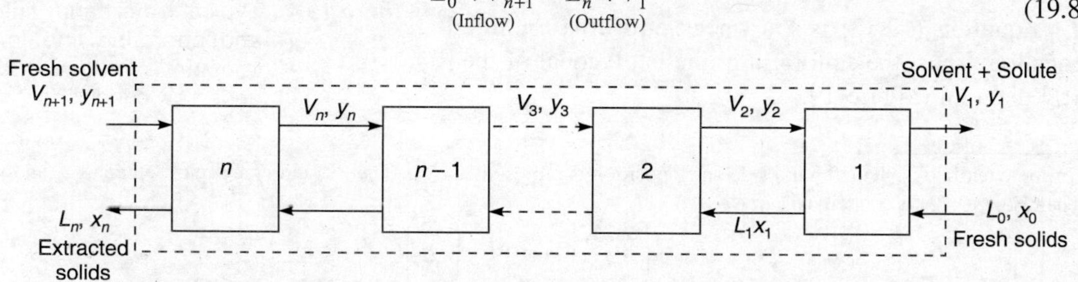

Figure 19.3 Schematic representation of counter-current extraction process.

The overall balance is applicable to the box in dotted lines. Similarly, mass balance on the solute gives:

$$L_0x_0 + V_{n+1}\, y_{n+1} = L_n\, x_n + V_1\, y_1 \qquad (19.9)$$

If a fresh solvent is used, $y_{n+1} = 0$. This indicates that the solute inflow is only through the fresh solids, whereas the solute leaves in the overflow along with the solvent, and it also leaves in the underflow adhering to the spent solids (extracted solids).

From Eq. (19.9), we can rite

$$y_{n+1} = \frac{\ln}{(V_{n+1})}(x_n) + \frac{V_1 y_1 - L_0 x_0}{V_{n+1}} \qquad (19.10)$$

Equation (19.10) is known as the **operating line equation**.

The value of V_{n+1} can be written as:

$$V_{n+1} = L_n - L_0 + V_1$$

and is replaced in Eq. (19.10)

$$y_{n+1} = \frac{L_n}{L_n - L_0 + V_1}(x_n) + \frac{V_1 y_1 - L_0 x_0}{L_n - L_0 + V_1}$$

Generally, the process conditions are such that equal underflows and equal overflows are used throughout.

Hence, $V_1 = V_2 = \ldots V_n = V_{n+1} = V$

and $L_0 = L_1 = \ldots L_n = L$ (19.11)

Combining Eqs. (19.10) and (19.11)

$$y_{n+1} = \left(\frac{L}{V}\right)(x_n) + y_1 - \left(\frac{L}{V}\right)x_0 \qquad (19.12)^{\dagger}$$

Equation (19.12) is operating line equation for constant overflow and constant underflow.

In each stage, it is assumed that the equilibrium reaches, and hence, the stage is known as **equilibrium stage**. In extraction process, the same outgoing solution is adhering to the outgoing solids. Hence,

$$y_n = x_n \qquad (19.13)$$

Equation (19.13) is known as **equilibrium line** equation. Thus, the equilibrium line is diagonal of the x, y diagram (Figure 19.4).

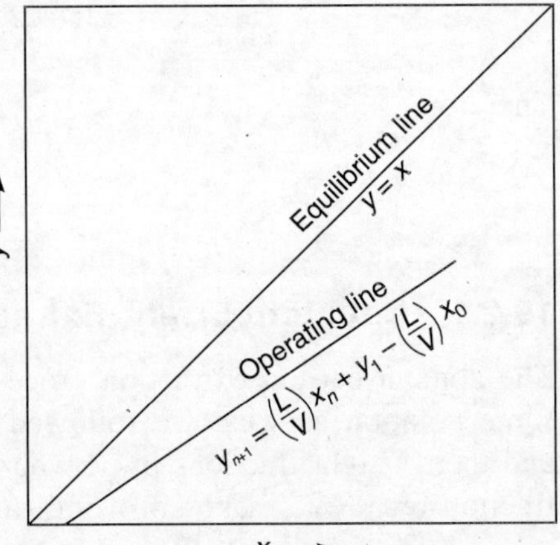

Figure 19.4 Typical equilibrium line and operating line for constant overflow and underflow conditions.

†In some textbooks, the left hand side entry position is represented by a. Then x_0 and y_1 are represented as x_a and y_a. Then Eq. (19.12) is written as:

$$y_{n+1} = \left(\frac{L}{V}\right)(x_n) + y_a - \left(\frac{L}{V}\right)(x_a)$$

Similarly, the right hand side exit position is represented by b.

The equilibrium line and operating line data are used to calculate the number of ideal stages in a counter-current extraction process. The procedure is explained here with Problem 19.2.

PROBLEM 19.2 A counter-current extraction process is used for extraction of oil using petroleum ether as solvent from oil seeds which initially contain 20 per cent oil by weight. It is desired to extract 90 per cent of the oil from the seeds, and the final concentration of oil in the exit solution is 50 per cent. The crushed seeds in the underflow carry with them 40 per cent of their weight of solution in each stage. Assuming that all the oil in the seeds is extracted in the first stage itself, and equilibrium is reached in each step, find the number of ideal stages. The overflows and underflows are assumed to be constant.

Solution The flow conditions are calculated by applying mass balance equation and the values are represented in Figure 19.5.

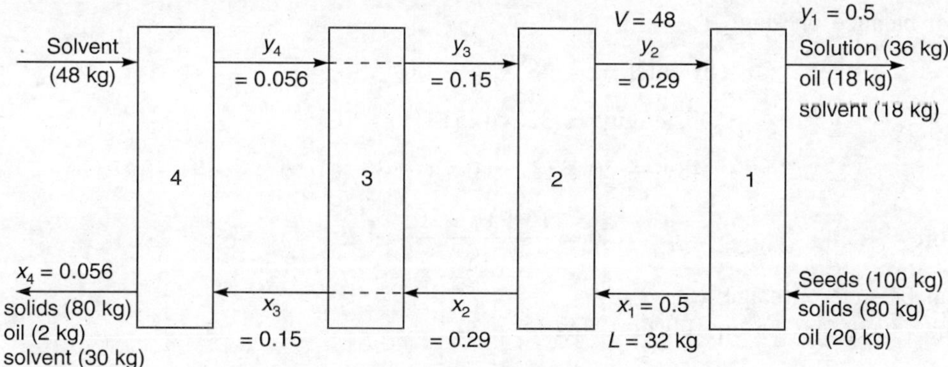

Figure 19.5 Material balance for Problem 19.2.

Basis: 100 kg of oil seeds.

The seeds contain 20 per cent of oil = 20 kg

Solids (inerts) = 80 kg

90 per cent of oil is to be extracted. Therefore,

$$\text{Oil in the final solution} = 20 \times 0.9 = 18 \text{ kg}$$

$$\text{Oil fraction in the final solution} = 50 \text{ per cent}$$

Hence,

$$\text{Solvent in the solution} = \frac{18}{50} \times 50 = 18 \text{ kg} \tag{1}$$

In the underflow, weight of solids = 100 – 20 = 80 kg

Solution retained by them = 40 per cent of their weight

= 80 × 0.4 = 32 kg

$$\text{Solvent in the final solution} = \text{Solution} - \text{Weight of oil in it}$$

$$= 32 - (20 \times 0.1) = 30 \text{ kg} \tag{2}$$

Therefore, total solvent entering = (1) + (2) = 18 + 30 = 48 kg

Total volume of final overflow = 18 + 18 = 36 kg

Mass balance:

Mass in (kg)		*Mass out* (kg)	
Overflow:		Overflow:	
Solvent:	48	Oil: 18 Solvent: 18	36
Underflow:		Underflow:	
Solids:	80	Solids: 80	
Oil:	20	Oil: 2	112
		Solvent: 30	
Total	**148**	**Total**	**148**

Material balance on stage 1:

$$\text{Oil entering} = 48 \times y_2 + 20$$

$$\text{Oil leaving} = 32 \times 0.5 + 36 \times 0.5$$

$$48y_2 + 20 = (32 \times 0.5) + (36 \times 0.5)$$

Therefore,
$$y_2 = \frac{16 + 18 - 20}{48} = 0.29$$

Material balance on stage 2:

Since stage 2 is an equilibrium stage,

$$x_2 = y_2 = 0.29$$

$$\text{Oil entering} = 48 \times y_3 + 32 \times 0.5$$

$$\text{Oil leaving} = 48 \times 0.29 + 32 \times 0.29$$

Therefore,
$$y_3 = \frac{(48 \times 0.29) + (32 \times 0.29) - (32 \times 0.5)}{48}$$

$$= \frac{13.92 + 9.28 - 16}{48} = 0.15$$

Material balance on stage 3:

$$y_3 = x_3 = 0.15$$

$$\text{Oil entering} = 48 \times y_4 + 32 \times 0.29$$

$$\text{Oil leaving} = 48 \times 0.15 + 32 \times 0.15$$

Therefore,
$$y_4 = \frac{(48 \times 0.15) + (32 \times 0.15) - (32 \times 0.29)}{48}$$

$$= \frac{7.2 + 4.8 - 9.28}{48} = 0.056$$

Since it is an ideal stage

$$y_4 = x_4 = 0.056$$

We require final x_n to be $\frac{2}{32} = 0.0625$

Since $x_4 < 0.0625$, only 4 stages are adequate.

19.2.2 McCabe–Thiele Method

The above method of calculating the ideal stages can also be done graphically by stepwise calculations using the McCabe–Thiele method by plotting the equilibrium line and operating line on *x–y* diagram. The method (McCabe and Thiele 1925) was originally proposed by McCabe and Thiele for stepwise calculation of number of ideal stages in a distillation column. The method can be better explained with Problem 19.3.

PROBLEM 19.3 Solve the Problem 19.2 by using McCabe–Thiele graphical method of stepwise calculation.

Solution The operating line and equilibrium line are drawn. The stepwise calculations method can be applied from 2nd stage onwards. In the first stage, it is assumed that all the oil from seeds is extracted; and hence, by applying material balance method we calculate x_1 and y_2.

$$x_1 = 0.5$$

$$y_2 = 0.29$$

Upto this point, the procedure is same as what we did earlier. The block in which the step-wise calculations are done is show in Figure 19.6.

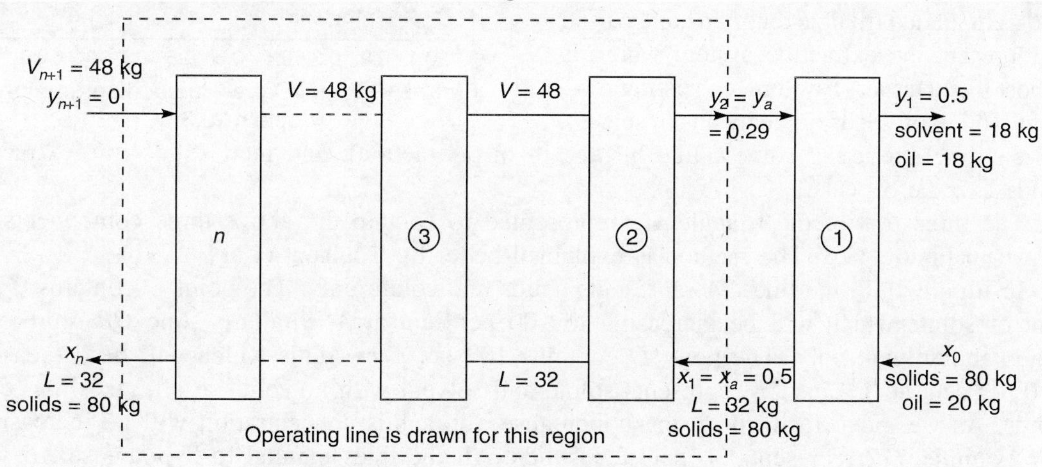

Figure 19.6 Representation of Problem 19.3.

Equation for equilibrium line $y_n = x_n$.
Equation for operating line:

$$y_{n+1} = \left(\frac{L}{V}\right)x_n + y_a - \left(\frac{L}{V}\right)x_a$$

$$L = 2 \text{ kg}$$

$$V = 48 \text{ kg}$$

$$\left.\begin{aligned} y_a &= y_2 = 0.29 \\ x_a &= x_1 = 0.5 \end{aligned}\right\}^{\dagger}$$ since we are considering from stage 2 onwards, and x_1 and y_2 are inlet and leaving conditions for stage 2.

Therefore,
$$y_{n+1} = \left(\frac{32}{48}\right)x_n + 0.29 - \left(\frac{32}{48}\right)0.5$$
$$= 0.667\ x_n - 0.04 \qquad (19.14)$$

The equilibrium line and operating line are drawn on Figure 19.7; and the stepwise procedure is shown.

$$x_4 = y_4 = 0.055$$

The final oil to be left out in the solids

$$= 20 \times (1.0 - 0.9)$$
$$= 2 \text{ kg}$$

Therefore, $x_n \le \dfrac{2}{32} \le 0.0625$

Since $x_4 < 0.0625$, four stages are adequate.

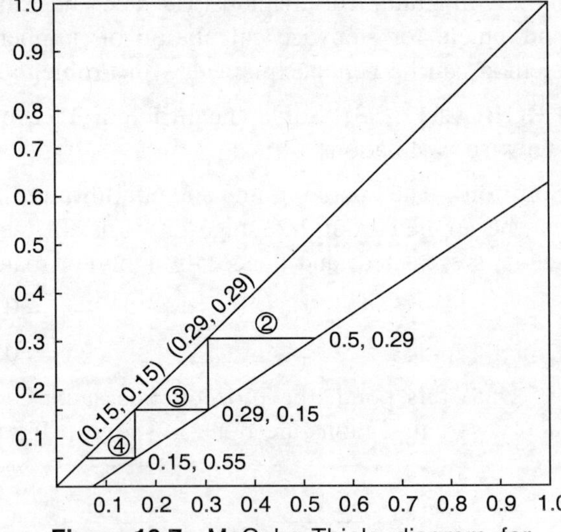

Figure 19.7 McCabe-Thiele diagram for Problem 20.3.

19.2.3 Right Angled Triangle Method

The right angled triangle diagram can be used to represent the extraction system which is almost like a ternary[‡] system, consisting of—solute (oil in oil seeds or oleoresins in spices), solvent (the extracting solvent like hexane in oil extraction), and inert solids (the extracted solids like the oil cake).

The three tips of the triangle are represented to contain the above three components as shown in Figure 19.8. The method is explained better by Coulson, et al., (1991).

In Figure 19.8, the line *OA* represents solids and solute only. The point *O* contains 0 per cent of solute which will be increasing to 100 per cent at *A*. Similarly, line *OB* represents solids and solvent only. The point *O* contains 100 per cent solids which will be decreasing to 0 per cent at *B*. Line *AB* represents solute and solvent with no solids at all, i.e., 0 per cent solids. As we move forward on the hypotenuse, the solids concentration will be increasing. For example, *QR* represents 30 per cent solids. All the lines parallel to hypotenuse *AB* will be containing the increased solvent as we go up. Similarly line *Q′R′* represents 70 per cent solids. Therefore, now any point *P* in the triangle represents 40 per cent solute, 30 per cent solvent and 30 per cent solids. Similarly, we can represent any composition of the ternary system in the triangle.

[†] y_a and x_a are the weight fraction of upper leaving stream and the weight fraction of the bottom entering stream, respectively.

[‡] The ternary system contains three components.

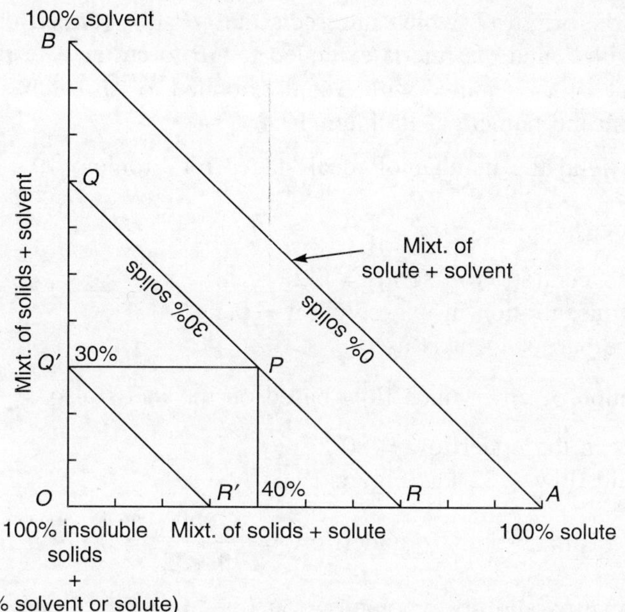

Figure 19.8 Representation of three component system. Line *OA* contains solids and solute. Solute is 0% at *O* and 100% at *A*; Solids are 100% at *O* and 0% at *A*. Line *OB* represents solids and solvent only.

Now, let us see how this method can be used to calculate the number of ideal stages. The method is also known as **difference point method**. For a constant underflow, the concentration of solids is fixed, and it can be represented by a line parallel to hypotenuse line *QR* or *Q′*R′.

Based on the conditions given in the problem, we will be knowing

$$x_0,\ y_1,\ x_n,\ y_{n+1}$$

We note x_0 on the *OA* (abscissa), and y_1 on the hypotenuse as the overflow does not contain any solids. In fact all the *y*'s will be lying on the hypotenuse *AB*. Generally, we use pure solvent for extraction. Hence, y_{n+1} will be at *B* (100 per cent solvent). We note down x_n on the underflow line since this point represents solvent, solute and inert solids. The difference point (Δ) is the point of intersection of extended (y_1, x_0) line and (y_{n+1}, x_n) line as shown in Figure 19.9. Once the Δ is located, further procedure is by stepwise calculation.

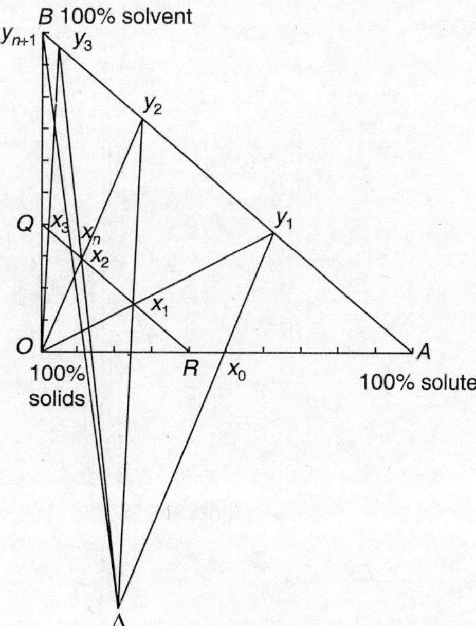

Figure 19.9 Graphical method of finding number of ideal stages using difference point method.

y_1 is joined to the origin O, which intersects the QR line (underflow) at x_1 (Figure 19.9). Now x_1 is joined with Δ and the line is extended to AB to cut at y_2. Subsequently y_2 is joined with O to cut QR at x_2. x_2 is joined with Δ and extended to AB to cut at y_3. The procedure is better explained with the numerical Problem 19.4.

PROBLEM 19.4 Find the number of ideal stages for Problem 19.2 using difference point method.

Solution

$x_a = x_0 =$ Oil concentration in solids = 0.2

$y_a = y_1 =$ Oil concentration in exit solution = 0.5

$y_{n+1} = 1.0$, since pure solvent is used.

$x_n =$ Concentration of oil in underflow based on the inerts also $= \dfrac{2}{112} = 0.0178$

The inert solids in the underflow = 80 kg

Underflow liquid flow = 32 kg (Figure 19.6)

$$\text{Therefore, fraction of inert solids in underflow} = \frac{80}{80+32} = 0.714 \cong 0.71$$

Line QR is drawn parallel to hypotenuse touching x-axis at $0.286 \cong 0.29$

Locating x_1, x_2, x_3, etc. on line QR is as described earlier.

The graphical solution is shown in Figure 19.10. x_4 is less than x_n. Hence, four equilibrium stages would be sufficient to achieve the desired extraction.

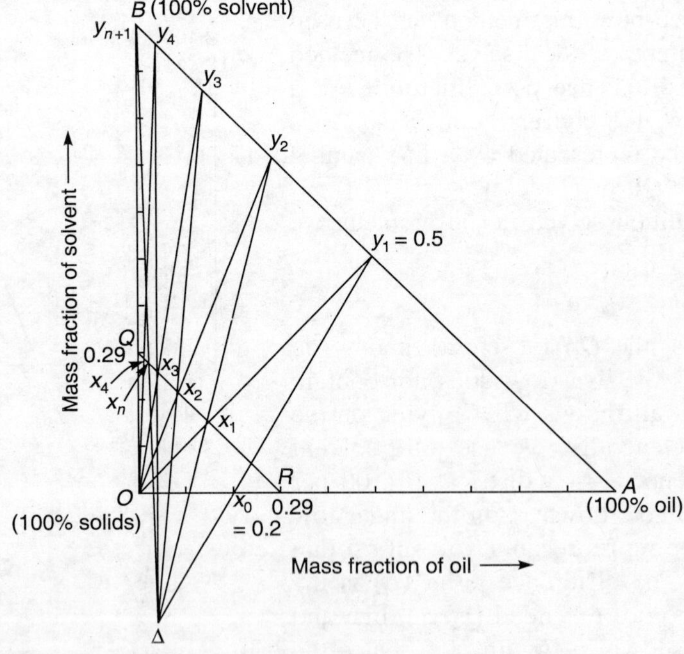

Figure 19.10 Graphical solution for Problem 19.4 by right angle triangle (difference point) method.

19.3 EXTRACTION EQUIPMENT

The extraction equipment for leaching active principles from the solids has to carry out the following operations:

- Bringing intimate contact between the solvent and solute.
- Leaching out the solute as quickly as possible by providing adequate opportunity for the solute to diffuse into the solvent from the porous structure of the solids.
- Leaving as little solute in the spent solids as possible.
- Extracting as much solute into the solution as possible.
- Leaving as little solvent in the spent cake (solids) as possible.

Some of the above operations are based on equipment characteristics, while others are based on the system parameters. The following are some of the system parameters:

- Diffusion of the solute from the solid matrix.
- Solubility of the solute in the solvent.
- Particle size of the solids to provide adequate contact area.

In the present section, we restrict ourselves to the various types of extraction equipment. The classification is shown in Figure 19.11.

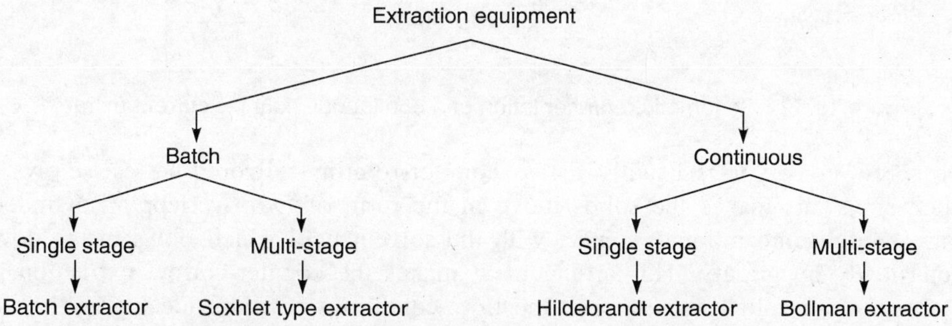

Figure 19.11 Classification of extraction equipment.

19.3.1 Batch Extractor

It has already been described in section 19.1.1 with Figure 19.1.

19.3.2 Continuous Counter-current Extractor

The advantages of a continuous counter-current extraction system have already been discussed in terms of

- improving the efficiency of extraction,
- reducing the solvent usage, and
- reducing the residual solute in the spent cake.

The extractor is slightly in a slanting position to allow the movement of the solvent fed at the upper end to flow continuously (Figure 19.12) by gravity. In fact, there is a constant liquid level maintained in the extractor. The solids are made into a suitable form (like pellets, flakes or granules, etc.) and fed from the lower end continuously. The solids are conveyed by means of a screw conveyor which transfers the solids through the liquid (solvent) pool in the upward direction. Heating jackets are provided to the extractor to heat the contents in case it is required.

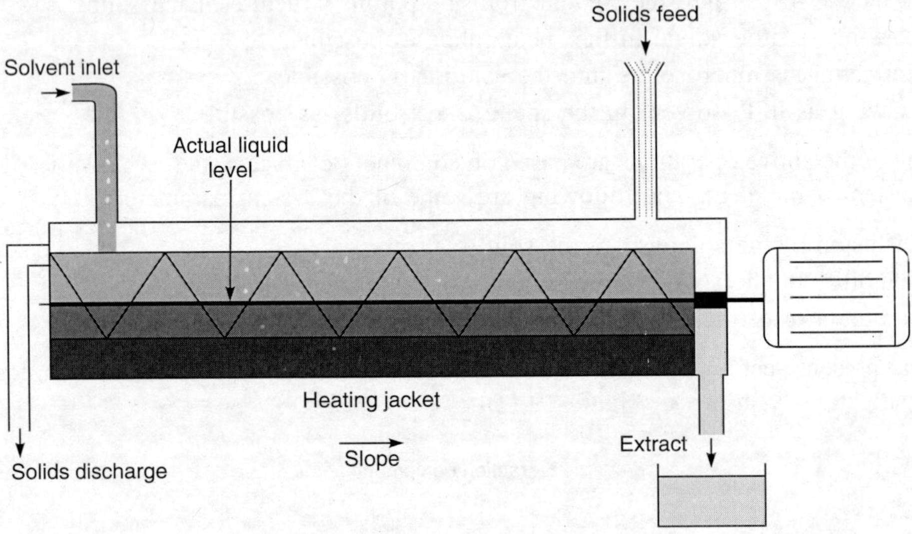

Figure 19.12 Schematic representation of a continuous counter-current extractor.

The screw conveyor is usually a two counter-rotating-mirrored-helical-screws with a controlled slip. This makes the solids move in the form of two overlapping cylinders. The solids are brought into intimate contact with the solvent pool which flows in the downward direction slowly by gravity. This arrangement makes the counter-current extraction process take place in a controlled manner with the flow conditions reaching ideal plug-flow in both the phases. The necessary residence time for the solids can be achieved by adjusting the rate of the solids. The flow rate of the liquid (solvent) can also be adjusted to give an optimum concentration of the extract.

The extract leaves at the bottom end. In some systems, self-cleaning filtration systems are arranged. The extracted solids are discharged at the upper end.

Hildebrandt extractor is one such arrangement shown in Figure 19.13 (Cofield 1951). It consists of a U-shaped screw conveyor with each limb having a separate screw conveyor with different rpm. The bottom leg also consists of a screw conveyor. The fresh solids in appropriate form are fed in one limb, and fresh solvent is fed in the other limb. Thus, the solids and the solvent move in counter-current direction. The fresh solids come in contact with the solution which is about to leave the extractor, whereas the fresh solvent comes in contact with the almost spent solids which are about to leave the extractor from the other end. The different speeds of the helices of the conveyors give enough compaction to the solids in the horizontal section.

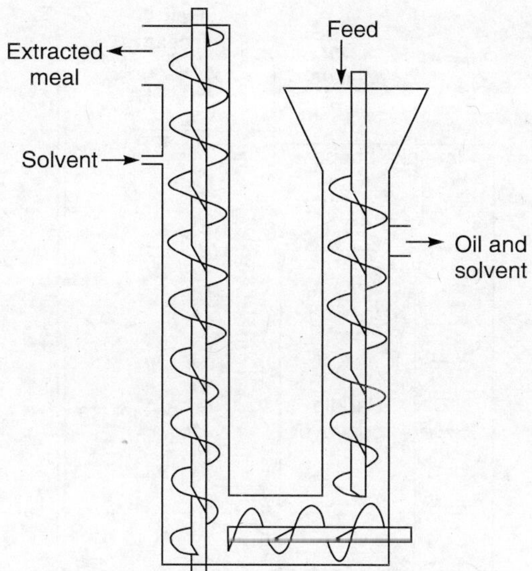

Figure 19.13 Hildebrandt extractor.

Source: Reprinted from Solvent extraction of oilseed, Cofield, Jr, E.P., *Chemical Engineering*, **58**(1), January, 1951, p. 129, Copyright (1951), with permission from Chemical Week Associates, New York.

19.3.3 Multi-stage Continuous Counter-current Extractor

It consists of a battery of extractors into which the fresh solids are fed at one end (to the first extractor) while the fresh solvent is added at the other end (to the last extractor). The extracted solution known as **full miscella** leaves after getting contact in the first extractor, whereas the spent solids (extracted solids) leave from the last extractor after getting in contact with the fresh solvent.

To facilitate the movement of solids and the half miscella, Bollman extractor is one such an arrangement, which is used industrially. The extractor is shown in Figure 19.14 which consists of bucket elevator (Cofield 1951). The buckets are perforated. At right hand side fresh solids are fed into each bucket which will be moving downwards until half the way. Later on the buckets will be moving upwards. Fresh solvent is fed on the left hand side continuously into the buckets containing almost spent solids. Later the buckets when they go to the top position, tilt upside-down, and the spent solids are discharged into the hopper of peddle conveyor and are taken away. The solvent by the time it reaches to the bottom of the left side leg of batteries, it extracts some solute from the solids, and is known as the half miscella. The half miscella is pumped to the top of the right side leg and is fed into the bucket into which the fresh solids are fed. Thus, half miscella comes in contact with the fresh solids and extracts more solute, and becomes full miscella by the time it reaches the bottom. Here the miscella and solids will be moving in co-current direction in the right leg. At the bottom of the right leg, the full miscella is discharged.

At the bottom of the extractor, the half miscella and full miscella are separated by a partition. In the left leg, the solvent and the solids will be moving in counter-current direction.

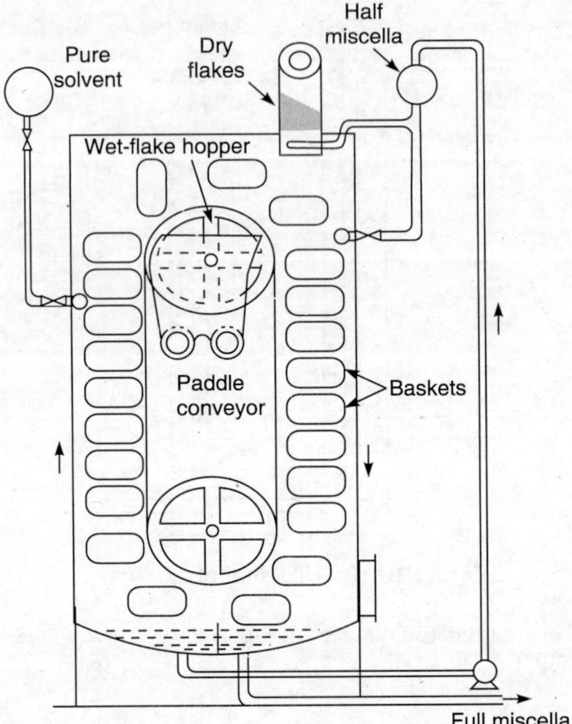

Figure 19.14 Bollman extractor.

Source: Reprinted from Solvent extraction of oilseed, Cofield, Jr, E.P., *Chemical Engineering*, **58**(1), January, 1951, p. 129, Copyright (1951), with permission from Chemical Week Associates, New York.

Thus, Bollman extractor has the advantage of both co-current and counter-current movement. Typical industrial units for extraction of soya bean oil operate in the capacity range of 50–500 tons of beans per 24 hours.

19.4 APPLICATIONS OF EXTRACTION IN FOOD PROCESSING

As has been mentioned, extraction finds extensive applications in food processing for the following:

19.4.1 Extraction of Oils and Fats

Most of the oils from oil seeds are extracted by solvent extraction, particularly if the oil content is less which is of the order of 15–20 per cent or even less. Edible grade hexane is used as solvent. Soyabean oil and rice bran oil are extracted invariably by solvent extraction only. Other oil seeds like cotton seed, mustard, peanut, gingily, etc. contain higher amounts of oils. Hence, major part of the oils is extracted by pressing in a screw press or *ghani.* When the residual oil content is brought down to 6–8 per cent in the cake, it is subjected to solvent extraction to remove the last traces of oil so that the cake is free from oil, and hence, can be stored without the problem of rancidity. The cake, after the oil is removed, becomes a rich source of protein and can be used for making protein concentrates and protein isolates.

Corn oil, which is a high value vegetable oil, is extracted from corn germs by solvent extraction even though the oil content in the germ is very high. Similarly fish oil can be extracted with the use of hexane, fish oil is nutritionally very rich.

19.4.2 Extraction of Oleoresins

Oleoresin is a mixture of oil and resin in the spices like pepper, ginger, turmeric and chilli. The oleoresin can be considered to be entirely the essential component in the spices. Extraction of oleoresins is done by solvent extraction of the spices. Usually hexane or diethyl ether is used to extract the oleoresin. Later on, extract is desolventized by distillation to recover pure oleoresin free of residual solvent, which is an essential quality attribute for oleoresins. In fact, much efforts go in reducing the residual solvent levels to few ppm. Central Food Technological Research Institute (CFTRI, Mysore) has developed technologies for extraction of oleoresins.

19.4.3 Extraction of Natural Food Colours

Natural food colours like annatto dye (norbixin), or red colour from paprika, or violet colour from Bangalore grapes, or dark red-cum-maroon colour from beet root, or golden yellow colour from marigold or yellow colour from turmeric or orange to red colour from coccinia, are all extracted by solvent/water extraction. Different solvents or different combinations of solvents are used. For example annatto dye is extracted from annatto seeds by using a dual solvent system initially to extract the oil portion by a non-polar solvent and subsequently by a polar solvent to extract the pigment (Satyanarayana, et al., 2003). Similarly, orange red colour can be extracted from ripened coccinia by using acetone solvent (Sunita, et al., 1999). CFTRI has developed technologies for a number of natural food colours.

19.4.4 Extraction of Caffeine

Caffeine is the active principle in coffee and tea which is responsible for stimulating action of the beverages. Due to various health reasons, it is also preferred to consume decaffeinated coffee by several consumers. Green (unroasted) coffee beans are decaffeinated by using hot water or some organic solvents. Later the brewing of coffee is done by extracting the roasted coffee beans using hot water.

19.4.5 Extraction of Coffee

Coffee and tea are extracted using hot water in the form of decoction. The decoction may be readily used for consumption, or may be concentrated upto good consistency (35–50 per cent solids) and later the concentrate is subjected to spray drying for preparing instant coffee or tea powders, which are shelf stable.

19.4.6 Extraction of Flavours and Pigments

Some of the undesired pigments from the foods are washed using solvent extraction. For example, the taint in cream is removed by stripping with steam to deodorize the cream. The vanillin flavour from vanilla beans is extracted by solvent extraction. Some of the volatile flavours present in small quantities in foods are extracted by *Simultaneous Distillation Extraction (SDE)* method using *Likens–Nickerson apparatus*. SDE method is generally used at laboratory scale only.

19.5 SUPER CRITICAL FLUID EXTRACTION (SCFE)

Hitherto the solvent extraction process we have studied is an excellent process except for the presence of residual solvents which are mostly organic solvents, and hence, may be carcinogenic. In this respect, a safe solvent like carbon dioxide under supercritical conditions could be a better alternative for extraction, particularly from natural products like coffee, tea, tobacco, cocoa, hops, spices and oil seeds to extract the active principles in them. This technique is popularly known as **Super Critical Fluid Extraction** (SCFE) method, and is a novel technique in extraction process which has its origin in early 1960s. In fact, the concept that compressed gases can be used to dissolve solids (solutes) was first demonstrated as early as 1870, in a publication in Proceedings of Royal Society (London). It has virtually taken eight decades to realize the utility of this concept.

Fluids beyond their critical conditions are known as **supercritical fluids**, and they exhibit behaviour and physico-chemical properties of both liquids and gases. For example, carbon dioxide (CO_2) which is mostly used as a supercritical fluid in food processing extractions is a gas at supercritical conditions (T_c = 31 °C and p_c = 73 atm.), and has the flowability of a gas, but has its density comparable to that of liquids, and hence, has the solubilizing power of a liquid. After extraction, the solvent (CO_2) can be removed simply by releasing the pressure when the CO_2 leaves the solute and escapes as gas. Thus, in general, fluids near their critical points have the dissolving power comparable to that of liquids, and have the transport properties of gas. This unusual combination of physico-chemical properties can be judiciously exploited to carry out separation operations by extraction which may be otherwise difficult by traditional extraction processes using organic solvents or water. In addition to it, SCFE

- is a quick extraction process,
- is an improved separation process,
- has lower operating costs,
- has no problem of residual solvents, and
- results in products which are relatively pure.

Various gases used for SCFE process are: CO_2, ethane, ethylene and N_2O.

CO_2 has unique advantage of having properties which are much different from those of water or non-polar organic solvents. In addition to it, CO_2 has the following advantages:

(i) CO_2 is most unobjectionable as a solvent from health point of view and scores over most of the organic solvents.
(ii) It is ubiquitous in nature. It is inexpensive and readily available in pure form.
(iii) It is inert and does not react with the solute during extraction.

(iv) It is neither inflammable nor toxic.

(v) It does not corrode the container in its native form or in combination with moisture.

Hence, CO_2 is a better candidate as SCFE fluid for extraction of natural products for edible purposes.

The SCFE process system consists of two units: extraction unit and expansion unit separated by an expansion valve. The schematic arrangement is shown in Figure 19.15.

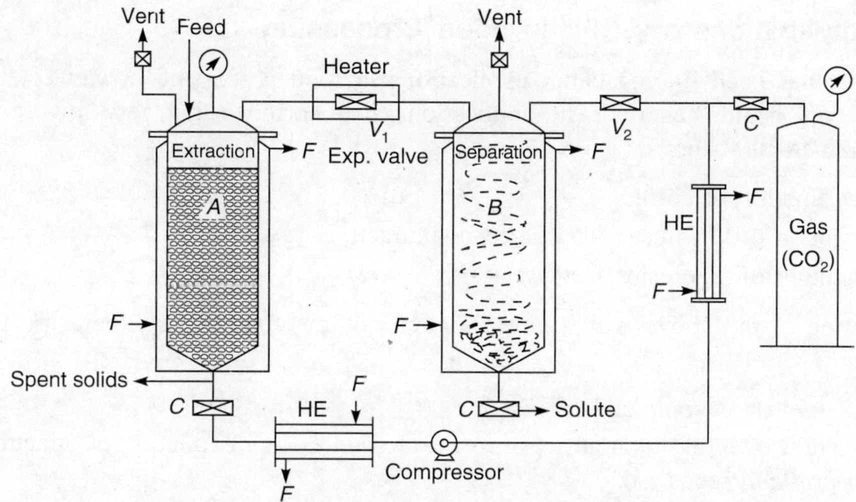

Figure 19.15 Schematic diagram of SCFE unit.
C: Valves
F: Heat exchanger fluid
HE: Heat exchanger

19.5.1 SCFE Process

Like any other batch extraction process, the material is fed into a cylindrical vessel *A* (known as **extraction section**). The vessel is filled with CO_2 or the SCFE solvent, and is pressurized using a compressor until the desired pressure is reached which is read on the pressure gauge. Usually the pressures are of the order of 75–300 atm (7.5 to 30 MN/m^2), and ideally for CO_2 the pressure is upto 150 atm, beyond which the increase in density of supercritical CO_2 is marginal. The preferred temperature range is 35–80 °C.

The system is allowed to equilibrate for sufficient time with the food solids and the supercritical fluid (CO_2). Indeed the diffusion of SCF into the solid matrix is rapid, but the solubility of solute into the solvent is slow which may require prolonged contact time for better extraction. The extracted CO_2 is taken into another chamber *B* (known as **separation section**) by opening valve V_1 and keeping valve V_2 closed. When the expansion valve V_1 is opened, care should be taken to see that the temperature does not fall because of Joule–Thomson effect. Sometimes the expansion valve is heated slowly to avoid cooling. Later on, the valve V_1 is closed. All extracted material with solvent at higher pressure is trapped in the expansion tank. Then the valve V_2 is slowly opened to release the CO_2 pressure. Once the pressure is released, CO_2 loses its supercritical nature; and hence, the solubility of the solute drastically reduces. Thus, the solute is separated in the chamber *B*. The CO_2 is either lost off or recycled into the compressor for next extraction (as shown in Figure 19.15). The spent solids are discharged from the extraction chamber subsequently to feed a fresh batch.

In some cases, an entrainer is used along with SCF to help better extractability. Various types of entrainers are in use.

19.5.2 Applications of SCFE in Food Processing

SCFE of late has been finding better applications in food processing in view of its obvious advantages with health benefits. The applications are enormous, but very few of them have reached commercial stature:

(i) Decaffination of coffee
(ii) Extraction of hops used in the beer manufacture
(iii) Extraction of oleoresins from spices

In addition to the above which have reached commercial success, there are many more applications:

- Extraction of flavours and aromas
- Extraction of food colourants from natural sources like extraction of annatto dye from *Bixa orellana* seeds
- Extraction and concentration of Eicosapentaenoic Acid (EPA) from fish
- Extraction of cholesterol from eggs
- Extraction of cholesterol from butter, lard and tallow
- Extraction of nicotine from tobacco
- Extraction of stimulants from tea, cocoa, etc.
- Extraction of edible oils from the oil seeds
- Extraction of soya bean oil
- Extraction of oil from corn, wheat germ, sunflower, safflower seeds and peanuts
- Extraction of natural products, viz., steroids, alkaloids and anticancer agents.

Some specific applications found in the literature based on an International Symposium on Supercritical Fluids, Societe Francaise De Chimia, Paris, 1985 (McHugh 1990) are as follows:

- Extraction of limonene and cineole
- Removal of cholestoral from butter, beef tallow
- Extraction of prime rose seeds
- Extraction of ginger and rosemary
- Extraction of pepper
- Extraction of some flavour components
- Extraction of fatty acids of oils from natural triglycerides
- Processing of fermentation broths by SCF extraction.

19.5.3 Constraints in SCFE Processing

The constraints faced in SCFE processing are two fold:

(i) Thermodynamic
(ii) Operational

The high pressure fluid phase equilibrium date for the solvent and the solute is generally lacking. Experimental measurements are very difficult in view of the very high pressures involved. Hence, we resort mostly to predictive equations based on equations of state. Another important parameter in SCFE is the polarity of the solute being extracted.

The operational constraints are based on the equipment. Generally, the SCFE equipment is operated in a batch manner. Since the process conditions are extreme especially in terms of pressure (imagine the vessels operating at 100 atm. pressure!), loading and unloading of the batch and making the equipment every time leak proof is a stupendous task. Design and mechanical fabrication of the vessels to withstand to such high pressures, welding joints for the flanges, etc. are all to be meticulously done. Setting up special requirements for the sealing elements is very essential. In addition to the above, the following are some operational constraints:

- Maintaining temperatures of the order of 100 °C which is of significance with regard to seals, etc.
- Meeting the food quality requirements in terms of corrosion-resistant equipment and piping.
- Pipelines for flowing supercritical fluids at high pressures, the bends and joints for the piping, etc.
- Valves to particularly arrest flow rate of fluids at supercritical conditions; frequent operations of valves should not give way for leaks.
- After every filling of the feed, quick closing of the lid, bolts and nut joints, etc. are to be ensured.
- The carry over problem of the feed solids, if the bulk densities are low (0.2–0.3 kg/lit) along with the SCF (whose density is approximately 0.9 kg/lit) into the separating chamber is a recurring nuisance.
- Channeling of the fluid in the extraction tank without adequately coming in contact with all the solids.

Thus, the operational difficulties restrict the commercialization of SCFE processes.

Symbols

A: area (m^2)
C: concentration (kg/m^3 or k mol/m^3)
C_A: concentration of A in bulk
C_{AS}: concentration of A at the interface
K: partition coefficient defined by Eq. (19.1)
k_1: mass transfer coefficient
L: underflow rates (kg/s or kg/h)
M: mass transferred (kg)
R: resistance for mass transfer
t: time (s or h)
V: overflow rates (kg/s or kg/h)
V_v: volume (m^3)
x: weight fraction of underflow
y: weight fraction of overflow

Subscripts

A: component A (usually solute)	1, 2: phases 1 and 2
s: interface	1, 2, ..., n: number of the stage
0: initial	

REVIEW QUESTIONS

19.1 Describe the process of solvent extraction.
19.2 Differentiate between extraction and leaching.
19.3 Differentiate between extraction and decoction.
19.4 What is meant by partition coefficient?
19.5 Write an expression for rate of leaching.
19.6 What are the factors that affect the rate of extraction?
19.7 What are various types of extraction processes? Classify them.
19.8 Describe Soxhlet extraction process.
19.9 Describe counter-current extraction process.
19.10 Describe counter-current multi-stage extraction process.
19.11 How do you describe the operating line equation for a multistage extraction process?
19.12 Describe McCabe–Thiele method to find out the number of equilibrium stages.
19.13 Describe the right angle triangle method to find out the number of equilibrium stages.
19.14 What are the basic requirements of an extraction equipment?
19.15 Describe a batch extractor with a neat diagram.
19.16 Describe a counter-current extractor with a neat diagram.
19.17 Describe the Bollman extractor with a neat diagram.
19.18 Describe the Hildebrandt extractor with a neat diagram.
19.19 What are various applications of extraction in food processing?
19.20 Describe the process of SCFE.
19.21 Why CO_2 is a better gas for SCFE?
19.22 What are the advantages of SCFE over conventional solvent extraction process?
19.23 What are applications of SCFE in food processing?
19.24 What constraints does one face in SCFE?

NUMERICAL PROBLEMS

19.1 10 grams of annatto seeds containing 2.1 per cent of dye by weight is extracted in a laboratory equipment using 500 ml of acetone. After 10 min of extraction, the spent seeds still retain 1.2 per cent of colour. The acetone has a saturated concentration of the dye to an extent of 5 g/l. If 2 kg of seeds are extracted in a pilot scale extractor of 150 l capacity filled with acetone upto (2/3) of the volume for ½ h, what will be the residual dye in the seeds. **(Ans:** 0.762 per cent)

19.2 Soyabeans containing 16 per cent oil is extracted with *n*-hexane in a counter-current extraction battery to recover 85 per cent of the oil. The final concentration of oil in the extract is 40 per cent. The crushed beans in the underflow carry with them 50 per cent solution in each stage. All the oil is extracted in the first stage itself. Equilibrium is reached in each stage. Find the number of ideal stages and x_n if the overflow and underflow rates are constant throughout. (**Ans:** $x_n = 0.0496$, No. of ideal stages = 4)

19.3 Calculate the number of ideal stages for numerical problem 19.2 by using McCabe–Thiele method.

19.4 Repeat numerical problem 19.2 by using graphical procedure of difference point method.

REFERENCES

Cofield, Jr., E.P. (1951), Solvent extraction of oil seed, *Chem. Engg.*, **58**(1), pp. 127–140.

Coulson, J.M., Richardson, J.F., Backhurst, J.R. and Harker, J.H. (1991), *Chemical Engineering*, Vol. 2, 4th ed., Pergamon Press, Oxford, pp. 391–426.

McCabe, W.L. and Thiele, E.W. (1925), Graphical design of fractionating columns, *Ind. Engg. Chem.* **17**(6), pp. 605–611.

McCabe, W.L., Smith, J.C. and Harriott, P. (1993), *Unit Operations of Chemical Engineering*, 5th ed., McGraw-Hill, New York, pp. 614–643.

McHugh, M.A. (1990), Supercritical fluid extraction in *Biotechnology and Food Process Engineering*, Schwartzberg, H.G. and Rao, M.A. (Ed.), Marcal Dekker, Inc., New York, pp. 203–210.

Satyanarayana, A., Prabhakara Rao, P.G. and Rao, D.G. (2003), Chemical processing and toxicology of annatto (*Bixa orellana*, L.) *J. Food Sci. Technol.*, **40**(2), pp. 131–141.

Sunita, M., Jyothirmayi, T. and Rao, D.G. (1999), A process for preparation of vegetable based natural food colour from *Coccinea indica*, Indian patent No. 1493/DEL/99.

CHAPTER

20

Filtration

We have seen so far most of the separation processes for separating one liquid from the other which are both miscible, and hence we had to resort to phase change from liquid to vapour and then condense the vapours to get a reasonably pure liquid component. Since phase change is caused by supply of heat, and diffusion of vapours from one phase to other is involved, these operations were categorized as simultaneous heat and mass transfer operations; the classical examples are drying, distillation, gas absorption, extraction, crystallization, etc. Now, we see such separation operations mainly for separation of solids from liquids, liquids from liquids or solids from solids which are essentially insoluble or immiscible; i.e., the solids are in a state of suspension in liquids, or one liquid is dispersed into the other liquid which is insoluble or one solid may be present as a separate entity in a solid mixture. Separation of such mixtures is categorized as only physical separation processes since no phase change is involved and no diffusional mass transfer is involved. The separation is based either on particle sizes or density differences. Hence, they are also known as **mechanical operations.** One such physical separation process for separation of suspended solids from a liquid is filtration which is the subject matter of this chapter.

Filtration is a physical separation process for separation of suspended solids from a liquid, the only criterion being that the solid particle size should be larger than the molecular size of the liquid. The liquid mixture is passed through a filtration medium which is usually a filter cloth or a polymeric material or a metallic screen. The solid particles which are bigger than the aperture size of the filter medium are retained on the top of the filter medium. The solid particles which remain on the top of the filter medium is known as **residue** or filter cake or simply cake, while the liquid that passes through is known as **filtrate**. Smaller size particles of course will pass through the filter medium.

The filter medium is chosen in such a way that the aperture size of the filter cloth is always less than the particle size. If the aperture size is very small, no doubt the filtration is efficient, but it causes a lot of resistance for the liquid to pass through. Hence, a proper choice of the filtration medium is desirable. Some criteria for choice of filtration medium are:

- The aperture size of the filtration medium should not be too small which causes higher pressure drops for the fluid flow, nor too big which will not be able to filter out the solids.

- It should be sufficiently strong to hold the materials and also be able to withstand the applied pressure or vacuum without being ruptured.
- It should not chemically react with the liquid or solids.
- It should not contaminate the solids or liquid.
- It should not get clogged frequently.
- It should be easily cleaned and replaced.
- Above all it should not be very costly.

Usually the filtration medium which we call as filter cloth is placed on a perforated metal screen to give sufficient strength for the cloth to hold and retain the filtered solids. The process is shown in Figure 20.1.

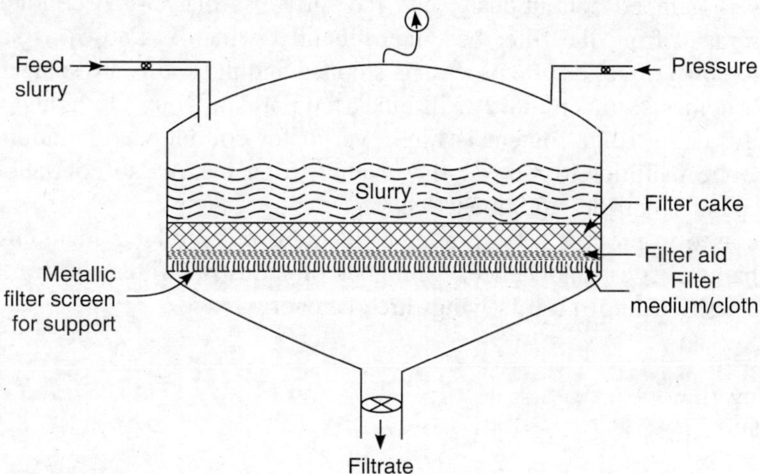

Figure 20.1 Schematic representation of filtration process.

In the initial stages of filtration, there will not be any bed thickness, and hence, the filtration is not efficient even though the rate of filtration is fast. As the filtration progresses, the cake builds up on the filter medium and causes some resistance for the flow of liquid as well as for the solid particles. This will make the filtration process efficient. Even though the filter bed reduces the filtration rate, it will make clear liquid to pass through by restricting the passage of finer particles.

To overcome this difficulty of less efficient filtration what we face in the initial stages of filtration, we add deliberately a foreign component known as **filter aid** to the slurry or spread the material uniformly on the filter cloth. Usually the filter aids are:

- Cellulose
- Diatomaceous earth
- Any inert chemical (such as *supercel*)

The filter aid should not chemically react with the slurry. It should not be too fine to block the pores of filter cloth. Thus, the basic criteria for the choice of filter aid are that it should be chemically inert and economically inexpensive. The addition of filter aid is both optional and conditional. It depends upon whether the filter cake is the desired product or the filtrate is the desired product. If latter is the case, filter aid can be safely used. In the former case (filter cake is the desired product) we do not use any filter aid; we pass the slurry initially until a sufficient height of filter bed builds up. The filtrate obtained during this period, which is not clear, is recycled for filtration once again. The initial stages of filtration will allow sufficient filter bed to build up, which does the job of filter aids.

The filtration rate through the filter cloth can be increased either applying pressure on the up stream side, or by applying vacuum on the down stream side to suck the filtrate. Usually pressure is applied on the upstream side either by continuously pumping the feed slurry by using a pump in continuous filtration, or by applying pressure by an inert gas in a batch filtration process. Industrially the continuous processes are preferred in view of higher through flows. The feed slurry is pumped continuously, and the filtrate is discharged continuously, whereas the solids are scraped from the filter bed intermittently, which is known as **semi-continuous** operation. For removal of solids, the process is stopped and the solids are scraped; otherwise the solid build-up becomes so much that it will block the pores of filter cloth, and reduce the flow rate considerably. If the solid content is high, we go for continuous operation. In continuous operations there are facilities to recover the residue (filter cake) also continuously. They will be described separately in the equipment section.

Filtration is used to process a large number of industrial products including food products with diverse characteristics, and hence, a number of industrial filters is available. Basically, the filters are classified into the following three categories:

- Cake filters
- Clarifying filters
- Cross-flow filters

Cake filters are mostly prominent, and is used for separation of solids from slurries having high solid content. The solids form a cake on the filter medium, and hence, the name. As the filtration progresses, the filter cake builds up and improves the efficiency of filtration. Clarifying filters are mostly used with very dilute slurries/suspension to clarify the suspensions as in the case of clarification of fruit juices. The cross-flow filters are used mainly for concentrating a dilute suspension. In this type of filters, there is no clear demarcation between the residue and filtrate. A dilute slurry passes over the filter medium in a cross-flow direction, and hence, the name of cross-flow filter. By the time the slurry passes over the filter, a part of the filtrate filters, and in the outlet we get a relatively concentrated slurry as shown in Figure 20.2. *Ultra filtration* is a classical example of cross-flow filtration.

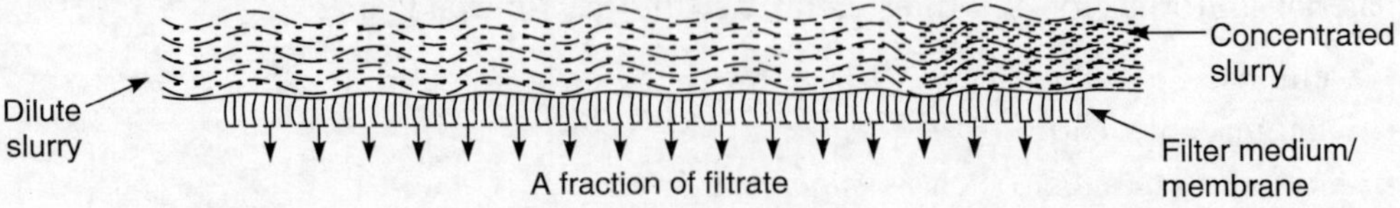

Figure 20.2 Schematic representation of cross-flow filtration.

20.1 THEORY OF FILTRATION

Any transfer process increases with the driving force and decreases with the resistance.

$$\text{Rate process} \propto (\text{driving force})$$

$$\propto \left(\frac{1}{\text{Resistance}}\right) \tag{10.10}$$

In addition to the above, the rate of filtration also increases with the cross-sectional area. Pressure differential is the driving force for filtration, and the resistance is due to the resistance for flow of filtrate through the filter bed. Hence, we can write

$$\text{Rate of filtration} \propto (\text{Pressure difference or pressure drop, } \Delta p)$$

$$\propto (\text{cross-sectional area})$$

$$\propto \left(\frac{1}{\text{Resistance to the flow}}\right) \tag{20.1}$$

Resistance for the flow of filtrate is due to:

- Viscosity of the liquid
- Filter bed/cake
- Filter medium (cloth) and filter aid, if any, etc.

Thus, we can write

$$R = \mu\alpha\,(L_c + L_m) \tag{20.2}$$

where R is the resistance for filtration, L_c is the cake thickness which will be increasing as the filtration progresses, L_m is a fictitious thickness which takes care of the filter medium, filter aid, supporting screen, etc., μ is the viscosity of the filtrate, which is usually taken as that of water, and α is the constant of proportionality and is known as **specific cake resistance**.

The specific cake resistance is defined as *the resistance offered to the flow of filtrate through filter cake, filter medium, etc. of unit thickness of them.* Hence, we combine the cake thickness with the filter medium, etc. in the parenthesis of Eq. (20.2) on the RHS. The units for specific cake resistance[†] is $1/m^2$.

Since we are combining L_m with L_c, L_m is customarily known as **equivalent cake thickness** which the filter medium, filter aid and filter screen would contribute for the resistance of flow like the filter cake, had it (cake) been of that thickness. Cake thickness (L_c) is the one which will be continuously changing with the volume (V) of the slurry. If w_s is the weight of solids present in unit volume of the slurry, w_sV would give the weight of the solids present in slurry in volume V[‡]. If ρ_s is the density of the solids we can write an expression for cake thickness is follows:

$$L_c = \frac{w_s V}{\rho_s}\left(\frac{1}{A}\right) \tag{20.3}$$

[†]The units for α are given as m/kg in some textbooks. It depends how w_s is defined.

[‡]In some textbooks, w_s is defined as the volume fraction of solids in the slurry, in volume of solids present per unit volume of the slurry.

in which A is the cross-sectional area of the filter bed. Combining Eqs. (20.2) and (20.3), we get

$$R = \mu\alpha\left(\frac{w_s V}{\rho_s A} + L_m\right) \tag{20.4}$$

Rate of filtration is measured by dV/dt

Therefore,

$$\frac{dV}{dt} = \frac{A(\Delta p)}{\mu\alpha\left(\frac{w_s V}{\rho_s A} + L_m\right)} \tag{20.5}$$

in which Δp is the pressure drop across the bed. Equation (20.5) is considered as the basic equation for filtration. Various terms involved in the equation can either be measured or evaluated or noted from the literature. This will help us find what is the specific cake resistance and what is the equivalent cake thickness. The rate equation is helpful for scale-up of the filtration process. We shall explain them with numerical exercises.

Equation (20.5) leads us to two possible conditions discussed as follows:

Constant rate filtration: In the early stages of filtration, generally the bed built-up is less, and hence, the resistance for flow is less; with the result that the rate of filtration is virtually constant. Integrating Eq. (20.5),

$$\frac{(V/A)}{t} = \frac{\Delta p}{\mu\alpha\,[(w_s/\rho_s)(V/A) + L_m]} \qquad (20.6)^{\dagger}$$

Equation (20.6) gives an expression for Δp. When once the cake built-up starts, the rate will be changing. To maintain constant rate, we need to adjust Δp by increasing the pressure on the upside flow, which is difficult. Hence, we generally would not like to follow the constant rate filtration; rather we prefer the constant pressure filtration.

Constant pressure filtration: In constant pressure filtration, we maintain Δp constant. The flow rate will be changing, which will be noted down. It is easy to maintain constant pressure rather than constant filtration rate.

Δp is constant in Eq. 20.5, which on rewriting yields

$$A\,(\Delta p)\,dt = \mu\alpha\left(\frac{w_s V}{\rho_s A} + L_m\right)dV \tag{20.7}$$

On integration

$$A\,(\Delta p)\,t = \mu\alpha\left(\frac{w_s}{2\rho_s}\left(\frac{V^2}{A}\right) + L_m V\right) \tag{20.8}$$

$$= \mu\alpha\left(\frac{w_s}{2\rho_s}\left(\frac{V}{A}\right) + L_m\right)$$

[†]Integration of Eq. (20.5) is not that simple as shown by Eq. (20.6), since Δp also varies with t.

$$\frac{At}{V} = \frac{\mu\alpha\, w_s}{2(\Delta p)\,\rho_s}\left(\frac{V}{A}\right) + \frac{\mu\alpha L_m}{\Delta p} \tag{20.9}$$

or

$$\frac{t}{(V/A)} = \frac{\mu\alpha w_s}{\Delta p\, 2\rho_s}\left(\frac{V}{A}\right) + \frac{\mu\alpha}{\Delta p} L_m \tag{20.10}$$

Equation (20.10) represents a straight line with (*V*/*A*) as independent variable and [*t*/(*V*/*A*)] as dependent variable. If we draw a plot between (*V*/*A*) and [*t*/(*V*/*A*)] on *x*-axis and *y*-axis, in which *V* is the volume of filtrate collected in a time period of *t*, we get

$$\text{Slope} = \frac{\mu\alpha}{\Delta p}\frac{w_s}{2\rho_s} \tag{2.11}$$

$$\text{Intercept} = \frac{\mu\alpha}{\Delta p} L_m \tag{20.12}$$

In Eq. (20.11), w_s is the weight of solids present in a unit volume of slurry which can be measured in the laboratory by taking a known volume of the slurry and evaporating it completely in an oven to find the weight of the solids gravimetrically. ρ_s is the density of the solids which can be measured independently. The pressure drop is maintained, and hence, can be noted from the reading of the pressure gauge. μ is the viscosity of the filtrate which is taken as that of water, since the filtrate is always a clear liquid which is mostly water. The cross-sectional area, *A* is measured by noting down the diameter of the filter bed. We collect the data on filtration in the laboratory by measuring the volume of the filtrate collected with time. We draw a plot between (*V*/*A*) and [*t*/(*V*/*A*)] as shown in Figure 20.3. The slope and intercept are measured. α can be measured from the slope by Eq. (20.11) with a knowledge of w_s, ρ_s, μ and Δp. Knowing α, the equivalent cake thickness (L_m) can be measured from the intercept by Eq. (20.12). This procedure is explained with Problem 20.1.

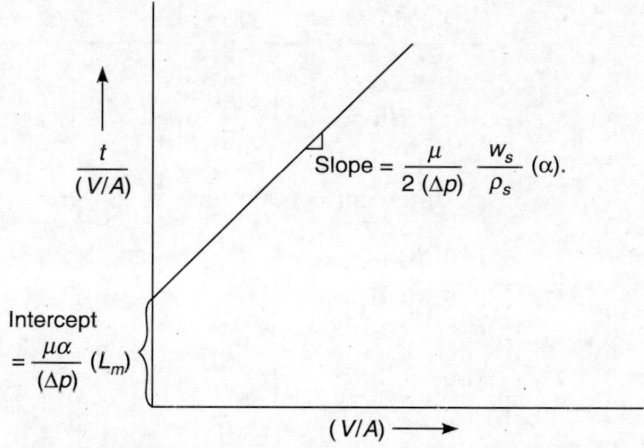

Figure 20.3 Method of evaluating α and L_m by graphical method.

PROBLEM 20.1 A fruit pulp is treated with pectinase enzyme to make a clear juice. Later on the juice is filtered in a plate and frame filter press† to clear it from any suspended particles and fibrous material. The total filtration area is 1 m^2. The following data on volume vs. time is obtained at 25 °C, keeping a pressure drop of 0.1 MN/m^2 (gauge).

Time, min	2	8	18.75	30.67	46	67.5
Volume, l	50	100	150	200	250	300

The slurry contains 25 g of solids per litre, and the density of solids is 900 kg/m^3. Calculate the specific cake resistance, and equivalent cake thickness.

Solution We note down all the data given to us

$$\rho_s = 900 \text{ kg/m}^3$$

$$w_s = 25 \text{ kg/m}^3$$

$$A = 1 \text{ m}^2$$

$$\Delta p = 0.1 \text{ MN/m}^2 = 1 \times 10^5 \text{ Pa}$$

The viscosity of water at 25 °C is noted from literature

$$\mu = 1 \times 10^{-3} \text{ Pa s}$$

From the data given to us, we calculate (*V*/*A*) and [*t*/(*V*/*A*)] as shown in Table 20.1 and are plotted in Figure 20.4.

Table 20.1 Data for Problem 20.1

t (min)	*V* (l)	*t* (s)	*V* (m^3)	*V*/*A* (m)	*t*/(*V*/*A*) (s/m)
2	50	120	0.05	0.05	2400
8	100	480	0.1	0.1	4800
18.75	150	1125	0.15	0.15	7500
30.67	200	1840	0.2	0.2	9200
46	250	2760	0.25	0.25	11,040
67.5	300	4050	0.3	0.3	13,500

$$\text{Slope} = \frac{11{,}000 - 600}{(0.25 - 0)} = 41{,}600 \text{ s/m}^2$$

$$\text{Intercept} = 600 \text{ s/m}$$

By Eq. (20.11)

$$\frac{\mu\alpha}{\Delta p}\frac{w_s}{2\rho_s} = 41{,}600$$

i.e.

$$\frac{1\times10^{-3}\,(\alpha)}{1\times10^5}\,\frac{25}{2\times900} = 41{,}600$$

†The plate and frame filter press is a filtration equipment which will be described in section 20.2.1.

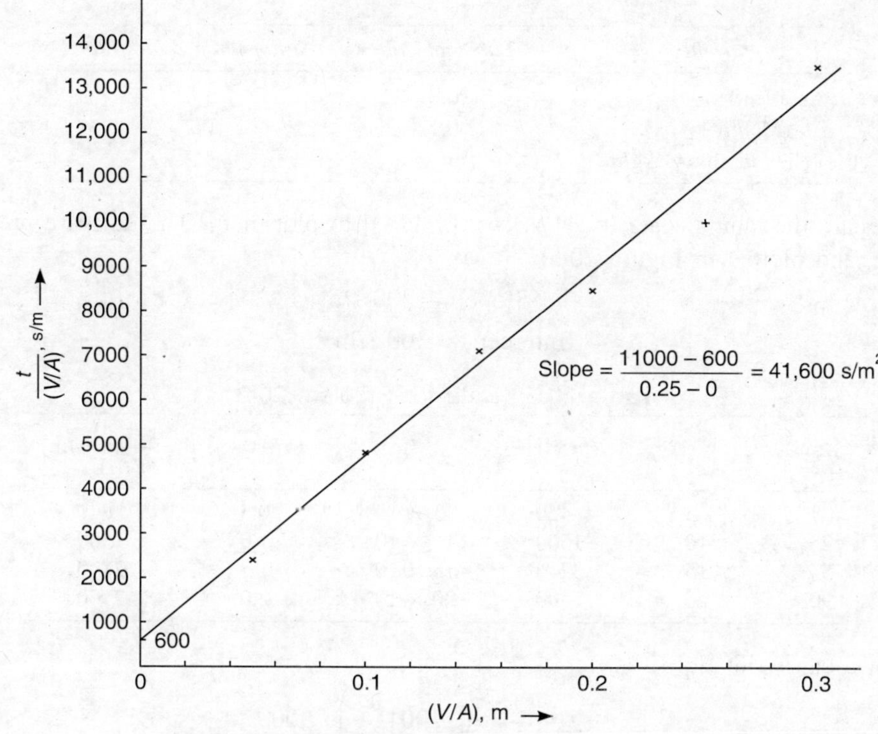

Figure 20.4 Plot of (V/A) vs. $\frac{t}{(V/A)}$ for Problem 20.1.

Therefore,

$$\alpha = 3 \times 10^{14} \text{ m}^{-2}$$

By Eq. (20.12),

$$\frac{\mu\alpha}{\Delta p} L_m = 600 \text{ s/m}$$

Therefore,

$$L_m = 600 \times \frac{1 \times 10^5}{3 \times 10^{14} \times 1 \times 10^{-3}} = 2 \times 10^{-4} \text{ m}$$

The rate equation [Eq. (20.10)] can also be used for predicting the performance of the filter press on scale-up. The procedure is explained with Problem 20.2.

PROBLEM 20.2 Sour grapes are crushed to extract the juice from the pulpy material by filtration in a laboratory filter press with 0.01 m^2 area under a constant pressure of 0.1 MN/m^2 (gauge). The following data on volume vs. time is recorded.

V, (l)	5	10	15	20
t, (min)	7	25	54	90

The data is used to predict the performance of a pilot plant scale filter press with 1 m^2 area under a pressure of 0.2 MN/m^2 to filter the same slurry, but having 20 per cent extra solids. Find the filtration rate in the pilot plant filter press in m^3/h.

Solution

Parameter	*Lab scale filter*	*Pilot plant scale filter*
Area (m^2)	0.01	1.0
Δp (MN/m^2)	0.1	0.2
Solids in slurry (kg/m^3)	w_s	1.2 w_s

We prepare the tabular data for (V/A) vs. $[t/(V/A)]$ to plot them. The data are presented in Table 20.2, and plotted in Figure 20.5.

$$\text{Slope} = 1200 \text{ s/m}^2$$

$$\text{Intercept} = 300 \text{ s/m}$$

Table 20.2 Data for Problem 20.2

t (min)	V (l)	t (s)	V (m^3)	(V/A) (m)	$\frac{t}{(V/A)}$ (s/m)
7	5	420	5×10^{-3}	0.5	840
25	10	1500	10×10^{-3}	1.0	1500
54	15	3240	15×10^{-3}	1.5	2160
90	20	5400	20×10^{-3}	2.0	2700

Now, we can write Eq. (2.10) as:

$$\frac{t}{(V/A)} = 1200\left(\frac{V}{A}\right) + 300$$

$$1200 = \frac{\mu\alpha}{0.1}\frac{w_s}{2\rho_s}$$

from which

$$\frac{\mu\alpha}{2\rho_s} = \frac{120}{w_s}$$

$$300 = \frac{\mu\alpha}{0.1}L_m$$

from which

$$\mu\alpha L_m = 30$$

Now, we write Eq. (20.10) for the pilot plant scale for which solid content in slurry is 1.2 w_s and Δp is 0.2 MN/m^2

$$\frac{t}{(V/A)} = \frac{\mu\alpha}{2\rho_s} \times \frac{1.2w_s}{0.2}\left(\frac{V}{A}\right) + \frac{\mu\alpha}{0.2}L_m$$

Replacing for $\frac{\mu\alpha}{2\rho_s}$ and $\mu\alpha L_m$ from the above

$$\frac{t}{(V/A)} = \frac{120}{w_s} \times \frac{1.2\,w_s}{0.2}\left(\frac{V}{A}\right) + \frac{30}{0.2}$$

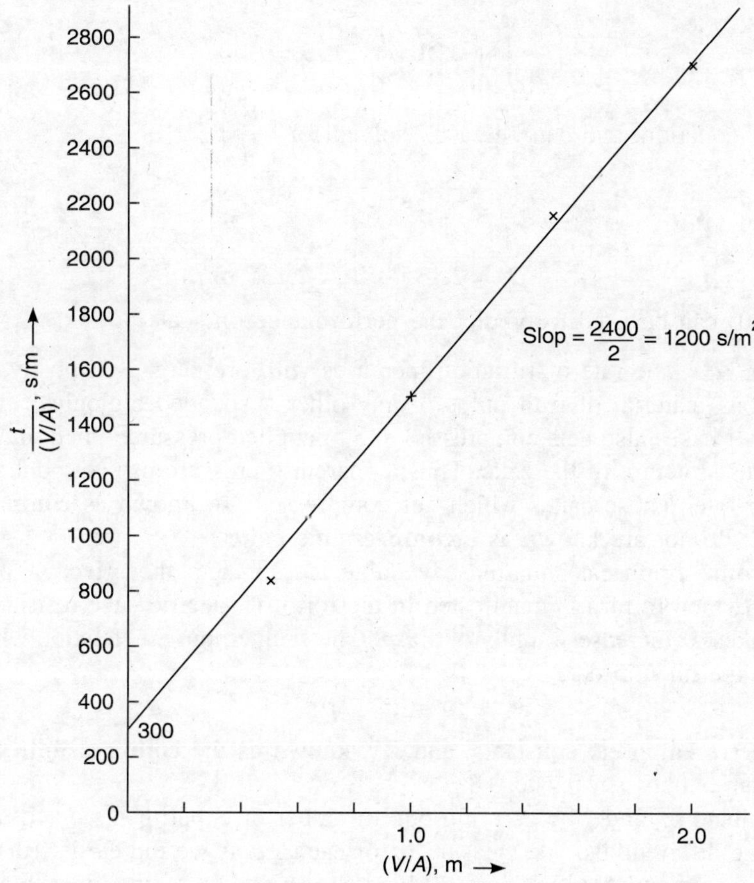

Figure 20.5 (V/A) vs. $\dfrac{t}{(V/A)}$ data for Problem 20.2.

$$= 720\left(\frac{V}{A}\right) + 150$$

Therefore,
$$t = 720\left(\frac{V}{A}\right)^2 + 150\left(\frac{V}{A}\right)$$

If we want to know the flow for one hour, $t = 3600$ s

Therefore,
$$3600 = 720\left(\frac{V}{A}\right)^2 + 150\left(\frac{V}{A}\right)$$

i.e.,
$$720 + \left(\frac{V}{A}\right)^2 + 150\left(\frac{V}{A}\right) - 3600 = 0$$

or $$\left(\frac{V}{A}\right)^2 + 0.21\left(\frac{V}{A}\right) - 5 = 0$$

It is a quadratic equation† and can be easily solved for $\left(\frac{V}{A}\right)$.

Hence, $$\frac{V}{A} = 2.135 \text{ m}$$

Therefore, $$V = 2.135 \times 1 = 2.135 \text{ m}^3\text{/h}$$

Thus, Eq. (20.10) can be used to predict the performance on scale-up.

Effect of Δp on α: The rate of filtration increases with pressure. But mostly, the filter cake consists of fibrous material of fruit pulps or any other suspended biological material. Hence, the material (filter cake) also gets compressed due to applied pressure which may in turn block/reduce the porous structure of the cake. Thus, apparently pressure has got contradicting effects on the filtration rate. Those cakes which get compressed are known as **compressible cakes**, and those which do not are known as **incompressible cakes**.

There are some empirical equations available to represent the effect of pressure on the cake resistance, which in turn is manifested in the form of specific cake resistance. For highly compressible cakes, α increases rapidly with Δp. One such empirical equation which reasonably represents the experimental data is:

$$\alpha = \alpha_0 \, (\Delta p)^s \tag{20.13}$$

where α_0 and s are empirical constants, and s is known as the **compressibility coefficient** of the cake, and varies between 0.2 to 0.8.

We can evaluate α_0 and s by carrying out a number of experiments on filtration studies at different pressure differentials. We measure α for each Δp as we did earlier (in Problem 20.1), and tabulate data on α and Δp. Equation (20.13) is linearized by taking logarithms on both sides

$$\ln \alpha = \ln \alpha_0 + s \ln(\Delta p) \tag{20.14}$$

Now, we can draw a plot of ($\ln \alpha$) vs. ($\ln \Delta p$) which results in the form of a straight line. We measure the slope and intercept of the line.

$$\text{Slope} = s \tag{20.15}$$

$$\text{Intercept} = \ln \alpha_0 \tag{20.16}$$

Thus, we can evaluate α_0 and s. The procedure is described in Problem 20.3.

PROBLEM 20.3 The enzymatically-treated fruit pulp of Problem 20.1 is clarified by filtration at three different pressures to assess the effect of differential pressure on specific cake resistance. The data on V vs. t obtained on a laboratory filter of 0.1 m^2 cross-sectional area at different pressures are presented in Table 20.3. Find what is the effect of pressure on α.

---

†Generalized quadratic equation is represented as $ax^2 + bx + c = 0$ for which the solution is

$$x = \frac{-b \pm \sqrt{(b^2 - 4ac)}}{2a}$$

Table 20.3 *V* vs. *t* Data for Different Δp for Problem 20.3

Δp = 148.4 Pa		Δp = 2981 Pa		$\Delta p = 1 \times 10^5$ Pa	
V (l)	*t* (min)	*V* (l)	*t* (min)	*V* (l)	*t* (min)
2	2	5	3.33	5	2
4	6	10	10	10	8
6	12.3	15	21.25	15	18.75
8	19	25	56.25	25	46
10	29	30	80	30	67.5

Solution Given data:

$$A = 0.1 \text{ m}^2$$
$$\rho_s = 900 \text{ kg/m}^3$$
$$w_s = 25 \text{ kg/m}^3$$
$$\mu = 1.0 \times 10^{-3} \text{ Pa s}$$

The laboratory data on *V* vs. *t* in Table 20.3 is used to calculate (*V*/*A*) and [*t*/(*V*/*A*)] at different Δp. The data are tabulated in Table 20.4 and plotted in Figure 20.6. α and L_m are measured by noting the slope and intercept for each pressure. The specific cake resistance at different Δps are measured using Eq. (20.11) and given in Table 20.5. Also noted data on ln α and ln (Δp). By using Eq. (20.14), a graph is plotted (Figure 20.7). From the graph, the slope and intercept are measured to be

$$\text{Slope} = 0.8$$
$$\text{Intercept} = 24$$

Table 20.4 Data for Problem 20.3 with $A = 0.1$ m^2

Δp = 148.4 Pa				Δp = 2981 Pa				$\Delta p = 1 \times 10^5$ Pa			
V (l)	*t* (min)	(*V*/*A*) (m)	$\frac{t}{(V/A)}$ (s/m)	*V* (l)	*t* (min)	(*V*/*A*) (m)	$\frac{t}{(V/A)}$ (s/m)	*V* (l)	*t* (min)	(*V*/*A*) (m)	$\frac{t}{(V/A)}$ (s/m)
2	2	0.02	6000	5	3.33	0.05	4000	5	2	0.05	2500
4	6	0.04	9000	10	10	0.1	6000	10	8	0.1	4800
6	12.3	0.06	12,300	15	21.25	0.15	8500	15	18.75	0.15	7500
8	19	0.08	14,300	25	56.25	0.25	13,500	25	46	0.25	11,040
10	29	0.1	17,440	30	80	0.3	16,000	30	67.5	0.3	13,500

Slope = 135,431 s/m^2
Intercept = 3900 s/m
$\alpha = 1.448 \times 10^{12}$ m^{-2}

Slope = 49,800 s/m^2
Intercept = 1076 s/m
$\alpha = 1.069 \times 10^{13}$ m^{-2}

Slope = 41,600 s/m^2
Intercept = 600 s/m
$\alpha = 3 \times 10^{14}$ m^{-2}

Table 20.5 Data for Calculating α_0 and s for Problem 20.3

Δp (Pa)	α (m^{-2})	$ln\ \Delta p$	$ln\ \alpha$
148.4	1.448×10^{12}	5.0	28.0
2981	1.069×10^{13}	8.0	30.0
1×10^5	3.0×10^{14}	11.51	33.33

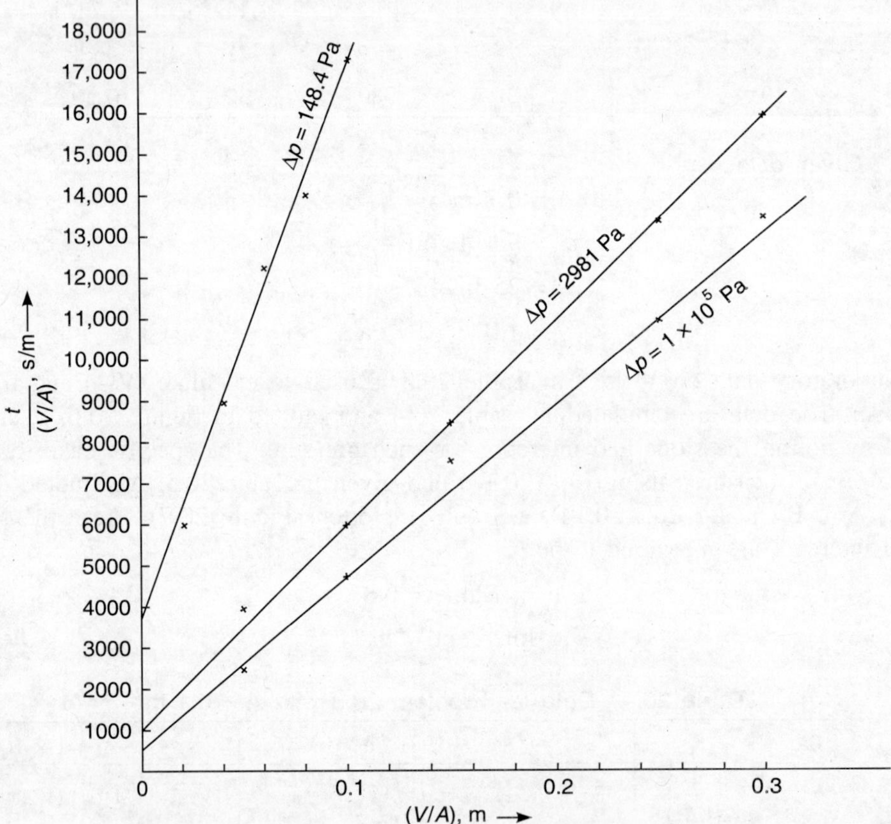

Figure 20.6 (V/A) vs. $\dfrac{t}{(V/A)}$ for various Δps for Problem 20.3.

By using Eqs. (20.15) and (20.16), we have

$$s = 0.8$$

$$\ln \alpha_0 = 24$$

Therefore,

$$\alpha_0 = 2.65 \times 10^{10}$$

By Eq. (20.13),

$$\alpha = 2.65 \times 10^{10}\ (\Delta p)^{0.8}$$

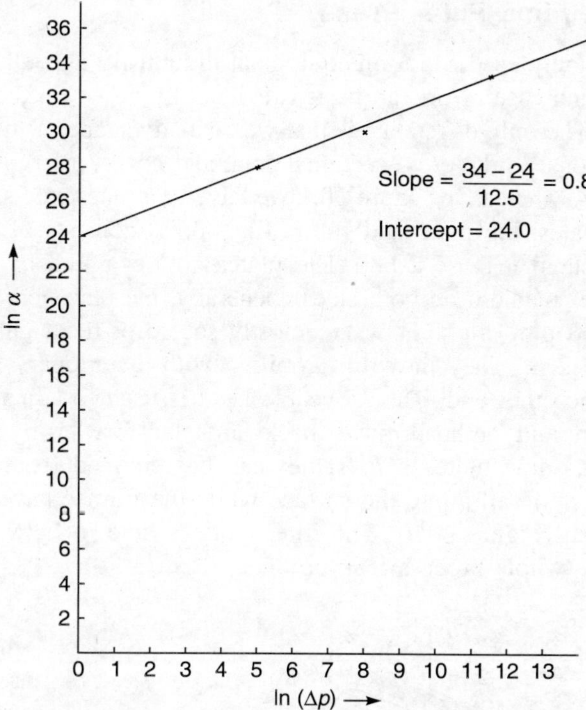

Figure 20.7 ln (Δp) vs. ln α data for Problem 20.3.

20.2 FILTRATION EQUIPMENT

In view of the industrial applications of filtration in chemical process industries and food processing, a good number of industrial filters is available in the market. They are batch (semi-batch) and continuous, using pressure on the upstream side or vacuum on the downstream side, small capacities and large capacities, both for laboratory use and for industrial use. Table 20.6 lists most of them. However, we describe here briefly a few of them, which have much applications in food processing.

Table 20.6 List of Industrial Filters

Plate and frame filter
Sand filter
Horizontal plate filter
Vertical leaf filter
Horizontal leaf filter
Sweetland pressure filter
Horizontal rotary vacuum filter
Rotary drum vacuum filter
Rotary disc vacuum filter
String discharge rotary drum vacuum filter
Continuous vacuum pre-coat filter
Horizontal belt filters
Horizontal pan filters
Larox automatic belt filters
Centrifugal filters

20.2.1 Plate and Frame Filter Press

It is one of the most widely used filtration equipment in industry as well as in the laboratory in view of its simplicity and ease of operation. It is essentially based on the principle of filtration shown in Figure 20.1. The only difference is that we take advantage of the fact that the filtration rate is directly proportional to the cross-sectional area, and hence, a large number of frames and plates are used in series, whereas in Figure 20.1 we have only one plate's surface area available.

As the name indicates, the plate and frame filter press consists of a series of plates and frames stacked on a stand (Figure 20.8). The plates are separated by hollow frames which have filter cloth or filter medium on both sides. Generally, the plates are 6 to 50 mm thick and the frames are 6 to 200 mm thick. They are mostly square or rectangular in shape with sizes varying from 15 cm to 2 m. They move on a railing with a stationary head on one side, and a moveable head on the other end. The moveable head is used to adjust the number of plates and frames in operation, and the head moves backwards and forward by a jack or screw. When the capacities are low, some plates and frames can be removed from the stack. The plates will have a small projection all along the surface while the frames have matching depressions along the circumference (Figure 20.9). The filter cloth is held in between the projection and depression to make the whole process leak proof.

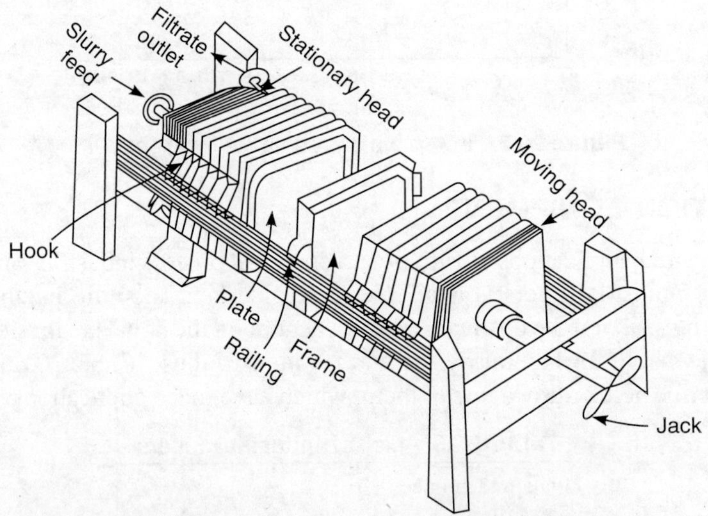

Figure 20.8 Plate and frame filter press.

The slurry is fed through the feed line by using a pump under pressure. Usually the pressure varies from 0.3 to 1.0 MN/m^2 (3 to 10 atm pressures). The feed line is usually on the top while the discharge line is usually in the bottom. The feed slurry enters into the hollow portion of the frame and passes through the filter medium into the plate groovings, and then enters into the discharge line. From the discharge line, the filtrate goes out. The filter cake or residue gets deposited on the filter cloth in the frame side.

As the filtration continues, the cloth gets clogged with the cake, and flow rate reduces considerably. It is time that the plates and frames are to be dismantled, the filter cloths to be removed, and the filter cake to be scraped. There will be an auxiliary provision in the press for

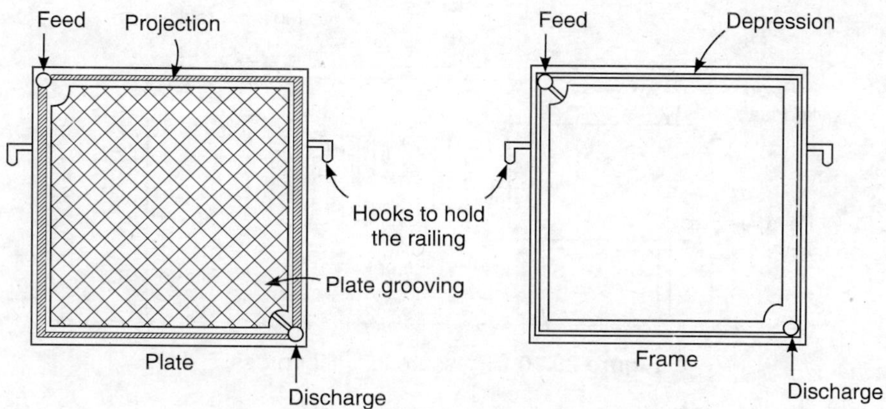

Figure 20.9 The plate and frame configuration.

a wash line through which wash liquid is passed to clean the cake, if the cake is the desired product. Sometimes even steam is passed which will clean and wash the cake. These operations are auxiliary, and are carried out only if the filter cake is the desired product. If the filtrate is the desired product, then the cake is simply discarded.

The plate and frame filter press has many advantages. It is simplest and economically cheapest filter press. It is a most versatile filter press to handle a wide variety of slurries with a provision to increase or decrease the cross-sectional area by increasing or decreasing the number of frames and plates. It requires very little floor space for a given capacity. Filter press can be easily dismantled and cleaned. There is also a facility to clean, wash and dry the cake if desired.

The greatest and serious disadvantage with the filter press is clogging or blinding of the filter medium by the solids which needs to dismantle the whole set up. The wear and tear on the filter cloth is more which make its life shorter. High labour costs are involved which sometimes overcompensate the low initial fixed costs. There are problems of leakage from the filter cloth which will make the process hall dirty. The workers are frequently exposed to the filtering materials which are not desirable in most of the chemical industries; and stringent statutory restrictions often prohibit such exposure of workers. Of course, this may not be a serious disadvantage in food processing operations.

20.2.2 Pressure Leaf Filter Press

It is also known as **shell and leaf filter**. It is similar in functioning to the plate and frame filter press, except that in this case the feed slurry enters from outside into the leaves under pressure, and clear filtrate is discharged through the leaves in a channel which is connected to the leaves. The filter cloth is kept outside the leaves on both sides. This filter press can be operated at a relatively higher pressure as compared to plate and frame filter press.

These filters are available in both horizontal and vertical directions, but the horizontal models are more popular. It consists of a horizontal cylindrical shell supported on a frame with a provision to feed the slurry (Figure 20.10). The filter leaves are usually square shaped, and are placed along the horizontal axis of the cylindrical shell on a movable rack with the roller. The filter leaf is a heavy screen or a groved plate hollow inside, over which the filter

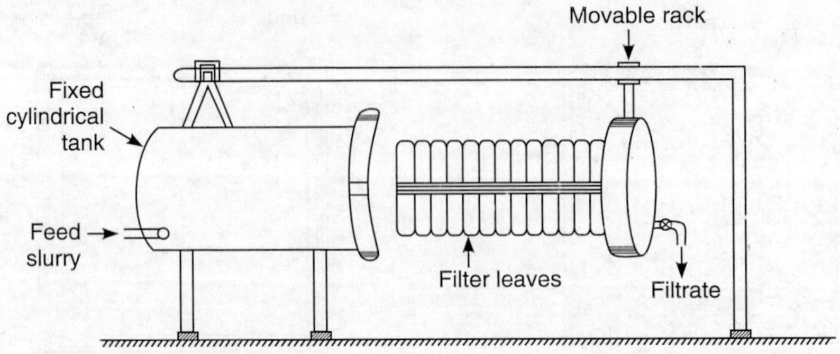

Figure 20.10 Pressure leaf filter press.

cloth or filter fabrics or fine mesh is fitted. Usually textile fabrics are more popular in view of their economical pricing and chemical inert nature. The textile fabrics are highly flexible, and hence, can be wrapped in the form of a bag around the leaf. The filter medium in the form of bags may be sewed, zippered, stapled or snapped around the filter leaves. Wire-screen cloths are sometimes used with the filter aid. The wire-screen is attached to the leaves by welding, riveting or bolting or may be clamped with a 180° bend in the wire cloth under tension (Figure 20.11). Somehow, the whole exercise is to see that the filter medium is wrapped around the filter leaves, and also it is essential to see that it is not sagged under the weight of the filter cake.

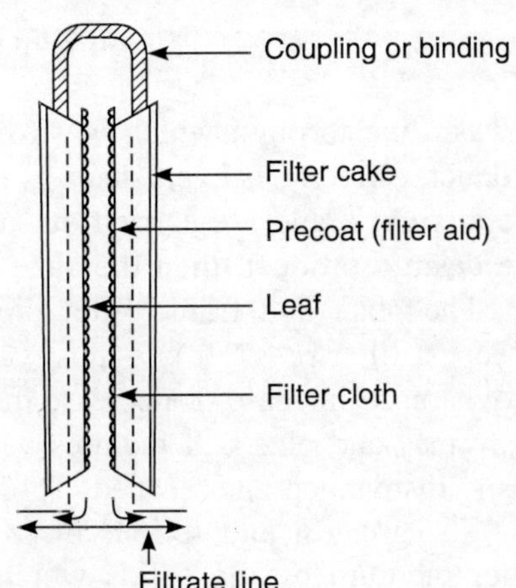

Figure 20.11 Filter leaf arrangement.

The leaves are supported usually on the top, and moves on the movable rack. Once the filter medium is fixed, the whole assembly of leaves is rolled inside the cylindrical vessel, and the lid is securely closed to avoid any leakage. The slurry is fed into the cylindrical vessel with a high pressure pump. The slurry is filled into the cylindrical vessel from which it filters into the hollow leaves through the filter medium, from where it is discharged through the filtrate discharge manifold. During filtration, slowly the filter cake coats over the filter medium, with the result the filtration rate falls. The leaf filters can be operated under constant pressure filtration or constant filtration rate. In the latter case, the pressure is adjusted to maintain constant flow rate. After a considerable amount of cake is formed which is indicated by reduced (low) flow rates of the filtrate, the filtration process is stopped, the leaves are disengaged, and the cake is removed by scraping or washing (in case cake is not the desired product). There are some arrangements where the cake is also discharged continuously by continuously scraping the cake and discharging the cake through a screw conveyor (not shown in figure).

The leaf filter is specially suitable where high pressure needs to be applied; and the contents should not come in contact with the surroundings. Assembly and dismantling can be done quickly.

20.2.3 Continuous Rotary Vacuum Filter Press

It is also known as the rotary drum filter since the filter press is in the form of a drum of 0.3 to 3 m diameter. The drum is 50–100 cm wide and is hollow inside. It is a continuous filter press unlike the earlier ones described. Instead of applying pressure, vacuum is applied in this from inside the drum. Since vacuum is applied, all the problems associated with maintaining vacuum are inevitable in this filter. Except for this one disadvantage, it is an excellent filter press where the solids (cake) are continuously scraped and removed. Hence, there is no problem of choking or blinding of the filter medium. The filtrate is also continuously discharged from the centre of the drum.

It consists of a cylindrical drum rotating very slowly at 0.1 to 2 rpm and is partly dipped into the feed tank (as shown in Figure 20.12). A firm metallic screen is fixed all along the circumference of the drum. In fact the drum is made up of two circular discs separated by a firm metallic screen. All along the screen, the filter cloth is securely tied with grip along the rim (Figure 20.13). Normally 30–35 per cent of the circumference of the drum is dipped into the feed tank into which the slurry continuously flows in. If the solids are less in the slurry, even upto 60 per cent of the drum can be dipped into the feed. Vacuum is applied inside the drum to an extent of 33 to 66 kPa. Hence, the slurry is sucked into the drum through the filter cloth which is firmly fixed to the rim of the cylindrical drum. Filter cake forms outside on the filter cloth, whereas the filtrate flows into the rotating drum, and is discharged from the centre of the drum through the filtrate discharge line provided axially at the centre. The cake is washed in the upper portion of the drum and is scraped at the other end with a doctor blade, and is discharged continuously.

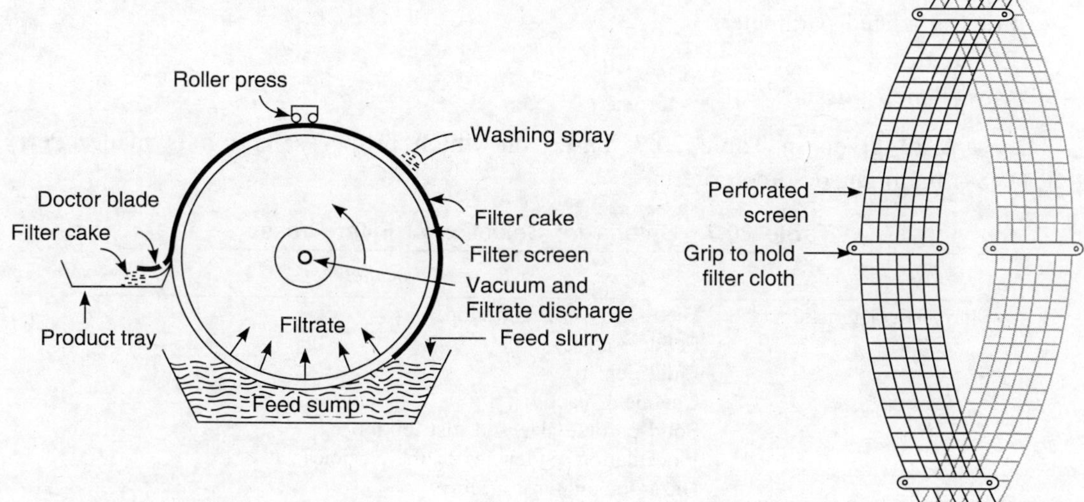

Figure 20.12 Rotary vacuum filter.

Figure 20.13 Perforated hollow cylindrical drum of the rotary vacuum filter press.

There are various types of arrangements available for the discharge of the cake. In some filters, in the upper segment of the drum, hot air is blown continuously from inside for drying the cake partially and also to bloat the filter cloth for easy scraping of the cake. However, at the time of applying hot air, the vacuum supply is stopped. In some designs, the cake is pressed from outside with a roller arrangement to remove the adhering filtrate.

The best advantage with this filter press is that it can be operated continuously, and both the cake and filtrate can be continuously discharged. No time is lost in stopping the filtration process to scrape the cake from the filter cloth. The filter cloth will have a longer life. Since most of the operations are continuous, and hence, are automatic, it hardly involves any expenditure for the labour costs.

20.2.4 Selection of a Filter Press

A variety of filtration equipment is available in the industry to meet every requirement of the processing industry. We have described very briefly only a few. For exhaustive information on the same, any standard textbook on unit operations may be consulted including Perry's Chemical Engineers' Handbook (Perry, et al., 1984). Proper choice of an equipment, for the desired purpose needs a special mention which is based on technical considerations, in addition to the cost factor.

The following are some of the guidelines on the basis of which the choice is made. The various factors are:

- Slurry characteristics
- Equipment features
- Process related parameters
- Economics
- Miscellaneous issues

They are all given in Table 20.7, based on which proper selection is made. Perry, et al. (1984) highlighted some criteria.

Table 20.7 Criteria for Selection of Filter Press

Basic factors	*Related parameters*
Slurry characteristics	Feed slurry concentration
	Fluid viscosity
	Fluid density
	Chemical reactivity
	Solid particle size and distribution
	Flocculation tendencies of the particles
	Hazardous nature of slurry
Equipment characteristics	Batch processing
	Continuous processing
	Filter media to be used
	– wire mesh
	– textile fabric
	Permissible materials of construction

(contd.)

Table 20.7 Criteria for Selection of Filter Press (*contd.*)

Basic factors	*Related parameters*
Process-related parameters	Amount of slurry to be handled
	Type of filter media to be used
	Whether filter aid is required
	Completeness of separation required
	Whether filtrate is desired or filter cake is desired
Economics	Absolute and relative values of filtrate and cake
	Installed costs
	Equipment life
	Operating costs
	Maintenance costs
	Labour costs
	Replacement of filter media
	Costs associated with product loss if any
Miscellaneous issues	Availability of floor area
	Whether the filter is to be tailor made or for multiple use
	Availability of any spare filters

20.3 APPLICATIONS OF FILTRATION IN FOOD PROCESSING

Applications of filtration in food processing are varied even though it is seldom used as a processing tool. For most of the clarification operations, filtration is used to clear the suspension of the fine suspended solids. Filtration of coffee decoction to make a clear extract of the coffee beans is a very common domestic and industrial application. Some of the major application of filtration in food processing may be classified as:

Clarified fruit juices: The fruit juices from non-pulpy fruits like citrus fruits, grapes, pineapple, etc. contain a lot of haze after extracting the juice in a fruit mill. Such fruit juices are filtered to remove the suspended solids before pasteurization or preservation.

Pulpy fruits like banana, mango, guava, papaya, etc. are processed in a pulper to get a thick pulp which on enzymatic treatment with a pectolytic enzyme gives clarified fruit juice on the top and the fibrous material at the bottom. The upper liquid is decanted, and still it contains some suspended solids. Such clarified fruit juices are further clarified by filtration by passing through a muslin cloth.

Sugar syrups are added to most of the fruit pulps or juices to adjust the acid sugar ratio. The sugar syrups, made by dissolving the commercial sugar in water, contain some undesirable foreign material which are filtered by passing through a muslin cloth.

Papaya latex, obtained from the raw papaya fruits from the trees by lancing on the surface, contains a lot of foreign materials. The latex is filtered before processing to make pappain.

Vegetable oil industry: Most of the vegetable oils are made by screw-pressing the oil seeds like ground nuts (peanuts), sesame, sunflower, dried coconut copra, etc. The screw pressed oil contains debris from the seed cake. Before marketing, such oils are invariably clarified by filtration by passing through a plate and frame filter press. Hence, plate and frame filter press is a very common equipment in any vegetable/edible oil industry.

Processing of plantation products: Plantation products like spices (chilli, turmeric) are processed by solvent extraction to extract the active principles. The extracts also contain the debris from the crushed spices. Such spice extracts are filtered for clarification before going for desolventization. Annatto seeds are extracted with an organic solvent for making crystalline edible dye. The extract is filtered for clarification before desolventization to get the crystalline dye, without which step (filtration) the crystalline dye contains solids and debris from the annatto seeds which would add unnecessarily to the weight of the dye without substantially contributing for the bixin content in the dye.

Sugarcane juice beverage is made by pasteurizing the sugarcane juice which is obtained by crushing the canes in open presses. The juice contains the debris from the canes. Such juice is clarified by filtration before pasteurization and bottling.

Symbols

A: cross-sectional area of the filter bed (m^2)
L_c: cake thickness (m)
L_m: equivalent cake thickness (m)
Δp: pressure drop (kPa or MN/m^2)
R: resistance for filtration
s: compressibility coefficient in Eq. (20.13)
t: time (s or h)
V: volume of the slurry (m^3)
w_s: weight of solids present in unit volume of slurry (kg/m^3)

Greek Symbols

α: specific cake resistance (1/m^2)
α_0: constant in Eq. (20.13)
μ: viscosity (Pa s)
ρ_s: density of solids (kg/m^3)

REVIEW QUESTIONS

20.1 Describe the process of filtration as a unit operation for separation of suspended solids.

20.2 How filtration is different from other separation processes like distillation or crystallization?

20.3 What are the desirable characteristics of a filter medium?

20.4 What is meant by filter aid? Why it is used in filtration process?

20.5 What are different types of filtration processes? Explain them briefly.

20.6 Derive an expression for constant rate filtration.

20.7 Derive an expression for constant pressure filtration.

20.8 What is meant by *specific cake resistance*? How do you measure it?

20.9 What is meant by *equivalent cake thickness*? How do you measure it?

20.10 What is the effect of pressure on specific cake resistance? How do you evaluate the effect?

20.11 What are various types of filters available?

20.12 Describe the plate and frame filter press with a neat diagram.

20.13 Describe pressure leaf filter press with a neat diagram.

20.14 Describe a rotary vacuum filter press with a neat diagram.

20.15 What are various parameters you need to consider for selection of a filter press?

20.16 What are various applications of filtration in food processing?

20.17 What are various applications of filtration in fruit processing?

20.18 What are various applications of filtration in processing of plantation products?

20.19 What are various applications of filtration in vegetable oil processing?

NUMERICAL PROBLEMS

20.1 A fruit juice is filtered using a filter press having a cross-sectional area of 0.03 m^2 under a gauge pressure of 0.02 MN/m^2. The solution has a solid content of 10 g/l whose density is 950 kg/m^3. The volumes of the filtrate collected with time are as follows:

t, min	1.5	11	21	33.33	50
V, liter	1.2	3	4.5	6	7.5

Find the specific cake resistance and equivalent cake thickness.

(**Ans:** $\alpha = 1.388 \times 10^{14}$ m^{-2}; $L_m = 4.32 \times 10^{-4}$ m)

20.2 A certain slurry was filtered in a laboratory filter press of cross-sectional area 0.5 m^2 under different pressure differentials. The solids concentration in the slurry is 52 kg/m^3. The density of solids is 970 kg/m^3. Find the effect of pressure on the filtration using the following data:

Δp, Pa	20.1	2980	22026	1.63×10^5
α, m^{-2}	3.6×10^9	4.82×10^{10}	7.2×10^{10}	1.96×10^{11}

(**Ans:** 1.32×10^9 $(\Delta p)^{0.45}$)

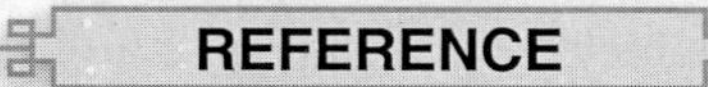

REFERENCE

Perry, R.H., Green, D.W. and Maloney, J.O. (1984), *Perry's Chemical Engineers' Handbook*, 6th ed., McGraw-Hill, New York, pp. 19.73–19.89.

CHAPTER

21

Sedimentation and Centrifugation

Sedimentation and centrifugation are physical separation processes of the substances in the immiscible phases; they could be:

- Solid-liquid,
- Liquid-liquid,
- Liquid-liquid-solid, or
- Gas-solid systems.

Mostly we refer to solid-liquid systems or two immiscible liquid systems. Sedimentation and centrifugation are often dealt as similar operations because in both the cases, the separation of the particles or globules from the other phase is affected by the relative differences in forces acting on the particles in upward and downward directions which ultimately determine the velocity of settling of the particle. The only difference is that the settling velocities are very slow in case of sedimentation and may take sometimes even days together, whereas in case of centrifugation the velocity of settling is faster and may complete the whole process of separation in an hour's time.

From this discussion, it is clear that these separation processes involve two immiscible phases: one is *dispersed phase* which is in small quantity, usually the solid particles are a dispersed phase; and the other is *continuous phase* which is in larger quantity, usually liquid phase or gas phase. In case of two immiscible liquid phases (like organic and aqueous phases), the one which is present in small quantity is called the **dispersed phase**, and the other the **continuous phase**. That is why, the two phases together (i.e., the system as a whole) are known as **dispersions** or **suspensions**. It is also clear form the above discussion that sedimentation and centrifugation are applied only if the phases are immiscible. If they are miscible and form into a single phase, they are known as **solutions**; and the separation is carried out by using such unit operations like distillation, extraction, crystallization or drying involving phase change with the application of simultaneous heat and mass transfer.

21.1 SEDIMENTATION

Sedimentation is a process of separation by the gravitational force of suspended solids in a liquid or dispersed liquid globules in a liquid. Hence, the separation process is very slow. When a solid particle is suspended in a liquid medium, the particle is pulled downwards by the gravitational force, and it will have an acceleration because of the gravity. As the body is moving downwards, its downward movement is obstructed by the buoyancy forces exerted by the liquid, and they are in the upward direction. Initially the particle moves with increasing velocity and ultimately it reaches a constant velocity which is known as **terminal velocity** of the particle. The various forces acting on a settling (moving) particle are shown in Figure 21.1. The gravitational pull is dependent upon the density of the particle, whereas the buoyancy force is dependent upon the density of the fluid. Hence, the velocity of the body is dependent upon the difference in the densities of the particle and fluid. If the difference is large, the velocity (rate of settling) is fast. If the difference is less, the settling (or terminal) velocity is low, and hence, separation is very slow. If the density difference is zero, separation never takes place; and the particles remain in suspension. If the difference is negative (i.e., the density of fluid is more than that of the particle), the particle never sinks; on the contrary, the particle goes upwards and stays on the top of the fluid as froth or as a layer.

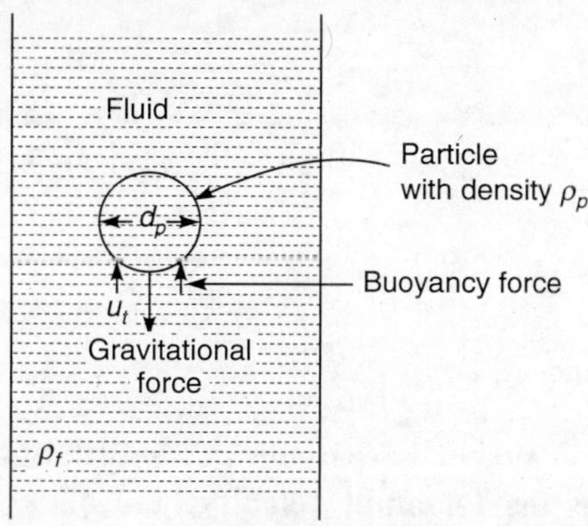

Figure 21.1 Forces acting on a particle moving in a fluid.

21.1.1 Principles of Sedimentation

A particle† at rest suspended in a fluid will be slowly moving downwards and it attains a constant velocity u_t known as **terminal velocity**. Obviously the movement of the particle is dependent upon:

- Size of the particle
- Shape of the particle
- Density of the particle
- Density of the fluid
- Viscosity of the fluid
- Gravitational pull (which is usually constant and is known as **acceleration** due to gravity)

The downward force (F_g) acting on the body (Earle 1983)

$$F_g = V_p g\,(\rho_p - \rho_f) \tag{21.1}$$

where V_p is volume of the particle, ρ_p and ρ_f are density of the particle and fluid respectively, and g is acceleration due to gravity.

$$g = 9.81 \text{ m/s}^2$$

†The discussion whatever is presented here for a solid particle is equally applicable to the liquid globule with similar diameter and density.

The drag force (F_d) acting on the particle is:

$$F_d = C_D \rho_f \frac{u^2}{2} A_p \tag{21.2}$$

where u is velocity of the particle (m/s), A_p is projected area of the particle (m^2), and C_D is drag coefficient which is obtained by *Stoke's law* as follows for streamline settling of spherical bodies:

$$C_D = \frac{24}{N_{Re,P}} \tag{21.3}$$

where $N_{Re,\,P}$ is particle Reynolds number

$$N_{ReP} = \frac{\rho_f d_p u}{\mu_f} \tag{21.4}$$

For spherical particles, the projected area is equal to a circle of equal diameter; hence

$$A_p = \frac{\pi d_p^2}{4}$$

and

$$V_p = \frac{\pi d_p^3}{6}$$

When the particle is freely falling in a liquid, it approaches a constant velocity known as the **terminal velocity**; and under these circumstances Eqs. (21.1) and (21.2) are equal. Substituting $N_{Re,\,P}$, A_p and V_p,

$$\frac{\pi d_p^3}{6} g\,(\rho_p - \rho_f) = 24 \mu_f \rho_f \frac{u_t^2}{2} \frac{\pi d_p^2}{4} \tag{21.5}$$

which on simplification, yields an expression for terminal velocity as follows:

$$u_t = \frac{d_p^2 g(\rho_p - \rho_f)}{18\,\mu_f} \tag{21.6}$$

Equation (21.6) is a very useful equation in designing the sedimentation tanks, particularly to find the surface area of thickeners. The equation is for the terminal velocity of the particle or liquid globule; and it also stands for the rise velocity of a gas bubble in a liquid medium. It indicates how fast a particle settles in a liquid medium. Higher is u_t, faster is the settlement rate of the particle; and hence, less will be the area of the settling tank for given conditions of operation. But as is evident from Eq. (21.6), u_t will be high if $(\rho_p - \rho_f)$ is high or if the particle diameter is larger or if the viscosity of the fluid is less. Obviously change of g is not possible. If the difference between the density of the particle and that of the fluid is less, less will be the rate of settlement, and hence, it takes longer time for the particles to settle down.

We shall notice that the Eq. (21.6) can be used to calculate the terminal velocity of solid particles in the form of a dust in air, or settling velocity of oil globules in an oil-water emulsion, etc.

PROBLEM 21.1 An emulsion of oil in water has oil droplets (globules) in the form of spheres of average diameter of 10 μ. The specific gravity of oil is 0.95. Find the raised velocity of oil globules.

Solution It is a straight case of application of Eq. (21.6)

Given data: $d_p = 10\ \mu = 10 \times 10^{-6}\ \text{m} = 1 \times 10^{-5}\ \text{m}$

ρ_f = Density of water = 1000 kg/m^3

ρ_p = Density of globule = Specific gravity × Density of water

$= 0.95 \times 1000 = 950\ \text{kg/m}^3$

μ = Viscosity of water = 1 cp

$= 1 \times 10^{-3}$ kg/m s

g = Acceleration due to gravity = 9.81 m/s^2

$$u_t = \frac{d_p^2 g (\rho_p - \rho_f)}{18\ \mu} = \frac{(10^{-5})^2 \times (9.81) \times (950 - 1000)}{18 \times 10^{-3}}$$

$$= -\ 2.725 \times 10^{-6}\ \text{m/s}$$

The negative sign value of u_t indicates that it is the raise velocity of the globules.

Now, we shall see how u_t is used for finding out the area of a thickener based on the experimental data of a batch sedimentation tank.

Batch sedimentation: A batch sedimentation experiment is often conducted in the laboratory to collect data which is useful for design of continuous industrial sedimentation tanks. When a liquid suspension is taken in a cylindrical container, it exhibits various stages as time progresses. Different zones of sedimentation can be noticed. Initially, the solids are equally distributed as shown in Figure 21.2(a). As the time progresses, the sedimentation process sets in. Particles start settling down. The whole slurry is divided into four zones as shown in Figure 21.2(b), viz., clear liquid on the top, followed by slurry, thick suspension, and the solids. The solid zone is at the bottom. The interface between the clear liquid and slurry will be moving slowly downwards until the completion of sedimentation. Finally when the sedimentation is completed, it results into only two zones as shown in Figure 21.2(c), viz., clear liquid on the top, and solids at the bottom. The solids are discharged, and the clear liquid is decanted.

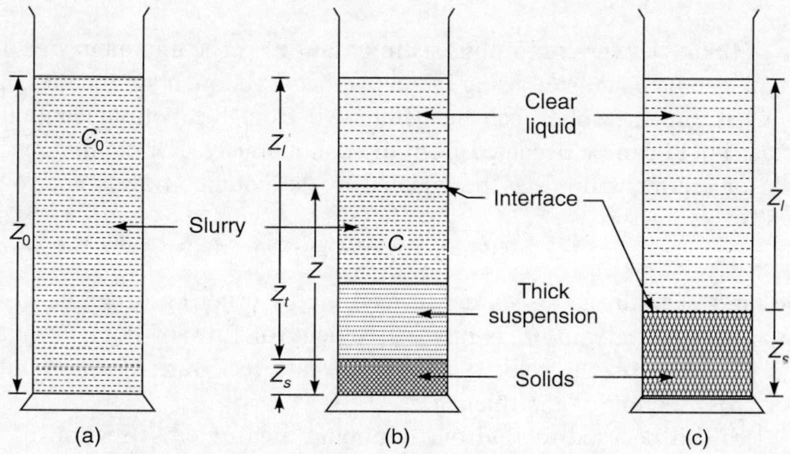

Figure 21.2 Batch sedimentation process.

The batch sedimentation process shown in Figure 21.2(b) is sometimes described in the form of only three zones at anytime, t (Leninger and Beverloo 1975):

- Clear liquid zone on the top, of height Z_l
- Thick suspension zone in the middle, of height Z_t
- Solid zone at the bottom, of height Z_s

The three zones are shown in Figure 21.3(a). In this, the boundary between the clear liquid and thick suspension (l–t zone) will be falling down, whereas the boundary between the thick suspension and the solids will be increasing as time progresses. They are shown in Figure 21.3(b). The two will merge after sometime as we reach to Figure 21.2(c). It is shown in Figure 21.3(b) as point A. Similarly the concentration of solids in l–t zone will be falling, whereas that in t–s zone will be building up. They will meet at point B. Beyond this point, virtually zone t disappears and we reach the situation as shown in Figure 21.2(c).

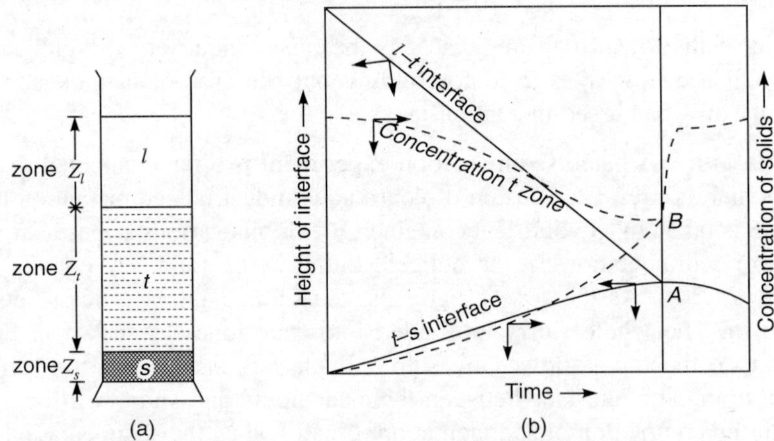

Figure 21.3 Position of interface (—) and variation of solids concentration (- - -) in the sedimentation process.

Minimum area of thickener for continuous sedimentation: The minimum area of a continuous sedimentation tank can be calculated using the above batch sedimentation data. This will enable us to know at what rate the solids can be withdrawn from the bottom continuously, and at what rate the clear liquid can be decanted from top continuously. For continuous sedimentation process, the rate of sedimentation can be equated to the counter flow velocity of the raising fluid (Earle 1983).

$$(W - L)\, f_s = A u_u \rho_f \tag{21.7}$$

where W is the ratio of liquid to solids in the feed, kg of water/kg of solids, L is the ratio of liquid to solids in the exit stream, u_u is upward velocity of flow of liquid, which in turn will be equal to the particle terminal velocity (u_t), f_s is mass feed rate of slurry in kg of feed/s, A is the area of cross-section of the thickener.

Equation (21.7) can be used to find out minimum area of cross-section of the thickener. We shall explain the utility of Eq. (21.7) with Problem 21.2.

PROBLEM 21.2 A slurry containing 20 per cent of sodium carbonate is to be concentrated (thickened) to 40 per cent solids in a continuous thickener. To find the area of continuous thickener for the above purpose, batch sedimentation data were collected in laboratory using different concentrations of slurries. The data are as follows:

Concentration of slurry (per cent solids)	20	22	25	29
Sedimentation rate (u_u), mm/s	0.12	0.094	0.065	0.05

It is desired to affect the separation from a feed rate of 0.85 kg/s. Calculate the minimum area of the thickener to affect the above sedimentation.

Solution We shall first convert the data in the table to our format, i.e., kg of water/kg of solids; and we shall also tabulate $(W - L)/u_u$.

The feed concentration contains 20 per cent solids.

Therefore, $W = \dfrac{80}{20} = 4$ kg of water/kg of solids.

Similarly, the final solution should contain 40 per cent solids.

Therefore, $L = \dfrac{100 - 40}{40} = 1.67$ kg of water/kg of solids.

L is constant through out. The data are prepared in Table 21.1. We have to design a thickener for minimum area. Minimum area is possible when $(W - L)/u_u$ is maximum. Hence, we design the thickener to address for the maximum value of

$$\frac{W - L}{u_u} = 2.046 \times 10^4$$

which we note from the last column in Table 21.1.

Table 21.1 Data for Problem 21.2 with $L = 1.67$

Slurry concentration (per cent)	W $\dfrac{\text{kg of water}}{\text{kg of solids}}$	$W - L$	u_u *Sedimentation rate*, (m/s)	$\dfrac{W - L}{u_u}$ (s/m)
20	4.0	2.33	1.2×10^{-4}	1.94×10^4
22	3.54	1.87	0.94×10^{-4}	2.0×10^4
25	3.0	1.33	0.65×10^{-4}	2.046×10^4
29	2.45	0.78	0.50×10^{-4}	1.56×10^4

By Eq. (21.7), we hav

$$A = \frac{(W - L) f_s}{u_u \rho_f}$$

$$\rho_f = \text{Density of water} = 1000 \text{ kg/m}^3$$

$$f_s = 0.85 \text{ kg/s}$$

Therefore,
$$A = \frac{2.046 \times 10^4 \times 0.85}{1000} = 17.39 \text{ m}^2$$

Batch sedimentation to calculate sedimentation rate and solids concentration: The batch data is collected on the height of the interface vs. time, and a plot is made. A typical plot is shown in Figure 21.4. By drawing tangents at different points of the curve, the slope dZ/dt is noted. This is the settling rate or sedimentation rate (R_s). Point A corresponds to a height of Z_L at time t_L. By drawing tangents at A which touches the y-axis at Z_i we can calculate

$$R_s = \frac{Z_i - Z_L}{t_L} \tag{21.8}$$

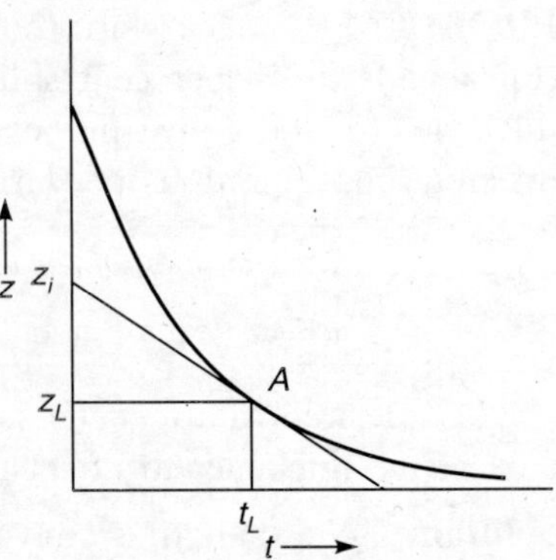

Figure 21.4 Batch sedimentation results.

We can also define a concentration C_L such that

$$C_L Z_i = C_0 Z_0 \tag{21.9}$$

i.e., Z_i is such height of the column where the solids will have a concentration of C_L.

The above procedure is explained with Problem 21.3.

PROBLEM 21.3 A batch sedimentation test was conducted on a slurry, the results of which are shown in Table 21.2. Find the settling rates of the solids and corresponding solid concentrations at different time intervals if the solid concentration in the slurry is 185 grams per litre.

Table 21.2 Test Data of Batch Sedimentation for Problem 21.3

Time t, h	*Height of interface, Z*, cm
0	45
0.2	43
0.4	40.5
0.6	39.0
0.8	36.0
1.0	34.5
1.5	30.5
2.0	27.5
3.0	21.0
4.0	16.5
6.0	11.0
7.0	9.5
8.0	8.5
9.0	7.5
10.0	6.5
12	6.0
15.0	5.0

Solution It is a straight case of application of Eqs. (21.8) and (21.9). To evaluate the data on Z_i at different t, we draw a plot of data in Table 21.2. The curve is shown in Figure 21.5. From the data given, $C_0Z_0 = 185 \times 45 = 8325$ gcm/l.

We will draw a tangent at any point say P on the curve. The tangent touches the y-axis at 32.5 cm which is equal to Z_i, where Z_L is 16.5 cm corresponding to t_L of 4 hours.

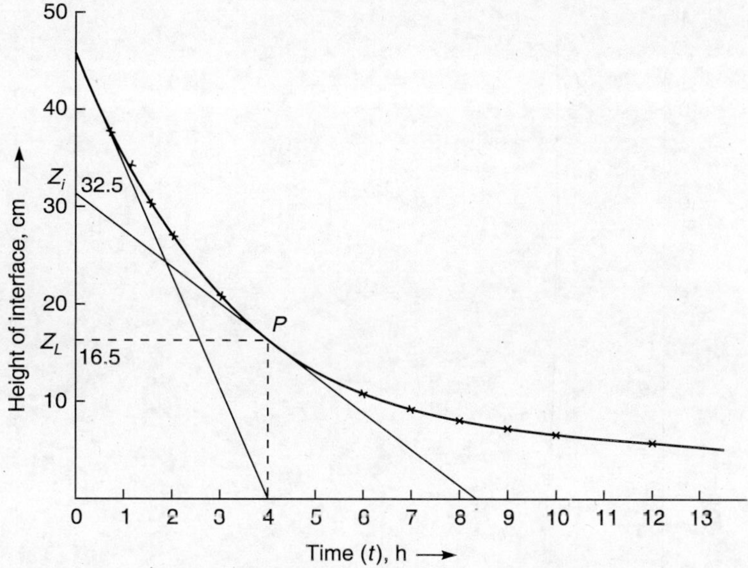

Figure 21.5 Method to calculate sedimentation rate for Problem 21.3.

Therefore,

$$R_s = \frac{Z_i - Z_L}{t_L} = \frac{32.5 - 16.5}{4} = 4 \text{ cm/h}$$

and

$$C_0 Z_0 = C_L Z_i$$

Therefore,

$$C_L = \frac{Z_0 C_0}{Z_i} = \frac{8325}{32.5} = 256.15 \text{ g/l}$$

Similarly, we can find out Z_i and C_L at different time periods. The data are shown in Table 21.3. A plot of R_s vs. C_L is shown in Figure 21.6.

Table 21.3 Settling Rate and Concentration Data at Different Time Periods for Problem 21.3

t_L, h	Z_L, cm	Z_i, cm	R_s, cm/h	C_L, g/h
0.5	39.5	45	11	185
1.0	34.5	44	8.5	189.2
1.5	30.5	41	7.0	203.0
2.0	27.5	40	6.25	208.1
2.5	24.0	39	6.0	213.5
3.0	21.0	36	5.0	231.25
4.0	16.5	32.5	4.0	256.15
5.0	13.5	26.5	2.6	314.15
6.0	11	22.0	1.83	378.4
7.0	9.5	19.5	1.43	426.9
8.0	8.5	17.0	1.06	489.7
10.0	6.5	12.5	0.6	666.0
12.0	6.0	8.0	0.166	1040.6

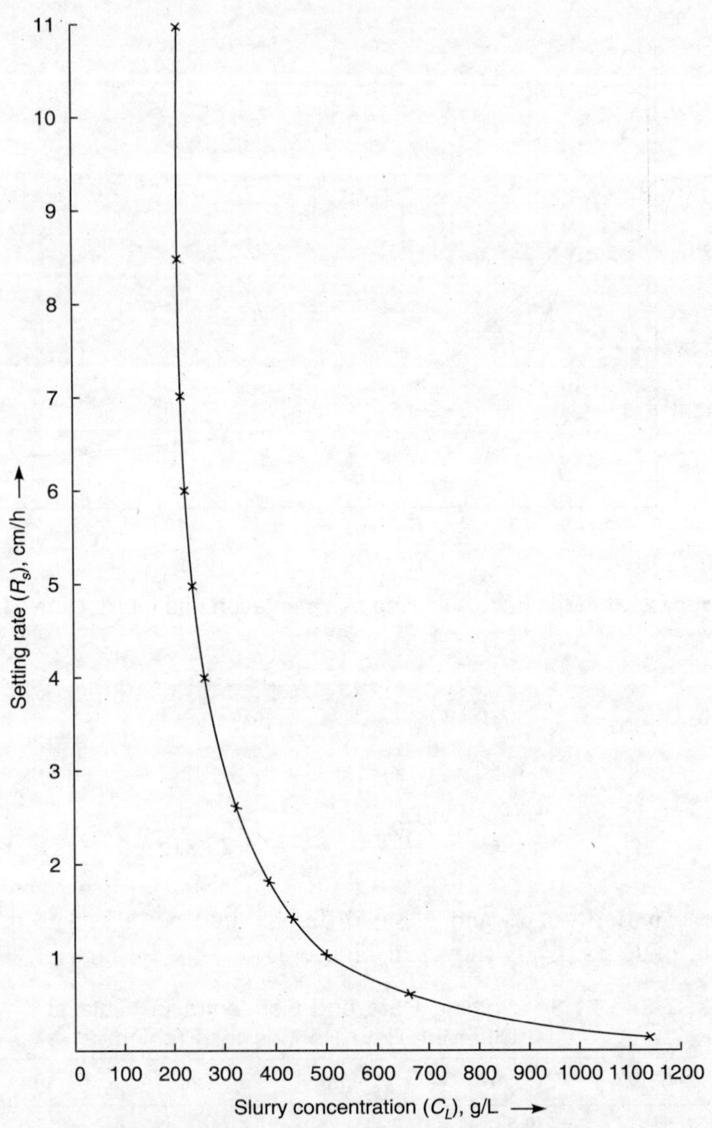

Figure 21.6 R_s vs. C_L data for Problem 21.3.

Thickener area by angle bisector method: Now, we describe a procedure to find the minimum thickener area by a method called **angle bisector method**.

From Figure 21.5, it is evident that in the early stages of settling, more or less the velocity of settling is constant and this zone is known as **free-settling** zone. Similarly, in the last stages of sedimentation, where the concentrations are high and settling velocities are low, we notice virtually that the slope is constant and horizontal. If we draw tangents in the initial stages and in the final stages, they intersect at a point P as shown in Figure 21.7. If we bisect the angle at P, the bisector will meet the settling curve at point Q. The concentration corresponding to Q is known as C_c and the time is t_c. It is at this time that the solids enter in the *compression zone.* The compression zone height Z_c is obtained by:

$$C_0Z_0 = C_cZ_c$$

where C_c is the concentration of solids in the compressed zone.

Generally, it is the compressed zone which goes as underflow. In some occasions, we fix the underflow conditions, particularly the concentration of solids in the underflow (C_U). If Z_U is height corresponding to this concentration, then we have

$$C_0Z_0 = C_UZ_U \qquad (21.10)$$

with the knowledge of C_0Z_0 and C_U, we can calculate Z_U from Eq. (21.10).

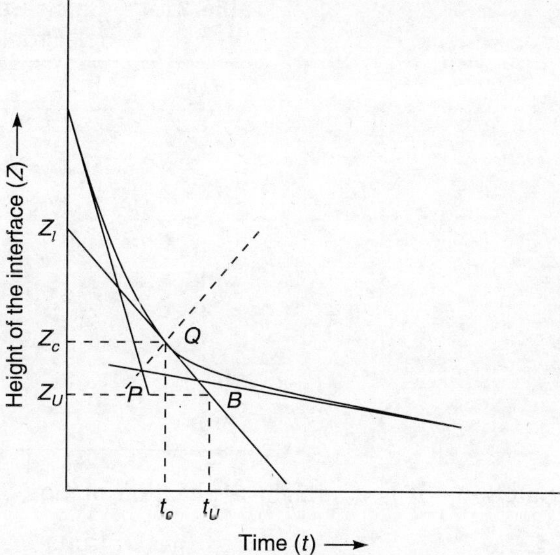

Figure 21.7 Graphical procedure to evaluate Z_U and t_U for finding A by angle bisector method.

Based on the inlet flow rate of the slurry (L_0) and concentration of solids in the inlet slurry (C_0), and final outlet concentration of the solids in the underflow (known as C_U), we can calculate the cross-sectional area of the thickener. Before doing this, however, we should also know the time parameter known as t_U corresponding to point B which is obtained by intersection of the horizontal line at Z_U and the tangent at Q. Once t_U is known, the cross-sectional area (A) can be found by Eq. (21.11).

$$A = \frac{L_0}{Z_0}t_U \qquad (21.11)$$

The theoretical basis of the procedure was described by Foust, et al. (1980).

PROBLEM 21.4 It is proposed to design a sedimentation tank for the wash waters coming from fruit washing plant of 150 tons of fruits/day plant which requires approximately 50 m^3 of water per ton of fruits. The initial load of dust and dirt in the wash waters is 2 g/l, and it is to be concentrated to 10 times of the solid concentration in the underflow of the sedimentation tank. Find the minimum area of the tank. The test data conducted in laboratory on a batch sedimentation set-up is given in Table 21.4.

Table 21.4 Batch Sedimentation Test Data for Problem 21.4

t (min)	Z (cm)
0	48
1	28
2	23.4
3	20.6
4	18.4

(contd.)

Table 21.4 Batch Sedimentation Test Data for Problem 21.4 (*contd.*)

t (min)	Z (cm)
5	16.8
6	15.7
7	14.8
8	14.0
10	12.8
12	11.8
14	11.2
16	10.6
18	10.4
20	10.2
26	10

Solution It is a straight application of Eq. (21.11) and Figure 21.7.

Given data:
$$L_0 = 150\,\frac{\text{tons of fruits}}{\text{Day}} \times 50\,\frac{\text{m}^3\text{ of water}}{\text{Ton of fruits}} = 7500\,\frac{\text{m}^3}{\text{Day}}$$

$$= 7.5 \times 10^3\ \text{m}^3/\text{day}$$

$$= \frac{7.5\times 10^3}{24\times 60} = 5.21\ \text{m}^3/\text{min}$$

$$C_0 = 2\ \text{kg/m}^3$$

$$C_U = 20\ \text{kg/m}^3$$

$$Z_0 = 48\ \text{cm} = 0.48\ \text{m}$$

Therefore,
$$Z_U = \frac{C_0 Z_0}{C_U} = \frac{2\times 0.48}{20} = 0.048\ \text{m}$$

$$= 4.8\ \text{cm}$$

The data in Table 21.4 is plotted in Figure 21.8. Point Q is located by angle bisector method. From the figure, by drawing tangent at Q, we get

$$Z_1 = 21.4\ \text{cm}$$

$$= 0.214\ \text{m}$$

By drawing horizontal line through Z_U, we locate point B which is intersection with the tangent at critical point Q. Corresponding to point B, $t_U = 17.4$ min.

By applying Eq. (21.11), we get

$$A = \frac{L_0 t_U}{Z_0} = \frac{5.21\times 17.4}{0.48} = 188\ \text{m}^2$$

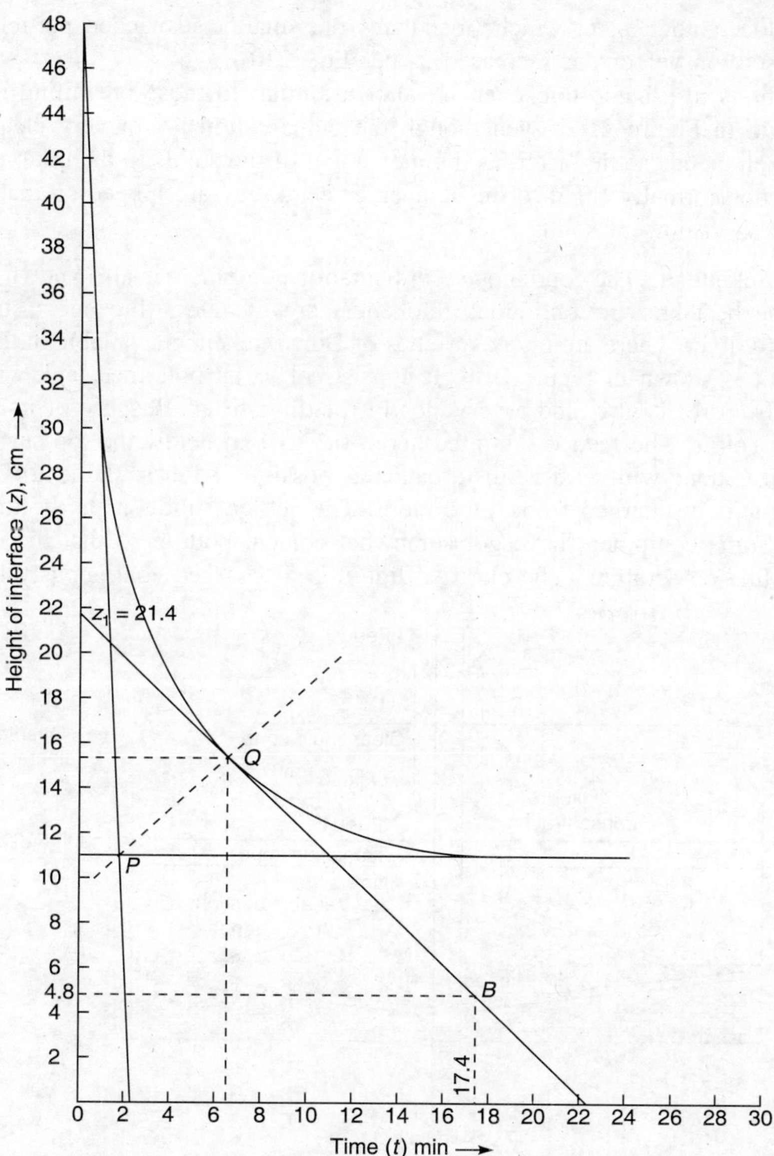

Figure 21.8 Graphical method to find *A* for Problem 21.4.

21.1.2 Sedimentation Equipment

The sedimentation equipment is mostly continuous or batch operated thickeners. Depending upon the throughputs, they are either operated in a batch manner or a continuous manner. A batch sedimentation tank is simply a tank with a slight conical bottom into which the suspension is filled and left for some time for settling to take place. Once the settling process is over, the supernatant clear liquid is decanted, and the thickened liquid (or thickened slurry) at the bottom is withdrawn separately. In some cases, the sedimentation is enhanced (or facilitated) by adding a flocculating agent which flocculates the smaller suspended particles into bigger lumps whose settling velocity is increased as per Eq. (21.6).

The conditions in a batch thickener are almost similar to those prevailing in a laboratory set-up as shown in Figure 21.2, even though the demarcation is not very clear. The bottom scraping will help solids settle faster, as the movement of the solids to the bottom is facilitated. The rakes are not normally fitted to the scraper; instead they are hinged so that they can fold in case of excess loading of solids.

Continuous thickener: The continuous sedimentation equipment is also known as **continuous thickener**. The best known continuous thickeners are the Dorr thickeners supplied by the company Dorr–Oliver. There are many varieties of Dorr thickeners available in the market. One such equipment is shown in Figure 21.9. It consists of a flat bottomed tank with a provision for feed inlet from the centre, and a very slowly rotating stirrer. Because of its very low rpm, it is known as **raker**. The feed is admitted about 0.3 to 1 m below the top surface layer. The feed admission is done with as little disturbance as possible. After settlement of the solids, the thickened liquor is discharged from the bottom. The settled solids at the bottom are scraped by the raker. Some equipment have got somewhat conical bottom so that the solids reach to the discharge line on scraping. The clarified liquid is discharged continuously from the top.

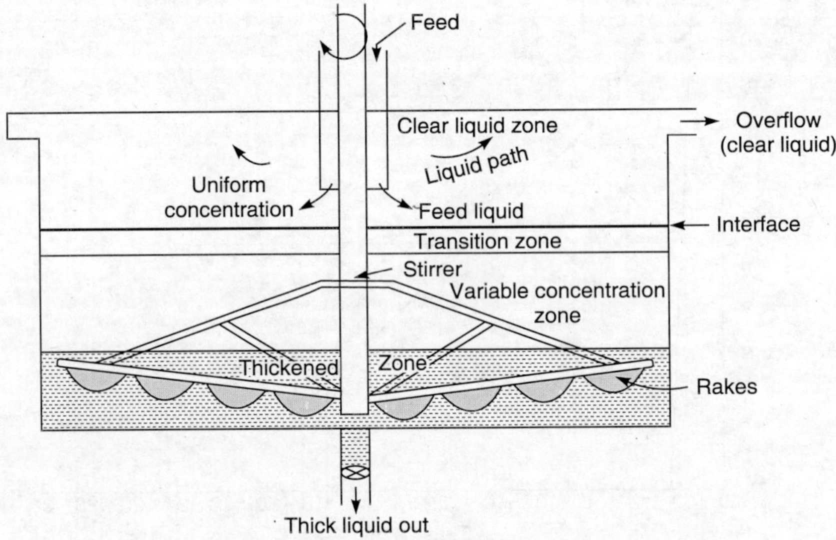

Figure 21.9 Continuous thickener.

The continuous thickeners are used mostly for clarification of industrial effluents where time is not a big constraint. Since the time taken for sedimentation is large, the thickeners are not normally used in the process line.

21.1.3 Applications of Sedimentation in Food Processing

As has already been mentioned, sedimentation is a very slow process, and hence, does not find much applications in processing line. Mostly the applications are restricted to desludging of process waters in the effluent treatment methods.

In case of fruit processing where large quantities of process waters are used to clean the fruits to remove dust and dirt adhering to the fruits, sedimentation tanks are used to clean the water for reuse.

Sedimentation is also used in the settlement of dirt from air, etc.

21.2 CENTRIFUGATION

Centrifugation is almost similar to sedimentation except that the centrifugal separation (settling) of particles is enhanced by centrifugal force as contrast to only gravitational force in case of sedimentation. Centrifugal force is generated in a material when it is rotated; the magnitude of centrifugal force on the material depends upon the radius at which the material is rotating from the centre of rotation, the mass of the body and the speed at which the body is rotating. In simple words, we can write

$$\begin{aligned} \text{Centrifugal force, } F_c &\propto \text{Radius } r \\ &\propto \text{Mass of the body } m \\ &\propto \text{Speed of rotation } N \end{aligned} \tag{21.12}$$

i.e.,

$$F_c = f\,(r, m, N)$$

Hence, it is evident that for a given system, the particles of higher weight will receive more centrifugal force, and hence, will have more r, i.e., they will move farther away from the centre, as compared to the lighter particles. This principle is made use of in the centrifuges for separation of heavier particles from the lighter particles. It could be for separation of

- Solid-liquid suspensions

  or

- Liquid-liquid dispersions

Mostly, the centrifuges are used for separations of very fine particles or suspensions for which the movement of the particles is very slow. Hence, it is reasonable to assume the application of Stoke's law in calculating the drag between the particles and the liquid. Centrifugal separations are reasonably faster and hardly take an hour's time as compared to sedimentation which takes even days together for separation to occur. This makes centrifugal separations an attractive feature; and hence, centrifugal separation processes are in vogue for more than a century in the industry, particularly in sugar industry and also for separation of cream in the dairy industry (Hsu 1986).

21.2.1 Principles of Centrifugation

One basic difference in sedimentation and centrifugation is that, the particles do not move downwards to settle in centrifugation, but they only move radially away from the centre of separation; and hence they never reach equilibrium velocity what is called as **terminal velocity** which occurs in case of sedimentation. However, the same equation for velocity can be used by replacing the term of the acceleration due to gravity with a term representing acceleration due to centrifugal forces.

$$u_t = \frac{d_p^2\, g(\rho_p - \rho_f)}{18\mu_f} \tag{21.6}$$

If the acceleration due to centrifugal force is a', then

$$u = \frac{d_p^2\, a'(\rho_p - \rho_f)}{18\mu} \tag{21.13}$$

If m is the mass of the particle rotating with an angular velocity of ω at a radius of r, the centrifugal force (F_c) generated in the particle is:

$$F_c = mr\omega^2 \tag{21.14}$$

and

$$\omega = 2\pi N \tag{21.15}$$

where N is number of revolutions per second.

Since acceleration is force divided by mass,

$$a' = \frac{F_c}{m} = r\omega^2 \tag{21.16}$$

we may also write,

$$\begin{aligned} F_c &= mr\omega^2 \\ &= mr\left(\frac{2\pi N'}{60}\right)^2 \\ &= 0.011\ mr\,(N')^2 \end{aligned} \tag{21.17}$$

where N' is rpm of rotation.

Combining Eqs. (21.13), (21.15) and (21.16)

$$\begin{aligned} u &= \frac{d_p^2\, r(\omega^2)(\rho_p - \rho_f)}{18\mu} \\ &= \frac{d_p^2\, r(2\pi N)^2\,(\rho_p - \rho_f)}{18\mu} \end{aligned} \tag{21.18}$$

Since u in Eq. (21.13) represents velocity of the particle, in case of centrifugation the movement is in radial direction, hence,

$$\frac{dr}{dt} = \frac{d_p^2\, r\omega^2\,(\rho_p - \rho_f)}{18\mu} \tag{21.19}$$

Centrifuge effect or g-number: It is a method of expressing how effective is the centrifuge as compared to its counter part, i.e., sedimentation equipment. The **g-number** of the centrifuge indicates that how many times, the acceleration in the centrifuge is improved over g which exists in sedimentation tank. The more is the number, the faster is the settlement. Dividing Eqs. (21.13) with (21.6) gives,

$$\frac{u}{u_t} = \frac{\text{Velocity in centrifugation}}{\text{Velocity in sedimentation}} = \frac{r\omega^2}{g} \tag{21.20}$$

The ratio (u/u_t) is also known as Z. Hence,

$$Z = \frac{r\omega^2}{g} \quad (21.21)$$

Sigma factor: Sigma factor, also written as Σ factor is an important factor usually associated with centrifuge. It is used to compare the performance of different centrifuges for the same duty of operation. It is also used as criteria for scale-up of centrifuges. It is defined as:

$$\Sigma = \frac{v}{2u_t} \quad (21.22)$$

where v is the volumetric feed rate in m^3/s, and u_t is usually the terminal velocity of the particle under gravitational force.

Σ has the units of m^2, and hence, it represents the cross-sectional area of gravity settler with same sedimentation characteristics.

PROBLEM 21.5 Find the g-number of centrifuge which spins at the rate of 5000 rpm at a maximum radius of 10 cm.

Solution It is a simple application of Eq. (21.21)

$$Z = g \text{ factor} = \frac{r\omega^2}{g}$$

where

$$\omega = 2\pi N = 2\pi\left(\frac{5000}{60}\right) = 523.6 \text{ s}^{-1}$$

$$r = 10 \text{ cm} = 0.1 \text{ m}$$

$$g = 9.81 \text{ m/s}^2$$

Therefore,

$$Z = \frac{0.1 \times (523.6)^2}{9.81} = 2794$$

The centrifuge generates 2794 g.

PROBLEM 21.6 A centrifuge is used for separation of an oil emulsion in water. The oil is dispersed in water in the form of fine globules of average diameter 50 μ, and the oil has a density of 900 kg/m^3. The centrifuge rotates at 2000 rpm. The effective radius at which the separation takes place is 40 mm. Calculate the velocity of oil globules in the water medium.

Solution It is a straight application of Eq. (21.18)

Given data:

$$d_p = 50\ \mu = 5 \times 10^{-5} \text{ m}$$

$$\rho_p = 900 \text{ kg/m}^3$$

$$r = 40 \text{ mm} = 0.04 \text{ m}$$

$$N = 2000 \text{ rpm} = \left(\frac{2000}{60}\right) \text{ rps}$$

Data taken from literature:

$$\rho_f = 1000 \text{ kg/m}^3$$

$$\mu = 1 \times 10^{-3} \text{Pa s}$$

$$u = \frac{d_p^2 r (2\pi N)^2 (\rho_p - \rho_f)}{18\mu}$$

$$= \frac{(5\times10^{-5})^2 \times 0.04\ (2\pi\ 2000/60)^2\ (900-1000)}{18\times10^{-3}}$$

$$= -\ 0.024 \text{ m/s}$$

The negative sign indicates that the oil moves towards the centre.
Now, let us verify whether application of Stoke's law is justifiable by finding the particle Reynolds number,

$$N_{Re,P} = \frac{\rho d_p u}{\mu} = \frac{900\ (5\times10^{-5})\times0.024}{1\times10^{-3}} = 1.1$$

The flow is streamlined, and hence, it is justifiable to use Stoke's law.
Therefore, velocity of oil globules through water phase = 0.024 m/s

PROBLEM 21.7 For Problem 21.6, if the flow rate of the feed is 100 lpm, find Σ factor of the centrifuge.

Solution It is also almost direct application of Eq. (21.22) after calculating u_t by Eq. (21.6)

$$u_t = \frac{d_p^2\, g(\rho_p - \rho_f)}{18\mu_f} \tag{21.6}$$

and
$$\Sigma = \frac{v}{2u_t}$$

$$g = 9.81 \text{ m/s}^2$$

$$v = 100 \text{ lpm} = \frac{100\times10^{-3}}{60} \text{ m}^3\text{/s}$$

$$= 1.67 \times 10^{-3} \text{ m}^3\text{/s}$$

$$u_t = \frac{(5\times10^{-5})^2 \times 9.81(1000-900)}{18\mu_f} = 1.3625 \times 10^{-4} \text{ m/s}$$

$$\Sigma = \frac{1.67\times10^{-3}}{2\times1.362\times10^{-4}} = 6.128 \text{ m}^2$$

Separation of liquids in a centrifuge: In some food processing operations, particularly in dairy processing, we come across the need for separation of one component in an immiscible liquid-liquid mixture in which one liquid is finely dispersed in the other liquid phase. In the milk processing, cream is finely dispersed into liquid milk, and needs to be separated. We generally use a tubular bowl centrifuge† into which milk is continuously fed at the centre, cream separates and is discharged through one outlet, whereas the skimmed milk is discharged from another outlet. The lighter material (cream) accumulates and gets discharged near to the central axis, whereas the milk is discharged from the outlet away from the centre. In such cases, it is essential to locate the neutral zone on one side of which has the lighter liquid, and on the other side has the heavier liquid. It is represented schematically in Figure 21.10.

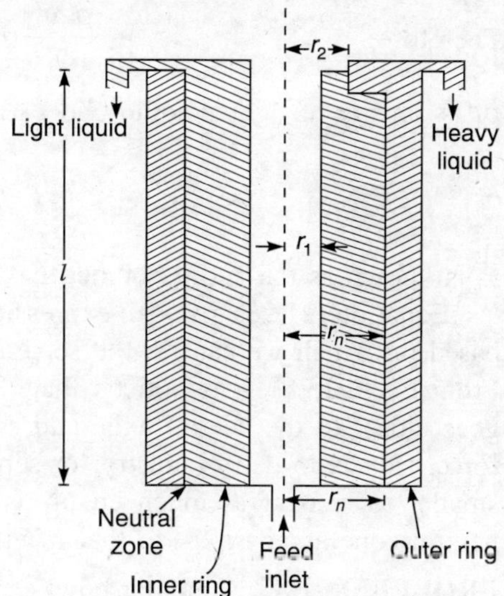

Figure 21.10 Liquids separation by bowl centrifuge.

In the liquid zone, if we consider an elemental concentric cylinder of fluid of thickness dr and mass dm, at a radius of r and of length l, the centrifugal force acting on it is given by Eq. (21.14).

$$dF_c = (dm)\ r\omega^2 \tag{21.23}$$

If ρ is the density of the fluid element

$$dm = \rho \times 2\pi rl\ (dr) \tag{21.24}$$

Pressure acting on the fluid element of area $2\pi rl$ is given by

$$\frac{dF_c}{2\pi rl} = dp = \rho\omega^2 rdr \tag{21.25}$$

On integration,

$$p = \rho\omega^2 \frac{r^2}{2}$$

Therefore, pressure at radius $r_1 = \rho\omega^2 \dfrac{r_1^2}{2}$

Similarly, pressure at radius $r_2 = \rho\omega^2 \dfrac{r_2^2}{2}$

The differential pressure acting on the small elemental cylinder of fluid is:

$$p_2 - p_1 = \Delta p = \frac{\rho\omega^2}{2}\ (r_2^2 - r_1^2) \tag{21.26}$$

If r_1 and r_2 are radii for discharge of lighter liquid and heavier liquid, respectively, and if r_n is the radius at which the separation is taking place, it is obvious that Δp on both sides of the cylinder of radius r_n must be same to keep the separating line in equilibrium,

†The tubular bowl centrifuge will be described in the next section.

Therefore, $$\frac{\rho_1\omega^2}{2}(r_n^2 - r_1^2) = \frac{\rho_2\omega^2}{2}(r_2^2 - r_n^2) \quad (21.27)$$

ρ_1 is density of lighter liquid and ρ_2 is density of heavier liquid.

or $$r_n^2 = \frac{(\rho_2 r_2^2 + \rho_1 r_1^2)}{(\rho_2 + \rho_1)} \quad (21.28)$$

r_n is known as the **radius of neutral zone**.

Equation (21.28) gives an expression to locate at what radius does the separation take place, based on which we can fix the separating plate, and accordingly radii r_1 and r_2 can be fixed. From Eq. (21.28), it is noticed that if r_2 is reduced, r_n is also reduced; i.e., the neutral zone goes closer to the central axis, and so the lighter liquid is exposed to only small centrifugal force. Hence, in dairy industry for separation of cream from the milk, the neutral zone is kept smaller to recover as much cream as possible. In such cases, we admit the feed as close to neutral zone as possible, so that it will enter into the system with least disturbance.

PROBLEM 21.8 A tubular bowl centrifuge is used for separation of cream from milk, which has the discharge radii of 5 cm and 7 cm. If the density of milk is same as that of water, and the cream density is equal to 900 kg/m^3, calculate the radius of the neutral zone.

Solution It is a direct application of Eq. (21.28).

Given data:

$$r_1 = 5 \text{ cm} = 0.05 \text{ m}$$

$$r_2 = 7 \text{ cm} = 0.07 \text{ m}$$

$$\rho_1 = 900 \text{ kg/m}^3$$

$$\rho_2 = 1000 \text{ g/m}^3$$

$$r_n^2 = \frac{(\rho_2 r_2^2 + \rho_1 r_1^2)}{(\rho_2 + \rho_1)}$$

$$= \frac{1000 \times (0.07)^2 + 900 \times (0.05)^2}{(1000 + 900)}$$

$$= 0.00389 \text{ m}^2$$

Therefore, $$r_n = 0.0624 \text{ m} = 6.2 \text{ cm}$$

The radius of the neutral zone = 6.2 cm.

21.2.2 Centrifugation Equipment

As has been mentioned earlier, the very basic purpose of centrifuge is to separate the solid suspensions or liquid emulsions. They are different from filtration equipment in the sense that filters work on the basis of differences in particle size, whereas the centrifuges work on density differences. They are like sedimenters except that the sedimentation (or settling!) rates are enhanced by application of centrifugal forces. The centrifuges enhance the acceleration values by a factor of $r\omega^2/g$ which is approximately equal to $4rN^2$ (Leninger and Beverloo 1975). Since N is squared, the centrifuges give good effect in boosting the centrifugal effect, which varies anywhere between 5000–20,000. Even ultracentrifuges work with $4rN^2 = 1{,}000{,}000$. Even though N is very high in these cases, the r values are quite small. The high speeds always result in some associated mechanical design problems such as bearings, mountings, etc.

Centrifuges are classified into two distinct categories (Coulson, et al., 1991). They are:

- Filtration centrifuges
- Sedimentation centrifuges

Some centrifuges are used only for liquid-liquid separation. They are also like sedimentation type centrifuges, but the system does not involve any solids. A broad classification of centrifugal equipment is shown in Figure 21.11. We shall describe a few of them.

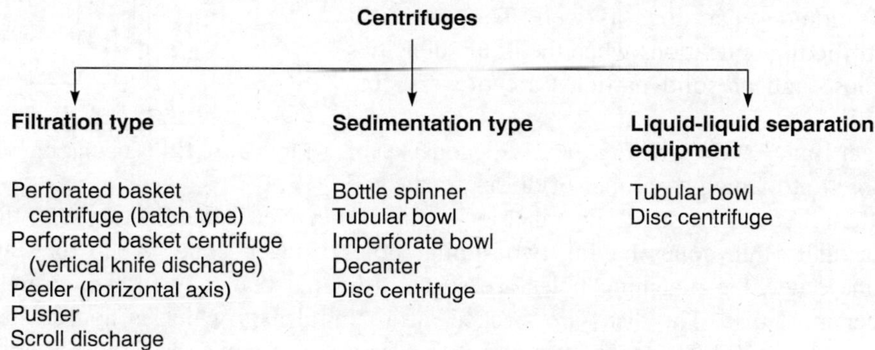

Figure 21.11 Classification of centrifugation equipment.

Tubular bowl centrifuge: They are also popularly known as **bowl centrifuges**, the main reason for this being their appearance. They are usually tall and narrow. The diameters are low and of the order of 10–15 cm. They rotate at high speeds, approximately 15,000 rpm. They are used for liquid-liquid separation, and rarely used for slurries. Even if they are used for slurries, the solid content in the slurry should be very low, and is of the order of 1 per cent with the specific gravity of the cake being less than 2. If the solid content is high, some adjustments are to be made in the internal arrangement. The solids' discharge is done batchwise or at the end of the operation while the liquids are continuously separated.

It consists of a tall and narrow bowl which rotates at high rpm. There is a provision for admitting feed from the bottom (Figure 21.12) which enters into the bowl and is subjected to centrifugation. As has been described in the earlier section, the feed separates into two phases, the lighter phase moving towards the centre, and the heavier phase going away from the central axis towards the walls. The radius of neutral zone is given by Eq. (21.28). The two phases slowly rise up in the bowl. An adjustable ring at the top is adjusted to suit to the neutral zone, and separate the phases. They are discharged through the respective outlets. If the feed contains suspended solids, they settle at the bottom (as shown in Figure 21.12) and slowly rise to the top, but away from the central axis, and almost touching the walls. They are removed intermittently by flushing water through a nozzle provided for the purpose. This arrangement is not shown in Figure 21.12. In fact the bowl centrifuges are not meant for separation of solids. They do not work efficiently with liquid-liquid-solid systems. This is particularly one of the disadvantages with the bowl centrifuges. For 3-phase systems, normally disc bowl centrifuges are preferred.

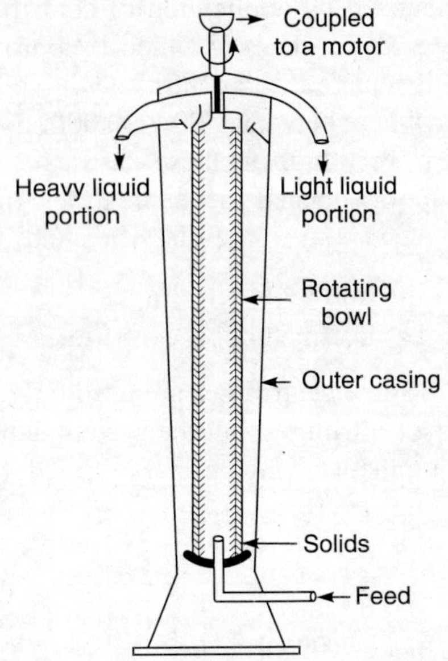

Figure 21.12 Tubular bowl centrifuge.

Disc bowl centrifuge: Disc bowl centrifuges are also known as **disc centrifuges** or disc stack centrifuges. They are ideal for a wide range of separations which involve less solid concentrations with smaller particle sizes. The advantage with these centrifuges is that they can be used for liquid-liquid separations and also for solid-liquid-liquid separations as well. They are best suited for difficult applications when the liquid densities are very close and or solid particles are of very fine suspensions.

The centrifuge, shown in Figure 21.13 consists of an outer bowl into which a number of discs are stacked one over the other with a small clearance of 3 mm in between each disc. In fact, the metallic discs are actually thin cones having two similar holes in each cone. When they are stacked one over the other, the matching holes are so co-ordinated that they form a straight path up to the upper most disc. The discs are stacked on a central axis which is housed in the bowl. The bowl is usually 20–50 cm diameter. The central axis is coupled at the bottom to a high speed motor, which rotates the bowl with the stack of discs at high rpm.

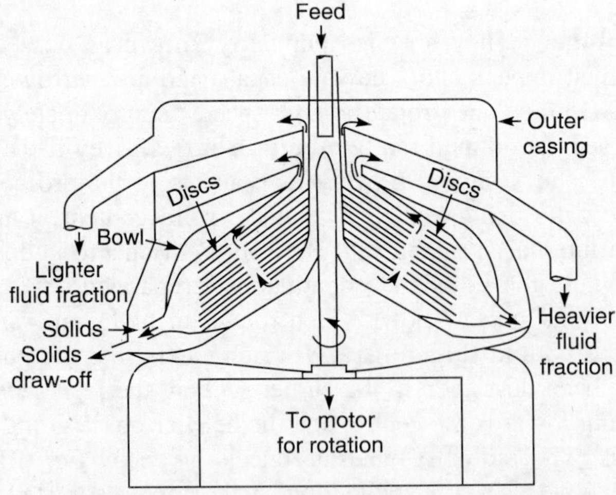

Figure 21.13 Continuous disc bowl centrifuge.

The feed liquid/suspension is fed from the top, and falls on the rotating central axis, and is thus, subjected to high centrifugal force. The heavier and lighter fractions get separated along with the solids if any. The lighter liquid fraction accumulates at the centre and rises up to be discharged through the outlet provided at the sides on the left hand side. The heavier liquid moves away from the axis, and rises up through the central-matching holes to the top to be discharged through the outlet provided at the sides on the right hand side. The solids if any will be accumulating at the bottom along the periphery of the bowl. If their quantity is low, they can be discharged at the end of centrifugation. In the nozzle-discharge centrifuge, a nozzle is provided at the bottom through which a jet of liquid can be introduced. The force of a small jet of liquid is sufficient enough to lift the whole stack of discs (which may be weighing few kilograms!). During centrifugation, the whole stack of discs becomes very light. The liquid jet washes the solids and discharges them through the outlet at the bottom.

These centrifuges are ideal for liquid-liquid dispersions with a small quantity of fine solid suspensions, as in the process of separation of coconut milk with a very fine solid suspensions.

Basket centrifuge: Basket centrifuge is one of the most frequently used centrifuges in industry for separation of solid suspensions from liquids. It is used only when the solid concentration is high, and the particle size is also big. Particularly if the solid suspensions are hazy or coagulated suspensions, basket centrifuges are ideally suited.

The basket centrifuge consists of a perforated basket rotating in an outer bowl. The basket is usually 75–120 cm in diameter and 45–75 cm deep. The basket is either hanged from the top and rotated, or supported from the bottom and coupled to a motor for rotation. The latter arrangement is shown in Figure 21.14. The whole basket rotates at a very high rpm. The feed is fed from the top. It falls on the knob of the rotating bowl, and gets centrifuged. The solids which are usually denser than the liquid will be thrown away from the centre to the walls of the basket. The lighter liquid will be accumulating at the bottom, and will be discharged through the outlet. The solids accumulate along the walls of the basket. The perforations help any adhering liquid to drip out of the solids. If the solid suspensions are very fine, a wet cloth is covered on the perforated basket from inside. In view of the centrifugal force, the cloth sticks to the basket. Inside the basket, there will be a provision to scrape the solids. The scraper is an adjustable loader knife. The solids can be discharged after stopping the centrifuge. The scraped solids can be washed with a jet of water or solvent. The washing process not only washes the solids but also helps to draw the solids from inside. The solids can be discharged only intermittently, and can only be done after stopping the centrifuge.

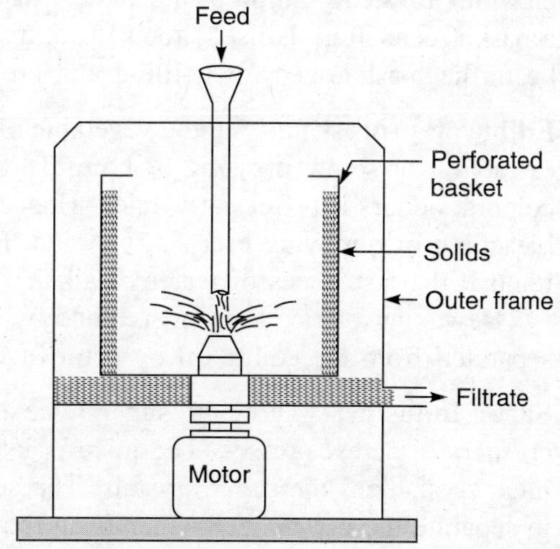

Figure 21.14 Basket centrifuge.

These centrifuges are very versatile, and are easy to operate irrespective of the solid concentration or particle size. They can only be used for solid-liquid suspensions; they cannot be used for separation of liquid-liquid dispersions. This is the only drawback with the basket centrifuges.

21.2.3 Applications of Centrifugation in Food Processing

As has been mentioned earlier, dairy processing is probably one of the biggest applications of centrifugation in food processing. There are many other applications in food processing, viz., clarification of fruit juices, separation of oil-water suspensions and also crystallized solids after crystallization.

Dairy processing: Centrifugation is used in dairy industry for separation of cream from the skimmed milk. The cream in turn is used for various applications. This also results in skimmed milk which can easily be spray dried into milk powder. Huge and domestic bowl centrifuges (Sharples) are used for this purpose.

Even in the case of wet processing of coconut, the coconut milk is centrifuged in a disc bowl centrifuge to separate cream, skimmed milk and some of the suspended solids that have escaped into the milk during screw pressing of coconut copra. The coconut cream is a high value product in the cosmetic industry.

Fruit beverage processing: Fruit juices obtained from some of the juicy fruits like oranges and pineapple obtained by screw pressing are centrifuged to obtain clarified fruit juices which can be further pasteurized. Fruit pulps are extracted from some of the pulpy fruits like guava or banana, and are subjected to enzymatic treatment using various types of commercial pectolytic enzymes to break the pulp into juice and fibre. Then it is subjected to centrifugation (usually a basket centrifuge is used) for separating the fibre to make a clarified fruit juice, which can be further pasteurized for bottling and preservation.

Edible oil processing: In the vegetable oil industry the oil is usually obtained either by solvent extraction or screw pressing or both. The oil after extraction contains some of the undesired colours, odours and free fatty acids. Hence, the oil is washed with water, alkali or charcoal for bleaching or removing Free Fatty Acids (FFA), etc. The oils are subjected to centrifugation to remove the last traces of water or alkali or charcoal.

Dewaxing of rice bran oil is done by enzymatic treatment. Later the gums and waxes are separated from the edible oil by centrifugation.

Sugar industry: The cane sugar juice is obtained from the sugarcanes by crushing in jaw crushers or screw presses. The juice is later concentrated by evaporation. The super saturated juice crystallizes the sugar crystals. The sugar crystals are removed in huge basket centrifuges to separate the crystallized sugar from the concentrated liquor. The mother liquor is known as **molasses** which is a feed material for a host of biotechnological products.

Savoury and snack foods: Mostly, the savoury and snack items are made by deep fat frying in oils. Such items contain (or retain) a good amount of oil adhering to them, probably of the order of 50 per cent by weight; classical examples are Indian traditional snacks like *bhujia and boondi.* The oil associated with snacks is not desired primarily because of rancidity problems during storage; and secondly the cost becomes prohibitive. Such oil traces are removed to a great extent in the tall basket centrifuges in which the solids stick to the perforated basket while the oil goes out of the basket and can be collected separately for reuse.

There could be a number of such applications in food processing for separation of solids from liquids or liquids from liquids.

Symbols

A: cross-sectional area (m^2)
A_p: projected area (m^2)
a': acceleration due to centrifugal force (m/s^2)
C: concentration, kg/m^3
C_D: drag coefficient given by Eq. (21.3)
C_L: concentration defined by Eq. (21.9)
C_U: concentration of solids in the compressed zone/underflow (kg/m^3)
d_p: particle diameter (m)
F_c: centrifugal force
F_d: drag force
F_g: downward force acting on the body
f_s: mass feed rate of slurry (kg/s)
g: acceleration due to gravity (9.81 m/s^2)
L: liquid content/kg solids in the exit stream
l: height of liquid path in Eq. (21.24)
m: mass of the body (kg)
N: rps (s^{-1})
$N_{Re,P}$: particle Reynold's number
N': rpm
p: pressure
Δp: pressure differential
R_s: sedimentation rate
r: radius (m)
t: time
u : velocity (m/s)
u_t: terminal velocity (m/s)
u_u: up-raise velocity of liquid (m/s)
V: volume (m^3)
V_p: volume of particle (m^3)
v: volumetric flow rate (m^3/s)
W: liquid content/kg of solids in the feed
Z: liquid height (m)
Z_L: liquid height at time t_L
Z_t: height of thick suspension
Z_U: height of compression zone

Subscripts

c: compression zone
f: fluid
l: liquid
n: neutral zone
p: particle
s: solids
t: anytime, t
U: underflow
0: initial
1, 2: light and heavy fluids

Greek Symbols

ρ: density (kg/m^3)
μ: viscosity (Pa s)
μ_f: fluid viscosity (Pa s)
ω: angular velocity (s^{-1})
Z: a factor defined by Eq. (21.21)
Σ: a factor defined by Eq. (21.22)

REVIEW QUESTIONS

21.1 Compare and contrast sedimentation and centrifugation.

21.2 Describe a sedimentation process.

21.3 Define terminal velocity of a freely falling particle in a liquid medium.

21.4 Show diagrammatically various forces acting on a freely falling particle in liquid medium.

21.5 Derive a mathematical expression for terminal velocity of a free falling particle in a liquid.

21.6 Describe various stages in a batch sedimentation process.

21.7 Describe systematically how do you use laboratory test data of batch sedimentation to evaluate the area of a continuous thickener.

21.8 Describe how do you evaluate minimum area of continuous thickener using angle-bisector method.

21.9 Describe a continuous thickener with a neat diagram.

21.10 What are various applications of sedimentation in food processing?

21.11 Describe a centrifugation process.

21.12 Derive a mathematical expression for the velocity of a particle in centrifugation process.

21.13 Define *g*-number, and write the expression for it.

21.14 What is meant by sigma factor of a centrifuge? What is its significance?

21.15 What is meant by radius of neutral zone in a centrifuge? Derive a mathematical expression for it.

21.16 How do you classify the centrifuges?

21.17 Describe the working of bowl centrifuge with a neat diagram.

21.18 Describe the working of disc bowl centrifuge with a neat diagram.

21.19 Describe the working of basket centrifuge with a neat diagram.

21.20 What are various applications of centrifugation in food processing?

NUMERICAL PROBLEMS

21.1 A coffee decoction has solids suspended in it. A sedimenter is used to settle the solids to decant the clear decoction. The suspended solids have an average diameter of 100 microns. Find their terminal velocity if the density of solids is 1100 kg/m^3 and the specific gravity of the coffee decoction is 1.06, and its viscosity is 1.1 centripoise.

(**Ans:** $u_t = 2.477 \times 10^{-4}$ m/s)

21.2 A continuous thickener is to be designed for treating a slurry containing 10 per cent solids. The final bottom concentration of solids should be 50 per cent with an average settling velocity of 0.62 kg/s. The test data collected in laboratory on a batch sedimentation is as follows. Calculate the minimum average area of the settling tank to meet the above requirements even at maximum load capacity.

Slurry Concentration (per cent solids)	10	15	20	25	30
Sedimentation Rate (mm/s)	0.15	0.142	0.1	0.06	0.047

(**Ans:** $A = 33$ m^2)

21.3 Determine how efficient is a centrifuge by determining its *g*-number at a radius of 6 cm. The centrifuge has an rpm of 2000. (**Ans:** $Z = 268.3$)

21.4 A centrifuge is used for separation of a coagulated protein from a solution fed at the rate of 30 lpm. The solids have an average particle size of 100 μ and their density is 1010 kg/m^3, while the density of mother liquor is 1000 kg/m^3, and viscosity is 1cp. Calculate the Σ factor of the centrifuge. (**Ans:** Σ factor = 0.45 m^2)

21.5 For the above problem, if the centrifuge spins at 5000 rpm, and if effective separation takes place at a radius of 5 cm, calculate the velocity of settling of the particle. (**Ans:** $u = 0.076$ m/s)

21.6 A bowl centrifuge is used to separate a solid suspension. The centrifuge has discharge radii of 8 cm and 10 cm. If the density of solids is 1050 kg/m^3, and the mother liquid is water, calculate the neutral zone so that the feed inlet may be designed. (**Ans:** $r_n = 9.07$ cm)

REFERENCES

Coulson, J.M., Richardson, J.F., Backhurst, J.R. and Harker, J.H. (1991), *Chem. Engg.*, Vol. 2, 4th ed., Pergamon Press, Oxford, pp. 365–389.

Earle, R.L. (1983), *Unit Operations in Food Processing*, 2nd. Ed., Pergamon Press, Oxford, pp. 143–159.

Foust, A.S., Wenzel, L.A., Clump, C.W., Maus, L. and Andersen, L.B. (1980), *Principles of Unit Operations*, 2nd ed., John Wiley & Sons, New York, pp. 622–637.

Hsu, H.W. (1986), Separations by liquid centrifugations, *Ind. Engg. Chem. Fundamentals*, **25**, pp. 588–593.

Leninger, H.A. and Beverloo,W.A. (1975), *Food Process Engineering*, D. Reinhold Pub. Co., Dordrecht (Holland), pp. 188–201.

CHAPTER

22

Mixing and Emulsification

So far we have seen the unit operations which are used for separation of one component from the other or one component from a multi-component mixture either with phase change or without phase change to recover the product in as pure form as possible. It could be a,

- liquid-liquid mixture,
- solid-liquid mixture,
- gas-liquid mixture, or
- liquid-liquid-solid mixtures.

The components in the mixtures may be in miscible form or in immiscible form.

In the present chapter we consider systems with various components to mix them as intimately as possible to make the products with good consistency. Hence, obviously, mixing is more concerned with the raw materials rather than the products, whereas all the earlier unit operations like drying, distillation, extraction or centrifugation are concerned with the product mix to recover the desired product in as pure form as possible. Another very distinct feature with mixing is that it is not only used for mixing of different components of raw material to make a homogeneous mixture, but it is also used to distribute the heat amongst various components of the mixture which in turn facilitates uniform temperature throughout the heating system. Thus, mixing helps better heat transfer either from the vessel to different components in the vessel, or better heat distribution amongst various components. The different phases involved in mixing could be gas, liquid and solid components in different combinations as mentioned earlier, and in addition to it we have solid-solid mixing.

Indeed, mixing of solid-solid components is more complicated than the mixing of other phases. Particularly if one of the solid components is in minor quantities, the task of mixing it evenly becomes more stupendous. For example, mixing a small quantity of yeast to maida to make dough for making bread, or mixing of vitamin premixes in ppm level to the cattle feed or poultry feed or to the dough in the bakery industry are some of the classical examples involving solid-solid mixing.

From the foregoing discussions, mixing may be defined as a unit operation which reduces or tends to reduce the gradients in composition or concentration or properties or temperature of materials in bulk. It is a method of dispersing one component throughout into the other. Dispersion need not essentially mean solubilizing. Even if the components are immiscible, what mixing does is only dispersing one component throughout into the other, and keeps it in suspension. Thus, mixing helps the components to be together, and hence, facilitate their flow or reaction in coexistence throughout. This makes mixing an important unit operation with a wide variety of applications in food processing. But many times, the extent of mixing is described qualitatively rather than quantitatively. We use many times phrases like *well mixed* or *reasonably mixed* or *poorly mixed*, etc. Quantification of mixing is rarely done; the main reason being that mixing is still a poorly understood science.

Now, let us go into the details of a few of them.

22.1 MIXING OF SOLIDS AND PASTES

Solids mixing is one of the very tricky operations, and becomes more complicated if one of the components is in a very small quantity. When all the solid components are taken and mixed in a container, what is expected under ideal conditions is that any sample randomly picked up from any corner of the container and analysed should contain the same proportion as what is taken initially. If the mixing is improper, or if the mixing time is insufficient, the samples taken will not have the same expected composition, some may be high while others may be less. Now, with this background, it is possible to define a perfect mixture as the one in which the components of samples are in proportions whose statistical chance of occurrence is equal to that of a statistically random dispersion of original components (Earle 1983).

From this discussion, it is evident that the sample analysis is obviously an effective tool for exercise. The importance of sample analysis can be understood from the following example. Let us say that we wish to add 0.1 per cent of vitamin premix for a food supplement; i.e., we add 1 kg of premix to 999 kg of food supplement to prepare 1000 kg food material. So far so good if all the 1000 kg bulk is sold or consumed at once as one sample. Suppose we want to sell the food supplement in unit packs of 1 kg each. Now, each pack should contain 1 gram of vitamin premix. Unless the mixing is proper, it is not possible. Some packs may contain less than one gram of the vitamin premix, while others may contain more than one gram. This means that the premix is randomly distributed in the unit packs; and such distributions are studied by quantifying a term known as **root mean square deviation** (σ), where

$$\sigma = \left[\frac{(x_1 - \bar{x})^2 + (x_2 - \bar{x})^2 + \cdots + (x_n - \bar{x})^2}{n-1}\right]^{1/2} \tag{22.1}$$

$$= \left(\frac{\sum (x_i - \bar{x})^2}{n - 1}\right)^{1/2}$$

$$= \left(\frac{\sum (x_i)^2 - \bar{x} \sum (x_i)}{n-1}\right)^{1/2} \tag{22.2}$$

where x_1, x_2, ..., x_n are compositions of different samples 1, 2, ..., *n*. σ is also known as the standard deviation, $\overline{x}$ is the average composition of all the samples and is equal to $\Sigma x_i/n$, *n* is the number of samples collected.

If the mixing process is perfect; then the deviation of *x* from $\overline{x}$ is minimum. Hence, $x_i - \overline{x}$ will be almost equal to zero. Then σ will be almost zero. In other words, we can say that smaller is the deviation, better are the mixing conditions. It is convenient to use σ^2 which is known as the **variance**.

$$\sigma^2 = \left(\frac{1}{n-1}\right)\left[(x_1 - \overline{x})^2 + (x_2 - \overline{x})^2 + \cdots + (x_n - \overline{x})^2\right] \tag{22.3}$$

Equation (22.3) can be simplified to

$$\sigma^2 = \frac{\sum (x_i)^2 - (\overline{x}) \sum x_i}{n-1}$$

We shall explain the concept with Problem 22.1.

PROBLEM 22.1 A vitamin premix of 1 kg is added to 999 kg of a food supplement. After mixing for some time in a blender, five samples each of 100 grams are collected and analyzed for the vitamin premix. The following are the results of each sample in grams:

0.11, 0.098, 0.09, 0.105, 0.12

Find the standard deviation and the variance.

Solution From the data given, we have

$$n = 5$$

$$x_1 = \frac{0.11}{100} = 0.0011$$

$$x_2 = \frac{0.098}{100} = 0.00098$$

$$x_3 = \frac{0.09}{100} = 0.0009$$

$$x_4 = \frac{0.105}{100} = 0.00105$$

$$x_5 = \frac{0.12}{100} = 0.0012$$

$$\Sigma x_i = 5.23 \times 10^3$$

$$\overline{x} = \frac{\sum x_i}{n} = 1.046 \times 10^{-3}$$

By Eq. (22.3), we have

$$\sigma^2 = \frac{\sum (x_i)^2 - (\bar{x}) \sum x_i}{n-1}$$

$$= \frac{1}{4} \times [(0.0011)^2 + (0.00098)^2 + (0.0009)^2 + (0.00105)^2$$

$$+ (0.0012)^2 - ((1.046 \times 10^{-3}) \times (5.23 \times 10^{-3}))]$$

$$= \frac{(5.5229 \times 10^{-6}) - (1.046 \times 10^{-3}) \times (5.23 \times 10^{-3})}{4}$$

$$= 1.308 \times 10^{-8}$$

Therefore, $\quad \text{variance} = \sigma^2 = 1.308 \times 10^{-8}$

$$\sigma = (1.308 \times 10^{-8})^{1/2} = 1.144 \times 10^{-4}$$

$$\text{Standard deviation} = \sigma = 1.144 \times 10^{-4}$$

22.1.1 Mixing Index for Granular Solids

Solids mixing is conveniently characterized by using the principles of probability theory. Let us say we have two components P and Q, and their relative fractional proportions are p and q. Obviously

$$p + q = 1 \tag{22.4}$$

In the unmixed state, each lump consists of either pure P or pure Q. If we randomly take a large number of samples from them, a proportion p of them would likely contain pure P, or a proportion q of them would likely contain only pure Q. Now, the variance (σ_0^2) for this probability of occurrence is given by,

$$\sigma_0^2 = p(1 - p) \tag{22.5}$$

Equation (22.5) is applicable for mixing of pastes, etc. Instead, if the two components are solid particles and mixed, and if we take any one particle, the probability that it is component P is p; and the probability that it is component Q is q. Instead of taking one particle, if we take N particles in one lot, the random variance (σ_r^2) is given by,

$$\sigma_r^2 = \frac{p(1-p)}{N} \tag{22.6}$$

Equation (22.6) is applicable for mixing of solids at zero time. Equations (22.5) and (22.6) can be used to quantify a term called **mixing index (I_s)**.

The performance of an industrial mixer is considered to be best when:

- time required for mixing is minimum,
- the product is excellent with least deviation from the average mixture composition, i.e., the standard deviation is the least, and
- the power load for mixing is also minimum.

Of these the first two parameters are described by the *mixing index*, which is defined as the ratio of standard deviation of various product samples during mixing to that at zero mixing. Mathematically, we can write it by combining Eqs. (22.2) and (22.5) as follows:

$$I_s = \frac{\sigma_r}{\sigma} = \left[\frac{(n-1)(p)(1-p)}{N\left(\sum x_i^2 - \bar{x}\sum x_i\right)} \right]^{1/2} \tag{22.7}$$

In case of mixing of pastes and doughs, the whole sample is considered as one lump only, and Eq. (22.7) is written as:

$$I_p = \frac{\sigma_0}{\sigma} = \left[\frac{(n-1)(p)(1-p)}{\sum (x_i - \bar{x})^2} \right]^{1/2} \tag{22.8}$$

I_s varies from 0 to 1; in which 0 indicates complete non-mixing, and 1 indicates fully mixed condition. It is seldom possible to get $I_s = 1.0$.

Rate of mixing: The rate of mixing under constant working conditions, viz., the mixer rotating at constant speed is quantified by using the index of mixing. Thus, the rate of mixing at any time t is proportional to the extent of mixing, and is given by

$$-\frac{dM}{dt} = K'M \tag{22.9}$$

in which M is a parameter which indicates extent of mixing and is something representative of variance, i.e., $M \propto \sigma^2$;

K' is a constant and represents the characteristic features of the mixing machine and conditions of mixing. Equation (22.9) indicates that rate of mixing falls down as the mixing time progresses.

Integrating Eq. (22.9) between the limits $M = M_0$ initially when $t = 0$ and $M = M$ at any time t, we get

$$\frac{M}{M_0} = e^{-K't} \tag{22.10}$$

Equation (22.10) can be used to predict the extent of mixing in a given time. It is explained with Problem 22.2.

PROBLEM 22.2 A fortified high protein dough is being made by adding 20 per cent soya flour to the maida flour. The two dry flours are mixed in a ribbon mixer to make the dough. After certain time, say 10 min., six samples were collected and analyzed for soya flour, the following are the fractional compositions:

0.2195, 0.22, 0.19, 0.185, 0.205, 0.191

Calculate mixing index, and standard deviations. Find how much time, it needs to be mixed for getting a variance of 1×10^{-4}.

Solution The mean composition of soya flour = 20 per cent = 0.2

$$\bar{x} = \frac{\sum x_i}{n} = \frac{(0.2195 + 0.22 + 0.19 + 0.185 + 0.205 + 0.191)}{6}$$

$$= \frac{1.2105}{6} = 0.20175$$

$$\sum x_i^2 = 0.2454$$

We calculate variance after 10 min. by Eq. (22.3)

$$\sigma_{10}^2 = \frac{0.2454 - (0.20175 \times 1.2105)}{(6-1)} = \frac{1.19287 \times 10^{-3}}{5}$$

$$= 2.38575 \times 10^{-4}$$

which may be taken as M_{10}

Therefore, $$M_{10} = 2.38575 \times 10^{-4}$$

Standard deviation, $$\sigma = \sqrt{(2.38575 \times 10^{-4})} = 0.0154$$

$$\sigma_0^2 = 0.2\ (1 - 0.2) = 0.16$$

which is also equal to $M_0 = 0.16$

$$\sigma_0 = 0.4$$

The Mixing Index (I_s) is given by Eq. (22.7)

Therefore, $$I_s = \frac{0.4}{0.0154} = 25.9$$

Now, we apply Eq. (22.10) to find mixing constant K', and hence, mixing time

$$\frac{M_{10}}{M_0} = e^{-K' \times 10}$$

i.e., $$\frac{2.38575 \times 10^{-4}}{0.16} = 1.491 \times 10^{-3} = e^{-10K'}$$

Therefore, $$K' = \frac{\ln(1.491 \times 10^{-3})}{(-10)} = 0.6508$$

If t is the time taken to reach desired mixing conditions

$$\frac{1 \times 10^{-4}}{0.16} = 6.25 \times 10^{-4} = e^{-0.6508 \times t}$$

Therefore, $$t = \frac{\ln(6.25 \times 10^{-4})}{-0.6508} = 11.33 \text{ min}$$

The mixing needs to be continued for another 1.33 min (11.33 – 10).

22.1.2 Mixing Equipment for Solids and Pastes

Mixers used for blending powders can also be used for mixing doughs and pastes. Hence, these equipments are described here together. The mixing is done by an impeller by

- tumbling,
- centrifugal smearing,
- impact,
- shearing,
- whipping,
- folding,
- stretching,
- compressing, or
- a combination of some of the above.

Different designs of impellers are available for industrial practice to ensure better mixing conditions and low energy consumption. In some designs the impellers rotate, whereas in some other designs the holding vessel rotates. Normally, the clearance between the rotor (impeller blades) and the vessel is kept minimum to ensure better mixing and to avoid stagnant zones since the solid particles and pastes are lethargic in moving unlike fluids.

Essentially the mixing equipment used with pastes and doughs are critical because of very high viscosity of the doughs. They resemble liquids, but their flowability is extremely poor. Here we describe some equipment used for mixing of solids, pastes and doughs which have extensive applications in food processing.

Planetary mixer: The planetary mixer consists of a vessel in which the agitator assembly is suspended through a planetary gear arrangement. The agitator consists of several vertical blades or fingers, and rotates on its own as well as it rotates within the vessel in a planetary motion, as shown in Figure 22.1. The blades may be bent or twisted to give better flowability to the pastes, and good mixing due to the planetary motion of the agitator. There is a provision for lifting up the whole agitator with planetary gears and the motor assembly, so that the vessel can be removed. For easy rolling of the vessel, it is provided with cart rollers. After the mixing is complete, the agitator is lifted, the blades are wiped and the vessel is removed. The contents are discharged from the vessel so that it can again be put into position for mixing another batch, or another similar vessel is replaced. Before feeding the next batch, the agitator is thoroughly cleaned.

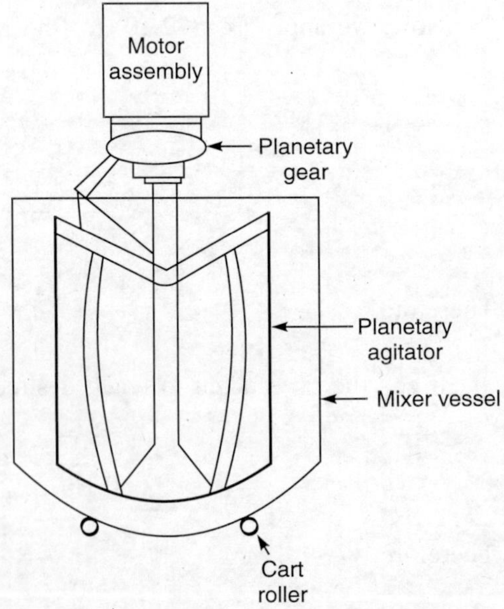

Figure 22.1 Planetary mixer.

The planetary motion enables the agitator blades to reach all parts of the vessel to mix the contents. The clearance between the agitator and vessel is kept to minimum to avoid any stagnant zones. Hence, planetary mixers are used to mix dry powders, particularly if one of the solid components is in minute quantities. They are also used for mixing slurries, thick doughs, etc.

Kneader: Kneading equipment is frequently used in bakery industry for mixing various components of the dough along with fat and make it a homogeneous mixture. Wherever, heavy materials are involved, or pasty and sticky materials are involved, kneaders are invariably used. The kneading equipment consists of two arms of special shape which rotate in opposite directions at different speeds as shown in Figure 22.2 in a container. Their rpm could be in the ratio of 3 : 2. The container could be of any shape and size. The design is mostly based on the curvature and concavity of the kneading blades. The container vessel is only meant for holding the contents and the kneading blades. Usually the vessel has a saddle-shaped bottom. The clearance between the blades and vessel walls is considerable. It is not like a typical scraping equipment where the clearance is the least.

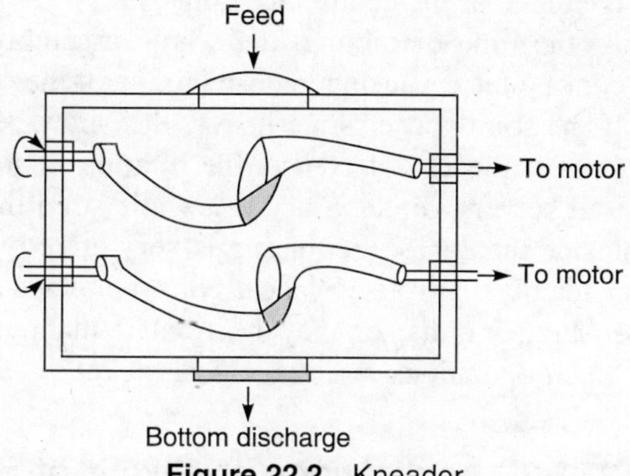

Figure 22.2 Kneader.

The kneader blades first squash the mass flat, and then roll it on itself, and squash once again. It tears the mass apart, and shear it in between the moving blades and the stationary surface walls. The mass becomes stiffy and rubbery during mixing, and hence, the energy consumption is quite large.

The effectiveness of mixing depends upon the kneader blade design. Sometimes they are made very rugged, and of different shapes other than those in Figure 22.2. The edges of the blades may be more serrated to give higher shredding action.

For heavy-duty operations, continuous kneaders are also available in the industry. But generally in food processing we use batch kneaders. In case of higher requirement of outputs, the units are multiplied by adding more batch units.

Ribbon mixer: Ribbon mixers are generally used for blending solids. The impeller or agitator is in the form of a ribbon and rotates horizontally in the tank (Figure 22.3). Since the ribbon consists of a large amount of empty space, the material slips through the empty cavity, and hence, the energy consumption is less, but still it ensures efficient mixing. The orientation of the ribbon blade ensures better mixing conditions.

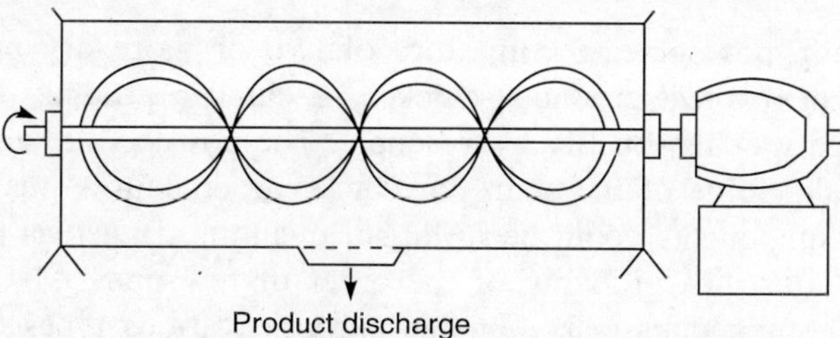

Figure 22.3 Ribbon mixer.

Double cone mixer: The double cone mixer is a simple and easy to operate equipment for mixing of small components, particularly if one of the components is in very small quantity, and the other components have bigger particle sizes. They are used to coat or sprinkle salt and spice mix to wafers and potato chips. They are also used for mixing different components of savoury mix items or snack items.

It consists of two cones joined at the centre with flat top and bottom as shown in Figure 22.4. The double cone is hinged at the centre, and rotates on a horizontal axis. At the time of mixing, the whole assembly of double cones rotates, causing intense mixing of the solids inside. If the solid particles are heavy, they may be sheared to smaller pieces. Of course, the rotation is not intense, it will be very gentle. Since the whole solid mass tumbles inside the cones, the mixing is very effective. The cone assembly may rotate by means of a motorized arrangement in bigger units, or may be rotated manually in small batch operations at cottage scale level.

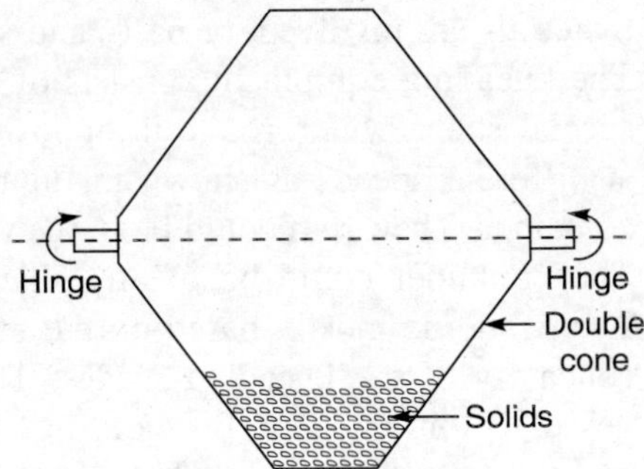

Figure 22.4 Double cone mixer.

22.1.3 Applications of Mixing of Solids in Food Processing

Applications of mixing of solids and pastes in food processing are:

Energy foods and weaning foods: Energy foods and weaning foods are a high-volume business industry with low-profit margins. There are continuous mixers which have capacities to process 1–2 tons per hour of dry powders. They are generally a mixture of

- carbohydrates (cereal)
- protein (pulse and oil seed cake)
- energy source (jaggery powder/sugar)

The cereals and pulses are separately roasted and ground to fine powders which are mixed with oil seed cake powder and jaggery powder or sugar. All the ingredients are mixed. Sometimes to fortify the energy food or baby foods, they are mixed with minerals and some micro-nutrients in ppm level. All these are thoroughly mixed in a double cone mixer or ribbon mixer depending upon the plant capacities.

Now, automatic weighing and dumping (feeding) units are available to mix and pack 50–100 tpd of products conveniently.

Soup powders: Soup powders are a mixture of two or more dry powders, made out of fruits, vegetables, meat or meat products, chicken or chicken products, or egg powders. The powders could be highly or reasonably hygroscopic. The powders are also mixed with spices/spice powders and salt. Some of the soup powders even contain a small quantity of vitamin premix. Some solid components could be in higher quantities, whereas the others could be in small or very small quantities probably of the order of few ppm. Mixing of all of them in even proportions is a stupendous task. For this purpose, various types of mixing equipments are used for proper and even mixing of various proportions of solids.

Chips and snacks industry: The chips and snacks industry of late, has grown into a big multi-product industry. Obviously in this, the snack items and chips are much bigger in shape as compared to the masala powders or natural flavour powders (viz., onion powder or tomato powder) to be mixed. In fact, the powders are sprinkled over the chips, and are rotated in big pans generally used for roasting except that the heat is not applied to the roasters. One convenience with this types of products is that strict adherence to uniform mixing conditions need not be there. A reasonable amount of mixing would do the job.

Poultry and cattle feeds: Poultry and cattle feeds are generally a mixture of various solid ingredients which are sources of carbohydrates and proteins. Animal based powders are also added in case of poultry feed to make them more palatable. The ingredients are seldom fine powders. Poultry feeds are made even by using extrusion techniques to give body and shape to the feed in the form of worms, so that the feed is relished by poultry. Any kind of usual ribbon mixers can be used since none of the ingredients are in smaller quantities. Most of them are of equal proportions; and hence, mixing is not a difficult task.

Bakery industry: Biscuit and bakery industry is one of the largest applications where the pastes and doughs of huge quantities are mixed before supplying to the leavening rooms for solid state fermentation to take place. Planetary mixers and kneaders are generally used in the bakery industry. Planetary mixtures are used in the bread making to knead the dough. Here the individual ingredients in the solid powder form are thoroughly mixed, and later they are mixed with water to knead the dough. Thus, the planetary mixer thoroughly mixes the solid powders and water so that the dough has even consistency throughout, which ferments better during leavening stage as the bakers yeast is also evenly distributed.

22.2 MIXING OF FLUIDS

We come across mixing of fluids in most of the food processing operations and fermentation processes. It could be mixing of

- two immiscible liquids to make emulsions,
- gas and liquid for physical absorption for making carbonated beverages,
- gas and liquid for physical absorption followed by chemical reaction as in dehydrogenation of oils and fats, and
- oxygen transfer in fermentation broths, etc.

Thus fluid-fluid mixing is an important unit operation in food biotechnology. We show a typical fermenter with all mixing accessories in Figure 22.5. Mixing becomes more complicated in case of viscous fluids like fermentation broths which consist of

- a nutrient medium,
- a substrate,
- cells in suspension, and
- gas bubbles.

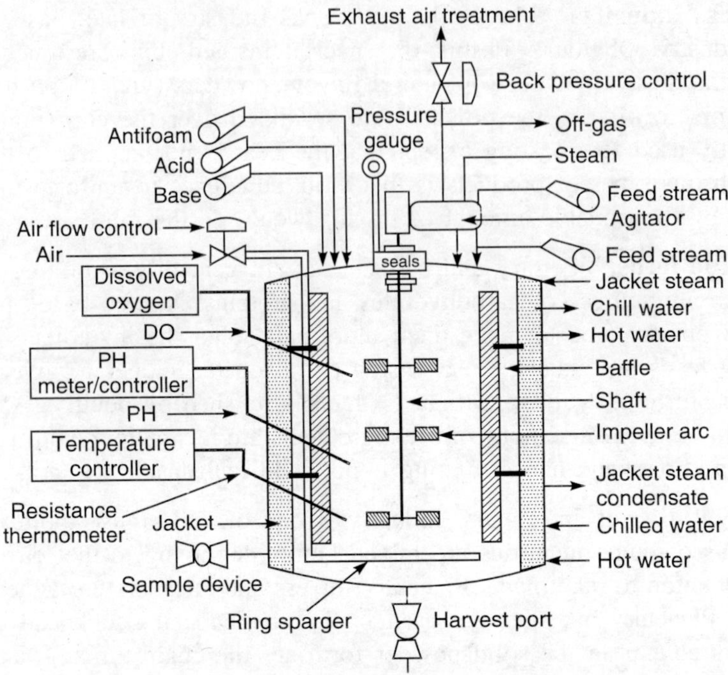

Figure 22.5 Representation of a jacketed fermenter.

Source: Reprinted from *Encyclopaedia of Chemical Technology*, Sengha, S.S. (Ed.) 4th ed., Kroschwitz, J.I., Vol. **10**, p. 373. Copyright (1993). This material is used by permission of John Wiley and Sons, Inc.

In all these processes, mixing serves two purposes:

- better mixing of different components evenly, and
- better heat transfer from the heating medium to various components in the mixture.

Various types of mixing equipments are used to achieve better degree of mixing. They all essentially contain a vessel (as shown in Figure 22.5) with an agitator/impeller to mix the components. The agitator is rotated using a motor at the top with a gear arrangement to achieve desired mixing conditions. For this purpose, various types of impellers are designed for better liquid-liquid mixing, and gas-liquid mixing.

Various types of impellers/agitators used in food processing operations are shown in Figure 22.6 (Rao 2005).

22.2.1 Power Required for Mixing

The entire theory of mixing revolves around the concept of power required for mixing or power required for rotating the impeller, which is rotated by an electric motor. Hence, it is the evaluation of wattage of power required by the motor for rotating the impeller when its blades impinge and overcome the viscous forces of the liquid broth that give resistance for the rotation of impeller blades. Overcoming of the shear resistance gives movement for the impeller blades. The power required by the impeller is thus, a function of

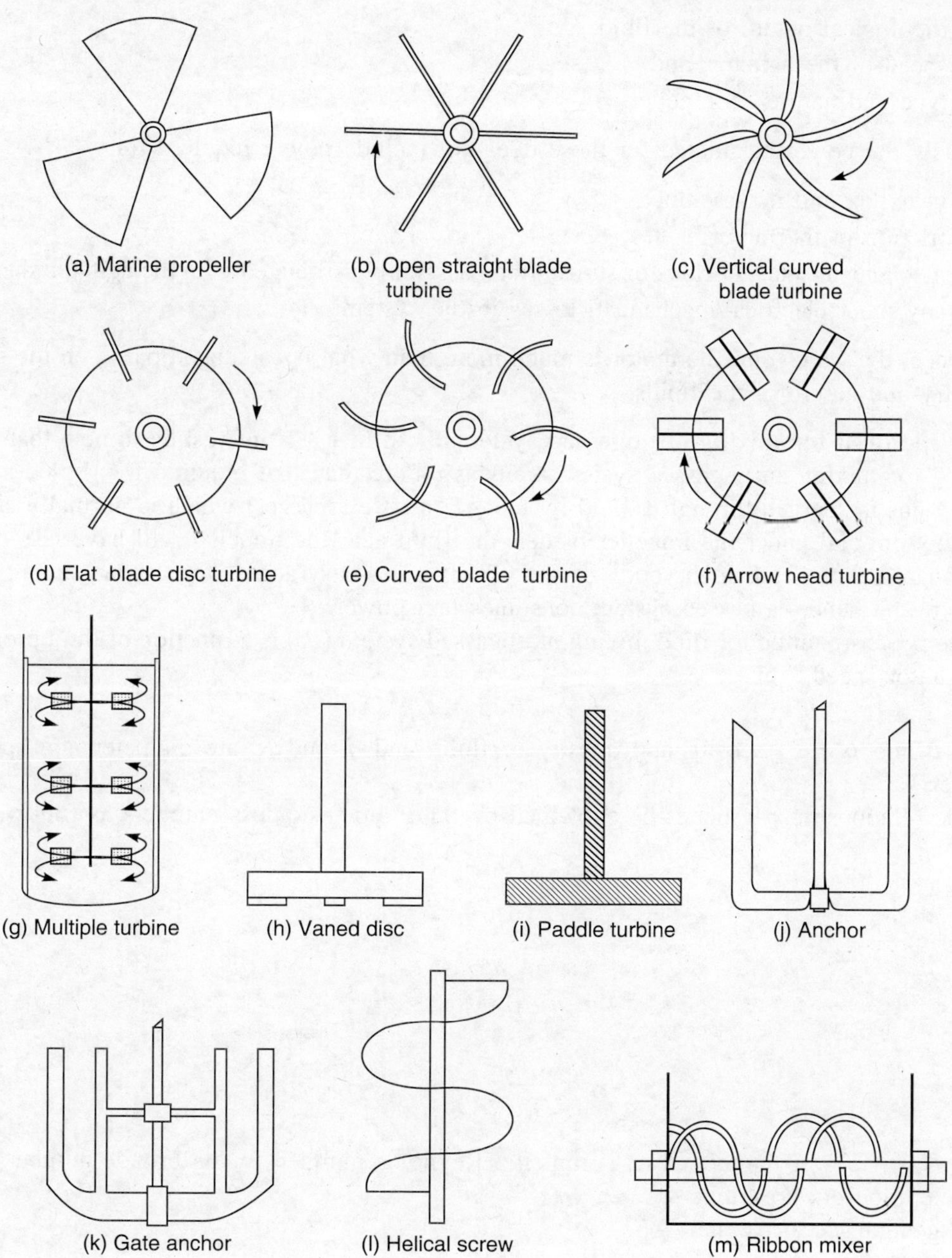

Figure 22.6 Various types of agitators.

Source: Reprinted from *Introduction to Biochemical Engineering*, Rao, D.G., Tata McGraw-Hill, New Delhi, p. 282. Copyright (2005). This material is used by permission of Tata McGraw-Hill.

- viscosity of the fluid,
- density of the fluid,

- rheological nature of the fluid,
- speed of the agitator, and
- type and nature of impeller.

Obviously the power calculated for these does not include power required for

- gear box and its assembly,
- friction in the motor,
- mechanical seals, O-rings or stuffing box assembly associated with the agitator shaft, and
- any other electrical/mechanical losses in the system.

Hence, the wattage of the motor is much more than what is calculated based on the system geometry and nature of the fluids.

Power required for mixing of ungassed systems: In fact it is interesting to note that power required for mixing an ungassed system is always more than that required for gassed system. When a gas is sparged through a fluid by using a nozzle (sparger) which is normally situated at the bottom and under the impeller blades, the fluid near the impeller will have less density as compared to the rest of the body of fluid. This gives easy access for the movement of the impeller, and hence, a gassed system consumes less power.

The power required for fluid mixing of ungassed system (P_0) is a function of fluid properties and agitator speed. Thus,

$$P_0 = f(\mu, \rho, N_i, d_i) \tag{22.11}$$

where μ and ρ are viscosity and density of fluid, and d_i and N_i are diameter and speed of impeller.

The relationship can better be explained by three dimensionless numbers as follows:

$$N_{P0} = \frac{P_0}{\rho N_i^3 d_i^5} \tag{22.12}$$

$$N_{Re,i} = \frac{\rho N_i d_i^2}{\mu} \tag{22.13}$$

$$N_{Fr} = \frac{d_i N_i^2}{g} \tag{22.14}$$

where N_{P0} is power number, $N_{Re,i}$ is impeller Reynolds number, N_{Fr} is Froude number, and g is acceleration due to gravity.

In general we can write,

$$N_{P0} = C(N_{Re,i})^{\alpha} (N_{Fr})^{\beta} \tag{22.15}$$

where C is a constant, and α and β are coefficients.

The Froude number [Eq. (22.14)] correlates the gravitational effect which is significant when the liquid is agitated by a propeller. If $N_{Re,i} < 300$, N_{Fr} is ineffective. Equation (22.15) becomes

$$N_{P0} = C\,(N_{Re,i})^{\alpha} \text{ if } N_{Re,i} < 300 \tag{22.16}$$

If $N_{Re,i} > 300$, the power number my be written as

$$N_{P0} = \left(\frac{P_0}{\rho N_i^3 d_i^5}\right)\left(\frac{1}{N_{Fr}}\right)^{\left(\frac{a - \log N_{Re,i}}{b}\right)} \tag{22.17}$$

where a and b are constants, and the values for common impellers were given by Skelland (1967), and are given in Table 22.1.

Table 22.1 Geometry of the Gas-liquid Systems Considered

Impeller dia/tank dia	1/3
No. of baffles	4 or 6
Baffle width/tank dia	1/10
Liquid height/tank dia	1
Tank height/tank dia	1.5
Various types of impellers	Rushton turbine (a = 1.0, b = 40, C_1 = 70, N'_{P0} = 5.0) Paddle (C_1 = 35, N'_{P0} = 2.0) Marine propeller (a = 2.1, b = 18.0, C_1 = 40, N'_{P0} = 0.35) Anchor (C_1 = 420, N'_{P0} = 0.35) Helical ribbon (C_1 = 1000, N'_{P0} = 0.35)

a and b values are from Skelland (1967)
C_1 values are from Doran (1995)
N'_{P0} values are from Rushton, et al. (1950)

The relationship between the power number and impeller Reynolds number has become a subject of great interest from the classical works of Rushton, et al. (1950) who presented the dependence of N_{P0} on $N_{Re,i}$ in the form of graphs for various types of impellers. Some details of various systems considered are presented in Table 22.1. The power correlation for propellers were reported by Bates, et al. (1966), and are shown in Figure 22.7 and that for Rushton turbine impeller designs in Figure 22.8 (Bates, et al., 1963). Rushton, et al. (1950) had also given power characteristics of mixing impellers. They compared the impellers, viz., propeller, flat blade and turbine, and presented the data in the form of graph drawn between power number and impeller Reynolds number (Figure 22.9). The plots are typical, and can be divided into the following three zones:

***Zone* 1.** In this zone where $N_{Re,i} \leq 10^2$, the power number continuously falls as $N_{Re,i}$ increases. This is considered as laminar flow zone. Here we notice

$$N_{P0} \propto \frac{1}{N_{Re,i}}$$

i.e.,

$$\frac{P_0}{\rho N_i^3 d_i^5} \propto \frac{\mu}{\rho N_i d_i^2} \tag{22.18}$$

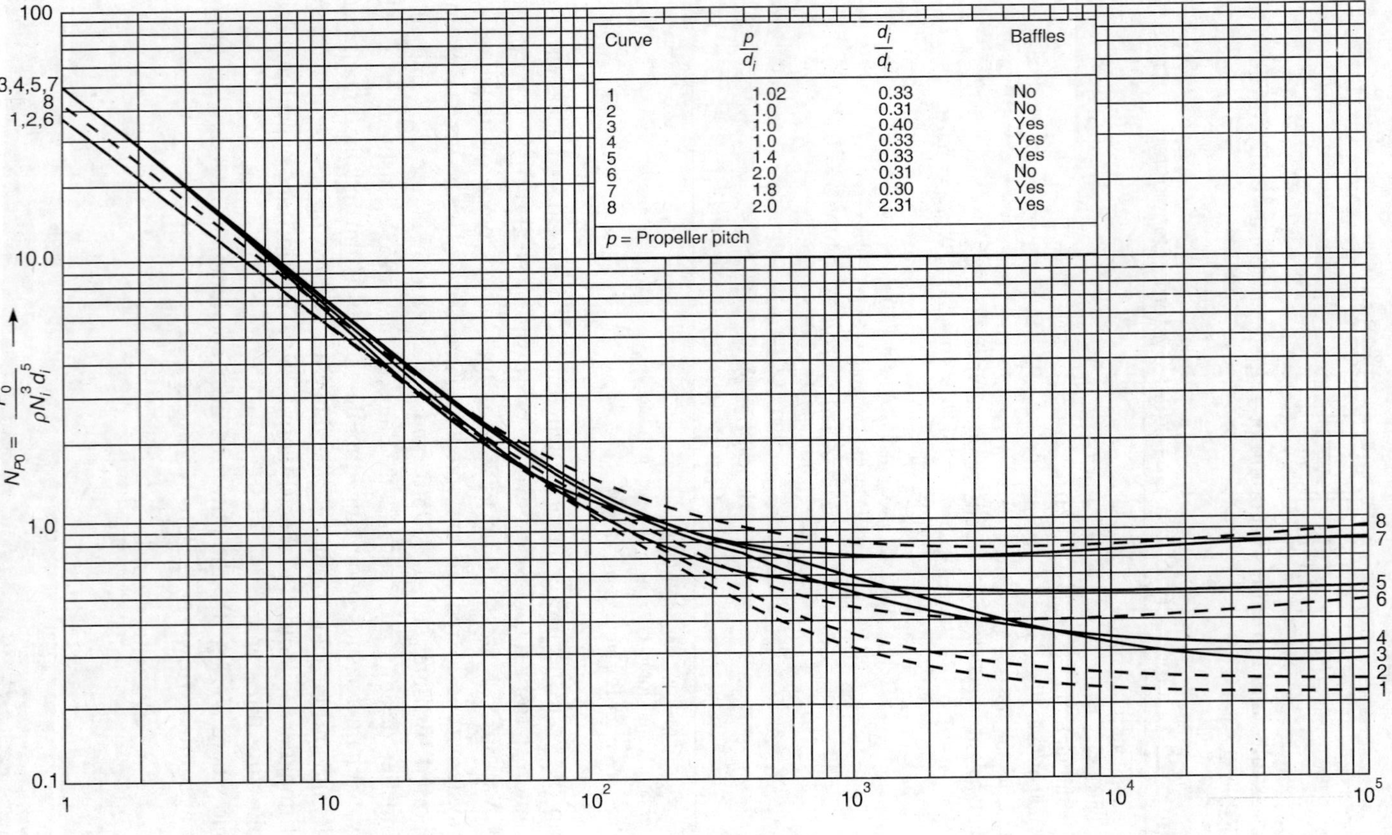

Figure 22.7 Correlation of power number for a three-blade marine propeller.

Source: This figure was published in *Mixing: Theory and Practice*, Vol. I, Uhl, V.W. and Gray, J.B. "Impeller Characteristics and Power," p. 130, Copyright Elsevier (1966).

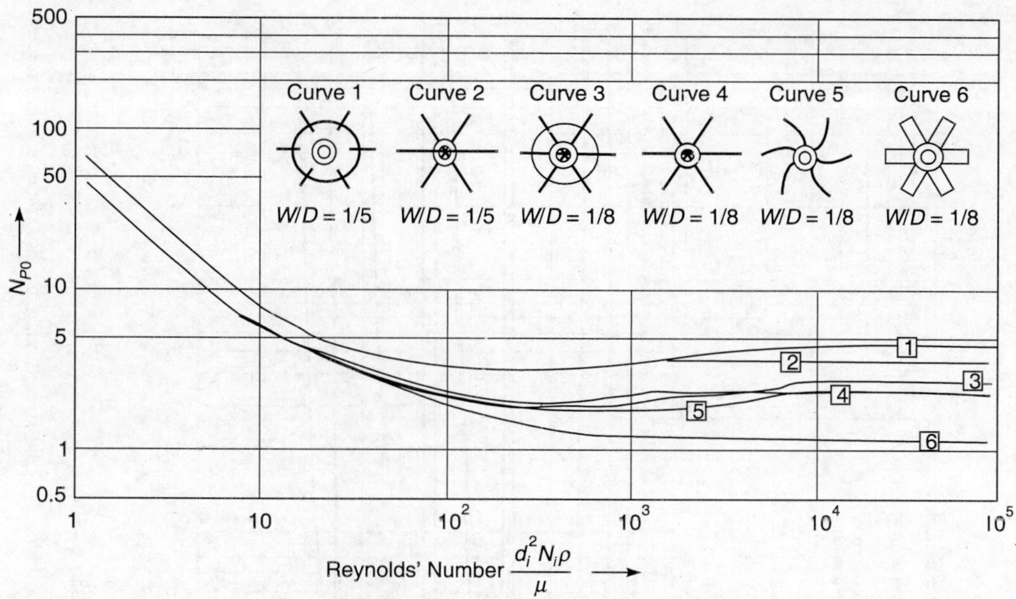

Figure 22.8 $N_{Re,i}$ vs. N_{P0} for various impellers.

Source: Reprinted from *An examination of some geometric parameters of impeller power*, Bates, R.L., Fondy, P.L. and Corpstein, R.R., p. 311, I&EC., Process Des. Develop., **2**(4), Copyright (1963), with the permission from American Chemical society.

which on simplification can be written as:

$$P_0 = C_1 \mu N_i^2 d_i^3 \tag{22.19}$$

where C_1 is the proportionality constant. Different values for C_1 were reported in literature (Doran, et al., 1995) and are shown in Table 22.1 for $N_{Re,i}$ in the range of 100. It is also interesting to note that the power input is not dependent on density in the laminar zone. It depends upon the viscosity of the fluid only.

***Zone* 2.** It is the transient zone in which the plots are curved. The zone extends from $N_{Re,i} > 10^2$ to $N_{Re,i} < 10^4$. It is in between laminar and turbulent zone; and the plots are curved. Power input depends upon viscosity and density of the fluid.

***Zone* 3.** In this zone, we notice that virtually $N_{Re,i}$ has no effect on power number, and the latter is constant. The $N_{Re,i} > 10^4$, and this zone is known as **turbulent zone.**

$$N'_{P0} = \text{Constant} = \frac{P_0}{(\rho N_i^3 d_i^5)} \tag{22.20}$$

Therefore,

$$P_0 = N'_{P0}\,(\rho N_i^3 d_i^5) \tag{22.21}$$

The approximate constant values of power number for various types of impellers were reported by Rushton, et al. (1950).

N'_{P0} = 5 for Rushton turbine

= 0.35 for propeller, anchor and helical ribbon

= 2 for paddle impeller

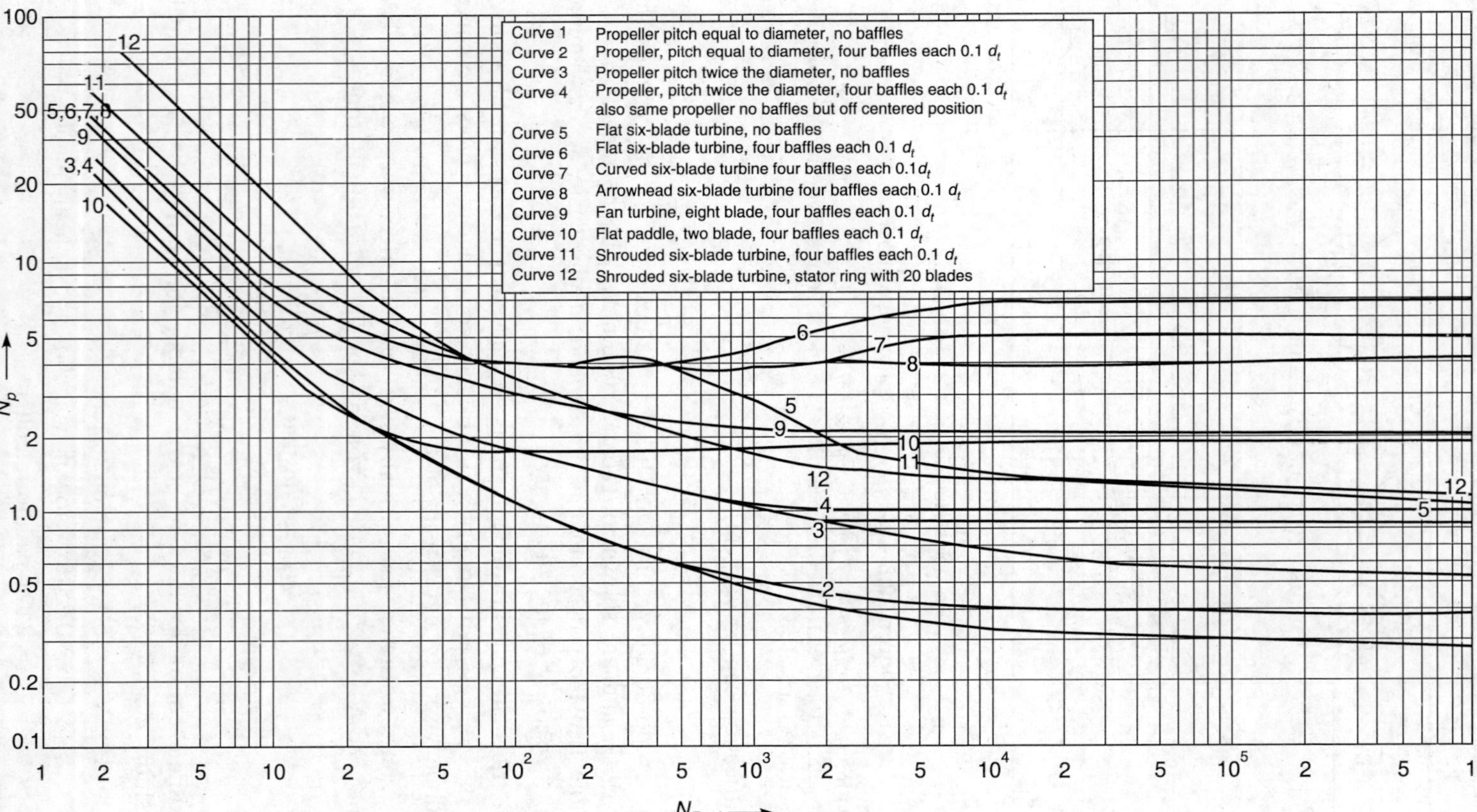

Figure 22.9 Comparison of impellers.

Source: Reprinted from *Power characteristics of mixing impellers: Part-II*, Rushton, J.H., Costich, E.W. and Everett, H.J., p. 470, Chem. Engg. Progress, Copyright (1950), with the permission from AIChE.

PROBLEM 22.3 A vitamin premix is blended with a fermentation broth in a 1 litre agitated tank and is agitated by a standard Rushton type turbine impeller. The vessel geometry is as per standard description. The impeller has six blades (curve 1, Figure 22.8), and is rotating at 10 rps. The fluid viscosity is 80 cp, and its density is 1050 kg/m^3. Calculate the power required to run the impeller.

Solution

Given data:

$$d_i = 3.6 \text{ cm} = 0.036 \text{ m}$$

$$N_i = 10 \text{ rps}$$

$$\mu = 80 \text{ cp} = 80 \times 10^{-3} = 8 \times 10^{-2} \text{ kg/ms}$$

$$\rho = 1050 \text{ kg/m}^3$$

Therefore,

$$d_t = 3d_i = 10.8 \text{ cm} = 0.108 \text{ m}$$

$$N_{Re,i} = \frac{\rho N_i d_i^2}{\mu} = \frac{1050 \times 10 \times (0.036)^2}{8 \times 10^{-2}} = 194$$

Since it is in the transient zone, we find N_{PO} by Figure 22.8.

$$N_{PO} = 3.5$$

Therefore,

$$3.5 = \frac{P_0}{\rho N_i^3 d_i^5} = \frac{P_0}{1050 \times (10)^3 \times (0.036)^5} = \frac{P_0}{0.063}$$

$$P_0 = 0.22 \text{ watts}$$

Power requirement for non-Newtonian fluids: So far we have studied the power requirements with Newtonian fluids. But incidentally most of the food materials and emulsions are non-Newtonian in nature. Let us see how we can use this concept for non-Newtonian fluids. As has been already mentioned in Chapter 9, the non-Newtonian fluids are characterized by the fact that they change their viscosity as the shear rate changes. Hence, we cannot use the term μ in finding impeller Reynolds' number by Eq. (22.13). Hence, we use a term called apparent viscosity (μ_a) in place of μ. The apparent viscosity is as defined earlier by Eq. (9.9).

$$\mu_a = K\gamma^{n-1} \tag{9.9}$$

where K is consistency index, n is flow behaviour index and γ is shear strain.

Now, we replace μ in Eq. (22.13) by μ_a as defined by Eq. (9.9). Hence,

$$(N_{Re,i})_N = (N_{Re,i})_{\text{non-Newtonian}} = \left[\frac{\rho N_i d_i^2}{K\gamma^{n-1}}\right] \tag{22.22}$$

Once the impeller Reynolds' number for the non-Newtonian fluid is calculated by Eq. (22.22), we can use the same in Figures 22.7 and 22.8 to measure the power number, etc. which in turn helps us calculate the power requirement for agitation by using various equations [Eqs. (22.18) to (22.21)].

The power number and Reynolds' number relationships for non-Newtonian fluids were well reported by Metzner, et al. (1961) in their classical works on non-Newtonian rheology. This work describes *agitation of viscous and Newtonian, and non-Newtonian fluids* by using a six-blade Rushton turbine impeller for a pseudo-plastic fluid in a baffled tank. The data for Newtonian and non-Newtonian fluids were compared. The data are shown in Figure 22.10. In the laminar zone when $N_{Re,i} < 10$, or $(N_{Re,i})_N < 10$, the results are virtually same. Similarly in the turbulent zone when $N_{Re,i} > 100$, the same trend continues. However, in the transition zone, when $N_{Re,i}$ or $(N_{Re,i})_N$ is in between 10 to 100, the power number for non-Newtonian fluids is slightly lower than that for Newtonian fluids. The data in Figure 22.10 is to be used with caution. Errors creep in during calculations mainly because of the difference in the flow pattern near the agitator and that in the rest of the mixer. For design purposes, however, the data in Figure 22.10 is adequate.

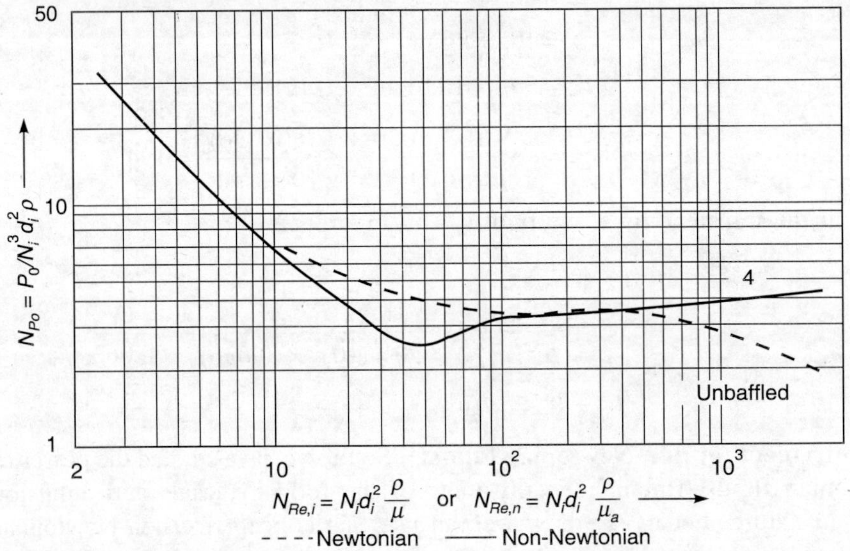

Figure 22.10 Agitation and mixing of liquids in unaerated mixing tank with Rushton turbine.

Source: Metzner, A.B., Feehs, R.H., Lopez Ramos, H., Otto, R.E. and Tuthill, J.D., *Agitation of viscous Newtonian and non-Newtonian fluids*, AIChEJ Vol. 7(1), 1961, p. 2, Copyright (1961), American Institute of Chemical Engineers. Reprinted by permission of John Wiley and Sons, Inc.

Power required for mixing of gassed systems: As has been already mentioned, the hydrodynamics in the vicinity of the sparger (and hence, in the vicinity of the impeller) for a gassed system will be different from the rest of the system because of the changes in density. The dissolved gas would reduce the density of the fluid around the impeller. Hence, it requires less power for stirring fluid. Thus, the power required for mixing a gassed system will be always less than that for ungassed system. For a turbine impeller, the power required by both the systems can be related as follows (Hughmark 1980):

$$\frac{P_g}{P_0} = 0.1\left(\frac{v_g}{N_i V}\right)^{-0.25}\left(\frac{N_i^2 d_i^4}{g w_i V^{2/3}}\right)^{-0.2} \tag{22.23}$$

where P_g is the power input for gassed system, v_g is volumetric gas flow rate, w_i is the width of the impeller, and V is the liquid volume (m^3).

Equation (22.23) is useful in finding out P_g/P_0 by knowing the dimensions of the impeller and system parameters.

P_0 can be found out by application of Eqs. (22.19) or (22.21). Hence, P_g can be determined using Eq. (22.23).

Thus, the power requirement of stirring a gassed system can be determined with a knowledge of P_0 (which is power required for stirring the ungassed system).

We define here dimensionless parameter known as **aeration rate** (N_a).

$$N_a = \frac{v_g}{N_i d_i^3} \tag{22.24}$$

Ohyama and Endoh (1955) correlated the experimental data on (P_g/P_0) vs. N_a for various types of impellers:

- flat blade turbine,
- vaned disc, and
- paddles.

Experimental data could be helpful for finding out P_g/P_0 with a knowledge of N_a. The values of N_a can be calculated using air flow rate, and diameter and speed of the impeller. Once if P_g/P_0 is known, P_g can be calculated with a knowledge of P_0. This method differentiates between various impellers, whereas the application of Eq. (22.23) does not differentiate the type of impeller used.

Van't Riet (1979) in a review article on non-viscous gas-liquid mass transfer in stirred vessels indicated that

(i) The type of stirrer had no effect in mass transfer.
(ii) The position of stirrer had no effect in mass transfer.
(iii) The number of stirrers had no effect in mass transfer.
(iv) The type of sparger had no effect in mass transfer.

But, it is not clear to what extent these observations are correct for viscous systems, and incidentally most of the food processing systems are viscous systems. However, it is difficult to believe that the type of stirrer had no effect on mixing characteristics in terms of mass transfer.

22.2.2 Mixing Equipment for Gas-liquid Systems

The mixing equipment essentially consists of a vessel to which the agitator is fitted. Hence, the description and design of mixing equipment are nothing but the choice of the proper impellers. Various types of impellers used for gas-liquid mixing are shown in Figure 22.6. The impellers *a–i* in Figure 22.6 are generally used for

- gas-liquid mixing, or
- liquid-liquid mixing

The impellers *j–m* are usually used for mixing of solids/pastes or for mixing very viscous liquids like corn starch, etc.

22.2.3 Applications of Gas-liquid Mixing in Food Processing

As has already been mentioned, the gas-liquid mixing systems are used in food processing for two purposes:

(i) for physical absorption of a gas in liquid, and
(ii) for absorption of a gas in a liquid for chemical reaction to take place.

Almost all the aeration and carbonation units come under the category of physical absorption. Particularly the absorption of CO_2 in water or in fruit juices is used for carbonation purposes. The carbonation helps in improving the storage life and in quenching the thirst. Normally this absorption is not done by mixing but by increasing the pressure of the absorption system as a whole. Thus, the absorption column is more a pressure column rather than a mixing column.

Chlorination of water is another classical example of physical absorption. Chlorinated waters are used in most of the process industries for disinfestation purposes. The chlorine gas in desired quantities is absorbed in liquid water by using some turbine impellers or a simple marine propellers.

Some of the dipping oil emulsions are used for easy dehydration of fruits/vegetables by giving a thin layer of coating on the fruit berries. Particularly grapes are dehydrated to make raisins. Before dehydration, the grape berries are dipped in a dipping oil emulsion. The dipping oil is made as an emulsion by mixing some fatty acid esters with 2.5 per cent aqueous potassium carbonate solution by using high speed marine propeller impellers (Giridhar, et al., 1999).

Ice creams are made by mixing air to an aqueous cream mixture with other ingredients of ice cream. Later the whipped mix with incorporated air is frozen. The whipping is done by a high speed homogenizer.

The other type of systems is absorption of a gas in a liquid system to enable the gas to react with the liquid, what we study under the category of mass transfer with chemical reaction. Oils and fats are hydrogenated by absorption of hydrogen gas into oils under high pressure in the presence of raney nickel as catalyst. Vaned disc type impellers are used for good gas-liquid contacting. The absorbed hydrogen in the oil reacts with the oils in the presence of nickel catalyst to saturate the unsaturated bonds in the oils. The saturation of chemical bonds by hydrogen makes the oil into a semi-solid fat which has got good applications in bakery and confectionary industry.

22.3 EMULSIFICATION

Food emulsifications have a great role to play in food industries. With increasing industrialization in the food sector, novel foods have been entering the market. Most of them need the use of emulsions in some form or other. Emulsions are formed by mixing of two immiscible liquids, usually one component in small quantities. Oil-in-water (O/W) emulsions are created by dispersing edible oils or fats in water. Milk is a classic example of a stable emulsion of fat in water.

Most of the ready to eat (RTE) foods and convenience foods contain a good amount of meat emulsions. The developments in new foods have given rise to quite a good number of emulsions. The use of monoglycerides in the margarine industry was in vogue for more than a century. Stabilized emulsions are needed to modify the structure and appearance of foods, or their rheological properties as per the choice of the customers or to increase the shelf-life of finished products or to reduce the fat content of a variety of foods. Still the emulsions remained more as an art than a science. Emulsification is a cumbersome process and is often assisted by an external agent known as emulsifier (Emulgent). In fact, most of the research in this area is towards devising various emulsifiers to keep the emulsions stable without separating into two layers.

The whole problem with emulsions is their thermodynamic instability since the interfaces formed between the immiscible liquids on a macroscale depend on a small but a positive energy for their existence. The emulsions are broken because of coalescence of droplets or their flocculation. This results in creaming, and finally the whole emulsion breaks into two separate phases (Figure 22.11). But the emulsions are expected to be stable (stay mixed) for a very long periods of time (sometimes even a year or so in case of emulsion paints, etc.). We have seen in the earlier section; the system need not be stable for very long periods. Once the reaction or process is over, the system need not be stable. Even in case of gas-liquid systems such as carbonated beverages, they are kept stable by applying external pressure. Once the pressure is released, they separate out into gas and liquid.

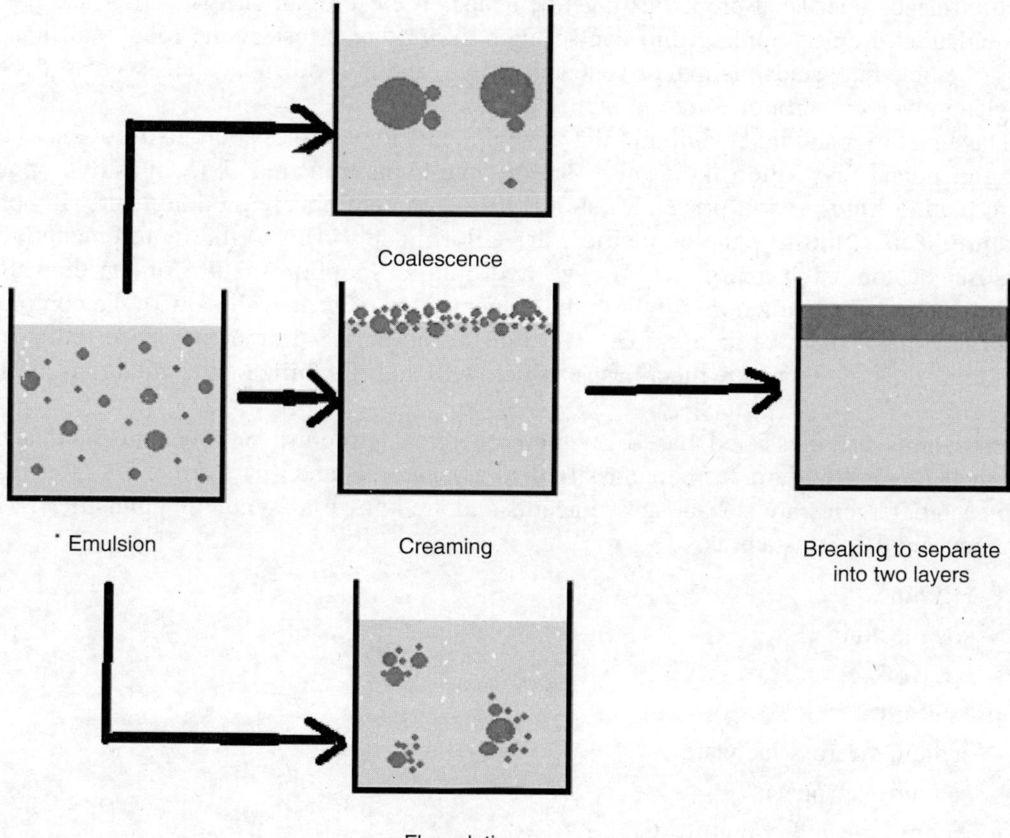

Figure. 22.11 Emulsion breaking process.

As a unit operation, emulsions are stable suspensions of two immiscible liquids, which are made stable by dispersing one liquid into the other in the form of fine droplets usually of the order of 80–250 nm sizes. The liquid in small quantities and is dispersed is known as *dispersed phase* and the other liquid in large quantities is known as *continuous phase.*

22.3.1 Emulsifiers

The two immiscible liquids do not remain stable on their own, however well we disperse the dispersed phase. The dispersed phase droplets will tend to coalesce and form bigger droplets which ultimately would settle down. Hence, in many instances the emulsions are made stable by adding a third component known as emulsifier (Emulgent) in very small quantities. The emulsifier (or emulsifying agent) is a surface-active component that stabilizes the emulsion by increasing its kinetic stability. It is adsorbed at the newly formed interface between the dispersed phase and continuous phase in the form of coating on the droplets. Thus, it prevents the immediate recoalescence of droplets. The recoalescence is the main culprit to result in separation of phases and destabilize the emulsion. Hence to make stable emulsions, substances which influence interfacial properties must be added. They form an energy barrier and prevent the emulsified droplets coming into contact with each other. Most of the food emulsions are esters of edible fatty acids (animal or vegetable origin) and polyvalent alcohols such as glycerol, propylene glycerol, sorbitol, sucrose, etc.

The emulsifier should be amphiphilic in nature, i.e., they possess both hydrophilic (water-loving or polar) lipophilic (liking oils). The lipophilic nature is also known as hydrophobic nature (hating water or non-polar). Most of the food emulsifiers are predominantly lipophilic in nature. Hence, most of the emulsifiers are tailor-made to suit to the requirements of the industrial customers (Friberg, 1976). The hydrophilic-lipophilic nature of the emulsifiers gives them an edge to anchor with both the dispersed droplets as well as with the continuous phase and make them remain stable. The interaction at the droplet interface reduces the interfacial tension between the phases which will help stabilize the emulsions without coalescing.

Emulsions are classified based on their chemical structure or mechanism of action. Chemical emulsifiers are further classified as (i) natural, and (ii) synthetic (Table 22.2). Soaps and detergents are surface active agents and are known as synthetic emulsifiers. There are many emulsifiers such as:

- lecithin
- soya lecithin
- egg yolk
- Mustard
- sodium stearoyl lactylate
- sodium phosphates
- carboxy methyl cellulose (CMC)
- diacetyl tartaric acid esters of monoglycerides and diglycerides (known as DATEM)

Table 22.2 Common Food Emulsifiers*

Common Name	*Label Declaration*	*Origin*	*Applications*	*Functionality*	*Direct Food Substances Affirmed as GRAS*	*Food Additives Permitted for Direct Addition to Food for Human Consumption*
Lecithin	Lecithin	Natural (phospho-lipids from egg yolk, soybean oil)	• Chocolate/ coatings/ confectionery • Baked goods • Margarines/ spreads • Beverages • Ice cream	• Rheology modification • Dough conditioner/ strengthener • Increased shelf-life	21CFR184.1400	
Mono-Di	Mono- and Diglycerides	Synthetic	• Baked goods/ cakes • Icings • Margarines • Shortenings	• Fat dispersion • Aeration • Emulsion stability	21CFR184.1505	
Distilled Mono	Mono-glycerides	Synthetic	• Breads/rolls • Batters • Cakes • Cake Mixes	• Crumb softener • Fat dispersion • Emulsion stability	21CFR184.1505	
Hydrated Mono	Mono-glycerides	Synthetic	• Breads/rolls • Batters • Cakes	• Instant functionality • Higher efficacy	21CFR184.1505	
LACTEM	Glyceryl-Lacto Esters of Fatty Acids	Synthetic	• Fillings/icings	• Aeration • Fat dispersion • Foam stabilization		21CFR172.852 GMP
Ethoxylated Mono	Ethoxylated Mono- and Diglycerides	Synthetic	• Breads/rolls	• Dough conditioner/ strengthener • Improves crumb softness		21CFR172.834
DATEM	DATEM	Synthetic	• Breads/rolls • Biscuits/ cookies	• Dough conditioner/ strengthener • Improves crumb softness	21CFR184.1101	
SSL	Sodium Stearoyl Lactylate	Synthetic	• Breads/rolls • Batters • Biscuits/ cookies • Fillings/icings • Dairy	• Dough conditioner/ strengthener • Improves crumb softness • Tightens cell structure • Fat reduction • Emulsion stabilization		21CFR172.846

(contd.)

Table 22.2 Common Food Emulsifiers* (*contd.*)

Common Name	*Label Declaration*	*Origin*	*Applications*	*Functionality*	*Direct Food Substances Affirmed as GRAS*	*Food Additives Permitted for Direct Addition to Food for Human Consumption*
PGME	Propylene Glycol Mono- and Diesters of Fats and Fatty Acids	Synthetic	• Bakery ingredients/ mixes • Cakes/ pastries/ sweet goods • Snack cakes • Frozen dairy	• Aeration/ emulsion stabilization • Fat reduction • Moisture retention • Increased shelf-life		21CFR172.856 GMP
Whey Protein	Whey Protein	Natural (milk proteins)	• Salad dressings • Baked goods • Beverages • Ice creams • Infant formula	• Emulsion stabilization • Increased shelf-life • Enhanced flavor/ texture/ processing stability	21CFR184.1979	
Casein	Casein	Natural (milk proteins)	• Dairy • Coffee creamers • Infant formula	• Emulsion stabilization • Thickening/ texturizing	N/A (food ingredient, not additive)	N/A (food ingredient, not additive)
Gum Arabic	Gum Arabic	Natural (glyco-proteins/poly-saccharides from acacia tree)	• Confectionery • Icings/fillings • Ice cream • Baked goods • Beverages	• Emulsion stabilization • Thickening/ texturizing	21CFR184.1330	

* Reprinted from INFORM, Vol. 27 (10), Cassiday, L. "Food emulsifier fundamentals", p. 12 (2016), with permission from The American Oil Chemists' Society

Mechanism of emulsifier action: Several mechanisms are proposed to hypothesize the concept of emulsion stability. They are (Avuchi, et al., 2019):

(i) *Surface tension theory*: It proposes that the emulsion stability is achieved by reducing the interfacial tension between two phases.

(ii) *Repulsion theory*: A film is created on the globules by the emulsifier which results in repulsion of various globules from agglomeration.

(iii) *Viscosity modification*: The increased viscosity results in preventing the globules from movement, and thus discourage their recoalescence. Some viscosity enhancing substances like gums (Acacia), glycerin, CMC increase the viscosity of the emulsions and keep them stable.

22.3.2 Factors Affecting Emulsion Stability

Stability of an emulsion is its ability to keep the dispersed phase in dispersion without being separated into two layers. That is, the system should be able to resist its physico-chemical properties to recoalescence. A few factors that affect the emulsion stability are:

- temperature
- pH
- viscosity

Temperature: Temperatures affect the stability of emulsions by altering their physico-chemical properties of oils, water, and interfacial films. Temperature also affects the solubility of the surfactant (emulsifier), and hence, may result in destabilization of the emulsion. Temperature effect on densities is well known. It may alter the densities of dispersed and continuous phases differently which results in destabilization of emulsion. Destabilization is accelerated with increased temperatures. Ostbring, et al. (2021) reported the effect of temperature on the emulsion stability with two emulsifiers. Rapeseed protein obtained from the oil pressed cake and soy lecithin at temperatures of 4° and 30°C on storage of six months. They reported that destabilization was accelerated with increasing temperature of storage.

pH: pH has a great role to play with emulsion stability. It is a matter of common experience that a droplet of lemon juice would break the stable milk emulsion into fat and water layers. pH effect cannot be directly quantified; but it depends on the type of buffer used. All emulsions are stable at neutral pH (pH of 7.0) Hunt and Dalgleish (1994) mentioned while working on the pH effect on the stability of emulsions made with whey protein isolate (WPI), citrate buffer could keep the emulsion stable at pH of 3. Soy lecithin was reported to stabilize the water-oil emulsions at lower pH up to 3 (Ostbring, et al., 2021).

Viscosity: Viscosity ensures higher stability by reducing the movement of dispersed droplets. If the continuous phase viscosity is increased using some emulsifiers. Higher viscosity reduces the terminal velocity of the droplets as indicated by Stoke's law (Eq 21.6). This in turn reduces the mobility of droplets and discourages recoalescence. If the continuous phase has higher initial viscosity (before introducing the second phase), the emulsions are likely to be more stable.

22.3.3 Challenges in Emulsification

The challenges in emulsification are typical in nature in not being stable. Some emulsions are expected to be stable for years as in the case of emulsion paints. The following are some challenges:

(i) *Lack of stability:* The dispersed phase droplets coalesce and separate out from the continuous phase. For emulsions to remain stable, enough energy is required to counter the surface tension forces. This is done by breaking the dispersed phase droplets into micro and nano sizes; smaller the particle size greater is the stability.

(ii) *High shearing* is required to break the dispersed phase droplets and keep them in suspension. Such dispersed particles may not be available for further processing as is generally required in case of suspension polymerization reactions, etc. However, we do not come across such cases in food processing.

(iii) *The high shear* generated by turbines or by sonometers will further add to the cost of processing.

22.3.4 Machinery and Equipment for Emulsification

Several methods are available to make macro and micro emulsions by using a wide range of equipment such as (i) static mixers, high-speed stirring vessels and injection equipment, (ii) high pressure systems, (iii) colloidal mills, (iv) homogenizers, and (v) ultrasound homogenizers. The main purpose of these equipment is to shear the dispersed (discontinuous) phase into fine droplets to prevent separation or coalescence. Invariably, emulsifiers or surfactants are used to help keep the emulsions stable. Most of the details of these equipment are dealt by the specific manufacturers who supply the equipment based on the customers' requirements and process demand. Essentially, they all contain a high-speed stirrer in a vessel. The shearing is such that the droplets size is brought down to 1μ or even less. The vessels are usually jacketed to heat up or cool down the system to the desired temperature of the homogenizer. The cooling is from –20°C to a heating of 80°C. Taha, et al. (2020) have tabulated various emulsification devices with the emulsification mechanism and their typical applications (Table 22.3).

Table 22.3 The Emulsification Mechanism and Applications of the Main Emulsification Devices*

Emulsification device	*Emulsification mechanism*	*Applications*	*References*
High shear mixers	The head of the mixer generates rotational, longitudinal and radial velocity. Gradients in the system, which breaks the larger droplets into smaller ones, disrupt oil/water interface and intermingles the liquids.	Coarse emulsion preparation and dissolution of powdered ingredients	McClemons, (2005)
Ultrasound homogenizers	The physical effects of ultrasound induced by acoustic cavitations can boost the disruption of oil droplets, facilitating the formation of stable o/w emulsions with small droplet size.	Emulsions and nano-emulsions preparation.	Awad, et al., (2012)
High pressure homogenizers	A coarse emulsion is forced to pass through a narrow homogenizer valve; then it undergoes extreme disruptive forces that disrupt oil droplets, forming fine emulsions.	Production of finely dispersed emulsions.	Stang, Schulmann and Schubert, (2001)
Microfluidizers	The fluids are brought together from two inlets at a high velocity (using pumping devices) that disrupt the oil droplets and produce fine emulsions after single pass.	Production of finely dispersed emulsions.	Jafari, et al. (2007)
Colloid mill	A colloidal mill typically consisted of two disks: a static disc and a rotating disc. The coarse emulsions are usually injected into the core area of the colloidal mill. The rotation speed of the rotating disk produces shear stresses in the gap between the disks that disrupt the oil droplets, producing emulsions.	Production of finely dispersed emulsions. Intermediate and high viscosity fluids homogenization.	McClemons, (2005)

(contd.)

Table 22.3 The Emulsification Mechanism and Applications of the Main Emulsification Devices* (*contd.*)

Emulsification device	*Emulsification mechanism*	*Applications*	*References*
Membrane emulsification	One of the immiscible liquids is forced (by pressure) to pass through a microporous membrane into the continuous phase. An emulsifier is usually dissolved in the continuous phase and adsorbed to the interface, thereby stabilizing the droplets against aggregation. The size of the droplets could be controlled by the size of the membrane spores.	Mainly used at the lab scale to produce emulsions with high dispersed phase fraction with specific dimensions and structures	Nazir, Schroen and Boom (2010)

* Reprinted from Trends in Food Science & Technology, Vol. 105, Taha, A., Ahmed, E., Ismaiel, A., Ashokkumar, M., Xu, X., Pan, S. and Hu, H.. "Ultrasonic emulsification: An overview on the preparation of different emulsifiers-stabilized emulsions", p. 364 (2020), with permission from Elsevier

(i) **High shear mixers:** They are a typical equipment in which the dispersed liquid is subjected to high shear stress to create high turbulence in a roto-stator assembly. The liquid is broken into fine droplets, and thus the emulsion becomes stable. Mostly it is used for making coarse emulsification. High mechanical force is applied to emulsify the liquid-liquid system. The speed of the homogenizer varies from 1,000–10,000 rpm which results in particle sizes as low as 2010–190 nm (Mulia, et al., 2019). A typical high speed shear mixer is shown in Figure 22.12 (Courtesy: Wahal Engineers, New Delhi).

Figure. 22.12 High speed shear emulsifier (With permission from WAHAL Engineers, New Delhi).

(ii) **Colloidal Mill:** Colloidal mill is used for making suspensions as well as emulsions. It contains a stator and a high-speed rotating rotor at 2,000–18,000 rpm. The dispersed material is subjected to high shear rates and results in droplet sizes as small as 1μ which are highly resistant for the emulsions to break. The gap between the stator and rotor can be adjusted to suit to particle size. Quite a good number of designs is available with the manufacturers, but they all operate on the same principle of a high-speed rotor rotating in a small gap between the rotor and stator and subject the fluids to high shearing. A schematic sketch of a colloidal mill is shown in Figure 22.13.

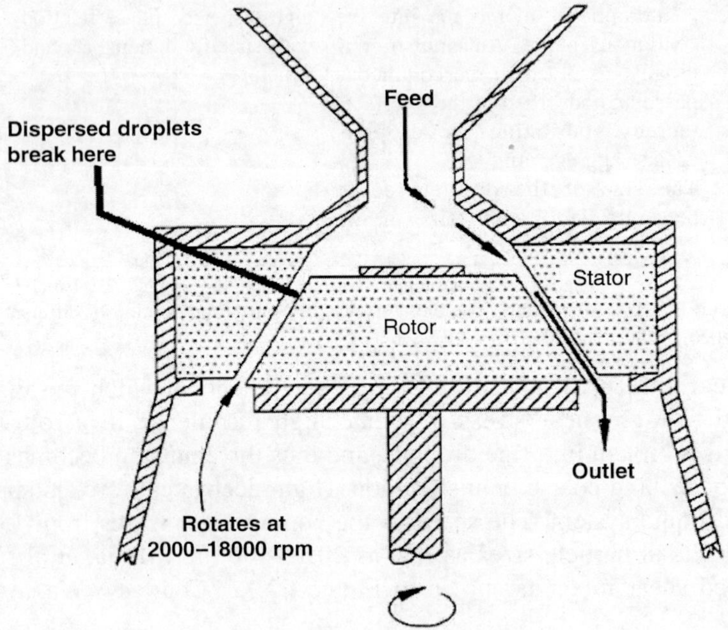

Figure. 22.13 Colloidal mill.

The feed material is not usually fed in the form of two different phases; it consists of a coarse emulsion of the dispersed and continuous phases. Colloid mills are generally used for homogenizing intermediates or high viscosity fluids such as peanut butter, fish or meat pastes, etc. Thus, colloidal mills are more suitable to be used for high viscosity fluids (0.1–1.0 Pa s).

(iii) **Sonicator or Ultrasound Homogenizer:** Use of ultrasound waves to create intense mixing is a usual phenomenon commonly employed in chemical reactors for multiphase reactions, extraction of phytochemicals, drying operations, processing of fruits and dairy products, etc. This technique is also used for emulsifications. The Sonicator is a machine that creates ultrasound frequencies. Sonicators apply ultrasound energy to agitate particles or droplets. Ultrasound waves are mechanical waves at frequencies greater than 20 kHz (Awad, et al., 2012). The ultrasound waves, when applied to the emulsion system, first compress the droplets, and then refract them. This continuous cyclic operations result in disrupting the droplets and collapsing them into finer droplets. A schematic representation of the phenomenon is shown in Figure. 22.14. The ultra-high frequencies are divided into High Intensity (HIU: 10–1000W/cm^2) or low intensity ultrasound waves (LIU $<$ 1W/cm^2).

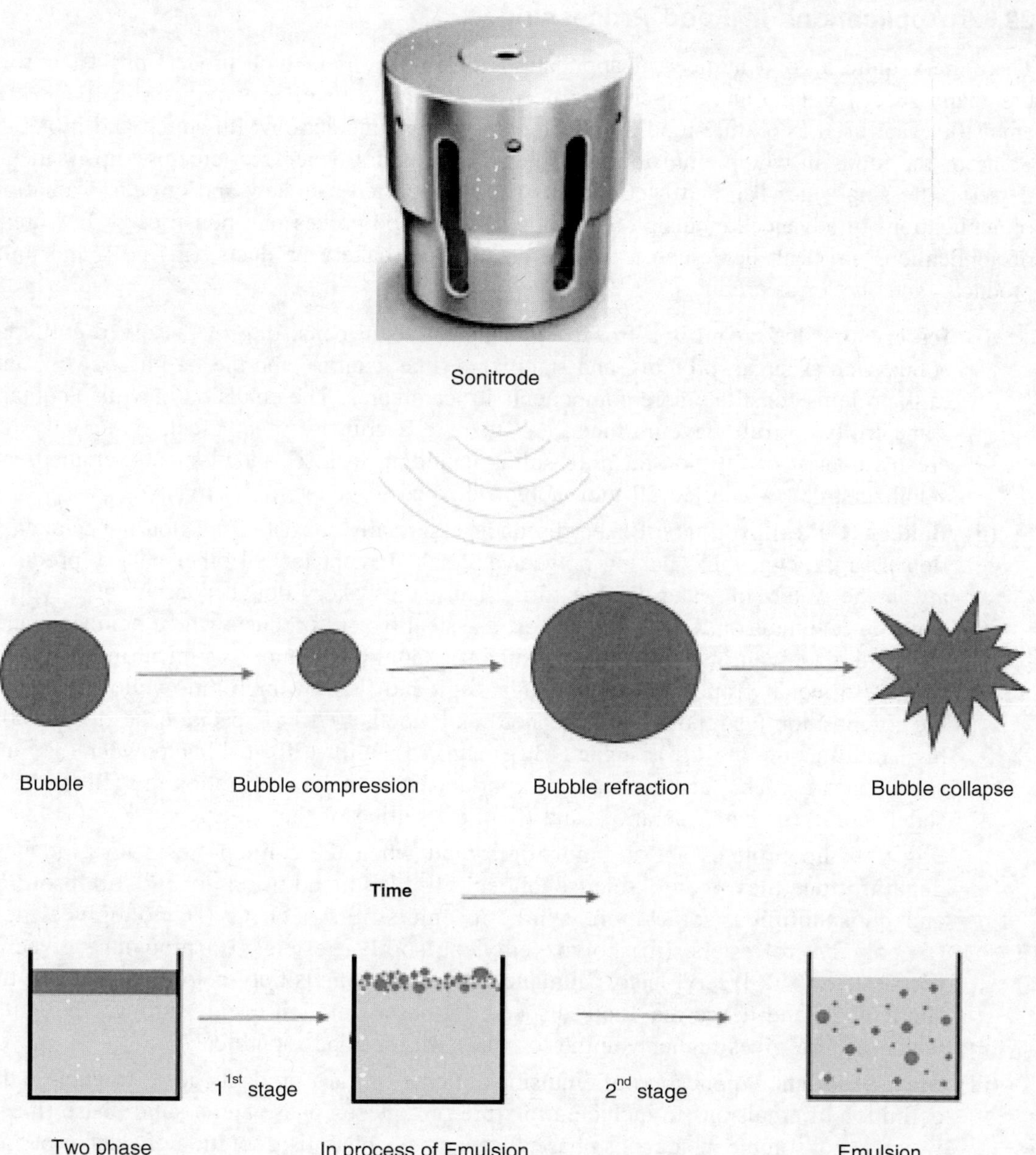

Figure. 22.14 Emulsion formation steps due to Accoustics.

Various applications of protein stabilized emulsion of HIU were listed by Taha., et al. (2020). The applications include whey protein concentrate and isolates, soya protein, sodium caseinate, rice bran protein, peanut protein concentrate, etc.

22.3.5 Applications in Food Processing

There are a quite a good number of applications of emulsions in food processing. Their role has many facets. Foams and suspensions need to be stable to give body for the food. The emulsifiers are used as texture-modifiers in foods by their interaction with starch and proteins. Some of the foods in which emulsion technology is used are beverages, creams, dips, sauces, desserts, dressings, etc. Tan and McClements (2021) reported a review and critical evaluation of applications of advanced emulsion technology in the food processing industries. A few areas of applications are dealt here such as (i) ice creams, (ii) bakery products, (iii) milk and milk products, and (iv) meat products.

(i) **Ice creams:** Ice cream is a frozen emulsion of foam consisting of dissolved colloidal solids such as sugar, proteins, and stabilizers (like gelatin), and the fat phase. Milk and milk proteins constitute core components in ice creams. The emulsion may also contain some fruits or fruit flavours, nuts, etc. to give a crunchy mouth feel. Essentially ice creams consist of fat, non-fat milk solids including lactose, added sugars, emulsifiers, stabilizers, etc. which are all thoroughly whipped to generate the foam.

(ii) **Bakery (Cakes) products:** Bakery products essentially consist of fat, flour (special cake flour), sugar, egg, milk, baking powder and salt. Except fat, all other bakery products are in the water- or water-soluble form. Thus, the bakery dough consists of a typical water-in-oil emulsion. Cakes are a most classical bakery products where emulsions and emulsifiers play an important role to impart a spongy structure. A typical modern cake consists of equal proportions of fat, sugar, egg and flour. A high ratio yellow cake has the composition (as reported by Shepherd and Yoell, 1976) of special cake flour (100), high ratio shortening (75), sugar (130), egg (95), milk (70), baking powder (3), and salt (2.5 parts). The factors that are important in the cake preparation are (i) emulsion stabilization, (ii) batter aeration, and (iii) heat setting of the batter.

Egg yolk lipoproteins act as emulsifiers; and when these lipoproteins undergo heat denaturation, they would release phospholipids. In addition to egg lipoproteins and phospholipids, some more synthetic emulsifiers such as (i) monoglycerides, (ii) polyglycerol esters, (iii) polyoxyethylene sorbitan esters, (iv) propylene glycerol monoesters, (v) glyceryl-lacto palmitate, etc. These emulsifiers help in dispersing the shortenings, and hence may help disperse fat-based aeration evenly through the batter which in turn gives higher volume to cakes with a good mouthfeel.

(iii) **Meat products:** Application of emulsions in case of meat products needs to enlarge the definition of emulsions to include a mixture of aqueous phase and a solid phase (meat) which is not soluble in aqueous phase. The aqueous phase is a solution of salts, proteins, and spices. This makes it a fit candidate to be called as a two-phase system in which the solid (meat particles) is dispersed into the aqueous phase and needs an intimate contact (Schut, 1976). Hence, it is called as meat emulsions by meat technologists, and is made by grinding meat and water. The water phase contains salt and spices. It forms a colloidal solution forming a matrix in which the animal fat is dispersed. It is essential that meat emulsions need to be stable with both high water holding capacity and high fat holding capacity by meat proteins. The animal protein itself acts as an emulsifying agent. In some cases, soya protein is added to an extent of 2% which also stabilizes the meat emulsions. Synthetic emulsifiers such as glycerine mono- and distearates and esters of fatty acids are also used as meat emulsifiers. However, mostly their use has always ended up with poor results. Polyglycerol fatty acid monoesters, monoacyl glycerol esters of dicarboxylic acids, sucrose esters of fatty acids, polyol monoesters of fatty acids and phospholipids are some emulsifiers used in preparing meat analogues. Phospholipids are also used to prepare filled-in meat products containing meat and texturized soya proteins.

Symbols

a: coefficient in Eq. (22.17)
b: coefficient in Eq. (22.17)
C: constant in Eq. (22.15)
C_1: proportionality constant in Eq. (22.19)
d_i: impeller diameter (m)
d_t: tank diameter (m)
g: acceleration due to gravity (m/s^2)
H_L: height of liquid in the tank (m)
I_p: mixing index for pastes
I_s: mixing index for solids
K: consistency index in Eq. (22.22)
K': constant in Eq. (22.9)
M: extent of mixing given by Eq. (22.9)
n: number of samples
n: flow behaviour index in Eq. (22.22)
N: number of particles in Eq. (22.6)
N_a: aeration rate defined by Eq. (22.24)
N_{Fr}: dimensionless Froude number
N_i: impeller speed in rps (s^{-1})
N_{P0}: dimensionless power number
N'_{P0}: dimensionless power number in the turbulent zone in Eq. (22.20)
$N_{Re,i}$: dimensionless impeller Reynolds' number
$(N_{Re,i})_N$: impeller Reynolds' number for non-Newtonian fluids
p: fraction of component P
P_g: power input for gassed system
P_0: power required for an ungassed system
q: fraction of component Q
t: time (s)
v_g: volumetric gas flow rate, (m^3/s)
V: volume (m^3)
w_i: width of the impeller (m)
x: fractional composition
$\bar{x}$: average fractional composition

Subscripts

i: component
1, 2, ..., n: samples

Greek Symbols

α: coefficient in Eq. (22.15)
β: coefficient in Eq. (22.15)
μ: viscosity (Pa s)
μ_a: apparent viscosity
γ: shear strain (s^{-1})
ρ: density (kg/m^3)
σ: rms deviation
σ^2: variance
σ_0^2: variance without mixing
σ_r^2: random variance

REVIEW QUESTIONS

22.1 Describe mixing as a unit operation with suitable examples.

22.2 Describe the process of mixing of solids and pastes. Explain how difficult it is to get homogeneous mixing with solids.

22.3 How do you characterize the solids mixing?

22.4 Define and explain the mixing index for granular solids.

22.5 Define rate of mixing. How do you characterize mixing rates?

22.6 Describe various operations taking place in a solid mixing unit.

22.7 Describe the working of a planetary mixer with a neat diagram.

22.8 Describe the working of a kneader mixer with a neat diagram.

22.9 Describe the working of a ribbon mixer with a neat diagram.

22.10 Describe the working of a double cone mixer with a neat diagram.

22.11 What are various applications of mixing of solids in food processing?

22.12 Describe the process of mixing of fluids.

22.13 How do you evaluate the power required for mixing of liquids?

22.14 Describe systematically how do you calculate the power requirement for a gassed system.

22.15 Define the following dimensionless parameters:

(i) power number

(ii) Froude number

(iii) impeller Reynolds' number

(iv) aeration rate

22.16 Describe the relationship for power number and impeller Reynolds' number.

22.17 How do you calculate the power requirement for mixing of non-Newtonian fluids?

22.18 What are various applications of gas-liquid mixing in food processing?

22.19 What is meant by emulsification?

22.20 What is an emulsifier? How does an emulsifier work?

22.21 What factors affect emulsion stability?

22.22 What are the challenges in emulsification?

22.23 Describe the machinery and equipment used for emulsification.

22.24 What are various applications of emulsification in food processing?

22.25 What are emulsification applications in meat industry?

22.26 What are emulsification applications in bakery?

REFERENCES

Avuchi,C.G., Igwe, V.S., and Echeta, K.K. (2019), The functional properties of foods and flours, *Int. J. Adv.Academic Research : Sci. Technol. & Eng*, **5**(11), pp. 139–160.

Awad, T.S., Moharram, H.A., Shaltout, O.E., Asker, D. and Youssef, M.M. (2012), Applications of ultrasound in analysis, processing and quality control of food: A review, *Food Research International*, 48(2), pp. 410–427.

Bates, R.L., Fondy, P.L. and Corpstein, R.R. (1963), An examination of some geometric parameters of impeller power, *Ind. Engg. Chem., Process Des. Develop.*, **2**(4), pp. 310–314.

Bates, R.L., Fondy, P.L. and Feni, J.G. (1966), Impeller Characteristics and Power, in *Mixing: Theory and Practice*, Uhl, V.W. and Gray, J.B. (Eds.), Academic Press, New York, pp. 111–178.

Cassiday, L. (2016), Food Emulsifier Fundamentals, *INFORM*, **27**(10), pp. 10–16.

Doran, P.M. (1995), *Bioprocess Engineering Principles*, Academic Press, London, p. 151.

Earle, R.L. (1983), *Unit Operations in Food Processing*, 2nd Ed., Pergamon Press, Oxford, pp. 166–172.

Freiberg, S. Ed. (1976), *Food Emulsions*, Marcel Dekkar Inc., New York.

Giridhar, N., Satyanarayana, A., Balaswamy, K. and Rao, D.G. (1999), A process for preparation of a formulation useful in drying of grapes, Indian Patent No. 804/DEL/99.

Hughmark, G.A. (1980), Power requirements and interfacial area in gas-liquid turbine agitated systems, *Ind. Engg. Chem., Process Des. Develop*, **19**, pp. 638–641.

Hunt, J.A. and Dalgleish, D.G. (1994), Effect of pH on the stability and surface composition of emulsions made with Whey Protein Isolate, *J. Agr. Food Chem.*, **42**, pp. 2131–2135.

Jafari, S.M., He, Y. and Bhandari, B. (2007), Production of sub-micron emulsions by ultrasound and micro fluidization techniques, *J Food Eng.*, **82**(4), pp. 478–488.

McClements, D. (2005), Food Emulsions: Principles, Practice and Techniques, *CRC series in Contemporary Food Science*, 3rd Ed., CRC Press.

Metzner, A.B., Feehs, R.H., Ramos, H.L., Otto, R.E. and Tuthill, J.D. (1961), Agitation of Viscous Newtonian and non-Newtonian Fluids, *AIChEJ.*, **7**(1), pp. 3–11.

Mulia, K., Safiera, A., Pane, I.F. and Krisanti, E.A. (2019), Effect of High Speed Homogenizer Speed on Particle Size of Polylactic Acid, SENTEN 2018 - Symposium of Emerging Nuclear Technology and Engineering. Novelty IOP Conf. Series: *Journal of Physics: Conf. Series* 1198, pp. 1–5.

Nazir, K. Shroen, K. and Boom, R. (2010), Premix emulsification, *J Membrane Sci.*, 362(1–2), pp. 1–11.

Ohyama, Y. and Endoh, K. (1955), Power characteristics of gas-liquid contacting mixers, *Chem. Engg., Japan*, **19** p. 2.

Ostbring, K.. Matos, M., Marefati, A., Ahlstrom, C. and Gutierrez, G. (2021), Effect of pH and storage temperatures on the stability of emulsions stabilized by Rapeseed proteins, *Foods*, **10**, pp. 1657–1673.

Rao, D.G. (2005), *Introduction to Biochemical Engineering*, Tata McGraw-Hill, New Delhi, p. 282.

Rushton, J.H., Costich, E.W. and Everett, H.J. (1950), Power characteristics of mixing impellers: Part I and Part II, *Chem. Engg. Progress*, **46**, pp. 395–404 and pp. 467–476.

Schut, J. (1976), Meat Emulsions, in Freiberg, S. Ed., *Food Emulsions*, Marcel Dekkar Inc., New York, pp. 386–453.

Shepherd, I.S. and Yoell, R.W. (1976), Cake emulsions, in Freiberg, S. Ed., *Food Emulsions*, Marcel Dekkar Inc., New York, pp. 221, 276.

Skelland, A.H.P. (1967), *Non-Newtonian Flow and Heat Transfer*, John Wiley & Sons, Inc., New York, pp. 27–49.

Stang, M., Schuchmann, H. and Schubert, H. (2001), Emulsification in high pressure homogenizers, *Engineering in Life Sciences*, 1(4), pp. 151–157.

Taha, A., Ahmed, E., Ismaiel,A., Ashokkumar, M., Xu, X., Pan, S. and Hu, H. (2020), "Ultrasonic emulsification: An overview on the preparation of different emulsifiers-stabilized emulsions", *Trends in Food Science & Technology*, 105, pp. 363–377.

Tan, C. and McClements, D.J. (2021), Application of Advanced Emulsion Technology in the Food Industry: A Review and Critical Evaluation, *Foods*, **10**, pp. 1–25.

Van't Riet, K. (1979), Review of measuring methods and results in non viscous gas-liquid mass transfer in stirred vessels, *Ind. Engg. Chem.*, Process Des. Develop., **18**(3), pp. 357–364.

CHAPTER

23

Size Reduction and Separation

Size reduction and size separation are two frequently used operations in food processing. Probably in the whole of processing industries, next to metallurgical industries, it is food industries which use size reduction operations of the raw food materials to make into various cookable, value-added and preservable products. God, who has been kind enough in bestowing a bounty of nature of food (raw) materials, has not provided them in a form ready to consume. Every raw material, including fruits and vegetables, needs to be brought down in size to bring it to consumable form. Even food grains, which are hardly of the size of a few millimeters to one millimeter and less, also need to be ground and milled before using. Milling is a form of size reduction, and results in:

- reduced particle size of the food material
- increased surface area

which have many obvious advantages which are given in Table 23.1.

Table 23.1 Advantages of Size Reduction

- Materials of definite sizes are desired for certain processing as in canning of fruits and vegetables or making of fruit bars.
- The increased surface area will help in enhanced heat and mass transfer; and hence, the rate processes are enhanced.
- The reduced particle size would enable accessibility to the interior of the food materials as in the leaching of spice oleoresins or oil from oil seeds.
- Intimate contact with various other ingredients is possible as in the case of preparation of soup mixes, etc.
- In preparation of certain soup mixes/gruel mixes/baby foods, etc. small particle sizes alone will work, and hence, size reduction/milling is not a choice but a compulsion.
- In preparation of certain baby foods/weaning foods where drum drying is used both for gelatinization of starch and drying of the prepared foods, reduced particle size is a necessity and not an option.

Most of the process operations related to size reduction are mostly applied to the solid foods. However, size reduction is also desired with liquids. Two immiscible liquids are broken into smaller fragments so that they can be mixed well to make an emulsion. This is known as **dispersion** of one immiscible liquid in the other.

In the present chapter, we restrict ourselves to mostly size reduction operations related to solids in view of their industrial importance. For example, making wheat grains into flour, *sooji*, semolina is an industrial operation used world wide for making various bakery products, pasta foods, extruded products, *chapattis* and various other Indian traditional foods.

23.1 THEORY OF SIZE REDUCTION

Size reduction is a highly energy inefficient process. When energy is applied in the form of a stress on the material, it breaks into two or more pieces resulting in generation of new surface area. Energy is used to break or overcome the intramolecular forces of the solid particles, so that they are broken into pieces and new area is created. The fracture mechanics of a particle is schematically represented by Figure 23.1. Grinding energy is one which is utilized to create new surface area. Remaining amount of energy is dissipated in the form of heat. Here, we may define a term called as **crushing efficiency** (η_c) which is the ratio of the energy utilized for creating new surface area to the energy absorbed by the solids for crushing or grinding (McCabe, et al., 1993).

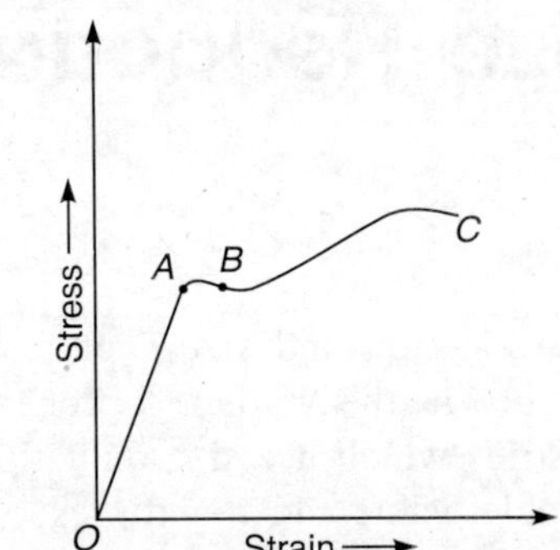

Figure 23.1 Stress-strain diagram of a food material. *A*: elastic limit, *B*: yield point, *C*: breaking point.

Thus, η_c is defined as:

$$\eta_c = \frac{E_s\,(A_p - A_f)}{E} \tag{23.1}$$

where E_s is surface energy of food material per unit area, A_f and A_p are areas per unit mass of the feed and product, respectively and, E is energy absorbed by unit mass of material.

Thus, the term $(A_p - A_f)$ indicates new surface area created per unit mass of food material. Even though there are some difficulties in measurement of various terms in Eq. (23.1) accurately from fracture physics data, it gives us an indicative relationship between the energy utilized for creating new surface area and the actual energy supplied for crushing or grinding. The efficiency of crushing is always poor, and is of the order of 1–3 per cent only (McCabe, et al., 1993: Prasher 1987). The remaining amount of supplied energy is wasted as heat, since grinding generates a lot of heat in the materials. Sometimes, the generated heat is so much that it is detrimental for product quality. For example, in the case of grinding of spices, the generated heat would volatilize the essential oils which are volatile in nature resulting in deterioration of the product. In such cases, we may need to go for *cryogenic grinding*.

23.2 VARIOUS TYPES OF SIZE REDUCTION

Size reduction is also known as **comminution** in engineering literature. It can be broadly classified into the following four methods:

23.2.1 Compression

The principle behind compression is application of compressive force or crushing to reduce the particle size. Nut cracker is a classical example. Crushing rolls or crushing mills are used industrially to bring down the particle size initially to some smaller sizes. Later they will be subjected to further processing to bring down the size.

23.2.2 Impact

The force applied at a time with a hammer to the material is known as **impact**. Hammering is an example of impact. Industrially, hammer mills/ball mills are used for size reduction. Hammering is also used initially to break the material into smaller pieces.

23.2.3 Shear

Shearing is an attrition-force or rubbing force applied to the material to bring down the particle size to finer sizes. Grinding and milling are attrition operations. Grinding stones, disc mills, *chakki* mill (stone mill), plate mills are some of the industrial/domestic units.

23.2.4 Cutting

Cutting is a generic term used to indicate size reduction. Usually cutting means, making a piece or particle into two pieces. Cutting with a knife or haksaw or saw into smaller pieces are used industrially. Chopping is also a term used to mean cutting. In the pulping industry, fruits are cut or chopped into pieces/slices before feeding to the pulper.

23.3 PRINCIPLES AND LAWS OF GRINDING

The grinding laws are based on the energy required to reduce the size of a material to create new surface. Since grinding is a highly energy inefficient process, calculation of it is very important, and holds significance in size reduction studies. Some theories have been put forward to calculate the energy required. All the theories are based on the assumption that the energy required to create a small change in the particle size is proportional to the size of the particle. Mathematically, in a differential form it can be expressed as:

$$-\frac{dE}{dL} = KL^n \tag{23.2}$$

where dE is the energy required to create a new length dL in a unit mass of particle of length L; K and n are constants. The negative sign on LHS indicates that E increases as L decreases. There are different laws which ascribe different valves for n. Most prominent amongst the laws are:

23.3.1 Rittingers' Law

Rittingers' theory was the first theory, originally proposed in the year 1867, and assumes the valve of $n = -2$ in Eq. (23.2) which means that an increment in surface is proportional to an increment in grinding energy. Thus,

$$-\frac{dE}{dL} = KL^{-2}$$

which on integration between the limits L_f and L_p, sizes of the feed and product,

$$E = K_R\left(\frac{1}{L_p} - \frac{1}{L_f}\right) \tag{23.3}$$

If P is the power required by the mill in terms of kW and $\dot{m}$ is mass flow rate, then

$$E = \frac{P}{\dot{m}} \tag{23.4}$$

Thus, the Rittingers' law states that the grinding energy is inversely proportional to the particle size, or grinding rate function should be proportional to the particle size. This is applicable for a fairly hard materials in a limited range of particle sizes (Prasher 1987).

23.3.2 Kicks' Law

Subsequently Kicks' law was proposed in the year 1885 which assumed that the work required to grind a given mass of material is constant for the same reduction ratio (L_f/L_p) which makes $n = -1$ in Eq. (23.2) Thus,

$$-\frac{dE}{dL} = KL^{-1} \tag{23.5}$$

On integrating Eq. (23.5), we get

$$E = K_K \ln\left(\frac{L_f}{L_p}\right) \tag{23.6}$$

in which K_K is Kicks' constant. This indicates that the grinding energy is proportional to the logarithm of reduction ratio. This was considered to be a fair representation of the grinding process. But unfortunately Eq. (23.6) shows that energy required to bring down the particle size from a very big particle to half its size is same as that required by a small particle to half its size, i.e., the energy required to bring down the particle size from 10 cm to 5 cm is same as the energy required to bring down a particle size from 1 mm to 0.5 mm which conceivably is irrational. Obviously the energy required in the latter case is much more than that required in the former case.

23.3.3 Bonds' Law

The Rittingers' and Kicks' laws did not seem to have been a fair representation of the grinding process in terms of the energy requirements because of their inherent inadequacies, and do not lead us to make any coherent conclusions to represent grinding energy requirements on a rational basis. Bond has proposed a third law in the year 1952 (Bond 1952), which assumed a value of $n = -3/2$ which is in between those of Rittingers' and Kicks'. Since during grinding, all particles will not have a uniform size, Bond assumed the final particle to be that size of the mesh[†] in which 80 per cent of the material passes. To dispel the discrepancy with Kicks' law

[†]Mesh size and evaluation of final particle size for a material which consists of various sized particles will be dealt in section 23.6.2.

for reduction ratio, Bond assumed the final particle size to be the one for which 80 per cent of the product passes through 100 micron size sieve. Thus, the final particle size is conceived to be 100 microns for calculation of Bonds' work index (W_i). Thus, Bonds' work index is defined as: *the gross energy required (in* kWh/ton*) by a large quantity of feed having a particle size which accounts for 80 per cent of the feed to reduce the final particle size to a value that 80 per cent of the product passes through 100 micron sieve.*
This leads us after integration of Eq. (23.2) to

$$E = \frac{P}{\dot{m}} = 10\ W_i \left(\frac{1}{\sqrt{L_p}} - \frac{1}{\sqrt{L_f}} \right) \tag{23.7}$$

In which W_i is Bonds' work index in kWh/ton, P is in kilowatts, $\dot{m}$ in tons/hour E is in kWh/ton, and L_p and L_f are in metres.

If L_p and L_f are expressed in millimeters, Eq. (23.7) becomes (McCabe, et al., 1993):

$$E = \frac{P}{m} = 0.3162\ W_i \left(\frac{1}{\sqrt{L_p}} - \frac{1}{\sqrt{L_f}} \right) \tag{23.8}$$

PROBLEM 23.1 Walde, et al. (1997) reported microwave drying and grinding of gum karaya samples to make a fine powder of gum karaya. The batch size they have taken was approximately 30 g and ground in a domestic grinder (555 W capacity). After grinding, the final particle size was evaluated by sieve analysis to be 0.55 mm. It took 20 s for grinding gum karaya of initial particle size 5 mm. Find the Bonds' work index (W_i)

Solution

Given:

Weight of the sample = 30 g

$L_f = 5$ mm

$L_P = 0.55$ mm

Wattage of grinder = 555 W

Time of grinding = 20 s

$$\text{Energy supplied for grinding 30 g of sample} = \frac{555}{1000} \times \frac{20}{3600}$$

$$= 3.08 \times 10^{-3} \text{ kWh}$$

$$E = 3.08 \times 10^{-3} \left(\frac{1000}{30} \right) = 0.103 \text{ kWh/kg}$$

$$E = 0.3162\ W_i \left(\frac{1}{\sqrt{L_p}} - \frac{1}{\sqrt{L_f}} \right) \tag{23.9}$$

i.e.,

$$0.103 = 0.3162\ W_i \left(\frac{1}{\sqrt{0.53}} - \frac{1}{\sqrt{5}} \right) = 0.285\ W_i$$

Therefore,
$$W_i = \frac{0.103}{0.285} = 0.36 \text{ kWh/kg}$$

23.4 SIZE REDUCTION EQUIPMENT

Various size reduction equipment used in the industry can be classified into the following (McCabe, et al., 1993), and are shown in Figure 23.2:

(i) Crushers
(ii) Grinders/fine grinders
(iii) Milling machines/attrition mills
(iv) Cutting machines

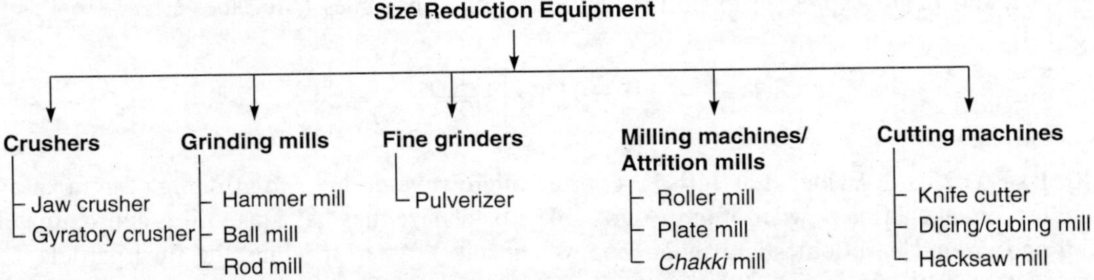

Figure 23.2 Classification of various size reduction equipment.

They are described briefly here,

23.4.1 Jaw Crushers

Jaw crushers are very versatile and are used to bring down the size of a big particle or a lump of particles into a reasonably smaller sizes, as in the case of crushing of several ores, lime stones, etc. It finds mostly application in mineral processing and less in food processing. It consists of typically two jaws, one of them is often stationary and the other is gyrating or moving on a pivot [Figure 23.3(a)]. The course material enters from the top. The moving jaw will be moving backwards and forwards over the pivot. The clearance between the two jaws is more on the top as compared to that at the bottom. The material gets crushed in between the two jaws, and the fine product is discharged from the bottom. The gyratory crusher also operates on the same principle except that the moving jaw will be actually rotating inside, whereas the outer jaw is fixed. The clearance between the two jaws will be decreasing as we move from top to bottom [Figure 23.3(b)].

23.4.2 Hammer Mill

Hammer mills are very commonly used in food industry for fine grinding of solids by impact. It consists of a centrally rotating shaft to which is attached a disc. The disc has a number of hammers swinging which are pivoted to the disc at the edge (Figure 23.4). The rotating disc turns at a very high speed inside a cylindrical casing. The casing is provided with a hard mesh

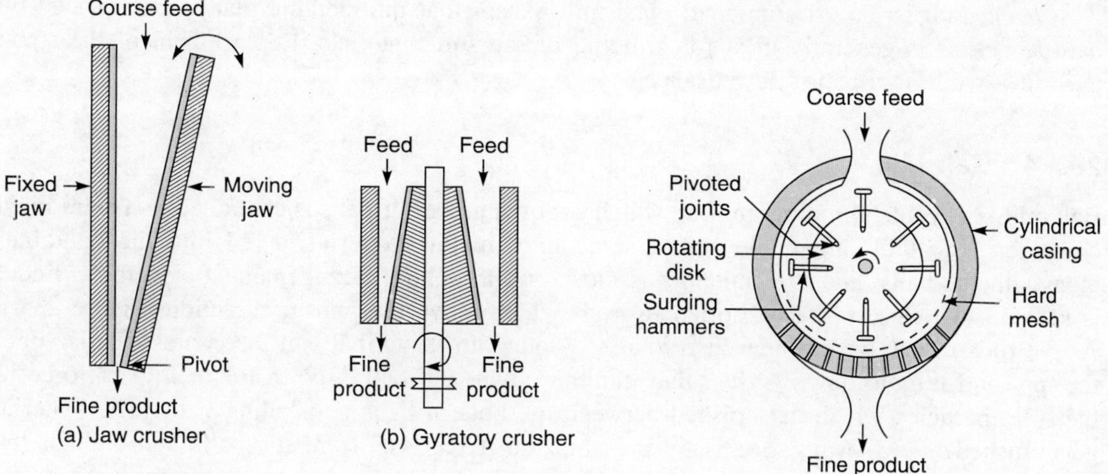

Figure 23.3 Crushers.

Figure 23.4 Hammer mill.

usually made out of stainless steel. The feed is subjected to beating by the swirling hammers until the material passes through the mesh. The mesh size is changed depending upon the fineness required.

23.4.3 Plate Mill

Plate mill, also known as **disc mill** is an attrition mill in which the feed is crushed between two hard metallic plates (Figure 23.5). Usually one plate is stationary and the other plate rotates. The stationary plate moves axially to adjust the clearance between two plates or discs. The plates will also have groovings to facilitate easy movement of the feed without slip. In some designs, both the plates will rotate, but in opposite directions. This will facilitate better grinding of the solids. If the groves between plates are worn off, they are dismantled, and fresh groves are carved. Otherwise, grinding becomes very inefficient.

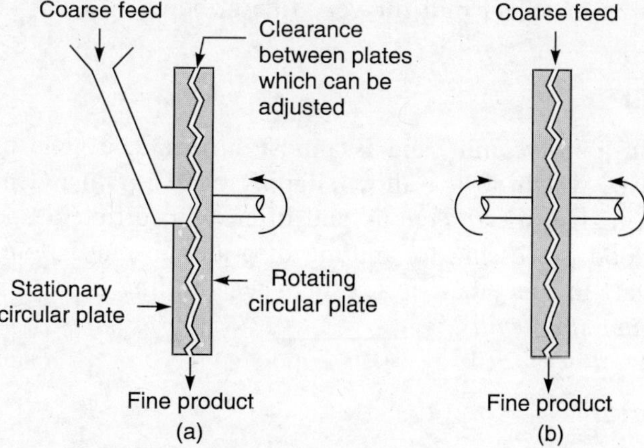

Figure 23.5 Plate mill.

Chakki mill is almost similar to plate mill except that the metallic plates are replaced by hard stones. It is generally used for grinding of the soft materials. Replacement of the stone once in a while is a recurring nuisance.

23.4.4 Ball Mill

Ball mill is a typical tumbling mill, in which a large number of balls is contained in a cylindrical container. The balls are made out of some hard metal. Material is fed into the container alongwith the balls, and the container is closed by a locking arrangement. Later, the cylinder is kept on two closely spaced horizontal rods which move in opposite direction (Figure 23.6). As the rods rotate, the cylinder also rotates. Alongwith it, the balls in the container raise upto the top, and tumble down. When they tumble down, the balls give a lot of impact force to the solid particles which get crushed between the bottom balls and falling balls. The material gets crushed in two ways, one is by the abrasion it receives from the balls and wall of the cylindrical container, and the other is by the impact force the material receives by the impinging (tumbling) balls. The material reduces to a fine powder.

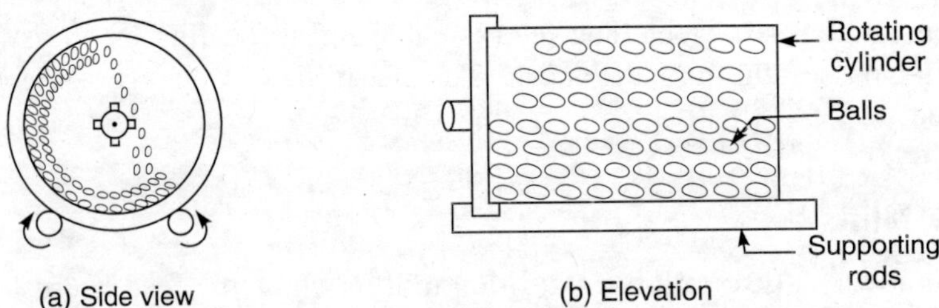

Figure 23.6 Ball mill.

It is usually a batch operation, in which the feed and balls are fed into the cylinder initially and closed. Later the cylinder is made to rotate on two horizontal rods. Because of the impact and abrasion, the material is ground into a fine powder. When the grinding is complete, the mill is stopped, the contents are discharged. Except for the noise it makes because of the tumbling balls, the ball mill is an excellent mill for very fine grinding.

23.4.5 Roller Mill

Roller mill is also an abrasive mill, and is almost similar to a plate mill or roller crushers. The rollers have groves which do not allow slipping of the grains. Generally both the rollers will be rotating but in opposite directions and often with different speeds. Roller mills are used extensively in food processing for milling of grains to make flour, *sooji*, and semolina, especially from wheat. In fact, their utility in wheat milling is so much that they become synonymous with *roller flour mills*.

23.4.6 Knife Cutter

Knife cutters are a typical cutting equipment used in food processing, especially for cutting and shredding of soft gelataneous materials like meat chunks, etc. It consists of a rotating plate with high speed, probably of the order of 200–900 rpm. The rotating plate has 2 to 12 oscillating knifes made of stainless steel. They cut the food material several times per minute against the stationary bed which is a hard metallic screen. The shredded material passes through a screen which has 5–8 mm openings. The knife cutters are extensively used for shredding meat to fill into the sausage casing.

23.5 APPLICATIONS OF SIZE REDUCTION IN FOOD PROCESSING

As has been mentioned earlier, food processing is a major user of size reduction operations in some form or other. All the four major types of size reduction operations, viz., cutting, grinding, impact and rubbing and sometimes a combination of them are used in food processing. Some of them are shown in Table 23.2. The list is not exhaustive, it is only indicative. Sometimes, even size reduction of mixed class of food materials like spices, nuts, oil-seeds, cereals, pulses or dry fruits and vegetables is also done.

Table 23.2 Applications of Size Reduction Operations in Food Processing

Major area	*Specific applications*
Fruits and vegetable processing	• Slicing of fruits before feeding to the pulper for making the fruit pulp • Slicing of potato chips • Cutting of raw mango, lemon and other sorts of acid fruits or vegetables for making pickles • Cutting of fruits and vegetables for usual preservation operations, viz., dehydration or canning, etc. • Cutting of citrus fruits to extract juice in a rosing machine • Shredding of fruits and vegetables in a fruit mill to dehydrate then to use in the manufacture of soup powders • Cutting of fruits and vegetables to remove the unedible portion before packaging or keeping for preservation
Cereals and pulses	• Flour milling of grains, especially wheat for making *maida,* wheat flour, *suji* and semolina; perhaps flour milling of wheat is the largest food processing operation; almost all the wheat all over the world is milled to flour for human consumption • Milling of minor millets like *ragi, bajra* to make into flour • Milling of brown rice to remove bran partially by abrasion technique; almost all the paddy grown all over the world is milled before it goes for human consumption as a staple food • Milling of most of the pulses to *dals* to make them ready for cooking • Grinding of some combinations of pulses, oilseeds and spices to make instant chutney powders • Wet grinding of black gram and green gram for making various traditional foods of India • Grinding of some non-edible portions of cereals and pulses or minor grains to make cattle feed
Spices and plantation products	• Most of the spices are used in the powder form either individually or in mixed form; the spices include chilli, pepper, turmeric, coriander, cloves, cinnamon, etc. • Coarse grinding of spices like chilli, turmeric, pepper and dry ginger to extract oleoresins

(contd.)

Table 23.2 Applications of Size Reduction Operations in Food Processing (*contd.*)

Major area	*Specific applications*
	• Wet grinding of ginger and garlic either to dry them as powders or to make ginger-garlic paste • Grinding of coconut to make copra for drying which is used as desiccated coconut powder • Wet rubbing of black gingelly seeds to remove the husk before drying
Animal products	• Virtually almost all the animals and birds are slaughtered by cutting operation before going for processing • Trimming of fish, prawns, crab and shrimp to remove the unedible portions before processing/packaging • Cutting of chicken into various edible portions like legs, wings, breast, etc. • Big chunks of meat pieces in frozen form are cut into smaller pieces before processing/packaging • Mincing of meat to fill into sausage casings • Grinding of dry fish and shrimp to make poultry feed

23.6 SIZE SEPARATION

When the materials are ground in any size reduction equipment, it is never possible to get all particle of the same size. Hence application of Eqs. (23.3), (23.5) and (23.6) is hypothetical. In fact, it is not possible to have uniform feed size as well. For example, when wheat is ground, we get a variety of products, viz., *maida*, flour, *sooji*, coarse semolina, fine semolina, and may be sometimes even unground grains. But many process operations require if not particles of uniform size, at least particles of a small range of sizes because often the process equipment are designed only to handle a certain range of particle sizes. Hence, the necessity to have equipment or methodology by which the particles should be segregated on the basis of size. This constitutes the subject matter of this section, and is applicable only for separation of solid particles.

23.6.1 Screening

Screening is a method of segregating or separating particles of different sizes into a range of particle sizes according to their size alone. When the feed material is dropped on a screen, all the particles (mostly) smaller than the size of the aperture of the screen pass through the screen. They are called **undersized** or **fines**; and are represented with – (negative) sign before the mesh size. All the particles above the size of the aperture of the screen will be retained on the screen. They are known **oversized** or **tails**; and are represented with + (positive) sign before the mesh size. Thus, one screen can separate particles into only two sizes; those which are undersized and those which are oversized. It does not give a correct representation of the size of the particles; because the undersized have no limit of their lower size; and similarly the oversized have no limit of their upper size.

Hence, to separate particles into a certain range of particle sizes, we use a series of screens one over the other. All the particles passing through one screen and retained on the other can be assigned a particle size which is the average of the aperture size of the two screens. Industrial screens are made out of woven wires, plastic clothes, or metal sheets with perforations or metallic bars. The screens are normally made out of stainless steel to be able to withstand the wear and tear. They are normally available in the range of 4″ to 400 mesh sizes[†].

23.6.2 Standard Screen Series

Average particle size is evaluated on the basis of standard screen analysis. This will help us segregate particles of sizes 3″ to 0.0015″. *Standard sieves*, also known as **test sieves**, are made out of woven wire screens, and are carefully standardized for the *mesh size*. The mesh openings are square in shape. The mesh size of the screen is identified by the number of meshes per inch. Hence, the actual size of the aperture is always less than the mesh number because of the thickness of the wires. Also it may be noted that as mesh size increases, the aperture size decreases. Thus, a 100 mesh sieve has an opening of 0.147 mm, whereas a 4 mesh size has 4.699 mm. There are two types of standard sieves available, viz.,

- BS (British Standard) mesh size, and
- Tyler series.

However, in the present book we follow mostly Tyler series. In the Tyler series, the screens are made out of mesh wire of 0.074 mm size. The size (area) of the aperture in one screen is exactly twice the size (area) in the lower screen in the series. Hence, the ratio of mesh size of one screen is equal to $\sqrt{2}$ (= 1.41) times of mesh size of the next lower screen. The Tyler standard sieve sizes are given in Appendix 11. For example, the clear opening for mesh size 4 is 4.699 mm, whereas that for mesh size 6 is 3.327. The ratio of 4.699/3.327 is 1.412. Sometimes, even intermediate screen sizes are also provided in which case the ratio of one mesh size to the next intermediate mesh size is $\sqrt[4]{2}$ (= 1.189). For example, the mesh size of 4 is 4.699 mm, whereas that for mesh size 5 is 3.962. The ratio of 4.699/3.962 is 1.186.

23.6.3 Average Particle Size

Average particle size is evaluated by carrying out *screen analysis* or *particle size analysis*. For doing this, a set of series of sieves is taken. The sieves are stacked with the smallest size at the bottom and the largest aperture size at the top. The bottom most is a blind sieve (i.e., the one without openings at all, also known as **pan**). The food material is fed to the top most sieve, and is closed with a lid. The stack is fixed in a gyratory moving arrangement. On the top of it, there will be an impinging hammer which hits on the top once in a while (Figure 23.7). This will facilitate to remove any particles that get entrapped in any of the apertures.

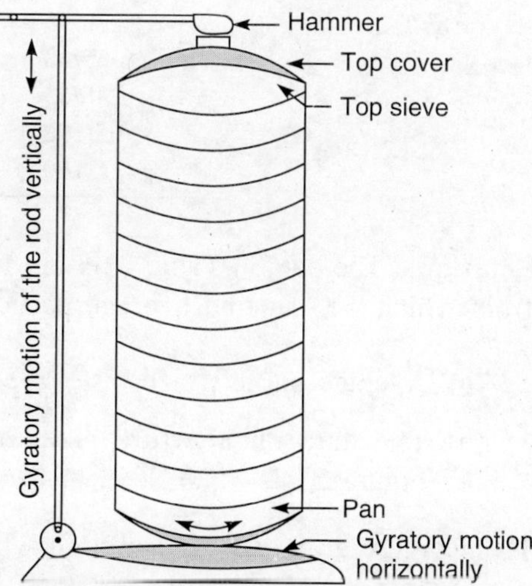

Figure 23.7 Stack of sieves with gyratory motion.

[†]See section 23.6.2.

The material collected in each sieve is taken and is weighed to find out the mass fraction of the material on each sieve (x_i). The particle size of material retained on any sieve i, is the mean of the screen opening of that sieve and that of the sieve above it [(i – 1) sieve]. Let us say that the material passes through mesh size 4 and is retained on mesh size 6, the average particle diameter retained on mesh 6 is equal to the mean of the screen openings of mesh 4 (4.699 mm) and that of mesh 6 (3.327).

Thus, the d_{pi} on mesh 6 is (4.699 + 3.327)/2 is 4.013 mm. This is known as **differential screen analysis.**

We take the average particle size of the whole feed on the basis of mass fractions, which is known as **mass mean diameter** ($\bar{d}_p$). Thus,

$$\bar{d}_p = \Sigma x_i d_{pi} \tag{23.10}$$

We shall explain the calculation of $\bar{d}_p$ with Problem 23.2.

PROBLEM 23.2 After grinding 25 g of wheat grains, the sieve analysis was carried out. The results are shown in Table 23.3. (a) Calculate average particle size, (b) Find the aperture size through which 80 per cent of material passes.

Table 23.3 Particle Size Analysis Data for Problem 23.2

Tyler mesh size	*Quantity retained*, g
10	1.977
10/20	11.693
20/28	4.508
28/35	1.199
35/48	2.428
48/65	0.63
65/100	1.095
100	1.47

Solution The data in Table 23.3 are written with the details of mesh sizes, etc. in Table 23.4, from which we can find average particle size.

(a) Average particle size = $\bar{d}_p = \Sigma x_i d_{pi}$ = 1.039 mm

(b) The data on aperture size, d_{pi} and the cumulative mass fraction are plotted in Figure 23.8.

From Figure 23.8, we get through which 80 per cent of the feed passes = 1.13 mm.

Table 23.4 Data for Problem 23.2

Mesh size	*Clear opening,* mm, (*Appendix* 11)	*Average aperture size,* mm	*Quantiy,* g	*Mass fraction* x_i	$x_i\ d_{pi}$, mm	*Cumulative mass fraction* Σx_i
10	1.65		1.977	0.079	0.13	1.0
		1.242	11.693	0.4677	0.581	0.9206
20	0.83					
		0.711	4.508	0.180	0.128	0.453
28	0.58					
		0.503	1.199	0.048	0.0241	0.273
35	0.41					
		0.356	2.428	0.09712	0.0345	0.225
48	0.29					
		0.251	0.63	0.0252	6.3252×10^{-3}	0.1278
65	0.20					
		0.178	1.095	0.0438	0.048	0.1026
100	0.14.					
		0.147	1.47	0.0588	0.0864	0.0588
			25.0	**0.999**	**1.039**	

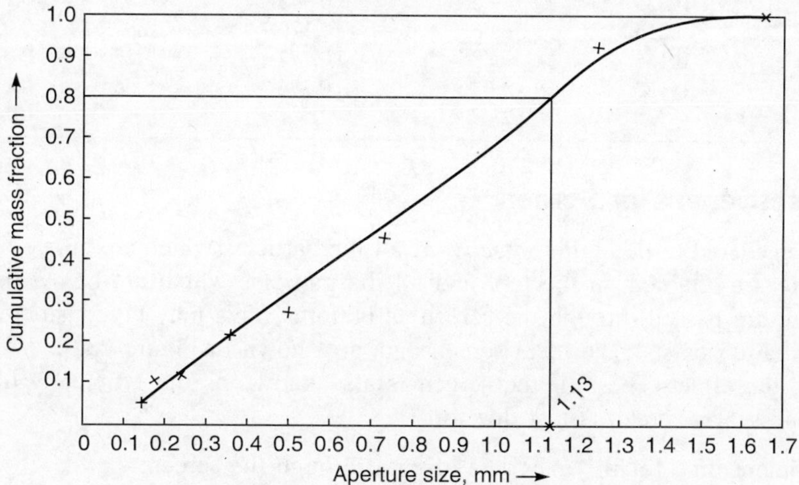

Figure 23.8 Solution for Problem 23.2.

PROBLEM 23.3 Chakkaravarthi and Rao (2002) reported on the grinding of garlic grits (28.2 g) which were earlier dried in a microwave oven. The particle size analysis was done using BS sieves. The data are presented in Table 23.5. Calculate the average particle size.

Solution The cumulative weights and weight fractions are shown in Table 23.6 Average particle size by Eq. (23.10) is:

$$\overline{d}_p = x_i d_p = 0.245 \text{ mm}$$

Table 23.5 Typical Sieve Analysis Data for Grinding of MicrowaveDried Garlic Grits for Problem 23.3

B.S. mesh size	*Quantity retained in* grams
10/20	–
20/30	0.28
30/40	1.37
40/60	9.68
60/80	3.65
80/100	4.86
100	8.36

Table 23.6 Cumulative Weight Fraction Data for Problem 23.3

B.S. mesh size	*Average aperture size* d_{pi}, mm *Appendix* 11	*Quantity retained,* gm	*Weight fraction* x_i	*Cumulative weight,* g	$x_i d_{pi}$, mm
10/20	1.265	–	–	28.6	–
20/30	0.675	0.28	0.01	28.2	6.78×10^{-3}
30/40	0.46	1.37	0.0485	27.92	0.022
40/60	0.335	9.68	0.343	26.55	0.115
60/80	0.215	3.65	0.129	16.87	0.028
80/100	0.165	4.86	0.172	13.22	0.028
100	0.15	8.36	0.296	8.36	0.044
		28.2	**1.0**		**0.245**

23.6.4 Effectiveness of Screen

A screen is considered to be highly *effective* if all the particles which are above the size of the screen aperture are retained on the top, and all the particles which are below the size of the screen aperture are passed through the screen as bottoms. This may be considered as an ideal screen which hardly exists. The ideal screen data are shown in Figure 23.9.

In a way, the effectiveness of the screen is also known as the **efficiency of the screen.** The effectiveness of a screen is lost due to

(i) insufficient time for the materials to pass through the screen,

(ii) the screen apertures may be elongated or contracted at certain positions,

(iii) some apertures may be blocked by the uneven shaped particles making the screen aperture size smaller, and

(iv) some wire meshes in the screen may be broken because of wear and tear due to long usage; this will make the aperture sizes much larger.

Thus, the efficiency or effectiveness of real screens is always less than 1.0.
The screen performance with the overflow and underflow is represented similar to that of a distillation column with the same notation as shown in Figure 23.10. Material balance on the screen yields

$$F = D + B \tag{23.11}$$

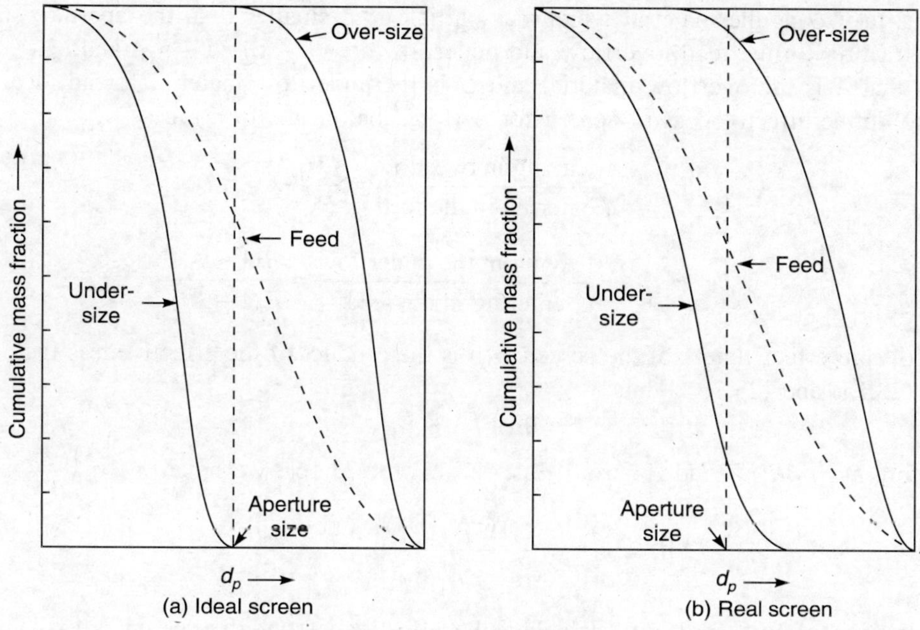

Figure 23.9 Analysis of screen data.

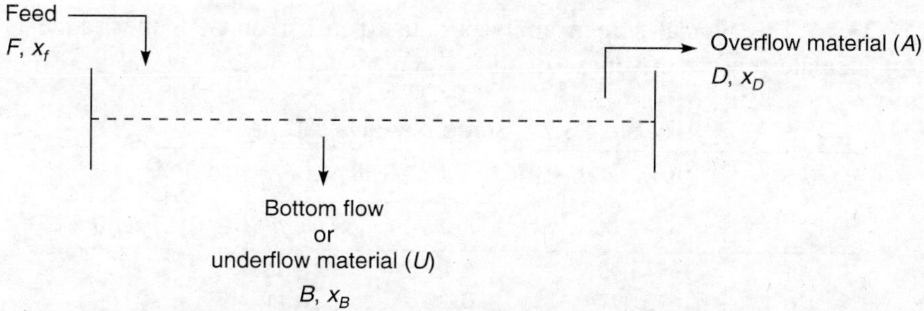

Figure 23.10 Material balance on a screen.

If we consider A is the material which has different particle sizes (both more than and less than the aperture size of the screen), and has its fractions in the feed, overflow and underflow as x_f, x_D, and x_B, then material balance yields:

$$\underset{(A \text{ in feed})}{Fx_f} = \underset{(A \text{ in overflow})}{Dx_D} + \underset{(A \text{ in bottomflow})}{Bx_B} \tag{23.12}$$

Eliminating by using Eqs. (23.11) and (23.12), we have

$$\frac{D}{F} = \frac{x_f - x_B}{x_D - x_B} \tag{23.13}$$

Similarly,

$$\frac{B}{F} = 1 - \frac{D}{F} = \frac{x_D - x_f}{x_D - x_B} \tag{23.14}$$

Let us take U as the material which has particle size smaller than the aperture size, then its mass fractions in the feed, overflow and underflow are $(1 - x_f)$, $(1 - x_D)$ and $(1 - x_B)$. This indicates that A is the overflow material and U is the underflow material for an ideal screen. We try to define effectiveness of screen for A (McCabe, et al., 1993) as:

$$E_A = \frac{A \text{ leaving in overflow}}{A \text{ entering in the feed}} = \frac{Dx_D}{Fx_f} \tag{23.15}$$

Similarly,

$$E_U = \frac{U \text{ leaving in the underflow}}{U \text{ entering in the feed}} = \frac{B(1-x_B)}{F(1-x_f)} \tag{23.16}$$

The overall effectiveness of the screen (η_s) is the product of the effectiveness factors given by Eqs. (23.15) and (23.16). Thus,

$$\eta_s = E_A \times E_U$$

Substituting D/F, B/F, E_A nd E_U from Eqs. (23.13) to (23.16), we have

$$\eta_s = \frac{(x_f - x_B)(x_D - x_f)\, x_D\, (1 - x_B)}{(x_D - x_B)^2\, x_f (1 - x_f)} \tag{23.17}$$

Application of Eq. (23.17) to calculate the effectiveness of the screen is explained with Problem 23.4.

PROBLEM 23.4 The typical screen analysis data for a screen of 8 mesh size is shown in Table 23.7. Calculate the effectiveness of the screen.

Table 23.7 Screen Analysis Data

Mesh size	*Aperture size* d_s, mm	*Cumulative Fraction of A*		
		Feed	*Overflow*	*Underflow*
4	4.699	0	0.02	
6	3.327	0.24	0.44	0
8	2.362	0.55	0.87	0.15
10	1.651	0.80	0.96	0.60
14	1.168	0.90	1.0	0.76
20	0.833	0.95		0.84
28	0.589	0.97		0.88
35	0.417	0.98		0.92
48	0.295	0.99		0.94
65	0.208	0.99		0.96
100	0.147	1.0		1.0

Solution The data in Table 23.7 are presented in Figure 23.11. From the data, corresponding to d_s = 2.362 mm

$$x_f = 0.55,\ x_D = 0.87,\ \text{and}\ x_B = 0.15$$

From Eq. (23.17), we have the screen efficiency

$$\eta_s = \frac{(x_f - x_B)(x_D - x_f)\,x_D\,(1 - x_B)}{(x_D - x_B)^2 \times x_f(1 - x_f)}$$

$$= \frac{(0.55 - 0.15)\,(0.87 - 0.55)\,0.87\,(1 - 0.15)}{(0.87 - 0.15)^2\;0.55\,(1 - 0.55)}$$

$$= 0.74$$

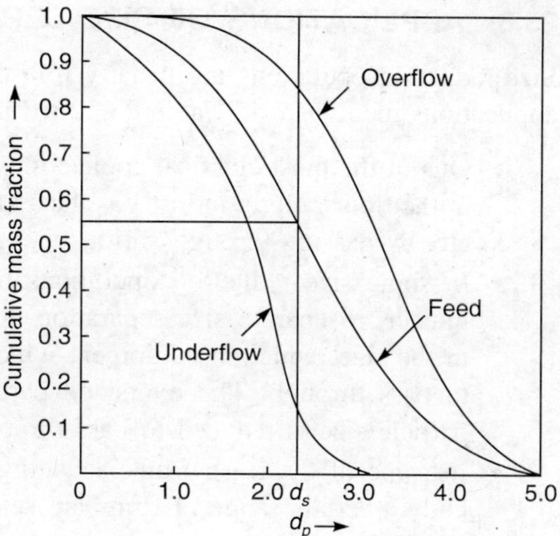

Figure 23.11 Solution for Problem 23.4.

23.7 SCREENING EQUIPMENT

The choice of screening equipment depends upon the purpose for which it is used and the value of the product (or feed) being processed. Broadly, they can be classified based on the movement of the screen, viz.,

Stationary screens: In case of stationary screens, the movement of particles is only by gravity. Hence, the screens are stranded in vertical position with some inclination to the ground. They are generally used for screening of dry solids of coarse nature. The separation need not be very rigid, as in the case of screening of sand or some grains/pulses before going for milling. The separation efficiency depends upon the inclination of the screen.

Vibratory screens: The screens vibrate with an electrical vibrator provided at the bottom of the screen in axial position, or at the end of the screen [Figure 23.12(a)]. The vibrations will facilitate easy movement of the material on the screen particularly, it is more helpful if the material is wet.

Gyratory screens: The gyratory screens are mostly used for industrial operations, where through-flow capacities are high, typically of the order of several tons per hour. The gyratory screen consists of two screens one over the other and inclined to horizontal by about 15°–30°. The gyrations are provided by an eccentric (operated with motor) provided at the middle [Figure 23.12(b)]. Typically, the gyrations are of the order of 600–1800 rpm. The feed material is fed at the upper edge of the upper screen. The feed gets segregated by the time it flows to the bottom edge of the screen. The oversized particles are separated from the top of the screen, and may be recycled to the size reduction equipment. The undersized particles are collected separately from the bottom of the screen. The fines are collected from the pan in a chute.

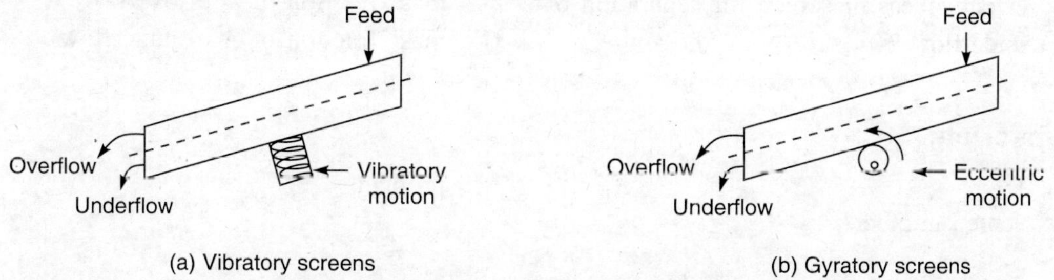

Figure 23.12 Schematic diagram of screen separation.

23.8 APPLICATIONS OF SIZE SEPARATION IN FOOD PROCESSING

Size reduction operations are usually followed by size separation operations. As such all the applications mentioned in Table 23.2 are applicable for size separations also.

- One of the most classical applications of size separation operations in food processing is in the flour milling industry. Wheat after grinding is separated into flour, *sooji*, semolina, etc. by passing through various sieves.
- In some size reduction operations like in a fruit mill or coconut copra unit or meat mincer, a separate size separation operation is not carried out, but a screen is built-in in the size reduction equipment which will allow only a particular size of the particles to pass through. The oversized particles are subjected to abrasions inside until their particle size is reduced to pass through the screen.
- In some of the fresh fruit packaging units, the fruits are segregated before packaging. Only a certain range of fruits are selected for packaging. Over sizes and undersizes are rejected.
- In some of the pulse-dehusking machines, the grains are segregated before milling. Only a certain size-range of grains is taken for processing. This will help adjust the clearance in the milling units.

Symbols

A: area
A_f: area per unit mass of feed
A_p: area per unit mass of product
B: bottom flow rate of the screen
D: upper flow rate of solids in the screen
d_{pi}: particle diameter on the ith screen
$\bar{d}_p$: average particle diameter
d_s: screen aperture size (mm)
E: energy required for grinding a unit mass of material (kJ/kg or kWh/kg)
E_A: effectiveness of screen for separation of overflow A
E_U: effectiveness of screen for separation of underflow U
E_S: surface energy per unit area
F: feed rate of solids to the screen
K: constant in Eq. (23.2)
K_K: Kick's constant
K_R: Rittengers' constant
L: length of the particle (m)
$\dot{m}$: mass flow rate (kg/s or tons/h)
n: constant in Eq. (23.2)
P: power (kW)
W_i: Bonds' work index
x_i: mass fraction of i
x_f: mass fraction of A in feed
x_D: mass fraction of A in overflow
x_B: mass fraction of A in underflow

Subscripts

f: feed
i: screen number
p: product

Greek Letters

η_c: crushing efficiency

η_s: screen effectiveness

REVIEW QUESTIONS

23.1 Write a brief note on size-reduction operations.

23.2 What is meant by crushing efficiency?

23.3 How do you classify size-reduction processes? Describe them briefly.

23.4 What are various laws of grinding? Explain them.

23.5 Compare and contrast various grinding laws.

23.6 Describe how Bonds' law scores over other two grinding laws.

23.7 Describe the working of a jaw crusher.

23.8 Explain the functioning of hammer mill with a neat diagram.

23.9 Write a note on ball mill.

23.10 What are various applications of size-reduction operations in food processing?

23.11 Write a brief note on size separation.

23.12 Describe the working of a screen separator.

23.13 Write a brief note on standard sieves.

23.14 How do you arrive at the average particle size after carrying out differential sieve analysis?

23.15 What is meant by average particle size?

23.16 What is meant by *effectiveness of screen*?

23.17 What are various screening equipment you use in food processing?

23.18 What are various applications of size separation/screening in food processing?

NUMERICAL PROBLEMS

23.1 Velu, et al. (2006) reported the grinding studies on maize grains (*Zea mays* L.) in a hammer mill operating on ½ hp motor. The initial particle size of the grains is 4.22 mm, and the desired final particle size is 0.4 mm. The Bonds' work index was reported to be 0.1172 kWh/kg. What should be the hourly flow rate of the feed grains?

(**Ans.** 9.2 kg/h)

23.2 Differential sieve analysis data during grinding of dry carrot grits are reported as follows. Calculate the

(a) final average particle size

(b) aperture size through which 80 per cent of the product can pass through.

Mesh size	*Average Aperture size* (microns)	*Quantity retained* (g)
10/20	1265	2.0
20/30	675	9.5
30/40	460	7.9
40/60	335	36.0
60/80	215	13.0
80/100	165	17.0
100/120	138	14.6

(**Ans.** (a) 322 microns, (b) 335 microns)

REFERENCES

Bond, F.C. (1952), The third theory of comminution, *AIME* Trans. **193**, pp. 484–494.

Chakkaravarthi, A. and Rao, D.G. (2002), Microwave drying and grinding characteristics of garlic (*Allium sativum*), *Indian Chemical Engineer*, **44**(3), pp. 180–182.

McCabe, W.L., Smith, J.C. and Harriott, P. (1993), *Unit Operations of Chemical Engineering*, 5th ed., McGraw-Hill, New York, pp. 960–1002.

Prasher, C.L. (1987), *Crushing and Grinding Process Handbook*, John Wiley and Sons (UK).

Velu, V., Nagender, A., Prabhakara Rao, P.G. and Rao, D.G. (2006), Dry milling characteristics of microwave dried maize grains (*Zea mays* L.), *J. Food Engg.*, **74**, pp. 30–36.

Walde, S.G., Balaswamy, K., Shivaswamy, R., Chakkaravarthi, A. and Rao, D.G., (1997), Microwave drying and grinding characteristics of gum karaya (*Sterculia urens*), *J. Food Engg.*, **31**, pp. 305–313.

CHAPTER

24

Material Handling and Transportation

We need to transport large quantities of raw materials or products or intermediate products in the capacity ranges of several tons per hour. The materials could be in the form of solids, liquids or slurries.

Generally, for transportation of liquids and slurries which are relatively easily flowable, we use different types of pumps (described in Section 8.9). When it comes to transportation of solids, we need a different methodology which we discuss in this chapter. Material handling is different from mass transport/mass transfer which we described in Chapter 16. The mass transfer operations are restricted to molecular level, and the transfer is essentially dependent upon the concentration gradient; whereas material handling and transportation normally deal with mega level, i.e., transfer of huge tonnage of materials. Here the concentration gradient has no role to play what-so-ever.

In food processing we come across a large number of applications of material handling (solids). A few of them are:

- transportation of grains for packaging/processing,
- hydraulic transportation of fruits and vegetables for cleaning and processing,
- pneumatic conveying of grains in the flour milling industry,
- lifting of grains, etc. to feed to the hoppers, and
- feeding of grains at controlled rate in the flour milling industry.

Various types of material transport: All these material handling operations can be classified into:

- Material (solids) transportation
  - over long distance
  - within the premises of plant for processing,
- Material feeding,
- Material lifting, etc.

The process materials include solids, liquids, gases, pastes, slurries, etc. However, we restrict ourselves in the present chapter only for handling of solid materials. Various material handling equipments are indicated in Table 24.1 which are useful for different purposes. The selection of a particular type of transportation equipment is dependent upon the material characteristics which are described in Table 24.2.

Table 24.1 Various Material Handling Equipment

Type of handling	*Equipment*
Material transportation over long distances	Belt conveyors Chain conveyors Hydraulic transportation
Material transportation within the premises/plant	Belt conveyors Screw conveyors Chain conveyors Pneumatic conveyors Bucket elevators Interlock transfer cranes Fork-lift trolleys Battery operated trucks Chain-pulley cranes
Feeding of materials	Mechanical shakers Oscillatory shakers Vibratory shakers Screw conveyors Bucket conveyors Pneumatic conveyors
Lifting of materials	Bucket elevators Flight elevators Pneumatic conveyors Hydraulic lifts Jockey lifts
Material transportation for processing	Chain-pulley conveyors Roller conveyors Belt conveyors Belt-roller conveyors Slat-belt conveyors

24.1 MATERIAL CONVEYING EQUIPMENT

As has been mentioned, material transport may be done over long distances after unloading from ships/trucks, or may be transported within the plant, which we describe as material handling and will be discussed separately. In this section we describe various material transport/conveying equipment.

24.1.1 Belt Conveyors

Belt conveyors are very much used in transportation operations such as in metallurgy, chemical processing, food processing, pharmaceutical industry and bulk drug industry, etc. Hence, they

Table 24.2 Material Characteristics of Various Bulk Solids

Parameter	*Material characteristics*	*Remarks*
Shape	Irregular, lumpy	Difficult to handle and transport
	Spherical	Easy to handle and transport
Size	Very fine	< 100 mesh
	Fine	100 mesh–3.18 mm
	Granular	3.18 mm–12.7 mm
	Lumpy	> 12.7 mm
Type	Lumpy	Cereal-based products
	Fibrous	Mostly animal-based products
	Stringy	Gelataneous materials
	Sticky	Containing high sugar
	Very light and fluffy	Some of the puffed grain products
	Fragile and shear sensitive	Needs care in handling material
Abrasiveness	Non-abrasive	Easy to transport
	Mildly-abrasive	
	Very abrasive	Difficult to transport
Flowability	Very free flowing	Angle of repose upto 30°
	Free flowing	Angle of repose 30°–45°
	Sluggish	Angle of repose > 45°
Special features	Contaminable	Affects the use/salability
	Hygroscopic	Needs to be protected
	Highly corrosive	Needs to protect the associated equipment
	Mildly corrosive	Needs to protect the associated equipment
	Gives off dust/fumes	Needs closed door transportation

are considered as most universal transportation equipment to carry heavy loads as high as 5000 tph, and the belt travels with a speed of 5 m/s. They can be used for transportation of solids to far off places upto few miles. They can also be used for short distance transportation and within the plant also. They are rarely used in processing operations.

A typical flat belt conveyor is shown in Figure 24.1. It is a short distance travel belt conveyor with a continuous belt. It just conveys the material from one end to the other. In some design, the plate support arrangement varies as shown in Figure 24.2. In some arrangements, the bottom support plates are not continuous sheets. They are idler plates with some degrees of inclination. Figure 24.2(a) shows 20° idler plates. Even 45° idler plates are also possible. Figure 24.2(b) shows a troughened belt on continuous plate. Some useful data of belt conveyors for troughened belts was given in Perry's Hand book (Perry, et al., 1984). In some equipment, the belt moves on rollers as shown in Figure 24.3.

Figure 24.1 Belt conveyor.

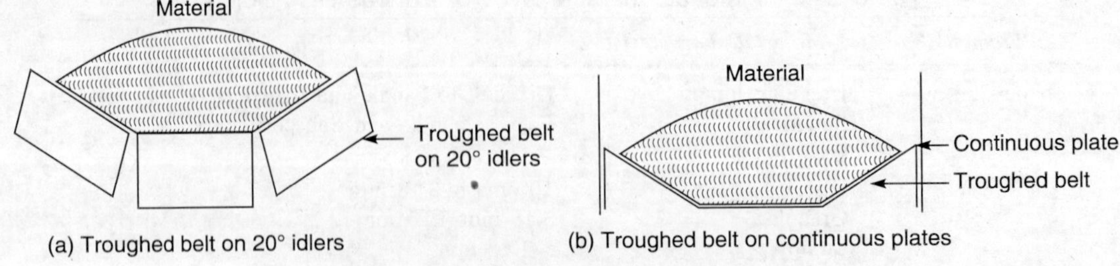

Figure 24.2 Typical belt conveyor idler arrangements.

Figure 24.3 Belt-roller conveyor.

It is normally not possible to change the direction of flow with belt conveyors. In case it is desired to change the direction of flow, we prefer to use another belt conveyor in the changed direction. In such a case, care must be taken to see that the material is properly handled on to the other belt drive without spillage. The flat belt conveyors with magnetic pulley arrangement are used to separate the magnetic materials (ferrous tramp metal) and iron pieces if any from the food material while transporting the material. Such an arrangement is shown in Figure 24.4.

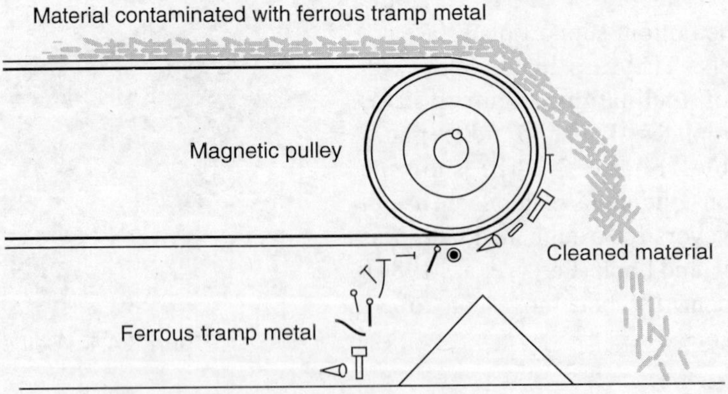

Figure 24.4 Magnetic pulley belt conveyor.

Belt conveyors are also used to transport the material not only in straight directions, but also in bent positions. In such cases, usually slat belts are used. A typical arrangement of a slat belt conveyor is shown in Figure 24.5. These conveyors comprise of wooden or metal slats, or sometimes plastic slats. The slats are arranged in such a way that they make a flat surface. They have good load-bearing capacity. Even in case of repair also, repairing/replacement of any one broken slat is much easier as compared to repairing or replacing of a whole belt. The slat conveyors are generally used for transporting packaged goods, cartons, cans or small bulky items like fruits, vegetables or fish.

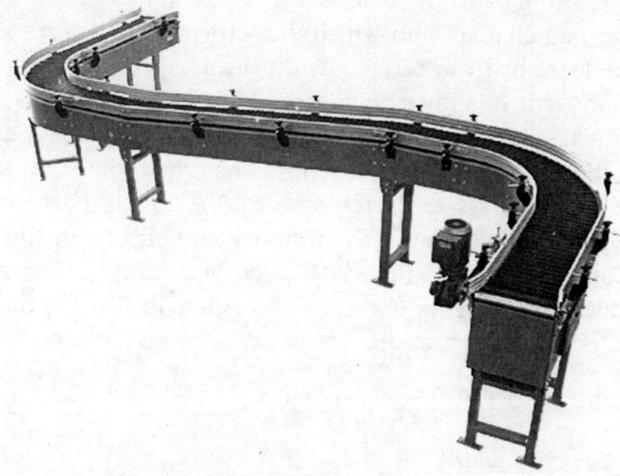

Figure 24.5 Slat belt conveyor.

The belt conveyors can be used to transfer the material in a slightly inclined direction also depending upon the flowability of the material. If the material is relatively sluggish, it can be transported even on an inclined belt conveyor. Figure 24.6 shows a typical belt conveyor in an industrial operation in inclined position.

Figure 24.6 Industrial belt conveyor.

In food processing, mostly belt conveyors are used for transporting grains to fill into silos or to fill into unit bags. Salt and powdered materials are transported on troughened belt conveyors.

24.1.2 Screw Conveyors

Screw conveyors are one of the oldest and most versatile conveyors used extensively in food processing and other allied industries for transportation of solids upto a maximum capacity of 4.7 m^3/min over a long or short distances. Preferably they are used for short distances within the process plant or for feeding of raw materials into the hopper at a predetermined rate. They consist of a trough or a long channel into which the screws in the form of helicoid flights (helix rolled from a flat steel sheet or from a flat steel bar) rotate by means of an electric motor[†]. A schematic representation of it is shown in Figure 24.7. It is similar to screw conveyor dryer shown in Figure 17.17, except that it does not have provision for passing steam for heating. The screw rotates horizontally with the help of a motor. Generally, the screws rotate with low rpm. The material is fed at one end. As the screw rotates, the flights push forward the material in the screw conveyor. Feed is fed continuously from one end either directly or from storage silos. The flow rate in the conveyor depends upon the pitch of the flight. One rotation will move the material forward by one pitch length. The rpm of shaft indicates the number of pitch lengths moved in one minute. Some specifications for a screw conveyor are shown in Figure 24.8.

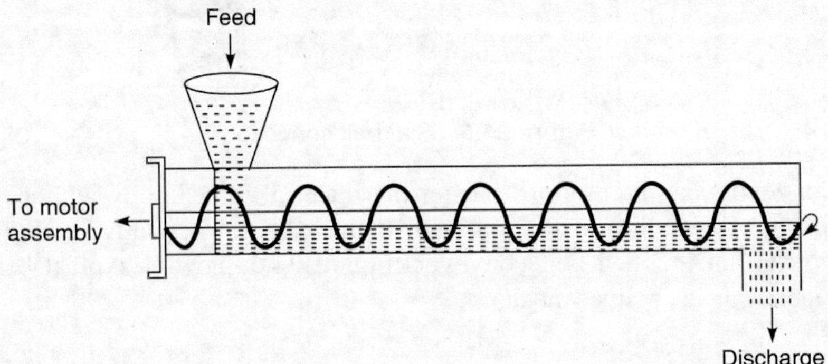

Figure 24.7 Schematic representation of screw conveyor.

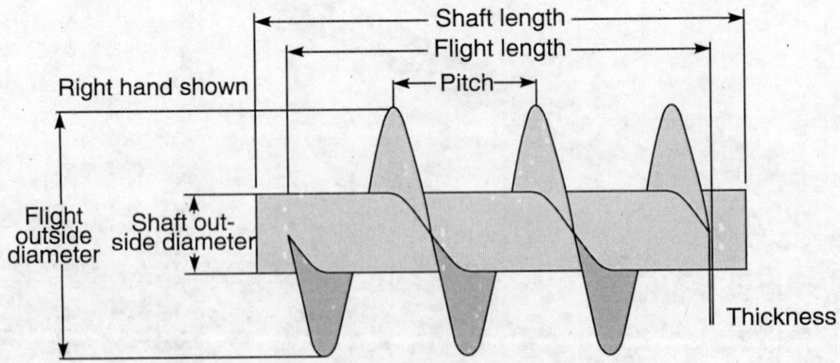

Figure 24.8 Conveyor screw specifications.

[†]Occasionally the screw is rotated manually especially for feed purpose upto a maximum distance of 30 cm or so.

They can also be used to lift the material and transport upto a maximum of 45° inclination depending upon the flowability of material. Figure 24.9 shows a typical industrial screw-conveyor arrangement. Table 24.3 shows some data of screw conveyors from Perry's Hand book (Perry, et al., 1984) for handling different capacities of material from 5–40 tph for a material whose density is 801 kg/m^3.

Figure 24.9 Industrial screw conveyor.

The screw conveyors are frequently used as feeders for feeding the raw material for the process plants, like feeding of wheat from the storage silo into grinding mill or for feeding of spices like chillis into the grinding unit to make powder, etc. Bulk dry materials (like sugar, flour or grains) and semi-liquid non-abrasive materials (like pulps or comminuted meat) can be transported over small distances by screw conveyors. They are also used for feeding the right quantity of materials by lifting them to a small height in an inclined screw conveyors.

Table 24.3 Some Useful Data for Screw Conveyors for Different Capacities for a Material with Density of 801 kg/m^3

Capacity (tph)	*Diameter of flights* (cm)	*Diameter of shaft* (cm)	*Speed* (rpm)	*Feed section diameter* (cm)	kW *of motor*† 4.6 m *maximum length*	9.2 m *maximum length*	13.8 m *maximum length*	18.5 m *maximum length*	22.9 m *maximum length*	*Max* kW *capacity at speed listed*
5	22.9	5.08	40	15	0.32	0.63	0.95	1.26	1.57	3.6
10	25.4	5.08	55	22.5	0.63	1.26	1.68	2.02	2.8	4.9
	25.4	5.08	80	22.5	0.95	1.68	2.52	2.94	3.67	7.15
15	30.5	5.08	45	25.4	0.95	1.68	2.52	2.94	3.67	4.0
	30.5	7.5			0.95	1.68	2.52	2.94	3.67	8.7
20	30.5	5.08	60	25.4	1.26	2.2	2.94	3.63	4.2	5.36
		7.5			1.26	2.2	2.94	3.63	4.2	11.62
25	30.5	5.08	75	25.4	1.58	2.8	3.67	4.2	4.9	6.7
		7.5			1.58	2.8	3.67	4.2	4.9	6.7
	35.5	7.5	45	30.5	1.58	2.8	3.67	4.2	4.9	8.7
30	35.5	7.5	55	30.5	1.68	2.94	3.76	5.0	5.6	10.65
35	35.5	7.5	65	30.5	1.95	3.41	4.4	5.2	6.5	12.6
40	40.6	7.5	50	30.6	2.2	3.35	5.0	6.0	7.45	9.7

†kilowatts listed are calculated for average conditions and are of the proper motor size with factors for length of conveyor, momentary overloads, etc. taken into consideration

Source: Perry, Green & Maloney, *Perry's Chemical Engineers' Handbook*, p. 7.7, 6th ed., Copyright (1984), McGraw-Hill, Singapore. The material is reproduced with the permission of The McGraw-Hill Companies.

24.1.3 Chain Conveyors

They are a type of continuous conveyors where the materials in the form of bulk packages are transported. The bulk packages/containers like crates, barrels or churns are placed over a moving chain with protruding lugs at the ground level. A typical chain conveyor (M/s Jorgensen conveyors, USA) is shown in Figure 24.10. Overhead chains are also used to transport wheeled trolleys or bulk couplings which move the meat carcass or chicken without being touched by hand in the meat or poultry industry. A horizontal conveying chain with side couplings or flights is shown in Figure 24.11. The movement is horizontal, but the railing can move on a overhead track. The material is hanged to the side flight which moves on the overhead railing along, and will be discharged when it reaches the end on to another conveyor. The chain can be stopped in between, and any processing if required can be done to the hanging material. For example, the deskinning of the carcass can be done by stopping the movement of the chain. Later on when the deskinning operation is over, the movement of the chain is started so that the material moves and gets discharged at the end.

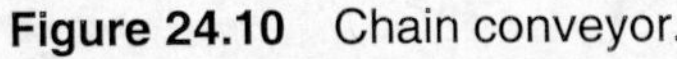

Figure 24.10 Chain conveyor.

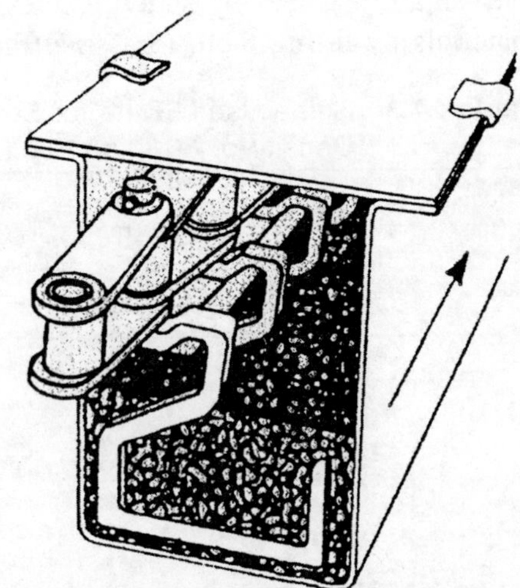

Figure 24.11 Horizontal conveyor with side-pull chain.

24.1.4 Roller Conveyors

Roller conveyors can be operated with or without power for movement. Usually they are powered with a chain drive for easy movement of material. The rollers are fitted with end sprockets which are in turn driven by an endless chain. The roller conveyors are invariably used for transportation of packages, cartons, pellets or any other packaged material. They are not normally used for unpacked food materials. They consist of a series of rollers fixed on either sides with ball bearings. The rollers are closely placed with a small clearance in between so that each roller rotates independently. While rotating, they transport the carton placed on them. A typical roller conveyor is shown in Figure 24.12 which moves horizontally only. Some roller conveyors are made to turn the movement of the material horizontally as shown in Figure 24.13.

Figure 24.12 Roller conveyor.

Figure 24.13 Roller conveyor with bending.

Figure 24.14 Roller conveyor in inclined position.

Normally, whenever we need to change the direction of movement of the packaged material/cartons, either roller conveyors or slat belt conveyors alone are preferred. Roller conveyors have an additional advantage that the moving material is never blocked as long as its size is bigger than the clearance between each roller. Roller conveyors can also be made for movement of the material in the inclined position also both in upward and downward direction (as shown in Figure 24.14). The inclination should be 10°–12° for gravity flow or power driven rollers. Higher degrees of inclination may result in slipping or scating of the packages. Chain driven rollers can be used for heavy duty operation where the speed of conveying could be as high as 2.5 m/s, whereas belt driven rollers are used only upto 0.25 m/s. They can be fitted with reversible drives so that the material can be moved in either direction. In some conveyors, provision is made for intermediate discharging of material also.

24.1.5 Pneumatic Conveyors

Pneumatic conveying is an excellent method for transportation of powders which are easily flowable. The conveying system uses a gas, usually air, to transport the material. Granular and powder foods like cereals, seeds, pulses, sugar, or any kind of ground products in the form of particulates or powders can be conveniently transported by pneumatic conveying systems.

The material is transported along with air flowing at high pressure or under vacuum. Usually air velocities are 15–30 m/s flowing in pipes of 5 to 40 cm diameters. A schematic representation of pneumatic conveyor is shown in Figure 24.15, where the material is fed from a storage silo through a rotary valve. The speed of the valve controls the flow rate of the feed. The material is kept in suspension along with air in the pipelines, and is transported through the pipelines upto the point of destination. The destination point could be the processing equipment. These conveyors are also used for unloading the materials from the ships to the place of storage, or into the silos. The distance of transportation can vary anywhere between 70–600 m. Usually air pressures are maintained at 700 kPa (7 atm). Higher air pressures upto 900–1000 kPa are also used, and materials can be transported upto 2–3 km (Leninger and Beverloo 1975).

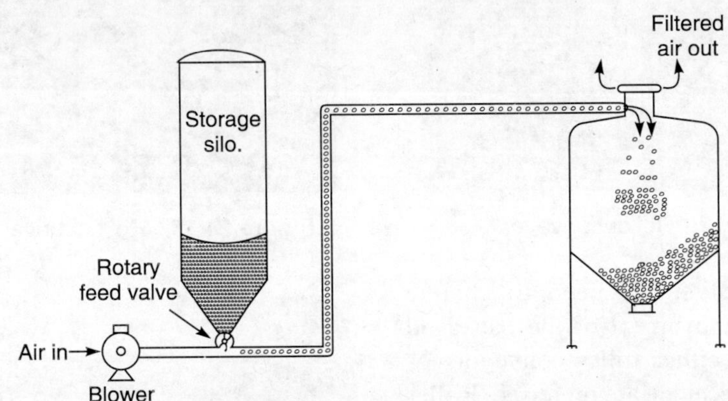

Figure 24.15 Pneumatic conveying system under pressure.

Pneumatic conveying can be done by the following methods also even though we have described only pressure conveying method.

(i) Pneumatic conveying under vacuum
(ii) Pneumatic conveying under pressure-vacuum combination
(iii) Pneumatic conveying by fluidization method

Just as the pressure systems blow the material, vacuum systems suck the material.

In the fluidization method, the material is kept in a state of fluidization by blowing air from the bottom. The material is transported either by means of entrainment (Section 17.4) or by the vibration of the bed which is known as vibro-fluidized bed. The fluidized bed system is shown in Figure 24.16.

In the pneumatic transportation systems, pressure and flow rate of air should be sufficient enough so that the material is kept in a state of suspension. If it is less, the material will start settling down. The pneumatic conveying process is an excellent method for transportation of powders, that too in closed containers (pipelines). This makes the transportation dust free which is an attractive feature to prevent atmospheric pollution. Another attractive feature is that the inlet and outlet of the pipelines can be moved with hose pipes which is very useful for unloading of material from the ships. High hygienic conditions can be maintained during transportation. The labour costs for unloading can be considerably reduced.

Only disadvantage with pneumatic conveying is high energy costs for pumping large quantities of air under high pressure. In case of transportation of powders, care should be taken to see that the material does not escape with the outlet air.

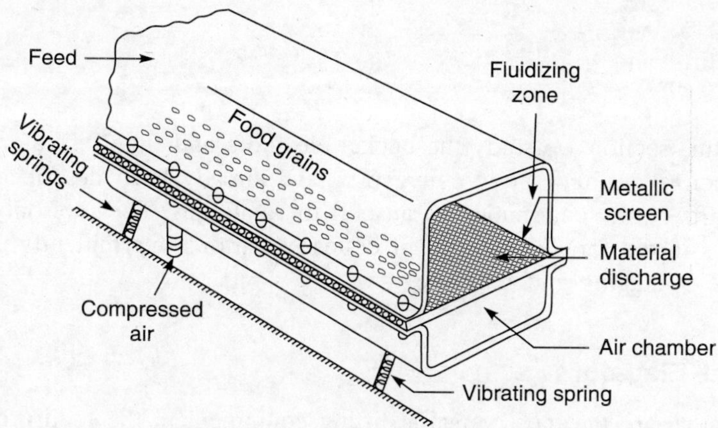

Figure 24.16 Vibro-fluidized conveyor.

24.1.6 Hydraulic Transportation

Hydraulic transportation/conveying is akin to pneumatic transportation except that the carrying fluid is a liquid, which is usually water. Since the particles with diameter less than 50 microns settle very slowly, they remain in a state of suspension in moving liquids, and hence, are amenable for hydraulic transportation. The fluid velocities are usually 1–5 m/s. Transportation is done by dragging the solid particles in a stream of water. Usually for transportation over a long distances (few kms), hydraulic transportation systems are used in the industrial practice. However, such applications are rare in food processing.

In food processing, usually hydraulic transportation is used in combination with rinsing or washing. For example, in the fruit processing units, the fruits are to be transported from the ripening sheds to the processing hall. They also have dust and dirt adhering to their bodies. Such fruits are loaded into a water channel (canal) in which there is a continuous flow of water. The fruits move along with the flowing water. During the process, the fruits tumble down and are cleaned. In some semi-automatic processes the workers will be standing on either side of the canal, and watch the fruits. The fruits which are spoiled or have blemishes or bruises will be separated and discarded manually. This serves the quality check for the processing of fruits.

The used water is collected at the end, and is recycled after filtration. The recycling water is pumped through the canal again from the feed end for transporting the fruits. This serves both the purposes of washing and transportation.

24.2 MATERIAL LIFTING EQUIPMENT

Under this section we describe various equipments for lifting the material vertically or in inclined position either for transportation or for further processing. The equipment used for lifting the material continuously are (Table 24.1):

- bucket elevators,
- flight elevators,
- pneumatic conveyors,

- screw conveyors,
- hydraulic lifts, and
- jockey lifts.

Of them, in this section we study the bucket elevators and flight elevators. The pneumatic conveyors and screw conveyors have already been described. The hydraulic lifts and jockey lifts do not continuously transport the materials. They are only used for lifting the food materials in the form of packets or crates or cartons intermittently to the processing height or into the packaging godowns, etc.

24.2.1 Bucket Elevators

The bucket elevators are the most versatile lifting equipment in the food processing industry for lifting the solid food materials either vertically or in an inclined position. It consists of a series of buckets connected to an endless belt or a chain. The belt or chain will be moving, and hence, the buckets also move upwards in one direction as shown in Figure 24.17. The bucket elevators carry the material in closed choots or in open form. The free flowing solids are fed at the bottom continuously by means of some feeding arrangement into the bucket under the feeding point. As the bucket moves upwards, the bottom bucket occupies the position of feeding point. Thus, the movement continues. The buckets may be closely placed or distantly placed (Figure 24.17 is a distantly spaced bucket elevator). The material discharge could be by way of centrifugal force or by simply dumping by tilting the buckets. Various forms of bucket elevators are available as follows (Perry, et al., 1984):

(i) Centrifugal discharge spaced buckets,
(ii) Positive discharge spaced buckets,
(iii) Continuous buckets,
(iv) Super capacity continuous buckets, and
(v) Spaced buckets with scooping arrangement.

All of these arrangements are characterized by bucket positioning, feeding method and discharge provisions. Figure 24.17 is a centrifugal discharge spaced buckets. In the positive discharge systems, the carrying chains are snubbed back as shown in Figure 24.18 (only discharge position is shown in the figure; rest of the arrangement is similar to that shown in Figure 24.17). This discharge method is particularly useful for wet or sticky materials. The flow rate of materials is controlled by the shape and spacing of the buckets and the speed of the conveyor (usually 15–100 m/min); continuous discharge is possible by closely spacing the buckets. Feeding is to be properly done by controlling the feed through a screw conveyor. Buckets are made of metal or plastic material. However, for longer shelf-life, metallic buckets are preferred.

Bucket elevators are commonly used for free flowing materials like sugar, beans, salt, cereals, etc. The bucket elevators are extensively used in wheat milling industry or paddy processing units for lifting the cereals from the ground level to feed into the mills at a higher level, and the products move downwards by gravity and get processed.

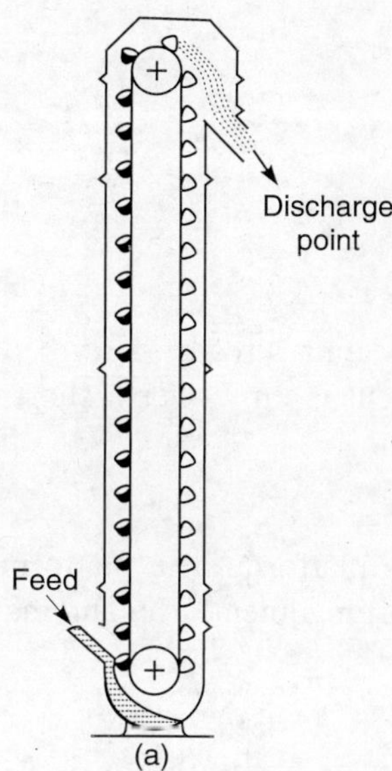

Figure 24.17 Bucket elevator.

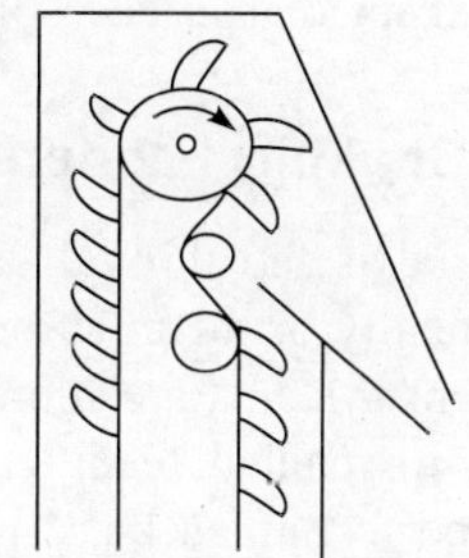

Figure 24.18 Positive discharge.

24.2.2 Flight Elevators

Flight elevators are particularly used for elevating granular/powdery material without being in contact with the environment. It consists of a casing in which the flights move on an endless chain pulley. The flights are specially designed to lift the material and deliver at the discharge end. A schematic representation of it is shown in Figure 24.19.

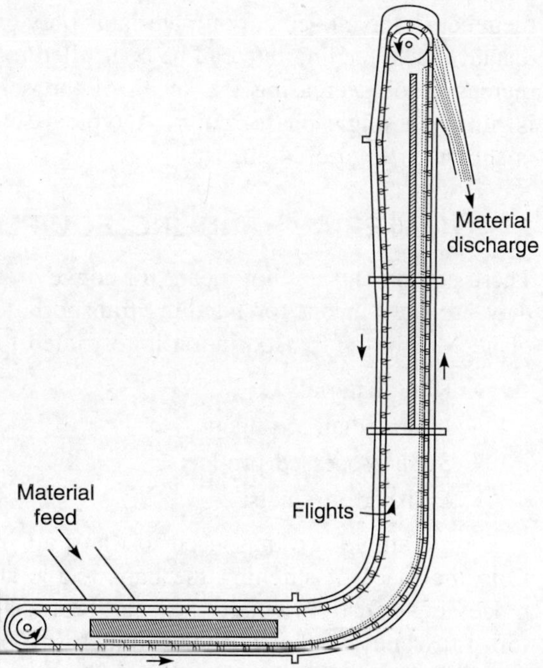

Figure 24.19 Flight elevator.

24.3 FEEDING EQUIPMENT

Feeding of known and desired quantity of materials is an essential requirement in any process industry including food processing industries. Feeding of known quantity of solids is more complicated (as contrast to liquids and gases whose flow rates can be controlled with valves) because of difficulty in flowing of solids. Different types of feeding equipments are:

(i) Shaking feeders
   - mechanical
   - oscillatory
   - vibratory

(ii) Bucket conveyor-cum-feeders

(iii) Screw conveyor-cum-feeders

The latter two feeders are similar to the conveying systems already described in earlier section, except that in case of feeders, the material movement is only over a short distance.

24.3.1 Shaking Feeders

The material is fed through a hopper on a shaking flexible platform. The platform is moved either manually or by an automatic oscillating or vibratory arrangement. The automatic feeding arrangement systems are more versatile and accurate as compared to manual feeding methods. Manual feeding is subjected to human bias.

24.3.2 Screw Conveyor-cum-feeders

They are also known as **screw feeders**, and are similar in construction and operation to the screw conveyors. In the feeders, the movement of material is only over a short distance. The feeding rate can be controlled/regulated by increasing or decreasing the speed of the screw which is run by a single phase motor. A typical screw feeder is shown in Figure 24.20.

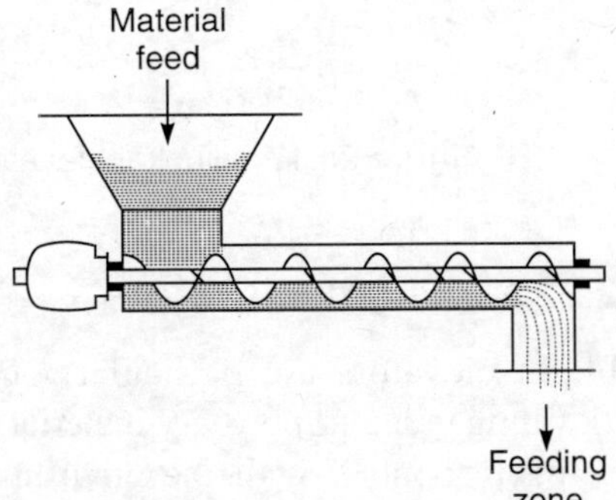

Figure 24.20 Screw conveyor-cum-feeder.

24.4 MATERIAL HANDLING EQUIPMENT

These equipments are not meant for conveying/transport of materials over long distances; rather they are only meant for handling/transportation of materials in the process hall or within the plant. Mostly the transportation is restricted to within the process hall. The materials could be:

- Raw materials,
- Intermediate products,
- Semi-processed products,
- Finished products.

For example, we may need to transport in a paddy processing plant the finished rice bags from the place of stitching the bags to the storehouse godowns or upto the point outside the processing hall from where they can be transported by trucks. We may prefer to use a simple fork-lift trolley. If the finished products are packed in the form of pellets, they are lifted from one place to other using fork-lift trolleys or by using hoist chains or cranes, etc. Thus, there are innumerable applications of transportation of materials within the process plant/within a small distance, and cannot be handled by human beings. This kind of mechanical transportation of materials by using various gadgets have several advantages:

- The handling is contamination free,
- The package is not distorted/deshaped,
- The transportation is economical and quick, and
- Any handling mishap may not result in injury to human beings (workers) in the area.

In view of some of the several advantages, a number of gadgets are used in food processing industries. They include:

(i) Fork-lift trolleys
(ii) Chain hoists
(iii) Chain pulley cranes
(iv) Interlock transfer cranes
(v) Miscellaneous light equipment
(vi) Belt conveyors
(vii) Slat-belt conveyors
(viii) Roller conveyors
(ix) Hydraulic lifts
(x) Jockey lifts, etc.

Thus, some of the equipment used for routine transportation like those mentioned in (vi) to (x) are also used within the process plant for material handling.

Fork-lift trolleys: Fork-lift trolleys are extensively used in food processing plants for shifting the packaged material in the form of cartons or pellets. The pellets will have a bottom slot. The fork-lift trolley shown in Figure 24.21 will have protruding forks which will enter into the slot of the pellet. Later the forks are slightly lifted upwards and pushed forward. Some trolleys have provision for lifting the forks slightly upwards by a hydraulic jack which is operated by the arms handle. The arrangement is not shown in Figure 24.21. Then the material (pellet) is moved by pushing the trolley forward manually or by a motorized engine operated by battery or by electric power or by petrol.

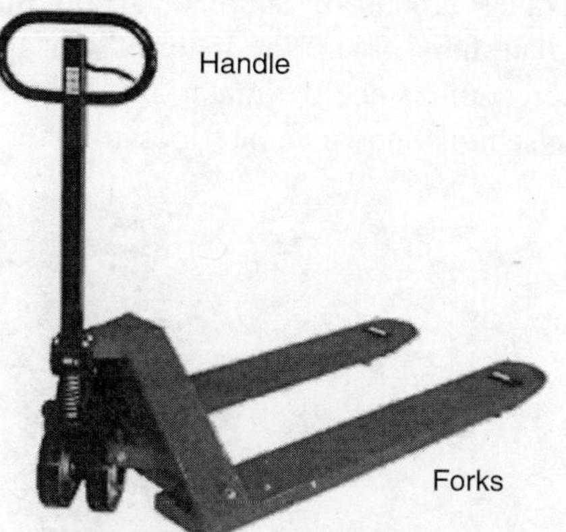

Figure 24.21 Fork-lift trolley.

Chain-hoist: The chain-hoist pulleys are generally used to lift the material so that it can be placed at desired position, or can be loaded on to the fork-lift trolleys or into trucks for transportation. The pulleys may be operated manually or by a motor. Weights upto a ton or more can be easily lifted manually by one man by pulling the endless pulley chain (Figure 24.22). It consists of two hooks, one at the top to fix the chain-hoist to the roof or to a tripod stand made with iron pipes. The bottom hook is attached to the package to be lifted. By operating (drawing) the chain slowly, the bottom hook moves along with the package to any height upto the bottom of the upper pulley.

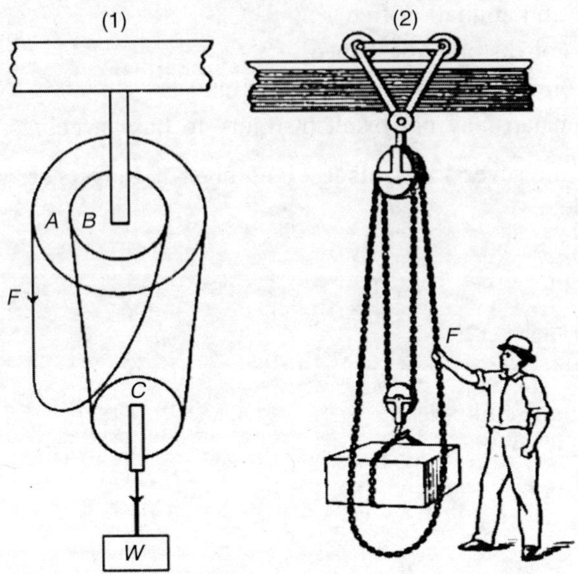

Figure 24.22 Chain-hoists.

Interlock transfer cranes: If the pellets/packages to be moved are heavy, interlock transfer cranes are used. Such an arrangement is shown in Figure 24.23. The lifting is similar to chain-hoist. Later the material can be moved backwards or forwards and also in a right-angled direction. Since the material also moves and is transported like in the case of cranes, the lifting system is known as interlock transfer crane. It is a combination of chain-hoist and transfer crane.

Figure 24.23 Interlock transfer cranes.

Source: Reproduced with permission from M/s. Gorbel Inc. Fishers, New York.

Miscellaneous lifting gadgets: A set of miscellaneous gadgets for lifting materials is shown in Figure 24.24. They are all manually operated, and are meant for lifting/transporting small packets of materials from the ground level to that of process tanks, or for stacking materials in the godowns/storehouses. In some gadgets, the material can be lifted and transported. Even though these miscellaneous gadgets do not find special mention in food engineering books, they are an integral part of any processing unit. They need to be regularly greased and serviced since they are all operated manually.

Some mechanically lifted cranes are shown in Figure 24.25.

Figure 24.24 Miscellaneous material handling equipment.

Figure 24.25 Miscellaneous material lifting equipment (mechanically lifted).

24.5 SELECTION OF MATERIAL HANDLING EQUIPMENT

Various types of material handling needs are highlighted in Table 24.1. These needs include:

(i) Transportation of bulk materials over very long distances, probably of the order of few kilometres.

(ii) Transportation of bulk materials over long distances which may be of the order of 1 km or so.

(iii) Lifting of bulk materials/packaged materials.

(iv) Feeding of materials to process plants.

(v) Handling of packaged materials within the process plant.

For (i), invariably hydraulic transportation or belt conveyors in closed conditions alone can be used. Much of such very long distance transportation is found in ore handling and metallurgy. We do not find much of this kind of application in food processing. Transportation over long distances as indicated in (ii) is found in unloading and delivery of raw materials from the ships like food grains or pulses, or loading of processed products in the form of packages into ships, etc. For this purpose also we use belt conveyors, roller conveyors or chain conveyors. For lifting of bulk raw materials like wheat, paddy, etc., we use bucket elevators or screw conveyors in inclined position. Pneumatic conveyors are also used, but only partially or in special conditions where the grain also needs to be dried before going for grinding. Pneumatic conveyors are excellent for transportation, but their prohibitive energy consumption discourages their utility/applications.

Brennan, et al. (1969) gave a categorization of material handling equipment and their specific applications in different areas of application (Table 24.4) which in fact can be taken as a guide.

Table 24.4 Categorization of Materials Handling Equipment

Classification	*Direction*					*Frequency*			*Location served*				*Service nature*			*Movement height*				*Material state*			
	Vertical up	*Vertical down*	*Incline up*	*Incline down*	*Horizontal*	*Continuous*	*Intermittent*	*Occasional*	*Point*	*Path*	*Limited area*	*Unlimited area*	*Permanent*	*Temporary*	*Non-fixed*	*Overhead*	*Working height*	*Floor level*	*Underfloor*	*Packaged*	*Bulk*	*Solid*	*Liquid*
Conveyors			×	×	×	×			×	×			×			×	×	×	×	×	×	×	
Elevators	×	×				×			×				×			×				×	×	×	
Cranes and hoists	×	×					×				×		×	×				×	×	×	×	×	
Trucks				×		×						×			×			×		×	×	×	
Pneumatic equipment	×	×	×	×	×	×				×			×			×	×				×	×	×

Source: Reprinted from *Food Engineering Operations*, Brennan, J.G., Butters, I.R., Cowell, N.D. and Lilly, A.E.V., p. 388, Copyright (1969), with permission from Elsevier.

24.6 APPLICATIONS OF MATERIAL HANDLING AND TRANSPORTATION IN FOOD PROCESSING

Applications of transportation, handling, lifting of solid materials are innumerable in food processing. The handling methods are required in view of the bulk of the materials (tonnage), and also to ensure hygiene. Some of them are discussed as follows:

24.6.1 Animal Products Processing

Handling techniques in animal products processing are mainly targeted towards hygiene.

- At the time of slaughtering, the animals are transported on roller conveyors for stunning and later they are slaughtered. The dead animal is hung on the chain-hoists and lifted to remove the skin and hides.
- The carcass is transported on chain conveyors for carrying out various process operations.
- Similarly the birds also after slaughtering and defeathering are transported on horizontal conveyors with side pull chain for carrying out various process operations with least intervention of human hands for maintaining high quality of hygiene.

- In marine processing, the fish or prawns are hydraulically transported from the storage place to the processing hall which also helps for cleaning and separation of unsuitable fish or prawns.
- In the IQF (Individual Quick Freezing) units, the frozen and packed prawns are transported on belt/chain conveyors to storage godowns.

24.6.2 Cereals and Pulses Processing

- Paddy in rice mills is lifted by bucket elevators.
- Wheat and other cereals in grinding mills are transported by bucket elevators or screw conveyors to the grinding rollers which are situated at high levels.
- In the energy food units, the raw materials (wheat and Bengal gram) are transported to roasters and then to grinders using bucket elevators and screw conveyors.
- In the bakery industry, the bread loaves/biscuits are transported on chain conveyors to the packaging units. After packaging, the packets are transported on belt conveyors.

24.6.3 Fruits and Vegetable Processing

- The fruits from the ripening godowns are hydraulically transported to processing hall for making the fruit pulp.
- The fruits after precleaning are transported by belt conveyors/roller conveyors to the pulping unit.
- The jam bottles after filling or cans after double seaming, are transported on roller conveyors.
- The squash bottles after processing, are transported on belt or chain conveyors to packaging unit.
- The packaged cartons are transported on lift trolleys or by cranes to the product storage godown.

24.6.4 Spices and Condiments Processing

- In the spice powder making, the individual spices after drying are transported by pneumatic conveyors to the grinding units.
- In the spice oleoresin industry, the ground spices are transported by screw conveyors or by flight elevators.

24.6.5 Confectionary and Convenience Foods

- In the chicklets/eclaire/toffee making units, the final products after wrap twisting are transported on belt conveyors to the packaging unit.
- Most of the convenience foods are in the form of powders, and hence, they are transported by screw conveyors or in some units the ground powders from silos are fed into the packaging units by screw feeders or some other mechanically vibrating feeders. The packages/cartons are transported by belt conveyors.

24.6.6 Miscellaneous Applications

- In any processing unit, the final packaged cartons are transported by belt conveyors or by roller conveyors.
- The feeding from the storage silos in most of the processing operations is done either by screw feeders or by rotary feed valves or by mechanical oscillatory feeders or by check plates.
- Various minor applications of material transportation in food processing using different miscellaneous equipment are innumerable, and hence, it is difficult to document all.

REVIEW QUESTIONS

24.1 How material handling/transportation is different from mass transfer?

24.2 What is the importance of material handling in food processing?

24.3 Classify various types of material handling/transportation systems.

24.4 Describe a belt conveyor with a neat diagram and its application in food processing.

24.5 Describe a screw conveyor with a neat diagram and its application in food processing.

24.6 Describe a chain conveyor with a neat diagram and its application in food processing.

24.7 Describe a roller conveyor with a neat diagram and its application in food processing.

24.8 Describe a pneumatic conveyor with a neat diagram and its application in food processing.

24.9 Describe a bucket conveyor with a neat diagram and its application in food processing.

24.10 Describe a flight conveyor with a neat diagram and its application in food processing.

24.11 Describe a hydraulic transport system and its importance in food processing.

24.12 What are various feeding systems you come across in food processing?

24.13 What are various lifting gadgets used in food processing? Describe them.

24.14 Describe a chain-hoist with a neat diagram and its applications in food processing.

24.15 Describe various material properties based on which you make a choice of the transportation system.

24.16 Explain the criteria for selection of material handling/transportation equipment based on material characteristics.

24.17 What are various applications of material handling/transportation systems in food processing industries?

24.18 What are various applications of material handling/transportation systems in animal products processing industries?

24.19 What are various applications of material handling/transportation systems in fruit processing industries?

24.20 What are various applications of material handling/transportation systems in cereal processing industries?

24.21 What are various applications of material handling/transportation systems in spice processing industries?

REFERENCES

Brennan, J.G., Butters, J.R., Cowell, N.D. and Lilly, A.E.V. (1969), *Food Engineering Operations*, Elsevier Pub. Co., Amsterdam (Netherlands), pp. 383–400.

Leninger, H.A. and Beverloo, W.A. (1975), *Food Process Engineering*, D. Reidel Pub. Co., Dordrecht (Holland), pp. 107–128.

Perry, R.H., Green, D.W. and Maloney, J.O. (1984), *Perry's Chemical Engineers' Handbook*, 6th ed., McGraw-Hill, New York, pp. 7.2–7.20.

CHAPTER

25

Fermentation

Fermentation is a biochemical process in which a substrate (reacting species) is converted to useful metabolic products or multiplied as a biomass under the action of microorganisms (bacteria, yeast and molds) or catalyzed by enzymes. The process develops a typical flavour or aroma or taste (mainly sour) and texture to the food products. The main food components such as cereals, pulses (which contain essentially carbohydrates) or fruit juices or sugars in the foods or the starchy materials in the foods act as substrates that undergo fermentation process. The enzymes present in the food materials or added externally catalyze the biochemical process. The fruit beverage products produced by fermentation and alcoholic drinks are in vogue from time immemorial (about 4000 BC). Preparation of curds by fermentation of milk or pickling of fruits and vegetables to extend the shelf-life are some of the classical applications of fermentation in food processing. Preservation of food and beverages resulting from fermentation has been an effective form of extending the shelf-life of foods. Fermented foods are made all over the world from various raw materials such as wheat, maize, sorghum, cassava, soybeans, vegetables, fish, milk and alcohol. Traditionally, preservation was through naturally occurring fermentation processes only. With industrialization for modern large-scale productions, the process is generally done by newly exploited (defined) strain starter systems. This would ensure consistent quality in the final products and enable large-scale production (mass production). Table 25.1 lists representative examples of fermented foods world over.

Table 25.1 Lactic Acid Bacteria (LAB) in Preservation of Various Foods

S. No.	*Fermentation of material*	*Product*	*Microorganisms used & Remarks*
1.	Fermentation of fresh vegetables such as cabbage	Sauerkraut, pickling of cucumbers Korean *Kimchi.*	Combination of LAB (Lactic Acid Bacteria).
2.	Fermentation of cereal (Maize)-yoghurt	Nigerian *Ogi*, Kenyan *Uji*	Lactic bacteria *Cephalosporium*, yeast and a number of organisms.

(contd.)

Table 25.1 (Lactic Acid Bacteria) in Preservation of Various Foods (*contd.*)

S. No.	*Fermentation of material*	*Product*	*Microorganisms used & Remarks*
3.	Fermentation of rice and black gram	South Indian *Idli* and *Dosa*	LAB, *Torulopsis, Candida, Trichosporon pullulans.* Idli is a steam-cooked spongy-cake consumed as a breakfast food.
4.	Non-baked bread like products without wheat	Philippines *Puto*	It is a generic term for steamed rice cakes made from fermented rice flour with yeast.
5.	Fermentation of milks	Yoghurts and cheese	LAB, *Propionibacterium shermanii.* Yoghurt is a good probiotic source.
6.	Fermented wheat-milk mixes	Egyptian *Kishk*, Greek *Trahanas*	*Lb. brevis, Lb. casei* and *Lb. plantarum.* It is made from sheep milk yoghurt and paraboiled wheat.
7.	Fermentation of soya beans	Indonesian *Tempe*	*Rhizopus oligosporus* Protein-rich vegetable meat substitutes.
8.	Pastes produced by fermentation of cereals and legumes	Soya sauce	Fungal fermentation by *Rhizopus oligosporus.* It is used as a condiment in several culinary preparations.
9.	Fermented cereal-fish-shrimp mix	Philippine *balao balao*	3% salt added to rice-shrimp mix, it ferments to 1.32% acidity in 4 days. Acid preserves it without any heat treatments.
10.	Fermentation of finger millet gruel	Indian *Ragi ambali*	*Lactobacillus*, and LAB. It is a rich source of vitamin B_{12} and probiotics.
11.	Fermentation of fruits and vegetables	Indian pickles	*Pediococcus cerevisiae, Lb plantarum.*
12.	Fermentation of white wheat flour	Indian *nan*	*Saccharomyces cerevisiae*, LAB. Consumed as a part of staple food.
13.	Fermentation of teff or sorghum	Ithopean *Injera*	Yeast, fungi such as *Pullaria sp, Aspargillus sp,* and some unidentified bacteria. It is a staple food of Ethiopea.

Some important bacteria used in food processing are:

- *Lactobacillacae* to produce lactic acid from carbohydrates,
- *Acetobacter* for fermentation of fruits and vegetables (which oxidize alcohol to vinegar),
- *Bacillus* for fermentation of legumes,
- Lactic acid bacteria (LAB).

Some important yeasts are:

- *Saccharomyces* (*S. cerevisiae*) for producing wine and alcoholic beverages, and for leavening of bread,
- Some of the yeasts like *candida* are useful for producing Single Cell Proteins.

The fermentation takes place in a container which is typically known as **Fermenter**, under some ideal process conditions such as temperature, pH, extent of mixing, type of microorganism/ enzyme, pressure, oxygen concentration, etc. It is imperative that fermentation proceeds with the intervention of living organisms. The presence of microorganisms to convert the substrate into useful metabolic products distinguishes the whole process from usual chemical processes in which no microorganisms are involved. Fermentation proceeds with the growth of microorganisms and depletion of the reactant (substrate). Thus, initially two competing processes go in fermentation concurrently. They are:

- growth of microorganisms
- depletion of substrate

For the initial growth of microorganisms, we require a nutrient medium. Once they substantially multiply, they survive by consuming the substrate in which process, the substrate gets converted into useful metabolic products.

Thus, for fermentation to proceed effectively we require the following:

- a microorganism/enzyme
- substrate
- nutrient
- a sterile medium
- ideal/suitable process conditions

25.1 ROLE OF FERMENTATION IN FOOD PROCESSING

Fermentation has several applications in food processing, viz.,

- preservation of foods by producing acids such as lactic acid, acetic acid, etc.,
- improving food safety by inhibiting pathogens and toxicants. *Bacillus subtilis* causes hydrolysis of proteins to amino acids and peptides releasing ammonia and thus increases alkalinity. Alkalinity makes the substrate unsuitable for growth of spoilage microorganisms,
- improving the nutritional status of foods by producing some vitamins,
- improving the taste of foods (organoleptic quality) by producing some food acids like acetic acid, lactic acid.

25.2 FERMENTATION GROWTH MODELS

The bacteria cell growth in fermentation broth has several stages, viz.,

- initially the cell growth is slow what we call as lag phase. In this phase, the cells utilize the nutrients and grow,
- later the cell growth is faster which is known as log phase until it reaches a stationary phase,
- the stationary phase continues for some time,
- finally, we have the death phase in which the cells decay due to the fermentation conditions and lead to the cessation of fermentation.

All the above four stages are shown in Figure 25.1. During the log phase, the cell growth is proportional to the number of cells present at that time; more is the number of cells, more is the growth rate. It can be represented by Malthu's law as follows:

$$\frac{dn}{dt} = \mu n \tag{25.1}$$

where n is the mass of cells per unit volume (cell density), μ is the constant of proportionality and is known as specific growth rate in $(h)^{-1}$ and t is the time in h.

However, as can be seen in Figure 25.1, the cell growth rate is not always increasing or decreasing, it takes different stages, viz., log phase, stationary phase and death phase. To account reasonably for all these stages, probably a better representation was proposed by Monod (1949) as follows:

$$\mu = (\mu_m\, s)/(K_s + s) \tag{25.2}$$

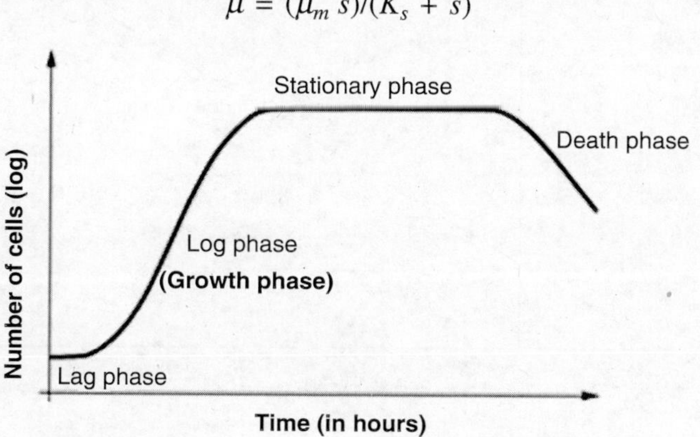

Figure 25.1 Cell growth curve in a batch fermentation.

where μ_m is the maximum achievable growth rate, s is the substrate concentration, and K_s is the limiting substrate concentration when the specific growth rate is equal to half of the maximum specific growth rate. Thus, when

$\mu = \mu_m/2$, as demonstrated in Figure 25.2, $s = K_s$

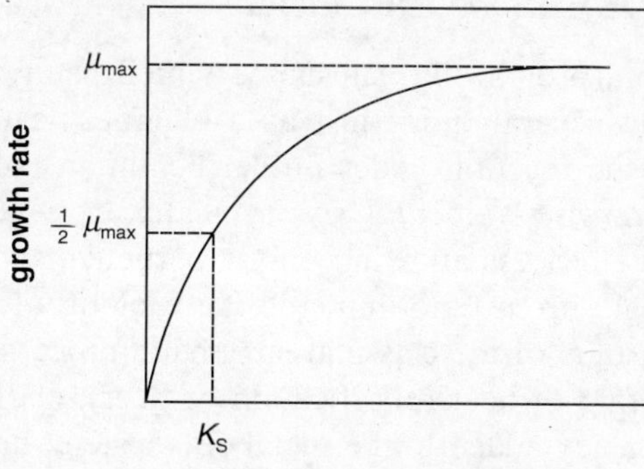

Figure 25.2 Determination of K_s at half specific growth rate.

If in Eq (25.2), μ is replaced by $\mu/2$, we get

$$\mu\frac{1}{2} = (\mu_m\, s)/(K_s + s) \tag{25.3}$$

In the above equations, it is always difficult to measure μ_m, we evaluate the same by linearizing the Monod equation as follows:

$$\frac{1}{\mu} = \frac{K_s}{\mu_m}\left[\frac{1}{s}\right] + \frac{1}{\mu_m} \tag{25.4}$$

Eq (25.4) is used to draw Figure 25.3 and measure μ_m and K_s by noting down intercept and abscissa values as shown in Figure:

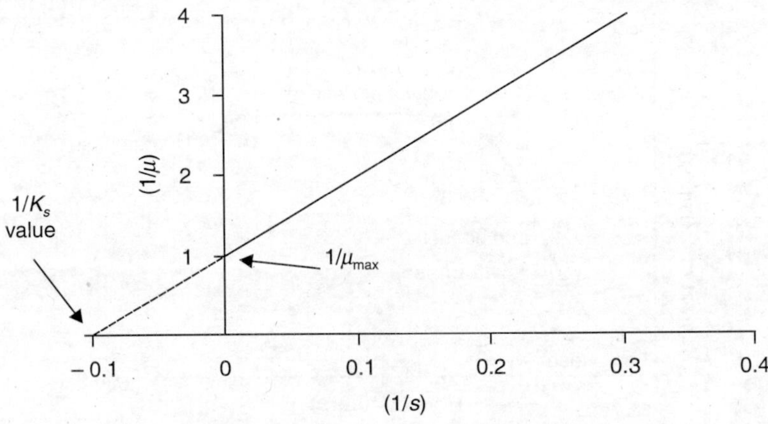

Figure 25.3 Linearized Monod Equation (Eq 25.4)

A detailed further discussion on cell growth kinetics and Monod equation with its limitations is available in several standard textbooks on Biochemical Engineering (Bailey and Ollis, 1986; Rao, 2010 (a)).

25.3 ENZYMATIC PROCESS MODELLING

An enzyme is a protein or a protein-like substance which catalyzes a biochemical process. It enhances the process rate several times just as a chemical catalyst does in the chemical process rate, or carries out the reaction under milder conditions of temperature and pressure. That is why enzymes are known as *biocatalysts*. Typically, enzyme activity is specific; i.e., it allows only a particular route process allowing only a particular type of reaction to take place. Enzyme reactions are in vogue from time immemorial. Making of cheese, leavening of bread, manufacture of vinegar are some of the classical enzymatic processes known to men. Enzymes are also used for chill proofing of beer (clarification of beer), clarification of fruit juices, etc.

Enzyme is an outside agency added to the metabolic system, unlike microorganisms which are ubiquitous and present naturally. Another peculiar difference in enzymatic processes (as contrast to microbial fermentation) is that they don't multiply in the reaction system unlike microorganisms. Their concentration remains the same during the process and remains with the system even after the reaction is over. In this respect, their behaviour is like chemical catalysts. Hence, the enzymatic reactions may be in:

- homogeneous phase (mixed with the system), or
- heterogeneous phase (does not mix with the system)

Heterogeneous systems are made by immobilizing the enzymes on a non-reactive support material.

An enzymatic reaction may be represented as follows:

$$E + S \leftrightarrow E.S \tag{25.5}$$

where E is an enzyme, S is a substrate and $E.S$ is the enzyme-substrate complex. It is a reversible reaction system in which $E.S$ complex forms and as well as it breaks depending on the process conditions. The reaction rate of such reversible enzymatic reactions is represented by Michaelis-Menten equation (Eq. 25.6).

25.3.1 Michaelis-Menten Equation

The rate expressions for homogenous catalytic systems are typically modeled by Michaelis-Menten (1913) equation as follows:

$$v = \frac{v_{max}[S]}{[S] + K_m} \tag{25.6}$$

where v is the enzymatic reaction rate, v_{max} is the maximum attainable rate, $[S]$ is the substrate concentration and K_{max} represents substrate concentration corresponding to reaction rate of $v_{max}/2$, i.e., when half of the maximum reaction rate is attainable as shown in Figure 25.4. It is similar to Figure 25.2.

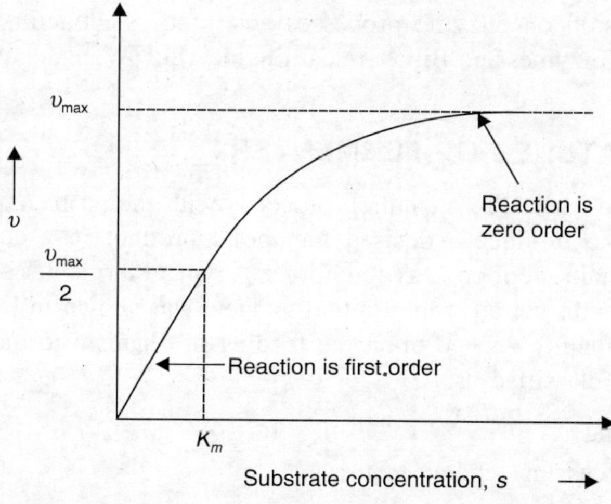

Figure 25.4 Michaelis-Menten plot to determine K_m.

The Michaelis-Menten constants are evaluated by linearizing the Michaelis-Menten equation by taking its reciprocal, as we did earlier for Monod equation.

$$\frac{1}{v} = \frac{K_m + [S]}{v_{\max}[S]} = \frac{K_m}{v_{\max}} \frac{1}{[S]} + \frac{1}{v_{\max}} \tag{25.7}$$

Eq (25.7) is drawn in the form of a plot which is known as Lineweaver-Burk plot (Figure 25.5). By noting the slope and intercept we can evaluate $K_{\max}$ and $v_{\max}$.

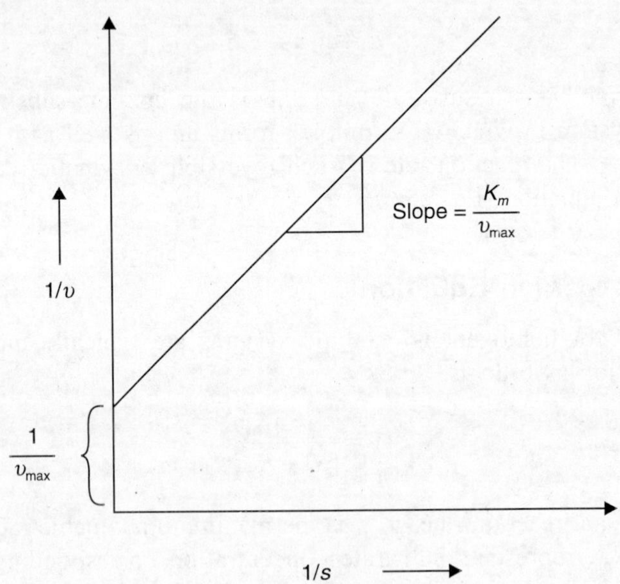

Figure 25.5 Lineweaver-Burk plot to determine K_m and v_{max}.

A detailed discussion on enzyme processes and process modeling are available in any standard textbooks on Enzymes and Biochemical Engineering (Wang, et al., 1979; Rao, 2010(b))

25.4 DESIGN FEATURES OF FERMENTER

As has been mentioned earlier, fermentation proceeds with the propagation of microorganisms from a raw material to produce a desired metabolic product in a container vessel, called **fermenter**, under a certain set of process conditions. A typical fermenter is shown in Figure 25.6. An exhaustive design with various input-output gadgets was shown in Figure 22.5. The design features of fermenter mainly depend upon the feeding mechanism to the fermenter. Basically, the feeding method is classified as:

(i) batch processing
(ii) continuous processing (chemostat)
(iii) semi-batch processing (fed batch)

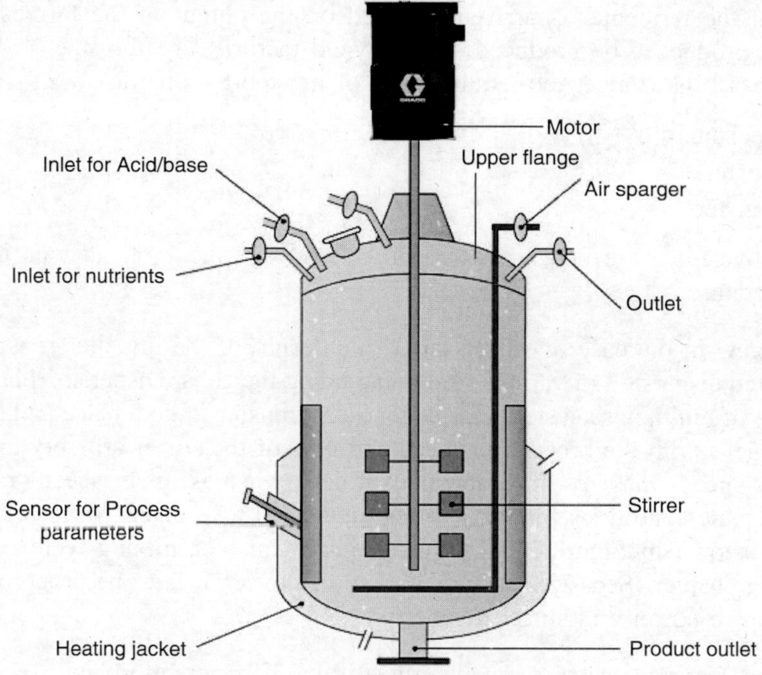

Figure 25.6 Typical Fermenter with all gadgets.

The above flow patterns are shown in Figure 25.7. Another basic difference lies in the process conditions, viz.,

- anaerobic fermentation (in the absence of air)
- aerobic fermentation (in the presence of air)

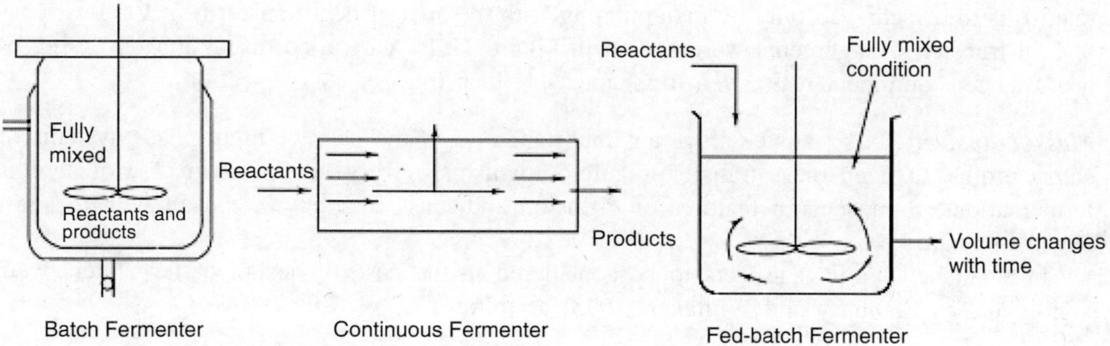

Figure 25.7 Various flow patterns of fermentation.

The design features vary based on the above. A very important aspect to be noted in case of design of fermenters is maintenance of strict aseptic conditions which is not the case with chemical reactors. Hence, knowledge of maintenance of asepsis is a desirable qualification for any firm manufacturing fermenters.

The size of the fermenter is arrived at based on the output of the fermentation, i.e., the quantity of the product to be produced and the yield factor. Once the size of the fermenter is fixed; the next task is to meet the requirements of transport properties, viz.,

- fluid mechanics
- heat transfer
- mass transfer

All the above three are dependent on the viscosity of the broth and its interrelationship with the temperature.

Fluid Mechanics: In the case of stirred tanks, the design of the impeller is very important to meet the required dispersion of various phases including the cells. Dispersion plays an important role in the case of multiphase systems for better mass transfer to take place and to achieve good contact of the cells with the broth. The speed and type of the stirrer are very important factors because higher speeds may result in abrasion of cells whereas lower speeds may not be able to provide adequate conditions for better mass transfer to take place. Various types of stirrers/ agitators were shown in Figure 22.6. Power requirements for mixing were comprehensively dealt in Mixing chapter (Sec. 22.2). The viscosity of the broth is to be properly accounted for while designing of stirrer and stirrer geometry.

Heat Transfer: Heat transfer also plays a crucial role in fermentation because we are handling here the living organisms which are more sensitive to heating. Generally, higher temperatures are not required in fermentations. The heat transfer rate (q) is given by Eq. (11.18).

$$q = U\,A\,(\Delta T) \tag{11.18}$$

Higher ΔTs are detrimental to the growth and survival of microorganisms while lower temperatures of the broth may not yield desired results. Maintaining uniform temperature throughout the fermentation broth is complicated by the high viscosity of the broth. Hence, desired heat transfer rate (q) is to be achieved only by manipulating with the area of heat transfer (A). Various types of heat transfer arrangements were shown in Figure 12.1. A detailed discussion on design of heat transfer equipment was given in Chapter 11 (heat transfer by convection).

Mass Transfer: Mass transfer plays an important role in the case of multi-phase systems or slurry broths. Oxygen mass transfer to microorganisms is very important in case of aerobic fermentations. Some design features of diffusion and mass transfer were dealt with to some extent in Chapter 16.

In a nutshell, various aspects to be considered in the case of design of fermenters were highlighted by Stanbury and Whitaker (1993) as follows:

- operating under aseptic conditions over a long period of time,
- the interactions of various transport properties are to be properly accounted for,
- in the case of aerobic fermentations, supply of adequate quantity of oxygen is to be ascertained,

- in the case of multi-phase systems/slurry broths, adequate amount of mixing is to be ensured,
- strict control of various process parameters is to be ensured,
- flexibility in operating fermenter is an essential design feature, and
- finally, the operating costs should be minimum without sacrificing any of the above factors including asepsis of the fermenter.

25.5 APPLICATIONS OF FERMENTATIONS IN FOOD PROCESSING

As has been mentioned, fermentation is essentially a food processing operation with the intervention of microorganisms, their growth, and their metabolic activity to produce useful products. Industrial applications of fermentation in food processing are relatively less as compared to fermentations in other biotechnological processes, medicines and vaccines, production of industrial enzymes, waste, and wastewater treatment. The major notable industrial applications are in alcohol production, alcoholic beverages, dairy and products of organic acids, leavening of bread dough in the bakery industry, hydrolysis of starch to produce high fructose corn syrup which is used in the confectionary industry, fruit and vegetable processing to produce pickles, sauerkraut by using Lactic Acid Bacteria (LAB), etc.

Basically, fermentation is employed in food processing to achieve the following objectives:

- enrichment of food products through nutrient enhancement,
- elimination of toxicants and anti-nutrients,
- improving the texture of foods through enhancement of organoleptic qualities,
- preservation of food raw materials (especially fruit and vegetables which are highly perishable) through lactic acid fermentation, alcoholic fermentation and acetic acid fermentation,
- enrichment of foods with proteins and vitamins,
- substantial saving in cooking time which in turn helps the economic viability of production process.

Here is described a few food processing applications.

25.5.1 Alcohol Production from Molasses

By and large, production of industrial alcohol (ethyl alcohol) or edible alcohol from molasses, and bioethanol from various starchy and sugar sources to be used as blends in gasoline is one of the major applications of fermentation process in food industry. Also, some of the fruit juices (including cashew fruit juice) are used to produce brandies and edible grade/pharmaceutical grade alcohol. When glucose is the starting material, invariably the microorganism used is *Saccharomyces cerevisiae.* If starch (carbohydrates) is the starting material, it is first hydrolyzed to glucose using any hydrolyzing enzyme such as glucoamylase and α-amylase; and later the resultant sugars are fermented to alcohol as follows:

$$C_{12}H_{22}O_{11} + H_2O \rightarrow 2\,C_6H_{12}O_6$$

$$C_6H_{12}O_6 \rightarrow 2C_2H_5OH + 2CO_2$$

Molasses is a dark-coloured viscous fluid and is a byproduct of the sugar industry after sugar crystals form from cane sugar juice by concentration and crystallization. Molasses is a byproduct of sugar industry when further crystallization of sugar is not economical. Still, it contains approximately 50% (40–50%) fermentable sugars. Maximum concentration of alcohol realized in fermentation is about 8–9%. Later the dilute alcohol is distilled to get 95.6% alcohol (known as rectified spirit).

Production Process: Initially the molasses is diluted to 14–16% consistency and saccharifying enzyme (*invertase*) is added to ferment molasses into fermentable sugars. In the fermentation stage, the mash is fermented using yeast in which step, the sugars are fermented at 27°C to 8-10% ethanol and carbon dioxide under anaerobic conditions. In the distillation step, the dilute alcohol is distilled to concentrated alcohol in which the maximum alcohol content is 95.6%. From one ton of sugar cane, 115 kg of sugar (at 11.5% recovery level) and 45 kg of molasses are produced. The 45 kg of molasses contain 18 kg of TFS (total fermentable sugars) which ferment to 10.8 liters of ethanol.

Industrial fermenters are usually of the sizes of 100 m^3 and diameters of 2.5 m and heights of 6.5 m and operate in semi-batch mode for 6–7 days under anaerobic conditions with continuous release of CO_2 gas (Rao, 2010 (c)).

25.5.2 Bakery

Various types of bakery products include:

- bread
- cakes
- buns
- pastries
- biscuits
- wafers
- cookies
- doughnuts
- crackers, etc.

Of the above, the bread and biscuits are the larger industries. Particularly, bread making is an important industrial activity with a world production of 160–180 million tons (158 mT in the year 2022). Hence, this process alone is described here briefly. Bread is a baked product of several ingredients as follows:

- wheat flour (100 parts),
- yeast (2–4 parts),
- salt (2 parts),
- sugar (6 parts),
- fats (2 parts), and
- water (60 parts).

There could be some variations in salt, sugar or fat content. The process involves the following five steps.

Step 1. *Mixing of all ingredients:* Mixing helps in proper distribution of all ingredients, and development of proteins (gluten). The dough is made from all the ingredients.

Step 2. *Rising (Fermentation):* After mixing of all ingredients, the dough is left to rise. This is the fermentation process and is known as *leavening* of bread. In this process, the dough becomes smooth with good gas holding properties. The yeast cells grow, alcohol and carbon dioxide are formed by breaking down of starch and sugars present in the wheat flour. The CO_2 formed in the process tries to escape out and results in rising of dough. Alcohol also escapes during the baking process.

Step 3. *Kneading and proofing:* Kneading is done typically by kneaders. During the process of stretching and folding, gluten develops well and releases the gas. The dough is allowed for second rising. This final rising is known as *proofing*. It takes place in about 30–50 min at 27–34°C at humidity of 85%.

Step 4. *Baking:* During the baking process in the oven, the temperature is of the order of 200°C. In this process, the dough transforms into easily digestible product. Heat releases any gases from the dough and makes the bread fluffy. The internal temperature of the bread reaches 98°C. Some weight is lost by evaporation of moisture. Browning reactions take place at above 160°C which make the bread crisp and form the colour.

Step 5. *Cooling and wrapping:* When the bread comes out of the oven at 200°C, the crust temperature is about 200°C, and the internal temperature of the bread is 98°C. It is quickly cooled to about 35°C; later it is sliced and wrapped.

Various stages in bread making process are shown schematically in Figure 25.8.

25.5.3 Dairy Industry

Cheese, curds, yoghurts and butter are some of the products made from milk in the dairy industry. Various fermented dairy products are shown in Figure 25.9 in a nutshell. Most of these products are made at domestic scale. However, cheese, yoghurts, butter are also made at industrial scale. Cheese is considered as a major product of dairy industry with a world production of 22.6 million tons. Various steps in cheese making process are:

- pasteurization of milk and cooling it to room temperature,
- inoculation of starter and non-starter bacteria and ripers,
- addition of rennet to form curd,
- cutting the curds and heating,
- draining whey and sending it to whey treatment plant,
- texturing the curd to extract whey,
- salting and ripening the cheese,
- packaging and distribution.

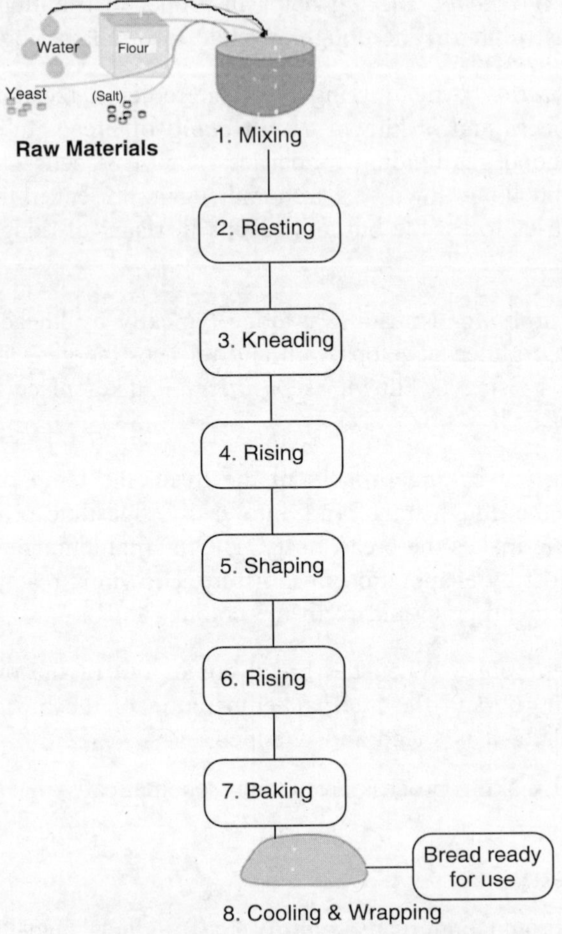

Figure 25.8 Bread making process.

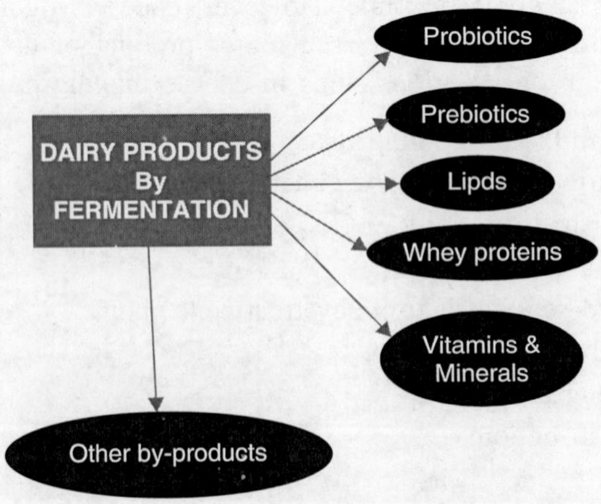

Figure 25.9 Various fermented dairy products.

Processing of various other dairy products is deliberately not discussed here as a number of standard textbooks is available on dairy industry.

25.5.4 Lactic Acid Fermentations

Lactic acid bacteria (popularly knowns as LAB) plays a tremendous role in the food fermentation in several processes such as those shown in Table 25.1. A few of them is discussed below (Steinkraus, 1992).

Sauerkraut: Vegetables like cabbage are preserved by lactic acid fermentation. After processing the raw vegetables, they are shredded and brined into 2% solution. The salt extracts the liquids (water) from the vegetables which in turn act as substrate for growth of lactic acid bacteria. Anaerobic conditions need to be maintained to avoid spoilage.

Kimchi (Korean): Kimchi is a relatively less acidic pickle and is made in the carbonated form. The vegetable (Mostly Chinese cabbage or radish) is salted in 5–7% brine for 12 hours. The contents are drained and rinsed. Fermentation continues at lower temperature (*ca* 10°C) until the lactic acid concentration reaches 0.4–0.8% pushing the pH to 4.2–4.5. Some minor ingredients such as garlic, onions, sugar, pepper, leaf mustard are added to flavour the pickle.

Pickled vegetables: Several vegetables are pickled by brining them. They include carrots, onions, ginger, cucumber, chilli, bamboo shoots, unripe small mangoes, papaya, gooseberry, lime, etc. The vegetables are brined in 5% salt solution for about 15 days. The duration of pickling depends on the local temperature. Acidity reaches up to 0.6–1.0% as lactic acid and pushes the pH to 3.4–3.6.

The pickled vegetables are not further used for cooking. They only constitute an adjunct for the main course of meal.

Nigerian Ogi: It is a sour porridge made of fermentation of corn, sorghum or millet. It has a typical flavour resembling the yoghurts. It has a solid content of 8%. The grains are cleaned and steeped for 3–5 days in water during which period the desired microorganisms are naturally grown. Hence, it is known as a naturally fermented drink. Later, the grains are ground in water and filtered to remove the course particles. Steeping is the only process to ferment the porridge to a pH of 4.3 in which the lactic acid concentration is about 0.65% and acetic acid concentration is 0.1%.

25.5.5 Soya Sauce

Soya sauce is an important all-purpose souring ingredient in most of the culinary preparations to give flavour and body to the dish. It is prepared from soya beans and roasted wheat, and fermented with *Aspargillus oryzae* solid cultures for two days to produce *Kozi*. The mixture of soya beans and wheat are hydrolyzed by enzymes (*Proteinases, Peptidases and Amylases*) to produce what is called *maromi*. When *maromi* is fermented, lactic acid is produced by *Pediococcus halophilus*. This results in lowering pH. Alcohol (2–3%) is also produced in the process by *Zycosaccharomyces rouxii*. Simultaneously some aroma compounds and phenolic compounds (by *Candida versatilis* and *Candida etchellsii)* are also produced. They all impart characteristic flavour and aroma to soya sauce. The process flowchart is shown in Figure 25.10.

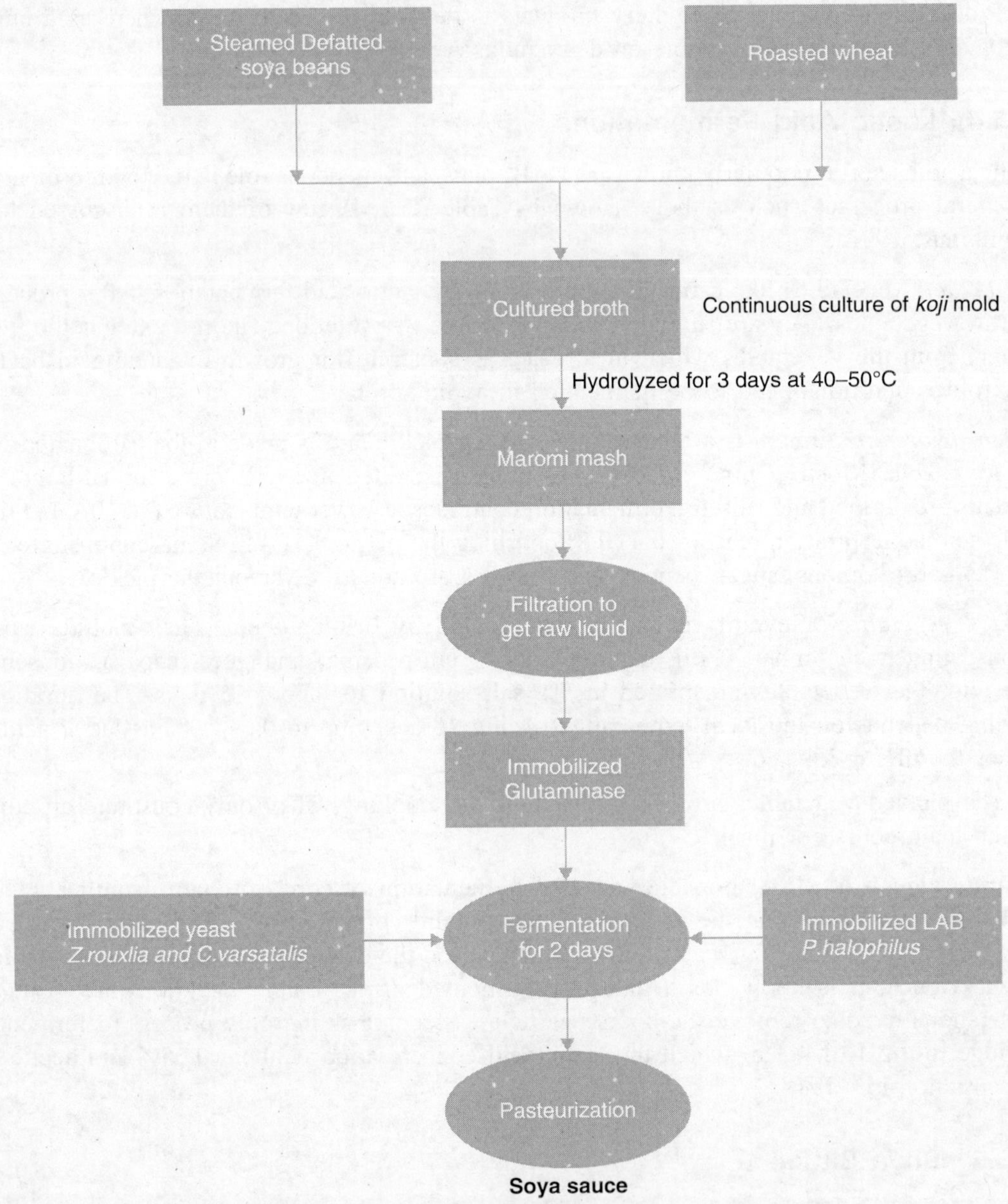

Figure 25.10 Industrial process for soya sauce in a bioreactor.

25.5.6 High Fructose Corn Syrup (HFCS)

High fructose corn syrup is an artificial sweetening agent and finds wide applications in food processing such as:

- beverage industry
- confectionary
- bakery

- breakfast foods
- dairy
- animal feeds
- pharmaceutical industry.

The world market of HFCS is of the order of 10^7 tons/year ($8.9 billion). It is made from any starchy materials such as corn, rice, potato, cassava, etc. The process essentially involves three steps (http://www.starchprojectsolution.com/faq/high_fructose_corn_syrup_produce_1170.html):

- liquefaction of starch by enzymatic hydrolysis/acid hydrolysis,
- saccharification,
- isomerization of glucose to fructose by using glucose isomerase.

All the essential steps are shown in Figure 25.11. In this process, glucose isomerization is the crucial step, because the glucose isomerase enzyme gets deactivated due to high temperature and acidic pH. HFCS is numbered based on the fructose percentage content in it, viz., to what extent glucose is isomerized to fructose. For example HFCS 90 means it contains 90% fructose and 10% glucose, similarly HFCS 42 contains 42% of fructose and 58% glucose.

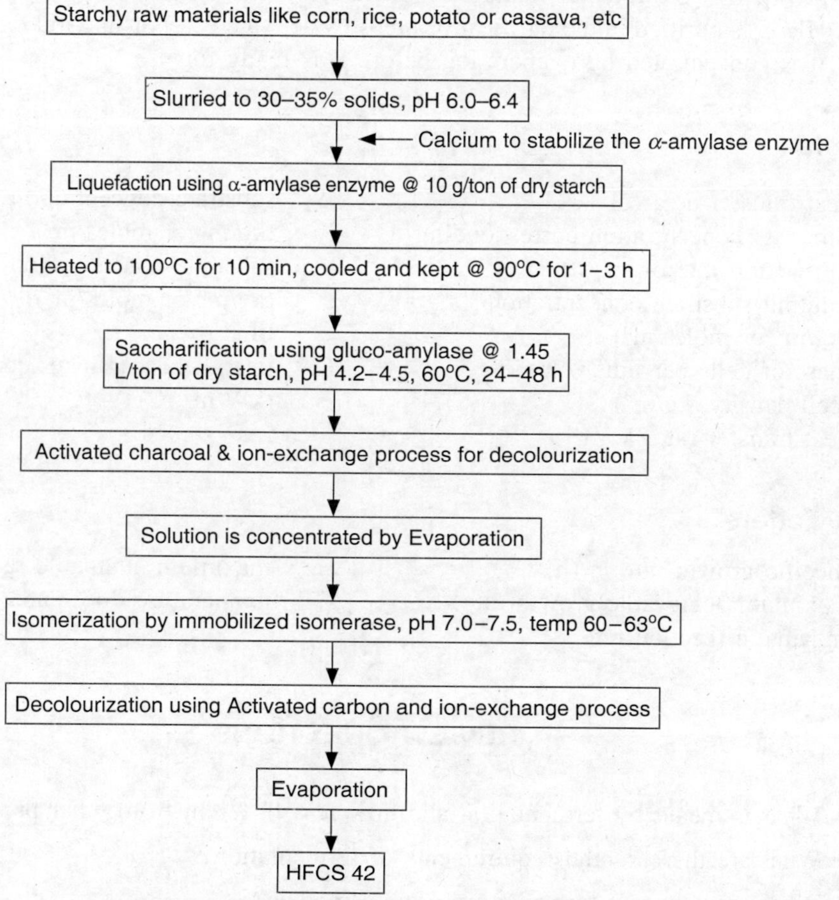

Figure 25.11 Flow chart for HFCS Process.

Liquefaction of starch: Starch is made into slurry to contain 30–35% of dry solids. The pH is adjusted to 6.0–6.4. Calcium is added to it to help stabilize the enzyme. α-amylase enzyme is added to an extent of 8–10 grams of enzyme/ton of dry weight of starch. This will initiate hydrolysis process. The contents are heated to 100°C for 10 minutes, later the temperature is brought down to 90°C, and kept at that temperature for 1–3 hours to complete the hydrolysis process.

Saccharification process: The next step in hydrolysis is saccharification process using an enzyme like *Amyl-glucosedase.* The enzyme is added to an extent of 1.45 L/ton of dry starch. The process is carried out at 4.2–4.5 pH, temperature of 60°C for 24–48 hours depending upon the speed of enzymatic reaction. This results in 95% glucose syrup.

Isomerization step: The above glucose syrup is purified and is taken for isomerization step in which the glucose is isomerized to fructose using an isomerase enzyme, usually in the immobilized form. The immobilized enzymes can be conveniently separated from the mother solution and can be reused until the enzyme loses its activity. During isomerization process, the pH is maintained at 7.0–7.5 and temperature at 60–63°C. The glucose concentration falls to 58% giving HFCS 42. This results in conversion rate of 42%.

The HFCS is purified in the purification steps by activated carbon and ion-exchange process. Finally, it is concentrated by evaporation before it is ready for use.

Symbols

A: heat transfer area (m^2)

K_{max}: substrate concentration corresponding to reaction rate of $v_{max}/2$

K_s: limiting substrate concentration (kg/m^3 or $moles/m^3$)

n: mass of cells per unit volume (cell density, kg/m^3)

q: heat transfer rate (kW)

$[S]$: substrate concentration (kg/m^3 or $moles/m^3$)

s: substrate concentration (kg/m^3 or $moles/m^3$)

t: time (h)

U: overall heat transfer coefficient ($kW/m^2\,°C$)

Greek Letters

μ: specific growth rate in $(h)^{-1}$

μ_m: maximum achievable growth rate

v: enzymatic reaction rate

v_{max}: maximum attainable rate

ΔT: temperature difference

REVIEW QUESTIONS

25.1 What is meant by fermentation and how it is different from other process operations?

25.2 What are the essential requirements of fermentation?

25.3 What are different fermentation growth models?

25.4 What are various applications of fermentation in food processing?

25.5 What are enzymes and how do they influence a food process?

25.6 What is Michaelis-Menten equation? Explain it.

25.7 What design features are important while designing an industrial fermenter?

25.8 What are three transport phenomena that are important while designing a fermenter?

25.9 Describe the production of alcohol from first principles.

25.10 What are various bakery products? Describe the production of bread with a flow chart.

25.11 Explain the production of cheese making?

25.12 How soya sauce is made? Explain various production steps with neat flow diagram.

25.13 What are various lactic acid fermentation process? Describe a few of them.

25.14 Describe how the high fructose corn syrup is made? What are its applications in food processing?

REFERENCES

Bailey, J.E. and Ollis, D.F. (1986), *Biochemical Engineering Fundamentals*, 2nd Ed., McGraw-Hill, Inc., New York.

Michaelis, L. and Menten, M.L. (1913), "The kinetics of invertin working", *Biochem. Z.* (in German), 49, p. 333.

Monod, J. (1949), The growth of bacterial cultures, *Ann. Rev. Microbiol.*, 3, pp. 371–394.

Rao, D.G. (2010), *Introduction to Biochemical Engineering*, 2nd Ed., McGraw–Hill Education, New Delhi, (a) pp. 151–169, (b) pp. 189–192, (c) pp. 107.

Ray, R.C. and Joshi, V.K. (2014), Fermented foods : Past, present and Future, in "*Microorganisms and Fermentation of Traditional Foods*", Ed by Ray. R.C. and Joshi, V.K., CRC Press, Boka Raton, pp. 1–36.

Stanbury, P.F. and Whitaker, A. (1993), *Principles of Fermentation Technology,* Pergamon Press, Oxford, p. 121.

Steinkraus, K.H. (1992), Lactic acid fermentations in "*Applications of Biotechnology in Traditional Fermented Foods*", National Academic Press, Washington DC., pp. 43–51.

Wang, D.I.C., Cooney, C.L., Demain, A.L., Dunnil, P., Humphrey, A.E. and Lilly, M.D. (1979), *Fermentation and Enzyme Technology,* John Wiley & Sons, New York.

Part IV

Food Industry Management

CHAPTER

26

Cleaning and Sanitation of Process Plants

Food processing is a typical modern manufacturing operation where hygiene and sanitation play a vital role. In the past, the food processing operations normally used to be taken up at cottage scale or small-scale level, and mostly in a batch manner. Of late, the modern processing operations take place in continuous manner with huge capacities. For example, the fruit processing industries of the order of 150–200 tpd are very common. Similarly, meat processing industries of the order of 100 tons of meat per day are very common. When such huge quantities of raw materials are being processed, it is quite likely that the hygiene of the place gets affected. Added to this, most of the food materials are biologically degradative. It is more so with the animal products. The raw materials get degraded very fast by the microorganisms.

The contamination problems are the same either in a batch process or in a continuous process, either in a small scale of production or on a large-scale production. But the contamination multiplies as the quantity of production increases. Hence, care should be taken to see that such contamination is totally done away with. First, let us see what are the possible sources for contamination to occur. It could be from

- raw materials if they are dirty,
- foreign bodies hopping on the food materials,
- enzymatic action causing degradation of the raw materials, and
- dirty production equipments.

Hence, the goals of cleaning and sanitation are to see that all these problems are addressed to. In fact, the concept of cleaning and sanitation should be made as an integral part of food processing.

26.1 BASIC PRINCIPLES OF HYGIENE AND SANITATION

Now, let us see the basic principles of hygiene and sanitation.

- It is advisable to take the raw materials in as pure form as possible. If the raw materials are dirty, they need to be cleaned at farm level or at the processing plant before they go for processing. Fruits, vegetables, roots, or nuts are naturally associated with dirt, and need to be cleaned in running water. The running water is recycled for further use. In such cases, care should be taken to see that the dirt is not recycled.
- All the materials and equipment parts coming in contact with the food materials should be inert and rust-free. The food materials should not pick up any toxic materials or pathogenic organisms from the contact parts.
- The surfaces coming in contact with the food materials should be smooth and polished. The pitches and crevices on any rough surface are a potential source for pathogenic organisms to harbour and thrive.
- Equipment should have provision for easy cleaning of the surface, and the surfaces should also be visible as far as possible so that visual check or observations can be made.
- All the equipment and machinery should be routinely cleaned at periodic intervals. For example, the milk processing line and equipment should be thoroughly sterilized every day after the processing is over.
- Generally bulk handling of solids/liquids is preferred to the small packs, tiny packs or unit packs. In case of latter we end up with a large number of bags or cans or packages or cartons which are potential source of contamination.
- The metallic wire screens of the sieves should never be made out of stainless steel. Even though stainless steel sieves are stronger and last longer, they have an inherent disadvantage that in case of breakage or rupture of screens, the debris of *SS* wire meshes escape undetected in the magnetic separators, and goes into the food material.
- The packaging materials are normally procured in bulk and stored. Hence, they should be properly disinfected before use in case of long storage periods, particularly with jute bags or cartons.
- The tanks and storage vessels should be designed with bottoms in such a way that the stagnant fluids drain-off through the pitched bottom or inclined bottom; i.e., the bottom surface should be designed for self draining. Otherwise, the stagnant fluids are an ineluctable source for toxic microorganisms to grow and thrive which will contaminate the food materials of next batch.
- The joints and welding surfaces should be as smooth as possible to avoid growth of any undesirable microorganisms.

26.2 VARIOUS METHODS OF CLEANING PROCESS

Cleaning of food process equipment routinely is a common practice in any food industry to maintain high levels of safety and hygiene. The cleaning can be done normally or by using automatic gadgets/equipment/accessories.

Depending upon the nature of the material and the nature of contamination, the cleaning process varies. Sometimes only one type of cleaning may be used; in some other times we may need to use a combination of cleaning processes. Table 26.1 presents a list of cleaning steps.

Table 26.1 Various Cleaning Steps

Cleaning step	*Cleaning process*	*Remarks*
Pre-cleaning operation	The equipment is rinsed with water primarily to remove adhering dust, dirt, loose soil, grass, etc.	This step may not be always required
Cleaning operation and rinsing step	Usual cleaning with water and detergents to remove any adhering contaminants after step 1. The adhering detergents and water are removed	–
Sanitizing step	This is the main cleaning process in which most of the microorganisms are destroyed either by heat sterilization or by application of chemicals or both	This is the main sanitizing operation
Final rinsing	The adhering chemicals (due to sanitization), etc. are removed by water rinsing	This is the final step of rinsing for removing any adhering chemicals, etc.

Essentially cleaning process may be carried out either by open cleaning methods or closed cleaning methods.

26.2.1 Open Cleaning Method

In the open cleaning method, various parts of the machinery and equipment are dismantled and cleaned. If the contamination is simple, the dismantled parts are cleaned in open place by soaking, washing and rubbing different parts with brushes, etc. This kind of open cleaning is known as **Cleaning Out of Place (COP)**. If the contaminants are hard stuck to the surface, they may be blasted with a jet of fluid (water) under high pressure or low pressures. The high pressure jets are of the order of 15–120 atmospheres, and the cleaning takes place mainly by the shear force applied by the jet, and not essentially based on the type of detergent used. Hence, this kind of cleaning process is known as **cleaning under high pressure**.

In the low pressure clearing process the pressures applied by the jets are of the order of 5 atmospheres or less. Hence, the cleaning is done not by the shear force, but by the type and quality of the detergent used. Another method of cleaning is by application of foam. In this method, the equipment parts are kept in contact with the foam of the detergent. Later they are rubbed with clean metallic brushes or rubber scrubbers. In this method, cleaning takes place based on the amount of shear force applied by rubbing, the contact time with the foam and the type and quality of detergent used.

26.2.2 Closed Cleaning Method

In case of closed cleaning methods, the cleaning takes place in the processing hall itself without dismantling the equipment. It is also known as **cleaning in place**. This can be done either by using the cleaning media only once or by repeated use.
Various methods of cleaning processes are shown in a nut shell in Table 26.2.

Table 26.2 Various Methods of Cleaning

Open cleaning	Cleaning out of place
	Cleaning with high pressure jets
	Cleaning with low pressure jets
	Cleaning with foam and rubbing
Closed cleaning	Single use CIP
	Multiple use CIP

26.3 CLEANING IN PLACE (CIP)

In case of manual cleaning/open cleaning, usually different parts are disassembled, and after cleaning are reassembled. Thus, manual cleaning is a time-consuming and labour-intensive process. In spite of that, some parts may not be accessible. Opening and cleaning of equipment may be harmful to employees and detrimental to product quality. Hence, the modern approach is to clean the equipment in place without dismantling the equipment. This procedure of cleaning in place is known in food technology terminology as **CIP**.

CIP has some of the obvious advantages, viz.,

- Cleaning time (production down time) can be considerably reduced.
- Labour involvement is eliminated.
- The CIP can be automated.
- It can give reproducible results in terms of product quality.

In the CIP system, the detergents and other cleansing solutions and steam are passed through the pipes and into the equipment. During the process of cleaning, the detergents and cleaning solutions may be used once or may be reused, based on which the CIP is classified as follows:

26.3.1 Single Use CIP System

In this method, the liquids and cleaning water are used only once, and later they are discarded (Figure 26.1). This method is simple and very effective except for the fact that it uses a large quantity of solutions and detergents. Since the solutions are not reused, the carry forward of the solids, dirt and dust, etc. is totally avoided. A standard cleaning protocol for a one tank cleaning system is shown in Table 26.3. A schematic flow chart is shown in Figure 26.2.

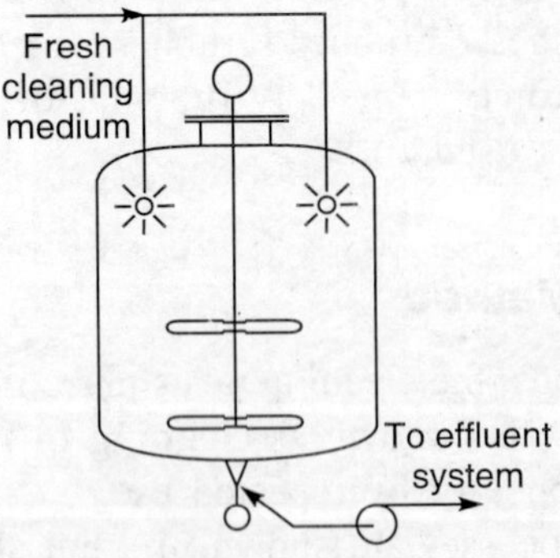

Figure 26.1 Concept of single use CIP systems.

Table 26.3 Cleaning Times for Each Step in a One Tank CIP System

Operation	*Time taken*	*Remarks*
Pre-rinses (3 times)	0.3 min with an interval of 0.67 min	They remove solid contamination
Injection of cleaning media and steam	Maintained for about 12 min	The spent lye is discharged to drain
Injection of water and organic acids	Circulation of cold solution for 3 min	The pH is lowered to 4.5 to 5 The cold solution is discharged to drain
Total time taken	20min	–

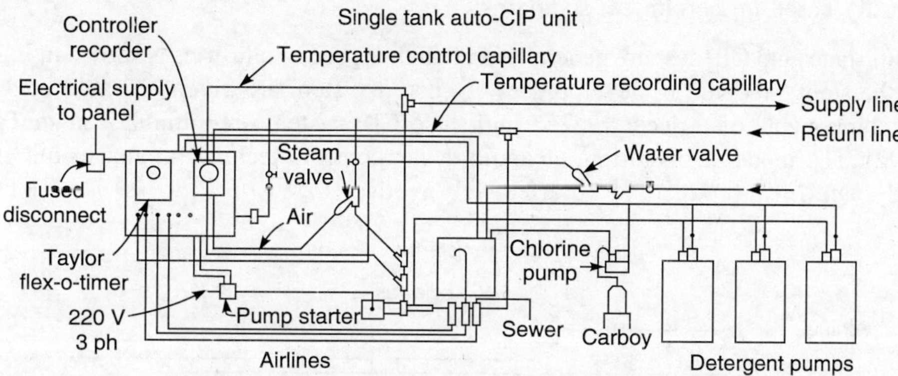

Figure 26.2 Single use Clean in Place (CIP) system.

26.3.2 Multi Use CIP System

In a multi use CIP system, the rinsing water, detergents, sanitizers are partly used depending upon the level of contaminants. It consists of a centralized reuse system from where the cleansing agents are drawn through pipes, solenoid valves and pumps. Afterwards they are returned to the centralized system, may be with a part of them replaced with fresh ones. It is schematically shown in Figure 26.3. There is a great saving in chemicals and detergents. The system can be automated with proper process control gadgets. In such systems, the quantity of cleaning media is reduced; and only the required quantities of cleaning media are used. This will reduce the consumption of rinsing water and steam to remove the cleaning media. Since exact quantities are used, even the cleaning media will be economized.

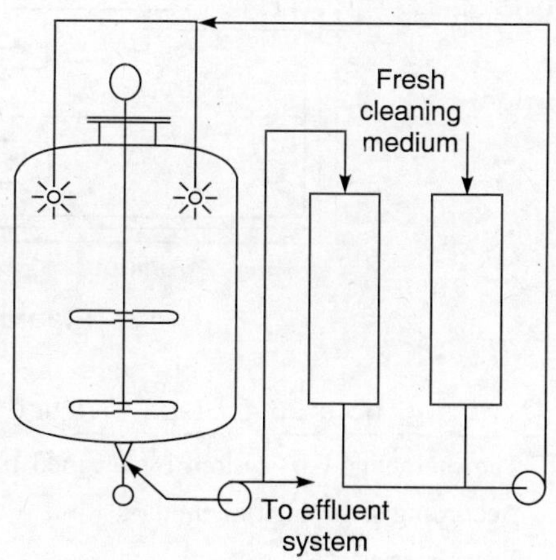

Figure 26.3 Concept of recirculation system with external tanks of multi use CIP.

This is an excellent system for cleaning provided the contamination levels are minimum. If the contamination loads are high, they will also recirculate, and the cleaning process will not be efficient. In some systems, the rinsing water is reused for precleaning operations, for which very clean water is not required.

Thus, a multi use CIP system has the following features:

- A cold dilute alkali solution, used to remove the dairy products or any other proteinaceous matters, will be generally the one which is obtained from earlier rinse solutions.
- Only approved sanitizers and hot water is recirculated; efforts are made to check the level of contamination picked up by the cleaning media, and then only decision is taken whether to recirculate the whole or part.
- The circulation time and temperatures of the alkali solutions and detergents are maintained strictly based on certain set standards.

A multi stage auto CIP system, generally used in the process industries is shown in Figure 26.4. Some of the steps involved in development of such a system are given in Table 26.4. Various principal components of a decentralized multi use CIP system are summarized in Table 26.5 (Plett 1992). The mode of operation, cleaning cycle, control mechanism, etc. would enable the rinse media and rinse products to be effectively recovered.

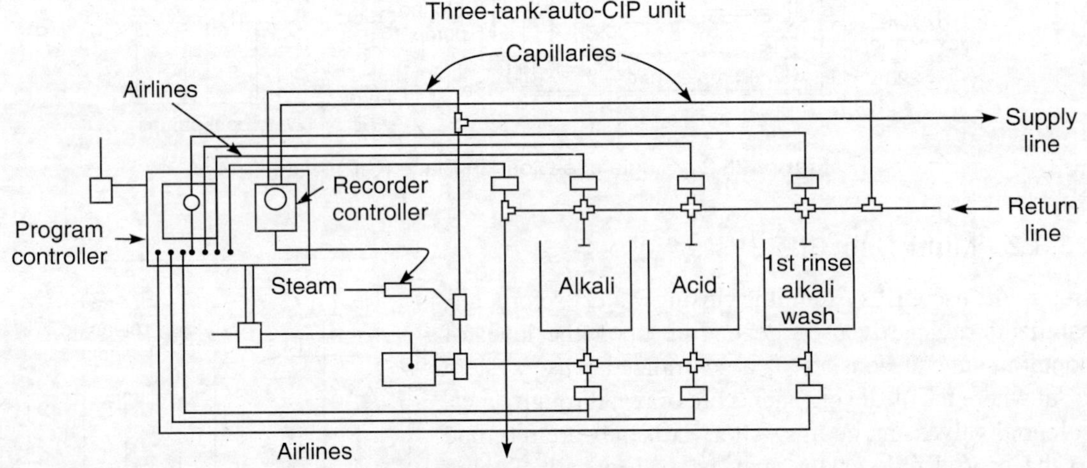

Figure 26.4 Multi use CIP system.

Table 26.4 Steps Involved in the Development of a Multi Use CIP System

The multi line CIP system is designed based on the process line requirements.
Accordingly the instrumentation for CIP system is developed.
The materials of construction are selected based on the system requirements.
The selection and installation of the recirculating equipment (pipes, valves, pumps, etc.) are finalized based on the system requirements.
Suitable mechanical spraying systems are designed so that the detergent/disinfectant reaches to all the desired and affected parts of the pipeline/equipment.

Table 26.5 Principal Components of a Multi Use CIP System

Control systems with valves, transmitters and controllers, etc.
Recirculation pumps with suitable valves.
Piping, valves, bends, elbows, etc. to connect the cleaning objects.
Various batch tanks to hold alkali, acid, detergents, disinfectant and other cleaning media.
Plate heat exchanger to quickly heat the acid solutions, alkali solutions or any other cleaning media to the desired temperature before they are used.
A concentrated acid tank and concentrated alkali tank are provided separately with provision for suitable dilution so that the desired diluted acid/alkali solutions can be made depending upon the need.
An air blow system (compressor) with valves to blow air at desired pressure and flow rates so that cleaning and/drying can be done effectively.
An overall control panel with all logic control units for carrying out the overall cleaning operations.

26.4 SANITATION METHODS

Sanitation is a method of cleaning in which chemicals (known as **sanitizers**) are used for destruction of microorganisms either by application of heat or without heating. In other words, sanitation is a chemical method of sterilization (as contract to thermal sterilization). The sanitizers can be used both for open cleaning and for in-plant cleaning. They are pumped in CIP process and sprayed/distributed on the affected parts. In open cleaning, the contaminated parts are kept for soaking. Various types of sanitizers are used and can be broadly classified into chlorine-based sanitizers, quaternary ammonium compounds, iodophores, and amphoteric bactericides.

Sanitizers are chosen based on the purpose of sanitization; i.e., whether sanitization needs to be done for removing the

(i) organic matter, viz., fats, proteins, vegetable matter, animal based material, etc.

or

(ii) bacteria or any other pathogenic microorganisms.

or

(iii) microorganisms associated with the organic matter.

Various items that are important in the selection of a sanitizer are given in Table 26.6. We shall discuss them briefly in the following sub-sections:

Table 26.6 Various Important Issues to be Considered for Selection of a Sanitizer

The nature of material to be sanitized and time of its exposure. For example, a high concentration of chlorine forms acidic solution and itches the metallic surfaces.
Allowable length of time of exposure of the contaminated parts with the sanitizer.
Temperature of the sanitizer solution. However, in case of chlorine-based sanitizers, higher temperatures are detrimental as the sanitizer gets dechlorinated at higher temperatures.
pH of the sanitizer solution.
The amount and quantity of organic material, viz., fats, proteins, vegetable-based materials or organic-based materials to be cleaned.
The quantity and the type of microorganisms or pathogens.
The amount and type of mineral deposits, milkstone or any other type of incompatible materials which obstruct the penetration of sanitizer to cleanse the bacteria.
Based on the phenol coefficient[†] of the sanitizer. Higher the phenol coefficient, more efficient is the sanitizer.
Cost of the sanitizer, which ultimately influences the choice.

[†]The phenol coefficient is obtained by comparing the sanitizer's activity with that of pure phenol under similar conditions of contact time, and known amount of bacterial count.

26.4.1 Chlorine-based Sanitizers

Chlorine when absorbed in water gives hypochlorous solution (HOCl) which is an effective sanitizing agent, and is frequently used for sanitizing the working hands and feet of the workers in the food processing plants.

Similarly, aqueous solutions of sodium hypochlorite are also frequently used as a sanitizing agent. The commercially available sodium hypochlorite solution contains 9–12 per cent of chlorine available. However, the frequently used liquid chloric cleaning medium will have a dosage of 0.15 per cent chlorine. For *SS* contact parts, the recommended level of chlorine is only 150–250 ppm. Even larger vessels are also sanitized by spraying 200 ppm chlorine solutions. Chlorine is known to have a rapid action against both gram-positive and gram-negative bacteria. Thus, chlorine based chemicals are effective for bactericidal action except in cases where the contaminant surfaces contain organic matter. Hence, it is essential that the contact surfaces are made free of organic detergent compounds before sanitation.

26.4.2 Quaternary Ammonium Compounds

Quaternary ammonium compounds are popularly known as **QUATS**. They are colourless, odourless and non-toxic in the concentration ranges they are used. They are a kind of sanitizers in which one or more or all of the hydrogen atoms in the ammonium radical $(NH_4)^+$ are replaced by alkyl or aryl groups, and at least one of them should be a long chain group. The inorganic anion is usually chloride or bromide; for example, alkyl trimethyl ammonium bromide or lauryl dimethyl benzyl ammonium chloride.

Since they are cationic in nature, they should not be mixed with anionic groups (detergents). Otherwise their activity gets reduced. Generally they are available as powders or pastes, and they are most effective in aqueous solutions. Only drawback with them is that they foam.

26.4.3 Iodophores

Iodophores are iodine-based sanitizers and they are non-ionic surface active agents. They can be used as a sterilant and detergent also. The aqueous solutions are effective even with hard water. They do not impart any stains on the surface, and they do not cause any objectionable odour or taste. They are excellent sanitizers except that they are costlier by 3 times as compared to chlorine-based sanitizers.

26.4.4 Amphoteric–Ampholytic Bactericides

They are organic compounds containing substituted amino acids or betaines. They can be both acidic and basic in nature, and exhibit cationic and anionic properties depending upon pH. They are good bactericides.

Table 26.7 highlights some information regarding various types of sanitizers.

Table 26.7 Comparative Details of Various Sanitizers[†]

Sr. No.	*Sanitizer (Other Name)*	*Optimum pH and temperature*	*Remarks*
1.	Chlorine-based (sodium hypochlorite)	Neutral-acidic pH and RT	Effective against a number of microorganisms. Corrosive against a number of materials. Unstable on long storage.
2.	Iodine-based (iodophores)	Acidic pH and RT	Highly stable and less corrosive. Less effective against bacteria. The cost is three times of (1).
3.	Quaternary Ammonium Compounds (QUATS)	Alkaline pH and RT	Stable and non-corrosive, but they are highly foaming. Costs double of (1).
4.	Peroxy acetic acid	Acidic pH and below RT (chilled conditions)	Effective against a number of organisms. It is biodegradable. But it is hazardous in concentrated form. Costs double of (1).
5.	Alcohols (isopropanol)	Neutral-slightly acidic pH and RT	Their action is fast, but they are not active against spores and limited effectiveness against viruses. Costs double of (1).
6.	Acidic-Anionic	Acidic pH. RT-hot	They are less corrosive; but limited microbial activity. Costs triple of (1).

RT: Room Temperature

[†]Chlorine-based sanitizers are taken as base for estimation of costs of others.

26.5 THERMAL STERILIZATION

Sterilization is a method of making the equipment and materials free of microorganisms by application of heat as contrast to sanitation by application of chemicals. Generally we come across the thermal sterilization of medium in biochemical fermentations, where we essentially want to deactivate the undesirable microorganisms. But in food processing, sterilization operations are restricted to sterilization of equipment, machinery, pipelines, etc. and the packaging materials like metallic cans, glass bottles, etc. However, the principles of thermal sterilization are the same in either processing activity, and are generally treated at length in any standard textbook of biochemical engineering [Rao 2010(a)].

Thermal sterilization is done by steam at 121 °C and 15 psig (0.1 MN/m^2, gauge) which is passed into the equipment, machinery, pipelines, valves, etc. and kept for 20 min. Normally at this temperature and pressure, the contact parts (viz., the condensers, thermowells, pressure transducers, analytical probe points, and any pits or crevices on any of these) get sterilized. It is one of the CIP methods. The condensate of the steam is allowed to drain off subsequently. This will cleanse/sterilize the whole process equipment and machinery. This kind of thermal sterilization is done routinely in the dairy industry.

After the sterilization is over, the equipment would be flushed with sterile air, and is kept under positive pressure. This will help prevent entry of unsterile air after cooling as the system would develop vacuum on cooling. The inside of the equipment should have minimum number of joints and all joint points should be as smooth as possible to avoid that they become potential points for harbouring contaminants. If by any chance, the feed or the packaging materials are separately sterilized in another place and brought back to the processing set-up, care should be taken to see that this process does not result in any foreign microorganisms entering the process line.

The thermal death kinetic studies of the inactivation of microorganisms, as we need to do in case of canning of fruit pulps or purees, are all extensively discussed in any textbook of biochemical engineering; and hence, are deliberately avoided here.

26.6 DETERGENTS

Detergents are water soluble inorganic or organic chemicals or some enzyme-based chemicals used essentially for wet cleaning of surfaces of food process equipment. They are essentially meant for removing the adhering dust and dirt or some of the oily and greasy materials from the surfaces. Most of the food processing equipment are constructed out of stainless steel.

The problems of corrosion are invariably absent with the use of detergents on *SS* surfaces. The detergents will emulsify any oil or fat globules. They have a good capacity for rinsing. Hence, they cleanse the surfaces by dislodging the dust or dirt or any of the fat globules. The detergents are of various types and are shown in Table 26.8.

Table 26.8 Various Types of Detergents Used in Food Processing

Type of detergent	*Sub-classification*	*Remarks*
Inorganic alkaline detergents	Sodium hydroxide	A very common detergent with excellent emulsification capacity. Known for its bactericidal properties. Extensively used in bottle washing machines.
	Sodium carbonate and bicarbonate	Used as water softners and cleaning agents. They are mainly used as buffering agents.
	Trisodium phosphate	Known for its emulsifying and dispersing properties. Used for water softening.
	Sodium meta silicate	Known for its good emulsifying, wetting and deflocculating properties. It is much less corrosive than NaOH.
	Sodium ortho silicate	It is a powerful saponifier. It is highly alkaline, and attacks proteins and greases.
Acid detergents	Hydrochloric acid Nitric acid Phosphoric acid Glutamic acid Citric acid Tartaric acid	Descaling of heat transfer equipment. Used extensively to remove milkstone in dairy industries. Corrosion of the plant material and personnel needs to be given consideration.

(contd.)

Table 26.8 Various Types of Detergents Used in Food Processing (*contd.*)

Type of detergent	*Sub-classification*	*Remarks*
Surface active agents	Anionic surfactants	The active ion is negatively charged. They have good wetting and dispersing power. Used for removal of fatty acids and inorganic soils.
	Cationic surfactants	Excellent bactericides. Hence, used mostly for sterilization.
	Non-ionic surfactants	They are not affected by hardness of water. They are good emulsifiers. Hence, used for removal of colloidal soils.
Complexing and sequestering agents	Trisodium phosphate Monosodium phosphate Disodium hydrogen phosphate Poly phosphates Ethylene Diamine Tetra Acetic Acid (EDTA) Gluconic acid	Mainly used to remove calcium and magnesium ions in solution. They loosen and disperse the soils.
Enzymes	Proteases	For cleaning greasy surfaces.
Miscellaneous compounds	Starch Sodium CMC Sodium sulphite Sodium silicate	They are good corrosion-inhibitors and antifoaming agents.

26.7 PEST PROOFING/FUMIGATION METHODS

Controlling of pests and their growth are an important embodiment of any food processing industry, and frequently referred to as pest proofing. Insects, weevils, worms, and larvae are all considered as pests. Even the rodents are also classified as pests. Various measures need to be taken, in the absence of which the pests grow and multiply, and make the stored materials unfit for human consumption. They mainly attack the raw materials in the godowns or the packaging material, particularly gunny bags kept in the inventory stores. Various methods of infestation control are in vogue to render the raw materials and packaging materials unaccessible for the pests.

One of the most effective methods of infestation control is to go for fumigation. Methyl bromide is a frequently used fumigant, but it is becoming obsolete as most of the countries are banning it. Methyl bromide is known to deplete ozone in the atmosphere. Hence, replacement to methyl bromide has been found. Phosphine, sulphuryl fluoride, and carbonyl sulphide are a few alternatives.

Fumigation is done to the raw materials before being processed. For example, in the manufacture of instant mixes of convenience foods like instant *dosa* mix, *vada* mix, *idli* mix, or *sambar* mix, all the ingredients like cereals, pulses, spices, etc. are fumigated before being processed. Similarly pest proofing of jute bags for packaging food grains is routinely done to avoid the entry of pests through the gunny bags into the stored material over a period of time during storage. Even the cereal grains like paddy or wheat, or pulses like red gram (toor dal) or black gram are stored in bins admixed with aluminium phosphate, etc. for future use.

In a bakery industry, the rooms where the dough is kept for fermentation, and other rooms where the baked products are stored are potential sources for pests and rodents to grow. Hence, such rooms are routinely fumigated with formalin (liquid formaldehyde) solution. Ethyl alcohol (rectified spirit) is frequently used to disinfect the chamber and containers before inoculation of microorganisms in any fermentation process.

Key steps for effective pest control: Pest control can be done effectively in any processing plant if the following key steps are followed:

(i) Identification of the pest causing the damage is the first step in any pest control measures. Once if the pest is identified, its biology, ecology and behaviour can be obtained; and hence, effective pest control measures can be initiated.
(ii) At low population of pests, most pest control methods work effectively. Hence, it is advisable to initiate early action so that the pest can be controlled.
(iii) The morphology and life cycle also matter in its control. Different pests can be effectively controlled at different stages of their growth.
(iv) Integrated Pest Management (IPM) strategies should be followed judiciously combining the use of chemical and non-chemical tactics.

26.8 WATER SUPPLY TO FPI

Water is one of the most important commodities and infrastructural facilities for any food processing industry. Probably compared to any other processing industry, food industry uses more water for various purposes, viz.,

(i) For general use like cleaning, rinsing, washing, etc.
(ii) For process operations.
(iii) For cooling water purposes in refrigeration and air-conditioning systems.
(iv) For boiler feed water for steam generation.

Water is used extensively in food processing for *general purposes* like cleaning of raw materials, equipment and machinery and processing hall and factory premises, in addition to drinking purpose by the plant personnel. The potable water should be clean, clear and free from any colour or odour, which is used not only for drinking, but also for cleaning of plant and machinery.

Water is used for *process operations* like washing of raw materials. Most of the fruits and vegetables which are dirty at the time of harvest will be washed with water as well as transported by flow of water. These waters are occasionally recirculated. If the water is hard, it needs to be softened. Otherwise they may affect the texture of fruits and vegetables during blanching operations. Water is also used for making sugar syrups in beverage industry or for making aqueous solutions of acids/alkali as a part of processing steps. In all these cases, the water should be clean, pure and free of any salts, etc. Any salts present in the water may result in deposition of scale on the walls of the equipment causing reduction in heat transfer coefficients.

The quality of water used for *cooling purposes* need not be very clean as long as it does not come in contact with the food materials. In food processing, cold waters are used sometimes for chilling purpose. If the contact of the chilled water is indirect, the quality of water does not matter. For refrigeration and air-conditioning purposes, cold water is used with indirect contact. In such cases, water should preferably be not hard, lest the scale formation would affect heat transfer characteristics.

Water used for *boiler feed* should not be hard, and should not contain any soluble solids. If the water is hard, it should be softened before being fed to the boilers. Hard water deposits the salts on the boiler tubes in the form of scales and reduces the heat transfer coefficient.

26.8.1 Source of Water

These are classified as:

(i) Natural water sources
(ii) Supplies from the municipality or state authorities.

The natural water sources may be:

- Surface waters in the form of tanks or lagoons
- Ground water sources in the form of wells/springs
- Underground water sources in the form of borewells.

26.8.2 Water Treatment Methods

Different types of impurities present in water are:

- Suspended solids/materials
- Dissolved mineral matter like iron, manganese
- Dissolved salts like carbonates or chlorides
- Organic matter which imparts colour, odour and taste
- Microorganisms
- Dissolved gases if any.

Various types of treatment methods to purify the water from the above impurities are shown in a nut shell in Table 26.9. Further details on the subject can be found in any standard textbooks on water purification including the one by Brennan, et al. (1969). Rao, et al. (2012) highlighted various wastewater treatment techniques by advanced processes and technologies. These techniques are equally applicable to process waters as well.

Table 26.9 Various Treatment Methods for Purification of Water

Impurity	*Treatment method*	*Remarks*
Suspended solids/matter	Settling	Done by gravity sedimentation in large tanks.
	Flocculation	Fine suspensions are agglomerated by adding flocculating agents, viz., aluminium sulphate, sodium aluminate, chitosan, etc.
	Filtration through sand beds	Water is passed through a filter bed consisting of fine sand supported on pebbles.

(contd.)

Table 26.9 Various Treatment Methods for Purification of Water (*contd.*)

Impurity	*Treatment method*	*Remarks*
Dissolved minerals like Fe and Mn in the form of bicarbonates	Aeration, settling and filtration	Sodium hexametaphosphate is added as sequestering agent. Mn is oxidized by aeration. The insoluble oxides are removed by filtration.
Dissolved salts in the form of sulphates, chlorides and bicarbonates of Na, Mg and Ca	Precipitation process	$Ca(OH)_2$ and Na_2CO_3 precipitate the salts in the form of insoluble hydroxides or sulphates. $Ca(HCO_3)_2 + Ca(OH)_2 \rightarrow 2CaCO_3 \downarrow + 2H_2O$ $CaSO_4 + Na_2CO_3 \rightarrow CaCO_3 \downarrow + Na_2SO_4$
	Zeolite processes	The cation exchanging ability of zeolites is exploited to remove the hardness of water. $Ca(HCO_3)_2 + Na_2Z \rightarrow 2NaHCO_3 + CaZ$ The zeolite can be regenerated by treatment with NaCl.
Organic matter which imparts colour due to presence of colloidal suspensions of ultramicroscopic particles	Coagulation, settling and filtration/ decanting	Coagulants like activated silica is used. Aeration and super-chlorination are also employed. Adsorption on activated carbon is also recommended by passing through a fixed bed of activated carbon.
Microorganisms in surface waters	Chlorination/treatment with chlorine derivatives	Common microorganisms are: diatoms, fungi, algae, protozoa, rotifers, nematodes and some pathogenic bacteria. Chlorination under optimum pH, temperature contact time and sufficient quantity would remove the microorganisms.
Dissolved gases	Coagulation, settling, filtration, softening, chlorination and degassing	The dissolved gases are usually CO_2, N_2, O_2 and H_2S (in sulphur waters). Sodium sulphite and hydrazine solutions are used for oxygen scavenging.

26.9 PROCESS PLANT WASTE MANAGEMENT

Food processing industries release a large amount of waste materials because they process the crude raw materials (viz., fruits, vegetables, animals, spices, condiments, cereals or pulses) into finished products. For example, the fruits and vegetables processing industries release 50 per cent of the weight of raw materials as waste products in the form of peels, stones or fibres. Similarly animal processing industries (slaughter houses) release a lot of waste products in the form of skins and hides, blood, fats, horns, hoofs, hairs, feathers, shells, bones, liver, intestines, etc. Table 26.10 lists different waste materials generating from various food processing industries. Some of them are of much use, whereas others are not. If they are left outside in the environment, they cause a lot of environmental pollution. They are called as **industrial wastes** and need to be addressed to for their safe disposal or conversion to some other useful byproducts. Various methods of treatment may be classified as physical, chemical, and biological. One or more of these combinations may be used depending upon the type of waste. The final choice depends upon the individual cases and local circumstances.

Table 26.10 Waste Products from Various Food Processing Industries[†]

Processing industry	*Waste materials*	*Remarks*
Animal products	Skins, hides, blood, fats, horns, hairs, bones, liver, intestines	Most of the materials are utilized to make various byproducts, or consumed as inferior foods. The blood spilling out and mixed with wash waters is a pollutant.
Poultry processing	Skin, blood, fats, hairs, feathers, bones, liver, intestines, wings, trimmed organs	Hair and feathers need to be disposed. Other products find some utility or other or mixed with poultry feed.
Marine products processing	Shells, roes, trimmed parts, pincers	Most of them go as poultry feed. The prawn shells are used for making chitosan.
Cereals and pulses processing	Husk, hulls, chaff, stalks	Mostly used as agro waste for generation of heat.
Fruits and vegetable processing	Skin, peels, stones, fibre, pith	Used for generation of biogas. Briquetted to use as fuel. Citrous peels are used to make pectin. Fruit processing waste is used as manure after purification.
Nuts	Shells, coir, pith	The coir is used for making a variety of fancy items. Shells are mostly used as fuel.
Spices and condiments	Hulls, stalks	Used as fuel.

[†]The list is not exhaustive.

26.9.1 Physical Treatment Methods

The physical methods of treatment include sedimentation or filtration or open evaporation, and these methods are used generally to remove the suspended solids. These methods also remove some of the suspended high BOD[†] materials like fats, oils or blood globules, etc. In case if the sedimentation velocities are low, some flocculating agent is used so that the suspended solids agglomerate and settle quickly.

[†]BOD stands for biological oxygen demand which is the oxygen required by waste materials to fully oxidize. In other words, this is the measure of oxygen required which the materials would take from the water in which they are suspended and deplete it (water) of the oxygen. Such waters cannot sustain aquatic life, and hence, are considered as polluted.

Open ponds or lagoons: Lagoons or open ponds are holding tanks. The waste materials are let into these tanks where they get degraded by natural means which can later be used as manure. In the desiccated coconut powder making unit, the coconut water is let into open tanks where the water simply evaporates.

Open yard drying: If adequate space is available for drying, the waste materials, particularly those from processing of fruits and vegetables or from animal-based processing industries can be dried. The dried material can be used either as poultry feed or cattle feed. Alternatively the wet material may be used as organic manure in agricultural fields.

Filtration through loose beds: This would enable small quantities of fats or blood globules, etc. to be retained by the bed. The clarified effluent waters can be let out.

26.9.2 Chemical Method

In chemical methods some of the flocculating agents are used to increase the sedimentation velocity, or the pH is adjusted. These methods are mostly used for treatment of effluents.

The citrus peels are subjected to various chemical reactions to extract the pectin from them which has a great utility in food and pharmaceutical industries as a gelling agent. Central Food Technological Research Institute (CFTRI), Mysore has developed a commercial process for extraction of pectin from citrous peels. Similarly the prawn waste from the prawn processing industry can be used by alkylation and acidification to make a polysaccharide called **chitosan** (a CFTRI, Mysore process) which is a very effective flocculating agent and is reported to be 200 times more active than alum for water treatment.

26.9.3 Biological Methods

The organic matter from food processing industries can be treated with microorganisms either in solid state fermentations or by submerged fermentation with suitable dilution. In the fermentation process, the microorganisms convert the organic matter into some useful products or may use it for generation of biogas by virtue of which the BOD is considerably reduced. The fermentation could be either anaerobic process[†] or aerobic process.

Anaerobic fermentation: Anaerobic fermentation which is carried out in the absence of air is an effective method for converting the fruit processing wastes by digestion to produce biogas which is a mixture of methane and carbon dioxide in the proportion of 1 : 3 to 1 : 1 with some amount of hydrogen gas. A number of microorganisms including *E. coli* can be used for generation of biogas. However, there are some selective microorganisms which can produce a good amount of biogas. It is a method of conversion of organic matter into CH_4 and CO_2 by means of:

- Hydrolysis
- Acetogenic process

[†]Anaerobic fermentation is the one which is carried out in the absence of air, whereas the aerobic process is the one which is carried out in the presence of air. Further details are available in chapter 25 on Fermentation.

- Homoacetogenic process
- Methanogenesis

Biogas generation: The fruit processing industry wastes contain:

Cellulose:	30–40 per cent
Hemi-cellulose:	20–40 per cent
Lignin:	20–30 per cent

which are incidentally the good source of carbon, and hence, can be converted into biogas by methnogenesis by using some of the following important strains of methanogens (Krishna Nand 1999):

Methanobacterium ruminantium
Methanothermo autotrophicum
Methanospirillium hungatti
Methanobacterium formicicum
Methanococcus vannieli
Methanosapcina barkeri
Methylomonas albus
Methylobacter capsulatus
Methylosinus trichosporium, etc.

The optimum temperatures for digestion are 35–40 °C, at which biogas can be produced after 20 days. Further details are available in Rao [2010(b)].

Bioconversion of waste by microalgae: This is one of the most effective methods of converting the food processing wastes, particularly fruit processing industry wastes, into fish-feed by using microalgae. Fruit processing wastes, even after extraction of pulp, are a rich source of organic matter which can be used for growing some blue-green algae as biomass. The biomass is a good source for fish-feed with which fish can be grown, which in turn is a good source of proteins for human beings. Thus, the bioconversion process is an indirect method of converting the food industry wastes into useful proteinaceous source in the form of fish for human consumption.

Sunita and Rao (2003) reported the method for bioconversion of wastes containing peels and fibre from a mango processing industry to a fish-feed by microalgae. The microalgae was isolated from soil into which the mango processing industry effluents were being discharged. The mango wastes were suitably macerated and diluted with water to make a solution containing 20 per cent solids. The solutions were treated with blue-green algae (mostly filamentous form), viz.,

Chlorella
Gleotrichia
Anabena
Oscillatoria
Nostoc

The trials were restricted to lab scale in 20 liter carboys. The harvested biomass after 15–20 days were fed to model fish *Tilapia mossambica* which have shown considerable increase in weight (of the order of 2.5 to 3.0 times) as compared to the control.

REVIEW QUESTIONS

26.1 Describe the importance of hygiene in food processing.

26.2 What are the basic principles of hygiene and sanitation?

26.3 What are various methods of cleaning practices?

26.4 Describe open cleaning method.

26.5 Describe closed cleaning method.

26.6 What is meant by CIP?

26.7 Describe a single use CIP system. What are its advantages and disadvantages?

26.8 Describe a multiple use CIP system. What are its advantages and disadvantages?

26.9 Describe various steps involved in the development of multi use CIP system.

26.10 What are the principal components of a multi use CIP system?

26.11 What are different sanitation methods?

26.12 What are the various important issues to be considered during selection of a sanitizer?

26.13 Describe chlorine-based sanitizers.

26.14 Describe sanitizers based on quaternary ammonium compounds.

26.15 What are iodophores?

26.16 Provide comparative details of various sanitizers.

26.17 What is meant by thermal sterilization?

26.18 What are various types of detergents used in food processing industries for cleaning purpose?

26.19 Describe a pest proofing method.

26.20 What are the key steps for effective pest control?

26.21 Describe the importance of water supply to food processing industry.

26.22 What are different methods of purification of water?

26.23 What are various process plant waste management methods?

26.24 Describe physical treatment methods for process plant waste management.

26.25 Describe chemical treatment methods for process plant waste management.

26.26 Describe biological treatment methods for process plant waste management.

26.27 Describe biogas generation process.

26.28 Describe a method of bioconversion of wastes by microalgae.

REFERENCES

Brennan, J.G., Butters, J.R., Cowell, N.D. and Lilly, A.E.V. (1969), *Food Engineering Operations*, Elsevier, Esser (UK), pp. 362–382.

Krishna Nand (1999), Biomethanation from agro industrial and food processing wastes, *Biotechnology: Food Fermentation*, Joshi, V.K. and Pandey, A. (Eds.), Vol. 2, Educational Publishers and Distribution, New Delhi, pp. 1349–1372.

Plett, E. (1992), Cleaning and Sanitation, in *Handbook of Food Engineering*, Heldman, D.R. and Lund, D.B. (Eds.), Marcel-Dekkar, Inc., New York, pp. 719–730.

Rao, D.G. (2010), *Introduction to Biochem. Engg.*, 2nd Ed., Tata McGraw-Hill, New Delhi, (a) pp. 116–134, (b) pp. 406–411.

Rao, D.G., Senthil Kumar, R., Byrne, J. and Feroz, S. (2012), Ed., *Wastewater Treatment: Advanced Processes & Technologies*, CRC Press, Taylor & Francis group, USA.

Sunita, M. and Rao, D.G. (2003), Bio conversion of mango processing waste to fish-feed by micro algae isolated from fruit processing industrial effluents, *J. Sci. Ind. Res.*, **62**, pp. 344–347.

CHAPTER

27

Food Process Economics

Process economics plays a vital role in any commercial operations. The ultimate success of any process, any technology, any R and D output is measured in terms of how much money (profits) it (the industry) would spin out. Hence, in this chapter we discuss some of the aspects related to process economics which would play an important role in any process, research and development, and commercial exploitation.

In food processing, the challenges are many. There is a continuous need to develop new products since the product life cycle of most of the food products is very short. Figure 27.1

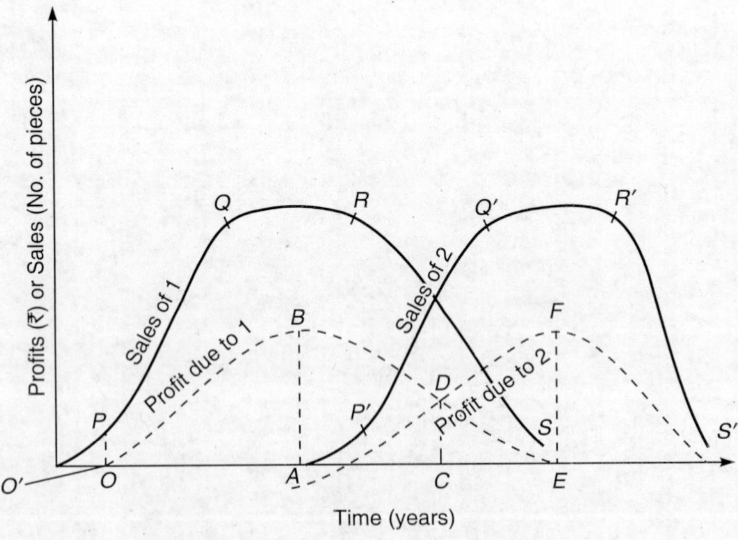

Figure 27.1 Life cycle of products (1 and 2).

shows a typical product life cycle, which is hardly of the order of 5–6 years. The reasons for it could be many:

(i) The changes of societal needs.
(ii) The changes in preferences of different demographic groups.
(iii) Replacement or rejuvenation of existing product that has declining life cycle.
(iv) Response to competitors.
(v) To increase the sales on the name of new products.
(vi) Repositioning of existing products.
(vii) The selected snobbish nature of consumers for esteem products.

In view of the above, there is a continuous need for R and D and diversions in a food processing sector. Figure 27.1 shows life cycle of two products concurrently to have continuous non-declining net profits. For example, *AB* is the profit due to product 1. Twice of *CD* is profit due to products 1 and 2 concurrently *EF* is net profit due to product 2 when product 1 has become obsolete in the market. *OP* is gestation period for product 1, *OO′* is expenditure during gestation period. *PQ* is growth phase, *QR* is stabilization period and *RS* is decay period for product 1. Similarly for product 2, *P′Q′* is growth period, *Q′R′* is stabilization period and *R′S′* is decay period. The development of a successful process has many objectives, which are tabulated in Table 27.1. Apparently all the objectives envisaged in the table are not achievable, but efforts should be made to see that as many of them as possible to be met within the given frame work.

Table 27.1 Basic Objectives in the Development of a Successful Process

The new process route should be economical.
The capital investment in the main equipment and ancillary machinery should be minimum.
The equipment and machinery should be flexible, so that they can be used for multiple purposes.
The new product should be replacement or rejuvenation of the existing product.
The new product should have additional health benefits or convenience or some niche quality to push through into the consumer market with boosted sales.
There should be some saving in labour; or the new process route should be amenable to automation which in turn reduces the recurring costs of production.
The recovery and purification of the product should be easier, simple and rapid.
The consumption of utilities should be as little as possible.
The effluent discharge should be minimum.
The floor space occupied should be minimum.
The new product should meet the statutory regulations of the local government.
It is desirable that the new product should be able to reap the benefits of the statutory concessions if any extended by the local governments.

27.1 PROCESS ECONOMICS

Every process/product has some peculiar characteristics associated with it. The process is described with a

- process flow concept,
- process equipment sizing,

- materials of construction,
- utility needs,
- costs of initial plant and operations,
- anticipated risk in the development,
- anticipated associated-costs in market development for the product, and
- estimate of Returns On Investment (ROI).

One of the important strategies to be considered in the process economics is to carefully visualize whether the developed product would attract/exploit any statutory concessions extended to the product by the local government[†].

These points are more related to the financial aspects of the process. Now, let us see how the economic considerations play a dominating role at various stages of plant design.

27.1.1 Various Stages in Plant Design

Various project stages in a plant design are important in terms of process economics. The first and the foremost important stage is inception of the project. The idea for conceiving a project may come from the experience of the sales personnel or a survey conducted by social scientists. Before conceiving the project, it may be worthwhile to make a preliminary financial estimation for the project. It is also known as market survey/preliminary evaluation of the market. It gives us an indication of the potential market for the product. Sometimes, even a market survey is also conducted with live product by preparing it in a small quantity in pilot plants or by modifying some of the existing facilities. If the results are satisfactory, efforts will be made to generate data for final design. The final design is based on the

(i) market survey, and
(ii) cost refinement.

Concurrently with the final design stage, we need to carry out financial analysis. It is also known as **final economic analysis** (Bailey and Ollis 1986). Various queries raised in the final economic analysis are shown in Table 27.2. If the answers for the queries are satisfactory, decision will be taken to take up the project, and detailed engineering design will be made. This stage is more technical in nature. Efforts would be initiated for consolidating information on

Table 27.2 Queries to be Made for Final Economic Analysis

What is the novelty of the proposed product?
Is the product likely to be a niche product?
Does the product fall into the charter of activities of the company? If so, what advantages the new product brings in? if not, what are likely complications?
Based on the market survey, how does the product compete in the market with the other existing products to generate the anticipated profits?
Whether the available man power and technology are adequate to meet the requirements of the new product?
Whether the proposed machinery and equipment are flexible to be used in some other process line?

[†]It will be discussed more in detail in the next chapter on plant location (Section 28.3).

the availability of machinery and equipment, their detailed specifications, piping and instrumentation, various control valves and techniques, a maze of other accessories and services, and availability of infrastructural facilities, etc. At this stage, even the plant and equipment layout, and details of building construction are also envisaged.

The next stage is to go for procurement of materials and starting of construction work with the help of some project engineering firm. After successful completion of the tasks, the last stage is to go for *start-up and trial production runs*. The trial production runs are taken up with the help of the consultant who provided the overall consultancy for the project.

In the overall project implementation, financial analysis ultimately plays a major role, and needs to be looked into:

27.1.2 Financial Analysis

Financial analysis deals with the capital investment in plant and equipment, and the returns on such investment. The primary aim of any such investments in financial ventures is to get profits, i.e., the returns on the investment should be more than the amount the investment would normally fetch by safe deposits and bank deposits. The returns can be expressed in many ways:

(i) Average annual return or investment,
(ii) Pay back period,
(iii) Returns On Investment (ROI), and
(iv) Net Present Value (NPV).

Since the investments cannot be infinite or limitless, it is always essential to make investments judiciously. This judicious decision is based on the following two questions:

(i) With the capital investment at our disposal, to what extent does it meet the minimum financial objectives of the company?
(ii) If there are several alternatives available for making the investment, which investment offers the best returns?

If there are several options available for a particular process to be carried out or a particular product to be made, financial analysis will help find the best option or technology on the basis of profits. The financial analysis is usually based on the following terms (Valentas, et al., 1991):

(i) *Taxable income*, which is the difference between the total earnings before payment of taxes and depreciation.
(ii) *Cash flow*, which is the total earnings (before the payment of taxes) minus the taxes.
(iii) *Book income*, which is the difference between the taxable income and taxes payable.
(iv) ROI

The cash flow diagram is shown in Figure 27.2.

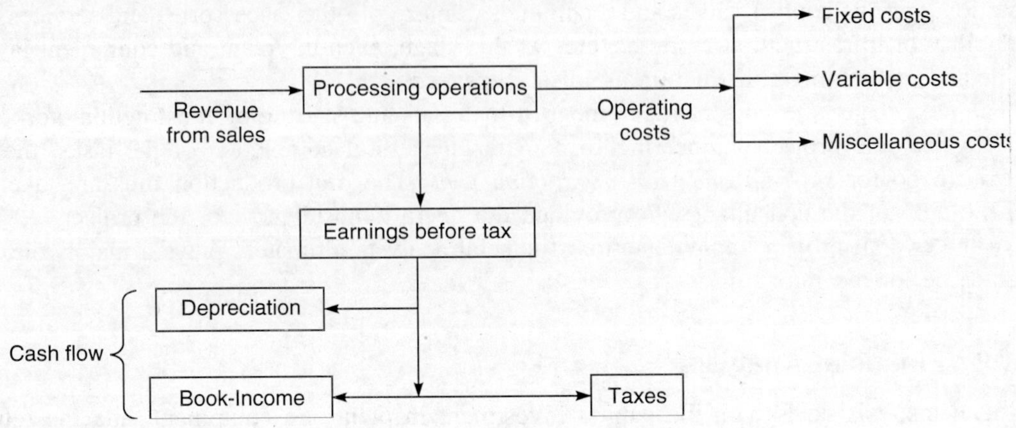

Figure 27.2 Cash-flow diagram.

The return on the investments, also known as ROI, is an important component in financial analysis. This is a way of finding out by investing how much capital initially, we get an assured cash flow of one rupee per year. We normally expect about 10 per cent earnings on our investments.

For the first year, the one rupee equivalent is 0.909 (i.e., 0.909 × 1.1 = 1.0). For the second year, the equivalent amount is ₹ 0.826 to get an earning of ₹ 0.909 (i.e., 0.826 × 1.1 = 0.909). Similarly, for the third year, it should be ₹ 0.751. For all the three years put together it becomes

$$0.909 + 0.826 + 0.751 = ₹\ 2.49$$

which is taken as the present value. That is, by investing an amount of ₹ 2.49 today, we get an assured amount of ₹ 1 each year for the coming 3 years. The ROI is 10 per cent on discounting basis (Valentas, et al., 1991). This works out to be

$$P_n = \frac{S_n}{(1+i)^n} \tag{27.1}$$

where P_n is the present value of principal amount in the nth year, S_n is the cash flow every year expected for n years, and i is the discount rate or ROI.
Therefore, the present investment is ΣP_n to get an assured income of S_n every year.

i.e., $$\text{Present value} = \Sigma P_n \tag{27.2}$$

and $$\text{the present value factor} = \frac{1}{(1+i)^n} \tag{27.3}$$

The present value factors $[1/(1+i)^n]$ at different discount rates are shown in Table 27.3.

Now, we work out in the reverse direction.
Initially we have invested ₹ 2.49,
At the end of first year it becomes an amount of 2.49 × 1.1 = ₹ 2.739, out of which we are taking out one rupee. The amount left out = 2.739 – 1.0 = 1.739.
Therefore, amount available at the beginning of second year = ₹ 1.739.
It would fetch 10 per cent earnings at the end of second year.

Table 27.3 Present Value Factors $[1/(1+i)^n]$ at Different Discount Rates

n	2%	4%	6%	8%	10%	12%	14%	16%	18%	20%	22%	24%	26%	28%	30%
1	0.98	0.962	0.943	0.926	0.909	0.893	0.877	0.862	0.847	0.833	0.82	0.806	0.794	0.781	0.769
2	0.961	0.925	0.89	0.857	0.826	0.797	0.769	0.743	0.718	0.694	0.672	0.65	0.63	0.61	0.592
3	0.942	0.889	0.84	0.794	0.751	0.712	0.675	0.641	0.609	0.579	0.551	0.524	0.5	0.477	0.455
4	0.924	0.855	0.792	0.735	0.683	0.636	0.592	0.552	0.516	0.482	0.451	0.423	0.397	0.373	0.35
5	0.906	0.822	0.747	0.681	0.621	0.567	0.519	0.476	0.437	0.402	0.37	0.341	0.315	0.291	0.269
6	0.888	0.79	0.705	0.63	0.564	0.507	0.456	0.41	0.37	0.335	0.303	0.275	0.25	0.227	0.207
7	0.871	0.76	0.665	0.583	0.513	0.452	0.4	0.354	0.314	0.279	0.249	0.222	0.198	0.178	0.159
8	0.853	0.731	0.627	0.54	0.467	0.404	0.351	0.305	0.266	0.233	0.204	0.179	0.157	0.139	0.123
9	0.837	0.703	0.592	0.5	0.424	0.361	0.308	0.263	0.225	0.194	0.167	0.144	0.125	0.108	0.094
10	0.82	0.676	0.558	0.463	0.386	0.322	0.27	0.227	0.191	0.162	0.137	0.116	0.099	0.085	0.073
11	0.804	0.65	0.527	0.429	0.35	0.287	0.237	0.195	0.162	0.135	0.112	0.094	0.079	0.066	0.056
12	0.788	0.625	0.497	0.397	0.319	0.257	0.208	0.168	0.137	0.112	0.092	0.076	0.062	0.052	0.043
13	0.773	0.601	0.469	0.368	0.29	0.229	0.182	0.145	0.116	0.093	0.075	0.061	0.05	0.04	0.033
14	0.758	0.577	0.442	0.34	0.263	0.205	0.16	0.125	0.099	0.078	0.062	0.049	0.039	0.032	0.025
15	0.743	0.555	0.417	0.315	0.239	0.183	0.14	0.108	0.084	0.065	0.051	0.04	0.031	0.025	0.02
16	0.728	0.534	0.394	0.292	0.218	0.163	0.123	0.093	0.071	0.054	0.042	0.032	0.025	0.019	0.015
17	0.714	0.513	0.371	0.27	0.198	0.146	0.108	0.08	0.06	0.045	0.034	0.026	0.02	0.015	0.012
18	0.7	0.494	0.35	0.25	0.18	0.13	0.095	0.069	0.051	0.038	0.028	0.021	0.016	0.012	0.009
19	0.686	0.475	0.331	0.232	0.164	0.116	0.083	0.06	0.043	0.031	0.023	0.017	0.012	0.009	0.007
20	0.673	0.456	0.312	0.215	0.149	0.104	0.073	0.051	0.037	0.026	0.019	0.014	0.01	0.007	0.005

Hence, amount at the end of the second year = 1.739 × 1.1 = ₹ 1.9129.
We take out one rupee from it.
Therefore, amount available at the beginning of third year = 1.9129 – 1.0 = 0.9129.
This amount would fetch an earning of 10 per cent for the third year.
Hence, amount available at the end of the third year = 0.9129 × 1.1 = 1.0.

We have one rupee available at the end of third year. By the end of third year, the whole investment is over. Thus, the financial analysis gives us an indication of investment and returns on the investment after discounting for an expected rate of earning.

27.1.3 Cost of Production

After having made financial analysis of the project in terms of investments and rates of returns, now let us try to see what constitutes cost of production. Any material or industrial activity when it is produced will have two types of costs

- fixed costs, and
- variable costs.

The former does not vary with the volume of production, whereas the latter varies directly with the volume of production. Table 27.4 shows the list of various items which constitute the fixed costs and variable costs. The actual cost of production of a material is:

$$\text{Total cost of production} = \text{Fixed costs} + \text{Variable costs} \quad (27.4)$$

Table 27.4 Components of Fixed Costs and Variable Costs of Goods

Fixed costs	*Variable costs*
Interest on fixed capital	Raw materials
Pay back of fixed assets[†] (in a given period: usually 5–6 years)	Utilities
Rent and rates	Wages to workers
Depreciation (usually 10 per cent on flat basis)	Sales expenses
R and D expenses	Transportation
Marketing costs	Packaging materials
Salaries	
Administrative expenses	
Factory overheads	
Property taxes payable to local municipality	

[†]In some cases, the pay back amount is not accounted.

Peters and Timmerhaus (1980) suggested a method to arrive at total product cost which was a combination of manufacturing costs and general expenses. Various costs involved in total product cost are shown in Figure 27.3.

27.1.4 Break Even Analysis

Break even analysis is a method to find out what should be the volume of production so that the industry reaps profits. Break Even Point (BEP) is that level of volume of production at

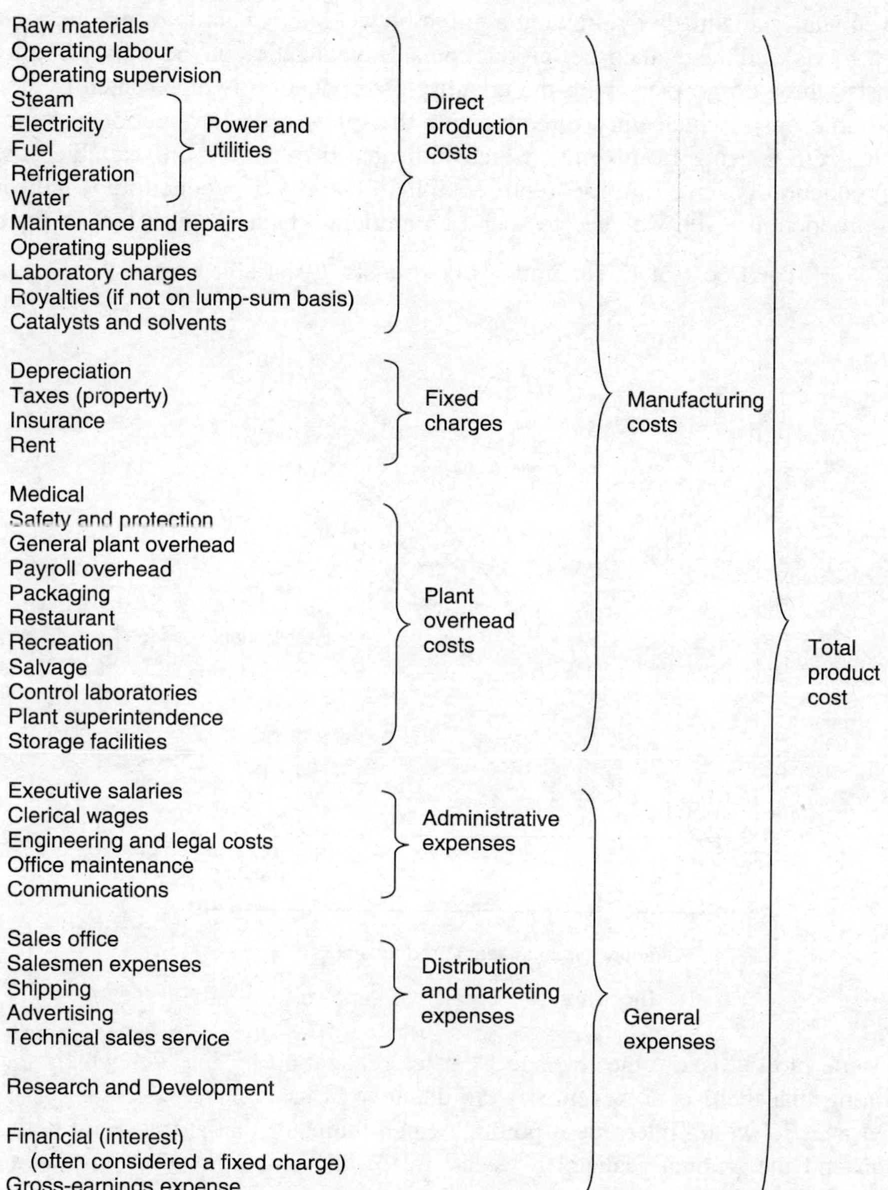

Figure 27.3 Costs involved in total product cost for a typical chemical process plant.

Source: Reprinted from, *Plant Design and Economics for Chemical Engineers,* 3rd ed., Peters, M.S. and Timmerhaus, K.D., p. 192, Copyright (1980), with permission from McGraw-Hill, New York.

which the cost of production of goods is just equal to the amount realized by selling the goods. This is a no-loss-no-profit situation. Any volume of production beyond BEP would result in profits, and any volume of production below the BEP would result in losses.

Break even analysis is made by drawing a graph between the quantity of production (No. of pieces) on the *x*-axis, and the total cost of production/sales realization on the *y*-axis, (Figure 27.4). The fixed costs which do not vary with the quantity of production is represented by horizontal line *BD*. Variable costs which vary directly with the quantity of production is represented by inclined line *OE* starting from origin which indicates that there is no variable cost if the quantity of production is zero. Line *BC* represents the total cost of production. It indicates that even at zero production level, we need to spend an amount which is equivalent to fixed costs.

$$\text{Total cost of production} = \text{Fixed costs} + \text{Variable costs}$$

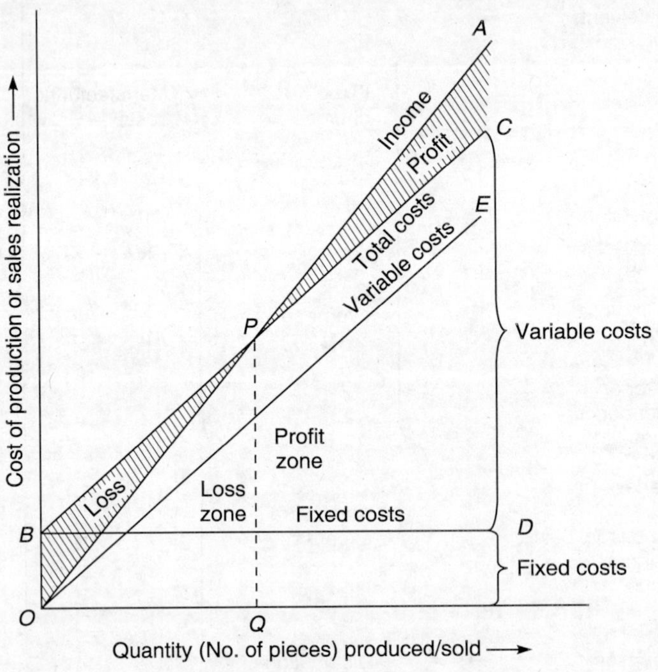

Figure 27.4 Break even analysis.

On the same plot we show the revenue by sales realization by line *OA* which starts from origin indicating that there is no revenue if the quantity of sales is zero.

Lines *OA* and *BC* would intersect at point *P* which indicates that at this point both the cost of production and the amount realized by sales are equal. This point *Q* is known as **Break Even Point** (BEP), and the point *Q* on the abscissa indicates the quantity corresponding to break even point. Any quantity of production less than *Q* results in losses and that beyond *Q* results in profits. The discussion in Figure 27.4 is restricted to the conditions prevailing for production costs and the sale price of the product. If there are any changes in it, obviously, the break even point also would shift. Accordingly the profits and losses would change. Profits would increase if the slope of sales line *OA* increases; i.e., the price of the product is high. Similarly the profits would also increase if the slope of the costs line *BC* is decreased, i.e., the production costs are reduced. It may not be always possible to increase the price of the product, since higher prices would reduce the market share of the product, unless the product is a monopoly item which rarely happens. Hence, the pragmatic approach would be to reduce the cost of production. Some of the possible alternatives for it could be by:

- using cheaper raw materials,
- cutting down the operating costs,
- increasing the efficiency of working personnel,
- reducing the consumption of utilities,
- reducing the processing time,
- reducing the inventory,
- effective realization of revenues, etc.

27.1.5 Business Perspective

Business perspective is a different subject which is neither purely management (business and economic matters) nor purely technical matters, but an amalgamation of both. In this the technical considerations are influenced by the management mood, and the management strategies are influenced by technological considerations. The business perspective becomes trickier in case of food processing, since food processing is a unique area where organoleptic qualities and human bias have a dominant role. As a food engineer, it is easy to maintain and manipulate the process parameters like temperature, pressure, degree of agitation, time of cooking, flow rates, etc. But when it comes to organoleptic qualities like taste, aroma, texture, likeness, colour, etc., it is difficult to gauge them and quantify them which enables one to manipulate them. Table 27.5 lists some non-measurable human parameters which make food processing a tricky activity without being amenable purely to technological issues nor managerial tactics.

Table 27.5 Some Non-measurable Human Parameters Related to Food Processing

Some peoples' belief that processed foods are not good for health.
In some parts of the globe, people like bland taste for foods, whereas in some other parts, people like spicy taste.
When it comes to food, peoples' choice goes with taste, whereas a bitter pill is also swallowed as medicine.
People are choosy and sensitive about the price of food products, whereas price does not matter in cosmetics, clothing and other fashion design produce.
Taste buds dominate the choice of foods, whereas the public image dominates in other spheres of life.

Let us also see to what extent the interaction between engineering and management is appropriate and adequate. The management's planning and outlook are more biased towards profits, whereas the engineering and technology are more biased towards science and scientific principles. Hence, a good management should hear to the technical and process information provided by the engineer, whereas a good engineer should heed to the financial implications and company's objectives. For successful running of the industry, management adopts

- A short range strategies to achieve quick results and financial gains.
- A long range strategies to achieve long range sustenance and steady growth.

The strategies are based on the

- seasonal supply and demand of product,
- supply and price of the raw materials,

- changes to be made in the capacity of production, and
- modify or change the structure of the product.

Almost akin to these, Rudd (1975) proposed a suitable industrial model which was more applicable to the development of intermediate chemical industry and is shown in Figure 27.5. Some of the strategic issues to be considered while taking a decision on the project are given in Table 27.6.

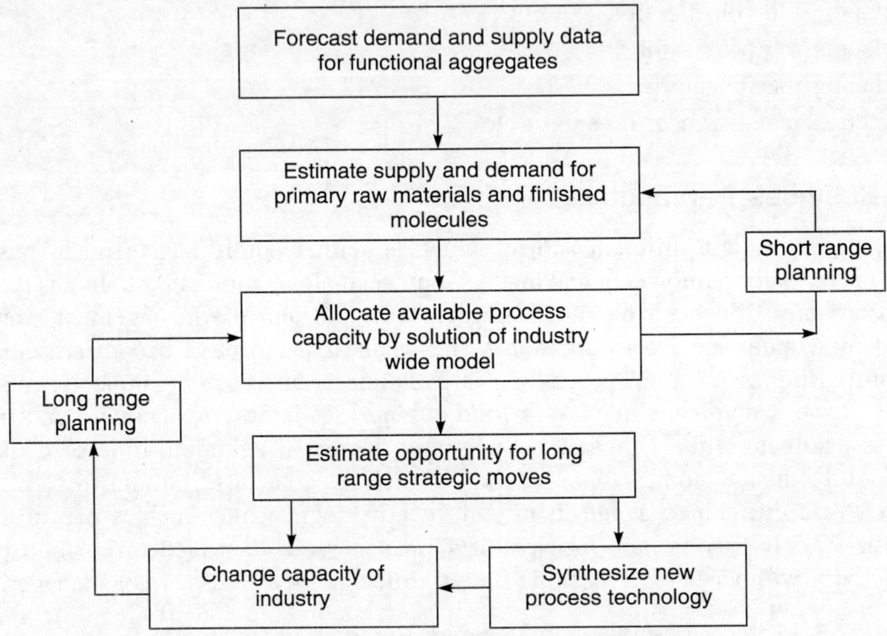

Figure 27.5 Model for strategic planning in industrial development.

Source: Reprinted from, Modelling the development of intermediate chemical industry, *The Chem. Engg. Journal*, Vol. 9, p. 7, Rudd, D.F., Copyright (1975) with permission from Elsevier.

Table 27.6 Issues to be Considered on a Project

What is the likely demand for the product? Is it increasing/decreasing?
Is the product going to be a niche product?
Whether the company has any prior experience in the new area of activity? Whether any related market outlets have been established?
Is it worth to go for internal production of the produce or to go for a franchisee/for some sub-contracting to go for contract work for another big company?
Whether the supply of raw materials is abundant? Whether the prices are stable?
How about the energy requirements and the availability of energy/power?
Whether the capital is available for construction work/is it needed to go for borrowing?
Is it necessary to go for renaming a product with some catchy words to explicitly spell out the distinct features of the product to reach niche market?
How serious are the environmental issues? What is the quantity of effluents? Is it easy to dispose off the effluents?
Whether the government notified concessions and intensives are likely to be available?
Whether any government regulations are likely to affect the production and marketing of the product?

From various issues considered above, it is evident that the success of food processing industry depends upon a judicious amalgamation of both technical and economic considerations. A project (food) engineer needs to identify the niche areas, and decipher the influence of personal emotions and bias from the economic viability of the product.

27.2 PROJECT ECONOMICS

For producing any commercial goods or for making any commercial ventures, we prepare the project proposals in which the project economics is an important component. It deals with various financial matters for the implementation of the project. It is something akin to estimating total investment for the project. For starting any project as a financial venture, we need to first develop the infrastructure for the project for carrying out the industrial activity. Once the infrastructure is created, we need to make a provision for production and operational expenditures. In some cases, we even need to make a provision for growth of the organization/ industry which will help it sustain for a longer duration. The types of capital requirement for any processing ventures may be classified as:

- fixed capital,
- working capital, and
- growth capital.

Let us discuss these things in more details in the following sections:

27.2.1 Fixed Capital

The capital required for meeting the expenses towards fixed assets for establishment of the industry is known as the **fixed capital**. These include capital for:

(i) land,
(ii) land development,
(iii) building,
(iv) machinery,
(v) equipment,
(vi) auxiliary equipment,
(vii) preoperative expenses,
(viii) accessories like computers, etc.

The capital investment invested in these fixed assets is a permanent asset, and does not change with the level of production. This cash does not come under the routine cash expenditure for the day to day production activities. Thus, the money invested in these fixed assets cannot be used for any other purpose. The repayment is contemplated over a period of time, usually of the order of 5–6 years.

The fixed assets are available as long as the industry runs, and also available for resale for some value even if the industry closes or becomes sick. Hence, the lending institutions take them as a part of security. The terms of payment are different. The banking interest rates will also be lower, and is usually charged on annual basis.

27.2.2 Working Capital

Working capital is used for routine running of the industry. It is the investment required to support the business activities on a short-term basis. Hence, the investment in this is also known as **circulating capital**. The amount will be changing day to day. The following are various items covered under working capital:

(i) Raw materials for the process
(ii) Auxiliary raw materials like acidulants, flavours, colours, preservatives, etc.
(iii) Packaging materials
(iv) Finished product inventories
(v) Handling and transportation charges
(vi) Marketing and sales activities
(vii) Salaries and wages of employees and workers
(viii) Taxes, insurance, security arrangements, etc.
(ix) Payment of telephone bills and bills for utilities, viz., water, electricity, diesel, furnace oil, coal, etc.
(x) Money to carry accounts receivable (e.g. credit extended to customers) minus accounts payable (e.g. credit extended by suppliers)
(xi) Additional cash required to carry out the business or process operations
(xii) Readily available cash to meet any emergencies (it varies from organization to organization).

From the list of the items covered above, it is evident that all these activities are required for day to day running of factory. Hence, the amount will be charging on day to day basis. For example, during the season, the amount required to procure the raw materials will be very high. Similarly in the first week of the month, cash is required for payment of salaries, wages and bills for various utilities. The lending institutions have a different pattern of charging interest. The interest rates are higher and will be charged on day to day basis.

Usually the working capital is of the order of 10–20 per cent of the total project cost (Peters and Timmerhaus 1980), but in case of food processing industries, major part of the working capital is required for purchasing the raw material during the season which will be used for processing for the rest of the year. Hence, the working capital requirements are usually large in food processing industries unlike in other processing industries. This is one of the major reasons for high mortality rate in food processing industries in small scale sector. This problem needs to be addressed tactfully for the success of food industry. Peters and Timmerhaus (1980) also suggested that it could be upto 50 per cent or more for companies producing products of seasonal demand because of large inventories.

27.2.3 Growth Capital

The growth capital is made keeping in view the future developments. It does not fluctuate like working capital based on seasonal fluctuations. The requirement of growth capital often arises if the existing industry plans for expansion or diversification. For example, let us take a fruit processing industry which is doing good business. If it has been buying the required pulp till recently, it may plan to go for expansion for starting a pulp making unit; such an industry requires capital for its growth.

The entrepreneur should keep all these three capital requirements separately in financial planning activity.

27.3 DEPRECIATION

Depreciation is the term used to mean decline in value/cost of the machinery and equipment. Usually in project economics, we make a provision for the depreciation of machinery and equipment, so that by the end of the depreciation period, the accumulated money can be used to replace the old and obsolete equipment. Eventhough the depreciation charges are included in the cost economics, practically there will not be any cash flow. From the total earnings, depreciation charges are deducted before the payment of taxes. It is presumed that this amount is deducted from the earnings and deposited separately in different account so that the amount is available for replacing the equipment as and when the need arises.

The depreciation rates are usually taken as 10–15 per cent of the cost of the equipment/machinery; i.e. in a period of 6–10 years, the equipment/machinery becomes obsolete. As has been mentioned and shown in Figure 27.2, the annual depreciation charge is deducted from the taxable income. The amount is only a book transaction, but reduces the tax payable. Hence, the percentages of depreciation charges have to be agreed upon by the competent tax authority of the government. There are two methods of calculating the depreciation:

(i) flat rate of depreciation, and
(ii) diminishing rate of depreciation.

In the former case, which is usually followed because of its simplicity, the fixed (or approved) depreciation per cent charges are deducted annually. If the cost of an equipment is ₹ 1.0 lakh and if the depreciation rate is 10 per cent, the annual depreciation charges are ₹ 10 thousand (10 per cent of ₹ 1.0 lakh). Hence, the value of the equipment becomes zero after 10 years.

In the diminishing rate of depreciation, the cost of the equipment never becomes zero. In this case, the depreciation percentage is based on the existing cost of the equipment. As in the earlier example, the depreciation charges at the end of first year is ₹ 10,000 and the cost of the equipment is ₹ 90,000. Hence, the depreciation amount in the next year is ₹ 9000 and the cost of the equipment is ₹ 81,000; and so on. The values are shown in Table 27.7. The cost of the equipment at the end of 10th year is ₹ 34,868 as contrast to zero in the former method.

Table 27.7 Depreciation and Cost of the Equipment Evaluated by Two Methods†.

Period (Years)	*Flat rate of depreciation* @ 10 per cent		*Diminishing rate of depreciation* @ 10 per cent	
	Depreciation amount (₹)	*Cost of the equipment* (₹)	*Depreciation amount* (₹)	*Cost of the equipment* (₹)
0	0	100,000	0	100,000
1	10,000	90,000	10,000	90,000
2	10,000	80,000	9000	81,000
3	10,000	70,000	8100	72,900

(contd.)

Table 27.7 Depreciation and Cost of the Equipment Evaluated by Two Methods (*contd.*)

Period (Years)	*Flat rate of depreciation* @ 10 per cent		*Diminishing rate of depreciation* @ 10 per cent	
	Depreciation amount (₹)	*Cost of the equipment* (₹)	*Depreciation amount* (₹)	*Cost of the equipment* (₹)
4	10,000	60,000	7290	65,610
5	10,000	50,000	6561	59,049
6	10,000	40,000	5905	53,144
7	10,000	30,000	5314	47,830
8	10,000	20,000	4783	43,047
9	10,000	10,000	4305	38,742
10	10,000	0	3874	34,868

†The values in the last two columns are adjusted to nearest one rupee.

27.4 GENERAL PROCESS ECONOMICS FOR CLARIFIED FRUIT JUICES

Fruits are generally valued for the vitamins, minerals and micronutrients present in them in addition to the dietary fibre. Pulpy fruits like banana, mango, guava or papaya give a thick pulp which is difficult/inconvenient to consume (drink). The fruit juice is held by the pectin to the fibrous materials. An envisaged methodology to make a clarified fruit juice (also known as **liquid fruits**) is to treat the fruit pulp with a pectolytic enzyme which will break the pectin and release the fruit juice from the fruit fibre. This results in a clear liquid fruit juice which can be centrifuged (or decanted) to separate from the fibre. Vani and Rao (2019) reported on the liquefaction of guava (*Psidium guaja L*) juice by using Pectinase enzyme (Pectinase CCM procured from Novozyme South Asia, Bangalore) immobilized on calcium alginate beads.

Central Food Technological Research Institute (CFTRI) at Mysore developed a commercial process for production of clarified fruit juices from various pulpy fruits.

The process operations include:

(i) Making of fruit pulp
(ii) Treating the pulp with enzyme
(iii) Decanting the clear liquid
(iv) Centrifuging the decanted solids and clear liquid
(v) Pasteurizing the clear liquid
(vi) Bottling and storage

The various process operations are shown in the form of a generalized flow sheet in Figure 27.6.

The process parameters to be standardized are:

(i) Method of extraction of pulp
(ii) Enzyme to pulp ratio
(iii) Whether to use enzyme in pure form or in immobilized form
(iv) Time of storage
(v) Standardization of centrifugation process
- time of centrifugation
- rpm of centrifuge

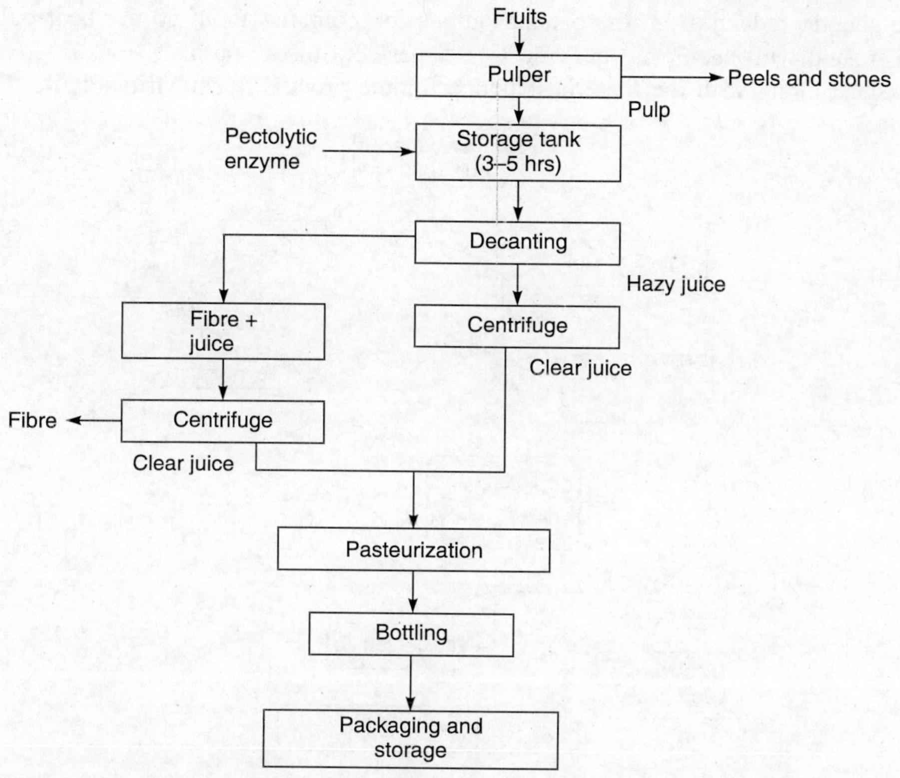

Figure 27.6 Generalized process flow sheet for making clarified fruit juices.

(vi) Packaging and bottling
- pasteurization and bottling, or
- asceptic packaging in tetrapacks

The decision on the use of immobilized enzyme needs to be taken based on the level of production which in turn is dependent upon the market survey. If the production capacities are high, it is better to go for immobilized enzymatic systems in which we can conserve the enzyme. Similarly depending upon the level of marketability and the ability to invest, decision is to be taken whether we should go for simple pasteurization and bottling or for asceptic packaging of the fruit juice in tetrapacks. Both the methods have their own advantages and disadvantages. Prospective planning is to be made based on the decision model for strategic planning shown in Figure 27.5.

27.5 CLARIFIED FRUIT JUICE FROM BANANA PULP

Banana is one of the most important fruit crops of India and is consumed on various occasions by various cross-sections of people. Banana is a highly perishable fruit and is very difficult to transport, because of which 35–40 per cent of the crop is lost. One of the best methods to reduce the post harvest losses is to go for marketing the pulp, and various other shelf-stable and value-added products from the banana pulp. Fruit contains about 50 per cent pulp, which is in turn treated with *pectinase*[†] enzyme @ 0.5 per cent of the pulp.

Flow sheet along with the material balance for the process is shown in Figure 27.7.

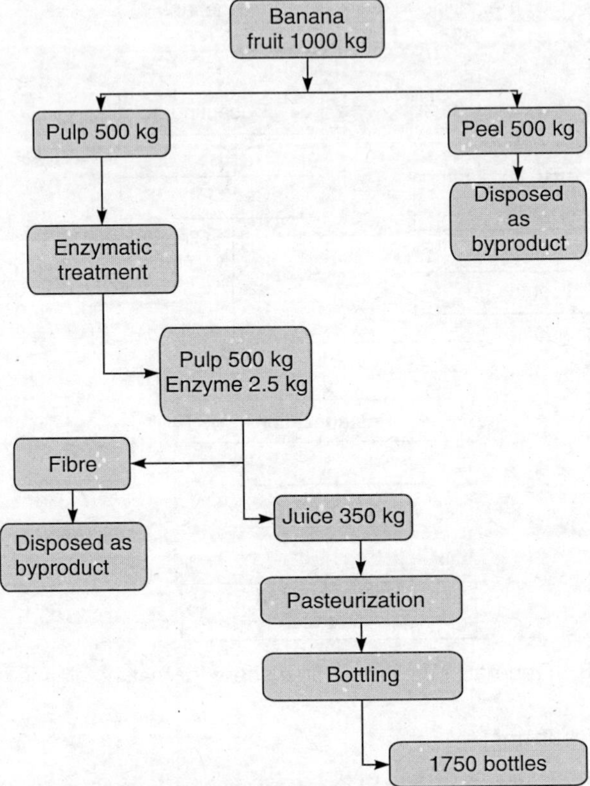

Figure 27.7 Flow sheet and material balance for clarified banana juice.

27.5.1 Estimation of Production Costs

Estimation of production costs is generally made by two ways:

(i) pre-design estimates, and
(ii) firm estimates.

The former is made with less amount of data, and hence, the information is not much authenticated. It gives an idea to the entrepreneur whether the project should be given further consideration to take up. Firm estimates need a lot of information and data. Sometimes, even experiments need to be carried out in the laboratory to get the desired data, particularly for the consumption of utilities. Hence, the information is authenticated and is more reliable. But it takes more time to make the estimates. Here we present only the former method of estimates.

[†]One of the major suppliers of the enzyme is Bangalore-based Biocon Industries.

Cost estimation[†]**:**

Fixed capital (₹):

Materials	*Cost (₹)*
Land	500,000
Buiiding	130,000
Land and building development	20,000
Equipment	500,000
Grand total	**1,150,000**

Equipment cost estimate:

Equipment	*Number*	*Cost (₹)*
Column	1	10,000
Storage tanks	2	90,000
Process tanks	1	70,000
Pulper	1	15,000
Pumps	4	10,000
Metering	1	10,000
Centrifuges	2	50,000
Heat exchanger	1	30,000
Reactor	1	200,000
Other instruments		15,000
Total		**500,000**

Working capital:

Operating cost per day:

Material	*Cost (₹)*
Raw materials	14,286
Packing material	7500
Enzymes required	7143
Salaries	2320
Wages	1000
Utilities (water, electricity, fuel, etc.)	483
Factory organization	166
Marketing and advertisement	5000
Interest on fixed asset	460
Depreciation on building and machinery	210
Pay back amount	639
Total	**39,157**

Cost of production: The cost estimation is done for 1 ton/day production of juices.

[†]The cost estimation is not authenticated. It is only given here as an example for easy comprehension of the procedure for arriving at the costs.

Basis:

1 day production
1 ton/day of juice
Pulp required = 1000/0.7 kg = 1428 kg
Fruits required = 1000/(0.7 × 0.5) = 2857 kg
Cost of fruits @ ₹ 5/kg = 2857 × 5 = ₹ 14,286 (A)
Packing (bottles) = 5000 bottles
Cost of bottles @ ₹ 1.50 = 1.5 × 5000 = ₹ 7500 (B)[†]
Enzyme cost @ ₹ 1000/kg
@ 0.5 per cent = 1428 × 0.5/100 = 7.143 kg
Cost of enzyme = 7.1 × 1000 = ₹ 7143 (C)

Wages:

10 persons @ 2500/month = 25,000
wage/day = 25,000/25 = ₹ 1000 (D)

Salaries per month:

1 food technologist	=	20,000
1 lab assistant	=	8000
2 diploma engineers	=	20,000
1 office assistant	=	6000
1 office boy	=	4000
Total (₹)		**58,000**

Salaries per day = 58,000/25 = ₹ 2320 (E)[‡]

Utilities per month (lumpsum):

Electricity = 10,000

Water = 3000

Total (₹) = 13,000

Utilities per day = 13,000/30 = ₹ 433 (F)

Marketing and advertisement (lumpsum) @ 10 per cent of sales (5000 bottles @ ₹ 10 = 50,000)

10 per cent of 50,000 = 50,000 × 0.1 = ₹ 5000 (G)

Interest component on fixed assets (@ 12 per cent) = 138,000/300 = ₹ 460 (H)

Factory organization: ₹ 50,000/annum = 50,000/300 = ₹ 166 (I)

Depreciation @ 10 per cent on building and machinery = 63,000/annum
= 63,000/300 = ₹ 210 (J)

Pay back amount in 6 years on fixed amount = ₹ 1,150,000

[†]Each glass bottle is presumed to have three recyclings, and hence, the cost of bottle is taken as: ₹ 4.5/3 = ₹ 1.5.
[‡]Even though salaries are made on the basis of 30 days per month, it is divided by 25 since the number of working days is 25, and we are trying to calculate the cost of production per day.

For one day = 1,150,000/6 × 300 = ₹ 639 (K)

Fixed costs = (E) + (G) + (H) + (I) + (J) + (K)

= 2320 + 5000 + 460 + 166 + 210 + 639 = ₹ 8795

Working capital required per day

= (A) + (B) + (C) + (D) + (E) + (F) + (G) + (H) + (I) + (J) + (K)

= 14,286 + 7500 + 7143 + 1000 + 2320 + 433 + 5000 + 460 + 166 + 210 + 639

= ₹ 39,157

Interest on working capital @ 16 per cent = 39,157 × 0.16 = ₹ 6265 (L)

Total cost of production per day = 39,157 + 6265 = ₹ 45,422

Out of this total cost of production, let us bifurcate the fixed costs and the variable costs.

Fixed costs per day	*Amount* (₹)
Interest on fixed capital	460
Depreciation	210
Payback amount [1,150,000/(12 × 300)]	639
Salaries	2320
Marketing and advertisement (lumpsum)	5000
Factory overheads	166
Total	**8,795**

It is estimated that each bottle of 200 ml can be sold for ₹ 10 to the wholesaler. The profit margins retained by the wholesaler and retailer is 30 per cent (Satyanarayana, et al., 2007) so that the product price in the market would be approximately ₹ 13 per bottle which is a reasonable price structure for pushing the product into the market.

Variable costs per day	*Amount* (₹)
Raw materials	14,286
Packaging material	7500
Enzymes	7143
Wages	1000
Utilities	433
Selling expenditures (@ 10 per cent of sales revenue)	5000
Bank interest on working capital[†]	1265
Total	**36,627**

[†]Bank interest is charged on the working capital @ 16 per cent annual interest rate on ₹ 39,157 which is ₹ 6265. it is presumed that working capital recycles 5 times in 1 year. Hence, the interest amount is 6265/5 = ₹ 1265. It is debatable whether the amount should be divided by 5. The decision is arrived at based on the fact that banks do not release all working capital lumpsum, but they release on instalment basis. They release only 1/5th working capital amount initially. Hence the interest is charged accordingly.

No. of bottles sold per day − 1000 lit/200 ml − 5000

Sales realization = 5000 × 10 = ₹ 50,000

Profit per day before taxes = 50,000 – 45,422 = ₹ 4578[†]

[†]The estimates are only pre-design estimates of Ali, et al. (2008), and hence, are not authenticated.

Based on these assumptions the break even analysis is done, and hence, break even production capacity (Q_{BEP}) is calculated.

27.5.2 Analysis of the Project

The project needs a fixed capital of ₹ 11.5 lakh. The working capital requirement per day is ₹ 39,157. Hoping that the working capital cycle is two months,

Working capital required = 39,157 × 50 days = ₹ 19.6 lakh
Fixed capital = ₹ 11.5 lakh
Total project cost = ₹ 31.1 lakh

Out of the total project cost, generally the equity participation is 1/3, and 2/3 would be procured from the lending institutions.

Therefore,
$$\text{equity share} = \frac{31.1}{3} = ₹\ 10.4 \text{ lakh} \tag{27.5}$$

Profits/day before paying tax = ₹ 4578
Annual profits before taxes = 4578 × 300 days = ₹ 13.7 lakh
Normally the taxation is upto 1/3 of the earnings.

$$\text{Hence, amount payable towards taxes} = \frac{13.7}{3} = ₹\ 4.6 \text{ lakh}$$

Therefore,
$$\text{Annual earnings} = 13.7 - 4.6 = ₹\ 9.1 \text{ lakh} \tag{27.6}$$

Comparing Eqs. (27.5) and (27.6), the project yields an earning of ₹ 9.1 lakh annually for an equity share of ₹ 10.4 lakh. Hence, the project proposal is attractive, and the *project seems to be viable.*

The likely constraints are:

(i) We have to make sure that 5000 bottles of clarified fruit juice can be sold at ₹ 13 per bottle of 200 ml volume.

(ii) The cost of the bottles (₹ 1.5 each) is arrived at on the assumption that the glass bottle has a life of 3 recycles, and each bottle costs ₹ 4.5.

(iii) It is presumed that the glass bottles would be recovered and can be recycled.

(iv) Provision needs to be made for preoperative costs and for the gestation period[†].

(v) The working capital cycle is taken as two months; efforts need to be put in to see that the working capital cycle is kept to the minimum.

Some additional and/attractive features are: annual profits are calculated without taking into consideration that the fixed assets are going to be repaid partly every year upto first six years; hence, the interest component will accordingly reduce for the first six years period. Once the fixed capital is repaid fully in the first six years, the profits would increase substantially after six years.

[†] The gestation period is the time period during which the product is not yet made, and hence, revenue by sales realization will not be there.

27.5.3 Break Even Point

Discussion on break even analysis was already presented in section 27.1.4. This is explained here with the specific example of clarified fruit juice from banana presented in the earlier section. Calculations are made for cost of production per day and the realization of sales per day.

The Fixed Costs (FC) = ₹ 8795

The variable costs (₹ 36,627) calculated in the earlier section is for a daily production of 5000 bottles. Hence, the variable costs for daily production of Q bottles are proportionately calculated as follows:

$$\text{Variable Costs (VC)} = \frac{36,627}{5000} \times Q \tag{27.7}$$

The data for various production capacities are given in Table 27.8, and it can be found at production capacities of 3000 and 3500 bottles/day the break even is occurring. The data are plotted in Figure 27.8. It can also be calculated

$$\text{Total cost of production} = \frac{36,627}{5000} \times Q_{\text{BEP}} + 8795 \tag{27.8}$$

Table 27.8 Costs and Sales Realization for Different Production Capacities/Day†

No. of bottles per day	*Variable cost (₹)*	*Total cost (₹)*	*Sales @ ₹ 10 per bottle*
0	0	8,795	0
500	3663	12,458	5000
1000	7326	16,121	10,000
1500	10,988	19,783	15,000
2000	14,652	23,447	20,000
2500	18,313	27,108	25,000
3000	21,987	30,773	30,000
3500	25,638	34,433	35,000
3288	24,086	32,881	32,880

†The fixed costs/day are ₹ 8795.

Similarly,

$$\text{Total sales revenue} = \frac{50,000}{5000} \times Q_{\text{BEP}} \tag{27.9}$$

At break even point, Eq. (27.8) is equal to Eq. (27.9)

i.e.,

$$7.3254\ Q_{\text{BEP}} + 8795 = 10\ Q_{\text{BEP}}$$

Therefore,

$$Q_{\text{BEP}} = \frac{8795}{(10 - 7.3254)} = 3288$$

Break even point occurs at a production capacity of 3288 bottles/day. The same is shown in Figure 27.8.

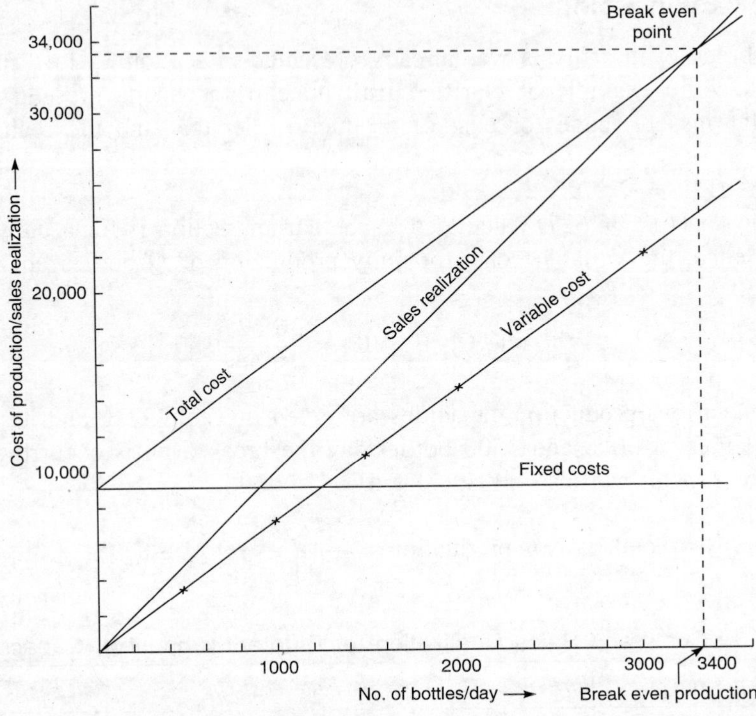

Figure 27.8 Break even analysis for a clarified fruit juice plant.

Symbols

BEP: Break Even Point
FC: Fixed Costs
i: discount rate or ROI
n: number of years
P_n: present value of principal amount in the *n*th year
Q: production capacity
Q_{BEP}: break even production capacity
ROI: Returns On Investment
S_n: cash flow expected every year for *n* years
VC: Variable Costs

REVIEW QUESTIONS

27.1 Describe the importance of food process economics.

27.2 How do you analyze a project in terms of economic returns?

27.3 What are the basic objectives in the development of a successful process route?

27.4 What is the importance of financial analysis?

27.5 What are various components of fixed costs?

27.6 What are various components of variable costs?

27.7 List some of the non-measurable human parameters related to food processing.

27.8 What are the issues to be considered while launching a project?

27.9 What components are involved in the cost of product?

27.10 What are the components involved in fixed costs?

27.11 What are the components involved in variable costs?

27.12 What is meant by growth capital?

27.13 What is meant by depreciation? How it is evaluated?

27.14 Draw a flow sheet for making clarified fruit juices.

27.15 What is break even analysis? What is its importance in any process operation?

27.16 Describe with a neat graph how the break even point is arrived at.

REFERENCES

Ali, M.S.S., Sreeman, E. and Kalyana Chakravarthy, S. (2008), Food enzymes for clarification of fruit juices, B.Tech. (BT) project report, Al-Ameer College of Engg. and Information Technology, Visakhapatnam, affiliated to JNTU, Kakinada.

Bailey, J.E. and Ollis, D.F. (1986), *Fundamentals of Biochem. Engg.*, 2nd ed., McGraw-Hill, New York, pp. 798–815.

Peters, M.S. and Timmerhaus, K.D. (1980), *Plant Design and Economics for Chemical Engineers*, 3rd ed., McGraw-Hill, Ch. 5, pp. 147–208.

Rudd, D.F. (1975), Modelling the development of the intermediate chemical industry, *The Chem. Engg.* J., **9**, pp. 1–20.

Sathyanarayana, A., Math, R.G., Jyothirmayi, T. and Rao, D.G. (2007), Novel concept of pilot model food processing centres in rural areas, *International conference on agriculture and food industries in developing countries: opportunities and challenges*, IIM-Lucknow, Aug. 10–12, 2007.

Valentas, K.J., Levine, L. and Clark, J.P. (1991), *Food Processing Operations and Scale Up*, Marcel Dekker Inc., New York, pp. 30–49.

Vani, M.M. and Rao, D.G. (2019), Effective diffusivity coefficients for degradation of pectin in guava (*Psidium guajava L.*) pulps using immobilized pectinase, *Applied food Biotechnology*, 6(2), pp. 119–126.

CHAPTER

28

Plant Design, Location and Equipment Layout

The principal activities of food engineers are to design, construct and operate food processing plants successfully. This is a continuous process in which the food engineer has to update his current knowledge in the latest developments in the processing sector. He gets this information from various sources including current publications. As compared to many other processing sectors, in fact, food processing is distinct because of the following reasons:

(i) Availability of raw materials is seasonable, and hence, availability of right quantity of raw materials at affordable cost is rather difficult.

(ii) Maintenance, hygiene and aseptic conditions are stringent.

(iii) Cost benefit analysis is rather influenced by social cost benefit analysis.

(iv) The project economics are price sensitive; the price of raw materials and products are always fluctuating which make the price dynamics rather complicated.

(v) Maintaining the inventory of raw materials and products is rather tricky, which needs a strategic approach.

These points weigh heavily in deciding the plant location and layout. In fact, many times, preliminary designs are made, and the details are further discussed with the skilled experts in the area of various engineering fields, viz., architectural, ventilating, electrical and civil. Of course, the technological details are discussed with the food technologists; and a pragmatic decision would be taken subsequently after weighing various issues and alternatives.

28.1 GENERAL PRINCIPLES

For design of food processing plant or for identifying the plant location, the food engineer or food technology consultant has to first carry out a feasibility survey which consists of various issues given in Table 28.1. Based on the survey report, we decide whether the plant is to be established.

Table 28.1 List of Items to be Considered in Making a Feasibility Survey

Availability of raw materials, their cost, and their seasonal variations.
Process technology in terms of yields, convertibility, selectivity and profits.
Availability of facilities and equipments, and how much of them are to be created.
Availability of market for the product, and need for its creation (the present market and potential market).
Safety aspects.
Any statutory requirements/restrictions.
Disposal of effluents/byproducts.
Materials of construction.
Transportation facilities and shipping restrictions/containers.
Cost of production, process economics and profit dynamics.

If the response is positive, we look into the process details, and the various equipment required are as follows:

(i) Vessels, their size, hold-up time, materials of construction.
(ii) Heat exchangers, their capacities, heat transfer areas, material of construction.
(iii) Pumps and compressors, different types of pumps, their capacities, pressure drops.
(iv) Process control instruments, various types of measuring instruments, special features of controlling instruments, sensitivity ranges.
(v) Special equipment, equipment required for process operations, viz., dryers, filters, centrifuges, mixers, grinders, etc.
(vi) Analytical facilities, analytical instruments and gadgets to check the product quality at different stages including the final product.

Subsequently, the details on the following are worked out before we go for design of the process plant:

- Specification for the raw materials and products
- Technology for the manufacturing process
- Material and energy balances
- Consumption of various utilities
- Materials of construction
- Plant site (location)

After going into the details, the preliminary design is almost complete. The next step is to go for process plant design.

28.2 PLANT DESIGN

Plant design operations are concerned with the

- type of the plant,
- volume of production,
- cost of product, etc.

In the earlier days, mostly the sizes of the plants were small, and hence, simple batch production processes were sufficient which did not require much attention on various issues. But the modern day production plants are very huge. Fruit processing plants of the order of 150–200 tpd are very common. Hence, the continuous processing operations are resorted to. Maintaining high level of hygiene and safety is very important. In such cases, the plant design is concerned with the following:

(i) Site selection
(ii) Design of buildings and housing the equipment in them
(iii) Plant and equipment layout

Of these items (i) and (iii) will be dealt in the next section. We deal with plant design in this section which will be discussed in the form of two sub-sections, viz.,

28.2.1 Design and Construction of Building

The building to house the food processing plant deserves special attention in view of the distinct nature of food materials.

Storage godowns: The raw materials are received in bulk and usually in contaminated form. For example, the fruits or vegetables after harvest are contaminated with dust, dirt, soil and sap; they need to be cleaned. They are also received in unripe form, and need to be ripened in the ripening sheds. They all require provision for cleaning and storage with good ventilation, etc. Hence, usually they are stored in huge godowns with not much of supporting pillars to effectively utilize the floor area.

Process hall: The process hall, in which all the processing operations take place, needs special attention with respect to (Brennan, et al., 1969):

- Floors
- Walls and ceiling
- Lighting
- Ventilation

The flooring of the process hall needs to be clean and should preferably be provided with tiles so that the large quantity of process waters can easily go into the drains. The floor inclination should be such, that water drains off to either end where the drainage is provided. Slip resistant, roughened tiles should be used to avoid damage to the personnel during processing, since the workers usually wear rubber shoes which are slippery in water. There should be adequate facility for water and water points. There is every possibility that the food materials may fall or splash on the floor during processing, and hence, needs to be cleaned. The stagnant pools of water should be avoided at any cost. The drains and drainage lines should invariably have screens to prevent rodents and cockroaches from entering into the processing areas.

Similarly, the walls and ceiling of the process hall should also be smooth, so that it is easy to clean with water. The walls also may be provided with smooth tiles, so that any food materials that may splash during processing can easily be cleaned. The joints between the walls and ceiling should not be sharp, rather they should be curved. The sharp joints would be difficult to clean, and hence, they harbour undesirable insects or microorganisms which are a potential hazard for the safety and hygiene of food processing.

False ceiling in the processing hall should never be permitted. False ceiling harbours the rodents, insects, etc. which will contaminate the processing operations. The process hall should also have adequate ventilation and light. Most of the food materials during processing release the volatiles and choke the place, with the result the operators feel suffocated. Particularly in case of grinding of chilli, the volatiles that come out are indeed suffocating. Similar is the case during processing of onions or garlic. Such processing should take place in well ventilated buildings where there is free flow of air and light. The ventilators should all be properly provided with screen meshes to avoid the inflow of flies and insects.

Poor ventilation also results in condensation of process vapours which may support the growth of microorganisms. It is also essential to carry out the processing in vessels or vats under hood cupboards so that the fumes or vapours are properly sucked out through the hoods, and their condensation may be avoided/reduced.

28.2.2 Functionality of the Building

The buildings should form an integral part of the overall design. The design of the main building and that of the auxiliary facilities should be in unison with each other. The auxiliary facilities include:

- Steam line
- Refrigeration plant
- Raw materials' entry point
- Raw materials' exit point
- Finished products' exit point
- Storage of packaging materials
- Quality control laboratory facility

For this purpose, the distribution line of various food processing operations follow any one of the following systems (Lopez–Gomez and Barbosa–Canovas, 2005):

- Straight line layout
- L-layout
- U-layout
- Z-layout

A straight line layout occupies more space in which the raw materials enter at one end, and finished product leaves at the other end as shown in Figure 28.1. This system has the advantage that none of the process steps interfere with each other. U-layout will have the entry of raw materials and the exit of finished goods from the same side as shown in Figure 28.1. In case of L-layout, the entry of raw materials and exit of products are in right angled direction to each other. For example, if the product is entering from the north side of the building the product will be leaving either from eastern side or western side of the building. Similarly, in case of Z-layout, the entry of raw materials and exit of products are in opposite directions just as in the case of straight line layout, but they are from different corners; i.e., if the raw materials are entering from the upper side of the western side, the product will be leaving from the bottom side of the eastern side. All these configurations are shown in Figure 28.1.

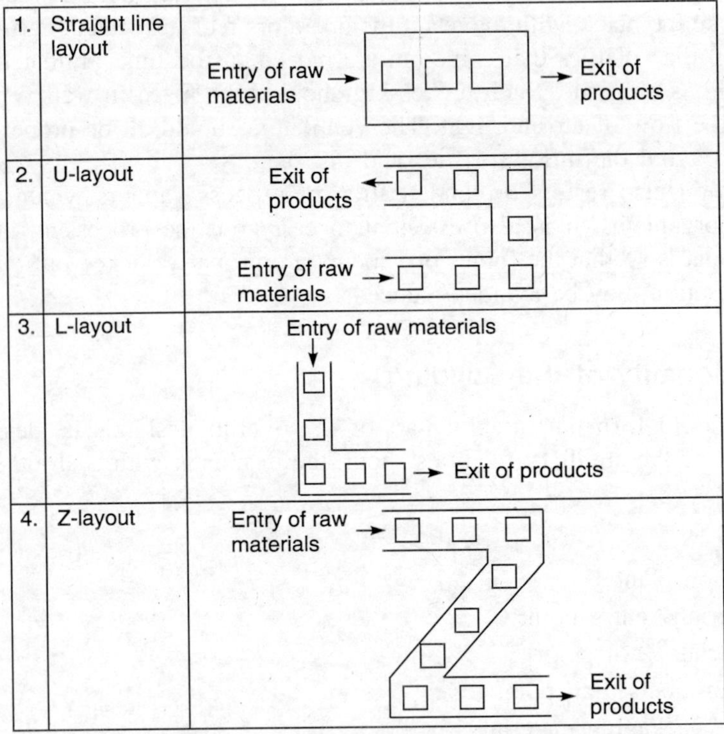

Figure 28.1 Pictorial description of various layouts.

The ceiling height of the building is based on the functional aspects going on in the building. As a rule of thumb, the building height should be atleast 2 m above the height of the tallest equipment in the building so that it facilitates good space for the movement of personnel on the top of the equipment during maintenance or repairs. Similarly, the height should be atleast one metre above the highest point to which the fork lift can reach.

Since food processing operations are always in a state of expansion, there should be a provision for space for expansion. Usually the processing operations start with procurement of raw materials in a semi-finished form. Later it would be felt economical to process the raw materials even to make the intermediate product also. For example, a fruit beverage industry may start with procurement of pulp for processing. Over a period of time, it may be felt economical to procure the fruits and make pulp within the factory for captive consumption. Thus, expansion is always possible; and hence, provisions for expansion forms an integral part of any functionality of the building. A typical pictorial description of various functional units is shown in Figure 28.2.

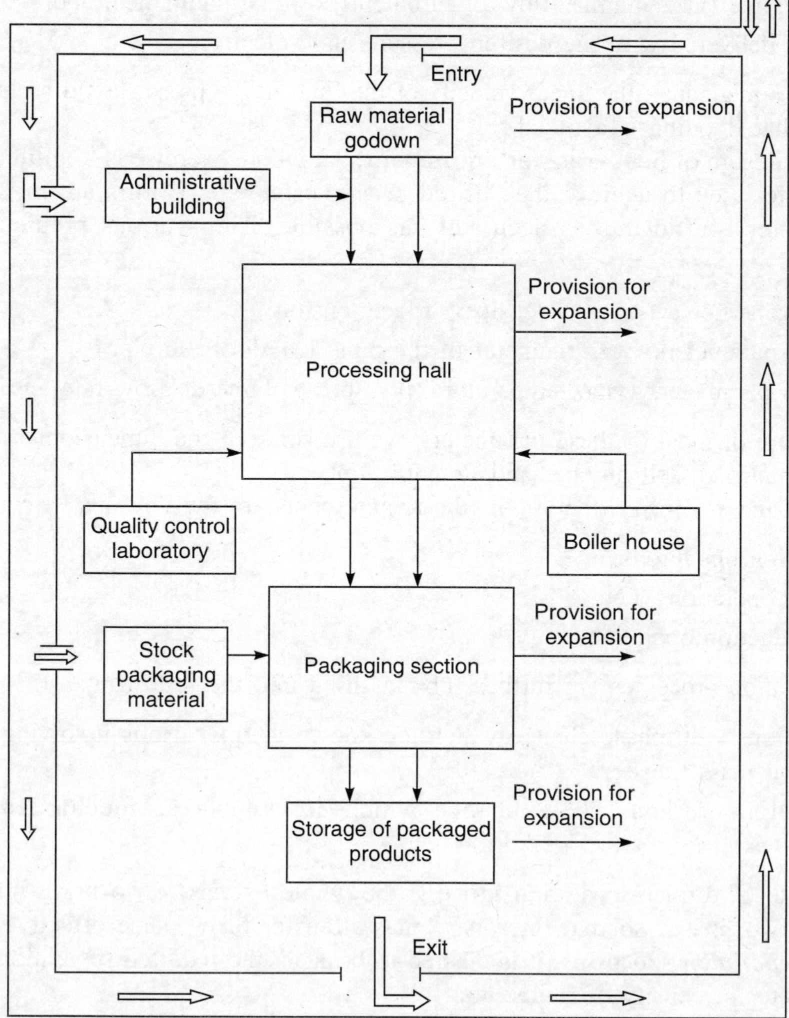

Figure 28.2 A typical plant layout showing various functional units.

28.2.3 Design and Fabrication of Equipment

The design and fabrication of equipment are usually done based on the need and throughput of processing. In food processing, special care needs to be taken for incorporating safety and hygiene in processing.

In *fluid flow* through pipes, the design features are to evaluate the diameter of the pipe based on the fluid properties like viscosity, density, etc. so that

- the energy required for pumping is the least,
- flooding is avoided,

- flow regime (viz., laminar flow or turbulent flow) is maintained, and
- fluid is delivered with the desired pressure and velocity.

All things being equal, the linear velocity of the fluid in the pipes should be less than 8′/sec (2.4 m/s) so that flooding is avoided.

In case of design of **heat transfer equipment**, the whole exercise is to ultimately evaluate the heat transfer area to achieve the desired heat transfer rate. Efforts are made to evaluate the heat transfer coefficients as accurately as possible using various predictive equations based on the

- fluid properties (viscosity, density, surface tension)
- system parameters (viz., diameter of the pipe, length of the pipe)
- process parameters (viz., temperature, pressure, volumetric flow rate, etc.).

To take care of most of these parameters, we use some of the dimensionless numbers like Reynolds number, Nusselt number and Prandtl number.

In case of *mass transfer* operations, the requirements are multifarious. It could be for

- separation operations,
- mixing operations, and
- size reduction operations.

The separation processes are further sub-classified into the following:

- separations with phase change which are known as **operations** involving simultaneous heat and mass transfer
- separations without phase change which are known as **mechanical separation processes**.

In the case of former separation process, the whole exercise is to bring intimate contact between the two phases so that interphase mass transfer takes place effectively. In case of *mechanical separation processes*, it is desired to bring in the required separation based on the particle diameter or density of fluids, etc.

In *mixing operations*, it is desired to bring in the required degree of mixing amongst different phases or particles so that a sample picked up from any corner of the container will have the same composition. It is also desired that the best degree of mixing be achieved with the consumption of least amount of energy for mixing. Mixing of gas-liquid systems or liquid-liquid dispersions is characterized by the fact that how best the dispersion takes place so that the maximum amount of interfacial area is realized which in turn results in maximum amount of mass transfer to take place.

For *size reduction operations*, it is always desired that the grinding energy is the least, and the grinding results in as uniform distribution of particle size as possible. The grinding mills, or cutters or choppers are designed to make the least amount of debris.

In *forming operations* like extrusion, moulding, etc. we look to maintain good consistency or structure of the food material so that its organoleptic qualities are enhanced.

Various aspects of design features of equipment are tabulated in a nutshell in Table 28.2.

Table 28.2 Design Features of Various Equipment

Equipment for	*Design features*	*Remarks*
Fluid flow	Diameter of pipe to avoid channeling and flooding	The design is based on the system parameters, fluid properties and process parameters.
Heat transfer	Heat transfer area	Heat transfer area should be adequate to transfer the desired amount of heat.
Mass transfer	Separation with phase change	Simultaneous heat and mass transfer are involved in operations like – drying – distillation – extraction – crystallization, etc.
	Separation without phase change	Mainly based on particle size or density in operations like – sedimentation – centrifugation – filtration
	Mixing operations	To achieve intimate contact amongst various phases, viz., – gas-liquid – liquid-liquid – liquid-solid – solid-solid – gas-liquid-solid
	Size reduction operations	To achieve the desired particle size utilizing as little grinding energy as possible
Forming operations	To get the desired shape with as good consistency as possible	Operations like – extrusion – forming – shaping – casing, etc.

28.3 PLANT LOCATION

The location of any processing plant plays an important role in the success of the project. Hence, a good amount of expertise is required by the project engineering firm which prepares the project details. An ill chosen plant location may fail the process economics, and unfortunately the discredit for failure is attributed to the technology. Similarly, it is true other way round also. A not-so-good technology installed in a strategic location may prove beneficial for the project. It is the experience and ingenuity of the project consultants which make or mar the project by choosing an appropriate plant location. Some general guidelines for plant location are tabulated in Table 28.3. Atleast a good number of them should be satisfied. Obviously all of them cannot be met since some of the requirements are contradictory for some other requirement. For example, availability of cheap labour is possible in rural areas where it is not possible to get and retain the qualified trained man power. Similarly, market is available in urban areas, whereas the raw materials at affordable price are available in rural areas. Transportation facilities are more available in places where human inhabitation is more, whereas we require remote areas for easy disposal of wastes and effluents.

Table 28.3 General Guidelines for Choosing Plant Location

Availability of major raw materials at affordable price throughout the year or atleast over a reasonable amount of period.
Infrastructural facilities like transportation by rail, road or sea; usually for imported raw materials or for exported products, the nearness of sea transport facility is a must.
Infrastructural facilities like availability of water, electricity, etc. at reasonable costs.
Availability of good quality water from internal sources (like borewells) rather than from municipal supply.
Availability of cheap labour, as most of the food processing operations are usually labour-intensive.
Availability of trained man power, and civil infrastructural facilities to retain the trained and qualified man power.
Adequate space/facilities for disposal of wastes/byproducts and residues/effluents.
Necessary storage facilities for storing either raw materials/finished products in the form of cold storage facilities to hire on contract basis.
Availability of packaging materials in right (desired) quantities at affordable price.
Market potentiality for the product in the vicinity.
Availability of any government incentives.
Above all, a favourable local environment, both climatic and governmental.

One of the most important considerations while choosing locations in case of food processing plants is to have adequate land for future expansion. The provision for excess land should be two to three times of the existing land. It is also important to consider whether processing plant should be close to the place where the raw materials are produced so that they can be processed immediately after harvest. It is truer with processing of fruits and vegetables; whereas it may not be that much serious in case of processing of grains or nuts. In some designs, the plant location is chosen based on the market availability. For example, a desiccated coconut plant could be established in areas close to the place where bakery industries are more so that it is easy to market the desiccated coconut which is used in the production of biscuits and other bakery products.

Since the food materials are perishable in nature and are highly labour intensive, it is advisable not to locate the plants in places where the labour union activities are vibrant. Frequent breakdowns may speak on the economic viability, and hence, the continuity of the project after it has been established. Similarly, are the local governmental policies. If the local state or federal governmental policies are not conducive for the industrial growth, it may also speak upon the economic viability of the project. Particularly adequate care has to be exercised in establishing animal-based product technologies or alcohol production. Some local governments are hostile for such industries on moral or ethical grounds. Sometimes even local people may not cooperate for some industries. It is not always possible to counter the sentiments of local people continuously.

General topography of the land and land level are also to be considered before choosing the location. Gentle topography and even slightly sloped lands will suit to the plant. Sloped land will help in maintaining water table near the plant location. Sometimes, the slope of the land is taken as an advantage for unloading the raw materials in the upside level, and the finished goods are delivered from the downside level. It is very essential to see that there are no stagnant water puddles/pools/ponds nearby. They may be a perennial source for insect proliferation leading to contamination.

The soil must possess adequate mechanical strength for the structures to stand erect permanently without getting sunk into the soil. If the soil is loose, it may result in sinking of the building structures over a period of time. This is a common precaution to be taken care while selecting land for any process industry irrespective of the processing operation.

28.4 COST–BENEFIT ANALYSIS

Cost-benefit analysis is an important component in any industrial operations/business ventures to gauge the profits vis-à-vis the costs involved. It is an important consideration in any project economics, and is an essential component in any project document. It is not simply the amount of sales realization minus the cost of production. The present value of the project cost, and its monetary benefits when the project is successfully implemented would enable the investor whether to go ahead versus not to go-ahead with the project. The cost-benefit analysis would be made by the experts using sophisticated financial analysis tools and softwares.

Another important component in it is *social cost benefit analysis*, which is not normally done by the individuals, but by the philanthropic organizations and government bodies.

28.4.1 Social Cost–Benefit Analysis

In the social cost-benefit analysis, profit alone is not the criteria, but the societal benefits the project would accrue if implemented. The social cost benefit analysis is more relevant in food processing as compared to any other processing sectors. By implementing a project, if it is going to

- Improve the living standards of the local people.
- Provide employment to locals.
- Prevent local produce from being spoiled for want of proper post harvest processing technologies.
- Provide remunerative prices to the local farmers for their agricultural produce.
- Encourage the local farmers to produce more in view of its consumption/utility.
- Improve the region economically which is otherwise backward.

The project would be implemented even though the benefits (profits) are not commensurate to the investment. The concept is explained with few examples.

The tribal people in hilly regions are engaged in the activity of collecting gum *karaya* from the wild grown trees. If a processing plant is established in such a region, it would provide not only remunerative prices for the local tribes, but it would discourage them from going away from national main stream into insurgency. Similarly, in an area where cashew plantation is there, it would be effective to start an alcohol making unit from cashew apple which would provide remunerative prices for the locals who live on collecting them. It generates mass employment to the locals. Establishment of plants in some remote areas would improve the area and bring-in a lot of benefits to the local people like a fruit processing plant in North Eastern Region of India. Like this, the instances are many.

28.5 PLANT LAYOUT

Food engineer has to play an important role in the plant layout (and in general, known as **plant and equipment layout**) of a food processing plant. One of the main reasons for it is that food processing is an important unique processing method where special expertise is required. Again, this is one area in which a number of unit operations is involved. Maintaining hygiene and aseptic conditions is an important characteristic feature of food processing.

There are essentially six basic objectives of any effective plant layout (Muther 1955) as follows:

1. *Overall integration* of all the various activities is to be done to integrate and coordinate the men, machinery, materials, movement and supporting activities in a satisfying manner.
2. *Movement* of materials within the processing hall should be minimum so that the overall productivity will increase.
3. *Flow* of materials should be in a sequential order as it comes in the flow sheet. Otherwise movement of materials will be zig-zag and unnecessarily hinder the work.
4. *Effective use of space* should be ensured preferably the process hall should not have any barricades or cubicles. The movement of materials should preferably be done by interlock transfer cranes/chain pulley cranes (Table 24.1).
5. *Safety and satisfaction* of the workers have to be ensured to have a long range strategy for improved productivity. After all, the workers are the intangible assets of the plant/factory.
6. *Flexibility* in layout is a very preferable approach, even though it is not always possible. All things being equal, it is better to have provision for adjusting and rearranging the layout with minimum amount of inconvenience.

The factors that would influence a plant layout are given in Table 28.4. A good plant layout with most of the above objectives will have a number of advantages which are tabulated in Table 28.5. A typical plant layout with a number of alternatives for a bottled milk plant is shown in Figure 28.3. A food engineer has a major role to play especially while planning a complete new layout. Before going for the design of layout for a new process plant, the food engineer takes the help of the technologists to identify what are the various process steps involved in the technology, based on which he can go for a rational layout. It is explained with an example for making a meat bun and a steamed bun. Various steps involved in the processing are shown in Table 28.6, based on which the flowchart is made and shown in Figure 28.4. Later the plant and equipment layout is made, and is shown in Figure 28.5.

Table 28.4 Factors Influencing a Typical Plant Layout

Materials movement should be minimum.
The accessibility to machinery should be fast.
Facilities to ensure better utilization of man power with good control.
Time factor should be such that the waiting time and delays should be minimum.
The building design should be such that it should ensure easy movement of raw materials and products; and the disposal of wastes, byproducts and effluents should be easy.

Table 28.5 Advantages in Terms of Costs of an Effective Plant Layout

A properly designed plant layout reduces the risk to health and safety of workers, which in turn improves their morale and satisfaction that in turn helps increase the productivity.
Reduced material handling and movement of men increase the output.
Better utilization of space and greater utilization of man power and machinery would reduce the production delays and help increase the output.
Less movement of materials would ensure shorter manufacturing times.
A good layout with cubicles for supervision and quality control personnel would ensure better quality check on the products as well on the performance of the workers.
Less congestion and confusion would ensure better material flow ensuring higher productivity.

In addition to these considerations, food processing is one activity where there is a strong need to keep a large amount of on-site space for future expansion. Hence, the building layout should be made in such a way. The raw materials received are usually dirty and need cleaning. Imagine the situation if the raw materials are animals in a slaughter house. The place will be usually dirty with blood, urine or dung, etc. The passage used for bringing in the raw materials should not be used for taking out the finished products, since the finished products may catch contamination from the unhygienic inlet passage. Hence, the outlet for the finished products can not be the same. Usually the outlet is made from another direction as shown in Figure 28.2. Also provision should be made for cleaning and sanitizing the trucks which carry the raw materials. Ramps need to be essentially provided near the raw material sheds to unload them, particularly more so in case of animals.

Table 28.6 Various Process Steps in Making a Meat Bun

Raw materials like wheat flour, sugar, water and yeast are taken in proper compositions and mixed together.
Vegetables are cut into desired sizes and dehydrated.
Meat is minced to desired size and mixed with the dehydrated vegetables.
The kneaded dough and the other filling materials are added to the filler hopper.
Filling and stuffing are carried out on to the dough belt.
Meat buns/filled bread are sent for proofing for fermentation.
The well fermented material is either sent to steaming for making steamed bread or to a baking oven to make the meat buns.
The materials are cooled and packed.
The packed material is sent to IQF (Individual Quick Freezing) unit or spiral freezer for deep freezing.
The frozen material is ready for sale.

Source: Courtesy: Anko Food Machinery Co. Ltd., Taiwan.

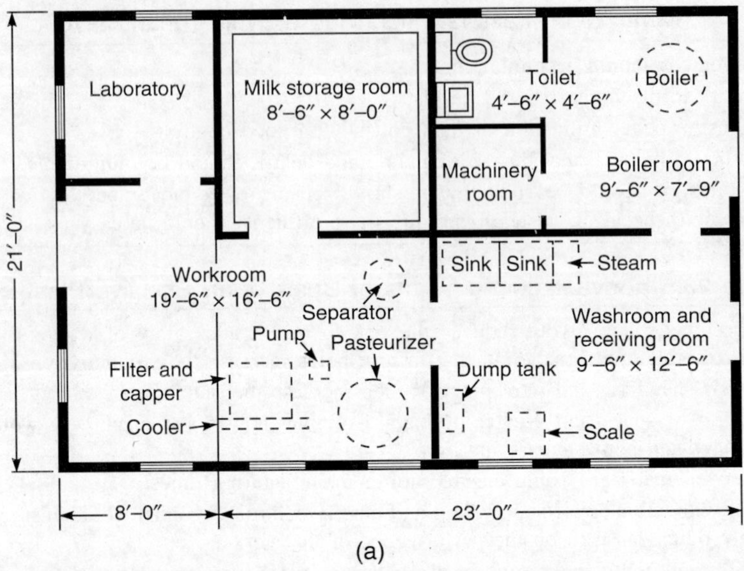

(a)

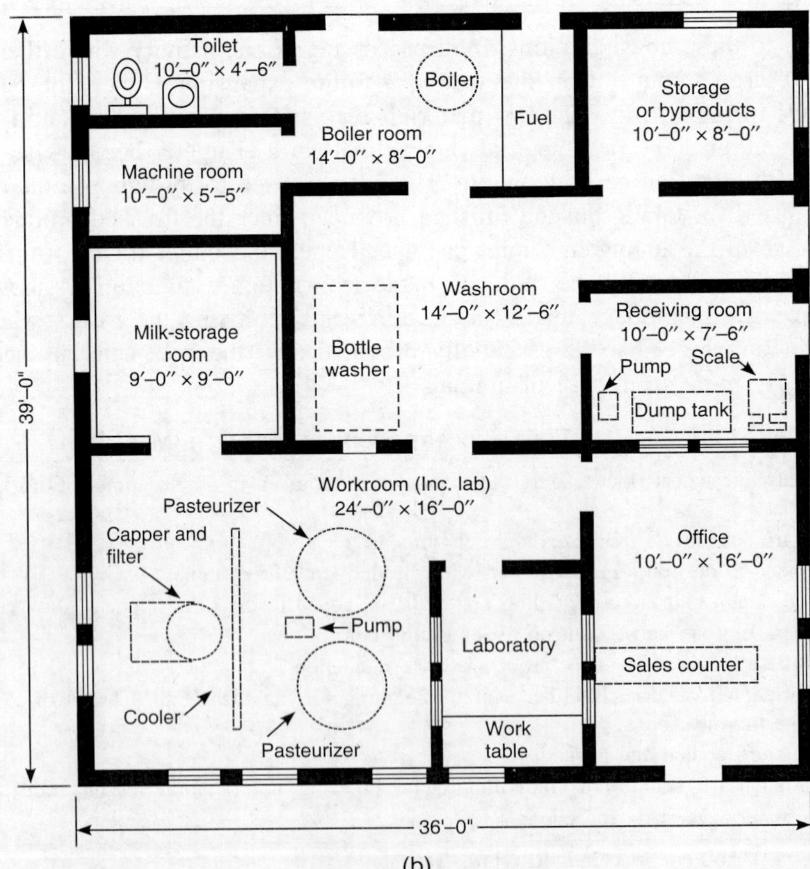

(b)

Figure 28.3 *(contd.)*

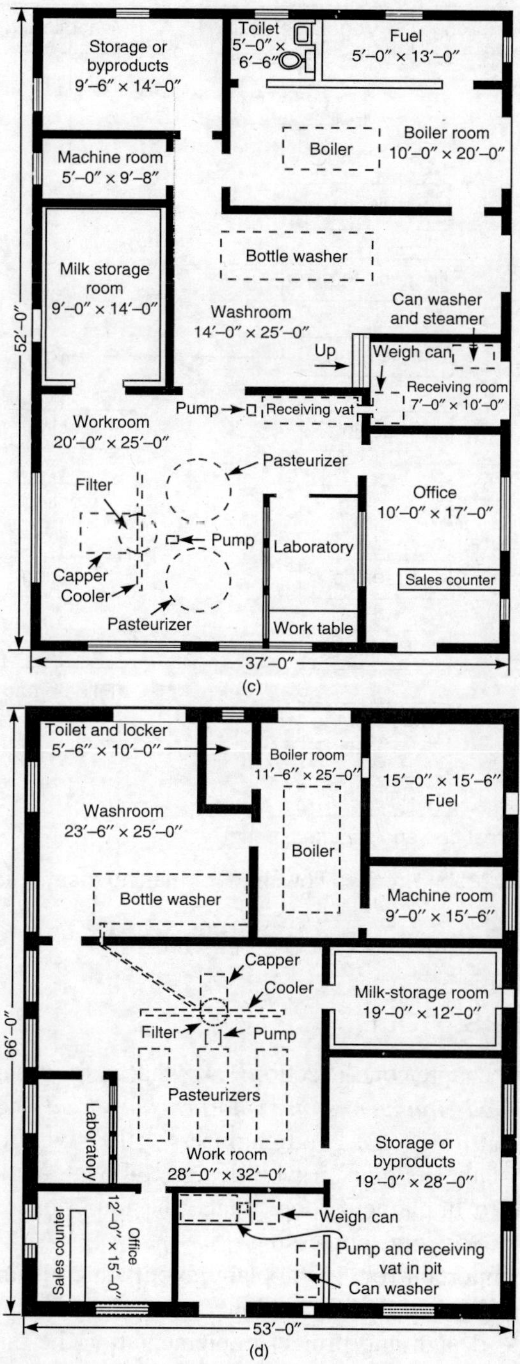

Figure 28.3 Plant layout of 4 different plants producing the same product (Bottled milk), but varying capacities.

Source: Reprinted from Muther (1955), *Practical Plant Layout*, McGraw-Hill, New York, Copyright 1955. With permission from McGraw-Hill.

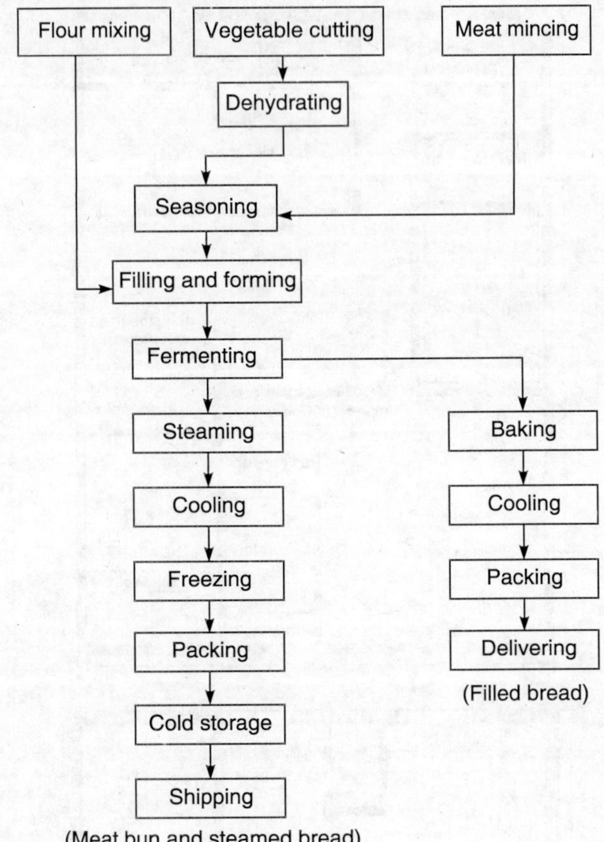

Figure 28.4 Process flowchart for making meat bread.

Source: Courtesy: Anko Food Machinery Co. Ltd., Taiwan; www.ankofood.com

28.6 FOOD ENGINEERS' ROLE

We have already defined in chapter one (Section 1.4) what is food engineering. Based on that concept, now *we define a food engineer is one who practises food engineering, irrespective of his/her basic engineering qualifications*. Food engineers' role is very critical during the design and erection of food plant. Other engineers who design, fabricate and establish process plants may overlook if they have to implement food plants, the importance of asepsis, hygiene and sanitation which are so vital for food processing.

He has also to play an important role in the plant layout which is unique for food processing operations as has been described in earlier section.

In the course of process design and project implementation, he may have to look for most optimum process route or technological path or choice of raw materials. Some of these are summarized in Table 28.7. A general survey from the published literature would often provide valuable information and pertinent data for the development of a design project. First approach

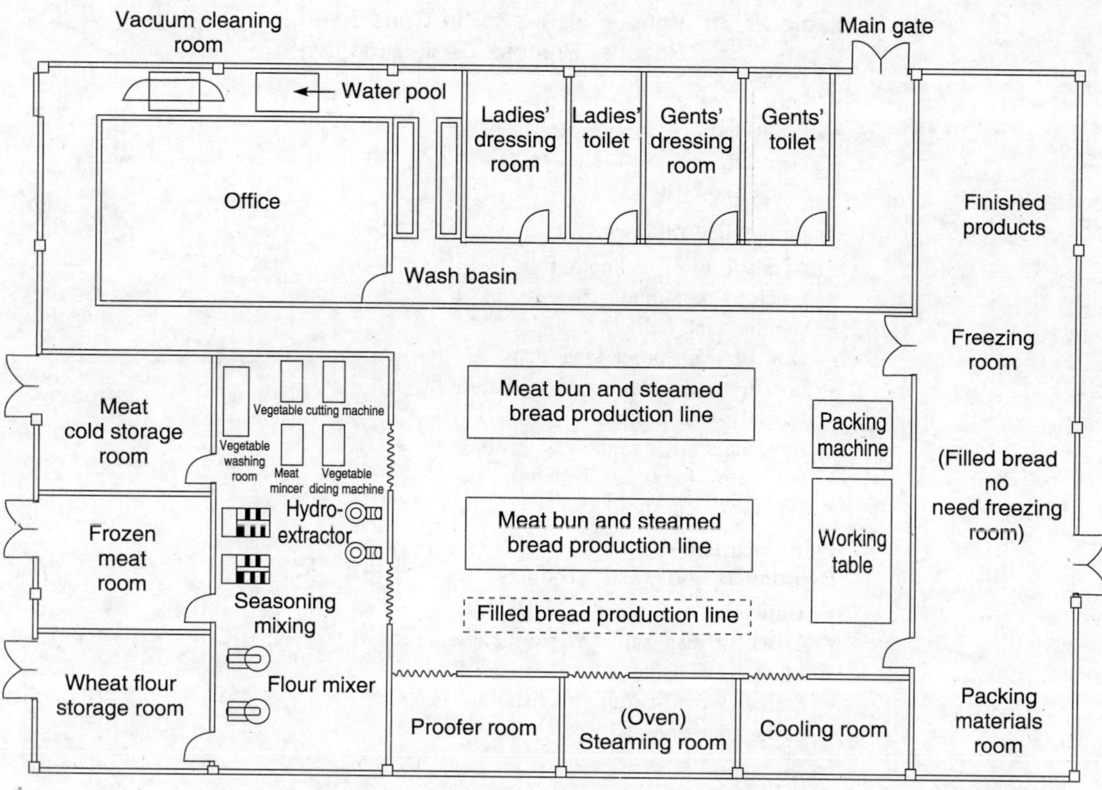

Figure 28.5 A typical plant layout for meat bun unit.

Source: With permission form M/s Anko Food Machinery Co. Ltd., Taiwan; www.ankofood.com

Table 28.7 Various Issues to be Considered in Rational Process Design

Raw materials
- Availability
- Seasonal variations
- Quality
- Storage considerations
- Susceptibility to pests, rodents, etc.
- Material handling problems

Equipment
- Availability of equipment
- Whether spare capacity equipment already available
- Initial cost vs. operating costs
- Maintenance and installation problems
- Man power to operate
- Any special design/alternate designs

Costs
- Raw materials
- Utilities

(contd.)

Table 28.7 Various Issues to be Considered in Rational Process Design (*contd.*)

- Labour costs
- Fixed and operating costs
- Depreciation
- Marketing costs

Technological matters

- Availability of technology
- Batch vs. continuous processing
- Commercial yields
- Down stream processing steps
- Energy requirements and availability
- Flexibility of process
- Raw materials common with other processes
- Possibility for future development
- Whether niche product

Plant location

Byproducts and waste products

- Utility for byproducts
- Costs for disposal of waste materials
- Method of disposal
- Statutory restrictions for disposal
- Effluents disposal

Time factor

- Market timeliness
- Deadlines for project completion
- Possible cost escalation if delayed

Market

- Availability of market for the product locally
- Monopoly market/open market
- Existing market/potential market/latent market
- Whether market promotion required
- Whether premium product
- Whether niche product
- Whether the product has export market

Statutory matters

- Any concessions by local government
- Environmental controls
- Any other local legislation
- Safety issues/hazards
- Legal restrictions and taxes

Miscellaneous issues

- General company objectives/charter
- Whether any moral or ethical issues involved
- Favourable climatic conditions

for it would be to go for survey of recent Food Technology abstracts† or FSTA‡ database search. Any recently published paper would give a lot of back references from which the whole literature search can be done. The physico-chemical properties of food materials, often required for design purpose can be seen from the book *Food Engineering Data Handbook* (Hayes 1987) or *Perry's Chemical Engineers' Handbook* (Perry, et al., 1984).

A primary source of information on all aspects of process design and equipment design can be obtained from Perry's Chemical Engineers' Handbook (Perry, et al., 1984) or Encyclopedia of Food Engineering (Hall and Farrall, 1986). Regular features about the industrial scenario, equipment costs and some contemporary issues are regularly published in Indian Food Industry Journal, Food and Beverage World Journal and Indian Food Packer Journal. Various important journals in food line are given in Table 28.8.

In the project implementation, food engineer plays a key role as a project manager, and bears ultimate responsibility for success or failure of the project. He will use some of the current state of the art project management tools such as PERT/CPM§ and probabilistic techniques to evaluate the progress of the project. The methodology is based on a five stage approach to project management.

Stage I-Project evaluation: It is done at the beginning of the project on the basis of cost-benefit analysis, and the envisaged difficulties in implementation of the project. The existing and potential markets are also assessed. It is a very crucial stage in deciding to take up the project. The whole work is done on paper in an office room.

Table 28.8 Various Important Journals Related to Food Science, Technology and Engineering

Journal	*Published by*
Beverage and Food World	Amalgamated Press, Mumbai
Bioresource Technology	Elsevier, USA
Bulletin of Grain Technology	Foodgrain Technologists' Research Association of India, Hapur
Cereal Chemistry	AACC (American Association of Cereal Chemists) International, USA
Critical Reviews in Food Science and Nutrition	Taylor and Francis Group, Amherst
Drying Technology	Taylor and Francis Group, Singapore
European Food Research and Technology	Springer, Germany
European Journal of Lipid Science and Technology	John Wiley and Sons, Germany
Flavours and Fragrance Journal	John Wiley and Sons, UK
Food Additives and Contaminants	Taylor and Francis Group, UK
Food and Bioproduct Processing	Elsevier, UK
Food Chemistry	Elsevier, UK
Food Manufacture	William Reed Business Media Ltd., UK

(contd.)

†*Food Technology Abstracts (FTA)* are regularly published by CFTRI, Mysore.

‡*Food Science and Technology Abstracts (FSTA)* are published bimonthly by IFIS (International Food Information Services), UK.

§PERT (Project Evaluation and Retrieval Techniques) and CPM (Critical Path Method) are important management tools.

Table 28.8 Various Important Journals Related to Food Science, Technology and Engineering (*contd.*)

Journal	*Published by*
Food Research International	Elsevier, Canada
Food Science and Technology International	Sage Publications, London
Food Technology	IFT, USA
Food Technology and Biotechnology	Faculty of Food Technology and Biotechnology, University of Zagreb, Croatia
Indian Food Industry Mag	AFST (I), Mysore
Indian Food Packer	AIFPA, New Delhi
Indian Journal of Dairy Science	Indian Dairy Association, New Delhi
International Journal of Dairy Technology	Blackwell Publishing, UK
International Journal of Food Engineering	www.bepress.com/ijfe
International Journal of Food Properties	Taylor and Francis Group, Muscat, Sultanate of Oman
International Journal of Food Science and Technology	Blackwell Publishing, UK
JAOCS	American Oil Chemists Society, USA
Journal of Agricultural and Food Chemistry	ACS Publications, California
The Journal of Associating Official Analytical Chemists International	Association of Official Analytical Chemists, USA
Journal of Chemical Technology and Biotechnology	Wiley Interscience, Hoboken, New Jersey
Journal of Food Biochemistry	Blackwell Publishing, Canada
Journal of Food Engineering	Elsevier, www.elsevier.com
Journal of Food Lipids	Blackwell Publishing, Canada
Journal of Food Process Engineering	Wiley Interscience, Hoboken, New Jersey
Journal of Food Processing and Preservation	Wiley Interscience, Hoboken, New Jersey, www.wiley.com/bw/journal
Journal of Food Protection	International Association for Food Protection, USA
Journal of Food Quality	Blackwell Publishing, Spain
Journal of Food Safety	Blackwell Publishing, USA
Journal of Food Science	IFT, USA
Journal of Food Science and Technology	AFST(I), Mysore by Springer
Journal of Lipid Research	The American Society for Biochemistry and Molecular Biology, Maryland
Journal of Texture Studies	Blackwell Publishing, USA
Journal of The Science of Food and Agriculture	Wiley Interscience, Hoboken, New Jersey
LWT (Lebensmittel–Wissenschaft-und Technologie)—Food Science and Technology	Elsevier, Switzerland
Packaging Technology and Sciences—International Journal	John Wiley and Sons, UK
Processed Food Industry	The Computype Media, New Delhi, www.pfionline.com
Times Food Processing Journal	Bennett, Coleman an Co. Ltd., Mumbai
Transactions of the American Society of Agricultural and Biological Engineers (ASABE)	American Society of Agricultural and Biological Engineers, Michigan (USA)
Trends in Food Science and Technology	Elsevier, UK

Stage II-Procurement: From now onwards, the project implementation starts. Efforts are put and funds are allocated for procurement of various items. This stage gives a clear definition for the problem and its implementation schedule with funds flow. Some of the key machinery and equipment are sourced, and purchase procedure is initiated. Sourcing and recruitment of trained man power are taken up.

Stage III-Implementation: The real project implementation starts in this phase. The whole project team would go to the field. Construction of the structures and buildings is taken up. The procured machinery will be installed in place. Efforts are also initiated to procure the utilities like power, water, etc.

Stage IV-Commissioning: This stage starts after successfully completing the above three stages. Now the plant is ready. Trial runs are taken with the feeding of raw materials and using the utilities. This stage gives confidence for the project team regarding the suitability of raw materials, quality and yield of product, quantity of the utilities required and the time schedules.

Stage V-Maintenance: This is the factory running stage, which comes after successful completion of the project. Regular production goes on. The operating procedures are finalized and the operating manuals are followed. The personnel are sent for training if required. The market tie-ups are strengthened and the product rolls out. Production goes on.

Food engineer is a happy man!

REVIEW QUESTIONS

28.1 How food processing is different from other processing activities in terms of plant design and location?

28.2 What are the general principles of plant design?

28.3 What are the design features of various equipments used for different unit operations?

28.4 What are the important issues you need to consider while making a feasibility survey for a food plant?

28.5 What factors do you consider for design and construction of a building for a food plant?

28.6 How the functionality of a plant building is important in the process activities?

28.7 What guidelines you need to follow for locating a food processing industry?

28.8 What is a plant layout?

28.9 Show a typical plant layout for any food processing activity of your choice.

28.10 What factors would influence a typical plant layout?

28.11 What advantages you envisage to reap by a good plant layout?

28.12 What is meant by cost-benefit analysis and how it is important in food processing?

28.13 What is social-cost-benefit analysis?

28.14 What is the role of a food engineer in the food plant implementation?

28.15 What are various issues a food engineer needs to consider for a rational process design?

REFERENCES

Brennan, J.G., Butters, J.R., Cowell, N.D. and Lilly, A.E.V. (1969), *Food Engineering Operations*, Elsevier, Amsterdam, pp. 342–345.

Hall, C.W. and Farrall, A.W. (1986), *Encyclopedia of Food Engineering*, AVI Pub. Inc., West Port, C.T.

Hayes, G.D. (1987), *Food Engineering Data Handbook*, Longman Group UK Ltd., Essex, UK.

Lopez–Gomez, A. and Barbosa–Canovas, G.V. (2005), *Food Plant Design*, CRC Press, Taylor & Francis Group, Boca Raton, FL (USA), pp. 351–361.

Muther, R. (1955), *Practical Plant Layout*, McGraw-Hill, New York.

Perry, R.H., Green, D.W. and Maloney, J.O. (1984), *Perry's Chemical Engineers' Handbook*, 6th ed., McGraw-Hill, New York.

APPENDIX

1A

Properties of Saturated Water and Steam

Temperature t, °C	Saturation pressure p, kN/m^2	Density ρ, kg/m^3		Specific volume υ, m^3/kg		Specific heat capacity C_p, kJ/kg K		Viscosity μ, mNs/m^2		Thermal conductivity k, W/mK		Prandtl number N_{Pr} $(C_p\mu/k)$ dimensionless		Surface tension σ, mN/m	Volume expansion coefficient $\beta^{\dagger}$, K^{-1} Water	Compressibility k, bar^{-1}
		Water	Steam	Water	Steam	Water	Steam	Water	Steam	Water	Steam	Water	Steam			
0.01	0.611	1000	0.00485	0.00100	206.2	4.217	1.854	1.755	0.0088	0.569	0.0173	13.02	0.942	75.6	0.060×10^{-3}	50.98×10^{-6}
10	1.227	1000	0.00940	0.00100	106.4	4.193	1.860	1.301	0.0091	0.587	0.0185	9.29	0.915	74.2	0.088×10^{-3}	47.89×10^{-6}
20	2.34	998	0.0173	0.00100	57.8	4.182	1.866	1.002	0.0094	0.603	0.0191	6.95	0.918	72.8	0.207×10^{-3}	45.91×10^{-6}
30	4.24	996	0.0304	0.00100	32.9	4.179	1.875	0.797	0.0097	0.618	0.0198	5.39	0.923	71.2	0.303×10^{-3}	44.75×10^{-6}
40	7.38	992	0.0513	0.00101	19.5	4.179	1.885	0.651	0.0101	0.632	0.0204	4.31	0.930	69.6	0.385×10^{-3}	44.22×10^{-6}
50	12.34	988	0.083	0.00101	12.05	4.181	1,899	0.544	0.0104	0.643	0.0210	3.53	0.939	67.9	0.458×10^{-3}	44.17×10^{-6}
60	19.92	983	0.130	0.00102	7.68	4.185	1.915	0.462	0.0107	0.653	0.0217	2.96	0.947	66.2	0.523×10^{-3}	44.50×10^{-6}
70	31.16	978	0.198	0.00102	5.05	4.190	1.936	0.400	0.0111	0.662	0.0224	2.53	0.956	64.4	0.584×10^{-3}	45.15×10^{-6}
80	47.36	972	0.293	0.00103	3.41	4.197	1.962	0.350	0.0114	0.670	0.0231	2.19	0.966	62.6	0.641×10^{-3}	46.10×10^{-6}
90	70.11	965	0.423	0.00104	2.36	4.205	1.992	0.311	0.0117	0.676	0.0240	1.93	0.976	60.7	0.696×10^{-3}	47.34×10^{-6}
100	101.3	958	0.598	0.00104	1.673	4.216	2.028	0.278	0.0121	0.681	0.0249	1.723	0.986	58.9	0.750×10^{-3}	48.90×10^{-6}
125	232.1	939	1.30	0.00107	0.770	4.254	2.147	0.219	0.0133	0.687	0.0272	1.358	1.047	–	–	–
150	476.0	917	2.55	0.00109	0.392	4.310	2.314	0.180	0.0144	0.687	0.0300	1.133	1.110	–	–	–
175	892.7	893	4.60	0.00112	0.217	4.389	2.542	0.153	0.0156	0.679	0.0334	0.990	1.185	–	–	–
200	1555.0	862	7.87	0.00116	0.127	4.497	2.843	0.133	0.0167	0.665	0.0375	0.902	1.270	–	–	–
225	2550	833	12.80	0.00120	0.0783	4.648	3.238	0.1182	0.0179	0.644	0.0427	0.853	1.36	–	–	–
250	3978	800	20.00	0.00125	0.0500	4.867	3.772	0.1065	0.0191	0.616	0.0495	0.841	1.45	–	–	–
275	5949	758	30.60	0.00132	0.0327	5.202	4.561	0.0972	0.0202	0.582	0.0587	0.869	1.56	–	–	–
300	8592	712	46.30	0.00140	0.0216	5.762	5.863	0.0897	0.0214	0.541	0.0719	0.955	1.74	–	–	–

(contd.)

(contd.)

Temperature t, °C	Saturation pressure p, kN/m^2	Density ρ, kg/m^3		Specific volume υ, m^3/kg		Specific heat capacity C_p, kJ/kg K		Viscosity μ, mNs/m^2		Thermal conductivity k, W/mK		Prandtl number N_{Pr} $(C_p\mu/k)$ dimensionless		Surface tension σ, mN/m	Volume expansion coefficient β, K^{-1}	Compressibility k, bar^{-1}
		Water	*Steam*	*Water*	*Steam*	*Water*	*Steam*	*Water*	*Steam*	*Water*	*Steam*	*Water*	*Steam*		*Water*	
325	12060	654	70.40	0.00153	0.0142	6.861	8.440	0.0790	0.0230	0.493	0.0929	1.100	2.09	–	–	–
350	16530	575	114	0.00174	0.0088	10.10	17.15	0.0648	0.0258	0.437	0.1343	1.50	3.29	–	–	–
360	18670	528	144	0.00190	0.0019	14.60	25.10	0.0582	0.0275	0.400	0.168	2.11	3.89	–	–	–
374.2 (Critical point)	22120	315	315	0.00317	0.0031	∞	∞	0.0450	0.0450	0.240	0.240	–	–	–	–	–

†The volume expansion coefficient, b, for water is at 101.3 kN/m^2 (1 atmosphere)

Source: Adapted from Chandrasekharan, K.D. and Venkateswarlu, D. (1974), *SI Units in Chem. Engg. Tech.*, Chem. Engg. Education Dev. Centre, IIT Madras.

APPENDIX

1B

Properties of Saturated Steam at Different Temperatures

Temperature t, °C	Pressure p, bar	Specific volume υ, m³/kg		Specific enthalpy, kJ/kg		Specific entropy, kJ/kg K	
		v_f, m³/kg	v_g, m³/kg	h_l	h_g	s_l	s_g
0	0.00611	0.0010002	206.2	–0.04	2501.6	–0.0002	9.1577
2	0.00706	0.0010001	179.9	8.39	2505.2	0.0306	9.1047
4	0.00813	0.0010000	157.3	16.80	2508.9	0.0611	9.0526
6	0.00935	0.0010000	137.8	25.21	2512.6	0.0913	9.0015
8	0.01072	0.0010001	121.0	33.60	2516.2	0.1213	8.9513
10	0.00123	0.0010003	106.4	41.99	2519.9	0.1510	8.9013
15	0.01704	0.0010008	77.98	62.94	2529.1	0.2243	8.7826
20	0.0237	0.0010017	57.84	83.86	2538.2	0.2963	8.6684
25	0.03166	0.0010029	43.40	101.77	2547.3	0.3670	8.5592
30	0.04241	0.0010043	32.93	125.66	2556.4	0.4365	8.4546
35	0.05622	0.0010060	25.24	146.56	2565.4	0.5049	8.3543
40	0.07375	0.0010078	19.55	167.45	2574.4	0.5721	8.2583
45	0.09582	0.0010099	15.28	188.35	2583.3	0.6383	8.1776
50	0.12335	0.0010121	12.05	209.26	2592.2	0.7035	8.0776
55	0.15741	0.0010145	9.579	230.17	2601.0	0.7677	7.9926
60	0.19920	0.0010171	7.679	251.09	2609.7	0.8310	7.9108
65	0.2501	0.0010199	6.202	272.02	2618.4	0.8933	7.8322
70	0.3116	0.0010228	5.046	292.97	2626.9	0.9548	7.7565
75	0.3855	0.0010259	4.143	313.94	2635.4	1.0154	7.6835
80	0.4736	0.0010292	3.409	334.92	2643.8	1.10753	7.6132
85	0.5780	0.0010326	2.829	355.92	2652.0	1.1343	7.5454
90	0.7011	0.0010361	2.361	376.94	2660.1	1.1925	7.4799
95	0.8453	0.0010399	1.982	397.99	2668.1	1.2501	7.4166
100	1.0133	0.0011437	1.673	419.06	12676.0	1.3069	7.3554
105	1.2080	0.0010477	1.419	440.17	2683.7	1.3630	7.2962
110	1.4327	0.0010519	1.210	461.32	2693.0	1.4185	7.2388

(contd.)

(contd.)

Temperature t, °C	*Pressure* p, bar	*Specific volume* v, m³/kg		*Specific enthalpy*, kJ/kg		*Specific entropy*, kJ/kg K	
		v_l, m³/kg	v_g, m³/kg	h_l	h_g	s_l	s_g
115	1.6906	0.0010562	1.036	482.50	2698.7	1.4733	7.1832
120	1.9854	0.0010606	0.8915	503.72	2706.0	1.5276	7.1293
125	2.3210	0.0010652	0.7702	524.99	2719.9	1.5813	7.0769
130	2.7013	0.0010700	0.6681	546.31	2719.9	1.6344	7.0261
135	3.131	0.0010750	0.5818	567.68	2726.6	1.6869	6.9766
140	3.614	0.0010801	0.5085	589.10	2733.1	1.7390	6.9284
145	4.155	0.0010853	0.4460	610.60	2739.3	1.7906	6.8815
150	4.760	0.0010908	0.3924	632.15	2745.4	1.8416	6.8358
155	5.433	0.0010964	0.3464	653.78	2751.2	1.8923	6.7911
160	6.181	0.0011022	0.3068	675.47	2756.7	1.9425	6.7475
165	7.008	0.0011082	0.2724	697.25	2762.0	1.9925	6.7075
170	7.920	0.0011145	0.2426	719.12	2767.1	2.0416	6.6630
175	8.924	0.0011209	0.2165	741.07	2771.8	2.0906	6.6221
180	10.027	0.0011275	0.1938	763.12	2776.3	2.1393	6.5819
185	11.233	0.0011344	0.1739	785.26	2780.4	2.1876	6.5424
190	12.551	0.0011415	0.1563	807.52	2784.3	2.2356	5.5036
195	13.987	0.0011489	0.1408	829.88	2787.8	2.2830	6.4654
200	15.549	0.0011565	0.1272	852.37	2790.9	2.3307	6.4278
205	17.243	0.0011644	0.1150	874.99	2793.8	2.3778	6.3906
210	19.077	0.0011726	0.1042	897.74	2796.2	2.4247	6.3539
215	21.060	0.0011811	0.09463	920.63	2798.3	2.4713	6.3176
220	23.198	0.0011900	0.08604	943.67	2799.9	2.5178	6.2817
225	25.501	0.0011992	0.07835	966.89	2801.2	2.5641	6.2461
230	27.976	0.0012087	0.07145	990.26	2802.0	2.6102	6.2107

Source: *Steam Tables: Thermodynamic Properties of Water including Vapour Liquid and Solid Phases Metric Units*, Keenam, J.H., Keyes, F.G., Hill, P.G. and Moore, J.G. Copyright (1969); Reprinted with permission of John Wiley and Sons, Inc.

APPENDIX

2

Approximate Heat Evolution of Fresh Fruits and Vegetables Stored at Different Temperatures

Commodity	Watts per Megagram, (W/Mg)[†]			
	0 °C	5 °C	10 °C	15 °C
Apple	10–12	15–21	41–61	41–92
Apricot	15–17	19–27	33–56	63–101
Artichoke, globe	67–133	94–177	161–291	229–429
Asparagus	81–237	161–403	269–902	471–970
Avocado	–	59–89	–	183–464
Banana, ripen	–	–	65–116	87–164
Bean, green or snap	–	101–103	161–172	251–276
Bean, lima (unshelled)	31–89	58–106	–	296–369
Beet, red (roots)	16–21	27–28	35–40	50–69
Blackberry	46–68	85–135	154–280	208–431
Blueberry	7–31	27–36	69–104	101–183
Broccoli, sprouting	55–63	102–474	–	514–1000
Brussel, sprouts	46–71	95–143	186–250	282–316
Cabbage	12–40	28–63	36–86	66–169
Cantaloupe	15–17	26–30	46	100–114
Carrot, topped	46	58	93	117
Cauliflower	53–71	61–81	100–144	136–242
Celery	21	32	58–81	110
Cherry, sour	17–39	38–39	–	81–148
Cherry, sweet	12–16	28–42	–	74–135
Corn, sweet	125	230	331	482
Cranberry	–	12–14	–	–
Cucumber	–	–	68–86	71–98
Fig	–	32–39	65–58	145–187
Garlic	9–32	17–29	27–29	32–81
Gooseberry	20–26	36–40	–	64–95
Grapefruit	–	–	20–27	35–38
Grape, American	8	16	23	47

(contd.)

(contd.)

Commodity	Watts per Megagram, (W/Mg)[†]			
	0 °C	5 °C	10 °C	15 °C
Grape, European	4–7	9–17	24	30–35
Honeydew melon	–	9–15	24	35–47
Horseradish	24	32	78	97
Kohlrabi	30	48	93	145
Leek	28–48	58–86	158–201	245–346
Lemons	9	15	33	47
Lettuce, head	27–50	39–59	64–118	114–121
Lettuce, leaf	68	87	116	186
Mushroom	83–129	210	296	–
Nuts, kind not specified	2	5	10	10
Okra	–	163	258	431
Onion, green	31–66	51–201	107–174	195–288
Onion	7–9	10–20	21	33
Olive	–	–	–	64–115
Orange	9	14–19	35–40	38–67
Peach	11–19	19–27	46	98–125
Pear	8–20	15–46	23–63	45–159
Pea, green (in pod)	90–138	163–226	–	529–599
Pepper, sweet	–	–	43	68
Plum, Wickson	6–9	12–27	27–34	35–37
Potato, immature	–	35	42–62	42–92
Potato, mature	–	17–20	20–30	20–35
Radish, with tops	43–51	57–62	92–108	207–230
Radish, topped	16–17	23–24	45–47	82–97
Raspberry	52–74	92–114	82–164	243–300
Rhubarb, topped	24–39	32–54	–	92–134
Spinach	–	136	327	529
Squash, yellow	35–38	42–55	103–108	222–269
Strawberry	36–52	48–98	145–280	210–273
Sweet potato	–	–	39–95	47–85
Tomato, mature green	–	21	45	61
Tomato, ripen	–	–	42	79
Turnip, roots	26	28–30	–	63–71
Water melon	–	9–12	22	–

[†]Conversion factor: (watts per megagram) × 74.12898 = Btu per ton per 24 hr.

Source: *ASHRAE Guide and Data Book*, 1961, Fundamentals and Equipment, Chapter 65, Table 4, physical properties of food materials, pp. 866–868. Copyright 1961, Reproduced with permission from American Society of Heating, Refrigenerating and Air-Conditioning Engineers, Inc., Atlanta, USA, www.ashrae.org

APPENDIX

3

Relative Humidity of Air from Wet and Dry Bulb Temperatures

Pressure range: 94.6 × 10^3 to 103.3 × 10^3 N/m², Dry bulb temperature: t °C, Wet bulb temperature: t_1 °C

	Relative humidity, $t-t_1$																																	
t	0.2	0.4	0.6	0.8	1.0	1.2	1.4	1.6	1.8	2.0	2.2	2.4	2.6	2.8	3.0	3.2	3.4	3.6	3.8	4.0	4.5	5.0	5.5	6.0	6.5	7.0	7.5	8.0	8.5	9.0	9.5	10.0	10.5	11.0
–10	93	87	80	74	67	61	54	48	41	35	28	22	16	9																				
–9	94	88	81	75	69	63	57	51	45	39	33	27	21	15	9																			
–8	94	88	83	77	71	65	60	54	48	43	37	32	26	20	15	10																		
–7	95	89	84	78	73	67	62	57	52	46	41	36	31	25	20	15	10	5																
–6	95	90	85	79	74	69	64	59	54	49	45	40	35	30	25	20	15	11	6															
–5	95	90	86	81	76	71	66	62	57	52	48	43	39	34	29	25	20	16	11	7														
–4	95	91	86	82	77	73	68	64	59	55	51	46	42	38	33	29	25	21	17	12														
–3	96	91	87	82	78	74	70	66	62	57	53	49	45	41	37	33	29	25	21	17	8													
–2	96	92	88	84	79	75	71	68	64	60	56	52	48	44	40	37	33	29	25	22	12													
–1	96	92	88	84	81	77	73	69	66	62	58	54	51	47	43	40	36	33	29	26	17	8												
0	96	93	89	85	81	78	74	71	67	64	60	57	53	50	46	43	40	36	33	29	21	13	5											
1	97	93	90	86	83	80	76	73	70	66	63	59	56	53	49	46	43	40	36	33	25	17	10											
2	97	93	90	87	84	81	78	74	71	68	65	62	59	55	52	49	46	43	40	37	29	22	14	7										
3	97	94	91	88	84	82	78	76	72	70	67	64	61	58	55	52	49	46	43	40	33	26	19	12	5									
4	97	94	91	88	85	82	79	77	74	71	68	65	62	60	57	54	51	48	46	43	36	29	22	16	9									
5	97	94	91	88	86	83	80	77	75	72	69	67	64	61	58	56	53	51	48	45	39	33	26	20	13	7								
6	97	94	92	89	86	84	81	78	76	73	70	68	65	63	60	58	55	53	50	48	41	35	29	24	17	11	5							
7	97	95	92	89	87	84	82	79	77	74	72	69	67	64	62	59	57	54	52	50	44	38	32	26	21	15	10							
8	97	95	92	90	87	85	82	80	77	75	73	70	68	65	63	61	58	56	54	51	46	40	35	29	24	19	14	8						
9	98	95	93	90	88	85	83	81	78	76	74	71	69	67	64	62	60	58	55	53	48	42	37	32	27	22	17	12	7					
10	98	95	93	90	88	86	83	81	79	77	74	72	70	68	66	63	61	59	57	55	50	44	39	34	29	24	20	15	10	6				

(contd.)

(contd.)

Relative humidity, $t–t_1$

t	0.2	0.4	0.6	0.8	1.0	1.2	1.4	1.6	1.8	2.0	2.2	2.4	2.6	2.8	3.0	3.2	3.4	3.6	3.8	4.0	4.5	5.0	5.5	6.0	6.5	7.0	7.5	8.0	8.5	9.0	9.5	10.0	10.5	11.0
11	98	95	93	91	89	86	84	82	80	78	75	73	71	69	67	65	62	60	58	56	51	46	41	36	32	27	22	18	13	9	5			
12	98	96	93	91	89	87	85	82	80	78	76	74	72	70	68	66	64	62	60	58	53	48	43	39	34	29	25	21	16	12	8			
13	98	96	93	91	89	87	85	83	81	79	77	75	73	71	69	67	65	63	61	59	54	50	45	41	36	32	28	23	19	15	11	7		
14	98	96	94	92	90	88	86	84	82	79	78	76	74	72	70	68	66	64	62	60	56	51	47	42	38	34	30	26	22	18	14	10	6	
15	98	96	94	92	90	88	86	84	82	80	78	76	74	73	71	69	67	65	63	61	57	53	48	44	40	36	32	27	24	20	16	13	9	6

t	0.5	1.0	1.5	2.0	2.5	3.0	3.5	4.0	4.5	5.0	5.5	6.0	6.5	7.0	7.5	8.0	8.5	9.0	9.5	10.0	10.5	11.0	11.5	12.0	12.5	13.0	13.5	14.0	14.5	15.0	16.0	17.0	18.0	19.0	20.0
16	95	90	85	81	76	71	67	63	58	54	50	46	42	38	34	30	26	23	19	15	12	8	5												
17	95	90	86	81	76	72	68	64	60	55	51	47	43	40	36	32	28	25	21	18	14	11	8												
18	95	91	86	82	77	73	69	65	61	57	53	49	45	41	38	34	30	27	23	20	17	14	10	7											
19	95	91	87	82	78	74	70	65	62	58	54	50	46	43	39	36	32	29	26	22	19	16	13	10	7										
20	96	91	87	83	78	74	70	66	63	59	55	51	48	44	41	37	34	31	28	24	21	18	15	12	9	6									
21	96	91	87	83	79	75	71	67	64	60	56	53	49	46	42	39	36	32	29	26	23	20	17	14	12	9	6								
22	96	92	87	83	80	76	72	68	64	61	57	54	50	47	44	40	37	34	31	28	25	22	19	17	14	11	8	6							
23	96	92	88	84	80	76	72	69	65	62	58	55	52	48	45	42	39	36	33	30	27	24	21	19	16	13	11	8	6						
24	96	92	88	84	80	77	73	69	66	62	59	56	53	49	46	43	40	37	34	31	29	26	23	20	18	15	13	10	8	5					
25	96	92	88	84	81	77	74	70	67	63	60	57	54	50	47	44	41	39	36	33	30	28	25	22	20	17	15	12	10	8					
26	96	92	88	85	81	78	74	71	67	64	61	58	54	51	49	46	43	40	37	34	32	29	26	24	21	19	17	14	12	10	5				
27	96	92	89	85	82	78	75	71	68	65	62	58	56	52	50	47	44	41	38	36	33	31	28	26	23	21	18	16	14	12	7				
28	96	93	89	85	82	78	75	72	69	65	62	59	56	53	51	48	45	42	40	37	34	32	29	27	25	22	20	18	16	13	9	5			
29	96	93	89	86	82	79	76	72	69	66	63	60	57	54	52	49	46	43	41	38	36	33	31	28	26	24	22	19	17	15	11	7			
30	96	93	89	86	83	79	76	73	70	67	64	61	58	55	52	50	47	44	42	39	37	35	32	30	28	25	23	21	19	17	13	9	5		
31	96	93	90	86	83	80	77	73	70	67	64	61	59	56	53	51	48	45	43	40	38	36	33	31	29	27	25	22	20	18	14	11	7		
32	96	93	90	86	83	80	77	74	71	68	65	62	60	57	54	51	49	46	44	41	39	37	35	32	30	28	26	24	22	20	16	12	9	5	
33	97	93	90	87	83	80	77	74	71	68	66	63	60	57	55	52	50	47	45	42	40	38	36	33	31	29	27	25	23	21	17	14	10	7	
34	97	93	90	87	84	81	78	75	72	69	66	63	61	58	56	53	51	48	46	43	41	39	37	35	32	30	28	26	24	23	19	15	12	8	5
35	97	94	90	87	84	81	78	75	72	69	67	64	61	59	56	54	51	49	47	44	42	40	38	36	34	32	30	28	26	24	20	17	13	10	7
36	97	94	90	87	84	81	78	75	73	70	67	64	62	59	57	54	52	50	48	45	43	41	39	37	35	33	31	29	27	25	21	18	15	11	8
37	97	94	91	87	84	82	79	76	73	70	68	65	63	60	58	55	53	51	48	46	44	42	40	38	36	34	32	30	28	26	23	19	16	13	10
38	97	94	91	88	84	82	79	76	74	71	68	66	63	61	58	56	54	51	49	47	45	43	41	39	37	35	33	31	29	27	24	20	17	14	11
39	97	94	91	88	85	82	79	77	74	71	69	66	64	61	59	57	54	52	50	48	46	43	42	39	38	36	34	32	30	28	25	22	18	15	12
40	97	94	91	88	85	82	80	77	74	72	69	67	64	62	59	57	54	53	51	48	46	44	42	40	38	36	35	33	31	29	26	23	20	16	14

APPENDIX

4A

Humidity Chart for Air-Water Vapour at 101 325 N/m²

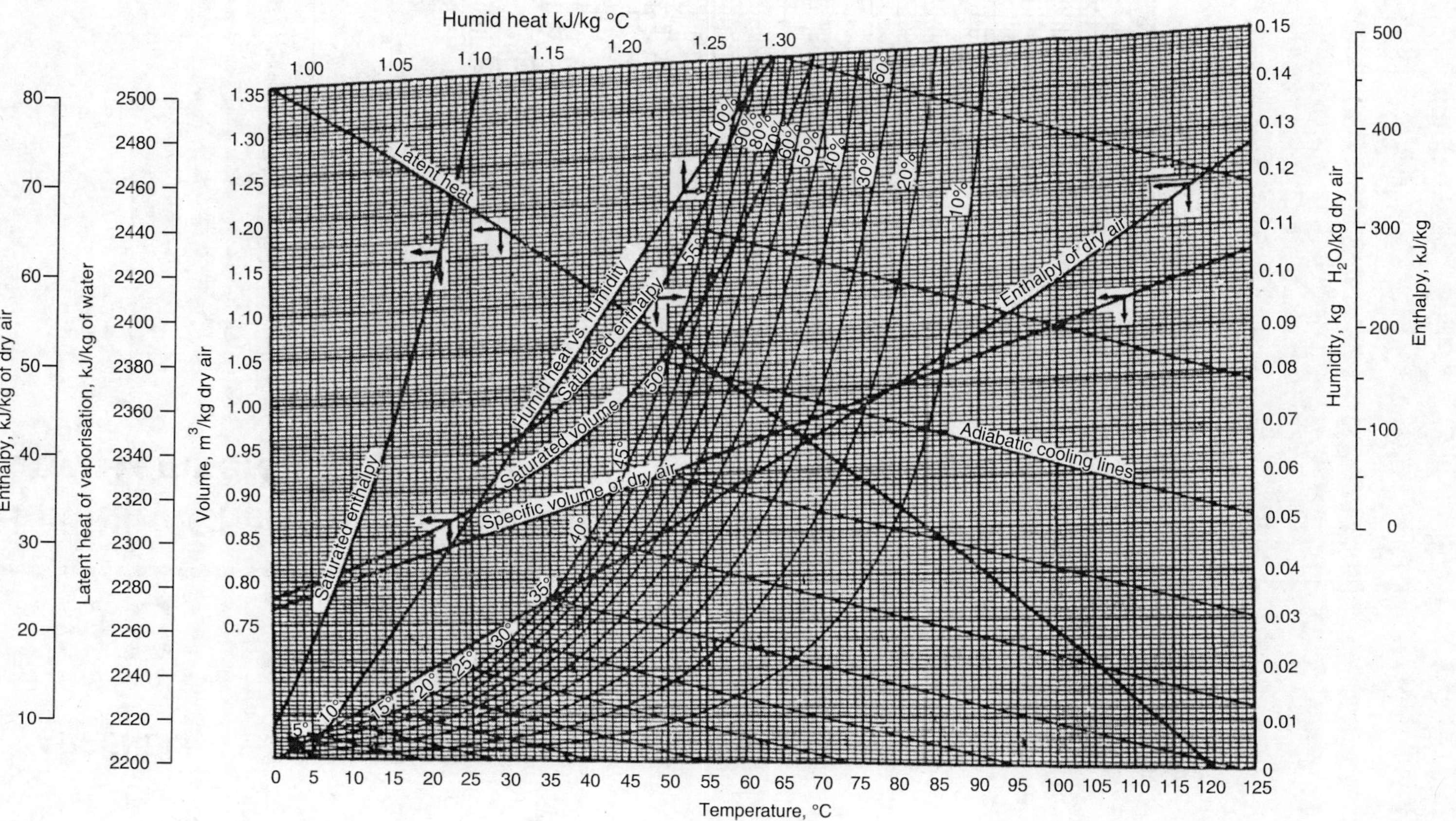

APPENDIX

4B

Humidity Chart for Air-Water-Vapour at 101 325 N/m²
(Low Temperature Range)

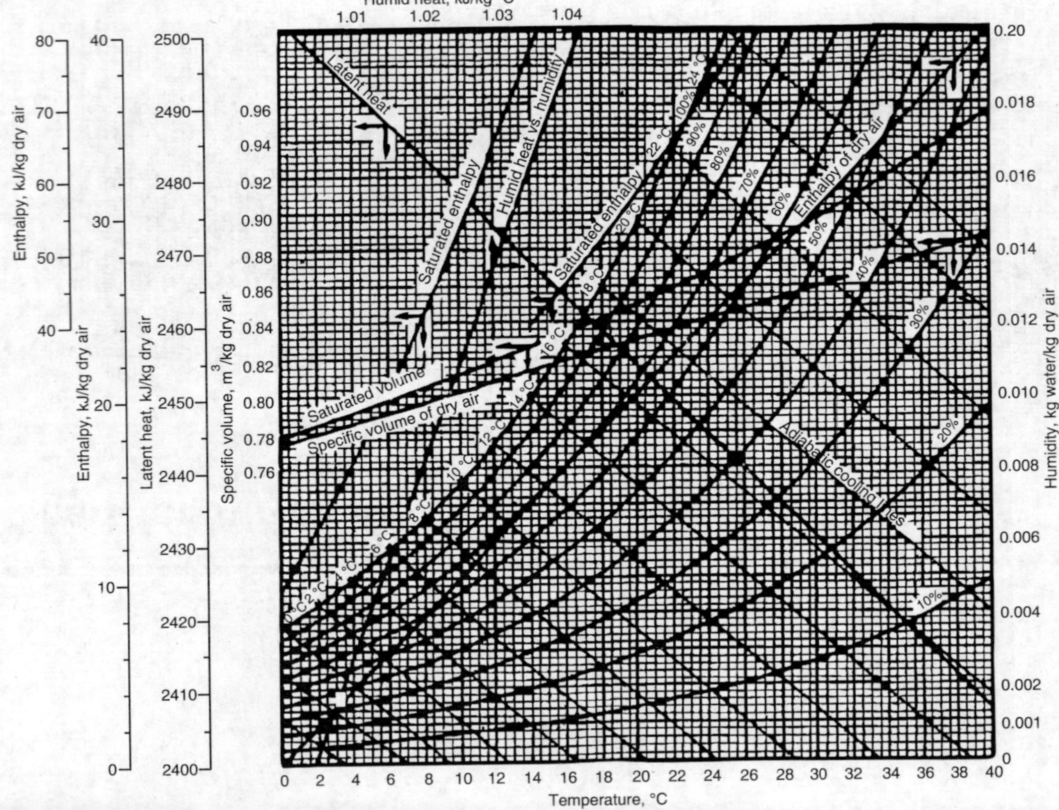

APPENDIX 5

Dimensions, Capacities, and Weights of Standard Steel Pipe

Nominal pipe size, in	Outside diameter, in	Schedule no.	Wall thickness, in	Inside diameter, in	Cross-sectional area of metal, in^2	Inside sectional area, ft^2	Circumference, ft or surface, ft^2/ft of length		Capacity at 1 ft/s velocity		Pipe weight lb/ft
							Outside	Inside	U.S. gal/min	Water lb/h	
1/8	0.40	40	0.068	0.269	0.072	0.00040	0.106	0.0705	0.179	89.5	0.24
		80	0.095	0.215	0.093	0.00025	0.106	0.0563	0.113	56.5	0.31
1/4	0.540	40	0.088	0.364	0.125	0.00072	0.141	0.095	0.323	161.5	0.42
		80	0.119	0.302	0.157	0.00050	0.141	0.079	0.224	112.0	0.54
3/8	0.675	40	0.091	0.493	0.167	0.00133	0.177	0.129	0.596	298.0	0.57
		80	0.126	0.423	0.217	0.00098	0.177	0.111	0.440	220.0	0.74
1/2	0.840	40	0.109	0.622	0.250	0.00211	0.220	0.163	0.945	472.0	0.85
		80	0.147	0.546	0.320	0.00163	0.220	0.143	0.730	365.0	1.09
1/4	1.050	40	0.113	0.824	0.333	0.00371	0.275	0.216	1.665	832.5	1.13
		80	0.154	0.742	0.433	0.00300	0.275	0.194	1.345	672.5	1.47
1	1.315	40	0.133	1.049	0.494	0.00600	0.344	0.275	2.690	1345	1.68
		80	0.179	0.957	0.639	0.00499	0.344	0.250	2.240	1120	2.17
1¼	1.660	40	0.140	1.380	0.668	0.01040	0.435	0.361	4.57	2285	2.27
		80	0.191	1.278	0.881	0.00891	0.435	0.335	3.99	1995	3.00
1½	1.900	40	0.145	1.610	0.800	0.01414	0.497	0.421	6.34	3170	2.72
		80	0.200	1.500	1.069	0.01225	0.497	0.393	5.49	2745	3.63

(contd.)

(contd.)

Nominal pipe size, in	Outside diameter, in	Schedule no.	Wall thickness, in	Inside diameter, in	Cross-sectional area of metal, in²	Inside sectional area, ft²	Circumference, ft or surface, ft²/ft of length		Capacity at 1 ft/s velocity		Pipe weight lb/ft
							Outside	Inside	U.S. gal/min	Water lb/h	
2	2.375	40	0.154	2.067	1.075	0.02330	0.622	0.541	10.45	5225	3.65
		80	0.218	1.939	1.477	0.02050	0.622	0.508	9.20	4600	5.02
2½	2.875	40	0.203	2.469	1.704	0.03322	0.753	0.647	14.92	7460	5.79
		80	0.276	2.323	2.254	0.02942	0.753	0.608	13.20	6600	7.66
3	3.500	40	0.216	3.068	2.228	0.05130	0.916	0.803	23.00	11500	7.58
		80	0.300	2.900	3.016	0.04587	0.916	0.759	20.55	10275	10.25
3½	4.000	40	0.226	3.548	2.680	0.06870	1.047	0.929	30.80	15400	9.11
		80	0.318	3.364	3.678	0.06170	1.047	0.881	27.70	13850	12.51
4	4.500	40	0.237	4.026	3.17	0.08840	1.178	1.054	39.6	19800	10.79
		80	0.337	3.826	4.41	0.07986	1.178	1.02	35.8	17900	14.98
5	5.563	40	0.258	5.047	4.30	0.1390	1.456	1.321	62.3	31150	14.62
		80	0.375	4.813	6.11	0.1263	1.456	1.260	57.7	28850	20.78
6	6.625	40	0.280	6.065	5.58	0.2006	1.734	1.588	90.0	45000	18.97
		80	0.432	5.761	8.40	0.1810	1.734	1.508	81.1	40550	28.57
8	8.625	40	0.322	7.981	8.396	0.3474	2.258	2.089	155.7	77850	28.55
		80	0.500	7.625	12.76	0.3171	2.258	1.996	142.3	71150	43.39
10	10.75	40	0.365	10.020	11.91	0.5475	2.814	2.620	246.0	123000	40.48
		80	0.594	9.562	18.95	0.4987	2.814	2.503	223.4	111700	64.40
12	12.75	40	0.406	11.938	15.74	0.7773	3.338	3.13	349.0	174500	53.56
		80	0.688	11.374	26.07	0.7056	3.338	2.98	316.7	158350	88.57

Source: Reprinted from ASME B36.1M-2004, by permission of The American Society of Mechanical Engineers.

APPENDIX

6

Thermo-Physical Properties of Some Solid Materials

	Density (g/cm³)	*Specific heat* kJ (kg°C)	*Thermal conductivity* (W/cm°C)
Non-metals			
Asbestos sheet	0.89	0.84	1.7×10^{-3}
Brick	1.76	0.92	7×10^{-3}
Cardboard	0.64	1.26	7×10^{-4}
Concrete	2.0	1.05	8.7×10^{-3}
Celluloid	1.4	1.55	2.1×10^{-3}
Cotton wool	0.08	1.26	4×10^{-4}
Cork	0.16	1.55	4.3×10^{-4}
Expanded rubber	0.072		4×10^{-4}
Fibreboard insulation	0.24		5.2×10^{-4}
Glass	2.24	0.84	5.2×10^{-3}
Ice	0.92	2.10	2.25×10^{-2}
Mineral wool	0.145		4×10^{-4}
Polyethylene	0.95	2.30	5.5×10^{-3}
Polystyrene foam	0.024		3.6×10^{-4}
Polyurethane foam	0.032		2.6×10^{-4}
Polyvinyl chloride	0.14	1.30	2.9×10^{-3}
Wood	0.7	2.5	2.8×10^{-3}
Metals			
Aluminum	2.64	0.87	2.21
Brass	8.65	0.38	0.97
Cast iron	7.21	0.42	0.55
Copper	8.9	0.38	3.88
Steel, mild	7.84	0.47	0.45
Steel, stainless	7.95	0.48	0.21

APPENDIX

7

Thermal Conductivity of Different Foods

Product	*Moisture content* (%)	*Temperature* (°C)	*Thermal conductivity* (W/m K)
Apple	85.6	2–36	0.393
Apple sauce	78.8	2–36	0.516
Beef, freeze dried			
1000 mm Hg pressure	–	0	0.065
0.001 mm Hg pressure	–	0	0.037
Beef, lean			
Perpendicular to fibres	78.9	7	0.476
Perpendicular to fibres	78.9	62	0.485
Parallel to fibres	78.7	8	0.431
Parallel to fibres	78.7	61	0.447
Beef fat	–	24–38	0.19
Butter	15	46	0.197
Cod	83	2.8	0.544
Corn, yellow dust	0.91	8–52	0.141
	30.2	8–52	0.172
Egg, frozen whole	–	–10 to –6	0.97
Egg, white	–	36	0.577
Egg, yolk	–	33	0.338
Fish muscle	–	0–10	0.557
Grape fruit, whole	–	30	0.45
Honey	12.6	2	0.502
	80	2	0.344
	14.8	69	0.623
	80	69	0.415
Juice, apple	87.4	20	0.559
	87.4	80	0.632
	36.0	20	0.389
	36.0	80	0.436
Lamb			
Perpendicular to fibre	71.8	5	0.45
	–	61	0.478

(contd.)

(contd.)

Product	*Moisture content* (%)	*Temperature* (°C)	*Thermal conductivity* (W/m K)
Parallel to fibre	71.0	5	0.415
	–	61	0.422
Milk	–	37	0.530
Milk, condensed	90	24	0.571
	–	78	0.641
	50	26	0.329
	–	78	0.364
Milk, skimmed	–	1.5	0.538
	–	80	0.635
Milk, nonfat dry	4.2	39	0.419
Olive oil	–	15	0.189
	–	100	0.163
Oranges, combined	–	30	0.431
Peas, black eyed	–	3–17	0.312
Pork			
Perpendicular to fibres	75.1	6	0.488
	–	60	0.54
Parallel to fibres	75.9	4	0.443
	–	61	0.489
Pork fat	–	25	0.152
Potato, raw flesh	81.5	1–32	0.554
Potato, starch gel	–	1–67	0.04
Poultry broiler muscle	69.1–74.9	4–27	0.412
Salmon			
Perpendicular	73	4	0.502
Salt	–	87	0.247
Sausage mixture	65.72	24	0.407
Soya bean oil meal	13.2	7–10	0.069
Strawberries	–	14–25	0.675
Sugars	–	29–62	0.087–0.22
Turkey breast			
Perpendicular to fibres	74	3	0.502
Parallel to fibres	74	3	0. 523
Veal			
Perpendicular to fibres	75	6	0.476
	–	62	0.484
Parallel to fibres	75	5	0.441
	–	60	0.452
Vegetable and animal oils	–	4–187	0.169
Wheat flour	8.8	43	0.45
	–	65.5	0.689
	–	1.7	0.542
Whey	–	80	0.641

Source: Reproduced from Singh, R.P. and Heldman, D.R., *Introduction to Food Engineering*, 2nd ed., Academic Press, Copyright 1993, with permission from Elsevier.

APPENDIX

8

Specific Heat Data of Raw Foods

Product	*Specific heat* (kJ/kg K)
Apple	3.759
Bacon	2.851
Beef, lean	3.579
Beef, roast	3.115
Blackberry, syrup pack	3.521
Blueberry, syrup pack	3.445
Butter	2.043
Carrot	3.864
Cheese	3.215
Cod, raw	3.697
Cucumber	4.061
Egg yolk	2.449
Fish	3.651
Milk, whole pasteurized	3.831
Milk dry, nonfat	1.520
Milk skim	3.935
Orange juice	3.822
Potato	3.483
Sardines	3.002
Shrimp	3.404
Starch	1.754
Veal	3.3483

APPENDIX

9

Thermal Diffusivity of Some Foodstuffs

Product	*Water content* (% weight)	*Temperature*[a] (°C)	*Thermal diffusivity* ($\times 10^7$ m^2/s)
Fruits, vegetables, and by-products, Apple, whole, Red Delicious	85	0–30	1.37
	37	5	1.05
	37	65	1.12
Applesauce	80	5	1.22
	80	65	1.40
	–	26–129	1.67
Avocado, flesh	–	24, 0	1.24
Seed	–	24, 0	1.29
Whole	–	41, 0	1.54
Banana, flesh	76	5	1.18
	76	65	1.42
Bean, baked	–	4–122	1.68
Cherry, tart, flesh	–	30, 0	1.32
Grape fruit, marsh, flesh	88.8	–	1.27
Grape fruit, marsh, albedo	72.2	–	1.09
Lemon, whole	–	40, 0	1.07
Lima bean, pureed	–	26–122	1.80
Pea, pureed	–	26–128	1.82
Peach, whole	–	27, 4	1.39
Potato, flesh	–	25	1.70
Potato, mashed, cooked	78	5	1.23
	78	65	1.45
Rutabaga	–	48, 0	1.34
Squash, whole	–	47, 0	1.71
Strawberry, flesh	92	5	1.27
Sugarbeet	–	14, 60	1.26

(contd.)

(contd.)

Product	*Water content* (% weight)	*Temperature*[a] (°C)	*Thermal diffusivity* ($\times 10^7$ m²/s)
Sweet potato, whole	–	35	1.06
	–	55	1.39
	–	70	1.91
Tomato, pulp	–	4, 26	1.48
Fish and meat products			
Cod fish	81	5	1.22
	81	65	1.42
Corned beef	65	5	1.32
	65	65	1.18
Beef, chuck[b]	66	40–65	1.23
Beef, round[b]	71	40–65	1.33
Beef, tongue[b]	68	40–65	1.32
Halibut	76	40–65	1.47
Ham, smoked	64	5	1.18
Ham, smoked	64	40–65	1.38
Water	–	30	1.48
	–	65	1.60
Ice	–	0	11.82

[a]Where two temperatures separated by a comma are given, the first is the initial temperature of the sample, and the second is that of the surroundings.

[b]Data are applicable only where juices that exuded during heating remain in the food samples.

Source: Reprinted with permission from, Thermal diffusivity of some food stuffs, Singh, R.P., *Food Technology*, **36**(2), 90, Copyright 1982, with the permission from Institute of Food Technologists, USA.

APPENDIX

10

Dielectric Properties Used for Critical Length Calculations

Material	f (MHz)	T (K)	K'	K''
Potato, freeze dried	3000	298.16	7.5	2.50
Potato, raw	3000	298.16	57.3	15.70
Aqueous nonfat dry milk	3000	308.16	63.3	16.00
Pizza stuffing	2800	323.16	10.1	3.10
Pizza baked dough	2800	323.16	4.6	0.60
Pizza baked dough	2800	373.16	6.2	0.90
Pineapple syrup	2800	323.16	67.8	11.60
Pineapple syrup	2800	373.16	57.1	9.20
Fish, cooked	2800	313.16	45.0	11.90
Fish, cooked	2800	353.16	42.6	12.70
Fish, cooked	2800	413.16	39.9	16.80
Gravy	2800	313.16	76.1	24.10
Gravy	2800	353.16	73.16	26.20
Gravy	2800	413.16	68.7	28.80
Water	2800	313.16	72.8	6.50
Water	2800	353.16	61.8	3.20
Water	2800	413.16	48.1	1.50
Potatoes, mashed	2800	313.16	60.6	17.40
Potatoes, mashed	2800	353.16	54.2	16.50
Potatoes, mashed	2800	413.16	44.7	15.00
0.1 M NaCl	2800	313.16	71.1	13.70
0.1 M NaCl	2800	353.16	63.5	14.80
0.1 M NaCl	2800	413.16	51.9	17.40
Beef, precooked 60°	2800	313.16	44.1	11.30
Beef, precooked 60°	2800	353.16	41.6	11.70
Beef, precooked 60°	2800	413.16	39.1	15.00
Beef, raw	2800	353.16	42.6	13.10
Beef, raw	2800	413.16	40.5	16.00
Beef, precooked 75°	2800	353.16	40.7	11.30

(contd.)

(contd.)

Material	*f* (MHz)	*T* (K)	*K'*	*K''*
Beef, precooked 75°	2800	413.16	38.7	14.80
Liver pate	2800	313.16	41.4	16.50
Liver pate	2800	353.16	38.9	18.40
Liver pate	2800	413.16	38.7	20.70
Pea, cooked	2800	313.16	60.8	12.60
Pea, cooked	2800	353.16	50.5	9.70
Pea, cooked	2800	413.16	41.2	8.90
Carrot, cooked	2800	313.16	70.1	11.80
Carrot, cooked	2800	353.16	61.8	12.50
Carrot, cooked	2800	413.16	41.5	11.00
Peanut butter	2800	323.16	3.1	4.10
Peanut butter	2800	373.16	3.5	5.00
Frankfurters	2800	323.16	39.0	26.90
Beef (ground) patties	2800	323.16	31.7	10.40
Beef (ground) patties	2800	373.16	31.7	12.60
Concentrated orange juice	2800	323.16	54.1	15.70
Beef steak, cooked	2800	323.16	37.0	10.60
Beef steak, cooked	2800	373.16	33.6	12.60
Ham	2800	323.16	66.6	47.00
Pork chop	2800	323.16	49.8	18.30
Pork chop	2800	373.16	44.5	19.40
Filet of turbot	2800	323.16	53.6	14.10
Ham salad	2800	323.16	38.7	26.80
Ham salad	2800	373.16	76.5	50.00
Potato salad	2800	323.16	56.4	22.70
Macaroni cheese	2800	323.16	54.5	20.70
Macaroni cheese	2800	373.16	59.2	27.40
Chicken a la king	2800	323.16	54.9	19.90
Sea food Newburg	2800	323.16	53.1	21.40
Sea food Newburg	2800	373.16	50.8	25.30
French fried potato	2800	323.16	12.4	4.60
French fried potato	2800	373.16	16.3	6.60
Pineapple pieces	2800	323.16	62.0	11.00
Water	2450	298.16	78.0	12.48
Beef, raw	2450	308.16	43.0	15.00
Beef, cooked	2450	308.16	30.5	9.60
Beef (cooked) juice	2450	308.16	62.6	21.90
Aqueous nonfat dry milk	1000	308.16	65.5	24.00
Potato, raw	1000	298.16	65.1	19.60
Beef, cooked	915	308.16	35.4	16.00
Beef, raw	915	308.16	43.0	40.00
Beef (cooked) juice	915	308.16	74.3	39.60
Beef, raw	900	313.16	49.2	25.60
Beef, raw	900	353.16	44.2	38.00
Beef, raw	900	413.16	37.6	72.10
Fish, cooked	900	313.16	52.0	27.40
Fish, cooked	900	353.16	47.0	37.50

(contd.)

(contd.)

Material	*f* (MHz)	*T* (K)	*K′*	*K″*
Fish, cooked	900	413.16	39.7	80.40
Gravy	900	313.16	51.5	52.20
Gravy	900	353.16	38.0	87.20
Gravy	900	413.16	35.0	144.00
Water	900	313.16	67.8	1.36
Water	900	353.16	56.8	0.69
Water	900	413.16	43.7	0.30
Potato, cooked	900	313.16	53.2	20.80
Potato, cooked	900	353.16	45.7	28.80
Potato, cooked	900	413.16	36.2	47.60
Water	450	298.16	78.5	1.80
Beef, raw	450	313.16	50.0	58.30
Beef, raw	450	353.16	42.0	87.30
Beef, raw	450	413.16	40.0	137.00
Fish, cooked	450	313.16	53.0	54.20
Fish, cooked	450	353.16	45.0	83.60
Fish, cooked	450	413.16	40.0	158.00
Gravy	450	313.16	50.0	126.00
Gravy	450	353.16	47.0	184.00
Gravy	450	413.16	42.0	303.00
Water	450	313.16	69.2	1.18
Water	450	353.16	58.4	0.83
Water	450	413.16	45.4	0.68
Potato, cooked	450	353.16	44.0	82.90
Potato, cooked	450	413.16	34.0	139.00
Beef, raw	300	413.16	43.0	120.00
Beef, cooked	300	308.16	38.0	47.00
Beef (juice), cooked	300	308.16	77.9	113.80
Potato, raw	300	298.16	80.0	47.80

Source: Reprinted from, Microwave heating: An evaluation of power formulations, *Chemical Engg. Science*, **46**, pp. 1010–1011, Ayappa, K.G., Davis, H.T., Crapiste, G., Davis, E.A. and Gordon, J., Copyright 1991, with the permission from Elsevier.

APPENDIX

11

Standard Sieves

Tyler Standard Screens		British Standard Screens	
Mesh size[†]	*Aperture* (mm)	*Mesh size*[†]	*Aperture* (mm)
3	6.680	5	3.353
4	4.699	6	2.8118
6	3.327	7	2.410
8	2.362	8	2.0574
10	1.651	10	1.6764
12[‡]	1.397	12	1.4046
14	1.168	14	1.204
16[‡]	0.991	16	1.0033
20	0.833	18	0.8534
24[‡]	0.701	22	0.6985
28	0.589	25	0.5994
32[‡]	0.495	30	0.5004
35	0.417	36	0.4216
42[‡]	0.351	44	0.3531
48	0.295	52	0.2946
60[‡]	0.246	60	0.2515
65	0.208	72	0.2108
80[‡]	0.175	85	0.1778
100	0.147	100	0.1524
115[‡]	0.124	120	0.1245
150	0.104	150	0.1041
170[‡]	0.088	170	0.0889
200	0.074	200	0.0762

[†]Meshes per inch

[‡]The ratio of diameter with the above mesh size is $1 : \sqrt[4]{2}$ instead of $1 : \sqrt{2}$

Index